Orthogonal Frequency Division Multiple Access Fundamentals and Applications

WIRELESS NETWORKS AND MOBILE COMMUNICATIONS

Dr. Yan Zhang, Series Editor
Simula Research Laboratory, Norway
E-mail: yanzhang@ieee.org

Broadband Mobile Multimedia: Techniques and Applications
Yan Zhang, Shiwen Mao, Laurence T. Yang, and Thomas M. Chen
ISBN: 978-1-4200-5184-1

Cooperative Wireless Communications
Yan Zhang, Hsiao-Hwa Chen, and Mohsen Guizani
ISBN: 978-1-4200-6469-8

Distributed Antenna Systems: Open Architecture for Future Wireless Communications
Honglin Hu, Yan Zhang, and Jijun Luo
ISBN: 978-1-4200-4288-7

The Internet of Things: From RFID to the Next-Generation Pervasive Networked Systems
Lu Yan, Yan Zhang, Laurence T. Yang, and Huansheng Ning
ISBN: 978-1-4200-5281-7

Millimeter Wave Technology in Wireless PAN, LAN and MAN
Shao-Qiu Xiao, Ming-Tuo Zhou, and Yan Zhang
ISBN: 978-0-8493-8227-7

Mobile WiMAX: Toward Broadband Wireless Metropolitan Area Networks
Yan Zhang and Hsiao-Hwa Chen
ISBN: 978-0-8493-2624-0

Orthogonal Frequency Division Multiple Access Fundamentals and Applications
Tao Jiang, Lingyang Song, and Yan Zhang
ISBN: 978-1-4200-8824-3

Resource, Mobility, and Security Management in Wireless Networks and Mobile Communications
Yan Zhang, Honglin Hu, and Masayuki Fujise
ISBN: 978-0-8493-8036-5

RFID and Sensor Networks: Architectures, Protocols, Security and Integrations
Yan Zhang, Laurence T. Yang, and JimIng Chen
ISBN: 978-1-4200-7777-3

Security in RFID and Sensor Networks
Yan Zhang and Paris Kitsos
ISBN: 978-1-4200-6839-9

Security in Wireless Mesh Networks
Yan Zhang, Jun Zheng, and Honglin Hu
ISBN: 978-0-8493-8250-5

Unlicensed Mobile Access Technology: Protocols, Architectures, Security, Standards, and Applications
Yan Zhang, Laurence T. Yang, and Jianhua Ma
ISBN: 978-1-4200-5537-5

WiMAX Network Planning and Optimization
Yan Zhang
ISBN: 978-1-4200-6662-3

Wireless Ad Hoc Networking: Personal-Area, Local-Area, and the Sensory-Area Networks
Shih-Lin Wu, Yu-Chee Tseng, and Hsin-Chu
ISBN: 978-0-8493-9254-2

Wireless Mesh Networking: Architectures, Protocols, and Standards
Yan Zhang, Jijun Luo, and Honglin Hu
ISBN: 978-0-8493-7399-2

Wireless Quality-of-Service: Techniques, Standards, and Applications
Maode Ma, Mieso K. Denko, and Yan Zhang
ISBN: 978-1-4200-5130-8

AUERBACH PUBLICATIONS

www.auerbach-publications.com
To Order Call: 1-800-272-7737 • Fax: 1-800-374-3401
E-mail: orders@crcpress.com

Orthogonal Frequency Division Multiple Access Fundamentals and Applications

Edited by
Tao Jiang ◆ Lingyang Song ◆ Yan Zhang

CRC Press
Taylor & Francis Group
Boca Raton London New York

CRC Press is an imprint of the
Taylor & Francis Group, an **informa** business
AN AUERBACH BOOK

CRC Press
Taylor & Francis Group
6000 Broken Sound Parkway NW, Suite 300
Boca Raton, FL 33487-2742

First issued in paperback 2019

ISBN-13: 978-1-4200-8824-3 (hbk)
ISBN-13: 978-0-367-38422-7 (pbk)

Library of Congress Cataloging-in-Publication Data

Orthogonal frequency division multiple access fundamentals and applications / editors, Tao Jiang,
　　Lingyang Song, and Yan Zhang.
　　　　p. cm. -- (Wireless networks and mobile communications)
　　Includes bibliographical references and index.
　　ISBN 978-1-4200-8824-3 (hard back : alk. paper)
　　1. Orthogonal frequency division multiplexing. I. Jiang, Tao, 1970 Jan. 8- II. Song, Lingyang. III.
Zhang, Yan, 1977-

TK5103.484.O765 2010
621.382--dc22

2009051226

Visit the Taylor & Francis Web site at
http://www.taylorandfrancis.com

and the CRC Press Web site at
http://www.crcpress.com

Contents

Preface vii

Editors ix

Contributors xiii

1   Introduction to OFDMA 1
    Li Zhang and Pengfei Sun

2   Radio Channel Modeling for OFDMA 11
    David W. Matolak, Jingtao Zhang, and Qiong Wu

3   Nonlinearity Analysis of OFDM-Based Wireless Systems 41
    Chunming Liu and Fu Li

4   Game Theory and OFDMA Resource Allocation 67
    Quang Duy La, Boon Hee Soong, Yong Huat Chew, and Woon Hau Chin

5   Resource Management Techniques for OFDMA 101
    Didier Le Ruyet and Berna Özbek

6   Scheduling and Resource Allocation in OFDMA Wireless Systems 131
    Jianwei Huang, Vijay Subramanian, Randall Berry, and Rajeev Agrawal

7   Spectrum Efficiency in OFDMA 165
    Thomas Magesacher

8   Resource Allocation in IEEE 802.16e Mobile WiMAX 189
    Chakchai So-In, Raj Jain, and Abdel Karim Al-Tamimi

9   Resource Management in MIMO-OFDM Systems 235
    Malte Schellmann, Lars Thiele, Thomas Wirth, Volker Jungnickel,
    and Thomas Haustein

10 Differential Space Time Block Codes for MIMO–OFDM 267
Benigno Rodríguez Díaz

11 Adaptive Modulation 301
Víctor P. Gil Jiménez and Ana García Armada

12 Training Sequence Design in Multiuser OFDM Systems 329
Jianwu Chen, Yik-Chung Wu, Tung-Sang Ng, and Erchin Serpedin

13 Fundamentals of OFDMA Synchronization 349
Romain Couillet and Merouane Debbah

14 Synchronization for OFDM and OFDMA 375
Thomas Magesacher, Jungwon Lee, Per Ödling, and Per Ola Börjesson

15 Multiuser CFOs Estimation in OFDMA Uplink Systems 397
Yik-Chung Wu, Jianwu Chen, Tung-Sang Ng, and Erchin Serpedin

16 Frequency Domain Equalization for OFDM and SC/FDE............................. 417
Harald Witschnig

17 MIMO Beamforming Schemes for Multiuser Access in OFDM–SDMA 453
Ahmed Iyanda Sulyman and Mostafa Hefnawi

18 Cooperative OFDMA in the Presence of Frequency Offsets 473
Zhongshan Zhang and Chintha Tellambura

19 Performance and Optimization of Relay-Assisted OFDMA Networks 507
Wern-Ho Sheen and Shiang-Jiun Lin

20 OFDM–MIMO Applications for High Altitude Platform Communications 537
Abbas Mohammed and Tommy Hult

21 OFDMA Systems and Applications 563
André Noll Barreto and Robson Domingos Vieira

22 OFDMA-Based Mobile WiMAX... 595
Jinsong Wu and Pei Xiao

Index 623

Preface

With the rapid increasing demand of modern society for high-speed reliable information exchanges anytime and anywhere, broadband wireless systems are envisioned for future wireless communication networks. It has been widely acknowledged that broadband wireless access faces some formidable technical challenges due to both the frequency selective multipath fading propagation environment and the underutilization of limited available radio spectra. Fortunately, it has also been accepted that orthogonal frequency division multiple access (OFDM/OFDMA) is a preferable and promising physical layer technology for the broadband wireless system, since it can efficiently mitigate frequency-dependent distortion across a wide frequency band and simplify equalization in a multipath fading environment. At the same time, much radio spectra can be saved via the spectra of individual subcarriers' orthogonal overlap. Moreover, OFDMA can effectively divide a common transmission medium among multiple users.

This book aims at providing a comprehensive picture of state-of-the-art research status, a representative sampling of important research results, and in-depth treatments of selected topics in the OFDM and OFDMA, which would provide enough background material and discuss advanced principles that could enable significant improvements in wireless network characteristics not realizable with current wireless infrastructures. Therefore, the book is suitable for graduate students, researchers, and practical engineers. The material could be used for an advanced special topic for undergraduate/graduate course in OFDM/OFDMA-based communications, and could also be used for independent study, for example, as an introduction for graduate students who wish to conduct research in the area. Practicing engineers will appreciate the up-to-date tutorial exposition to timely topics. A large number of illustrative figures, cross-references, and comprehensive references for readers interested in more details are provided.

The book is organized as follows. Chapter 1 takes a comprehensive look at OFDMA and its history, principles, fundamental theories, and recent development. Chapter 2 is tailored for radio channel modeling in OFDMA-based wireless communication systems, focusing on the outdoor environment. Chapter 3 takes us on a comprehensive analysis based on the spectrum approach in OFDM wireless local area network (WLAN) systems. Chapters 4 through 9, in particular, cover the various techniques of effective resource management for OFDM/OFDMA-based wireless communication systems, including allocation of radio resources (frequency channels, power, etc.) and utilization of scarce spectra. Chapter 10 describes how one can combine the OFDM transmission technique and the multiple antennas–multiple input–multiple output to guarantee the high data rate transmission for OFDMA wireless communication systems. Chapter 11 gives a concise overview of the concepts and key techniques for adaptive modulation in OFDMA wireless communication systems, for example, adaptive bits and power loading. Chapters 12 through 15 are devoted to channel estimation and synchronization in OFDM/OFDMA systems. Specifically, Chapter 12 provides a review on the training sequence designed for channel and frequency estimations in multiuser OFDM systems. Chapter 13 begins with a review of different types of offsets, then introduces the basics of estimating and correcting timing offsets and carrier-frequency offsets in a single-user OFDM/OFDMA communication system, and finally outlines the synchronization procedure for an operational OFDMA system. Chapter 14 details the complete synchronization steps needed for practical OFDMA-based systems to rapidly enter the proper data exchange phase. Chapter 15 discusses how one can achieve good frequency offset estimation in multiusers' OFDMA uplink systems. Chapter 16 focuses on the detection strategies of frequency-domain equalization and their advantages over time-domain equalization for broadband communications, while Chapter 17 provides extensive treatment of adaptive MIMO beamforming suitable for multiuser access in OFDM–SDMA systems.

Chapter 18, in particular, evaluates how the performance of an OFDMA can be improved with cooperative relaying to obtain much spatial diversity gain, followed by Chapter 19 that aims to investigate the theoretical performance of the relay-assisted OFDMA network in the multicell environment with optimized system parameters. To increase the capacity of high-altitude platform communication links, in Chapter 20 a novel combination of OFDM and the orbital angular momentum of the electromagnetic field is proposed and compared with the conventional OFDM system. Chapter 21 gives a comprehensive overview of the main wireless communications technologies that employ OFDM/OFDMA, both mature and emerging, and Chapter 22 focuses on a detailed discussion of the aspects of Mobile WiMAX in a TDD OFDMA-based physical layer and a multiple access control layer, including OFDMA frame structure and subchannelization, power saving mechanisms, and handover approaches.

We acknowledge all contributors for their efforts and time to make this book worthwhile with their high-quality contributions. All contributors were extremely professional and cooperative, and did a great job in the production of the book. We also acknowledge the advice and comments from our anonymous reviewers, which greatly improved the quality of the book. Moreover, we express our sincere thanks to Amy Blalock from CRC Press, a Taylor & Francis Company, for the continuous support, patience, and professionalism from the beginning to the final stage. We owe a great deal to our families and friends—it would have been impossible to maintain our spirit and work habits without their continuous love and support. Last but not least, we thank our contributing authors.

MATLAB® and Simulink® are registered trademarks of The MathWorks, Inc. For product information, please contact:

The MathWorks, Inc.
3 Apple Hill Drive
Natick, MA 01760-2098, USA
Tel: 508 647 7000
Fax: 508-647-7001
E-mail: info@mathworks.com
Web: www.mathworks.com

Tao Jiang
Lingyang Song
Yan Zhang

Editors

Tao Jiang is a professor in the Department of Electronics and Information Engineering at Huazhong University of Science and Technology, Wuhan, People's Republic of China. He received his BS and MS degrees in applied geophysics from China University of Geosciences, Wuhan, People's Republic of China, in 1997 and 2000, respectively. In June 2004, he received his PhD degree in information and communication engineering from Huazhong University of Science and Technology, Wuhan, People's Republic of China. He was an academic visiting scholar at Brunel University, UK, from August 2004 to 2005. He worked in the United States as a postdoctoral researcher at universities, including the University of Michigan from October 2006 to December 2007. His research focuses on nano-networks, areas of wireless communications based on OFDM and MIMO, especially cognitive radio networks, cooperative communications, wireless sensor networks, free space optical communication, and corresponding digital signal processing. He has authored or co-authored over 60 technical papers in refereed journals, conference proceedings, and book chapters in these areas. He served or is serving as a symposium technical program committee member of many major IEEE conferences, including INFOCOM, VTC, ICC, GLOBCOM, WCNC, etc. Dr. Jiang has been invited to serve as TPC Symposium Chair for IWCMC 2010. He served or is serving as an associate editor of technical journals in communications, including Wiley's *Wireless Communications and Mobile Computing (WCMC) Journal* and Wiley's *International Journal of Communication Systems*. He is a member of IEEE, IEEE ComSoc, and IEEE Broadcasting.

Lingyang Song received a BS degree in communication engineering from Jilin University, People's Republic of China, in 2002 and a PhD in differential space–time codes and MIMO from the University of York, UK, in 2007, where he received the K. M. Stott Prize for excellent research. From January to September 2003 he worked as a software engineer in Hwasun Tomorrow Technology, Beijing, People's Republic of China. He was a postdoctoral research fellow at the University of Oslo, Norway, until rejoining Philips Research UK in March 2008. Now, Dr. Song is with the School of Electronics Engineering and Computer Science, Peking University, People's Republic of China. He is a co-inventor of a number of patents and an author or co-author of over 70 journal and conference papers. He is currently on the editorial board of the *International Journal of Communications, Network and System Sciences, Journal of Network and Computer Applications*, and the *International Journal of Smart Homes*, and a guest editor of *Elsevier Computer Communications* and *EURASIP Journal on Wireless Communications and Networking*. He serves as a member of the Technical Program Committee and is co-chair for several international conferences and workshops. He is a member of IEEE and IEEE ComSoc.

Yan Zhang received a BS degree in communication engineering from the Nanjing University of Post and Telecommunications, People's Republic of China; an MS degree in electrical engineering from the Beijing University of Aeronautics and Astronautics, People's Republic of China; and a PhD degree from the School of Electrical & Electronics Engineering, Nanyang Technological University, Singapore.

He is an associate editor or on the editorial board of Wiley's *International Journal of Communication Systems (IJCS)*; *International Journal of Communication Networks and Distributed Systems (IJCNDS)*; Springer International's *Journal Ambient Intelligence and Humanized Computing (JAIHC)*, *International Journal of Adaptive, Resilient and Autonomic Systems (IJARAS)*; Wiley's *Wireless Communications and Mobile Computing (WCMC)*, *Security and Communication Networks*, *International Journal of Network Security, International Journal of Ubiquitous Computing, Transactions on Internet and Information Systems (TIIS)*, *International Journal of Autonomous and Adaptive Communications Systems (IJAACS)*, *International Journal of Ultra Wideband Communications and Systems (IJUWBCS)*, and the *International Journal of Smart Home (IJSH)*.

Dr. Zhang is currently the book series editor for the *Wireless Networks and Mobile Communications* book series (Auerbach Publications, CRC Press, Taylor & Francis Group). He serves as guest co-editor for Wiley's *Wireless Communications and Mobile Computing (WCMC)* special issue for best papers in the conference IWCMC 2009; ACM/Springer's *Multimedia Systems Journal* special issue on "Wireless Multimedia Transmission Technology and Application"; Springer's *Journal of Wireless Personal Communications* special issue on "Cognitive Radio Networks and Communications"; Inderscience's *International Journal of Autonomous and Adaptive Communications Systems (IJAACS)* special issue on "Ubiquitous/Pervasive Services and Applications"; EURASIP's *Journal on Wireless Communications and Networking (JWCN)* special issue on "Broadband Wireless Access"; IEEE's *Intelligent Systems* special issue on "Context-Aware Middleware and Intelligent Agents for Smart Environments"; and Wiley's *Security and Communication Networks* special issue on "Secure Multimedia Communication"; guest co-editor for Springer's *Wireless Personal Communications* special issue on selected papers from ISWCS 2007; guest co-editor for *Elsevier Computer Communications* special issue on "Adaptive Multicarrier Communications and Networks"; guest co-editor for Inderscience's *International Journal of Autonomous and Adaptive Communications Systems (IJAACS)* special issue on "Cognitive Radio Systems"; guest co-editor for *The Journal of Universal Computer Science (JUCS)* special issue on "Multimedia Security in Communication"; guest co-editor for Springer's *Journal of Cluster Computing* special issue on "Algorithm and Distributed Computing in Wireless Sensor Networks"; guest co-editor for EURASIP's *Journal on Wireless Communications and Networking (JWCN)* special issue on "OFDMA Architectures, Protocols, and Applications"; and guest co-editor for Springer's *Journal of Wireless Personal Communications* special issue on "Security and Multimodality in Pervasive Environments."

He is co-editor for several books: *Resource, Mobility and Security Management in Wireless Networks and Mobile Communications*; *Wireless Mesh Networking: Architectures, Protocols, and Standards*; *Millimeter-Wave Technology in Wireless PAN, LAN, and MAN*; *Distributed Antenna Systems: Open Architecture for Future Wireless Communications*; *Security in Wireless Mesh Networks*; *Mobile WiMAX: Toward Broadband Wireless Metropolitan Area Networks*; *Wireless Quality-of-Service: Techniques, Standards, and Applications*; *Broadband Mobile Multimedia: Techniques and Applications*; *Internet of Things: From RFID to the Next-Generation Pervasive Networked Systems*; *Unlicensed Mobile Access Technology: Protocols, Architectures, Security, Standards, and Applications*; *Cooperative Wireless Communications*; *WiMAX Network Planning and Optimization*; *RFID Security: Techniques, Protocols, and System-On-Chip Design*; *Autonomic Computing and Networking*; *Security in RFID and Sensor Networks*; *Handbook of Research on Wireless Security*; *Handbook of Research on Secure Multimedia*

Distribution; *RFID and Sensor Networks*; *Cognitive Radio Networks*; *Wireless Technologies for Intelligent Transportation Systems*; *Vehicular Networks: Techniques, Standards, and Applications*; *Orthogonal Frequency Division Multiple Access (OFDMA)*; *Game Theory for Wireless Communications and Networking*; and *Delay Tolerant Networks: Protocols and Applications*.

Dr. Zhang serves as Program Co-Chair for IWCMC 2010, Program Co-Chair for WICON 2010, Program Vice Chair for CloudCom 2009, Publicity Co-Chair for IEEE MASS 2009, Publicity Co-Chair for IEEE NSS 2009, Publication Chair for PSATS 2009, Symposium Co-Chair for ChinaCom 2009, Program Co-Chair for BROADNETS 2009, Program Co-Chair for IWCMC 2009, Workshop Co-Chair for ADHOCNETS 2009, General Co-Chair for COGCOM 2009, Program Co-Chair for UC-Sec 2009, Journal Liaison Chair for IEEE BWA 2009, Track Co-Chair for ITNG 2009, Publicity Co-Chair for SMPE 2009, Publicity Co-Chair for COMSWARE 2009, Publicity Co-Chair for ISA 2009, General Co-Chair for WAMSNet 2008, Publicity Co-Chair for TrustCom 2008, General Co-Chair for COGCOM 2008, Workshop Co-Chair for IEEE APSCC 2008, General Co-Chair for WITS-08, Program Co-Chair for PCAC 2008, General Co-Chair for CONET 2008, Workshop Chair for SecTech 2008, Workshop Chair for SEA 2008, Workshop Co-Organizer for MUSIC'08, Workshop Co-Organizer for 4G-WiMAX 2008, Publicity Co-Chair for SMPE-08, International Journals Coordinating Co-Chair for FGCN-08, Publicity Co-Chair for ICCCAS 2008, Workshop Chair for ISA 2008, Symposium Co-Chair for ChinaCom 2008, Industrial Co-Chair for MobiHoc 2008, Program Co-Chair for UIC-08, General Co-Chair for CoNET 2007, General Co-Chair for WAMSNet 2007, Workshop Co-Chair FGCN 2007, Program Vice Co-Chair for IEEE ISM 2007, Publicity Co-Chair for UIC-07, Publication Chair for IEEE ISWCS 2007, Program Co-Chair for IEEE PCAC'07, Special Track Co-Chair for "Mobility and Resource Management in Wireless/Mobile Networks" in ITNG 2007, Special Session Co-organizer for "Wireless Mesh Networks" in PDCS 2006, a member of the Technical Program Committee for numerous international conferences, including ICC, GLOBECOM, WCNC, PIMRC, VTC, CCNC, AINA, ISWCS, etc. He received the Best Paper Award at the IEEE 21st International Conference on Advanced Information Networking and Applications (AINA-07).

From August 2006, he has been working with Simula Research Laboratory, Norway. His research interests include resource, mobility, spectrum, data, energy, and security management in wireless networks and mobile computing. He is a member of IEEE and IEEE ComSoc.

Contributors

Rajeev Agrawal
Department of Advanced Networks
  and Performance
Motorola
Arlington Heights, Illinois

Abdel Karim Al-Tamimi
Department of Computer Science
  and Engineering
Washington University
St. Louis, Missouri

Ana García Armada
Department of Signal Theory
  and Communications
University Carlos III de Madrid
Leganés, Madrid, Spain

André Noll Barreto
Department of Electrical Engineering
University of Brasília
Brasília, Brazil

Randall Berry
Department of Electrical Engineering
  and Computer Science
Northwestern University
Evanston, Illinois

Per Ola Börjesson
Department of Information Technology
Lund University
Lund, Sweden

Jianwu Chen
Department of Electrical and
  Electronic Engineering
The University of Hong Kong
Hong Kong, People's Republic of China

Yong Huat Chew
Department of Modulation
  and Coding
Institute for Infocomm
  Research
Singapore

Woon Hau Chin
Toshiba Research Europe
  Limited
Bristol, United Kingdom

Romain Couillet
ST-Ericsson
Sophia Antipolis, France

Merouane Debbah
Alcatel-Lucent Chair on
  Cognitive Radio—
  Supélec
Gif sur Yvette, France

Benigno Rodríguez Díaz
Telecommunications
  Department
University of the Republic
Montevideo, Uruguay

Thomas Haustein
Department of Broadband Mobile
  Communication Networks
Fraunhofer Institute for
  Telecommunications, Heinrich
  Hertz Institute
Berlin, Germany

Mostafa Hefnawi
Department of Electrical and Computer
  Engineering
Royal Military College of Canada
Kingston, Ontario, Canada

Jianwei Huang
Department of Information
  Engineering
The Chinese University of Hong Kong
Hong Kong, People's Republic of China

Tommy Hult
Department of Signal Processing
Blekinge Institute of Technology
Ronneby, Blekinge, Sweden

Raj Jain
Department of Computer Science
  and Engineering
Washington University
St. Louis, Missouri

Víctor P. Gil Jiménez
Department of Signal Theory
  and Communications
University Carlos III de Madrid
Leganés, Madrid, Spain

Volker Jungnickel
Department of Broadband
  Mobile Communication
  Networks
Fraunhofer Institute for
  Telecommunications, Heinrich
  Hertz Institute
Berlin, Germany

Quang Duy La
School of Electrical and Electronic
  Engineering
Nanyang Technological University
Singapore

Didier Le Ruyet
Electronics and Communications Laboratory
Conservatoire National des Arts et Métiers
Paris, France

Jungwon Lee
Marvell Semiconductor, Inc.
Santa Clara, California

Fu Li
Department of Electrical and
  Computer Engineering
Portland State University
Portland, Oregon

Shiang-Jiun Lin
Department of Electrical Engineering
National Chiao Tung University
Hsinchu, Taiwan

Chunming Liu
Department of Electrical and
  Computer Engineering
Portland State University
Portland, Oregon

Thomas Magesacher
Department of Electrical Engineering
Stanford University
Stanford, California

David W. Matolak
School of Electrical Engineering
  and Computer Science
Ohio University
Athens, Ohio

Abbas Mohammed
Department of Signal Processing
Blekinge Institute of Technology
Ronneby, Blekinge, Sweden

Tung-Sang Ng
Department of Electrical and
  Electronic Engineering
The University of Hong Kong
Hong Kong, People's Republic of China

Per Ödling
Department of Information Technology
Lund University
Lund, Sweden

Berna Özbek
Department of Electrical and
    Electronics Engineering
Izmir Institute of Technology
Izmir, Turkey

Malte Schellmann
Department of Broadband Mobile
    Communication Networks
Fraunhofer Institute for
    Telecommunications, Heinrich
    Hertz Institute
Berlin, Germany

Erchin Serpedin
Department of Electrical and
    Computer Engineering
Texas A&M University
College Station, Texas

Wern-Ho Sheen
Department of Information and
    Communication Engineering
Chaoyang University of Technology
Taichung, Taiwan

Chakchai So-In
Department of Computer Science
    and Engineering
Washington University
St. Louis, Missouri

Boon Hee Soong
School of Electrical and Electronic
    Engineering
Nanyang Technological University
Singapore

Vijay Subramanian
Hamilton Institute
National University of Ireland
    Maynooth
Maynooth, County Kildare, Ireland

Ahmed Iyanda Sulyman
Department of Electrical Engineering
King Saud University
Riyadh, Saudi Arabia

Pengfei Sun
School of Electronic and
    Electrical Engineering
University of Leeds
Leeds, United Kingdom

Chintha Tellambura
Department of Electrical and
    Computer Engineering
University of Alberta
Edmonton, Alberta, Canada

Lars Thiele
Department of Broadband Mobile
    Communication Networks
Fraunhofer Institute for
    Telecommunications, Heinrich
    Hertz Institute
Berlin, Germany

Robson Domingos Vieira
Nokia Technology Institute
Brasília–DF, Brazil

Thomas Wirth
Department of Broadband Mobile
    Communication Networks
Fraunhofer Institute for
    Telecommunications, Heinrich
    Hertz Institute
Berlin, Germany

Harald Witschnig
NXP Semiconductors
Gratkorn, Austria

Jinsong Wu
Department of Electrical and
    Computer Engineering
Queen's University
Kingston, Ontario, Canada

Qiong Wu
School of Electrical Engineering and
    Computer Science
Ohio University
Athens, Ohio

Yik-Chung Wu
Department of Electrical and
    Electronic Engineering
The University of Hong Kong
Hong Kong, People's Republic of China

Pei Xiao
Institute of Electronics, Communications
    and Information Technology
Queen's University Belfast
Belfast, Northern Ireland, United Kingdom

Jingtao Zhang
School of Electrical Engineering and
    Computer Science
Ohio University
Athens, Ohio

Li Zhang
School of Electronic and Electrical
    Engineering
University of Leeds
Leeds, United Kingdom

Zhongshan Zhang
Department of Electrical and
    Computer Engineering
University of Alberta
Edmonton, Alberta, Canada

Chapter 1

Introduction to OFDMA

Li Zhang and Pengfei Sun

Contents

1.1 Wireless Multiple Access Techniques..1
1.2 OFDM Basics 3
1.3 OFDMA..4
    1.3.1 General Introduction..4
    1.3.2 Subcarrier Allocation...5
    1.3.3 Transmitter...5
    1.3.4 Receiver...6
1.4 Issues Addressed in the Book...7
References...7

1.1 Wireless Multiple Access Techniques

To improve spectrum efficiency, a wireless system is required to accommodate as many users as possible with least possible degradation in the performance of the system by effectively sharing the limited bandwidth [1]. The multiple access method allows multiple users to share a common communication channel to transmit information to a receiver. There are several ways of sharing radio resources among multiple users, by allocating regions in frequency, time, and space to different users [2]. Examples include classic methods such as frequency division multiple access (FDMA), time division multiple access (TDMA), and code division multiple access (CDMA), and more recently developed methods such as space division multiple access (SDMA) and orthogonal frequency division multiple access (OFDMA).

FDMA was one of the earliest multiple access techniques for cellular systems. In this technique, the available bandwidth for a given service is divided into a number of subchannels, which are allocated to individual users on request. Since users exclusively occupy their subchannels all the time, FDMA does not require synchronization, and hence it is simple to implement. However, to minimize adjacent channel interference, unused frequency slots, that is guard bands, are introduced between neighboring subchannels. This leads to a waste of bandwidth. In addition, when continuous transmission is not required for a user, bandwidth goes wasted since its allocated bandwidth is not in use. FDMA is used as the principal multiplexing technique in radio and television broadcast and in the first-generation (1G) cell phone systems, such as the advanced mobile phone system (AMPS) and wireless metropolitan area network (WMAN) [2].

The TDMA scheme is to subdivide the frame duration to nonoverlapping, equal-length time slots, which are assigned to different users when they need to transmit information. Each user is allowed to transmit throughout the entire bandwidth for a finite period of time (time slots). TDMA requires careful time synchronization since users share the bandwidth in the time domain. In TDMA, spacing in time between time slots is required as guard time to minimize the interference between users. Compared to FDMA, the guard time between time slots is considerably smaller. However, in cellular communications, when a user moves from one cell to another there is a chance that the user could experience a call loss if there are no free time slots available. TDMA is widely used in wireless applications such as global systems for mobile communications (GSM) and IEEE 802.16 wireless MAN standards.

FDMA and TDMA are sometimes used together, such as in the second-generation cellular systems [Interim Standard 54 (IS-54), GSM] and in the 2.5G wireless system (enhanced data rates for GSM evolution-EDGE). In such systems, the available spectrum is divided into frequency subchannels, which are divided into time slots. Each user is then given a subchannel and a time slot during a frame. Different users can use the same frequency in the same cell, except that they must transmit at different times.

CDMA techniques allow users to simultaneously access all of the available frequency and time resources. Typically, in CDMA systems, each user is assigned a channel by being assigned a unique spreading code sequence, which spreads the information signal across the assigned frequency band. The code sequences of different users are orthogonal to each other. This allows the signals that are transmitted from multiple users to completely overlap both in time and in frequency domains. If the spreading codes of different users are not exactly orthogonal, it will cause multiple access interference (MAI). The receiver is required to know the spreading code used by the transmitter in order to separate signals from different users by cross correlation of the received signal with each of the possible user spreading code sequences. Unlike TDMA, CDMA does not require time synchronization between the users. CDMA is the basis of the principal 3G cellular phone systems, such as CDMA2000 [3] and WCDMA [4]. It is also used in the second-generation cellular standard IS-95.

SDMA provides another degree of freedom to conventional TDMA-, FDMA-, or CDMA-based multiplexing access. It utilizes the spatial separation of users in order to share the frequency spectrum [5]. Conventionally, SDMA was to reuse the same frequency in different cells in a cellular network. This requires the cells to be sufficiently separated and hence limits the frequency reuse factor. The more advanced approach utilizes antenna arrays smart antenna technique to steer the antenna pattern in the direction of the desired user and place nulls in the direction of the interfering signals [6]. A different set of beam forming weights (leading to a different pattern) is applied to receive the signal from a different user while eliminating signals from other users. In this way, the frequency can be reused within the cell as long as the spatial separation between users is sufficient. As a result, a direct increase in the system capacity is achieved. SDMA has been studied for the wireless local area network (WLAN) [7] and IEEE 802.16 [8] systems. In a practical cellular environment, it is likely that multiple transmitters fall within the receiver beam width. In such a scenario, SDMA can be used in conjunction with other multiple access techniques, such as TDMA and CDMA.

Recently, the OFDMA technique has been introduced based on the popular OFDM digital modulation scheme [9]. In this technique, distinct users are assigned with mutually exclusive groups of subcarriers for simultaneous transmission. This multiplexing technique provides significant advantages in terms of high spectrum efficiency, robustness against multipath fading channels, resistance to multiuser interference, simplified equalization, and so on. These features have led to a wide range of adoption of OFDMA in future wireless communication systems, such as the cable TV (CATV) network [9], digital terrestrial television [10], the emerging IEEE 802.16e worldwide interoperability for microwave access (WiMAX) [11], and the fourth-generation cellular networks. OFDMA has been viewed as one of the most promising multiple access technologies for current and future wireless networks. This chapter will therefore concentrate on the introduction to OFDMA in general, starting from the fundamental principle of the OFDM technique.

1.2 OFDM Basics

OFDM is a bandwidth-efficient signaling scheme for wideband digital communications. The development of the OFDM technique dates, back to the 1950s [12]. The concept of parallel data transmission and frequency division multiplexing (FDM) was suggested by Chang [13] and Saltzberg [14] in the 1960s. In 1970, an OFDM patent was issued in the United States. The idea of this technique was to use parallel data transmission and FDM with overlapping subchannels to provide significant advantages in terms of simplified equalization, robustness to multipath distortion, and considerably increased spectrum efficiency [15]. The initial application of OFDM was to high-frequency signaling in military communications, such as KINEPLEX [12] and ANDEFT [16]. However, practical implementation of OFDM using a large number of very accurate oscillators and filters is unreasonably expensive and complex [15], and for this reason, OFDM could not come into use rapidly, particularly for a large number of subchannels. The development of digital signal processing techniques [17] made it possible to perform modulation and demodulation using inverse fast Fourier transform (IFFT) and FFT, respectively. The advances in very large scale integration (VLSI) technology made high-speed, large-size FFT chips readily available, and then OFDM became practical to implement, leading to its current popularity.

When OFDM was initially adopted by digital audio broadcasting [18] in the 1980s and then for digital video broadcast (DVB) [19], it started to attract a lot of attention. In the 1990s, OFDM was exploited for wideband data communications over mobile radio [20], ADSL, HDSL, and VHDSL [21,22]. In the past few years OFDM has been studied and implemented for digital television, HDTV terrestrial broadcasting, and digital audio terrestrial broadcasting [18,23–25]. More recently, it became established as the transmission technique of choice for broadband wireless systems, being used in several emerging wireless standards, such as 802.11 WiFi, 802.16 WiMAX, and 3GPP LTE-2 [26–28].

In OFDM, the overall signal bandwidth is divided into a number of closely spaced narrow subchannels. This converts a wideband frequency-selective fading channel into a series of narrowband and relatively flat subchannels, which require simpler equalization. The difference between OFDM and the traditional FDM system is that the spectra of the individual subchannel can overlap, thanks to the orthogonality among them, which is achieved by carefully selecting the carrier spacing to be exactly equal to the symbol rate [13], as illustrated in Figure 1.1. The orthogonality allows simultaneous transmission on subcarriers in a tight frequency space without interfering with each other. At the receiver, the orthogonal signals can be separated by correlation techniques or by using a conventional matched filter.

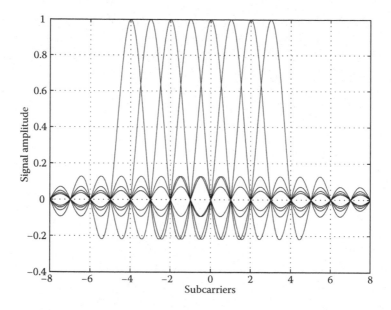

Figure 1.1   OFDM spectrum with eight subcarriers.

In practical implementation, the incoming data sequence is divided into a number of parallel lower rate data streams, each of which is modulated onto one of the subcarriers by IFFT. A cyclic prefix (CP) is inserted to extend the transmitted signal beyond the nominal symbol period to create a cyclically extended guard interval, so that both the intersymbol interference (ISI) and the intercarrier interference (ICI) can be completely eliminated as long as the CP is longer than the maximum delay spread of the channel [15].

Considering the significant advantages offered by OFDM, extending the OFDM concept to multiuser communication scenarios attracts intensive research interest. Multiple access, such as TDMA, FDMA, CDMA, and SDMA, can be implemented in OFDM systems in the same way as in a single carrier system [29–31]. Among them, the combination of FDMA and OFDM results in OFDMA, which has been widely considered as one of the most promising multiple access schemes for next-generation systems [32].

1.3 OFDMA

1.3.1 General Introduction

OFDMA is a straightforward extension of the OFDM into a multiuser environment. The concept of OFDMA was first introduced by Sari and Karam [33], who investigated the application of OFDMA to send upstream information from subscriber premises to the cable head-end in CATV networks. This scheme was then adopted in the standard of digital terrestrial television (DVB-RCT) in the uplink of the interaction channel [34]. OFDMA became popular in 2002, when it was adopted as the air interface for emerging IEEE 802.16e standards for wireless metropolitan area networks [8]. The purpose was to replace the "last mile" connection by a wireless link and provide wireless broadband network access. It promises ubiquitous broadband access even to rural or developing areas where broadband is not available due to the lack of cabled infrastructure [35]. The appealing advantages of OFDMA, including its scalability, intrinsic protection against MAI, flexibility in resource management, and those attractive features inherited from OFDM, such as robustness to multipath channels and simplified equalization, have recently attracted intensive research interest. It has been widely accepted as one of the most promising multiple access techniques and is thus adopted in a wide range of systems, such as the IEEE 802.20 [36], which is supposed to be the true mobile wireless technology with high capacity and high mobility, ETSI broadband radio access networks [37], multiuser satellite communications [38], and the fourth-generation cellular network [39].

The OFDMA subcarrier structure can support a wide range of bandwidths, by adjusting FFT size to channel bandwidth while fixing the subcarrier frequency spacing. In this way, the basic unit of physical resource is fixed and the impact to higher layers is minimized. This scalability is one of the most important advantages offered by OFDMA and allows flexibility in deployment [40].

1.3.2 Subcarrier Allocation

The principle of OFDMA is to divide the available subcarriers into several mutually exclusive groups (i.e., subbands) according to the subcarrier allocation strategies. Then each group of subcarriers is assigned to one user for simultaneous transmission [32]. The orthogonality among subcarriers ensures that users are protected from MAI.

Three commonly used carrier allocation schemes (CAS) are illustrated in Figure 1.2. In the subband CAS (SCAS) systems, all the subcarriers of each user are grouped together and thus the signals can be easily separated by filter banks. However, this scheme does not exploit frequency diversity and a deep fade can affect a substantial number of subcarriers of a given user. The interleaved CAS (ICAS) solves this problem by allocating subcarriers with uniform spacings. However, it still has restriction on resource allocation. The most flexible and desirable method is called generalized CAS, which allows users to select the best available subcarriers to transmit their information, and thereby fully exploits channel frequency diversity and provides high flexibility in resource allocation [32].

1.3.3 Transmitter

At the transmitter of the OFDMA system, the overall N subcarriers are allocated to K users according to a specific CAS. The data sequence of each user is divided into blocks of length M. The M data symbols in each block are modulated onto their corresponding subcarriers.

For downlink transmission, other subcarriers in the length N block are modulated with data from other users. The resultant length N block is sent to the conventional OFDM modulator, that is the N-point IFFT. A single CP is then appended to the front of the OFDM symbol before

Subband CAS   Interleaved CAS   Generalized CAS

▬▬▬ User 1   ▬ ▬ ▬ User 2   ▬▬▬ User 3

Figure 1.2 Three subcarrier allocation schemes: subband CAS; interleaved CAS; generalized CAS.

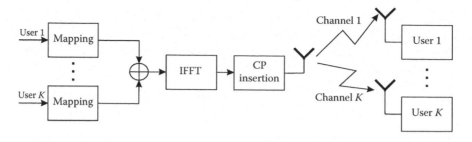

Figure 1.3 Block diagram of OFDMA downlink transmission.

it is transmitted over K channels to K different users. A diagram of the downlink transmitter is demonstrated in Figure 1.3.

Unlike the downlink transmitter, each uplink transmitter left the other subcarriers empty. The resultant length N block with M nonzero entries is sent to an N-point IFFT, followed by the insertion of a distinct CP before it is transmitted through a distinct channel to the base station (BS). The diagram of the OFDMA uplink transmitter is shown in Figure 1.4.

1.3.4 Receiver

At the receiver, after removing the CP, the remaining N samples are passed to an N-point FFT. One of the notable advantages of OFDMA is that it effectively converts a wide frequency-selective fading channel into a set of parallel flat-fading channels. Thus, channel equalization can be easily performed using a bank of one-tap multipliers, one for each subcarrier. The output of the equalizer is then used for data detection. For downlink transmission, each user only picks up the M data symbols transmitted over its allocated subcarriers for channel equalization and data detection. For uplink, the received signal at the BS is the superposition of signals from all active users, which pass through different channels and experience different frequency offsets and propagation delays. Signals from different users can only be separated after correcting the offsets of individual users. Thus, all the parameters have to be estimated jointly and this makes the synchronization of OFDMA uplink a very difficult problem. However, it has to be performed accurately before the data detection can be carried out.

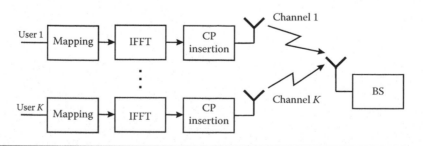

Figure 1.4 Block diagram of OFDMA uplink transmission.

1.4 Issues Addressed in the Book

Despite the above appealing advantages, OFDMA has some weaknesses. For example, similar to OFDM, OFDMA is extremely sensitive to timing errors and carrier frequency offsets (CFO). The presence of CFOs destroys the orthogonality among subcarriers and results in massive ICI and MAI. Timing errors result in inter-block interference, which will lead to serious performance degradation. In addition, an OFDM signal is the sum of a number of independently modulated subcarriers, and thus has a large peak to average power ratio. These problems have to be resolved in practical implementation. Other important issues, including CP design, channel estimation, optimized subcarrier and power allocation, interference control solutions, and so on, must also be carefully considered during system design. This book will investigate different issues in an OFDMA system and provide a complete survey of recent developments on OFDMA technology.

The book is organized as follows. Chapter 2 describes wireless channel modeling for OFDMA, with a focus on terrestrial (outdoor) settings, for which most current OFDM/A systems are being designed, but some discussion is more broadly applicable. Chapter 3 discusses interference control techniques for OFDMA. Chapter 4 is concerned with resource management techniques for OFDMA systems. This includes an overview of the resource management in general OFDMA systems, resource allocation and scheduling techniques for OFDMA systems, resource allocation using game theoretical approaches, and analysis on the spectrum efficiency of OFDMA. Chapter 5 examines transmission techniques for OFDMA, including differential space time block codes and the adaptive modulation in OFDMA systems. Chapter 6 investigates channel estimation techniques for OFDMA, particularly the training sequence design. Chapter 7 looks at the important synchronization issue in the OFDMA system, and provides an overview on the available synchronization techniques for multiple-input multiple-output (MIMO) OFDMA systems. Chapter 8 covers detection techniques, focusing on frequency domain equalization strategies. Chapters 9 through 11 discuss the application of the OFDMA system in conjunction with other techniques, such as cooperative and cognitive networks. Finally, Chapter 12 examines current applications of OFDMA systems, such as in high attitude platform communications and mobile WiMax systems.

References

1. ITU-R M.1645, Framework and overall objectives of the future development of IMT-2000 and systems beyond IMT-2000, 2003.
2. X. Wang, H. V. Poor, *Wireless Communication systems*, Prentice Hall, Upper Saddle River, NJ 07458.
3. V. K. Garg, IS-95 and CDMA2000, Prentice Hall, Upper Saddle River, NJ, 2000.
4. T. Ojanpera, R. Prasad, Wideband CDMA for third generation mobile communications, Artech House, Norwood, MA, 1998.
5. K. Winand, A Smart Indoor Receiver, M.S. thesis, University of Calgary, September 1998.
6. F. B. Gross, *Smart Antennas for Wireless Communications with MATLAB®*, McGraw-Hill Education, New York, 2005.
7. P. Vandenameele, *Space Division Multiple Access for Wireless Local Area Networks*, Kluwer Academic Publisher, The Netherlands, 2001.
8. Draft Amendment to IEEE Standard for Local and Metropolitan Area Networks, Part 16: Air Interface for Fixed Broadband Wireless Access Systems-Amendment 2: Medium Access Control Modifications and Additional Physical Layer Specifications for 2-11GHz, IEEE P802.16a/D3-2001, March 2002.
9. H. Sari, G. Karam, Orthogonal frequency division multiple access and its application to CATV networks, *Eur. Trans. Commun.*, 45, 507–516, 1998.

10. Interaction Channel for Digital Terrestrial Television (RCT) Incorporating Multiple Access OFDM, ETSI DVB RCT, March 2001.
11. IEEE Standard for Local and Metropolitan Area Networks, Part16: Air Interface for Fixed and Mobile Broadband Wireless Access Systems Amendment 2: Physical and Medium Access Control Layers for Combined Fixed and Mobile Operation in Licensed Bands, IEEE Std. 802.16e, 2005.
12. R. R. Mosier, R. G. Clabaugh, A bandwidth efficient binary transmission system, *AIEE Transactions*, 76, 723–728, 1958.
13. R. W. Chang, Synthesis of band-limited orthogonal signals for multichannel data transmission, *Bell System Technical Journal*, 45, 1775–1796, 1966.
14. B. R. Saltzberg, Performance of an efficient parallel data transmission system, *IEEE Transactions on Communications Technology*, COM-15, 805–811, 1967.
15. W. Y. Zou, Y. Wu, COFDM: an overview, *IEEE Transactions on Broadcasting*, 41, (1), 1995.
16. G. C. Porter, Error distribution and diversity performance of a frequency-differential PSK HF modem, *IEEE Transactions on Communicaitons Technology*, Com-16, 567–575, 1968.
17. S. Weinstein, S. Weinstein, P. Ebert, Data transmission by frequency-division multiplexing using the discrete fourier transform, *Communications, IEEE Transactions on Communications*, 19, (5), 628–634, 1971.
18. B. Le Floch, R. Lassalle, D. Castelain, Digital sound broadcasting to mobile receivers, *IEEE Transactions Consumer Electronics*, 3, 493–503, 1989.
19. European Standard EN 300 744 V1.1.2, Digital Video Broadcasting (DVB); Framing Structure, channel coding and modulation for digital terrestrial television.
20. E. F. Casas, C. Leung, OFDM for data communication over mobile radio FM channels–Part I: Analysis and experimental results, *IEEE Transactions of Communications* COM-39, (5), 1991.
21. P. S. Chow, J. C. Tu, and J. M. Cioffi, Performance evaluation of a multichannel transceiver system for ADSL and VHDSL services, *IEEE Journal on Selected Areas in Communications*, SAC-9, 909–919, 1991.
22. P. S. Chow, J. C. Tu, J. M. Cioffi, A discrete multitone transceiver system for HDSL applications, *IEEE Journal on Selected Areas in Communications*, SAC-9, 895–908, 1991.
23. J. F. Helard, B. Le Floch, Trellis-coded orthogonal frequency division multiplexing for digital video transmission, Proc. of Globecom, December 1991.
24. B. Sueur, D. Castelain, et al., Digital Terrestrial Broadcasting of Audiovisual Signals, Spectrum 20/20 Conf, Toronto, September 1992.
25. European Telecommunication Standard, Radio broadcast systems: digital audio broadcasting (DAB) to mobile, portable and fixed receivers, ETSI final draft pr ETS 300 401, November 1994.
26. Supplement to IEEE Std. 802.11, Wireless LAN media access control (MAC) and physical (PHY) specications: High-speed physical layer in the 5GHZ band, IEEE Std. 802.11a, 2001.
27. IEEE Standard for Local and Metropolitan Area Networks, Part 16: Air inter-face for fixed and mobile broadband wireless access systems amendment 2: Physical and medium access control layers for combined fixed and mobile operation in licensed bands, IEEE Std. 802.16e, 2005.
28. 3GPP TS 36.300 Evolved Universal Terrestrial Radio Access (E-UTRA) and Evolved Universal Terrestrial Radio Access Network (E-UTRAN); Overall description.
29. T. Muller, K. Bruninghaus, H. Rohling, Performance of coherent OFDM-CDMA for broadband mobile communications, *Wireless Personal Communications*, 2, 295–305, 2005.
30. H. Rohling, R. Grunheid, Performance comparison of different multiple access schemes for the downlink of an OFDM communication system, VTC'97, Phoenix, 1365–1369.
31. P. Vandenameele, L. Van Per Perr, et al., A combined OFDM/SDMA approach, *IEEE Journal on Selected Areas in Communications*, 18, (11), 2000.
32. M. Morelli, C. C. Jay Kuo, M.-O. Pun, Synchronisation techniques for orthogonal frequency division multiple access (OFDMA): A tutorial review, *Proceedings of the IEEE*, 95, (7), 2007.
33. H. Sari, G. Karam, Orthogonal frequency-division multiple access and its application to CATV networks, *European Transactions on Communications*, 45, 507–516, 1998.
34. Interaction Channel for Digital Terrestrial Television (RCT) Incorporating Multiple Access OFDM, ETSI DVB RCT, March 2001.

35. A. Ghosh, D. R. Wolter, et al., Broadband wireless access with WiMax/802.16: current performance benchmarks and future potential, *IEEE Communications Magazine*, 43, 129–136, 2005.
36. IEEE 802.20 WG, Mobile broadband wireless access systems 'Five criteria': Vehicular mobility, IEEE 802.20 PD-03, November 13, 2002.
37. Broadband Radio Access Networks (BRAN), Inventory of Broadband Radio Technologies and Techniques, Eur. Telecommun. Standards Inst. (ETSI), Sophia Antipolis, France, ref. DTR/BRAN 030 001, 1998.
38. L. Wei, C. Schlegel, Synchronization requirements for multiuser OFDM on satellite mobile and two-path Rayleigh fading channels, *IEEE Transactions on Communications* 43, 887–895, 1995.
39. 3GPP, TR25.814(V7.1.0), Physical layers aspects for Evolved UTRA, October 2006.
40. H. Yin, S. Alamouti, OFDMA: A broadband wireless access technology, Sarnoff Symposium, 2006 IEEE.

Chapter 2

Radio Channel Modeling for OFDMA

David W. Matolak, Jingtao Zhang, and Qiong Wu

Contents

2.1 Introduction... 12
    2.1.1 Frequency Bands and Environments.. 12
    2.1.2 Importance of Channel Modeling.. 13
    2.1.3 Some Unique Features of OFDM/A Channel Modeling....................... 14
    2.1.4 Chapter Scope.. 15
2.2 Statistical Channel Characterization Overview.. 16
    2.2.1 Remarks on Small-Scale versus Large-Scale Fading, and
           Deterministic Models.. 16
    2.2.2 Channel Impulse Response and Transfer Function............................. 17
    2.2.3 CIR and CTF Correlation Functions... 20
    2.2.4 Correlated/Uncorrelated Scattering and WSS.................................. 21
2.3 Characterization Functions and OFDM/A Channel Models........................... 23
    2.3.1 OFDM/A Signal Model.. 23
    2.3.2 Time versus Frequency Domain Modeling.................................... 24
    2.3.3 Cyclic Extensions.. 26
    2.3.4 Modeling Frequency and Time Offsets.. 27
    2.3.5 Time Variation and Doppler Spreading....................................... 29
    2.3.6 Channel Parameters and OFDM/A Signal Parameters........................ 30
2.4 Detailed OFDM/A Channel Models... 31
    2.4.1 A Canonical OFDM/A Channel Model.. 31
    2.4.2 Practical OFDM/A Channel Models... 32
    2.4.3 Nonstationary Channel Modeling... 34

2.4.4 Extensions to MIMO and MB Models... 35
2.4.5 Channel Estimation.. 36
2.4.6 Open Issues.. 37
2.5 Summary and Conclusions... 37
Abbreviations.. 37
References... 38

2.1 Introduction

This section provides background information for the rest of the chapter. We introduce some definitions, discuss the settings we consider, and also describe the chapter's scope. To motivate the discussion, we provide a brief description of the importance of channel modeling in general, and also some remarks specific to orthogonal frequency division multiplexing/orthogonal frequency division multiple access (OFDM/A).

2.1.1 Frequency Bands and Environments

The use of OFDM/A as a modulation and multiple access scheme is growing for both indoor and outdoor applications. This includes commercial communication systems such as wireless local area networks (WLANs), military communication systems, and ad hoc, mesh, and peer-to-peer networking [1]. One reason for this growth is the flexibility and performance achievable by such a "two-dimensional" modulation scheme. Our discussion here does not address these advantages, but focuses on the channel encountered when using an OFDM/A signaling scheme. Also worth noting is that from the channel perspective, OFDM and OFDMA are equivalent—in OFDMA different system user signals occupy a portion of the spectrum, whereas in OFDM a single user's transmission occupies the entire spectrum.

One of the most prominent OFDM systems is the IEEE 802.11a WLAN [2]. This OFDM version of the 802.11 standard systems (which are often termed "Wi-Fi") is primarily used indoors, in the 2.4 and 5 GHz bands, but can also be used outdoors. For indoor systems, mobility is generally limited so that channel dynamics are slow and of little concern. Indoor channel dispersion due to multipath propagation can be significant though, particularly for the wider values of channel bandwidth, so our discussion on this phenomenon applies equally to indoor and outdoor channels. Although in principle OFDM/A could be used in nonterrestrial environments such as aeronautical and satellite systems, our focus here is on the outdoor setting for terrestrial communications, since most current attention is being given to these types of systems. We also limit ourselves to channels for conventional populated areas such as urban, suburban, and some rural environments; hence, we do not consider locations such as wilderness areas or maritime locations. Also worth pointing out is that OFDM/A can and is used in wired applications, for example [3], where it is often termed discrete multitone modulation. We do not consider such wired channels here.

Also true is that, in principle, OFDM/A could be used in any arbitrary frequency band, but as with other systems, regulatory constraints and technology limitations narrow the choices. For this chapter we restrict ourselves to frequency bands below 10 GHz, and generally above a few hundred MHz in the ultra high frequency (UHF) band. These bands have the desirable properties of moderate propagation loss and efficient antennas of small size that are needed for mobile, portable communications. This upper frequency limit also applies to the newer OFDM/A standard IEEE 802.16 (often termed "WiMax") [4]. We consider primarily mobile cases, where at least one platform—transmitter or receiver (Tx, Rx, respectively)—is in motion. Some implementation parameters of

Table 2.1 System Parameters Comparison of Wi-Fi and WiMax

Parameters	Wi-Fi	Fixed WiMax	Mobile WiMax
IEEE standards	802.11a/g/n	802.16-2004	802.16e-2005
Carrier frequency band (GHz)	2.4 and 5	3.5 and 5.8 initially	2.3, 2.5, and 3.5 initially
Channel band-width(s) (MHz)	20 for 802.11a/g; 20/40 for 802.11n	3.5 and 7 in 3.5 GHz band; 10 in 5.8 GHz band	3.5, 7, 5, 10, and 8.75 initially
Modulation	BPSK, QPSK, 16QAM, and 64QAM	QPSK, 16QAM, and 64QAM	QPSK, 16QAM, and 64QAM
Transmission scheme(s) (M = no. of subcarriers)	$M = 64$ ODFM for 20 MHz	Single carrier, $M = 256$ or $M = 2048$ OFDM	Single carrier, $M = 256$ OFDM or scalable OFDM with $M = 128, 512, 1024,$ or 2048
Gross data rate (Mbps)	6, 9, 12, 18, 24, 36, 48, or 54; up to 600 for 802.11n	1–75	1–75

Source: Based on Andrews *et al. Fundamentals of WiMAX: Understanding Broadband Wireless Networking,* Prentice Hall PTR, Upper Saddle River, NJ, 2009.

Wi-Fi and WiMax systems are compared in Table 2.1 [12]. Newer applications such as vehicle-to-vehicle communications are also of interest [5], and a version of the 802.11 standard is being developed for this application [6], but that is beyond the scope of our treatment here. The 802.16 standard was developed as an improvement to the 802.11 standard in terms of higher data rates and greater mobility (i.e., higher platform velocities), and also has options for more sophisticated signal processing for generally better system performance. Worth noting is that nearly all of our theoretical and modeling results can be usable in frequency bands outside those noted above. The actual channel parameters for any model are, however, generally frequency-band- and environment-dependent. Also, generally speaking, as channel bandwidth increases, model complexity also increases, in terms of the number of parameters required to accurately specify the channel.

For this discussion on channel models, we consider only the effects of the propagation medium itself, which means that we do not consider important effects such as noise or intra- or inter-system interference. Noise is generally modeled as additive, spectrally white, and Gaussian, and techniques for estimating system performance in the presence of this impairment are well known [7]. The effects of interference can also be accounted for using known analytical techniques [8]; worth noting is that the propagation of such interference can be addressed by the models described in this chapter. We also do not consider the effect of nonlinear devices (e.g., power amplifiers), whose effect must be properly compensated in multicarrier schemes such as OFDM/A.

2.1.2 Importance of Channel Modeling

Regardless of the modulation scheme used, some knowledge of the wireless channel characteristics is vital to the design and performance of any communication system [9]. For engineering purposes, *mathematical* channel characterization results are most desirable, as these lend themselves to both analysis and simulations, which are extensively used in communication system physical layer design and development. Mathematical models for the wireless channel are often used to compare contending schemes or to evaluate the performance of the chosen scheme over a given channel. For example, features of the modulation scheme such as transmission bandwidth, and higher layer properties such as packet durations, can be optimized using good knowledge of the channel. A good model for the wireless channel for the environment of interest allows fair comparisons for any such study.

A set of good channel models also enables estimation of expected system performance [bit error ratio (BER), latency, etc.], which can lead to the design of remedial measures to counteract channel effects. Popular channel-counteracting techniques include the use of multiple diversity antennas, interleaving and coding, and equalizers. As with many areas of study in the field of communication systems, models are invaluable for determining critical features and inspiring creative design. These models are typically "block diagram" in form, and the wireless channel model represents a key block in the cascade of models that represents the entire communication system. In modern research and development of such systems, computer simulations are used extensively, and these provide a natural "environment" to use mathematical channel models, even models that are entirely empirical and consist of stored measured channel data.

Nearly all modern communication systems are sophisticated from the perspective of the user. One element of sophistication is the systems' ability to adapt to changing conditions, and OFDM/A systems are no exception. These conditions include a change in the number of system users or a change in a user's communication requirements, but often the most dynamic and stressful changes are those imposed by the wireless channel itself. Hence even if the system is adaptive, impairments caused by the channel can significantly degrade system performance if the system is not prepared to counteract them. These channel-induced degradations have been widely studied, for example, in Refs. [8,9]. Some example effects that can result from improperly modeling the wireless channel are a BER "floor" and a large delay. The BER floor refers to the phenomenon where error probability reaches a lower limit despite any increase in received signal strength. A large delay, or latency, can yield inefficiencies in data throughput, and for some protocols can yield a "link outage," in which the communication connection between transmitter and receiver is (temporarily) severed. One channel effect that can cause a large latency is a deep channel fade in which signal strength drops substantially (e.g., by several orders of magnitude). For modern systems that employ error correction and re-transmission, the data lost during the fade may need to be re-sent, yielding an increase in latency and an effective reduction in throughput. This effect also typically degrades subjective system performance as viewed by the system user. Thus, these channel fading characteristics should be modeled as accurately as possible for use during system research, design, and development, since they affect both physical layer (PHY) and protocol performance. Ultimately, accurate channel models are essential for wireless system optimization, and this is true of OFDMA systems as well.

Table 2.2 lists a number of important channel parameters and the signal design parameters they directly affect [10]. The signal design parameters refer mostly to the physical and data link layers, but, as noted, have impact directly upon higher layers. The parameters are defined subsequently.

2.1.3 Some Unique Features of OFDM/A Channel Modeling

The OFDM modulation scheme has been known for decades [11], but only with the advent of inexpensive and rapid signal processing (primarily the fast Fourier transform, FFT) has this technique gained widespread use. In general, the technique employs a number of narrowband, partially overlapping subcarriers. For most cases the subcarrier bandwidth is selected to be small, so

Table 2.2 Channel Parameters and Corresponding Signal/System Parameters They Affect

Channel Parameters	Affected Signal/System Design Parameters
Multipath delay spread T_M Coherence bandwidth B_c	Signal and subcarrier bandwidths, symbol rate
Channel attenuation α	Transmit power P_t, link ranges, modulation/detection type, data rate R_b
Doppler spread f_D Coherence time t_c	Data block or packet size, FEC type and strength transceiver adaptation rates, duplexing method
Spatial/temporal correlations ρ_s, ρ_t	Diversity method, FEC type, multiplexing method

that upon transmission, channel dispersion—roughly, the length of the channel impulse response (CIR)—is less than the reciprocal of the subcarrier bandwidth, denoted by the symbol time T_s. This means that in the frequency domain, the channel amplitude response varies only slightly, if at all, over the bandwidth of any single subcarrier. For modeling purposes, this means that if one models in the frequency domain, a sufficient description of the channel amplitude response at any given time is a vector of amplitude samples, one per subcarrier.

Since adjacent subcarrier spectra overlap when the (inverse) FFT (IFFT) is used for modulation at the transmitter, orthogonality can be destroyed by Doppler spreading due to motion. This is more severe for an OFDM/A signal than for a single-carrier signal of the same total bandwidth; hence, good models for Doppler spreading are more important for channel models for OFDM/A than for single-carrier schemes.

Finally, as with most modern systems, transmissions are typically in blocks, or packets of symbols. For an OFDM system, each packet of N time-domain symbols consists of NM total symbols, where M is the number of transmitted subcarriers per time-domain symbol. The time-domain packet of duration NT_s is denoted as a frame. Again when at least one of the platforms (Tx or Rx) is mobile, channel time variations over a frame can occur, and these variations must be tracked. The most common way to do this in OFDM/A is the use of known, or pilot, symbols, which are interspersed among the multiple subcarriers and in time. When mobile velocity becomes large enough, the channel can change significantly, requiring either a larger fraction of pilot symbols (reducing useful throughput) or a shorter frame duration. Although single-carrier systems must also contend with this effect and appropriately select frame durations and pilot symbols, the complexity of pilot allocation is larger in OFDM/A systems.

2.1.4 Chapter Scope

As stated, we limit our discussion to terrestrial wireless channels with frequency in the range of a few hundred MHz to 10 GHz. Based upon currently implemented and planned systems, we also limit ourselves to channel bandwidths of 20 MHz or less [12]. (Section 2.1.1 addresses extension beyond these conditions.) In Section 2.2, we provide an overview of statistical channel characterization and the most popular type of channel model based on this characterization, the tapped-delay line (TDL). Section 2.3 provides a description of the next step in modeling: translating channel characterization functions to usable channel models. In that section, we consider time- versus frequency-domain modeling, cyclic extensions, and several other practical effects for OFDM/A signaling. Some detailed channel models are provided in Section 2.4. This section discusses a popular model seen in the literature, and then turns to more realistic and accurate models and what this means for the model user. We also briefly describe how these models can be extended for use in multiple antenna systems (multiple-input/multiple-output or MIMO), and comment on methods for practical channel estimation and open issues. The last section concludes the chapter with a summary.

2.2 Statistical Channel Characterization Overview

This section describes some fundamentals of statistical channel characterization. We discuss large- versus small-scale fading. Brief comments on deterministic modeling are also given. We then describe the CIR and its Fourier transform, the channel transfer function (CTF), and the appropriate correlation functions of these two functions for statistical modeling.

2.2.1 Remarks on Small-Scale versus Large-Scale Fading, and Deterministic Models

We focus only on small-scale fading, which usually arises due to the destructive interference from multiple replicas of the transmitted signal arriving at the receiver with different delays. This effect is termed multipath propagation—the signal from Tx to Rx travels multiple paths, each of which can have a different amplitude, delay, and phase. Small-scale fading is observed on spatial scales on the order of one-half wavelength, the distance required for a propagating wave's phase to change by π radians. In contrast, for frequency bands of current interest for OFDM/A, large-scale fading, also termed shadowing or obstruction [13], occurs on scales of many wavelengths (e.g., 20 or more [9]).

Attenuation or propagation path loss can be treated along with large-scale fading. Strictly, this loss is not a fading phenomenon, but rather due to signal spatial spreading as the electromagnetic wave propagates. Path loss does not vary that rapidly with distance, and is generally compensated by specifying adequate transmit power and antenna gains. Large-scale shadowing varies over moderate distances (e.g., on the size of buildings) so that fading margin and/or power control can compensate. Other types of transceiver processing are needed to counteract small-scale fading. For mobile communication, all fading types can occur, and in complicated environments (e.g., urban), these fading effects are most compactly modeled stochastically.

As is widely accepted, the functions that describe variation of the channel's amplitude, phase, and delay are very well modeled as random. This is invariably true for complicated environments. One could, in principle, compute the resulting electromagnetic field at any point in space distant from the transmitting antenna given all the electrical, geometric, and kinematic parameters of objects in the environment, but in most practical situations these parameters are unknown or their estimates are insufficiently accurate. Even if accurate parameter estimates were available, significant computational resources are required to solve the electromagnetic field equations or trace "rays." Computation time can also constrain how fast field strengths can be estimated, and the consequence of this is an upper limit for platform velocity in the model. Ultimately, in complicated environments with mobility, we usually do not need the exact value of field strength at a specific point, but only the average and range of values over a small spatial extent.

In both principle and practice, deterministic channel models are essentially site specific, and to be accurate they are computationally intensive in comparison to statistical models. This is one advantage of statistical models. Statistical models do not even attempt to provide exact estimates of a channel's small-scale fading characteristics, but rather they aim to faithfully reproduce channel *variation*.

2.2.2 Channel Impulse Response and Transfer Function

Before we introduce the CIR and CTF, we provide a summary of several widely used channel parameters, including multipath delay spread and Doppler spread. With knowledge of just these two parameters plus channel attenuation, a communication system designer can estimate not only the detrimental effects the channel will have on any given signaling scheme, but also the need for, and complexity of, remedial measures to counteract these effects.

The *multipath delay spread* is essentially the duration of the CIR. It is reciprocally related to *coherence bandwidth*, which is a measure of the channel's frequency selectivity. The coherence bandwidth expresses the width of the contiguous frequency spectrum over which the channel affects a signal roughly equally; specifically, at each frequency within the coherence bandwidth, the channel's effect on any signal transmitted through the channel is approximately the same. The *Doppler spread* is essentially the range of frequencies over which a transmitted tone is spread as a result of transiting the channel. Doppler spread is reciprocally related to *coherence time*, which is a measure of the rate of channel time variation. Coherence time is roughly the time over which the channel does not change appreciably. These parameters are more precisely defined, in statistical terms, in a subsequent section.

Table 2.3 presents a brief summary of some key channel parameters [10], along with two additional "parameters." These additional "parameters" are the number of channel "taps" in the channel model (L) and the parameter probability density function [$p_z(x)$]. More detail will be provided on these subsequently, but, in brief, the number of taps L (a positive integer) represents the length of the CIR relative to the signal symbol duration T_s. The probability density function $p_z(x)$ is used to describe numerous parameters, but most importantly describes the channel's amplitude distribution.

The wireless channel is modeled as a time-varying linear filter, and its multipath CIR, in complex baseband form, can be expressed as follows:

$$
\begin{aligned}
h(\tau, t) &= \sum_{k=0}^{L(t)-1} z_k(t)\alpha_k(t) \exp\{j[\omega_{D,k}(t)(t - \tau_k(t)) - \omega_c(t)\tau_k(t)]\}\delta[\tau - \tau_k(t)] \\
&= \sum_{k=0}^{L(t)-1} z_k(t)\alpha_k(t)e^{j\phi_k(t)}\delta[\tau - \tau_k(t)], \quad (2.1)
\end{aligned}
$$

where $h(\tau, t)$ represents the response of the channel at time t to an impulse input at time $t - \tau$. This representation employs "discrete impulses" via the Dirac deltas, which implies that the channel imposes specific discrete attenuations, phase shifts, and delays upon any signal transmitted. This approximation is very good for signal bandwidths of tens of MHz or more [14].

The variable α_k in Equation 2.1 is the amplitude of the kth resolved component, the argument of the exponential $\phi_k(t)$ is the kth resolved phase, and τ_k is the delay of the kth path. The radian carrier frequency is ω_c, and the kth resolved Doppler frequency is $\omega_{D,k}(t) = 2\pi v(t)f_c \cos[\theta_k(t)]/c$, with $v(t)$ the relative velocity between Tx and Rx, $\theta_k(t)$ the aggregate phase angle of all components arriving in the kth delay "bin," c the speed of light, and $f_c = \omega_c/(2\pi)$ the carrier frequency in Hz.

Table 2.3 Channel Parameters and Definitions

Channel Parameter	Definition (Units)	Comments
T_M	*Multipath delay spread*: Extent in delay, of the CIR, usually weighted by energy (seconds)	• Most often specified statistically via r.m.s. value σ_τ; maximum and minimum values also of interest • Typically account for all impulses within some threshold (e.g., 25 dB) of "main" impulse
B_C	*Coherence bandwidth*: Bandwidth over which a channel affects a signal equally (Hz)	• Reciprocally related to T_M, often estimated as a/T_M, a = small constant>0 • Precisely, values of frequency separation at which channel amplitude correlation falls to some value, e.g., 0.5, 0.1 • Measure via Fourier transform of power delay profile, correlation of spectral components
f_D	*Doppler spread*: Maximum value of Doppler shift incurred by signal (Hz)	• Can be estimated analytically via classical physics, i.e., $f_D = v\cos(\theta)/\lambda$, v = maximum relative Tx–Rx velocity, θ = angle between propagation vector and velocity vector, λ = wavelength • Measure via Fourier transform of spaced-frequency, spaced-time correlation function, at fixed delay
t_C	*Coherence time*: Time over which channel remains approximately constant (seconds)	• Reciprocally related to f_D • As with B_C, desire values of time separation at which channel correlation falls to some specified value • Measure via spaced-time correlation function, or compute from f_D
α	*Attenuation*: Power loss, function of frequency and distance (unitless, dB)	• Analytically "estimatable" via traditional physics, e.g., free-space ($20\log(4\pi d/\lambda)$ dB). "Plane-earth" models • Multiple models available in software
L	*Impulse response length*: Length, in signal elements, of CIR (unitless integer)	• Depends upon signal element (bit, symbol, chip) duration • Estimate as $\lceil T_M/T \rceil$; T = smallest signaling duration, with $\lceil x \rceil$ = smallest integer $\geq x$
$p_z(x)$	*Probability density function of random variable z* (unitless)	• Random variable can be varying in time, frequency, space • Common amplitude distributions: Rayleigh, Ricean, Nakagami, Weibull • For phase, common distributions are uniform, Gaussian

The width of the delay bin is approximately equal to the reciprocal of the signal bandwidth, for example, for a 20 MHz signal, the bin width is 50 ns, which means that components separated in delay by an amount smaller than 50 ns are "unresolvable."

The CIR form of Equation 2.1 is generalized from that typically used [7], and allows for a time-varying number of transmission paths $L(t)$; a "persistence process" $z(t)$ accounting for the finite "lifetime" of propagation paths, and explicit time variation of carrier frequency to model transmitter oscillator variations and/or carrier frequency hopping.

The CTF corresponding to Equation 2.1 is

$$H(f,t) = \sum_{k=0}^{L(t)-1} z_k(t)\alpha_k(t)e^{j2\pi f_{D,k}(t-\tau_k(t))}e^{-j2\pi f_c\tau_k(t)}e^{-j2\pi f\tau_k(t)}, \qquad (2.2)$$

where we have suppressed time variation in $f_{D,k} = \omega_{D,k}(t)/(2\pi)$ for brevity, and assume the carrier frequency f_c is constant. The last exponential expresses the frequency dependence. The second exponential, which typically dominates small-scale fading since f_c is usually *much* larger than $f_{D,k}$, can change significantly with small changes in delay $\tau_k(t)$ when f_c is large, for example, nanosecond delay changes can cause 2π shifts in this exponential argument when $f_c = 1$ GHz. The first exponential term embodies slower time variation associated with $f_{D,k}$.

The CIR equation shows variation in both delay τ and time t. Bello termed this function the input delay spread function in his classic treatment of random channels [15]. In Figure 2.1 we show a conceptual CIR with both fading in time and variation of impulse energy ($\sim\alpha_k^2$) with delay. This type of CIR can be used in analysis, simulations, or hardware using the popular TDL model. The TDL is a linear, finite impulse response filter, shown in Figure 2.2, with input symbols x_k, output symbols y_k, and $h_k(t) = z_k(t)\alpha_k(t)e^{j\phi_k(t)}$. Most often in practice, the CIR essentially decays (ends) in delay τ long before any appreciable change of the components in time t. The multipath persistence process $z_k(t)$ can be added to Figure 2.2 as a binary {0,1} process that multiplies $\alpha_k(t)e^{j\phi_k(t)}$.

As previously noted, the CIR is often modeled as random, and the most significant quantity for communication system performance is amplitude r [aggregate amplitude of $h(\tau, t)$ or amplitude of

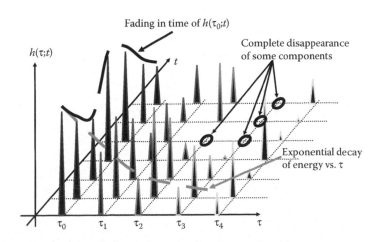

Figure 2.1 Conceptual illustration of time-varying CIR. (From D. W. Matolak, *IEEE Commun. Mag.*, 46(5), 76–83, 2008. ©2008 IEEE, reprinted with permission.)

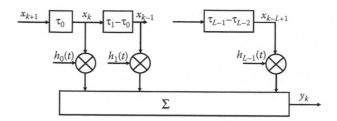

Figure 2.2   TDL channel model with $h_k(t) = z_k(t)\alpha_k(t)e^{j\phi_k(t)}$.

one component, $\alpha_k(t)$]. In Table 2.4 we summarize the most common probability density functions used for modeling amplitude fading. The mean-square value is Ω for all except the lognormal distribution.

2.2.3 CIR and CTF Correlation Functions

For random channels the CIR and CTF are stochastic processes; hence a complete specification would require joint probability density functions of all orders. We can never attain these in practice

Table 2.4   Commonly Used Fading Amplitude Probability Density Functions

Distribution Type	Probability Density Function	Comments
Rayleigh	$p_R(r) = \frac{2r}{\Omega} \exp\left[-\left(\frac{r^2}{\Omega}\right)\right]$	• Widely used for ease of analysis • Derived from Central Limit Theorem arguments
Ricean	$p_{Ri}(r, K) = \dfrac{2r(1 + 10^{K/10})}{\Omega}$ $\times \exp\left[\dfrac{-r^2(1 + 10^{K/10})}{\Omega} - 10^{K/10}\right]$ $\times I_0\left(2r10^{K/20}\sqrt{\dfrac{1 + 10^{K/10}}{\Omega}}\right)$	• Rice or "K-factor" is equal to $k = \dfrac{\text{(LOS power)}}{\text{(Scattered power)}}$, $K = 10\log(k)$ (dB) • For $k \to 0$, pdf $\to$ Rayleigh • For $k \to$ large, pdf $\to$ nonfading • $I_0 =$ modified Bessel function of first kind, zero order
Nakagami-m	$p_N(r, m) =$ $\dfrac{2m^m r^{2m-1}}{\Gamma(m)\Omega^m} \exp\left[\dfrac{-mr^2}{\Omega}\right]$	• $m \geq 0.5$ • For $m = 2$, equals the Rayleigh • $\Gamma =$ Gamma function
Weibull	$p_W(r) = \dfrac{b}{a^b} r^{b-1} \exp\left[-\left(\dfrac{r}{a}\right)^b\right]$	• $b =$ shape factor, determines fading severity ($\sim$ Ricean k, Nakagami m) • $a =$ scale parameter, $a = \sqrt{\Omega/\Gamma[(2/b) + 1]}$
Lognormal	$p_L(r, \mu, \sigma) = \dfrac{10}{r\ln(10)\sqrt{2\pi\sigma^2}}$ $\times \exp\left\{-[10\log(r) - \mu]^2/(2\sigma^2)\right\}$	• For $r =$ received power, $w =$ power in dBW, pdf of w is Gaussian, with mean $= \mu$, standard deviation $= \sigma$

Source: Adapted from D. W. Matolak, Wireless channel characterization in the 5 GHz microwave landing system extension band for airport surface areas, Final Project Report for NASA ACAST Project, Grant Number NNC04GB45G, May 2006.

and fortunately do not need this comprehensive a description. The short summary here uses the functions of Bello [15], and, as in Ref. [15], we limit our attention to second-order statistics. The CIR correlation function is*

$$R_{hh}(\tau_0, \tau_1, t_0, t_1) = E[h(\tau_0, t_0)h^*(\tau_1, t_1)].$$ (2.3)

We use correlation instead of covariance functions because for many cases of interest, the CIR is at least approximately zero mean, and for stationary or wide-sense stationary (WSS) cases, the correlation and covariance differ only by a constant. This zero mean condition does not hold when there is a dominant (often LOS) component, with a slowly varying or constant phase, as in the well-known Rician channel [7]. The meaning of Equation 2.3 is that R_{hh} quantifies the correlation of the resolved multipath component (MPC) at delay τ_0, time t_0, with that of the MPC at delay τ_1, time t_1; see Figure 2.1. A larger delay difference ($|\tau_0 - \tau_1|$) generally means a lower correlation, since widely differing delays usually correspond to distinctly separate propagation paths.

The CTF correlation function is

$$R_{HH}(f_0, f_1, t_0, t_1) = E[H(f_0, t_0)H^*(f_1, t_1)],$$ (2.4)

and has meaning analogous to that for R_{hh}: R_{HH} measures the correlation between the CTF evaluated at frequency/time "points" (f_0, t_0) and (f_1, t_1). These two correlation functions are related by a double Fourier transform (F):

$$R_{HH}(f_0, f_1, t_0, t_1) = \iint R_{hh}(\tau_0, \tau_1, t_0, t_1)e^{-j2\pi f_0\tau_0}e^{-j2\pi f_1\tau_1}d\tau_0\,d\tau_1.$$ (2.5)

Bello also defined the delay-Doppler correlation function:

$$R_{SS}(\tau_0, \tau_1, \nu_0, \nu_1) = \iint R_{hh}(\tau_0, \tau_1, t_0, t_1)e^{-j2\pi\nu_0 t_0}e^{-j2\pi\nu_1 t_1}\,dt_0\,dt_1,$$ (2.6)

which describes the correlation between the MPCs at delay-Doppler shift pairs (τ_0, ν_0) and (τ_1, ν_1).

Bello's last correlation function $R_{DD}(f_0, f_1, \nu_0, \nu_1) = F\{F\{R_{SS}\}\}$ (with respect to τ_0, τ_1) also equals $F\{F\{R_{HH}\}\}$ (with respect to t_0, t_1). R_{DD} is the correlation function of the output Doppler-spread function $H_O(f, \nu)$, but neither H_O nor R_{DD} is as widely used as the other functions.

2.2.4 Correlated/Uncorrelated Scattering and WSS

If MPCs at different delays are uncorrelated, the channel is an uncorrelated scattering (US) channel. In the US case, the CIR and CTF correlation functions become

$$R_{hh}(\tau_0, \tau_1, t_0, t_1) = R_{hh}(\tau_0, t_0, t_1)\delta(\tau_0 - \tau_1),$$ (2.7)

$$R_{HH}(f_0, f_1, t_0, t_1) = R_{HH}(\Delta f, t_0, t_1),$$ (2.8)

---

* For complex baseband functions, a factor of $1/2$ actually multiplies the expectation in the correlation functions, to account for the complex envelope scaling [7], but we ignore this factor here.

so R_{hh} is nonzero only when $\tau_0 = \tau_1$, and with $\Delta f = |f_0 - f_1|$, the CTF correlation function depends only upon the difference between f_0 and f_1. This condition on the CTF is termed WSS in frequency, so US(delay) $\Leftrightarrow$ WSS(frequency). With US, the delay-Doppler correlation function is $R_{SS}(\tau_0, v_0, v_1)\delta(\tau_0 - \tau_1)$.

A stochastic process is WSS if its mean is constant, and its autocorrelation depends only on the time difference and not on absolute times [16]. With $\Delta t = |t_0 - t_1|$, a WSS CIR means

$$R_{hh}(\tau_0, \tau_1, t_0, t_1) = R_{hh}(\tau_0, \tau_1, \Delta t). \tag{2.9}$$

The CTF autocorrelation becomes

$$R_{HH}(f_0, f_1, t_0, t_1) = R_{HH}(f_0, f_1, \Delta t) \tag{2.10}$$

and the delay-Doppler correlation function becomes

$$R_{SS}(\tau_0, \tau_1, v_0, v_1) = \delta(v_0 - v_1)P_S(\tau_0, \tau_1, v_0), \tag{2.11}$$

with $P_S(\tau_0, \tau_1, v_0)$ the delay-Doppler cross power spectral density, and Equation 2.11 means that the contributions to the CIR from different scatterers are uncorrelated if the scatterer Doppler shifts are different; in short, WSS(time) $\Leftrightarrow$ US(Doppler).

If the channel is both US(delay) and WSS(time), the channel is termed WSSUS, yielding

$$R_{hh}(\tau_0, \tau_1, t_0, t_1) = R_{hh}(\tau_0, \Delta t)\delta(\tau_0 - \tau_1), \tag{2.12}$$

$$R_{HH}(f_0, f_1, t_0, t_1) = R_{HH}(\Delta f, \Delta t), \tag{2.13}$$

$$R_{SS}(\tau_0, \tau_1, v_0, v_1) = \delta(\tau_0 - \tau_1)\delta(v_0 - v_1)P_S(\tau_0, v_0), \tag{2.14}$$

with R_{HH} termed the spaced-frequency, spaced-time (SFST) correlation function, and $P_S(\tau, v)$ the scattering function.

Setting $\Delta t = 0$, $R_{hh}(\tau, 0) = \psi_h(\tau)$, the power delay profile (PDP), equal to the channel's average power output versus delay, and $\mathbf{F}\{R_{hh}(\tau, 0)\} = \mathbf{R_{HH}}(\Delta f, 0) = \psi_{H_f}(\Delta f)$, the spaced-frequency correlation function, which quantifies the correlation between channel effects at frequencies separated by Δf. The width of $\psi_h(\tau)$, the multipath delay spread T_M, is reciprocally related to the width of $\psi_{H_f}(\Delta f)$, the coherence, or correlation bandwidth B_c.

Similarly, with $\Delta f = 0$ in R_{HH}, $R_{HH}(0, \Delta t) = \psi_{H_t}(\Delta t)$, the spaced-time correlation function. This quantifies the correlation between channel effects Δt apart in time. The scattering function $P(\tau, v) = \mathbf{F}\{\mathbf{F}\{R_{HH}(\Delta f, \Delta t)\}\}$, and $P_S(0, v) = P_S(v)$ is the Doppler spectrum, whose width is the Doppler spread f_D. Doppler spread is reciprocally related to the width of $\psi_{H_t}(\Delta t)$, which is the coherence, or correlation time t_c.

Summarizing these relationships, we have $T_M =$ width of $\psi_h(\tau) \sim 1/B_c$, and $f_D =$ width of $P_S(v) \sim 1/t_c$. Delay spread T_M quantifies the amount the channel spreads an input impulse in delay and, analogously, Doppler spread f_D quantifies the amount the channel spreads an input tone in frequency. The coherence bandwidth B_c measures the channel's frequency selectivity, and the coherence time t_c measures the channel's time selectivity (time rate of variation). The most common measure for T_M is the root-mean-square delay spread (RMS-DS)σ_τ.

Regarding stationarity, any real channel can be statistically stationary for only a limited time; hence, statistical wireless channel models for mobile channels can only be truly WSS [15]. A primary reason for the appeal and use of the WSS (and WSSUS) assumption(s) is their relative mathematical simplicity. Yet this simplicity has inherent limits. For example, cellular [17], LAN [18,19], metropolitan area networks [20,21], and vehicle-to-vehicle channel models [6] often incorporate nonstationarity (NS) without explicit models by providing distributions of channel delay spreads. Only a few NS models presently exist. Some examples are the COST 259 model [22], and the recent UMTS model [23], which employs a set of two distinct models that are "switched between" as in Ref. [24]. Additional NS models are contained in the author's work, for example, [5,25], and work on the general topic of NS modeling is growing [26]. In practice one would like to incorporate "just enough" nonstationarity to capture the most significant effects, without unduly complicating the model. NS models for effects at the network layer are also being developed [27].

The phenomenon of correlated scattering has been documented to a lesser degree, for example, [28]. In most cases, this phenomenon appears to be less detrimental to communication system performance than does NS.

2.3 Characterization Functions and OFDM/A Channel Models

Despite the fact that real wireless channels are NS if observed long enough, and also exhibit correlated scattering, sample PDPs and CIRs can of course still be measured, and delay spreads, coherence bandwidths, Doppler spreads, and coherence times can be estimated for environments of interest. Generally, we will not have full knowledge of the channel correlation functions R_{hh} or R_{HH}, or scattering function $P(\tau, \nu)$, but representative CIRs and CTFs can be analyzed statistically.

To completely specify the TDL channel model, we must first specify the number of taps (where a tap is a resolvable MPC). This number is obtainable from T_M (or B_c). We must also specify the taps' time rate of change (based upon f_D or t_c), and a statistical model for the tap amplitude random processes.* Last, we also require the relative energy of each tap from the PDP. Typically, the longer the delay, the weaker the MPC, so an exponentially decaying energy versus delay characteristic is most often employed, for example

$$E_h(k) = c_1 e^{-c_2 k} - c_3, \qquad (2.15)$$

for tap index $k = 1, 2, \ldots$, and the c's appropriate constants (see Figure 2.1) [5].

In this section, we describe the OFDM/A signal model, time- versus frequency-domain modeling, cyclic extensions, time variation and Doppler spread, and explicit relationships between the channel parameters and OFDM/A signaling parameters.

2.3.1 OFDM/A Signal Model

The OFDM signal (baseband) for the nth symbol can be expressed as

$$x_{un}(k) = \sum_{m=-M/2}^{M/2-1} X_{un}(m) e^{j2\pi km/M} \qquad (2.16)$$

where $x_{un}(k)$ is the kth transmitted time-domain sample of the uth user, and $X_{un}(m)$ is the transmitted frequency domain symbol of the uth user on the mth subcarrier, with $k = 0, 1, \ldots, M-1$,

---

* Tap phases are usually assumed to be uniformly distributed on $[0, 2\pi)$, except when there is a dominant component, in which case that tap phase is modeled as having a narrow distribution about its mean phase.

and M the number of subcarriers. Equation 2.16 represents the IFFT operation at the transmitter. The frequency domain symbols can be expressed by

$$X_{un}(m) = \begin{cases} d_{un}(m), & m \in I_{ud}, \\ p_{un}(m), & m \in I_{up}, \end{cases} \tag{2.17}$$

where I_{ud} and I_{up} are the uth user's index sets for data ($d_{un}(m)$) and pilot ($p_{un}(m)$) subcarriers, respectively. For OFDMA, symbols of the form of Equation 2.17 from multiple users are grouped to form the entire length-M vector of symbols that is used in the IFFT of Equation 2.16.

2.3.2 Time versus Frequency Domain Modeling

Since any signal can be analyzed in either the time or the frequency domain, the effects of channel fading on transmitted signals can be naturally studied in either domain. As noted, the wireless channel is generally modeled as a linear time-varying filter, completely described by either the CIR $h(\tau, t)$ or the CTF $H(f, t)$. The CIR is a time-domain function, and the CTF a frequency-domain function. Any channel model based on generating CIRs is viewed as a time-domain channel model. Likewise, any channel model based on generating CTFs is a frequency-domain channel model.

Historically, wireless communication systems have evolved from narrowband applications to wideband applications; therefore, wireless channel models have had to correspondingly change from narrowband to wideband. Different channel features are of interest for narrowband and wideband channel models. For narrowband models, the received power statistics are critical, whereas detailed time-varying multipath characteristics are more important for wideband channel models. Because of platform mobility and consequent Doppler effects, narrowband signals can see significant spectral broadening whereas wideband signals are proportionately less affected, but suffer much more severe frequency selectivity. Current and future OFDM/A systems will aim at providing high data rate transmissions and therefore our OFDM/A channel models belong to the wideband class.

As noted, the time-domain channel model $h(\tau, t)$ can be completely represented by the well-known TDL model of Equation 2.1 [29]. To facilitate the digital signal processing of a signal with fixed bandwidth ($W/2$ Hz at baseband), the CIR is modeled as a sampled version, so that $\tau_k = k/W$ is the delay for the kth path. As indicated by our tap persistence process, there may be some taps with zero energy for some time. In the popular "Stanford University Interim" (SUI) models [30] for example, three taps with different delays are modeled. Typically, in modeling we set the average channel energy to unity: $\sum_{k=1}^{L-1} E(|h_k(t)|^2) = 1$, where $E(.)$ denotes expectation. The kth MPC $h_k(t)$ with fixed delay $\tau_k = k/W$ is still usually the aggregate of many physical paths ("subpaths") with nearly the same delay (within the "delay bin"). For a WSSUS channel assumption, the MPCs are also assumed to be uncorrelated.

Assuming the channel MPCs are essentially time invariant over at least several symbols, the sampled frequency-domain model corresponding to Equation 2.2 is

$$\begin{aligned} H(f, t) &= \sum_{k=0}^{L-1} z_k(t)\alpha_k(t)e^{j2\pi f_{D,k}(t-\tau_k)}e^{-j2\pi f_c\tau_k}e^{-j2\pi f\tau_k} \\ &= \sum_{k=0}^{L-1} h_k(t)e^{-j2\pi f\tau_k} = \sum_{k=0}^{M-1} h_k(t)e^{-j2\pi fk/W} = \sum_{m=0}^{M-1} H(f_m, t), \end{aligned} \tag{2.18}$$

where we assume $h_k = 0$ for $k = L, L + 1, M - 1$ since M is usually much larger than L. For a system with sampling rate W, we are actually interested in a discrete frequency-domain model, obtained by the sampled CTF ($H(mW/M, t), m = 0, 1, \ldots, M - 1$). For notational simplicity, we drop the t parameter in the CIR and CTF, and express $h_k(t)$ and $H(mW/M, t)$ as h_k and H_m, respectively. The relationship between these two can be expressed as

$$H_m = \sum_{k=0}^{L-1} h_k e^{-j2\pi mk/M} = \sum_{k=0}^{M-1} h_k e^{-j2\pi mk/M}. \tag{2.19}$$

The above equation allows representation of the sampled CTF as a vector $\mathbf{H} = (H_0, H_1, \ldots, H_{M-1})^T$, the discrete Fourier transform (DFT) of the CIR vector $\mathbf{h} = (h_0, h_1, \ldots, h_{M-1})^T$, that is, $\mathbf{H} = \text{DFT}\{\mathbf{h}\}$. Note that the number of DFT points, M, determines the frequency resolution. To have a sufficient frequency resolution, M should be larger than some minimum value, determined by the actual system requirements; we address this in a subsequent section.

A frequency-domain model has to contain M independent or correlated time-varying variables. Whether analytical or based upon actual frequency-domain channel measurements, model complexity is proportional to channel bandwidth and frequency resolution. This is one reason why wideband channel measurements are often taken in the time domain, yet an equivalent frequency-domain model is readily available using Equation 2.19.

The prevalence of the statistical time-domain model has historical as well as technical reasons. Channel measurements and modeling have focused more on the time domain ever since the study of wireless narrowband channels began. Standards setting bodies also usually specify a time-domain model for system performance evaluation such as the SUI models [30] for nonmobile applications of the IEEE standard 802.16. This is not only because most engineers are more familiar with time-domain models, but also because they are easy to incorporate into time-domain-based analyses and simulations. For time-domain channel models with a very small number of taps, an equivalent frequency-domain channel model will require a large number of coefficients when M is large. Often though researchers are more interested in frequency-domain models when most of the signal processing occurs in the frequency domain, and this is especially true for OFDM/A systems. With a frequency-domain channel model, the frequency response of each resolvable frequency/subcarrier can be easily controlled.

With a frequency-domain channel model, the time-domain convolution of the transmitted signal with the CIR can be simplified to a frequency-domain multiplication of the transmitted signal on each subcarrier with a multiplicative channel coefficient, and the resulting received frequency-domain vector $\mathbf{R}$ is

$$\mathbf{R} = H\mathbf{X} + \mathbf{W}, \tag{2.20}$$

where H is a diagonal matrix, with elements from the discrete CTF, $H = \text{diag}(\mathbf{H})$, vector $\mathbf{X} = (X_0, X_1, \ldots, X_{M-1})^T$ is the transmitted data symbol vector in one OFDM symbol, and $\mathbf{W}$ is the noise vector for each of the M subcarriers.

A convenient property of the DFT is that multiplication in the frequency domain corresponds to circular convolution in the time domain. To use Equation 2.20, a circular convolution has to be constructed, while only a linear convolution obtains when a signal passes through a linear channel [12]. This circular convolution can be constructed at the transmitter by using a cyclic extension (cyclic prefix (CP), cyclic suffix, or both), while still retaining the useful signal part in the OFDM symbol.

2.3.3 Cyclic Extensions

Guard intervals as shown in Figure 2.3 are reserved to avoid ISI between adjacent OFDM symbols. No energy is transmitted in such guard intervals. As long as the guard interval duration is longer than the channel delay spread T_M, the delayed multipath energy from the previous OFDM symbol will not "smear into" the current symbol and thus ISI can be completely avoided. The use of this guard interval will decrease the bandwidth efficiency.

One significant reason for the popularity of OFDM is the use of the FFT, which has low computational complexity. Unlike the general Fourier transform, the FFT (or DFT) of the circular convolution ($\otimes$) of two signals equals the product of their individual FFTs (or DFTs), as in

$$\text{FFT}\{x_k \otimes h_k\} = \text{FFT}\{x_k\} \times \text{FFT}\{h_k\}. \tag{2.21}$$

However, when signals go through a wireless channel, the received signal is the linear convolution ($*$) ($F\{x_k * h_k\} \neq F\{x_k\} \times F\{h_k\}$) of the transmitted signal with the channel filter. To use the FFT in OFDM systems, a circular convolution has to be constructed. Thus the cyclic extension is introduced to replace the guard intervals. Three different types of cyclic extension are illustrated in Figure 2.3. The extension copies each OFDM symbol's ending part of duration T_{pre} and places it before the OFDM symbol; the cyclic postfix extension copies each OFDM symbol's beginning part of duration T_{post} and places it after the OFDM symbol, while the CP *and* postfix extension does both.

Since the duplicated signal part is actually transmitted during the cyclic extension period, the use of any cyclic extension, as with a guard interval, decreases spectrum efficiency. It is also not power efficient since it consumes part of the transmitted energy. Nonetheless use of a cyclic extension provides more benefits than drawbacks. The cyclic extension can also be used for signal detection and symbol synchronization since there is a fixed delay between the cyclic extension and the same replica in the useful part of the OFDM symbol. The cyclic extension works the same as a guard interval to avoid ISI. When the cyclic extension is larger than T_M, it provides some timing synchronization margin, which relaxes timing synchronization requirements. If, however, the cyclic extension is shorter than T_M, ISI cannot be avoided. An example of this is shown in Figure 2.4, where two adjacent OFDM symbols are shown: one OFDM symbol consists of five samples and the CP, only

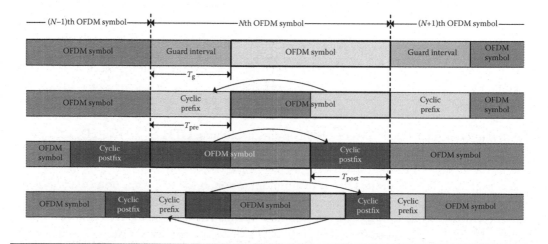

Figure 2.3   OFDM signal with guard interval and different cyclic extensions.

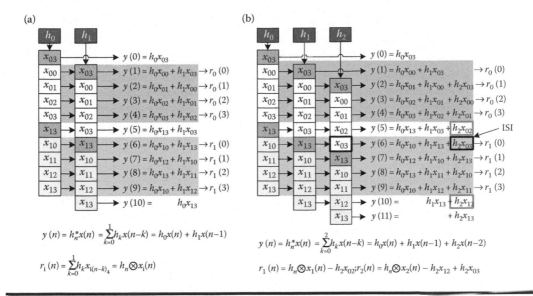

Figure 2.4   Example OFDM symbols with CP: (a) sufficient CP; (b) insufficient CP.

one sample, equals 1/4 of the useful signal part. For a two-tap channel, the useful part of the received time-domain signal can be written as the circular convolution of the transmitted signal with the CIR. Therefore the transmitted signals on different subcarriers are weighted by a multiplier, which is only determined by the channel. There is no ISI or ICI. For the three-tap channel in Figure 2.4, the cyclic extension is shorter than the channel delay spread. The third MPC introduces ISI to the next OFDM symbol and the received signal no longer has the cyclic convolution format.

2.3.4 Modeling Frequency and Time Offsets

A frequency offset (or error) f_e between the Tx and Rx can result from imperfect frequency tracking at the receiver, frequency drifts of oscillators at either or both Tx and Rx, or from a Doppler shift caused by the channel. It is not essential to model frequency offset explicitly in the channel model, but since it can have a significant impact on OFDM system performance, it should be considered in OFDM system design.

In the previous section, although the discrete CTF vector is modeled as the DFT of the CIR vector assuming the channel is time invariant, real channels will change with time, and this is particularly true when mobile velocity is large. The time-varying channel effect can be modeled in a general form as Ref. [31]

$$r(t) = \zeta(t)x(t), \tag{2.22}$$

where $x(t)$ is the transmitted signal in the time domain, $r(t)$ is the received signal, and $\zeta(t)$ is the time-varying channel effect. When $\zeta(t)$ is not constant, it will generally introduce inter-channel (inter-subcarrier) interference (ICI).

The frequency offset can be modeled as a deterministic or random process. Within several OFDM symbols it can often be modeled as a constant, and thus $\zeta(t)$ will be a deterministic function

$$\zeta(t) = e^{j2\pi f_e t}. \tag{2.23}$$

This frequency offset effect can also be incorporated into the OFDM channel vectors, **h** and **H**. The elements of **h** will have a linear phase rotation, that is, $h_k \rightarrow h_k e^{j(2\pi f_e k/W + \phi_0)}$ Here ϕ_0 is an initial phase, which is the same for all samples in one OFDM symbol. If the channel is carefully modeled, the effect of ϕ_0 can be included in the h_k coefficients, but this also means that **h** has to be updated each OFDM symbol. Thus the corresponding CTF vector will be changed as

$$H'_m = \sum_{k=1}^{M-1} h_k \, e^{j2\pi f_e k/W} \, e^{-j2\pi nk/M} = \sum_{k=0}^{M-1} h_k \, e^{-j2\pi(m-n)k/M} \, e^{j2\pi \mu k/M}. \qquad (2.24)$$

The frequency offset can be expressed as $f_e = (n + \mu)\Delta f = (n + \mu)W/M$, where Δf is the frequency resolution or the subcarrier spacing, n is an integer and μ is a real number with $|\mu| < 0.5$. When $\mu = 0$, the frequency offset is $f_e = n\Delta f$, $H'_m = H_{m-n}$, which means the elements in **H** will be cyclically shifted. When $\mu \neq 0$, **H** will be completely changed. For both cases, **H** has to be updated for each OFDM symbol because of the different initial phase. Whether in the time or frequency domains, the effect of a frequency offset on the channel model is to complicate the model. Thus frequency offset effects, although treatable as part of channel effects, are typically modeled separately (e.g., at either Tx or Rx) and often in the time domain.

Note that the two vectors **h** and **H** are assumed constant within one OFDM symbol. By introducing the frequency offset effect, some time variation is actually introduced. This is easy to accommodate in system simulations. Therefore frequency offsets of a fixed value are usually separately modeled, and, once estimated, they can be easily removed. Practical OFDM systems usually have dedicated processing modules to estimate frequency offset and remove its effect in system synchronization.

The frequency offset f_e can also be modeled as a random process, which takes values in some range $[-f_L, f_L]$ with a certain distribution, for example, uniform or Gaussian. The offset f_e is not likely to change rapidly and can typically be updated only once every several frames. Usually f_e is assumed to take its maximum value when evaluating its effect on performance.

A frequency offset distorts subcarrier orthogonality and causes ICI, which can be very detrimental to OFDM system performance. The ICI energy is proportional to the square of the ratio of frequency offset to subcarrier bandwidth [32]. Therefore to reduce ICI, the subcarrier bandwidth should be as large as possible, or equivalently for a fixed system bandwidth the number of subcarriers (M) should be chosen as small as possible. Note that this must be a compromise with the desire for small subcarrier bandwidth to ensure flat fading. Other techniques are also used to further reduce ICI, such as time-domain Gaussian windowing, which can trade ISI with ICI by a single parameter adjustment [32].

Timing offset is caused in the timing synchronization process when the optimal timing instant (t) is not employed and a timing error t_e relative to this optimum position moves the timing window back ($t_e < 0$) or forth ($t_e > 0$). A received OFDM symbol (of duration T_s) as in

$$x'(t) = x(t + t_e), \quad 0 \leq t \leq T_s \qquad (2.25)$$

can include part of the energy from the prior or subsequent OFDM symbol if there is no guard interval, which means ISI will occur. Generally, a pure delay introduces a phase shift to the desired signal and also some additive inference, which depends on the OFDM frame structure. Different effects for systems with a null interval, and with or without a cyclic extension, are shown in Ref. [31].

Since there is usually a cyclic extension between OFDM symbols, the timing accuracy can be relaxed. If the timing offset is within the timing synchronization margin [12], that is, $t_e \leq T_M - T_G$, where T_G is the CP duration and T_M is the maximum channel delay spread, there is only a fixed phase shift on different subcarriers, which can be taken as part of the channel effect and removed with equalization. If the timing offset is beyond this timing synchronization margin, not only is useful OFDM signal energy lost, but the additive effect of ISI is introduced. Thus the signal-to-noise ratio (SNR) is decreased according to the following approximation [12]:

$$\Delta \mathrm{SNR}(t_e) \approx -2\left(\frac{t_e}{T_s}\right)^2 = -2\left(\Delta f\, t_e\right)^2. \tag{2.26}$$

A longer OFDM symbol duration or a narrower subcarrier bandwidth is more robust to the timing offset, which is opposite to the guideline for frequency offsets. So to avoid SNR degradation and conflicts with the frequency offset requirement, a sufficient cyclic extension is employed.

2.3.5 Time Variation and Doppler Spreading

The Doppler effect causes a change in frequency of the received signal due to the relative motion between Tx and Rx. A Doppler frequency shift observed at the receiver can be expressed as $f_d = v\cos(\theta)/\lambda$, where λ is the wavelength of the received signal, v is the magnitude of the relative velocity vector between the receiver and transmitter, and θ is the angle of the incoming wave relative to the direction of the velocity vector. Note that the phase shift of individual incoming waves over time can be expressed as a fixed frequency offset of f_d Hz, and for an OFDM system this can be modeled in both time and frequency domains using Equations 2.23 and 2.24. To obtain an estimate of the Doppler spread of a radio channel, we can simply transmit a sinusoid at a desired frequency and Fourier transform the channel output signal to generate a Doppler spectrum estimate. This requires of course that our measurement equipment samples the received signal fast enough to capture the highest Doppler shift. Alternatively, if available, we can simply Fourier transform the SFST correlation function with $\Delta t = 0$.

In a typical multipath environment, the received signal travels over many reflected/scattered paths, each with a different Doppler shift. The resulting received spectrum is spread, and the Doppler spread f_D quantifies the total amount by which the channel spreads a transmitted tone in frequency. This Doppler spectrum is $P_S(v)$ and $f_D \sim 1/t_c$ is its width.

If all the MPCs arrive independently with the same amplitude and uniformly distributed phases and angles of arrival (isotropic scattering), the classic Clarke model [8] applies, with a "U-shaped" Doppler spectrum,

$$D(f) = \frac{1}{2\pi f_D}\left[1 - \left(\frac{f}{f_D}\right)^2\right]^{-1/2}, \quad |f| \leq |f_D|. \tag{2.27}$$

In many practical cases a more conventional lowpass response better describes the actual Doppler spectrum, for example, a Gaussian [34] or "rounded" spectrum [30]. Generally, the shape of the Doppler spectrum depends on the specific channel environment, but the maximum Doppler shift f_D is determined by the largest velocity.

In OFDM systems, if we assume flat fading for any subcarrier signal and the delay of this signal is approximately constant, then for any subcarrier signal $x(t)$, the received signal is $r(t) = h_k(t)$

$\delta(t - \tau_k) * x(t) = h_k(t)x(t - \tau_k)$; thus the subcarrier experiences multiplicative fading. Correspondingly, in the frequency domain we have $R(f) = H_k(f) * X(f)$ (ignoring the exponential phase term that comes from the delay). Convolution in the frequency domain hence implies spreading of the subcarrier spectrum. Thus the time-varying impairments can be modeled as a multiplicative distortion $\zeta(t)$ [31], where $\zeta(t)$ is equal to $|h_k(t)|$, and this can often be modeled as a zero-mean WSS stochastic process with spectral density given by the Doppler spectrum. It has been shown [31,35] that the ICI power resulting from Doppler spread can be upper bounded as follows:

$$P_{\text{ICI}} \leq \frac{\kappa}{12}(2\pi f_D T_s)^2 \leq \frac{1}{12}(2\pi f_D T_s)^2, \tag{2.28}$$

where κ is a constant depending on the Doppler spectrum shape.

If $f_D T_s$ is very small, that is, $f_D \ll B_s$, the ICI due to Doppler spread will be negligible. However, as $f_D T_s$ becomes larger (even ~ 0.01), the orthogonality between subcarriers is compromised, leading to performance degradation. A large Doppler spread also makes the channel parameters vary rapidly from one block to another, making channel estimation and synchronization more challenging.

Through the inverse relationship between Doppler spread and coherence time, we can consider the Doppler spread to be the typical fading rate of the channel. By accurately modeling the time variation in the time-domain model [i.e., in $h_k(t)$] we will naturally induce Doppler spreading, similarly in the frequency-domain model [i.e., in $H_m(t)$]. The filtered additive Gaussian noise (AGN) method [8] or the sum of sinusoids method [34] can be used to generate such time variation in simulations.

2.3.6 *Channel Parameters and OFDM/A Signal Parameters*

One of the main advantages to employing a multicarrier scheme instead of a single carrier scheme is the ability of the multicarrier technique to circumvent multipath delay spread, and convert a time dispersive channel into a set of nondispersive channels, one for each subcarrier. A channel is said to be dispersive if the delay spread (typically σ_τ) is on the order of the symbol duration or larger. Thus to ensure frequency nonselective fading on each of the M subcarriers, the symbol duration T_s on each subcarrier should be much larger than the delay spread, or $T_s \gg \sigma_\tau$. Via the serial-to-parallel conversion of the high rate data sequence at rate R symbols/second to M lower-rate sequences of rate R/M, the original symbol duration $T = 1/R$ is increased to $T_s = M/R$. Hence knowledge of the channel delay spread is critical to selection of M given the desired data rate R. For channels with a range of delay spreads, one can conservatively use the maximum value. The condition $T_s \gg \sigma_\tau$ also implies that the channel coherence bandwidth B_c must be greater than the subcarrier bandwidth B_s; if $T_s \gg \sigma_\tau$, then by the reciprocal relation between T_s and B_c, we have $B_c \gg B_s$.

As noted, most OFDM/A systems employ a CP to prevent channel dispersion across blocks of M symbols. To do this, the CP length J (samples) should also be larger than the delay spread. Thus after the IFFT in the transmitter, the last J ($< M$) samples of the block are added to the beginning of the sequence to be transmitted, and the resulting "total symbol time" is $T_T = T_s + J/R = T(M + J) = T_s + T_G$—the CP duration JT is sometimes called a guard time (T_G).

There is a tradeoff between the CP duration and spectral efficiency, since for a given OFDM symbol time T_s, a longer CP means a smaller portion of T_T is allocated to user data. In the 802.16 standard for Example [4], the CP is allowed to be one of four values, given as fractions of T_s: 1/32, 1/16, 1/8, and 1/4. Different FFT sizes can have the same actual CP duration, but different values of J.

In addition to time delay dispersion, the rate of channel time variation should also be known and used in OFDM/A design. For this we typically employ the coherence time t_c. For coherent detection, one must have $t_c > T_T$, and for so-called "slow fading" we typically have $t_c \gg T_T$ (also $t_c \gg T_s$). In the frequency domain, this also means the subcarrier bandwidth should be much larger than the Doppler spread ($1/T_s = \Delta f \gg f_D = 1/t_c$). The slow fading condition is also important in selecting the frame duration T_f, where a frame consists of a block of N consecutive OFDM/A symbols; hence $T_f = NT_T$. For accurate channel estimation, we typically desire the coherence time to be longer than a frame so that channel estimates based upon pilot symbols are accurate over the entire frame.

In summary, the main rules are as follows:

- $T_T \gg \sigma_\tau$ to circumvent channel dispersion and get flat fading on each subcarrier
- $T_G > \sigma_\tau$ to circumvent inter-block ISI
- $t_c > T_T$ to enable coherent detection
- $t_c > T_f$ to yield slow fading for accurate channel estimation

Note that for a given channel and pilot symbol pattern, the third rule requires $T_T = T_s + T_G$ to be small, but the first requires T_T to be large, so there is a compromise between spectral efficiency and channel estimation accuracy.

2.4 Detailed OFDM/A Channel Models

In this section we describe a common form for OFDM/A channel models we denote a canonical model, provide examples of more realistic models and how to incorporate channel nonstationarity, and then extend these models to MIMO and multiband (MB) cases. The section concludes with a brief discussion on channel estimation.

2.4.1 A Canonical OFDM/A Channel Model

Based upon the fact that with a large number of subcarriers, the per-subcarrier bandwidth is generally small, the flat fading approximation for each subcarrier will typically be very good. With this approximation, in most terrestrial environments where the number of MPCs will be large in the nonopen areas in which receivers are located, the Central Limit Theorem approach that yields Rayleigh fading on any individual subcarrier is generally a very good one. Thus in terms of Equation 2.1, each $|h_k(t)|$ is well modeled as a Rayleigh random process. If a dominant LOS component is present, the Ricean model is appropriate. As an example, the initial 802.16e OFDMA "profile" employs 512 subcarriers and a symbol duration of just over 100 μs in a bandwidth of 5 MHz [12]. This yields a subcarrier bandwidth $B_s \cong 9.8$ kHz, for which the flat fading assumption holds for delay spreads up to approximately 20 μs when we use the relation $B_c \sim 1/(5\sigma_\tau)$ [36]. Delay spreads of this magnitude are quite rare [37]; hence the flat fading assumption is very good. If the CIR Rayleigh/Ricean MPCs are assumed uncorrelated, then the channel amplitude coefficients in the frequency domain ($|H(f_n, t)|$) can also be shown to be Rayleigh/Ricean.

For most OFDM/A symbol rates of current interest, this Rayleigh or Ricean fading will be slow for a substantial range of terrestrial velocities. Using again the initial 802.16e OFDMA "profile," for a carrier frequency of 3.5 GHz, and a mobile velocity of 14 m/s ($\sim$ 31 miles/h), the maximum Doppler spread will be $f_D \cong 163$ Hz, yielding a coherence time of $t_c \cong 6.1$ ms. Hence $t_c/T_T \cong$ 60, clearly enough for coherent detection, and slow enough for "slow fading" over packets of at least a few tens of symbols. For smaller velocities and/or smaller carrier frequencies, fading is even slower.

When the coherence bandwidth is much larger than the subcarrier bandwidth, adjacent subcarriers will be correlated. The amount of correlation can be quantified by the SFST correlation function $R_{HH}(\Delta f, 0)$, or in practice by estimates of this function based upon measurements, using the technique in Ref. [38]. A reasonable estimate of correlation can be obtained by assuming that the correlation decays linearly with frequency separation Δf from a value of 1 at $\Delta f = 0$ to a value of 0.5 at approximately $\Delta f = 0.5/\sigma_\tau$ [39], that is, correlation $\rho(\Delta f) \cong 1 - \sigma_\tau \Delta f$.

Finally, in the case of very slow mobile velocities, Doppler spreading on each subcarrier will be negligible. For moderate velocities, as in the 802.16e example here, where $f_D/B_s \cong 0.015$, orthogonality between subcarriers will be affected, but for this small amount of Doppler spreading, it can be shown that the ISI that results will be tens of dB below the desired signal energy [32].

In summary, a canonical time-domain OFDM/A channel model for built-up terrestrial environments can employ the following conditions:

- Rayleigh/Ricean amplitude coefficients ($|h_k(t)|$)
- Number of taps (L) equal to $\lceil \sigma_\tau/T_{samples} \rceil = \lceil M\sigma_\tau/T_s \rceil$
- Uniform tap phases [$\phi_k(t)$]
- An exponential PDP, with decay constant determined by environment
- Slow fading on each tap based upon the ratio of coherence time to OFDM symbol time
- Correlated amplitude fading on subcarriers with correlation approximated by $\rho(\Delta f) \cong 1 - \sigma_\tau \Delta f$
- (Resulting) flat fading on each subcarrier.

2.4.2 Practical OFDM/A Channel Models

Due to the wide variety of possible operational environments of OFDM/A systems, it is not possible to provide channel models for all settings. Here we describe some representative models for several environments of current interest.

First we describe the often-cited SUI models, for fixed wireless applications with link distances less than 7 km, and base station and receiver antenna heights of 30 and 6 m, respectively. We do not repeat these models in full here but show three representative examples. Worth noting is that for fixed applications, directional antennas can be used to limit received multipath energy from wide angles, but for brevity we list only the omni-directional models here. Table 2.5 shows three models, all of which contain only three MPCs.

Table 2.5   Representative SUI Models for Fixed Wireless OFDMA Application

Model	Tap Delays (μs)	Tap Relative Powers (dB)	Ricean k-Factors (Linear)	Terrain Type
SUI-1	0,0.4,0.9	0,−15,−20	4,0,0	C
SUI-3	0,0.4,0.9	0,−5,−10	1,0,0	B
SUI-5	0,4,10	0,−5,−10	0,0,0	A

Source: Institute of Electrical and Electronics Engineers, IEEE Broadband Wireless Access Working Group document IEEE 802.163c-01/29 Channel Models for Fixed Wireless Applications, July 2001, http://wirelessman.org/tg3/contrib/802163c-01_29r4.pdf

The SUI terrain types were taken from Ref. [40], with type "C" denoting flat terrain with light tree density, type "A" denoting hilly terrain with moderate to heavy tree density, and type "B" denoting either flat terrain with moderate to heavy tree density or hilly terrain with light tree density. The Ricean k-factor of zero denotes Rayleigh fading, and all the k-factors are specified to be at or above the values listed with probability 0.9. For all these models, the Doppler spreads are less than 2.5 Hz—given that the models pertain to nonmobile channels, Doppler effects presumably emanate from motion of scatterers in the environment. Clearly these are simplified models, easily generated in computer simulations and easily translated to frequency-domain models.

A frequency-domain model for one of these channels could be readily constructed. For example, suppose we choose the SUI-5 model, and a channel bandwidth of $W = 5$ MHz. This means our quantization in delay is $1/W = 0.2$ μs. The three taps, with normalized energies $[0.7061, 0.2233, 0.0706]$, would then be used to construct a vector (of length 49 to span the 10 μs delay) with energies $E_h = [0.7061, \underline{0}_{18}, 0.2233, \underline{0}_{28}, 0.0706]$, with $\underline{0}_k$ denoting a vector of k zeros. Given the desired value of M (e.g., 512), we then generate three complex Gaussian samples with energies as specified by E_h, and take a length-M FFT of the E_h-length vector (padded with zeros), to obtain the M frequency-domain coefficients for one OFDM symbol. Since temporal correlation is most easily constructed in the time domain, we would typically generate correlated samples in time for each of the taps, and FFT each vector to obtain the frequency-domain coefficients, which will also be correlated in time. If US does not apply, correlation among taps can also be created by using correlated complex Gaussian variables for the taps.

In contrast to the simplicity of the SUI models, the 3GPP channel models [41] are complex [12]—given this complexity, we do not reproduce these models here. Allowing for MIMO operation, the 3GPP model CIR is defined in terms of a sequence of impulses, each consisting of multiple "subpaths," where subpaths are unresolvable, and similar to the "cluster" concept made popular in indoor channel modeling in Ref. [42]. The PDP is of exponential form, with the number of MPCs from 1 to 20, depending on environment. Each subpath has a random phase and a gain that is a function of the Tx and Rx antenna gains, distance, and the angles of arrival and departure from the Tx and Rx, respectively. A lognormal shadowing parameter is also included for each MPC.

Another comprehensive set of models is that of the WINNER project [43]. Like the 3GPP models, the WINNER models are very detailed, and are "ray-based" models capable of being used for MIMO applications. The models, which apply to both indoor and outdoor settings, are based on analyses and data from the existing literature, with model parameters obtained from a number of measurement efforts. These models also incorporate lognormal shadowing and, in addition, account for correlations among several large-scale parameters such as delay spread and shadowing standard deviation. The PDPs employ an exponential shape with decay factors (parameter c_2 in Equation 2.15) ranging from approximately $0.3/\sigma_\tau$ to $0.6/\sigma_\tau$ in suburban and urban areas. Urban models have 20 MPCs, and all models use a cluster of 10 "subpaths" per MPC. All MPCs are assumed to be Rayleigh in the urban models.

In some instances, fading can be more severe than Rayleigh. This result has been ascribed to multiple physical effects, including multiple scattering, too few "subpaths" per cluster to invoke the Central Limit Theorem, and rapid channel transitions wherein average received power changes abruptly and frequently [44]. The severe fading phenomenon has been noted for decades, but its occurrence is rarer than the Rayleigh case. For highly accurate models though, even these rare cases need to be modeled [45]. We present one model from our own work that exhibits severe fading in Table 2.6 [46]. This eight-tap model for a channel bandwidth of 5 MHz (tap spacing 100 ns) is based upon measurements in the 5 GHz band, in a campus environment with base antenna height approximately 20 m and receiver antenna height 1.5 m. The Weibull probability density function

[16] used to model amplitude fading is roughly analogous to the Ricean, with the Weibull "shape factor" b analogous to the Ricean k-factor. For $b = 2$, the Weibull is the Rayleigh density, and for $b < 2$, fading is more severe than Rayleigh. These results pertain to pedestrian velocities (~ 2 m/s or less), yielding maximum Doppler spreads of approximately 33 Hz. We address the last three columns of Table 2.6 in the next section.

2.4.3 Nonstationary Channel Modeling

As noted in Section 2.2, real wireless channels can only be statistically stationary for some limited time. In addition to the NS models cited previously, several indoor NS models have been proposed [47,48]. These models (indoor and outdoor) often employ Markov chains to emulate NS effects. This is embodied in the CIR of Equation 2.1 by the "persistence process" $z_k(t)$, where $z_k(t)$ takes binary values zero or one, with one representing the presence of the MPC and zero its absence. As noted in Ref. [49], this persistence can be viewed as a "medium-scale" or "mesoscale" [50] fading phenomenon with fading rate slower than the small-scale multipath fading but faster than typical shadowing. The persistence process can be likened to a birth/death process [16]; some standards, for example, the 3GPP, even incorporate such channel processes into their testing specifications [51].

First-order homogeneous Markov chains are the simplest to implement, and to specify these we need the steady state probability vector S and the transition probability matrix P. For our two-state persistence process, these are

$$P = \begin{bmatrix} P_{00} & P_{01} \\ P_{10} & P_{11} \end{bmatrix}, \quad S = \begin{bmatrix} P_0 \\ P_1 \end{bmatrix} \tag{2.29}$$

where P_{ij} denotes the probability of going from state i to state j, and P_i denotes the long-term probability of being in state i. As noted, state "0" denotes MPC "off," and state "1" denotes MPC "on." We also have $P_0 = 1 - P_1$; $P_{01} = 1 - P_{00}$; $P_{10} = 1 - P_{11}$. Table 2.6 shows these probabilities for the campus channel model. As expected, the longer-delay, weaker taps are "on" for the smallest

Table 2.6   Campus Area NLOS Channel Model Parameters, 5 MHz Bandwidth

Tap Index	Weibull Shape Factors (b)	Tap Energy	P_1	P_{10}	P_{00}
1	2.68	0.559	1	0	NA
2	1.7	0.123	0.963	0.036	0.051
3	1.5	0.089	0.909	0.085	0.145
4	1.5	0.061	0.853	0.130	0.243
5	1.5	0.051	0.805	0.169	0.299
6	1.5	0.044	0.778	0.197	0.309
7	1.5	0.038	0.750	0.228	0.312
8	1.5	0.035	0.722	0.255	0.335

Source: G. Pai et al., *Proc. 16th Virginia Tech Symp. on Wireless Pers. Comm.*, Blacksburg, VA, June 7-9, 2006.

fraction of time. Translating this persistence into a frequency-domain model is easy enough via FFT, but its frequency-domain interpretation, and accurate specification of the rate of this medium-scale effect, are subjects for future work.

Naturally, employing NS processes in the channel model complicates the modeling process. This is at the expense of greater modeling accuracy. For the most accurate communication system performance evaluation, we have found that employing MPC persistence does indeed matter. As an example, for 802.16e OFDMA systems in vehicle-to-vehicle channels, not incorporating this effect can yield BER estimates that are optimistic by 1/2 to 1 order of magnitude [52]. One can also incorporate NS at a slower time scale to model the change of environment over time (e.g., from urban to suburban) [10].

2.4.4 Extensions to MIMO and MB Models

The use of MIMO techniques for communication system performance enhancement has received much attention in the literature [53,54]. These techniques can increase link capacity, increase data throughput, and improve BER performance [55]. To maximize these improvements, the MIMO channel should ideally be a set of parallel, independent channels among the set of N_T transmit and N_R receive antennas. MIMO options exist in most OFDM/A standards, including some of the IEEE 802.11 and 802.16 standards.

In practice, independence—or at least uncorrelatedness—among the $N_T \cdot N_R$ pairwise channels is difficult to attain unless separation between antenna elements is large (e.g., more than a few wavelengths). The MIMO channel model can be represented by a matrix extension of the CIR in Equation 2.1, where $h_{ij}(\tau, t)$ represents the CIR between Tx antenna $i \in \{1, 2, \ldots, N_T\}$ and Rx antenna $j \in \{1, 2, \ldots, N_R\}$. In the frequency domain we have $H_{ij}(f_n, t)$.

In most academic works addressing OFDM, for example, [56], each subcarrier frequency response $H_{ij}(f_n, t)$ is assumed to have Rayleigh (or Ricean) fading, and for simplicity these responses are often assumed to be uncorrelated. This flat Rayleigh assumption is also true for some established standards such as [57] for 802.11 systems. This 802.11 standard does allow for correlation among the different antennas, using various models based upon results for the power angular spread from Ref. [58]. As with the 3GPP and WINNER models, the CIRs contain several multipath "clusters."

In OFDMA systems, the signal can be easily controlled to transmit on selected subcarriers. This high degree of scalability not only allows such a system to use only part of the available bandwidth, but also allows OFDMA to work in separate bands simultaneously, that is, to work in a MB channel. A MB channel consists of a number of dis-contiguous frequency bands, with each of these bands, termed a subband, having a moderate bandwidth. The subband bandwidths might be between 1 and 20 MHz, for example. When the center frequencies of the subbands are close enough, the MB channel is essentially a traditional wideband channel, with the only difference being that the entire available bandwidth is not used. There is no clear definition on how far apart these subbands must be before being termed an MB channel, but in principle the channel is not MB if the subbands are not far enough apart to have distinctly different effects on transmitted signals.

In Ref. [43], the so-called MB, OFDMA system employs multiple disjoint subbands within a total bandwidth range of 100 MHz. This kind of MB channel can still use the traditional wideband channel model. But for other MB channels with subbands far separated from each other, the different subbands have to be modeled separately. The first factor that must be considered is the large-scale or medium-scale fading difference between subbands. These fading differences must be treated in link budget analysis by treating the different subbands separately, possibly using different transmitted power levels or different antenna gains. For simplicity, these large-scale and medium-scale fading differences can be modeled together, either deterministically or randomly. In Ref. [59], a constant fading difference was modeled for a two-band MB channel where the carrier frequencies differ by a factor of approximately 5.5. Different receiver qualities can also be modeled, for example, a 3-dB noise figure difference is incorporated in Ref. [59]. For small-scale fading, a different fading rate has to be used for each subband. For coherent detection, this requires the channel estimation module to update channel information at different rates in the two bands. Finally, we may also need to use different correlation values to model the correlations in frequency across subcarriers in the multiple bands.

2.4.5 Channel Estimation

Channel estimation in OFDM/A is almost always done using some known (pilot) symbols, and what is typically estimated is the frequency-domain coefficient, one for each subcarrier, that is, the elements of the discrete CTF **H** in Equation 2.20. The pilot symbols are distributed in both time and frequency among the N symbols and M subcarriers of each frame. Most standards, for example, [4] have explicit formats for these distributions. The pilot symbols should be dense enough in the frequency domain to enable good estimates of the subcarrier channel coefficients between those subcarriers containing pilots, and the pilot symbols should be frequent enough in the time domain to ensure that multiple pilots are received within the channel coherence time. In other words, we would like to "sample" with pilot symbols at the Nyquist rate [7] in both the time and frequency domains.

Given the pilot symbols, there are multiple methods for estimating the (frequency domain) channel coefficients, and these techniques are similar to those used in single-carrier systems [60]. First the receiver must extract the channel-corrupted pilot symbols and then perform some filtering. The channel-corrupted pilot symbols are given by $\mathbf{R} = H\mathbf{X} + \mathbf{W}$ in Equation 2.20 hence to find the channel estimate for the kth subcarrier, we typically divide by the known pilot symbol X_k: $\hat{H}_k = R_k/X_k = H_k + W_k/X_k$. Two-dimensional Wiener filtering [61] of these estimates is optimal for estimating the channel coefficients on the data subcarriers, but this is significantly more complex than separate time- and frequency-domain techniques such as averaging over the pilot symbols (in either domain) or interpolation (in either domain). Often the minimum mean-square error (MMSE) is used to judge the quality of channel estimation methods (analytically or in simulations) [24], and MMSE filtering techniques can also be used to update channel estimates; some details on several of these approaches appear in Ref. [31].

As with equalization, channel estimation can also be done in a decision-directed mode, where estimates of actual (unknown) data symbols are used in the same way as the known pilot symbols. This can yield a larger number of "effective" pilot symbols per frame, at the expense of a larger estimation error (and consequent BER) when the detected symbol sequence contains errors. Matrix techniques for MIMO channel estimation are also addressed in Ref. [31].

Finally, channel estimates at an Rx must be conveyed back to the sending Tx, and this is often done using one or more subcarriers of an OFDMA signal. The channel model for this "return" channel takes the exact same form as that of the "direct" channel, but must also account for the inherent delay in estimation. For time-division duplexing schemes, which usually imply short link distances, the "return" channel is in the exact same frequency band, whereas for frequency-division duplexing schemes, the return channel is in a different band, and hence is often uncorrelated with the "direct" channel. Again though, the "return" channel model takes the same form as described in this chapter.

2.4.6 Open Issues

We have noted several challenging areas in OFDM/A channel modeling, including modeling NS sufficiently but without increasing model complexity significantly, and tracking rapid time variations. Nonstationary channels mostly arise when Tx and/or Rx and/or scatterers are fast moving, for example, V2V cases on highways or when aeronautical platforms are used. For any applications involving safety, reliable communication is critical, and accurate channel models are vital to system design. Modeling of nonstationary effects is also required for slower variations that take place as a mobile unit moves between regions with different characteristics (e.g., from urban to suburban). These are "large-scale" or "long-term" NS effects that are more easily modeled, but are nonetheless critical for seamless communications over long terms and/or wide areas.

Other issues deserving future work are efficient models for MIMO and MB channels, and estimation schemes for these channels. In the MIMO case, with more than a few antennas and a large number of subcarriers, channel model complexity can be significant; hence any means to simplify models are important for researchers and system developers. The final section of Ref. [62] provides a good summary of MIMO open research issues.

Our treatment limited the channel bandwidth to 20 MHz, so it is worth noting that a wider bandwidth channel model may require some differences in modeling to account for a large frequency range. One obvious issue here is the additional complexity of the time-domain model: as bandwidth increases, the number of taps in the time-domain model may become large (e.g., 50–70 for a 50 MHz bandwidth in Ref. [25]). In these cases, the complexity of frequency-domain models becomes relatively less; hence more use of frequency-domain modeling may become popular for such wide bandwidth channels.

2.5 Summary and Conclusions

In this chapter we described radio channel modeling for OFDM/A systems. A brief summary of channel characterization in terms of the CIR and its transform, the CTF, was provided, using the correlation functions of these responses. Our focus was on statistical models for small-scale fading. We described the relationships between the set of parameters that approximately describe the channel correlation functions (delay spread, coherence bandwidth, Doppler spread, and coherence time) and the OFDMA/signal parameters, and how the signal parameters should be selected given this channel knowledge. Time- and frequency-domain channel models were described. The topics of cyclic extensions, and time and frequency offsets were briefly addressed. A "canonical" OFDM/A channel model was also described, as were several realistic models, including some that exhibit NS and severe fading. The topics of modeling for MIMO and MB channels, as well as channel estimation, were briefly reviewed.

The most common models for practical channels are time-domain models, but these can typically be converted to frequency-domain models, which are particularly convenient for OFDM/A analysis and simulations.

Abbreviations

BER	Bit error ratio
CIR	Channel impulse response
DFT	Discrete Fourier transform
FFT	Fast Fourier transform

IEEE   Institute of Electrical and Electronics Engineers
LOS   Line of sight
MAC   Medium access control
MIMO   Multiple-input/multiple-output
MPC   Multipath component
NS   Nonstationary
PDP   Power delay profile
PHY   Physical layer
RMS-DS   Root-mean-square delay spread
Rx   Receiver
SFST   Spaced-frequency spaced-time
TDL   Tapped-delay line
Tx   Transmitter
US   Uncorrelated scattering
WLAN   Wireless local area network
WSS   Wide-sense stationary

References

1. I. F. Akyildiz, X. Wang, and W. Wang, Wireless mesh networks: A survey, *J. Comput. Networks*, 445–487, 2005.
2. Institute of Electrical and Electronics Engineers, IEEE Wireless Local Area Networks Working Group website, http://ieee802.org/11/, April 2008.
3. S. Baig and N. D. Gohar, A discrete multitone transceiver at the heart of the PHY layer of an in-home power line communication local-area network, *IEEE Commun. Mag.*, 41(4), 48–53, 2003.
4. Institute of Electrical and Electronics Engineers, IEEE Broadband Wireless Access Working Group website, http://grouper.ieee.org/groups/802/16/, 2008.
5. I. Sen and D. W. Matolak, Vehicle-vehicle channel models for the 5 GHz band, *IEEE Trans. Intell. Transp. Syst.*, 9(2), 235–245, 2008.
6. Institute of Electrical and Electronics Engineers (2008), IEEE Wireless Access in Vehicular Environments website, http://grouper.ieee.org/groups/802/11/Reports/tgp_update.htm, 2008
7. J. G. Proakis, *Digital Communications*, 4th ed., McGraw-Hill, Boston, MA, 2001.
8. G. L. Stuber, *Principles of Mobile Communication*, 2nd ed., Kluwer Academic Publishing, Boston, MA, 2001.
9. J. D. Parsons, *The Mobile Radio Propagation Channel*, 2nd ed., Wiley, New York, 2000.
10. D. W. Matolak, Wireless channel characterization in the 5 GHz microwave landing system extension band for airport surface areas, Final Project Report for NASA ACAST Project, Grant Number NNC04GB45G, May 2006.
11. R. W. Chang, Synthesis of band-limited orthogonal signals for multi-channel data transmission, *Bell Syst. Tech. J.*, 46, 1775–1796, December 1966.
12. J. G. Andrews, A. Ghosh, and R. Muhamed, *Fundamentals of WiMAX: Understanding Broadband Wireless Networking*, Prentice Hall PTR, Upper Saddle River, NJ, 2007.
13. E. Lutz, D. Cygan, M. Dippold, F. Dolainsky and W. Papke, The land mobile satellite communication channel-recording, statistics, and channel model, *IEEE Trans. Veh. Tech.*, 40(2), 375–385, May 1991.
14. R. C. Qiu, A study of the ultra-wideband wireless propagation channel and optimum UWB receiver design, *IEEE J. Sel. Areas Comm.*, 20(9), 1628–1637, December 2002.
15. P. Bello, Characterization of random time-variant linear channels, *IEEE Trans. Commun.*, 11, 360–393, December 1963.
16. A. Papoulis and U. Pillai, *Probability, Random Variables, and Stochastic Processes*, 4th ed., McGraw-Hill, New York, 2001.
17. COST 231 Final Report, http://www.lx.it.pt/cost231/final_report.htm, 2007.

18. Institute of Electrical and Electronics Engineers, IEEE 802.16.3c-01/29r4, Channel models for fixed wireless applications, 2007.
19. J. Medbo, H. Hallenberg, and J. E. Berg, Propagation characteristics at 5 GHz in typical radio-LAN scenarios, *Proc. IEEE Vehicular Tech. Conf.*, 1, 185–189, April 1999.
20. Institute of Electrical and Electronics Engineers, IEEE 802.20-03/48, Channel Models for IEEE 802.20 MBWA system simulations, http://grouper.ieee.org/groups/802//20/Contribs/C802.20-03-48.pdf, 2005.
21. Institute of Electrical and Electronics Engineers, IEEE 802.20 channel models (V1.0), IEEE 802.20 PD-08, http://grouper.ieee.org/groups/802//20/P_Docs/IEEE_802.20-PD-08.doc, 2005.
22. L. M. Correia (ed.), *Wireless Flexible Personalised Communications* (COST 259 Final Report), Chichester, UK, Wiley, 2001.
23. European Telecommunications Standards Institute (ETSI), Universal Mobile Telecommunications System (UMTS): Selection procedures for the choice of radio transmission technologies of the UMTS (UMTS 30.03, ver. 3.1.0), TR 101 112, v3.1.0, Section B.1.4.2, 1997.
24. D. Schafhuber and G. Matz, MMSE and adaptive prediction of time-Varying channels for OFDM systems, *IEEE Trans. Wirel. Commun.*, 4(2), 593–602, April 2005.
25. D. W. Matolak, I. Sen, and W. Xiong, The 5 GHz airport surface area channel: Part I, measurement and modeling results for large airports, *IEEE Trans. Veh. Tech.*, 57(4), 2014–2026, July 2008.
26. G. Matz, On non-WSSUS wireless fading channels, *IEEE Trans. Wirel. Commun.*, 4(5), 2465–2478, Sept. 2005.
27. A. Konrad, B. Y. Zhao, A. D. Joseph, and R. Ludwig, A Markov-based model algorithm for wireless networks, *Wirel. Networks*, 9, 189–199, 2003.
28. D. W. Matolak, I. Sen, W. Xiong, and R. D. Apaza, Channel measurement/modeling for airport surface communications: mobile and fixed platform results, *IEEE Aerosp. Electron. Mag.*, 22(10), 25–30, October 2007.
29. F. Adachi, and T. T. Tjhung, Tapped delay line model for band-limited multipath channel in DS-CDMA mobile radio, *IEE Electron. Lett.*, 37(5), March 2001.
30. Institute of Electrical and Electronics Engineers, IEEE Broadband Wireless Access Working Group document IEEE 802.163c-01/29 Channel Models for Fixed Wireless Applications, July 2001, http://wirelessman.org/tg3/contrib/802163c-01_29r4.pdf
31. Y. Li, and G. L. Stuber, *Orthogonal Frequency Division Multiplexing for Wireless Communications*, Springer, New York, NY, 2006.
32. A. Goldsmith, *Wireless Communications*, Cambridge University Press, New York, NY, 2005.
33. A. R. S. Bahai, and B. R. Saltzberg, *Multi-Carrier Digital Communications: Theory and Applications of OFDM*, Kluwer Academic Publishers, New York, 2002.
34. M. Patzold, *Mobile Fading Channels: Modelling, Analysis, and Simulation*, Wiley, New York, NY, 2002.
35. Y. G. Li, and L. J. Cimini, Jr., Bound on the interchannel interference of OFDM in time-varying impairments, *IEEE Trans. Commun.*, 49(3), 401–404, March 2001.
36. T. S. Rappaport, *Wireless Communications: Principles and Practice*, 2nd ed., Prentice-Hall, Upper Saddle River, NJ, 2002.
37. T. K. Sarkar, Z. Ji, K. Kim, A. Medouri, and M. Salazar-Palma, A survey of various propagation models for mobile communication, *IEEE Antennas Propag. Mag.*, 45(3), 51–82, June 2003.
38. R. J. C. Bultitude, Estimating frequency correlation functions from propagation measurements on fading channels: A critical review, *IEEE J. Sel. Areas Commun.*, 20(6), 1133–1143, August 2002.
39. I. Sen, 5 GHz channel characterization for airport surface areas and vehicle-vehicle communication systems, Ph.D. Dissertation, School of Electrical Engineering and Computer Science, Ohio University, August 2007.
40. V. Erceg et al., An empirically based path loss model for wireless channels in suburban environments, *IEEE J. Sel. Areas Comm.*, 17(7), 1205–1211, July 1999.
41. Third Generation Partnership Project, web site http://www.3gpp.org/, October 2008.
42. A. Saleh, and R. Valenzuela, A statistical model for indoor multipath propagation, *IEEE J. Sel. Areas Commun.*, 5(2), 128–137, February 1987.
43. Wireless World Initiative New Radio project, web site http://www.ist-winner.org/, October 2008.

44. D. W. Matolak, I. Sen, and W. Xiong, Wireless channels that exhibit "worse than rayleigh" fading: Analytical and measurement results, *Proc. MILCOM 2006*, Washington, DC, October 23–25, 2006.

45. M. A. Taneda, J. Takada, and K. Araki, A new approach to fading: Weibull model, *Proc. Int. Symposium Personal Indoor Mobile Radio Communication*, pp. 711–715, Osaka, Japan, September 4, 1999.

46. G. Pai, D. W. Matolak, I. Sen, W. Xiong, and B. Wang, 5 GHz campus area wireless channel characterization, *Proc. 16th Virginia Tech Symp. on Wireless Pers. Comm.*, Blacksburg, VA, June 7–9, 2006.

47. T. Zwick, C. Fischer, and W. Wiesbeck, A stochastic multipath channel model including path directions for indoor environments, *IEEE J. Sel. Areas Commun.*, 20(6), 1178–1192, August 2002.

48. C.-C. Chong, C.-M. Tan, D.I. Laurenson, S. McLaughlin, M.A. Beach, and A.R. Nix, A novel wideband dynamic directional indoor channel model based on a Markov process, *IEEE Trans. Wirel. Commun.*, 4(4) 1539–1552, July 2005.

49. D. W. Matolak, Channel modeling for vehicle-to-vehicle communications, *IEEE Commun. Mag., Special Section on Automotive Networking*, 46(5), 76–83, May 2008.

50. G. Calcev, D. Chizhik, B. Göransson, S. Howard, H. Huang, A. Kogiantis, A. F. Molisch, A. L. Moustakas, D. Reed, and H. Xu, A wideband spatial channel model for system-wide simulations, *IEEE Trans. Vehicular Tech.*, 56, 389–403, March 2007.

51. Third Generation Partnership Project, UMTS TS 25.101 v8.0, web site http://www.3gpp.org/, October 2008.

52. B. Wang, I. Sen, and D. W. Matolak, Performance evaluation of 802.16e in vehicle to vehicle channels, *Proc. IEEE Fall Vehicular Tech. Conf.*, Baltimore, MD, October 1–3, 2007.

53. D. Gesbert, H. Bolcskei, D. A. Gore, and A. J. Paulraj, Outdoor MIMO wireless channels: Models and performance prediction, *IEEE Trans. Commun.*, 50(12), 1926–1934, December 2002.

54. P. Almers, et al., Survey of channel and radio propagation models for wireless MIMO systems, *EURASIP J. Wirel. Commun. Networking*, vol. 2007, doi:10.1155/2007/19070, 2007.

55. H. Taoka, K. Dai, K. Higuchi, and M. Sawahashi, Field experiments on MIMO multiplexing with peak frequency efficiency of 50 bit/second/Hz using MLD based signal detection for OFDM high-speed packet access, *IEEE J. Sel. Areas Commun.*, 26(6), 845–856, August 2008.

56. Y. Q. Bian, A. R. Nix, E. K. Tameh, and J. P. McGeehan, MIMO-OFDM WLAN architectures, area coverage, and link adaptation for urban hotspots, *IEEE Trans. Veh. Tech.*, 57(4), 2364–2374, July 2008.

57. IEEE 802.11 Wireless LANs, *TGn Channel Models*, IEEE Std. 802.11-03/940r4, May 2004.

58. L. Schumacher, K. I. Pedersen, and P. E. Mogensen, From antenna spacings to theoretical capacities C guidelines for simulating MIMO systems, *Proc. PIMRC Conf.*, vol. 2, pp. 587–592, September 2002.

59. J. Zhang and D. W. Matolak, FG-MC-CDMA system performance in multi-band channels, *Proc. IEEE CNSR 2008*, Halifax, Nova Scotia, Canada, pp. 132–138, May 2008.

60. Y. G. Li, L. C. Cimini, and N. R. Sollenberger, Robust channel estimation for OFDM systems with rapid dispersive fading channels, *IEEE Trans. Comm.*, 46(7), 902–915, July 1998.

61. A. Molisch (ed.), *Wideband Wireless Digital Communications*, Prentice Hall, Upper Saddle River, NJ, 2001.

62. P. Almers, E. Bonek, A. Burr, N. Czink, M. Debbah, V. Degli-Esposti, H. Hofstetter, P. Kyosti, D. Laurenson, G. Matz, A. F. Molisch, C. Oestges, and H. Ozcelik, Survey of channel and radio propagation models for wireless MIMO Systems, *EURASIP J. Wirel. Commun. Networking*, 2007, 1–19, 2007.

## Chapter 3

Nonlinearity Analysis of OFDM-Based Wireless Systems

Chunming Liu and Fu Li

Contents

3.1 Introduction.. 41
    3.1.1 OFDM in Wireless Broadband Communication Systems......................... 42
    3.1.2 OFDM Challenges.. 43
        3.1.2.1 Spectrum Efficiency in Wireless Broadband Communication Systems. 43
        3.1.2.2 Nonlinearity in RF Power Amplifier................................... 44
3.2 IEEE 802.11a Systems 44
3.3 Nonlinearity of RF Amplifier in WLAN OFDM Systems............................ 48
3.4 Model Description 49
    3.4.1 802.11a OFDM Signal Equivalent Mathematical Model...................... 49
    3.4.2 High-Power Amplifier's Mathematical Model 55
3.5 PSD of Amplified 802.11a OFDM Signal.. 57
3.6 Design Example and Comparison with Simulations............................. 63
References... 64

3.1 Introduction

Marconi's innovative perception of the electromagnetic waves for radio transmission in 1897 was the first milestone on the important road to shared use of the radio spectrum. But only after almost a century did mobile wireless communication start to take off. The communication world in the late 1980s was rapidly becoming more mobile for a much broader segment of communication, users than ever before. Today, the spectacular growth of video, voice, and data communications over the Internet and the equally rapid growth of mobile telephony justify great expectations for mobile multimedia. Research and development are taking place all over the world to define the next generation of wireless broadband communication systems. These may create a "global information village," which consists of various components at different scales ranging from global to cellular. Present communication systems are primarily designed for one specific application such as speech on a mobile telephone or high-rate data in a wireless local area network (WLAN). Supporting such large data rates with sufficient robustness against radio channel impairments requires careful choosing of modulation techniques. The most suitable modulation choice seems to be orthogonal frequency division multiplexing (OFDM).

3.1.1 OFDM in Wireless Broadband Communication Systems

OFDM was proposed for digital cellular systems in the mid-1980s [1]. OFDM has been shown to be effective for digital audio broadcasting (DAB) and digital video broadcasting-terrestrial (DVB-T) at multi-megabit rates in Europe and was standardized by the European Telecommunications Standards Institute (ETSI) [2]. OFDM has also been incorporated into Fourth Generation Mobile Communication (4G) technology [3–5]. The key feature of the physical layer (PHY) in new WLAN standards including IEEE 802.11a, 802.11g, and high performance local area network type 2 (HiperLAN/2) is that OFDM has been selected as the modulation scheme [6–8], owing to its good performance on highly dispersive channels [9]. Single-carrier modulation cannot efficiently support high bit rates. This is an important factor since these wireless broadband communication systems are required to support much higher bit rates.

In single-carrier wireless communication systems, the bandwidth of the carrier signal must be high enough to transmit data at a high rate. OFDM, however, sends a high-speed data stream by splitting it into multiple lower-speed streams and transmitting it in parallel over lower-bandwidth subcarriers. In normal frequency division multiplexing (FDM) systems, the channels are nonorthogonal and so guard bands are required to prevent interference between channels. In OFDM, the subcarriers are orthogonal so that they do not interfere with each other: they can be made to overlap such that the main lobe of each subcarrier lies on the nulls of the other carriers, thereby increasing the spectral efficiency. The generation of the multiple orthogonal subcarriers in OFDM is accomplished by using the discrete Fourier transform (DFT).

OFDM can also be used to overcome the multipath-fading problem. In a single-carrier FDM system, the radio signal transmitted to a receiver may be reflected off other objects in the vicinity. The combination of all the signals arriving at the receiver causes the modulated signal to be distorted. In the worst case, the entire signal can be lost. One way to eliminate such multipath signal distortion is to use an equalizer at the receiver of the single-carrier system. However, this requires increasing the complexity of the receiver. OFDM can overcome the multipath fading problem since it transmits the subcarrier signals at a much lower data rate than a single-carrier system. Thus, the delayed signals from reflections are late by only a small fraction of a symbol time.

Five attractive features make OFDM suitable for next-generation mobile networks: (a) OFDM can be very efficiently implemented due to recent advances in very large scale integration (VLSI) technology that make high-speed fast Fourier transform (FFT) chips commercially available. OFDM is much less computationally complex than FDM with an equalizer; (b) OFDM uses digital signal processing (DSP) techniques to efficiently use the available radio frequency (RF) spectrum. This is done through adaptive modulation and power allocation across the subcarriers to match the changing channel conditions. Thus, OFDM provides a bandwidth-on-demand technology and improves spectral efficiency; (c) narrowband interference is not a major problem for OFDM since it affects only a small fraction of the subcarriers; (d) the operation of OFDM does not require contiguous bandwidth, unlike that in existing single-carrier third generation mobile communication (3G) technologies; and (e) with OFDM, it is possible to have single-frequency networks. This is especially useful in broadcasting applications.

3.1.2 OFDM Challenges

3.1.2.1 Spectrum Efficiency in Wireless Broadband Communication Systems

The Federal Communication Commission's 1985 ruling (rule modified in 1990) allowed unlicensed spread spectrum use of the three industrial, scientific, and medical (ISM) frequency bands. This ruling encouraged development of a number of wireless technologies. Today, many unlicensed wireless broadband communication products occupy all three ISM bands and create a level of interference. In the next 5–10 years, even more ISM applications will emerge, creating more interference. The bandwidth available for a worldwide bandwidth allocation is severely constrained and, even if such an allocation were feasible, the bandwidth resources would be very limited. Thus much more efficient utilization of the spectrum will be required.

These wireless broadband communication systems are expected to provide substantially higher data rates to meet the requirements of future high-performance multimedia applications. For example, the minimum target data rate for WLAN systems is expected to be 6 Mbps and maximum bit rates can be up to 54 Mbps. To provide such data rates, different modulation schemes must be developed to achieve much higher spectral efficiencies.

On the negative side, compared to single-carrier systems, OFDM has a larger peak-to-average power ratio (PAPR) [1]. This is OFDM's biggest drawback. It makes OFDM systems very sensitive to nonlinearities in RF power amplifiers, causing severe spectrum regrowth in WLAN systems. There is a special requirement for limiting spectrum regrowth in IEEE 802.11a standard. The standard is very clearly shown in Figure 3.1 that spectrum regrowth should not grow beyond the specially designed mask.

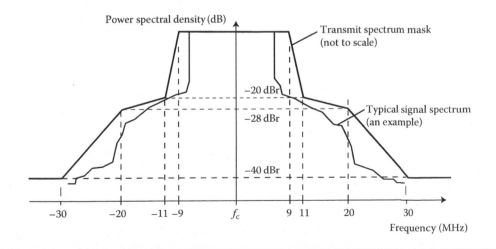

Figure 3.1   Spectrum regrowth control specified in IEEE 802.11a.

In general, modulation schemes that pass wave-packets with nonconstant envelopes through an amplifier in compression or saturation are limited. In such cases, the amplifier causes the spectrum to regrow, conflicting directly with the goal of high spectrum efficiency and adding extra cost to the RF power amplifier. The effect limits the application of these OFDM-based wireless broadband communication systems, such as DAB, DVB-T, and WLAN systems.

3.1.2.2 Nonlinearity in RF Power Amplifier

RF power amplifier cost is a major factor in wireless communication system cost, and the RF power amplifier itself is a major contributor to power supply requirements, heat management, and equipment size. Because no device has perfect linearity, the nonlinearity of an RF amplifier is inherent [10]. On the other hand, the OFDM signal has a relatively large PAPR and, hence, is prone to suffer from nonlinearities in RF power amplifiers. The direct result is (a) degraded OFDM-based wireless broadband communication system spectrum efficiency, (b) increased bit error rate, and (c) interference with adjacent channels. Alternative strategies to avoid severe spectrum regrowth, that is, reducing the power amplifier efficiency, will increase the cost of the power consumed by the transmitter amplifier. Therefore, the efficiency–linearity tradeoff is a critical design issue to be considered when attempting to limit spectrum regrowth in OFDM-based systems.

Accurately representing a typical digitally modulated signal requires multitones (about 500 tones) [11], making conventional methods unsuitable for simulating the spectrum regrowth caused by RF power amplifier nonlinearity. To the best of our knowledge, there has been no quantitative expression relating spectrum regrowth to power amplifier nonlinearity in OFDM-based communication systems. The lack of such a relation poses difficulties for developers when designing OFDM-based wireless broadband communication systems. This problem is generic in the design of RF power amplifiers for nonconstant envelope digital modulations. Since OFDM has a relatively larger PAPR as compared with single-carrier modulation, spectrum regrowth is more severe. Further, experiments and analyses revealed that, in some situations, only using the amplifier's third-order intermodulation is not sufficient to describe the spectrum regrowth, especially when the fifth-order intermodulation is relatively high compared to the third-order intermodulation. In previous studies, such as the work reported in reference [12], it was assumed that the effects of the fifth- and higher-order intermodulations could be ignored. However, if the output power is high and the signal bandwidth is wide, the out-of-band emission power levels caused by the fifth-order intermodulation can be significant even in single-carrier systems [13–16].

3.2 IEEE 802.11a Systems

Since the beginning of the nineties, WLAN for the 900 MHz, 2.4 GHz, and 5 GHz ISM bands have been available, based on a range of proprietary techniques. In June 1997, IEEE approved an international interoperability standard [7]. The standard specifies both medium access control (MAC) procedures and three different PHYs. There are two radio-based PHYs using the 2.4 GHz band. The third PHY uses infrared light. All PHYs support a data rate of 1 Mbps and optionally 2 Mbps. The 2.4 GHz frequency band is available for license-exempt use in Europe, the United States, and Japan.

User demand for higher bit rates and the international availability of the 2.4 GHz band has spurred the development of a higher speed extension to the 802.11 standard. In July 1998, a proposal was selected for standardization, which describes a PHY providing a basic rate of 11 Mbps and a fall back rate of 5.5 Mbps. This PHY can be seen as a fourth option to be used in conjunction with the MAC that is already standardized. Practical products, however, are expected to support both high-speed 11 and 5.5 Mbps rate modes and 1 and 2 Mbps modes.

In this fourth PHY option, OFDM was selected by the 802.11a task group of the IEEE Standards Committee as its high-speed extension to the 802.11 standard. This decision was followed by the approval of IEEE 802.11a standard in September 1999 by the IEEE Standards Board. The RF WLAN system is initially aimed for the 5.15–5.25, 5.25–5.35, and 5.725–5.825 GHz Unlicensed National Information Infrastructure (U-NII) bands, as regulated in the United States by the Code of Federal Regulations, Title 47, Section 15.407. The OFDM system provides a WLAN with data payload communication capabilities of 6, 9, 12, 18, 24, 36, 48, and 54 Mbps. The system uses 52 subcarriers that are modulated using binary/quadrature phase shift keying (BPSK/QPSK), 16-quadrature amplitude modulation (QAM), or 64-QAM. Forward error correction (FEC) coding (convolutional coding) is used with a coding rate of 1/2, 2/3, or 3/4 [6].

This new standard is the first one to use OFDM in packet-based communications. Following IEEE 802.11 decision, the ETSI and Multimedia Mobile Access Communications (MMAC) promotion association within the Association of Radio Industries and Broadcasting (ARIB) in Japan also adopted OFDM for their PHY standards [8,17]. The three bodies have worked in close cooperation since then to make sure that differences among the various standards are kept to a minimum, thereby enabling the manufacturing of equipment that can be used worldwide [1]. The spectrum allocation of HiperLAN/2 in Europe, ARIB MMAC in Japan, and IEEE 802.11a in the United States is shown in Table 3.1.

Table 3.2 lists the main OFDM parameters of IEEE 802.11a standard [6]. A key parameter that largely determined the choice of the other parameters is the guard interval of 800 ns. This guard interval provides robustness to root-mean-square delay spreads up to several hundreds of nanoseconds, depending on the coding rate and modulation used. In practice, this means that the modulation is robust enough to be used in any indoor environment, including large factory buildings. It can also be used in outdoor environments, although directional antennas may be needed in this case to reduce the delay spread to an acceptable amount and increase the range.

To minimize the signal-to-noise ratio (SNR) loss caused by the guard time, it is desirable to have the symbol duration much larger than the guard time. It cannot be arbitrarily large, because larger symbol duration means more subcarriers with smaller subcarrier spacing, a larger implementation complexity, and more sensitivity to phase noise and frequency offset [18], as well as an increased

Table 3.1   HiperLAN/2, MMAC, and IEEE 802.11a Frequency Band at 5 GHz

Geographical Area	Approval Standards	Frequency Range (GHz)	Approval Authority
USA	IEEE 802.11a	5.15–5.25 5.25–5.35 5.725–5.825	IEEE
Europe	HiperLAN/2	5.15–5.30 5.470–5.725	ETSI
Japan	MMAC	5.15–5.25	ARIB

Table 3.2 Main OFDM Parameters of IEEE 802.11a Standard

Parameter	Value
Data rate	6, 9, 12, 18, 24, 36, 48, 54 Mbps
Modulation	BPSK, QPSK, 16-QAM, 64-QAM
Coding rate	1/2, 2/3, 3/4
Number of data subcarriers (N_{SD})	48
Number of pilot subcarriers (N_{SP})	4
Number of subcarriers, total (N_{ST})	52($N_{SD} + N_{SP}$)
Channel spacing	20 MHz
Subcarrier frequency spacing (Δf)	0.3125 MHz (20 MHz/64)
IFFT/FFT period (T_{FFT})	3.2 μs(1/Δf)
OFDM symbol duration (T_s)	4 μs($T_{GI} + T_{FFT}$)
Guard interval duration (T_{GI})	0.8 μs(T_{FFT}/4)
Training symbol guard interval duration (T_{GI2})	1.6 μs(T_{FFT}/2)

PAPR [19,20]. Therefore, a practical design choice is to make the symbol duration at least five times the guard time, which implies a 1 dB SNR loss because of the guard time.

To limit the relative amount of power spent on the guard time to 1 dB, the symbol duration chosen is 4 μs. This also determines the subcarrier spacing at 0.3125 MHz, which is the inverse of the symbol duration minus the guard time. By using 48 data subcarriers, uncoded data rates of 12–72 Mbps can be achieved by using variable modulation types from BPSK to 64-QAM. In addition to the 48 data subcarriers, each OFDM symbol contains an additional four pilot subcarriers that can be used to track the residual carrier frequency offset that remains after an initial frequency correction during the training phase of the packet.

To correct for subcarriers in deep fades, FEC across the subcarriers is used with variable coding rates, giving coded data rates from 6 to 54 Mbps. Convolutional coding is used with the industry standard rate 1/2 with generator polynomials (133, 171). Higher coding rates of 2/3 and 3/4 are obtained by puncturing the rate 1/2 code. The 2/3 rate is used together with 64-QAM only to obtain a data rate of 48 Mbps. The 1/2 rate is used with BPSK, QPSK, and 16-QAM to give rates of 6, 12, and 24 Mbps, respectively. Finally, the 3/4 rate is used with BPSK, QPSK, 16-QAM, and 64-QAM to give rates of 9, 18, 36, and 54 Mbps, respectively.

Figure 3.2 shows the channelization for the lower and middle U-NII bands. Eight channels are available with a channel spacing of 20 MHz and a guard spacing of 30 MHz at the band edges in order to meet the stringent Federal Communications Commission (FCC) restricted band spectral density requirements. The FCC also defined an upper U-NII band from 5.725 to 5.825 GHz, which carries another four OFDM channels, shown in Figure 3.3. For this upper band, the guard spacing from the band edges is only 20 MHz, as the out-of-band spectral requirements for the upper band are less severe than those of the lower and middle U-NII bands.

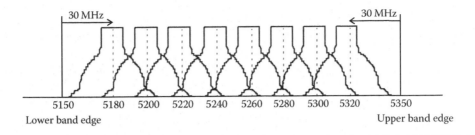

Figure 3.2   Lower and middle U-NII bands: eight carriers in 200 MHz/20 MHz spacing.

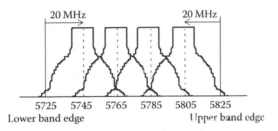

Figure 3.3   Upper U-NII bands: four carriers in 100 MHz/20 MHz spacing.

The general block diagram of the baseband processing of an OFDM transceiver is shown in Figure 3.4. In the transmitter path, binary input data are encoded by a standard rate 1/2 convolutional encoder. The rate may be increased to 2/3 or 3/4 by puncturing the coded output bits. After interleaving, the binary values are converted into QAM values. To facilitate coherent reception, four pilot values are added to each 48 data values. The total 52 QAM values of an OFDM symbol are modulated onto 52 subcarriers by applying the IFFT. To make the system robust to multipath propagation, a cyclic prefix is added. Further, windowing is applied to attain

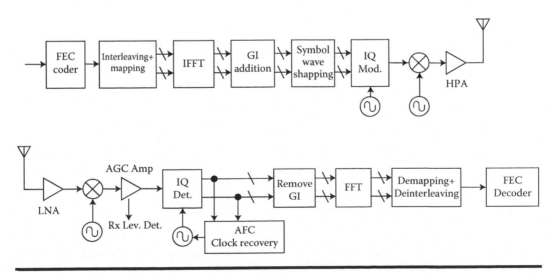

Figure 3.4   Transmitter and receiver block diagram for the OFDM PHY in IEEE 802.11a system.

a narrower output spectrum. After this step, the digital output signals can be converted to analog signals, which are then up-converted to the 5 GHz band, amplified, and transmitted through an antenna.

The OFDM receiver basically performs the reverse operations of the transmitter, together with additional training tasks. First, the receiver has to estimate frequency offset and symbol timing, using special training symbols in the preamble. Then it can do an FFT for every symbol to recover the 52-QAM values of all subcarriers. The training symbols and pilot subcarriers are used to correct for the channel response as well as the remaining phase drift. The QAM values are then demapped into binary values, after which a Viterbi decoder can decode the information bits.

3.3 Nonlinearity of RF Amplifier in WLAN OFDM Systems

One of the most important problems of OFDM modulation, the multicarrier modulation scheme, is that the transmit signals have a nonconstant power envelope. This means that any nonlinearity in the transmitter will cause in-band noise and spectral splatter—causing degraded performance and adjacent channel interference. The peak power of an OFDM signal is, in theory, N_{ST} times higher than the average power, where N_{ST} is the number of subcarriers [1]. The transmit power amplifier in an OFDM system has to be backed off by a large factor in some situations even after using a complementary coding scheme across the carriers to control the PAPR of the transmit signal [21], and this has a major impact on battery life in a mobile system. Furthermore, as part of IEEE 802.11a standard, special requirements have been set for the control of nonlinearity of RF amplifiers used in WLAN systems [6]. The nonlinearity is also called spectrum regrowth. Figure 3.5 shows the typical power spectrum of a transmit signal in a WLAN system, as well as amplified by a nonlinear

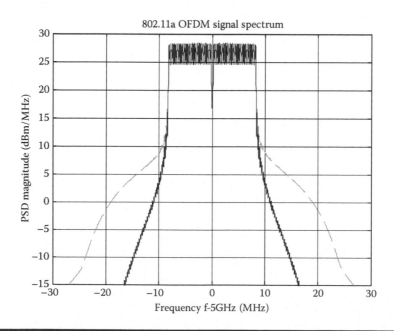

Figure 3.5 PSD of amplified WLAN OFDM signal (solid line: without regrowth; dashed line: with regrowth due to amplifier nonlinearity).

power amplifier. A frequency slot in the WLAN is 20 MHz, so it is clear that the distorted signal would cause severe adjacent channel interference. It is therefore very important for WLAN system designers to know the relationship between the spectrum regrowth and parameters of the system power amplifier.

Traditionally, the nonlinearity of an RF amplifier is described by using third-order interception point (IP$_3$) or by using the 1 dB compression point [22,23]. In simulations and analyses it was discovered that, in some cases, using IP$_3$ alone is not accurate enough to describe the spectrum regrowth, especially when OFDM is used as the modulation scheme. Quantitatively, to the best of our knowledge, there is no explicit relationship or expression between the out-of-band emission level and the traditional amplifier nonlinearity description for WLAN OFDM signal amplification. The lack of such a relationship makes it difficult for RF power amplifier designers to choose components.

In our early effort, we analyzed the nonlinear effect of an RF power amplifier on CDMA, TDMA, and Motorola iDEN systems. Expressions were developed for estimated out-of-band emission levels for signals in these systems in terms of (a) the power amplifier's intermodulation coefficients IP$_3$ and IP$_5$ (where IP$_5$ is a parameter defined in a similar manner as IP$_3$ to describe the fifth-order intermodulation quantitatively), (b) the signal power level, and (c) the signal bandwidth [13–16]. Continuing this past effort into the current work, we develop the spectrum analysis approach for WLAN OFDM signals. This chapter (a) presents the analysis of spectrum regrowth caused by the nonlinear effects of an RF power amplifier in OFDM-based WLAN systems, (b) presents the relationship between spectrum regrowth levels and the RF power amplifier's nonlinearity parameters (IP$_3$ and IP$_5$), (c) proposes a theoretical method and a model to predict the spectrum regrowth of WLAN systems, and (d) derives expressions relating the out-of-band power emission levels of an amplifier to its nonlinearities. The results enable OFDM-based 802.11a system designers to specify and measure spectrum regrowth using simple RF power amplifier intermodulation descriptions effectively. The results enable spectrum administrators to manage and plan spectrum allocation efficiently. The expressions turn out to be simpler and easier to use for the case where IP$_5$ is ignored. In addition, a spectrum comparison between the simulated and predicted results is presented.

3.4 Model Description

3.4.1 802.11a OFDM Signal Equivalent Mathematical Model

An OFDM symbol in the 802.11a system, $b_n(t)$, is constructed as an IFFT of a set of $d_{k,n}$, which can be defined as transmitted data, pilots, or training symbols in the 802.11a system. The mathematical model of an OFDM symbol can be described as [6]

$$b_n(t) = w_{\mathrm{T}}(t) \sum_{\substack{k=-N_{\mathrm{ST}}/2 \\ k \neq 0}}^{N_{\mathrm{ST}}/2} d_{k,n} e^{j2\pi k \Delta f(t-T_{\mathrm{GI}})}. \tag{3.1}$$

The parameters in Equation 3.1 are described in Table 3.2, where N_{ST} is the total number of subcarriers and is equal to 52 for the IEEE 802.11a system, and Δf is the subcarrier frequency spacing, chosen as 0.3125 MHz. Therefore, the resulting waveform of an OFDM data symbol is periodic with a period of $T_{\mathrm{FFT}} = 1/\Delta f$, with shifting time T_{GI}, which is a guard interval time to

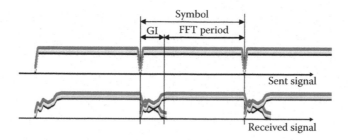

Figure 3.6 Sent and received signals with a guard interval time.

create the circular prefix used in OFDM to avoid the intersymbol interference (ISI) from the previous symbol. Three kinds of guard interval time are defined: for the short training sequence (0 μs), for the long training sequence ($T_{GI2} = 1.6$ μs), and for data OFDM symbols ($T_{GI} = 0.8$ μs). The sent and received signals with a guard interval time are shown in Figure 3.6.

The boundaries of the OFDM symbol are set by multiplication by a time-windowing function, $w_{T \text{subframe}}(t)$, which is defined as a rectangular pulse $w_T(t)$ of duration T, accepting the value T_{subframe}. The time-windowing function, $w_T(t)$, depending on the value of the duration parameter T may extend over more than one period T_{FFT}. In particular, window functions that extend over multiple periods of the FFT are utilized in the definition of the preamble. Figure 3.7 illustrates the possibility of extending the windowing function over more than one period, T_{FFT}, and additionally shows smoothed transitions by application of a windowing function, as exemplified in Equation 3.2:

$$
w_T(t) = \begin{cases}
\sin^2\left[\dfrac{\pi}{2}\left(0.5 + \dfrac{t}{T_{\text{TR}}}\right)\right] & \text{for } \dfrac{-T_{\text{TR}}}{2} < t < \dfrac{T_{\text{TR}}}{2}, \\[2ex]
1 & \text{for } \dfrac{T_{\text{TR}}}{2} \le t < T - \dfrac{T_{\text{TR}}}{2}, \\[2ex]
\sin^2\left[\dfrac{\pi}{2}\left(0.5 - \dfrac{t-T}{T_{\text{TR}}}\right)\right] & \text{for } T - \dfrac{T_{\text{TR}}}{2} \le t < T + \dfrac{T_{\text{TR}}}{2},
\end{cases}
\tag{3.2}
$$

where T is the duration time of the symbol, equal to T_s chosen as 4 μs for an OFDM data symbol, and T_{TR} is the transition time, about 100 ns. It is shown in Equation 3.2 that in the case of vanishing T_{TR}, the windowing function degenerates into a rectangular pulse of duration T. The smoothed transition is required in the 802.11a standard in order to reduce the spectral side lobes of the transmitted waveform [6]. In implementation, higher T_{TR} is typically implemented in order to smooth the transitions between the consecutive subsections. This creates a small overlap between them, of duration T_{TR}, as shown in Figure 3.7.

Hence, the concatenation of these OFDM symbols can now be written as

$$
\begin{aligned}
b(t) &= \sum_{n=-\infty}^{\infty} b_n(t - nT_s) \\
&= \sum_{n=-\infty}^{\infty} w_T(t - nT_s) \sum_{\substack{k=-N_{\text{ST}}/2 \\ k \neq 0}}^{N_{\text{ST}}/2} d_{k,n} e^{j2\pi k \Delta f (t - T_{\text{GI}} - nT_s)},
\end{aligned}
\tag{3.3}
$$

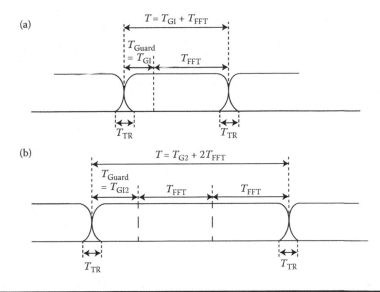

Figure 3.7 Illustrations of the OFDM frame with cyclic extension and windowing: (a) single reception; (b) two receptions of the FFT period.

where $b(t)$ is the transmitted baseband OFDM data signal, $d_{k,n}$ is the modulated transmitted data in the nth OFDM symbol and the kth subcarrier. The OFDM subcarrier shall be modulated by using binary phase shift keying (BPSK), QPSK, 16-QAM, or 64-QAM modulation, depending on the transmitted data rate requested. All the eight modulations with different coding rates are shown in Table 3.3. In this chapter, 16-QAM is chosen to analyze the nonlinear effect of a high-power amplifier due to the amplifier's nonlinearity for OFDM signals.

Table 3.3   PHY Modes with Different Coding Rates and Modulation Schemes

Data Rates (Mbps)	Modulations	Coding Rate	Coded Bits per Subcarrier	Coded Bits per OFDM Symbol	Data Bits per OFDM Symbol
6	BPSK	1/2	1	48	24
9	BPSK	3/4	1	48	36
12	QPSK	1/2	2	96	48
18	QPSK	3/4	2	96	72
24	16-QAM	1/2	4	192	96
36	16-QAM	3/4	4	192	144
48	64-QAM	2/3	6	288	192
54	64-QAM	3/4	6	288	216

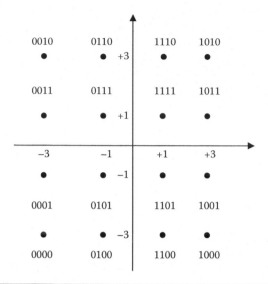

Figure 3.8 16-QAM constellation bit encoding.

In the 16-QAM modulation mode chosen in the 802.11a system, the encoded and interleaved binary serial input data are divided into groups of 4 bits and converted into complex numbers representing 16-QAM constellation points. The conversion shall be performed according to Gray-coded constellation mapping, illustrated in Figure 3.8. c_0 is the earliest input bits in the stream. The input bits $c_0 c_1$ determine the *in-phase* (*x*) value and $c_2 c_3$ determine the *quadrature* (*y*) value, as illustrated in Table 3.4. The output values, $d_{k,n}$, are complex numbers, formed by multiplying the resulting $x_{k,n} + jy_{k,n}$ value by a normalization factor K_{MOD}, described as

$$d_{k,n} = (x_{k,n} + jy_{k,n})K_{MOD} \tag{3.4}$$

and the equivalent form

$$d_{k,n} = (R_{k,n}e^{j\theta_{k,n}})K_{MOD}, \tag{3.5}$$

where $R_{k,n} = \sqrt{x_{k,n}^2 + y_{k,n}^2}$ and $\theta_{k,n} = \tan^{-1}(y_{k,n}/x_{k,n})$. The normalization factor K_{MOD} is chosen as $1/\sqrt{10}$ for 16-QAM modulation mode. The purpose of the normalization factor is to achieve the same average power for all mappings. In practical implementations, an approximate value of

Table 3.4   16-QAM Encoding Table

Input bits ($c_0 c_1$)	x-out	Input bits ($c_2 c_3$)	y-out
00	−3	00	−3
01	−1	01	−1
11	1	11	1
10	3	10	3

the normalization factor can be used, as long as the device conforms with the modulation accuracy requirements described in the 802.11a standard [6].

Substituting Equation 3.5 into Equation 3.3, the transmitted baseband OFDM data signal $b(t)$ can also be given by

$$
\begin{aligned}
b(t) &= \sum_{n=-\infty}^{\infty} w_{\mathrm{T}}(t - nT_{\mathrm{s}}) \sum_{k=-N_{\mathrm{ST}}/2}^{N_{\mathrm{ST}}/2} \left(R_{k,n} e^{j\theta_{k,n}} K_{\mathrm{MOD}} \right) e^{j2\pi k \Delta f(t - T_{\mathrm{GI}} - nT_{\mathrm{s}})} \\
&= \sum_{n=-\infty}^{\infty} \sum_{\substack{k=-N_{\mathrm{ST}}/2 \\ k \neq 0}}^{N_{\mathrm{ST}}/2} R_{k,n} K_{\mathrm{MOD}} w_{\mathrm{T}}(t - nT_{\mathrm{s}}) e^{j[2\pi k \Delta f(t - T_{\mathrm{GI}} - nT_{\mathrm{s}}) + \theta_{k,n}]}.
\end{aligned}
\tag{3.6}
$$

Furthermore, the mathematical model of the actual transmitted signal $s(t)$ can be presented by its baseband envelope as

$$
\begin{aligned}
s(t) &= \sum_{n=-\infty}^{\infty} \sum_{\substack{k=-N_{\mathrm{ST}}/2 \\ k \neq 0}}^{N_{\mathrm{ST}}/2} R_{k,n} K_{\mathrm{MOD}} w_{\mathrm{T}}(t - nT_{\mathrm{s}}) \cos[2\pi k \Delta f(t - T_{\mathrm{GI}} - nT_{\mathrm{s}}) + \theta_{k,n}] \cos(2\pi f_c t) \\
&\quad - \sum_{n=-\infty}^{\infty} \sum_{\substack{k=-N_{\mathrm{ST}}/2 \\ k \neq 0}}^{N_{\mathrm{ST}}/2} R_{k,n} K_{\mathrm{MOD}} w_{\mathrm{T}}(t - nT_{\mathrm{s}}) \sin\left[2\pi k \Delta f(t - T_{\mathrm{GI}} - nT_{\mathrm{s}}) + \theta_{k,n}\right] \sin(2\pi f_c t) \\
&= \sum_{n=-\infty}^{\infty} \sum_{\substack{k=-N_{\mathrm{ST}}/2 \\ k \neq 0}}^{N_{\mathrm{ST}}/2} R_{k,n} K_{\mathrm{MOD}} w_{\mathrm{T}}(t - nT_{\mathrm{s}}) \cos\left[2\pi k \Delta f(t - T_{\mathrm{GI}} - nT_{\mathrm{s}}) + \theta_{k,n} + 2\pi f_c t\right] \\
&= \mathrm{Re}\left\{ \left[\sum_{n=-\infty}^{\infty} \sum_{\substack{k=-N_{\mathrm{ST}}/2 \\ k \neq 0}}^{N_{\mathrm{ST}}/2} R_{k,n} K_{\mathrm{MOD}} w_{\mathrm{T}}(t - nT_{\mathrm{s}}) e^{j[2\pi k \Delta f(t - T_{\mathrm{GI}} - nT_{\mathrm{s}}) + \theta_{k,n}]} \right] e^{j2\pi f_c t} \right\} \\
&= \mathrm{Re}\left\{ r(t) e^{j2\pi f_c t} \right\},
\end{aligned}
\tag{3.7}
$$

where $\mathrm{Re}\{\cdot\}$ denotes the real part of $\{\cdot\}$, and $r(t) = \sum_{n=-\infty}^{\infty} \sum_{\substack{k=-N_{\mathrm{ST}}/2 \\ k \neq 0}}^{N_{\mathrm{ST}}/2} R_{k,n} K_{\mathrm{MOD}} w_{\mathrm{T}}$

$(t - nT_{\mathrm{s}}) e^{j[2\pi k \Delta f(t - T_{\mathrm{GI}} - nT_{\mathrm{s}}) + \theta_{k,n}]}$ is the baseband envelope of the actual transmitted signal $s(t)$, and f_c denotes the carrier frequency of the WLAN system.

Furthermore, $r(t)$ can be expressed as

$$r(t) = \sum_{n=-\infty}^{\infty} \sum_{\substack{k=-N_{ST}/2 \\ k\neq 0}}^{N_{ST}/2} R_{k,n} K_{MOD} w_T(t - nT_s) e^{j[2\pi k\Delta f(t-T_{GI}-nT_s)+\theta_{k,n}]}$$

$$= \sum_{\substack{k=-N_{ST}/2 \\ k\neq 0}}^{N_{ST}/2} \left\{ \sum_{n=-\infty}^{\infty} R_{k,n} K_{MOD} w_T(t - nT_s) e^{j[2\pi k\Delta f(-T_{GI}-nT_s)+\theta_{k,n}]} \right\} e^{j2\pi k\Delta f t} \qquad (3.8)$$

$$= \sum_{\substack{k=-N_{ST}/2 \\ k\neq 0}}^{N_{ST}/2} g_k(t) e^{j2\pi k\Delta f t},$$

where $g_k(t) = \sum_{n=-\infty}^{\infty} R_{k,n} K_{MOD} w_T(t - nT_s) e^{j[2\pi k\Delta f(-T_{GI}-nT_s)+\theta_{k,n}]}$, which is a pulse shaped nonreturn-to-zero (NRZ) function. In Ref. [22], the general expression for the power spectrum density (PSD) of a digital signal $s(t) = \sum_{n=-\infty}^{\infty} a_n f(t - nT_s)$ is presented as $P_s(f) = \left(|F(f)|^2/T_s\right) \sum_{n=-\infty}^{\infty} R(k) e^{j2\pi kf T_s}$, where $F(f)$ is the Fourier transform of the pulse shape $f(t)$, and $R(k)$ is the autocorrelation of the data. By using this result and considering that $R_{k,n}$ is uniformly distributed in 16-QAM constellation, the PSD of $g_k(t)$ can be obtained as [24]

$$P_{g_k}(f) = R_s |W(f)|^2, \quad k = \pm 1, \pm 2, \ldots, \pm \frac{N_{ST}}{2}, \qquad (3.9)$$

where $R_s = 1/T_s$ is the symbol rate, and $W(f)$ is the Fourier transform of the time-window function, $w_T(t)$, of the form:

$$W(f) = T_s \times \text{sinc}(T_s f) \times \frac{\cos(\pi T_{TR} f)}{1 - 4T_{TR}^2 f^2} e^{-j\pi T_s f}, \qquad (3.10)$$

where $\text{sinc}(x)$ is defined as $\text{sinc}(x) = (\sin(\pi x)/\pi x)$. It is clearly shown in Equation 3.10 that the PSD of the ODFM subcarrier is affected only by the symbol shaping window and symbol rate, no matter what the other parameters are in this system.

Since the spectrum of a bandpass signal is directly related to the spectrum of its baseband envelope [22], the PSD of $r(t)$ can be expressed as

$$P_r(f) = \frac{R_s}{4} \sum_{\substack{k=-N_{ST}/2 \\ k\neq 0}}^{N_{ST}/2} \left[|W(f - k\Delta f)|^2 + |W(-f - k\Delta f)|^2 \right], \qquad (3.11)$$

where $k = \pm 1, \pm 2, \ldots, \pm N_{ST}/2$.

3.4.2 High-Power Amplifier's Mathematical Model

Generally speaking, a practical amplifier is only a linear device in its linear region, meaning that the output of the amplifier will not be exactly a scaled copy of the input signal when the amplifier works beyond the linear region. Considering an amplifier as a functional box, it can be modeled by a Taylor series [22,23]. Using the OFDM signal equivalent mathematical model $s(t)$ in Equation 3.7, the output of an amplifier generally can be written as

$$y(t) = O\{s(t)\} = F[r(t)]\cos\{2\pi f_c t + \Phi[r(t)]\}, \tag{3.12}$$

where $O\{\cdot\}$ denotes the operation of the amplifier, $F[\cdot]$ is amplitude-to-amplitude conversion (AM/AM), and $\Phi[\cdot]$ is amplitude-to-phase conversion (AM/PM). The functions $F[\cdot]$ and $\Phi[\cdot]$ are dependent on the nonlinearity of the amplifier and modeling type.

Since we are, generally, interested only in the output band near the carrier frequency f_c, the phase distortion in the band is negligible using a Taylor series model, that is, $\Phi[r(t)]=0$ [11]. Therefore, Equation 3.12 becomes

$$y(t) = O\{s(t)\} = F[r(t)]\cos(2\pi f_c t). \tag{3.13}$$

Assuming $\tilde{y}(t) = F[r(t)]$, the Taylor expansion of $O\{s(t)\}$ can be used to determine $\tilde{y}(t)$. Generally, the Taylor model of a high-power amplifier can be written as

$$y(t) = \sum_{i=0}^{\infty} a_{2i+1} s^{2i+1}(t). \tag{3.14}$$

Here, only the odd-order terms in the Taylor series are considered; since the spectra generated by the even-order terms are at least f_c away from the center of the passband, the effects from these terms on the passband are negligible. Furthermore, as a linear amplifier, the third- and fifth-order terms dominate in Equation 3.14 for distortion. Therefore, in this analysis, the following model is used for a high-power amplifier:

$$y(t) = a_1 s(t) + a_3 s^3(t) + a_5 s^5(t). \tag{3.15}$$

Here, the coefficient a_1 is related to the linear gain G of the amplifier, and the coefficients a_3 and a_5 are directly related to IP$_3$ and IP$_5$, respectively. For an amplifier with gain compression ($a_3 < 0$), it can be proven after a lengthy derivation that the expression for these coefficients becomes [25]

$$a_1 = 10^{G/20}, \quad a_3 = -\frac{2}{3}10^{((-\mathrm{IP}_3/10)+(3G/20))}, \quad a_5 = -\frac{2}{5}10^{(-(\mathrm{IP}_5/5)+(G/4))}. \tag{3.16}$$

Substituting the input passband signal $s(t) = r(t)\cos(2\pi f_c t)$ into $y(t)$ of Equation 3.15, we can write $y(t)$ as follows if the components of the passband are ignored:

$$y(t) = \tilde{y}(t)\cos(2\pi f_c t), \tag{3.17}$$

where

$$\tilde{y}(t) = \tilde{a}_1 r(t) + \tilde{a}_3 r^3(t) + \tilde{a}_5 r^5(t) \tag{3.18}$$

with

$$\tilde{a}_1 = a_1, \quad \tilde{a}_3 = \frac{3}{4}a_3, \quad \tilde{a}_5 = \frac{5}{8}a_5. \tag{3.19}$$

From Equations 3.16 through Equation 3.19, it can be seen that an amplifier's output $y(t)$ is a function of G, IP_3, IP_5, and the input signal $s(t)$. Consequently, using Equation 3.17 and the PSD of $s(t)$, the PSD of $y(t)$ can be calculated and the power emission levels can be determined. Therefore, all of the nonlinear effects of the high-power amplifier with the 802.11a OFDM signals can be evaluated.

Before proceeding further, it is worth examining the method for measuring the nonlinear parameters IP_3 and IP_5 for a given transistor or amplifier. For most power transistors, the normally obtained IP_3 parameters are listed in the data books. The actual IP_3 of an amplifier is usually measured using a two-tone test [26] as shown in Figure 3.9. According to this test, IP_3 is calculated by

$$IP_3 = P_t + \frac{IM_3}{2}, \tag{3.20}$$

where P_t is the power of the original tone signals at the output. This expression is derived directly from the geometric relation shown in Figure 3.9. In order to measure IP_3 accurately, the tone signal's power P_t should be chosen low enough so that the fifth-order intermodulation IM_5 can be ignored at the output.

Unfortunately, IP_5 is usually not provided in the data books. However, it may also be measured by the two-tone test. Similarly to Equation 3.20, IP_5 can be determined by

$$IP_5 = P_t + \frac{IM_5}{4}. \tag{3.21}$$

The IM_5 measurement is shown in Figure 3.10. In this test, the power level of P_t is higher than the power level of the IM_3 measurement so that IM_5 can be measured reliably.

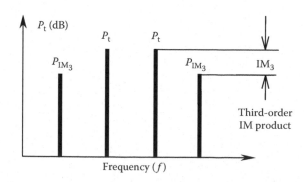

Figure 3.9 Two-tone test for the third-order intermodulation levels.

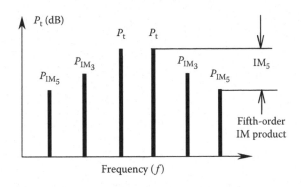

Figure 3.10 Two-tone test for the fifth-order intermodulation levels.

3.5 PSD of Amplified 802.11a OFDM Signal

Since $y(t) = \tilde{y}(t) \cdot \cos(2\pi f_c t + \theta)$, the PSD of $y(t)$ can be determined by the PSD of $\tilde{y}(t)$ as [22]

$$P_y(f) = \frac{1}{4}\left[P_{\tilde{y}}(f - f_c) + P_{\tilde{y}}(-f - f_c)\right] \tag{3.22}$$

and the PSD of $\tilde{y}(t)$ can be derived by the Wiener–Khintchine theorem as [22]

$$P_{\tilde{y}}(f) = \int_{-\infty}^{\infty} \phi_{\tilde{y}}(\tau) e^{-j2\pi f \tau}\, d\tau = F\{\phi_{\tilde{y}}(\tau)\}, \tag{3.23}$$

where $F\{\cdot\}$ is the Fourier transform of $\{\cdot\}$ and $\phi_{\tilde{y}}(\tau)$ is the autocorrelation of $\tilde{y}(t)$. By definition, $\phi_{\tilde{y}}(\tau)$ can be expressed as

$$\phi_{\tilde{y}}(\tau) = E\{\tilde{y}(t)\tilde{y}(t + \tau)\}, \tag{3.24}$$

where $E\{\cdot\}$ is the mathematical expectation of $\{\cdot\}$.

Since $\tilde{y}(t) = \tilde{a}_1 r(t) + \tilde{a}_3 r^3(t) + \tilde{a}_5 r^5(t)$, $P_{\tilde{y}}(f)$ can be expressed as

$$
\begin{aligned}
P_{\tilde{y}}(f) &= F\left\{\phi_{\tilde{y}}(\tau)\right\} = F\left\{E\left\{\tilde{y}(t)\tilde{y}(t + \tau)\right\}\right\} \\
&= F\left\{E\left\{\left[\tilde{a}_1 r(t) + \tilde{a}_3 r^3(t) + \tilde{a}_5 r^5(t)\right] \times \left[\tilde{a}_1 r(t + \tau) + \tilde{a}_3 r^3(t + \tau) + \tilde{a}_5 r^5(t + \tau)\right]\right\}\right\} \\
&= F\left\{E\left\{\tilde{a}_1^2 r(t) r(t + \tau)\right\} + E\left\{\tilde{a}_1\tilde{a}_3 r(t) r^3(t + \tau)\right\} + E\left\{\tilde{a}_1\tilde{a}_5 r(t) r^5(t + \tau)\right\}\right. \\
&\quad + E\left\{\tilde{a}_1\tilde{a}_3 r^3(t) r(t + \tau)\right\} + E\left\{\tilde{a}_3^2 r^3(t) r^3(t + \tau)\right\} + E\left\{\tilde{a}_3\tilde{a}_5 r^3(t) r^5(t + \tau)\right\} \\
&\quad \left. + E\left\{\tilde{a}_1\tilde{a}_5 r^5(t) r(t + \tau)\right\} + E\left\{\tilde{a}_3\tilde{a}_5 r^5(t) r^3(t + \tau)\right\} + E\left\{\tilde{a}_5^2 r^5(t) r^5(t + \tau)\right\}\right\}.
\end{aligned}
\tag{3.25}
$$

Obviously, the first term in the Fourier transform of Equation 3.25 can be easily expressed as

$$E\{\tilde{a}_1^2 r(t) r(t + \tau)\} = \tilde{a}_1^2 E\{r(t) r(t + \tau)\} = \tilde{a}_1^2 \phi_r(\tau), \tag{3.26}$$

where $\phi_r(\tau)$ is the autocorrelation of $r(t)$.

After a lengthy and tedious derivation [27], the left-hand side terms in the Fourier transform of Equation 3.25 can also be described as functions of $\phi_r(\tau)$:

$$E\{\tilde{a}_1 \tilde{a}_3 r(t) r^3(t + \tau)\} = E\{\tilde{a}_1 \tilde{a}_3 r^3(t) r(t + \tau)\} = \tilde{a}_1 \tilde{a}_3 [3\sigma_r^2 \phi_r(\tau)], \tag{3.27}$$

$$E\{\tilde{a}_1 \tilde{a}_5 r(t) r^5(t + \tau)\} = E\{\tilde{a}_1 \tilde{a}_5 r^5(t) r(t + \tau)\} = \tilde{a}_1 \tilde{a}_5 [15\sigma_r^4 \phi_r(\tau)], \tag{3.28}$$

$$E\{\tilde{a}_3 \tilde{a}_5 r^3(t) r^5(t + \tau)\} = E\{\tilde{a}_3 \tilde{a}_5 r^5(t) r^3(t + \tau)\} = \tilde{a}_3 \tilde{a}_5 [45\sigma_r^6 \phi_r(\tau) + 600\sigma_r^2 \phi_r^3(\tau)], \tag{3.29}$$

$$E\{\tilde{a}_3^2 r^3(t) r^3(t + \tau)\} = \tilde{a}_3^2 E\{r^3(t) r^3(t + \tau)\} = \tilde{a}_3^2 [9\sigma_r^4 \phi_r(\tau) + 6\phi_r^3(\tau)], \tag{3.30}$$

$$E\{\tilde{a}_5^2 r^5(t) r^5(t + \tau)\} = \tilde{a}_5^2 E\{r^5(t) r^5(t + \tau)\}$$
$$= \tilde{a}_5^2 [255\sigma_r^8 \phi_r(\tau) + 600\sigma_r^4 \phi_r^3(\tau) + 120\phi_r^5(\tau)], \tag{3.31}$$

where σ_r is the standard deviation of $r(t)$.

Substituting Equations 3.26 through 3.31 into Equation 3.25, $P_{\tilde{y}}(f)$ is written as

$$P_{\tilde{y}}(f) = F\{\phi_{\tilde{y}}(\tau)\}$$
$$= F\{\tilde{a}_1^2 \phi_r(\tau) + 2\tilde{a}_1 \tilde{a}_3 [3\sigma_r^2 \phi_r(\tau)] + 2\tilde{a}_1 \tilde{a}_5 [15\sigma_r^4 \phi_r(\tau)]$$
$$+ \tilde{a}_3^2 [9\sigma_r^4 \phi_r(\tau) + 6\phi_r^3(\tau)] + 2\tilde{a}_3 \tilde{a}_5 [45\sigma_r^6 \phi_r(\tau) + 600\sigma_r^2 \phi_r^3(\tau)]$$
$$+ \tilde{a}_5^2 [255\sigma_r^8 \phi_r(\tau) + 600\sigma_r^4 \phi_r^3(\tau) + 120\phi_r^5(t)]\}$$
$$= (\tilde{a}_1^2 + 6\tilde{a}_1 \tilde{a}_3 \sigma_r^2 + 30\tilde{a}_1 \tilde{a}_5 \sigma_r^4 + 9\tilde{a}_3^2 \sigma_r^4 + 90\tilde{a}_3 \tilde{a}_5 \sigma_r^6 + 225\tilde{a}_5^2 \sigma_r^8) \times F\{\phi_r(\tau)\}$$
$$+ (6\tilde{a}_3^2 + 120\tilde{a}_3 \tilde{a}_5 \sigma_r^2 + 600\tilde{a}_5^2 \sigma_r^4) \times F\{\phi_r^3(\tau)\} + (120\tilde{a}_5^2) \times F\{\phi_r^5(\tau)\}. \tag{3.32}$$

Since $F\{\phi_r(\tau)\} = P_r(f)$, where $P_r(f)$ is the PSD of $r(t)$, we can have

$$F\{\phi_r^3(\tau)\} = F\{\phi_r(\tau)\} \otimes F\{\phi_r(\tau)\} \otimes F\{\phi_r(\tau)\} = P_1 \otimes P_1 \otimes P_1 = P_3, \tag{3.33}$$

$$F\{\phi_r^5(\tau)\} = F\{\phi_r(\tau)\} \otimes F\{\phi_r(\tau)\} \otimes F\{\phi_r(\tau)\} \otimes F\{\phi_r(\tau)\} \otimes F\{\phi_r(\tau)\}$$
$$= P_1 \otimes P_1 \otimes P_1 \otimes P_1 \otimes P_1 = P_5, \tag{3.34}$$

where $P_1 = P_r(f)$, $P_3 = P_1 \otimes P_1 \otimes P_1$, and $P_5 = P_1 \otimes P_1 \otimes P_1 \otimes P_1 \otimes P_1$, in which $\otimes$ denotes the convolution operator.

Therefore, $P_y(f)$ is finally expressed as a function of the PSD of $r(t)$, $P_r(f)$ as

$$P_{\tilde{y}}(f) = \left(\tilde{a}_1^2 + 6\tilde{a}_1\tilde{a}_3\sigma_r^2 + 30\tilde{a}_1\tilde{a}_5\sigma_r^4 + 9\tilde{a}_3^2\sigma_r^4 + 90\tilde{a}_3\tilde{a}_5\sigma_r^6 + 225\tilde{a}_5^2\sigma_r^8 \right) \times P_1$$
$$+ \left(6\tilde{a}_3^2 + 120\tilde{a}_3\tilde{a}_5\sigma_r^2 + 600\tilde{a}_5^2\sigma_r^4 \right) \times P_3 + \left(120\tilde{a}_5^2 \right) \times P_5. \tag{3.35}$$

However, this expression is not easy to use because it is not tied with IP_3 and IP_5, and the linear output power P_o of the amplifier yet. Since we have

$$\tilde{a}_1 = a_1 = 10^{G/20}, \quad \tilde{a}_3 = \frac{3}{4}a_3 = -\frac{1}{2}10^{(-(IP_3/10)+(3G/20))},$$
$$\tilde{a}_5 = \frac{5}{8}a_5 = -\frac{1}{4}10^{(-(IP_5/5)+(G/4))}, \tag{3.36}$$

and

$$P_o = \frac{a_1^2\sigma_r^2}{2} = \frac{a_1^2 N_{ST}}{4}, \tag{3.37}$$

where $\sigma_r^2 = N_{ST}/2$; after substituting Equations 3.36 and 3.37 into Equation 3.35 and inserting the result into Equation 3.22, we can obtain the final result of the power spectrum $P_y(f)$ of $y(t)$ in terms of the amplifier nonlinear parameters IP_3 and IP_5, and the linear output power P_o of the amplifier:

$$P_y(f) = \left(2P_o - 12P_o^2 10^{(-IP_3/10)} - 60P_o^3 10^{(-IP_5/5)} + 18P_o^3 10^{(-IP_3/5)} \right.$$
$$\left. + 180P_o^4 10^{(-IP_3/10)(-IP_5/5)} + 450P_o^5 10^{(-2IP_5/5)} \right) \frac{P_1(f - f_c)}{N_{ST}}$$
$$+ \left(48P_o^3 10^{(-IP_3/5)} + 960P_o^4 10^{((-IP_3/10)-(IP_5/5))} + 4800P_o^5 10^{(-2IP_5/5)} \right) \frac{P_3(f - f_c)}{N_{ST}^3}$$
$$+ \left(3840P_o^5 10^{(-2IP_5/5)} \right) \frac{P_5(f - f_c)}{N_{ST}^5}, \tag{3.38}$$

where f_c is the carrier frequency of the WLAN OFDM system, $P_1 = P_r(f) = (R_s/4) \sum_{\substack{k=-N_{ST}/2 \\ k \neq 0}}^{N_{ST}/2}$
$\left[|W(f - f_k)|^2 + |W(-f - f_k)|^2 \right]$, $P_3 = P_1 \otimes P_1 \otimes P_1$, and $P_5 = P_1 \otimes P_1 \otimes P_1 \otimes P_1 \otimes P_1$, in which $f_k = k\Delta f$, $k = \pm 1, \pm 2, \ldots, \pm N_{ST}/2$.

The derived spectrum $P_y(f)$ is shown in Figure 3.11 as a solid line with output power as 10 W, $IP_3 = 57$ dBm, and $IP_5 = 52$ dBm.

If IP_5 is ignored, Equation 3.38 will become

$$P_y(f) = \left(2P_o - 12P_o^2 10^{-IP_3/10} + 18P_o^3 10^{-IP_3/5} \right) \frac{P_1(f - f_c)}{N_{ST}} + 48P_o^3 10^{-IP_3/5} \times \frac{P_3(f - f_c)}{N_{ST}^3}. \tag{3.39}$$

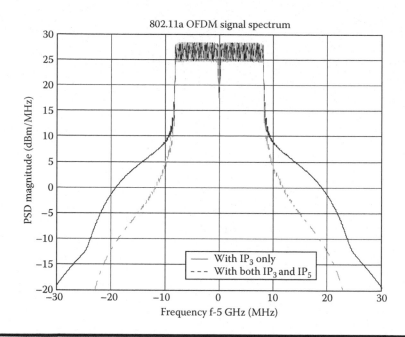

Figure 3.11 The theoretical PSD of amplified 802.11a signal (solid line: with both IP$_3$ and IP$_3$; dashed line: with IP$_3$ alone).

Thus, Equation 3.39 is a special case of Equation 3.38. It provides a simpler and easier result to use for the case when IP$_5$ is ignored. The $P_y(f)$ of Equation 3.39 is shown in Figure 3.11 as a dashed line. It is clearly shown that when IP$_5$ is relatively high compared to IP$_3$, such as IP$_3$ = 57 dBm and IP$_5$ = 52 dBM in this example, the out-of-band emission power levels caused by fifth-order intermodulation are significant.

The result can turn out to be a closed form if the following model is chosen to represent $P_r(f)$, which is the PSD of baseband signal $r(t)$:

$$P_1(f) = P_r(f) = \begin{cases} N_0 & \text{for } |f| \le B, \\ N_0 e^{-\alpha(|f|-B)} & \text{for } |f| > B, \end{cases} \quad (3.40)$$

where α (α ranging from 0^+ to ∞) is like a roll-off factor in this model, and B is the signal bandwidth. When $\alpha \to \infty$, the model of $P_r(f)$ degenerates into a rectangular pulse with bandwidth B.

After a long and tedious deviation, we get

$$P_3(f) = P_1(f) \otimes P_1(f) \otimes P_1(f)$$

$$= \frac{3N_0^3}{4\alpha^2} \left[e^{-\alpha(f+B)} + e^{\alpha(f-B)} \right] + 3N_0^3 B^2 - N_0^3 f^2 + \frac{6}{\alpha} B N_0^3 \quad \text{for } |f| \le B$$

$$\frac{N_0^3}{4\alpha^2} \left\{ 3e^{-\alpha(|f|+B)} + e^{-\alpha(|f|-B)} \left[-6\alpha \left(|f| - B \right) - 21 \right] \right.$$

$$\left. + 2 \left\{ \left[\alpha \left(|f| - 3B \right) - 3 \right]^2 + 3 \right\} \right\} \quad \text{for } B < |f| \le 3B$$

$$\frac{N_0^3}{4\alpha^2} \left\{ 3e^{-\alpha(|f|+B)} + e^{-\alpha(|f|-B)} \left[-6\alpha\left(|f| - B\right) - 21 \right] \right.$$
$$\left. + 2\left\{ [\alpha(|f| - 3B) + 3]^2 + 3 \right\} e^{-\alpha(|f|+3B)} \right\} \quad \text{for } 3B < |f|$$

$$(3.41)$$

and

$$P_5(f) = P_1(f) \otimes P_1(f) \otimes P_1(f) \otimes P_1(f) \otimes P(f)_1$$
$$= \frac{N_0^5}{48\alpha^4} \left\{ 15e^{\alpha(f-3B)} + [60\alpha(f - B) - 390]e^{\alpha(f-B)} + \alpha^4(12f^4 - 120f^2B^2 + 460B^4) \right.$$
$$+ \alpha^3(1840B^3 - 240f^2B) + \alpha^2(2160B^2 + 240f^2) + 480\alpha B + 960$$
$$\left. - [60\alpha(f + B) + 390]e^{-\alpha(f+B)} + 15e^{-\alpha(f+3B)} \right\}$$

for $|f| \le B$.

$$\frac{N_0^5}{48\alpha^4} \left\{ 15e^{-\alpha(|f|+3B)} - [60\alpha\left(|f| + B\right) + 390]e^{-\alpha(|f|+B)} \right.$$
$$+ \left[60\alpha^2(|f| - B)^2 + 720\alpha\left(|f| - B\right) + 2490 \right]e^{-\alpha(|f|-B)}$$
$$- \alpha^4\left(8f^4 - 80|f|^3B - 80|f|B^3 + 240f^2B^2 - 440B^4 \right)$$
$$+ \alpha^3\left(80|f|^3 - 480f^2B + 240|f|B^2 + 1760B^3 \right)$$
$$\left. - \alpha^2\left(480f^2 - 1440|f|B - 1440B^2 \right) + \alpha(1440|f| - 960B) - 1920 \right\}$$

for $B < |f| \le 3B$.

$$\frac{N_0^5}{48\alpha^4} \left\{ 15e^{-\alpha(|f|+3B)} - [60\alpha\left(|f| + B\right) + 390]e^{-\alpha(|f|+B)} \right.$$
$$+ \left[60\alpha^2\left(|f| - B\right)^2 + 720\alpha\left(|f| - B\right) + 2490 \right]e^{-\alpha(|f|-B)}$$
$$- \left[20\alpha^3\left(|f| - 3B\right)^3 + 330\alpha^2\left(|f| - 3B\right)^2 + 2130\alpha\left(|f| - 3B\right) + 5265 \right]e^{-\alpha(|f|-3B)}$$
$$+ \left[2\alpha^4\left(|f| - 5B\right)^4 - 40\alpha^3\left(|f| - 5B\right)^3 + 360\alpha^2\left(|f| - 5B\right)^2 \right.$$
$$\left. \left. - 1680\alpha\left(|f| - 5B\right) + 3360 \right] \right\}$$

for $3B < |f| \le 5B$.

$$\frac{N_0^5}{48\alpha^4} \left\{ 15e^{-\alpha(|f|+3B)} - [60\alpha\left(|f| + B\right) + 390]e^{-\alpha(|f|+B)} \right.$$
$$+ \left[60\alpha^2\left(|f| - B\right)^2 + 720\alpha\left(|f| - B\right) + 2490 \right]e^{-\alpha(|f|-B)}$$

$$-\left[20\alpha^3\left(|f|-3B\right)^3+330\alpha^2\left(|f|-3B\right)^2+2130\alpha\left(|f|-3B\right)+5265\right]e^{-\alpha(|f|-3B)}$$

$$+\left[2\alpha^4\left(|f|-5B\right)^4+40\alpha^3\left(|f|-5B\right)^3+360\alpha^2\left(|f|-5B\right)^2\right.$$

$$\left.+\,1680\alpha\left(|f|-5B\right)+3360\right]\times e^{-\alpha(|f|-5B)}\Big\}$$

for $5B<|f|$. (3.42)

Several observations are made by inspecting Equations 3.39, 3.41, 3.42, and the data plot in Figure 3.11:

a. In the passband $|f-f_c|\le B$, the first term $(2P_o/N_{ST})\cdot P_1(f-f_c)$ corresponds to the linear output power density; the remaining terms in the passband are caused by the nonlinearity. In other words, these remaining terms are due to the cross-modulation. For a linear amplifier, the cross-modulation usually is much lower than the linear output power. Therefore, the cross-modulation does not affect the passband spectrum significantly.

b. In the band $B<|f-f_c|\le 3B$, the nonzero PSD is generated by the third-order as well as the fifth-order intermodulation. This result shows that the out-of-band spectrum density is determined completely by the intermodulation. These out-of-band frequency components are usually called spectrum regrowth. This region contains the most harmful out-of-band emission.

c. In the band $3B<|f-f_c|\le 5B$, the nonzero power density is generated by the fifth-order intermodulation. In practice, if the output power P_0 is low, the effect of the fifth-order intermodulation is considered negligible. The spectrum regrowth in this band may be so small that it is covered by noise.

d. In the band $5B<|f-f_c|$, the power density is still nonzero. This result is obtained because the signal was assumed to be band-unlimited in the derivation. The emission in this band will very likely be covered by noise.

Furthermore, with the explicit power spectrum of the output 802.11a OFDM signal, the out-of-band spurious emission power may be calculated in a particular frequency band. It is this power that is used in the 802.11a standard to specify the limit for the out-of-band control. To keep the result easy to use, only IP_3 is considered here.

Let a frequency band be defined by f_1 and f_2 outside the passband. Using the results from $P_y(f)$ of Equation 3.39, the emission power level within the band (f_1,f_2), denoted as $P_{IM_3}(f_1,f_2)$, can be determined easily by

$$P_{IM_3}(f_1,f_2)=\int_{f_1}^{f_2}P_y(f)\,df=\left(2P_o-12P_o^2 10^{-IP_3/10}+18P_o^3 10^{-IP_3/5}\right)\times\frac{\int_{f_1}^{f_2}P_1(f-f_c)df}{N_{ST}}$$

$$+48P_o^3 10^{-IP_3/5}\times\frac{\int_{f_1}^{f_2}P_3(f-f_c)\,df}{N_{ST}^3}. \tag{3.43}$$

Equation 3.43 can be also expressed as

$$C_1\times 10^{-IP_3/5}+C_2\times 10^{-IP_3/10}+C_3=0, \tag{3.44}$$

where

$$C_1 = \frac{48P_o^3}{N_{ST}^3} \int_{f_1}^{f_2} P_3(f - f_c)\, df + \frac{18P_o^3}{N_{ST}} \int_{f_1}^{f_2} P_1(f - f_c)\, df,$$

$$C_2 = -\frac{12P_o^2}{N_{ST}} \int_{f_1}^{f_2} P_1(f - f_c)\, df,$$

$$C_3 = \frac{2P_o}{N_{ST}} \int_{f_1}^{f_2} P_1(f - f_c)\, df - P_{IM_3}(f_1, f_2). \qquad (3.45)$$

In most design procedures, a designer is concerned with the required IP_3 for a given out-of-band emission level. To obtain the desired IP_3, Equation 3.44 is solved for IP_3 with the given $P_{IM_3}(f_1, f_2)$, which yields

$$IP_3 = -10 \times \log_{10}\left(\frac{-C_2 + \sqrt{C_2^2 - 4C_1 C_3}}{2C_1} \right), \qquad (3.46)$$

where C_1, C_2, and C_3 are described in Equation 3.45.

This result provides a direct relationship between the out-of-band emission power of a WLAN OFDM signal power amplifier and its IP_3.

3.6 Design Example and Comparison with Simulations

In this example, the result shown in Equation 3.46 is used to design a 10 W amplifier, which complies with the out-of-band emission level control requirement proposed for IEEE 802.11a standard. The out-of-band emission level controls required in 802.11a are given as follows:

The transmitted spectrum shall have a 0 dBr (dB relative to the maximum spectral density of the signal) bandwidth not exceeding 18 MHz, −20 dBr at 11 MHz frequency offset, −28 dBr at 20 MHz frequency offset, and −40 dBr at 30 MHz frequency offset and above. The transmitted spectral density of the transmitted signal shall fall within the spectral mask, as shown in Figure 3.1.

Therefore, for this amplifier, $P_o = 10$ W, and for the ($f_c + 11$ MHz) to ($f_c + 20$ MHz) band, the corresponding maximum $P_{IM_3}(f_1, f_2)$ is expressed as

$$P_{IM_3}(f_1, f_2) = 10 \times 10^{-28/10} = 0.0158 \text{ W}. \qquad (3.47)$$

For the worst case, f_1 and f_2 are assumed at the lower edge of $\left[f_c + 11 \text{ MHz}, f_c + 20 \text{ MHz} \right]$, that is, $f_1 = f_c + 11$ MHz and $f_2 = f_c + 20$ MHz.

Then, from Equation 3.46, the required IP_3 becomes $IP_3 = 57$ dBm. For the band described above, in order to meet 802.11a requirement, this amplifier must have an IP_3 of at least 57 dBm.

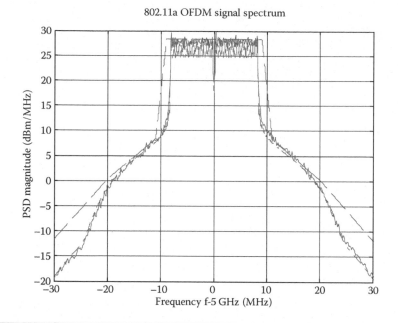

Figure 3.12 PSD of amplified IEEE 802.11a OFDM signal (solid line: signal spectrum generated by simulation; dash-dotted line: derived signal spectrum; dashed line: spectral mask given in IEEE 802.11a standard).

As mentioned before, IP_5 is not given in the data book. Fortunately, IP_5 could be measured through a two-tone test [26]. Therefore, without loss of generality, IP_5 can be assumed as 52 dBm at the same output power level.

In simulation, the WLAN OFDM signals were generated in accordance with IEEE 802.11a standard [6]. Figure 3.12 shows the derived power spectrum given by Equation 3.38 of this example compared to the spectrum generated by simulation with the spectral mask given in IEEE 802.11a standard [6]. The simulated RF amplifier spectrum agrees with the analytically predicted spectrum in both the in-band and the shoulder area.

References

1. R. van Nee and R. Prasad, *OFDM for wireless multimedia communications*, Artech House, Boston, 2000.
2. W. Y. Zou and Y. Y. Wu, "COFDM: an overview," *IEEE Transactions on Broadcasting*, 41, 1–8, 1995.
3. A. Bria, F. Gessler, O. Queseth, R. Stridh, M. Unbehaun, J. Wu; J. Zander, and M. Flament, Fourth-generation wireless infrastructures: scenarios and research challenges, *IEEE Personal Communications*, 8, 25–31, 2001.
4. J. Z. Sun, J. Sauvola, and D. Howie, Features in future: 4G visions from a technical perspective, in *Proc. IEEE GLOBECOM*, 2001, vol. 6, pp. 3533–3537.
5. S. Ohmori, Y. Yamao, and N. Nakajima, The future generations of mobile communications based on broadband access technologies, *IEEE Communications Magazine*, 38, 134–142, 2000.
6. IEEE, Supplement to standard for information technology—telecommunications and information exchange between systems—local and metropolitan area networks—specific requirements. Part 11: Wireless LAN medium access control (MAC) and physical layer (PHY) specifications: high speed physical layer in the 5 GHz band, Std. 802.11a, September 1999.

7. IEEE, Information technology—telecommunications and information exchange between systems—local and metropolitan area networks—specific requirements—Part 11: Wireless LAN medium access control (MAC) and physical layer (PHY) specifications, Std. 802.11, June 1997.

8. ETSI, "Broadband radio access networks (BRAN); HIPERLAN Type 2; System overview, TR 101 683 V1.1.1, February 2000.

9. B. R. Saltzberg, Performance of an efficient parallel data transmission system, *IEEE Transactions on Communications*, COM-15, 805–813, 1967.

10. G. Gonzalez, *Microwave Transistor Amplifiers: Analysis and Design*, 2nd Ed., Prentice Hall, Upper Saddle River, NJ, 1996.

11. M. C. Jeruchim, P. Balaban, and K. S. Shanmugan, *Simulation of Communication Systems*, Plenum Press, New York and London, 1992.

12. Q. Wu, M. Testa, and R. Larkin, Linear RF power amplifier design for CDMA signals, *IEEE International Microwave Symposium Digest*, 2, 851–854, 1996.

13. C. M. Liu, H. Xiao, Q. Wu, and F. Li, Spectrum design of RF power amplifier for wireless communication systems, *IEEE Transactions on Consumer Electronics*, 48, 72–80, 2002.

14. Q. Wu, H. Xiao, and F. Li, Linear RF power amplifier design for CDMA signals: a spectrum analysis approach, *Microwave Journal*, 41, 22–40, 1998.

15. C. M. Liu, H. Xiao, Q. Wu, and F. Li, Spectrum modeling of an RF power amplifier for TDMA signals, *Microwave Journal*, 44, 88–109, 2001.

16. C. M. Liu, H. Xiao, Q. Wu, and F. Li, Spectrum modeling of RF power amplifier for MIRS M-16 QAM signals, *International Journal of Electronics*, 89, 135–146, 2002.

17. J. Kruys, Standardization of wireless high speed premises data networks, Wireless ATM Workshop, Espoo, Finland, September 1996.

18. T. Pollet, M. Van Bladel, and M. Moeneclaey, BER sensitivity of OFDM systems to carrier frequency offset and wiener phase noise, *IEEE Transactions on Communications*, 43, 191–193, 1995.

19. M. Pauli and H. P. Kuchenbecker, Minimization of the intermodulation distortion of a nonlinearly amplified OFDM signal, *Wireless Personal Communications*, 4, 93–101, 1997.

20. C. Rapp, Effects of HPA-nonlinearity on a 4-DPSK/OFDM signal for a digital sound broadcasting system, *Proc. of the Second European Conference on Satellite Communications*, 1991, pp. 179–184.

21. R. van Nee, OFDM codes for peak-to-average power reduction and error correction, *IEEE Global Telecommunications Conference 1996 (GLOBECOM 1996)*, London, November 1996, vol. 1, pp. 18–22.

22. L. W. Couch II, *Digital and Analog Communication Systems*, Prentice Hall, Upper Saddle River, NJ, 1996.

23. T. S. Rappaport, *Wireless Communication Principles and Practice*, Prentice Hall, Upper Saddle River, NJ, 1996.

24. C. M. Liu and F. Li, Spectrum modeling of OFDM signals for WLAN, *IEE Electronics Letters*, 40, 1431–1432, 2004.

25. H. Xiao, Spectrum modeling for linear RF power amplifier design for digital cellular communication signals, Ph.D. dissertation, Portland State University, May 1999.

26. H. Xiao, Q. Wu, and F. Li, Measure a power amplifier's fifth-order interception point, *RF Design*, pp. 54–56, April 1999.

27. C. M. Liu, Spectral modeling and nonlinear distortion analysis of OFDM based wireless LAN signals, Ph.D. dissertation, Portland State University, June 2005.

Chapter 4

Game Theory and OFDMA Resource Allocation

Quang Duy La, Boon Hee Soong, Yong Huat Chew,
and Woon Hau Chin

Contents

4.1 Introduction: Efficient Resource Management 68
    4.1.1 Principles 68
    4.1.2 Dynamic Resource Allocation and OFDMA 69
4.2 Game Theoretical Framework 69
    4.2.1 Literature Review and Motivations 70
    4.2.2 Noncooperative Games: Concept of Nash Equilibrium 71
    4.2.3 Cooperative Games: Bargaining Solutions 72
4.3 OFDMA System Model 74
    4.3.1 Single-Cell Scenario 74
    4.3.2 Multi-Cell Scenario 75
4.4 Noncooperative Game Solutions 76
    4.4.1 Game Formulation 76
    4.4.2 Iterative Best-Response Noncooperative Game 77
    4.4.3 Noncooperative Game with Virtual Referee 79
        4.4.3.1 Power Minimization Game 80
        4.4.3.2 Virtual Referee Game: Rate Adjustment and Channel Removal 81
    4.4.4 Numerical Results 83
        4.4.4.1 Nash Equilibrium Convergence for the Iterative Best-Response Game. 84
        4.4.4.2 Power and Rate Allocation for the Virtual Referee Game 85
4.5 Cooperative Game Solutions 85
    4.5.1 Game Formulation 86

4.5.2 The Pair-Bargaining with Coalition Algorithm.................................... 88
    4.5.2.1 The Overall Multiuser Algorithm 88
    4.5.2.2 Grouping Strategies... 89
    4.5.2.3 Two-User Bargaining Algorithm....................................... 90
4.5.3 Round-Robin Carrier Assignment Algorithm 92
    4.5.3.1 Theoretical Evaluation of Bargaining Solutions......................... 92
    4.5.3.2 The Algorithm.. 93
4.5.4 Comparison of the Four Bargaining Solutions and Numerical Results........... 94
4.6 Concluding Remarks... 98
4.6.1 Summary.. 98
4.6.2 Open Issues... 99
References... 99

4.1 Introduction: Efficient Resource Management

In this chapter, issues of resource allocation in wireless communications are first discussed. We emphasize the need for adaptive, dynamic allocation algorithms in resource allocation problems, which are often treated as mathematical optimization. Next, we review the basics of game theory and the motivation to use it as a tool for allocating resources in communication networks.

4.1.1 Principles

Wireless communication systems suffer from the difficult and unforeseeable behaviors of wireless radio channels. More importantly, the use of radio frequency resources (channels and bandwidth) always poses a challenge to system managers as they are of a much more limited amount than those in cable or fiber optic communication networks. This may become a serious problem in the near future as the demand for more traffic is growing at so fast a pace that the availability of new radio resources is inadequate.

In 2001, the first IEEE 802.16 standard was approved, followed by several amendments years afterwards [1–5]. The IEEE 802.16 specifies a non-line-of-sight environment for the frequency band of 2–11 GHz; this allows OFDM/OFDMA (orthogonal frequency division multiple access) to be included (WirelessMAN-OFDM and WirelessMAN-OFDMA), which can combat multipath propagation. However, the resource allocation schemes and algorithms remain an open issue [6]. In short, the OFDMA system proves to be a promising technique, but its deployment was not fully utilized up to the present and a lot of research is focusing on the resource management aspects.

This urges us to revisit the prevailing principles of resource management in the present and next generations of wireless networks, as mentioned by Pietryk [6]. Nowadays, wireless communication networks must support connectivity for not a single user but for multiple users, which can be technically referred to as *multiple access*. In multiple-access systems, several users who are connected to the network will have to share a common pool of system resources. Familiar examples include time division multiple access (TDMA), frequency-division multiple access (FDMA), and code division multiple access (CDMA), where users are assigned different timeslots, frequencies, or codes. More advanced technologies include OFDMA, where the common radio resource is a set of orthogonal frequency channels.

Sharing of resources does not always mean that we divide the "cake" into different pieces and give one to each user. Resources (especially a spectrum) can also be *reused* to increase the utilization efficiency and accommodate more users, provided that careful designs are taken to ensure that conflict does not take place. Here, more than one user can share the same resource (e.g., frequency band) simultaneously. Obviously, they would cause interference to each other, but the effect of this interference can be carefully controlled. Owing to this constraint, a performance limitation exists for every system.

Another closely coupled issue in resource management is the need to support individual *quality of service* (QoS). This need stems from the fact that data traffic is becoming more and more complex, diversified, and application dependent. We refer to the OSI seven-layer hierarchical model: on top lies the Application layer, which consists of various user application programs such as messaging, multimedia streaming, Internet browsing, and social blogging. Each type of traffic has its own profile, and supporting the various traffic profiles is known as provisioning of QoS. The evolution of these various traffic types makes it hard for us to predict the dominant ones in the future. That diversified trend of many emerging new traffic services will continue to grow. Future wireless networks, however, must be able to accommodate any type of application service, requiring a flexible and customizable network capable of quickly responding to market demand. In order to achieve this, it is extremely crucial to provide sufficient QoS to users and services.

4.1.2 Dynamic Resource Allocation and OFDMA

Having laid out the basic principles for an effective resource allocation mechanism, we emphasize the importance of dynamic resource allocation. Early systems such as first-generation mobiles employed *fixed allocation*, in which channels or timeslots are permanently assigned to users without overlapping. This is a simple design and does not require knowledge of the environment, but can result in a waste of resources. Unlike fixed resource allocation, *dynamic allocation* enables users to adapt parameters for the appropriate use and reuse of resources to minimize interference and increase overall network performance. In other words, it allocates resources according to users' need and quality. The requirement for complex algorithms and detailed knowledge of the environment may prove to be a heavy burden, but can be overcome by today's advances in hardware and computer technology. The types and nature of such complex algorithms are very diversified.

Many contemporary works have recognized the importance of dynamic resource allocation for OFDMA, for examples [7–10] and references therein. It must be pointed out that, in whatever allocation schemes proposed, all schemes seek an *optimum allocation point*. As a matter of fact, they are all trying to solve a *mathematical optimization problem*.

In this chapter, we will look at this optimization problem from one particular angle, which is the game theoretical viewpoint. The next section offers a closer look at this approach.

4.2 Game Theoretical Framework

Game theory [11–13] has long been exploited in microeconomics to deal with competition among selfish, intelligent decision makers. It is a useful tool for studying interactive behaviors individually and collectively under conflict of interest. Recently, it has been adopted by the communication engineering community to solve certain optimization problems, for example, CDMA power control [14], cognitive radio [15], and OFDMA resource allocation, as we shall see later on. Firstly, we review the problem of game theory on OFDMA and its motivations. This is followed by a general game theoretical framework for both noncooperative and cooperative situations.

4.2.1 Literature Review and Motivations

The issue of efficient resource management in OFDMA systems was addressed back in the 1990s and several algorithms have been developed since then. However, studies of the game theoretical method and its application to OFDMA resource allocation have been carried out only recently. In MacKenzie and Wicker's preliminary but useful paper [16], the authors laid out the game theoretical framework and concepts applicable to communication systems. They explained the importance and potential of game theory in solving resource allocation problems in distributed networks. However, the authors did not apply their methods to the OFDMA system. Later, OFDMA works based on game theory began to emerge and can be divided into two categories: *noncooperative* and *cooperative* *approaches.*

The first noncooperative game theoretical approach to OFDMA was introduced by modeling the traditional power minimization problem as a game among multiple cells, according to Zhu Han et al. [17,18]. The significant aspect of the papers was existence of a final allocation point and improvement of the results over traditional convex optimization methods. It was shown that total power consumption was decreased using the proposed algorithm. Kwon and Lee [19] attempted to solve the rate optimization of separate users in multicell OFDMA also with their noncooperative game. They introduced a concept of "net utility" with power pricing. The proposed method was a greedy algorithm for assigning channels to users, coupled with best-response waterfilling for power control. Thus, their channel assignment allowed less complexity in computation such that no looping was required. Nevertheless, the minor issue of fluctuation in allocation results when the equilibrium point did not exist needed to be addressed. The same idea as Kwon and Lee's was adopted by a few others. For example, Wang et al. [20] also looked at multicell OFDMA but with an alternative utility function. The net utility concept was also used by Yu et al. for their OFDMA-relay network [21] and by Chen et al. in an OFDM power distribution game [22].

Other authors have also looked at the cooperative game theoretical approach. Among them, the optimal rate allocation for multiple users in single-cell OFDMA was associated with the Nash bargaining solution [23–25] and the Kalai–Smorodinsky bargaining solution [25]. Zhu Han et al. [23,24] proposed a novel allocation scheme for two users and expanded it for multiple users using the concept of coalition. The computational complexity of this algorithm was, however, reasonably high at degree $O(RK^2N \log N + RK^4)$. Chee et al. [25] suggested a reduced complexity to degree $O(KN + K^2)$. A comparison for different fairness schemers was later obtained in the work by Ibing and Boche [26], in which four criteria were weighted against each other, namely *utilitarian, egalitarian, Nash bargaining,* and *Kalai–Smorodinsky bargaining solutions.* From the above work, some insights to the question of which solution is the fairest and most efficient were offered.

The results from all the above works imply that the game theoretical approach to OFDMA resource allocation is a new but promising area of research that can contribute to the broader topic of efficient spectrum management for future wireless-access networks. It is worth mentioning the most attractive features of game theory that motivate researchers:

- *Rational decision making*: Intelligence and rationality are two basic assumptions in game theory. Ironically, human decision making hardly satisfies the two assumptions in the strict sense. However, if we consider a game among various computer stations, which are programmed to always aim at maximizing a predefined objective function, then we can expect the two assumptions to hold.
- *Distributed optimization*: As communication systems grow larger and larger in scale, the traditional centralized allocation mechanism (in which a central authority monitors all the tasks)

becomes more and more difficult and computationally infeasible. So one option is to allow each party to participate in the allocation process. Noncooperative game theory that specializes in solving conflict in a distributed context can be very useful in this situation.

■ *Inherent risks and uncertainties*: On very few occasions the outcome of a game can be predicted with exact certainty due to randomness and imperfect information; hence players have to face the risks of loss in utility. The same argument applies very well to wireless communication systems where users cannot have full knowledge of the channel conditions of others to avoid interference.

In Section 4.2.2 we will cover some basic concepts in game theory. The presentation is limited to a minimal level that is sufficient for readers to understand the discussion later in the chapter. For thorough and in-depth treatments of game theory, it is suggested that readers refer to the dedicated texts of Tirole and Fudenberg [12] or Myerson [13] or of Thomson [27], which deals with cooperative bargaining models.

4.2.2 Noncooperative Games: Concept of Nash Equilibrium

A game can always be established if its three fundamental elements are clearly defined:

■ *The set of players*
■ *The strategies associated with the players*—All the actions that a player can possibly select from
■ *The utilities (payoffs) for the players*—A function/rule that governs the payoff that a player will be awarded for taking a certain move given the other players' moves

We restrict the discussion to games in *strategic form*, which is widely used. In this form the actions of every player happen simultaneously (i.e., no sequential course of action) and so the full set of outcomes can be represented in tabular form.

Formally, let $\mathcal{G}$ be a game. Then,

$$\mathcal{G} = [\mathcal{N}, \{S_i\}_{i \in \mathcal{N}}, \{U_i\}_{i \in \mathcal{N}}]. \tag{4.1}$$

Here $\mathcal{N} = \{1, 2, \ldots, n\}$ is the set of n participating players. The strategy set of the ith player is

$$S_i = \{s_i | s_i \quad \text{is a strategy for the } i\text{th player}\}. \tag{4.2}$$

Further, we denote the *strategy space* as $\mathbb{S} = S_1 \times S_2 \times \cdots \times S_n$. Then, each element $\sigma = [s_1, s_2, \ldots, s_n] \in \mathbb{S}$ is said to be a *strategy profile*. Sometimes, we are only interested in one player (say, the ith player), so we rewrite $\sigma = [s_i, s_{-i}]$ where s_{-i} is the (joint) strategy adopted by its opponents. Then the utility function for each user i is a function with a real value of σ, that is, $U_i = f_i(\sigma) : \mathbb{S} \to \mathbb{R}$, where $\mathbb{R}$ is the set of real number. The notion $U_i(\sigma)$ will be frequently used.

The strategy set can contain discrete actions (e.g., rock/scissors/paper) or continuous actions (e.g., power level from an interval). In the discrete case (known as finite strategy game), it may be necessary to distinguish *pure strategy* from *mixed strategy*. A pure strategy requires the player to play a certain move with absolute certainty, that is with probability 1. Meanwhile, in a mixed strategy, a portfolio of strategies is defined with a predetermined probability assigned to each component. For simplicity, we only consider a pure strategy.

Nash equilibrium is a crucial concept in predicting a game's outcome. By definition, a Nash equilibrium is a strategy profile in which given that the others' strategies remain unaltered, no player would be tempted to change course from its current strategy. Mathematically, σ is a Nash equilibrium if and only if

$$U_i(\sigma_i, \sigma_{-i}) \geq U_i(\sigma_i', \sigma_{-i}) \quad \forall i \in \mathcal{N}. \tag{4.3}$$

It is perhaps the most important concept in game theory. At Nash equilibrium no user is able to gain by deviating from the point. Thus it is seen as a kind of "stable operating point" for the system point of view. As a celebrated result in game theory, the existence of Nash equilibrium in certain games is already established in the following theorems [12]:

THEOREM 4.1

A game has a Nash equilibrium if for all i $\in \mathcal{N}$, *the sets* S_i *are nonempty, convex, and a compact subset of a Euclidean space, and the utility function is continuous and quasi-concave in each* s_i.

THEOREM 4.2

Every finite strategy game has a Nash equilibrium in mixed strategy.

Another interesting concept is *Pareto optimal* strategies. They refer to strategies from which no deviation can mutually improve the payoffs of all players.

4.2.3 Cooperative Games: Bargaining Solutions

Cooperation permits players to exchange information and come to agreements with each other. In the context of this chapter, we are only interested in a special problem, which is the bargaining game. In such a problem, the players act together to divide a certain amount of resources among each other until an agreement is made. A bargaining problem is usually denoted as a pair $\mathcal{B} = (\mathbf{S}, \mathbf{d})$. Here, $\mathbf{S} \subseteq \mathbb{R}^n$ is a closed and convex set called the *set of feasible outcomes*, and $\mathbf{d} \in \mathbf{S}$ is referred to as the *status quo* or *disagreement point*. This is the default outcome when no agreement is reached.

A bargaining solution σ results from some rules that can find a unique feasible agreement point to the bargaining problem, that is $\sigma = f(\mathbf{S}, \mathbf{d})$. Bargaining theory can show that without cooperation, players should expect the noncooperative outcome, which is actually utilities at status quo, and a bargaining agreement should help all players to mutually improve their own utilities [13]. Thus, it requires that $\mathbf{S}$ must contain some points that dominate $\mathbf{d}$, that is $\exists \mathbf{s} \in \mathbf{S}$, $\mathbf{s} \succeq \mathbf{d}$, for the game to be nontrivial. Moreover, a solution may be characterized axiomatically by making them satisfy a collection of axioms representing some fairness criteria. A simplified explanation for four important bargaining solutions will be presented with respect to a *two-player* bargaining game. Generalization for more players can be easily carried out, but we will leave that to interested readers. Generally, let $\mathbf{s} = (s_1, s_2) \in \mathbf{S}$ where $\mathbf{S} \subseteq \mathbb{R}^2$; $\mathbf{d} = (d_1, d_2)$. Let $\sigma = (\sigma_1, \sigma_2) \in \mathbf{S}$ be a solution. A few axioms are proposed as follows:

Axiom 1 (Pareto Optimality): If $\mathbf{s} \succeq \sigma$, then $\mathbf{s} = \sigma$, $\forall \mathbf{s}$.
Axiom 1' (Weak Pareto Optimality). $\nexists \mathbf{s}: \mathbf{s} \succ \sigma$.
Axiom 2 (Individual Rationality): $\sigma \succeq \mathbf{d}$.

Axiom 3 (Scale Invariance):   For any linear transformation H, $H(\sigma)$ is a solution to $H(\mathcal{B}) = (H(\mathbf{S}), H(\mathbf{d}))$.

Axiom 4 (Symmetry):   If $d_1 = d_2$ and $\{(s_2, s_1) | (s_1, s_2) \in \mathbf{S}\} = \mathbf{S}$, then $\sigma_1 = \sigma_2$. (That is, two players in a symmetrical situation would expect the same payoffs.)

Axiom 5 (Independence of Irrelevant Alternatives):   If $\mathbf{P} \subseteq \mathbf{S}$ and $\sigma, \mathbf{d} \in \mathbf{P}$ for any closed and convex set $\mathbf{P}$, then σ is also the solution of the bargaining game $\mathcal{B}' = (\mathbf{P}, \mathbf{d})$.

Axiom 6 (Individual Monotonicity):   Let $m_i = \max\limits_{s \in \mathbf{S}, s \succeq \mathbf{d}} s_i$ and $\varphi_i(s_j) = \max\{s_i | (s_i, s_j) \in \mathbf{S}\}$ for $i, j = 1, 2$ and $i \neq j$. For two games $(\mathbf{S}, \mathbf{d})$ and $(\mathbf{P}, \mathbf{d})$, if $m_i(\mathbf{S}) = m_i(\mathbf{P}) = M$ and $\varphi_j(s_i, \mathbf{P}) \leq \varphi_j(s_i, \mathbf{S})$ for all $s_i \in [d_i, M]$, then $\sigma_j(\mathbf{P}) \leq \sigma_j(\mathbf{S})$ and vice versa.

Several bargaining solutions can be thereby established making use of certain combinations of the above axioms. Four most significant solutions are as follows.

Utilitarian solution: This solution satisfies axioms 1′, 2–4, and the condition below [13]. Intuitively, it maximizes the sum of all payoffs. For the two-player case,

$$\sigma = \arg\max_{s \in S}(s_1 + s_2). \tag{4.4}$$

Egalitarian solution: This solution gives every player equal gain from his/her status quo. It satisfies axioms 1′, 2–4, and the condition below [13]. Specifically,

$$\begin{cases} \sigma = \arg\max\limits_{s \in S}(\mathbf{s}) \\ \text{s.t.} \quad s_1 - d_1 = s_2 - d_2. \end{cases} \tag{4.5}$$

Nash bargaining solution: This is a unique solution that satisfies all the axioms from 1 to 5 (excluding 1′) [13]. It maximizes the Nash product:

$$\sigma = \arg\max_{s \in S}(s_1 - d_1)(s_2 - d_2). \tag{4.6}$$

Kalai–Smorodinsky solution: This solution yields gains $(\sigma_i - d_i)$ that are proportional to their respective maximum values $(m_i - d_i)$. It is a unique solution that satisfies axioms 1–4 (excluding 1′) and 6.

$$\begin{cases} \sigma = \arg\max\limits_{s \in S}(\mathbf{s}) \\ \text{s.t.} \quad \dfrac{s_2 - d_2}{s_1 - d_1} = \dfrac{m_2 - d_2}{m_1 - d_1}. \end{cases} \tag{4.7}$$

In this section, we have introduced the necessary game theoretical framework. In the next section, we will model the OFDMA system and address the resource allocation problem.

4.3 OFDMA System Model

We describe a generic OFDMA network modeled after existing work in the literature, from which some most typical and significant resource allocation problems are stated [17,19,23,25]. We subdivide our system into single-cell and multicell scenarios.

4.3.1 Single-Cell Scenario

Our OFDMA cell consists of a single base station (BS) and K users or mobile stations (MSs) randomly located around the cell area. There are N available orthogonal subcarriers over the whole system bandwidth B. We need to assume that the bandwidth of each subcarrier B/N is smaller than the coherent bandwidth, so as to have flat-fading channels. Each MS can be assigned a subset of the available subcarrier pool in order to transmit data to the BS (i.e., uplink communication) or receive data from the BS (i.e., downlink communication). We let $\mathbf{A} \subset \mathbb{R}^{K \times N}$ be the subcarrier assignment matrix whose element $a(k, n) \in \{0, 1\}$ is the relationship indicator between the kth MS and the nth subcarrier. Each element is set to 1 if the carrier is assigned to the MS, and 0 otherwise. In single-cell OFDMA, one carrier can only be used by a single user; thus $\sum_{k=1}^{K} a(k, n) \leq 1, \forall n$.

Another assumption is the perfect estimation of the channel condition over each subcarrier for each user. In order to clarify this, first we denote $\mathbf{G} \subset \mathbb{R}^{K \times N}$ as the channel gain matrix whose element $g(k, n)$ represents the channel gain between the BS and the kth MS via the nth subcarrier. Channel gains should take into account the effects of path loss, shadowing, and flat fading. The BS knows the exact value of $\mathbf{G}$; however, each kth MS has full knowledge of only the kth row of $\mathbf{G}$ but not the other rows. This is a reasonable assumption as the BS can apply the necessary measurements and feedback the values to the MS. We also assume that the BS and MS are perfectly synchronized.

Now between BS and one MS there exist N possible channels and the transmission power has to be distributed among those N channels. Thus, we also define the power matrix $\mathbf{P} \subset \mathbb{R}^{K \times N}$ whose elements $p(k, n)$ represent the transmission power for the link between BS and the kth MS via the nth subcarrier (typically valid for the uplink case). Obviously, elements of $\mathbf{P}$ are non-negative. Occasionally we may want a maximum value $P_{\max}$ on the total transmitted power for a single MS– BS link or for the entire network. For example, the maximum total power constraint for a single link is

$$\sum_{n=1}^{N} p(k, n) \leq P_{\max} \quad \forall k. \tag{4.8}$$

The signal-to-noise ratio (SNR) between BS and the kth MS via the nth subcarrier will then be determined as $\gamma(k, n) = p(k, n)g(k, n)/\sigma^2$, where σ^2 is the noise variance from additive-white Gaussian noise (AWGN) channels that is assumed to be uniform. We can then approximate the transmission rate for each link between BS and the kth MS. On each nth subchannel, the approximate achievable rate is

$$R(k, n) = \frac{B}{N}\log_2\left(1 + \frac{\gamma(k, n)}{\Gamma}\right), \tag{4.9}$$

where $\Gamma = -\ln(5 \times \text{BER})/1.5$ is a function of the required bit error rate (BER), often referred to as the SNR (and later SINR) gap [17,19,23,25].

Hence, the total rate for the link between BS and the kth MS is

$$R_k = \sum_{n=1}^{N} a(k, n) \cdot R(k, n). \tag{4.10}$$

Note that the rate is a function of power and assignment parameters.

We have established a generic single-cell OFDMA system. The above discussion can be applied to both downlink and uplink scenarios. In Section 4.5, we shall introduce the two optimization problems raised in Refs. [23,24] and [25] based on the single-cell model as well as their game theoretic formulation and algorithms.

4.3.2 Multi-Cell Scenario

Modeling for the multicell scenario can be obtained in a straightforward manner via modification of the single-cell scenario. We now have a total of K MSs distributed uniformly around M cells, each with a BS at the center. For each cell m, define Ω_m as the set of MSs that belongs to that cell. There are still N available subcarriers over a total bandwidth B. The same assumptions about channel conditions and users' knowledge of them still hold. Within a cell, a subcarrier cannot be reused but it can be used in other cells, that is the cellular network has a reuse factor of 1. The assignment matrix $\mathbf{A}$ is now expanded to a three-dimensional matrix of $\mathbb{R}^{M \times K \times N}$ whose element $a(m, k, n) \in \{0, 1\}$ suggests if the nth subcarrier is used between the mth BS and the kth MS. The same explanation can be applied to elements $g(m, k, n)$ of $\mathbf{G}$. For power, however, only the transmitted power matrix is defined. If communication is uplink, $\mathbf{P} \subset \mathbb{R}^{K \times N}$ is a $K \times N$ matrix defined for the MSs; if downlink, $\mathbf{P} \subset \mathbb{R}^{M \times N}$ is an $M \times N$ matrix defined for the BSs. A maximum power constraint can also be imposed.

Unlike the single-cell situation, now a subcarrier can be used by more than one user; hence a link will experience co-channel interference from signals from neighboring cells. The signal-to-noise-and-interference (SINR) for a link between the mth BS and the kth MS via the nth subcarrier is defined as

$$\gamma(m, k, n) = \frac{g(m, k, n)p(m, n)}{\sum_{i=1, i \neq m}^{M} g(i, k, n)p(i, n) + \sigma^2}. \tag{4.11}$$

It then follows that the achievable rate for the (m, k, n) tuple is

$$R(m, k, n) = \frac{B}{N} \log_2 \left(1 + \frac{\gamma(m, k, n)}{\Gamma} \right). \tag{4.12}$$

The total rate for each mobile user k can be determined:

$$R_k = \sum_{n=1}^{N} a(m, k, n) R(m, k, n), \tag{4.13}$$

which is also a function of power and assignment parameters.

The formulation is valid for both downlink and uplink scenarios. In Section 4.4, we will establish two optimization problems raised in Refs. [17,18] and [19] for the multicell model as well as their game theoretic formulation and algorithms.

4.4 Noncooperative Game Solutions

In the earlier works of Zhu Han et al. [17,18] and Kwon and Lee [19], noncooperative game models for the multicell OFDMA were proposed. Although the two approaches differ in certain ways, a unifying treatment is still possible. We will first present the formulation of the game. Two adaptations and algorithms and numerical results will follow.

4.4.1 Game Formulation

Consider the multicell OFDMA system described in Section 4.3.2. We further limit the case to a *downlink* scenario (i.e., transmitters are the BSs) as treatment is nearly the same for the uplink case. The power matrix is now an $M \times N$ matrix and $p(m, n)$ is the power transmitted by the mth BS via the nth subcarrier. Every optimization problem must have an optimizing objective. In the case of maximization, the objective function is also known as utility function. The same concept applies to a game theoretical problem where each player has its own payoff (utility). We will define the same utility function for each BS that is viewed as a player.

Rate as utility: The most basic maximizing objective for the OFDMA system is the transmission rate. For each cell m we investigate the following utility function $U1_m$:

$$U1_m = \sum_{k \in \Omega_m} R_k \quad \forall m, \tag{4.14}$$

which is the sum of data rates for all the users in that cell, or the total data rates that the BS simultaneously transmits.

Net utility with rate award and power cost: The net utility concept (mentioned in [19]) is an improvement from the previous rate utility. Net utility is basically "reward" minus "cost." In noncooperative games, utility function normally represents the preference of a selfish user, who wants to maximize its own reward. A player in an OFDMA game may somehow modify its transmission parameters such as power in such a way that increases not only its own reward (i.e., the previous rate utility), but also interference caused to others. This produces a negative overall effect as users deliberately neglect the harm it does to others. Hence there may be a need to introduce pricing into utility function to reduce that negative effect. Pricing will commonly be represented by a negative term added to the utility. This mechanism encourages a certain level of cooperation among users, which is what motivates us to modify $U1$ into the function $U2$:

$$U2_m = \sum_{k \in \Omega_m} R_k - \alpha \sum_{n=1}^{N} p(m, n) \quad \forall m, \tag{4.15}$$

where α is the cost factor in units of bps/W, as the cost is the total power transmitted by the BS. Here, we implicitly employ a linear pricing scheme in terms of power.

Power minimization objective: In Refs. [17,18] a power minimization game was mentioned where the objective is to minimize the total transmitted for each user. It is equivalent to maximizing the negative power sum. Such a game can be interpreted as having the utility $U3$:

$$U3_m = -\sum_{n=1}^{N} p(m, n) \quad \forall m. \tag{4.16}$$

After defining the utility function, we can begin to formulate our resource allocation game. We let $\mathcal{M} = \{1, 2, \ldots, M\}$ be the set of players. Clearly, the M BS are the distributed participants in the resource allocation process, trying to allocate subcarriers and power as well as monitoring rates for the MS.

Next, to identify the strategies, note that the players have the options of modifying the transmission power and the channel assignment parameters. Let $\mathbf{p}_m$ be the mth row vector of $\mathbf{P}$; then it is the power vector associated with the mth BS and that BS has total control over this quantity. Thus, we define

$$S1_m = \{\mathbf{p}_m | \quad \mathbf{p}_m \succeq 0; \sum_{n=1}^{N} p(m, n) \le P_{\max}\} \quad \forall m \tag{4.17}$$

as the *power strategy set* for BS m. Also, let $\mathbf{a}_m$ be the tow-dimensional (2D) matrix extracted from $\mathbf{A}$ as $a(m, :, :)$ (we abuse a MATLAB® notation here). The mth BS also has total control over $\mathbf{a}_m$. Thus, we define the *assignment strategy set* for each BS as

$$S2_m = \{\mathbf{a}_m | \quad a(m, k, n) \in \{0, 1\}; \sum_{k \in \Omega_m} a(m, k, n) \le 1\} \quad \forall m. \tag{4.18}$$

A generic joint power and subcarrier allocation game is formulated. Using the game theoretical notation in Section 4.2.2,

$$\mathcal{G} = [\mathcal{M}, \{S1_m \times S2_m\}_{m \in \mathcal{M}}, \{U_m\}_{m \in \mathcal{M}}], \tag{4.19}$$

where $\mathcal{M}, S1_m$, and $S2_m$ are all defined as aforementioned. U_m can be chosen from $U1, U2$, or $U3$. Note that individually, the sets $S1$ are continuous sets that are closed and bounded; meanwhile the sets $S2$ are discrete-valued. Analysis techniques for the two types of sets differ; hence many works tend to break them down and treat them separately, or consider only one set, rather than $S1 \times S2$ as a whole, which is more difficult.

In the sections that follow, we will document the two adaptations of the proposed generic game as well as the algorithms in Refs. [17–19].

4.4.2 Iterative Best-Response Noncooperative Game

One approach in Ref. [19] attempted to break down the generic game in Equation 4.19 into two steps. The net utility function $U2$ was employed. The details can be summarized as follows:

- Initially, assume constant power. Each BS then allocates subcarriers to mobile users greedily, reducing the game into a utility maximizing game with power strategy.
- Using the channel allocation results from before, each player now plays this subgame using its best-response strategy, which is optimal power distribution determined from the Karush–Kuhn–Tucker (KKT) condition.

In the first stage, assume that the power matrix $\mathbf{P}$ is constant and known. Each BS follows the greedy rule and assigns each subcarrier to the user associated with it that yields the largest achievable data rate. Consequently, the total data rate in that cell will achieve the maximum value. The optimal allocation is given by

$$
a^*(m, k, n) = \begin{cases} 1 & \text{if } k = k^*_{mn} = \arg \max_k (R(m, k, n)), \\ 0 & \text{otherwise}, \end{cases} \tag{4.20}
$$

where the notation k^*_{mn} suggests that the value of k should depend on m and n. The final result is the optimal assignment matrix $\mathbf{A}^*$. After the subcarriers have been assigned according to Equation 4.20, the utility function $U2_m$ (i.e., the net utility perceived by a BS) becomes

$$
U2_m = \sum_{n=1}^{N} \{R(m, k^*_{mn}, n) - \alpha p(m, n)\} \quad \forall m, \tag{4.21}
$$

which can now be understood as *the sum over all subcarriers of all the maximum achievable rates, less the respective power cost*. The original game is reduced to

$$
\mathcal{G}' = [\mathcal{M}, \{S1_m\}_{m \in \mathcal{M}}, \{U2_m\}_{m \in \mathcal{M}}]. \tag{4.22}
$$

which is identical to the distributed maximization problem

$$
\begin{cases} \max_{\mathbf{P}_m} U2_m = \sum_{n=1}^{N} \{R(m, k^*_{mn}, n) - \alpha p(m, n)\} \\ \text{s.t.} \quad p(m, n) \geq 0 \text{ and } \sum_{n=1}^{N} p(m, n) \leq P_{\max} \end{cases} \quad \forall m. \tag{4.23}
$$

Using game theory, we can prove the existence of Nash equilibrium for $\mathcal{G}'$. The strategy space $S1_1 \times S1_2 \times \cdots \times S1_M$ with each $S1_m$ defined previously satisfies the closure, convexity, and compactness condition. Meanwhile from Equations 4.12 and 4.23, each utility function $U2_m$ is now a quasi-concave function in $\mathbf{p}_m$ when all other parameters are constant. Thus the second condition in Theorem 1 (Section 4.2.2) is also satisfied. It follows that the existence of Nash equilibrium is verified. We state that as a theorem.

THEOREM 4.3 (*Existence*)

Nash equilibrium exists for the game in Equation 4.22.

Now we express our utility function as a function of the player's own strategy and its opponents' strategies: $U2_m = U2_m(\mathbf{P}) = U2_m(\mathbf{p}_m, \mathbf{P}_{-m})$. Note that $\mathbf{p}_m \in \mathbb{R}^N$ and $\mathbf{P}_{-m} \in \mathbb{R}^{(M-1) \times N}$. Given that the opponents, strategies are fixed, a player's *best-response* strategy $\mathbf{p}^*_m$ is defined as one that maximizes its own utility:

$$
\mathbf{p}^*_m = f(\mathbf{P}_{-m}) = \max_{\mathbf{P}_m} U2_m(\mathbf{p}_m, \mathbf{P}_{-m}), \quad \text{where } f(.) : \mathbb{R}^{(M-1) \times N} \to \mathbb{R}^N. \tag{4.24}
$$

Since $U2_m$ is a quasi-concave function in $\mathbf{p}_m$ when all other parameters are constant, Equation 4.23 is a concave optimization problem [28]. Therefore, the best response is equivalent to the optimal solution for Equation 4.23, which can be found from the KKT condition [28]. The final result is given in the following lemma.

LEMMA 4.1

The KKT best response for BS m is

$$
\begin{cases}
p^*(m,n) = \max\left(0, \dfrac{B}{N}\dfrac{1}{(\alpha+\mu_m)\ln 2} - \dfrac{\Gamma\left(\sum_{i=1,i\neq m}^{M} g(i,k_{mn}^*,n)p(i,n)+\sigma^2\right)}{g(m,k_{mn}^*,n)}\right), \\[4mm]
\mu_m\left(\displaystyle\sum_{n=1}^{N} p^*(m,n) - P_{\max}\right) = 0,\ \mu_m \geq 0.
\end{cases}
\tag{4.25}
$$

In reality, the problem of finding all best responses $\mathbf{p}_m^*$ simultaneously cannot be solved easily. However, in Ref. [19], the difficulty was avoided by relying on an iterative algorithm, in which the best-response determination of power is carried out using the values of $\mathbf{P}_{-m}$ from the *previous round* of iteration in order to find the *next* value of $\mathbf{p}_m^*$. That, coupled with the greedy channel assignment, is repeatedly done so that the results from one step are fed to the next step, until the whole process comes to a stable outcome, which is obviously a Nash equilibrium since no player now wants to change strategies. Table 4.1 demonstrates the algorithm.

Initially, for convenience we set constant power for all players. Now as every player is playing its best response given the others' strategies, in the long term such a repeated game will converge to a Nash equilibrium. Numerical results and convergence issues of this algorithm will be discussed in Section 4.4.4.

4.4.3 Noncooperative Game with Virtual Referee

A different approach for the multicell OFDMA game is documented in Refs. [17,18]. The system in Refs. [17,18] is further relaxed so that each cell only contains one BS and one MS, that is $M = K$ and $\Omega_m = \{m\}$. Thus, we now define a player as a pair consisting of the mth BS and the mth MS.

Table 4.1 Iterative Best-Response Algorithm

I. Initialization   Each BS sets equal power $P_{\max}/N$ over each subcarrier
II. Greedy Channel Assignment   Each BS performs greedy channel assignment according to Equation 4.20 to determine its optimal power matrix $\mathbf{a}_m^*$
III. Best-response Power Estimation   Each BS determines its best-response power vector $\mathbf{p}_m^*$ according to Equation 4.25 using the updated network carrier assignment matrix $\mathbf{A}^*$ from Step II and the previous round opponents' power $\mathbf{P}_{-m}$
IV. Iteration   Repeat Steps II and III until convergence is obtained

For power and channel gain, the notations $p(m, n)$ and $g(m, k, n)$ still have the same meaning. However, the carrier assignment parameter is now reduced to a 2-D $a(m, n)$. The rate achieved by a player m is now $R_m = \sum_{n=1}^{N} a(m, n)R(m, n)$, where

$$R(m, n) = \frac{B}{N}\log_2\left(1 + \frac{\gamma(m, n)}{\Gamma}\right) \text{ and } \gamma(m, n) = \frac{g(m, m, n) \cdot p(m, n)}{\sum_{i=1, i \neq m}^{M} g(i, m, n) \cdot p(i, n) + \sigma^2}. \quad (4.26)$$

Furthermore, Refs. [17,18] require a *fixed-rate* condition for all players, that is $R_m = $ const, $\forall m$. The modified power strategy sets $S1'_m$ and $S2'_m$ are

$$\begin{cases} S1'_m = \{\mathbf{p}_m | \quad \mathbf{p}_m \succeq 0; \sum_{n=1}^{N} p(m, n) \leq P_{\max} \text{ and } R_m = \text{const}\} \\ \\ S2'_m = \{\mathbf{a}_m | \quad a(m, n) \in \{0, 1\} \text{ and } R_m = \text{const}\} \end{cases} \quad \forall m. \quad (4.27)$$

Note that as R_m is dependent on both power and assignment parameters, the fixed-rate constraint affects both strategy sets. The fixed total rate constraint is a nonlinear constraint in $\mathbf{p}_m$. Also, thereafter R_m will be referred to as a constant.

4.4.3.1 Power Minimization Game

A game with power minimization as the objective (i.e., utility function $U3$) is first investigated. Each user in the network now wishes to minimize its transmission power over all subcarriers to obtain a fixed transmission rate. This is essentially the following game:

$$\mathcal{G}_1 = [\mathcal{M}, \{S1'_m \times S2'_m\}_{m \in \mathcal{M}}, \{U3_m\}_{m \in \mathcal{M}}]. \quad (4.28)$$

The game above is an alternative statement of the optimization problem

$$\begin{cases} \min \sum_{n=1}^{N} p(m, n) \\ \\ \text{s.t.} \quad \sum_{n=1}^{N} R(m, n) = R_m; \mathbf{p}_m \succeq 0 \text{ and } \sum_{n=1}^{N} p(m, n) \leq P_{\max}. \end{cases} \quad (4.29)$$

We first assume that the problem is feasible, that is the minimum power allocation satisfies both the fixed rate and maximum power constraint. Also, since every subcarrier can be used, we assume that in this game each player first transmits over all subcarriers (i.e., $\mathbf{A}$ is an all-one matrix). Then, the above is a convex problem that can be solved by applying the KKT condition (in fact, it is a dual problem of the waterfilling problem [17,18]). A detailed discussion on waterfilling can be found in Boyd and Vandenberghe's text [28]. We omit the details. The solution is given as

$$p(m, n) = \max\left(0, \mu_m - \frac{\Gamma\left(\sum_{i=1, i \neq m}^{M} g(i, m, n) \cdot p(i, n) + \sigma^2\right)}{g(m, m, n)}\right), \quad (4.30)$$

where μ_m is introduced as Lagrangian multipliers in obtaining the KKT condition.

Similar to the iterative best-response game in Section 4.4.2, it can also be shown by the same argument that the strategy space for the game (4.28) is convex and compact, and the utility function $U3$ is obviously quasi-concave with respect to each $p(m, n)$. Thus the existence of Nash equilibrium is confirmed by Theorem 1. Moreover, it is shown in Refs. [17,18] that the Nash equilibrium can coincide with the global optimum in certain cases. The following theorem results:

THEOREM 4.4 (*Existence and Optimality*)

Nash equilibrium exists for the game in Equation 4.28. If the global minimum of (4.29) occurs, which satisfies the constraints as well as $R(m, n) > 0, \forall a(m, n) > 0$, then it coincides with Nash equilibrium.

4.4.3.2 Virtual Referee Game: Rate Adjustment and Channel Removal

The previous power minimization game is sufficient to solve the resource allocation problem when the channel condition is good and interference among neighboring cells is low. In this case, all BS–MS pairs use up all subcarriers and consume power within the maximum limit. The obtained optimum point will also coincide with a Nash equilibrium point, so optimality is achieved from both the system's and the individual's perspectives. However, this is hardly the case. It is observed [17,18] that within the power limit, the required rate could hardly be achieved; moreover, the severe co-channel interference might cause the system to end up at a point below optimality. Thus it might be necessary either for the users to sacrifice the rate requirement to reserve power consumption, or to remove some channels from usage when that channel condition is too bad. By following the preventive actions, system performance is expected to be improved. Thus, an improved scheme with iterative rate adjustment followed by channel removal is implemented. The "game referee" is a virtual concept that serves to monitor channel removal. The algorithm is shown in Table 4.2.

Rate adjustment can be done exclusively on the player's part without cooperation. When a player needs to reduce its rate, it is understood that there is no feasible solution for the previous power minimization game to stabilize the allocation result for that player. Thus, it gives up its fixed-rate requirement, and instead tries to get the highest rate possible with its power budget. Such a player will obtain a reduced rate but the amount of adjustment is player-dependent so there is no need for a coordinator to monitor this action. Under these circumstances, this user is said to play another rate maximization game (i.e., using utility function $U1$). Since the fixed-rate constraint is removed, the power strategy is now back to $S1_m$ and $S2_m$ defined in Equations 4.17 and 4.18. For the subcarrier assignment strategy, because rate adjustment and channel removal are done iteratively, at the jth iteration a number of channels may already be removed; thus we denote $S2_{m,(j)}$ as the modified subcarrier set at the jth iteration for player m, wherein some elements of a_m have become zeros. Note that then at the jth iteration, a similar $S2'_{m,(j)}$ is used for $\mathcal{G}_1$. The rate maximization game can now be formally expressed as

$$\mathcal{G}_2 = [\mathcal{M}, \{S1_m \times S2_{m,(j)}\}_{m \in \mathcal{M}}, \{U1_m\}_{m \in \mathcal{M}}] \tag{4.31}$$

or in terms of mathematical optimization as

$$\begin{cases} \max_{\mathbf{P}_m} R_m = \sum_{n=1}^{N} a(m, n) \frac{B}{N} \log_2 \left(1 + \dfrac{p(m, n)g(m, m, n)}{\Gamma \left(\sum_{i=1, i \neq m}^{M} g(i, m, n).p(i, n) + \sigma^2 \right)} \right) \\ \text{s.t.} \quad \sum_{n=1}^{N} p(m, n) \leq P_{\max} \text{ and } p(m, n) \geq 0 \quad \forall n \end{cases} \quad \forall m. \tag{4.32}$$

Table 4.2 Virtual Referee Game Algorithm

I. Initialization
■ Every player loads data over all subcarriers ■ Use initial rate requirement
II. Power Minimization Game
■ Using current subcarrier strategy set and rate requirement, each plays the game $\mathcal{G}_1$ in Equation 4.28
III. Rate Maximization Game For players whose Step II cannot reach equilibrium:
■ Switch to game $\mathcal{G}_2$ in Equation 4.31 ■ Set new rate requirement at optimal rate obtained from $\mathcal{G}_2$ and go to Step IV
For others, remain at Step II
IV. Channel Removal by Virtual Referee
■ The referee decides the new subcarrier strategy for users from Step III using Equation 4.34 ■ Go back to Step II
V. Iteration Iterate from Step II until the system performance stabilizes

Problem (4.32) is an exact waterfilling problem, similar to (4.29). Hence a unique optimal solution for each $p(m, n)$ can be derived. The solution is given by

$$
p(m, n) =
\begin{cases}
\max\left(0, \mu_m - \dfrac{\Gamma\left(\sum_{i=1, i \neq m}^{M} g(i, m, n) \cdot p(i, n) + \sigma^2\right)}{g(m, m, n)}\right) & \text{if } a(m, n) = 1, \\
0 & \text{if } a(m, n) = 0.
\end{cases}
\tag{4.33}
$$

Any player who needs to play the second game is expected now to have achieved an undesirable outcome, since some of the subcarriers might prove to be too costly. It may need to refrain from using a certain subcarrier, and since other players also have to remove some subcarriers, interferences present in its remaining channels are to decrease and so the player is tempted to reduce power to still achieve the same rate from the second game. For that reason the game allows an iteration in which players can return to the first game.

Now we turn to the subcarrier removal issue. It is worth recalling that, in general, a player does not know which are the other users currently occupying a certain subcarrier. The player itself lacks the necessary information to select which channel to drop. The decision needs to be done from another party who is aware of the situation. Therefore the referee is introduced as a coordinator who monitors the game. It is a virtual entity that can be set up at one of the BSs, in the system. A simple rule is to use the SINR as the main criterion. For any player m who has to go through the removal stage, the referee will search through the players who share some subcarriers with player

m, and one of the other players (say, player i) will remove a shared subcarrier for which the SINR perceived by it is the lowest. Moreover, it is required that every player has at least one remaining subcarrier, and no subcarrier is left unoccupied. Formally,

$$a(\hat{i}, \hat{n}) \to 0 \text{ if } (\hat{i}, \hat{n}) = \arg \min_{i,n} \gamma(i, n) \text{ s.t. } \begin{cases} a(m, n) = a(i, n) = 1, & m \neq i, \\ \exists l, \, l \neq \hat{n} : a(i, l) = 1 & \forall i, \\ \exists j, \, j \neq \hat{i} : a(j, n) = 1 & \forall n. \end{cases} \qquad (4.34)$$

One final remark on the virtual referee scheme is that the distributed degree of the game has been lessened by including a rather centralized decision maker; hence the game is not totally noncooperative. The introduction of some cooperation to the game is one way of improving system performance from the inefficiency of noncooperative Nash equilibrium. The numerical results for this algorithm will be discussed in Section 4.4.4.

4.4.4 Numerical Results

In this section we discuss some important findings from our simulation of the schemes in Sections 4.4.2 and 4.4.3. We simulated the OFDMA system of seven cells ($M = 7$) with hexagonal structure. One instant of the simulation is depicted in Figure 4.1, showing cells with the BSs at the center and several MSs randomly distributed around the area.

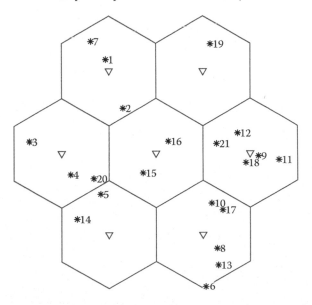

Figure 4.1 Graphical representation of the system.

4.4.4.1 Nash Equilibrium Convergence for the Iterative Best-Response Game

The convergence of the iterative best-response algorithm was the main concern in Ref. [19]. We verified the convergence of this algorithm by simulating two different scenarios: one with low interference and one with severe interference. A few global parameters were set as follows: bandwidth per subcarrier $B/M = 100\,\text{kHz}$; cell radius 1 km; target BER 10^{-5}; total maximum transmission power for each cell $P_{max} = 5\,\text{W}$; noise power density $-174\,\text{dBM/Hz}$; and cost factor $\alpha = 0.3\,\text{Mbps/W}$ for all users. The channels were modeled with Rayleigh fadings and a path loss exponent of 3.

In the first case, a low-interference scenario was simulated, where there are $K = 21$ users (i.e., about three users per cell) competing for $N = 15$ available subcarriers (i.e., five subcarriers per user on average). The simulation shows that the operating point of each player, in terms of its final value of utility function, stabilized after very few iterations. In Figure 4.2, the utility value of one random cell, normalized by its final equilibrium value, was recorded throughout the iteration. Clearly, in a case of abundant resource, the algorithm reaches a Nash equilibrium point where all the players of the noncooperative game achieved satisfied outcomes. Moreover, a Monte Carlo simulation of 1000 runs for this particular setting showed that the convergence rate is over 95%, which is sufficiently high.

Another scenario was generated where $K = 30$ mobile users compete for $N = 8$ subcarriers. With a relatively reduced sharing ratio per subcarrier, the network was shown to experience severe interference and as a result a desirable Nash equilibrium outcome may not be obtained. The utility of players might fluctuate as shown in Figure 4.3, and the convergence rate was reduced to only 60%. In Ref. [18] the same phenomena were briefly mentioned and attributed to the deficiency of the game's feasible region, but no results were shown nor explained.

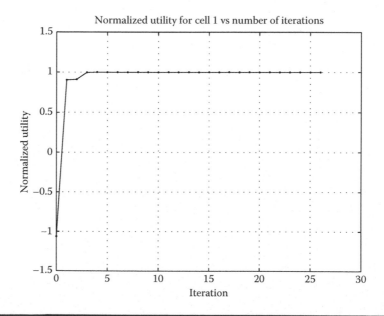

Figure 4.2 Utility for the randomly chosen cell converges quickly.

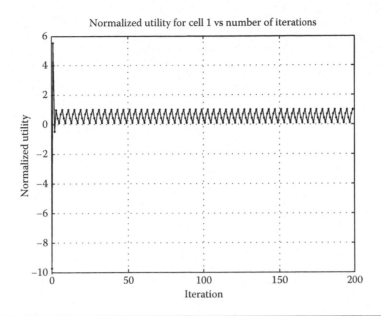

Figure 4.3 A case with oscillating utility.

4.4.4.2 Power and Rate Allocation for the Virtual Referee Game

By using an OFDMA system as in Figure 4.1, simulation studies of the game with a virtual referee for rate adjustment and channel removal were conducted. The system settings were adjusted similarly to Ref. [18], with $M = 7$ pairs of BS–MS; $N = 32$ subcarriers; maximum power constraint for each cell $P_{max} = 10$ mW; required BER $= 10^{-3}$; total bandwidth $W = 6.4$ MHz; thermal noise -70 dBm; using Rayleigh fading channels with path loss exponent 3.5; and cell radius $R = 0.1$ km. Moreover the initial fixed-rate requirement for all players was $R_m = 12$ Mbps. In Refs. [17,18], the important parameters (the eventual power and rate of players) were measured against the distance between two neighboring BSs when two cells were located further apart. The improved game with channel removal and rate adjustment $\mathcal{G}_2$ (labeled "VRG") was compared with the original power minimization game $\mathcal{G}_1$ (labeled "G1 only") in Figure 4.4.

It can be shown that when the distance between BSs increased, interference was reduced so that performance for both games could improve. The virtual referee game was shown to have slightly better performance than the original power game: slightly higher rate at less power.

Noncooperative games have been implemented for the uplink OFDMA scenario, and the studies of two algorithms have indicated the feasibility of the approach. The applicability of cooperative games to OFDMA resource allocation will be studied next.

4.5 Cooperative Game Solutions

The bargaining problem from cooperative game theory was employed for OFDMA resource allocation in the single-cell scenario [23–25]. In this section, we will formulate this resource allocation problem as a bargaining game and apply the four bargaining solution concepts (Section 4.2.3) to

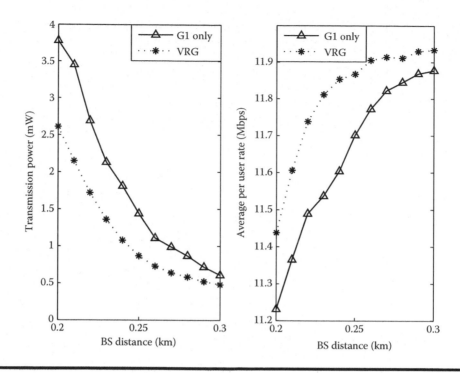

Figure 4.4 Power and rate allocation for power game and VRG.

define fairness criteria. Next we discuss two game adaptations [23–25] as well as their proposed algorithms and important numerical results.

4.5.1 Game Formulation

We shall revisit the single-cell OFDMA system described in Section 4.3.1. As a problem of distributed manner, the various mobile users need to participate in the decision making process, so we assume an uplink situation. The set of players is then $\mathcal{K} = \{1, 2, \ldots, K\}$. The available subcarriers and transmission power are taken as strategies. The players compete to maximize the same objective, which is the transmission rate. Hence we can define $S1_k \times S2_k$ as the strategy set where

$$S1_k = \{\mathbf{p}_k | \quad \mathbf{p}_k \succeq 0; \sum_{n=1}^{N} p(k, n) \leq P_{\max}\} \quad \forall k \tag{4.35}$$

and

$$S2_k = \{\mathbf{a}_k | \quad a(k, n) \in \{0, 1\}; \sum_{k=1}^{K} a(k, n) \leq 1\} \quad \forall k, \tag{4.36}$$

with $\mathbf{p}_k$ and $\mathbf{a}_k$ being respectively the rows of $\mathbf{P}$ and $\mathbf{A}$ corresponding to the kth MS. We can also define $U_k = R_k$, $\forall k$ as in Equation 4.10 as the utility function. Thus we have a proper game

$$\mathcal{G} = [\mathcal{K}, \{S1_k \times S2_k\}_{k \in \mathcal{K}}, \{U_k\}_{k \in \mathcal{K}}]. \tag{4.37}$$

From a different perspective, we treat the above problem as a bargaining game of transmission rate. Let $\mathbf{F} \subseteq \mathbb{R}^K$ be the set of feasible bargaining outcome (before cooperation) where

$$\mathbf{F} = \{\mathbf{r} = (R_1, R_2, \ldots, R_K) | \mathbf{r} \in \mathbb{R}^K, \mathbf{r} \succeq 0, \text{ is constrained by } S1_k \text{ and } S2_k \ \forall k\}. \tag{4.38}$$

Note that the set $\mathbf{F}$ is not identical to the strategy space; hence we avoid using the symbol $\mathbf{S}$ here. It is bounded due to the constraints imposed on $S1_k$ and $S2_k$; however, it may not be convex. Nonetheless, cooperative game theory can verify that when the players come to work together, they can enlarge their feasible outcome set into a closed, convex set $\mathbf{F}'$, which is the *convex hull* of $\mathbf{F}$ [11].

In order to complete the bargaining game, all we need is to define the disagreement point. Now, suppose that every player k has a minimum rate requirement $R_k^{\min}$ that it can achieve via no cooperation. It now wishes to improve its rate by acting together with other users sharing the same resource. This is indeed a reasonable assumption. So, define $\mathbf{d} = (R_1^{\min}, R_2^{\min}, \ldots, R_K^{\min})$. as the status quo and assume $\mathbf{d} \in \mathbf{F}'$ such that it is strictly dominated by other feasible outcome, which is highly likely. The bargaining game is fully defined as

$$\mathcal{B} = (\{\mathbf{r} | \mathbf{r} \in \mathbf{F}'\}, (R_1^{\min}, R_2^{\min}, \ldots, R_K^{\min})). \tag{4.39}$$

At this point, we can proceed to characterize the bargaining solution using the four fairness criteria discussed in Section 4.2.3. Let σ be a possible bargaining solution. The four possibilities are presented as follows.

Utilitarian solution: As suggested from Equation 4.4, the utilitarian solution aims to maximize the total rates for all players, that is a network throughput maximizer.

$$\sigma_{\text{util}} = \arg \max_{\mathbf{r} \in \mathbf{F}'} \sum_{k=1}^{K} R_k. \tag{4.40}$$

Egalitarian solution: Theoretically, the egalitarian solution resulting from Equation 4.5 is written as

$$\begin{cases} \sigma_{\text{egal}} = \arg \max_{\mathbf{r} \in \mathbf{F}'} \mathbf{r} \\ \text{s.t.} \quad R_i - R_i^{\min} = R_j - R_j^{\min}, \ \forall i \neq j \in \mathcal{K}. \end{cases} \tag{4.41}$$

The true egalitarian bargaining solution gives every user equal rates, the value of which is maximized, if we further assume that everyone has identical minimum rate requirements ($R_i^{\min} = R_j^{\min}$ for any $i \neq j$). However, in reality it is often replaced by a "most egalitarian" [23,26] fairness criteria namely the "max–min fairness" where all users get the maximum out of all the minimum rate requirements:

$$\sigma_{\text{max-min}} = \arg \max_{\mathbf{r} \in \mathbf{F}'} \{\min_i R_i\}. \tag{4.42}$$

Nash bargaining solution: The Nash bargaining solution is obtained as a maximizer of the Nash product. It is a highly favorable solution as it ensures proportional fairness [23,24]. Mathematically,

$$\sigma_{\text{NBS}} = \arg \max_{\mathbf{r} \in \mathbf{F}'} \prod_{i=1}^{K} (R_i - R_i^{\min}). \tag{4.43}$$

Kalai–Smorodinsky bargaining solution: With the Kalai–Smorodinsky solution, any user i now needs to consider its maximum achievable rate $R_i^{\max}$. This value is seen as the rate that this user obtains when it is the only user that transmits in the system using all the subcarriers, which is now clearly determined. The Kalai–Smorodinsky solution is then

$$\begin{cases} \sigma_{KS} = \underset{\mathbf{r} \in \mathbf{F}'}{\arg \max} \ \mathbf{r} \\[2ex] \text{s.t.} \quad \dfrac{R_i - R_i^{\min}}{R_i^{\max} - R_i^{\min}} = \dfrac{R_j - R_j^{\min}}{R_j^{\max} - R_j^{\min}}, \quad \forall i \neq j \in \mathcal{K}. \end{cases} \tag{4.44}$$

The following theorem is a reaffirmation of previous results:

THEOREM 4.5 (*Existence and Uniqueness*)

The four aforementioned bargaining solutions exist and are unique for the bargaining game $\mathcal{B}$ in Equation 4.39.

Although the four bargaining solutions were useful concepts in cooperative game implementation, finding the exact bargaining outcomes in a strict mathematical sense is still a difficult problem. Some potential difficulties are the nonlinear and discrete nature (e.g., integer constraints in assignment parameters) of the optimization problem, and the assumption of the convexity of our feasible set, which is true only if we consider mixed strategies and correlated strategies [11]; hence it is probable that the exact desired outcome (e.g., the Nash bargaining solution) is unattainable in pure strategy. Another difficulty is that the cost of the search for realistic outcomes closest to the theoretical solution can be tremendous. Thus, the common approach is to resort to a low-cost suboptimal algorithm for the problem, or a relaxation of the original problem. The focal tasks are still to work out schemes for *subcarrier assignment* and *power allocation*, which are our game strategies, so that together they approximate the four bargaining solutions best. We document in the following sections two adaptations for the OFDMA cooperative game approach in Refs. [23–25].

4.5.2 The Pair-Bargaining with Coalition Algorithm

In Refs. [23,24], a novel scheme is proposed to solve the bargaining problem (4.39). Several smaller two-player games are formed within the original game in which each pair acts as a *coalition* (i.e., a nonempty subset of the player set, usually with more than one, created for the purpose of cooperation [11]) and bargaining is done within those smaller games. Coalition can be regrouped several times until no further improvement can be achieved. Furthermore, this algorithm only considers two specific bargaining solutions, namely the utilitarian and the Nash solutions, for reasons that will be made clear later.

4.5.2.1 The Overall Multiuser Algorithm

In this subsection, the "master" algorithm for the larger multiuser game is discussed. Table 4.3 outlines the steps of this algorithm.

The steps of this algorithm are straightforward to explain. More can be discussed about the initial channel assignment. Every user is first guaranteed at least one subcarrier, which serves as its "capital" to "invest." The greedy assignment of the remaining subcarriers essentially gives the

Table 4.3 Overall Multiuser Algorithm

I. Initialization
■ Each user first grabs one subcarrier and tries to obtain minimal rate ■ The remaining subcarriers are greedily assigned to the users
II. Coalition Grouping
■ Users are grouped into coalition pairs based on a specified grouping scheme ■ If the number of users is odd, create a dummy user and group as usual. The dummy user's game strategy is to use zero power and no subcarriers so the network is not affected
III. Bargaining in Pairs
■ In each pair, the two users play the bargaining game according to the two-user algorithm
IV. Iteration
■ Repeat Steps II and III until there is no more improvement

subcarrier to the user who makes the best use of it in terms of the highest rate, in a similar manner to what was discussed in Section 4.4.2. Note that the power constraint may be violated in this initial stage to first satisfy the minimum rate requirement. Then after one round of iteration, each coalition between players i and j is expected to find the desirable bargaining objective function Q_{ij}, [i.e., the utilitarian sum of rates $R_i + R_j$, or the Nash product of rates $(R_i - R_i^{\min})(R_j - R_j^{\min})$]. The overall criterion for comparison between different rounds is the total of every user's rate, or the Nash product of the whole system. This is why utilitarian and Nash solutions are the two suitable candidates for this algorithm. Meanwhile, two main problems should be addressed: the grouping strategy and the algorithm for the two-user game. In the following sections, the two problems will be tackled.

4.5.2.2 Grouping Strategies

An efficient grouping method should aid the overall game to quickly converge, but on the other hand should not carry high computational cost. Two alternatives are considered [23,24]:

■ *Random method:* Pairs are randomly formed. No computational complexity is required. However, in terms of quick convergence it is of little effect [23,24] because a user's preference for a channel leads to it being selective about whom to pair with.

■ *Hungarian method:* The grouping issue is modeled after the assignment problem, in which the Hungarian algorithm [29] is applied to find the optimal assignment outcome. The method is suitable for obtaining quick convergence. The only drawback is its complexity of $O(K^4)$ [29], so when the number of users K gets larger this may become a limitation.

The Hungarian algorithm* specific to this game is discussed in detail. One particular form of this problem is that *given K workers and K tasks and assigning a worker to a task resulting in some*

---

* The algorithm was invented by Harold W. Kuhn in his original paper from 1955 [29]. It stemmed from previous works of two Hungarian mathematicians Kőnig and Egerváry, and hence was named this way by Kuhn. Since then it has become an important prototype method for problems in combinatorial optimization.

Table 4.4 Hungarian Algorithm

1. For each row of **B**, subtract the entire row by its minimum entry
2. Now for each user (i.e., row), there is at least one zero entry representing the most beneficial partner(s). If zero pairing results in every user having one and only one partner, exit the algorithm. Else, form as many pairs as possible and go to the next step
3. We now want to mark ■ All rows without assignment ■ All columns with unassigned zeros ■ From each column mentioned above, mark the row with assignment Do until no more can be marked
4. Now suppose the rows and columns of **B** can be colored. We color ■ All *marked* columns ■ All *unmarked* rows
5. From the uncolored elements ■ Find the minimum element and subtract it from the rest ■ Add the minimum to the elements at the intersection of a colored column and a colored row ■ Go back to (2), discard the coloring, and repeat

benefits, one would like to find an assignment that maximizes total benefits. Here the benefit b_{ij} of pairing player i to player j is defined as

$$b_{ij} = \max(\hat{Q}_{ij} - \bar{Q}_{ij}, 0), \tag{4.45}$$

where $\hat{Q}_{ij}$ represents the outcome of that two-player game after bargaining and $\bar{Q}_{ij}$ is the initial outcome without bargaining. Clearly, $b_{ii} = 0$ and $b_{ij} = b_{ji}$. The symmetric benefit table **B** can then be built with adjusted elements $B_{ij} = \max(b_{ij}) - b_{ij}$, upon which the Hungarian algorithm is applied. The outline of a simplified algorithm is summarized in Table 4.4.

The Hungarian algorithm can guarantee an optimal assignment. Note that the matrix **B** changes after each iteration; hence each time the algorithm is applied, we do not obtain similar results. The grouping strategy has now been specified. Lastly, we investigate the bargaining game within the pair.

4.5.2.3 Two-User Bargaining Algorithm

At the beginning of the two-user game, each player starts with its own initial resource (subcarriers), and wishes to exchange with the other player so as to gain mutual benefits. A two-band partition selection algorithm [30] was originally proposed, which can produce near-optimal subcarrier allocation under high SNR, and was adopted in Refs. [23,24]. The algorithm is tabulated in Table 4.5.

When the game starts, the union set of the subcarriers of the two is established and sorted based on information from both players. The sorted list is arranged in such a way that the *first player's preference for a subcarrier is decreasing along the list while that of the second player is increasing.* Then, by considering several ways of partitioning the list as in Step III, we select the one with the best

Table 4.5 Two-User Bargaining Algorithm

I. Initialization: Set the initial ϱ_i parameters: ■ For utilitarian objective, $\varrho_i = 1, \forall i$ ■ For Nash bargaining, set ϱ_i according to Equation 4.48
II. Sorting 　The set of subcarriers available for the two users $\{n \mid n \in$ subset of $\{1, 2, \ldots N\}\}$ is sorted in descending order according to $h_{1n}^{\varrho_1}/h_{2n}^{\varrho_2}$, which is defined in Equations 4.49. Let $\{1, 2, \ldots, L\}$ be the sorted list.
III. Two-band Calculation: For $n = 1$ to $L - 1$, ■ The first user occupies subcarriers 1 to n in the sorted list ■ The second user occupies subcarriers $n + 1$ to L ■ In every situation, a user always allocates power by waterfilling in Equation 4.47 ■ Calculate the bargaining objective function $Q_{ij}(n)$ in each case
IV. Two-band Selection ■ Select the best partition: $n^* = \arg\max_{ij} Q_{ij}(n)$
V. Decision ■ In the utilitarian case, exit after one iteration ■ In the Nash case, update $\varrho_i = 1/(R_i - R_i^{\min})$ for $i = 1, 2$ ■ If no more improvement from the system objective function can be obtained, exit; else, go back to (II) and iterate

outcome. After each round, the decision parameters ϱ_i are updated and the process continues until a stable allocation result is reached.

We now look at the problem of distribution of power over the subcarriers, which a player encounters in Step III. Fortunately, in each and every situation, the user knows exactly which channels to transmit and how good they are in terms of SNR. Let Δ_i be the set of available subcarriers for player i. (Here we use player i in the context of the two-user game, while in the context of the multiuser game, it is denoted k.) The user's goal is to obtain the maximal rate out of the available power. Thus, similar to the discussion in Section 4.4.3 this is exactly a waterfilling problem:

$$
\begin{cases}
\max_{P_i} R_i = \displaystyle\sum_{n \in \Delta_i} a(i, n) \frac{B}{N} \log_2\left(1 + \frac{p(i, n)g(i, n)}{\sigma^2 \Gamma}\right) & \\
\text{s.t.} \quad \displaystyle\sum_{n \in \Delta_i} p(i, n) \leq P_{\max} \text{ and } p(i, n) \geq 0 \quad \forall n \in \Delta_i &
\end{cases}
\quad \forall i = 1, 2, \qquad (4.46)
$$

where the solution is given by

$$
p(i, n) =
\begin{cases}
\max\left(0, \mu_i - \dfrac{\sigma^2 \Gamma}{g(i, n)}\right) & n \in \Delta_i \\
0 & \text{otherwise}
\end{cases}
\quad \forall i = 1, 2. \qquad (4.47)
$$

Here μ_i is the Lagrangian multiplier (also the water level), which will be computed by computer during the waterfilling algorithm. With that the power allocation problem is completely solved.

Next we tackle the principles behind the two-band partitioning algorithm. In such a scheme where the channels are sorted in the previously described manner, it was proven in Ref. [30] that optimizing the weighted sum of rates is a concave problem, in which both first-order and second-order optimality conditions [28] are satisfied. So at the optimal partitioning, the utilitarian sum of rates (of equal weight one) is maximized. Therefore, by simply letting weights $\varrho_i = 1$, the utilitarian solution is obtained in just one iteration. For the Nash bargaining solution, the objective is to maximize $Q_{ij} = (R_i - R_i^{\min})(R_j - R_j^{\min})$. In Refs. [23,24], similar results for the Nash solution were derived by applying the KKT condition to Q_{ij} above. Basically, an equivalent weight ϱ_i is defined as

$$\varrho_i = \begin{cases} \dfrac{1}{R_i - R_i^{\min}} & R_i > R_i^{\min} \\ 1/\epsilon & \text{otherwise} \end{cases} \quad \forall i = 1, 2, \tag{4.48}$$

where ϵ is a small positive number. This is to ensure a large weight for a player with a rate less than the minimum requirement. Assuming that the power has been determined from waterfilling in Equation 4.47, we define

$$h_{in} = g(i, n)/\Gamma\sigma^2 \quad \forall i = 1, 2; \quad \forall n. \tag{4.49}$$

Then $h_{1n}^{\varrho 1}/h_{2n}^{\varrho 2}$ becomes the subcarrier preference indicator for the two-player game as described before. So, sorting the subcarriers in descending order of $h_{1n}^{\varrho 1}/h_{2n}^{\varrho 2}$ makes the problem similar to that of Ref. [30], and by applying the KKT condition, optimality can be proven. We skip the mathematical manipulation since the scope of this chapter is limited.

We end this subsection with a discussion of the algorithm's complexity. For the two-user game, proper sorting methods can have the complexity $O(N \log N)$, and since it was claimed that convergence was obtained within only two or three rounds [23,24], the overall complexity approaches $O(N \log N)$ for each two-user game. Thus in Step III of the "master" scheme, the number of two-user games that can be played approaches $O(K^2)$, so the complexity of Step III is $O(K^2 N \log N)$. In Step II, if grouping is done using the Hungarian algorithm, the complexity in this stage is $O(K^4)$. The total complexity for the multiuser game is thus $O(RK^2N \log N + RK^4)$, where R is the average number of iterations needed.

4.5.3 Round-Robin Carrier Assignment Algorithm

This approach [25] attempts to separate carrier assignment and power allocation into two stages. In the first stage, the player sets an initial constant power level and selects channels in a round-robin iterative manner. In the second stage, power will be adjusted to further improve the rate outcome. The main goal is to approximate the Nash and Kalai–Smorodinsky solutions.

4.5.3.1 Theoretical Evaluation of Bargaining Solutions

Another condition is added to the general single-cell OFDMA system to help obtain a closed form for the Nash and Kalai–Smorodinsky solutions. It is now assumed that the *total achievable rate* of the whole network R_T is known and the equality constraint $\sum_{k=1}^{K} R_k = R_T$ is enforced. (Note that $R_T > \sum_{k=1}^{K} R_k^{\min}$ should be satisfied.)

With the addition of the equality condition, the maximum Nash product can easily be found as a maximizer of the product of K positive quantities whose sum is constant. Maximum product occurs when all quantities are of equal value, which is their sum divided by K. (Mathematical derivation can be given by applying Cauchy's mean theorem.) The theoretical Nash solution is

$$R_{k(\text{NBS})}^* = \frac{R_T - \sum_{j=1}^{K} R_j^{\min}}{K} + R_k^{\min} \quad \forall k. \tag{4.50}$$

For the Kalai–Smorodinsky solution, we can derive the formula by first applying the following normalization:

$$\hat{R}_k = \frac{R_k - R_k^{\min}}{R_k^{\max} - R_k^{\min}} \quad \forall k \tag{4.51}$$

and thus $\sum_{k=1}^{K} \hat{R}_k = \hat{R}_T$. The Kalai–Smorodinsky solution now governs that $\hat{R}_i^* = \hat{R}_j^* = \hat{R}_T / K$, $\forall i, j$. Thus,

$$
\begin{aligned}
\hat{R}_k^* &= \frac{R_1^* - R_1^{\min}}{R_1^{\max} - R_1^{\min}} = \cdots = \frac{R_K^* - R_K^{\min}}{R_K^{\max} - R_K^{\min}} \\
&= \frac{\sum_{j=1}^{K} R_j^* - \sum_{j=1}^{K} R_j^{\min}}{\sum_{j=1}^{K}(R_j^{\max} - R_j^{\min})} \\
&= \frac{R_T - \sum_{j=1}^{K} R_j^{\min}}{\sum_{j=1}^{K}(R_j^{\max} - R_j^{\min})}.
\end{aligned}
\tag{4.52}
$$

Thus, from Equations 4.51 and 4.52, we obtain the final result as

$$R_{k(KS)}^* = R_k^{\min} + \frac{(R_T - \sum_{j=1}^{K} R_j^{\min})(R_k^{\max} - R_k^{\min})}{\sum_{j=1}^{K}(R_j^{\max} - R_j^{\min})} \quad \forall k. \tag{4.53}$$

The above formulae serve as theoretical results for the bargaining solutions. In the next section, an algorithm is used to allocate subcarriers and power in order to obtain outcomes close to these solutions.

4.5.3.2 The Algorithm

The round-robin carrier assignment algorithm is presented in Table 4.6.

Carrier and power allocation constitute the two steps of the algorithm. Carrier assignment is done based on a round-robin manner in which each user takes turn to select one suitable subcarrier. The moderation process where users exchange subcarriers is a practical implementation of cooperative play where users carry on bargaining until agreement is reached. Here since the exact Nash or Kalai–Smorodinsky solutions may not be reached, an acceptable region of 95% the desirable rate is applied.

Table 4.6   Round-Robin Carrier Assignment Algorithm

Step I: Carrier Assignment
1. Initialization
a. Each user sets total transmission power at $p(k, n) = \overline{P} = P_{\max}/N$ b. For any user k, sort its SNR $\gamma(k, n)$ in descending order

2. Carrier assignment

a. For any user k, select the best channel according to the sorted SNR list in a round-robin manner. If the channel is already taken, skip to the next available on the list
     b. Compute the obtained rate R_k and the ratio $\eta_k = R_k/R_k^*$ where R_k^* is either $R_{k(NBS)}^*$ or $R_{k(KS)}^*$
     c. If $\eta_k > 0.95$ then user k stops; else wait for the next turn
     d. If no more subcarriers are available and not all users achieve 95%, go to (3)
     e. If all users achieve 95% and there are still available subcarriers, assign each subcarrier to the user that generates the highest rate (greedy assignment) and go to Step II

3. Moderation

a. Sort η_k in descending order; thus divide the users into two groups based on $\eta_k > 0.95$ and $\eta_k \leq 0.95$
     b. Process from the bottom of the sorted list: a user from group 2 borrows a subcarrier from one in group 1 when this subcarrier is best on its sorted SNR list and the group 1 user has more than one subcarrier
     c. Repeat (b) until $\eta_k > 0.95$ for all users or break after a number of rounds. Go to Step II

Step II: Power Estimation   For each user k, set $P(k, n)$ according to Equation 4.54.

In Step II, estimation of the transmission power to improve the rate outcome is done according to the following [25]:

$$
p(k, n) = \begin{cases} \overline{P} + \dfrac{1}{L} \sum_{i=1}^{K} \sum_{j=1}^{N} \dfrac{1}{\gamma(i,j)} - \dfrac{1}{\gamma(k,n)} & \text{if } a(k, n) = 1, \\ 0 & \text{otherwise,} \end{cases} \tag{4.54}
$$

where $L = \sum_{j=1}^{N} a(k, j)$ is the total number of subcarriers user k is assigned.

The algorithm's complexity can be shown to be considerably lower compared to the previous pair bargain with coalition algorithm. In Part 2 of Step I, loops of $O(KN)$ can occur. In Part 3 of Step I, $K(K-1)/2$ exchanges can occur [25]. The sorting algorithm can be chosen so that it is less than $O(K^2)$. Thus in the long term, the algorithm approaches $O(KN + K^2)$, which is clearly less than that of the pair-bargaining with coalition algorithm.

4.5.4 Comparison of the Four Bargaining Solutions and Numerical Results

In this section, we present an illustrative comparison of the four bargaining solutions in question as well as some numerical results for the two algorithms.

Figure 4.5 illustrates a geometrical representation of a two-player bargaining game. Along the x-axis and y-axis are the payoffs (e.g., rates) for the first and second players, respectively; a closed, convex region representing the feasible set is formed, the boundary of which is called Pareto optimal curve $\mathcal{P}$ where bargaining solutions should rest on since axiom 1 is satisfied (see Section 4.2.3).

The utilitarian solution is the point where the line $x + y = a$ is tangential to $\mathcal{P}$, which also maximizes the sum $x + y$. By definition, the egalitarian solution must be the intersection of the line $y = x$ and $\mathcal{P}$. The Kalai–Smorodinsky solution satisfies $y/x = y_{max}/x_{max}$, thus lying on the line joining the origin and point (x_{max}, y_{max}). The Nash solution uniquely maximizes the product $xy = C$, so C is the limiting value for the hyperbola $y = C/x$ to just touch $\mathcal{P}$.

Now we compare each solution against others in terms of two criteria: *efficiency* [23,26] in terms of total throughput $x + y$ and proportional *fairness* [25,26] represented by (x/x_{max}) versus (y/y_{max}) (assuming $x_{min} = y_{min} = 0$). As Figure 4.5 suggests (and true for general cases), the Kalai–Smorodinsky and Nash solutions maintain good proportional fairness among players as the ratios at these points are equal or nearly similar. In terms of total throughput, the utilitarian solution always achieves optimal values, and the egalitarian solution often yields results far from optimum; the Nash and Kalai–Smorodinsky solutions are normally close to optimum (reflected by how far the line $x + y = a$ is from the origin). That is why these two solutions are usually preferred due to good performance in both criteria. Some actual experimental results for the OFDMA system were documented in Ref. [26], which confirms our discussion. In Figure 4.6 we show the total throughput and average rate per user for the four solutions against the number of users (here the "max–min" (most egalitarian) solution (4.42) was used). The results agreed with the two-player example. (More details on the experiment can be found in Ref. [26].)

In Refs. [23,24] computer simulation was carried out to test the performance of the pair bargaining with coalition algorithm. Also, the Nash bargaining solution obtained from the algorithm

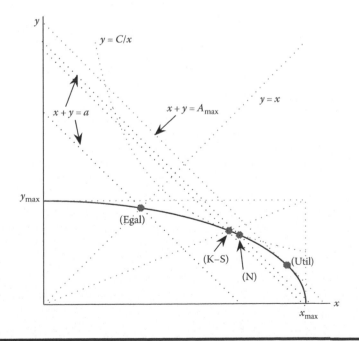

Figure 4.5 Locations of four bargaining solutions in a two-player example.

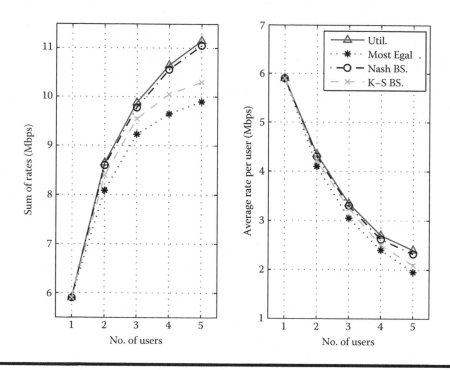

Figure 4.6 Comparison of four bargaining solutions for the experiment in Ref. [26].

was compared with the utilitarian solution and the minimal-rate scheme. The minimal rate turned into the "max–min" solution since all the rate requirements were similar. An OFDMA single-cell system was set up with $K = 128$ subcarriers over the total bandwidth $B = 3.2$ MHz, noise level $\sigma^2 = 10^{-11}$ W, maximum power constraint $P_{max} = 50$ mW, cell radius 0.2 km, and users' minimum rate $R_i^{min} = 25$ kbps.

The total rates for all users were plotted versus the number of mobile users in Figure 4.7. The "most egalitarian" solution performance seemed to be inferior to the other two. The utilitarian solution achieved the highest rate, but was "extremely unfair" [24] and when the fairness criteria discussed above were taken into account. The Nash bargaining solution calculated by the pair-bargaining with coalition algorithm was shown to achieve total throughput close to the utilitarian solution (see Figure 4.7). Basically the results agreed with previous examples regarding the strength and weakness of certain solutions. In addition, it was added that the Hungarian algorithm helped speed up convergence considerably compared to the random method [24].

The performance of the round-robin channel assignment algorithm was evaluated in Ref. [25]. The settings involved an OFDMA single-cell system with $K = 10$ users and $N = 256$ subcarriers. The average SNR over the network was set up to vary from 10 to 30 dB. The total rate was again an important measure, and was studied under different SNR conditions of the environment. The Nash and Kalai–Smorodinsky solutions obtained from the algorithm were also compared with the utilitarian solution. Another resource allocation scheme, which is the fixed channel assignment scheme, was also compared. The results are shown in Figure 4.8. As expected, the fixed channel assignment scheme exhibited the worst performance due to no resource reuse and optimization. The other bargaining solutions were able to achieve high total rates, with the utilitarian solution at the highest and the other two slightly below.

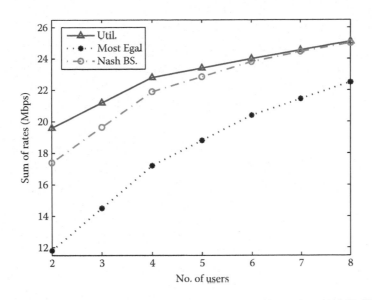

Figure 4.7 Comparison from the pair bargaining with coalition game. (From Z. Han, Z. Ji, and K. J. R. Liu, *IEEE Transactions on Communications*, **53, 8, 2005, pp. 1366–1376. With permission.)**

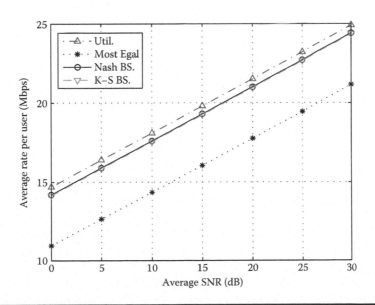

Figure 4.8 Comparison from the game using the round-robin carrier assignment algorithm. (From T. K. Chee, C.-C. Lim, and J. Choi, in *IEEE Singapore International Conference on Communication Systems*, **vol. 10, October 2006, pp. 1–5. With permission.)**

4.6 Concluding Remarks

4.6.1 Summary

In this chapter we have addressed the relatively new topic of using game theory to adaptively allocate power and bandwidth resources in an OFDMA system in order to obtain optimized transmission rates as well as efficient power consumption. After a general review of the noncooperative and cooperative realms of game theory, we showed how the OFDMA single-cell and multicell scenarios with distributed users competing for resources fitted into the game theoretical models. Existing game theoretical solutions in the literature were documented and assessed.

In the multicell scenario, players are the BSs of the various cells, who need to distribute the OFDMA subcarriers and power to their MSs scattered around the cell and at the same time try to avoid the potentially high co-channel interference from neighboring cells. The joint carrier-power assignment problem is a generic noncooperative game equivalent to a hard optimization problem, and is often broken into smaller games of power minimization and rate maximization in consequence. We introduced two algorithms, which tackled the problem in different ways. The iterative best-response algorithm relies on the best-response nature of individuals' decision making to play a repeated game, which may eventually reach a Nash equilibrium point. Another merit of this approach is the concept of "net utility," which reflects the award and cost from decision making. The other algorithm proposes a refereed game, which is an improved version of an original purely noncooperative game. Although the concept of a "referee" (a game monitor) is only a virtual implementation, it helps to allow a certain level of coordination among noncooperative players so that the inefficient noncooperative Nash equilibrium outcome can be improved upon. As the algorithm is of iterative nature, convergence is one of the main concerns. From the numerical results, convergence can be obtained unless the system resources are too stringent and restrictive.

Similarly, the single-cell scenario was dealt with using the cooperative game theoretical approach. Distributed mobile users within the same cell now act together to find the most efficient and fairest rate allocation, and the BS may serve as a network hub for users to share information and carry out their resource bargaining. Four well-known bargaining solutions in game theory are then investigated: the utilitarian, egalitarian, Nash, and Kalai–Smorodinsky bargaining solutions. In the literature, the true egalitarian solution is often replaced by a max–min scheme (i.e., "most egalitarian"). Preliminary studies showed that the utilitarian solution always brought optimal throughput, although resource usage among users is largely unfair. The "most egalitarian" scheme gave every user the same rate, which unfortunately is inferior to the other solution; however the Nash and Kalai–Smorodinsky solutions can somehow balance the two criteria of fairness and optimality, as suggested from the numerical results. In order to obtain the desirable bargaining outcome, we studied two algorithms: the pair-bargaining with coalition algorithm and the round-robin carrier assignment algorithm. The first algorithm enables cooperation by letting users form pairs and bargaining for resources repetitively. A Hungarian pair grouping method was proposed to further enhance the effectiveness of the algorithm, but at the expense of computational complexity. The round-robin carrier assignment offers a lower complexity level. It allows cooperation among users by letting players with enough rates and subcarriers transfer their "wealth" to other players in need of resources.

Due to the game theoretical framework, novel viewpoints were offered into the traditional optimization problem associated with resource allocation in wireless-access systems like OFDMA. However, game theory alone does not solve the optimization problem. Instead it provides rather theoretical solution concepts such as Nash equilibrium allocation points and the different bargaining solutions, and thus makes way for optimization techniques and algorithms, which are still the backbone of the problem in question, to compute those solutions effectively.

4.6.2 Open Issues

A major conclusion after investigating OFDMA game theoretical algorithms is that they have already become hot research topics despite numerous work done. Several issues are open for further improvement, and the following points are what we believe are key research directions.

First, we need more vigorous mathematical analysis to completely understand the strategy and utility space, as well as more insights into the issue of uniqueness versus multiplicity of Nash equilibrium. Nash's theorem itself only guarantees the existence of such points, and although efforts can be spent to numerically study the problems, we hope at least to be able to characterize the equilibrium set and to bring the game to the 'most optimal' equilibrium.

Second, it should be pointed out that the resource allocation problem dealt with so far in this chapter is restricted to power and channel assignment, neglecting the adaptive modulation issue [6]. Incorporating the bit loading matrix into the strategy sets of users poses great challenges to game modeling as well as analysis of solutions.

Next, we may look towards the continuing development of the new IEEE 802.22 standard [31]. This new standard employs OFDMA for the PHY layer with cognitive radio technology in the MAC layer, thus opening the new topic of radio resource management for OFDMA-based cognitive radio. The use of game theory for cognitive radio systems where smart nodes can sense the environment and adaptively adjust their transmission parameters has also been a fast emerging topic [15,32,33], with new game-theoretically several related concepts like the supermodular game and the potential game [32], and games based on auction theory [34]. We believe that combining the available techniques for OFDMA and cognitive radio will open new doors to satisfying the demanding need for efficient spectrum management.

References

1. IEEE Standard 802.16-2004 (2004), http://standards.ieee.org/getieee802/download/802.16-2004.pdf
2. IEEE Standard 802.16e (2005), http://standards.ieee.org/getieee802/download/802.16e-2005.pdf
3. IEEE Standard 802.16f (2005), http://standards.ieee.org/getieee802/download/802.16f-2005.pdf
4. IEEE Standard 802.16g (2007), http://standards.ieee.org/getieee802/download/802.16g-2007.pdf
5. IEEE Standard 802.16k (2007), http://standards.ieee.org/getieee802/download/802.16k-2007.pdf
6. S. Pietrzyk, *OFDMA For Broadband Wireless Access*. Boston/London: Artech House, 2006.
7. W. Rhee and J. M. Cioffi, Increase in capacity of multiuser OFDM system using dynamic subchannel allocation, in *IEEE Vehicular Technology Conference*, 2000, pp. 1085–1089.
8. H. Yin and H. Liu, An efficient multiuser loading algorithm for OFDM-based broadband wireless systems, in *IEEE Globecom*, 2000, pp. 103–107.
9. D. Kivanc and H. Liu, Subcarrier allocation and power control for OFDMA, in *Conference on Signals, Systems and Computers*. vol. 1, 2000, pp. 147–151.
10. Y. H. Chew and K. Zhou, Heuristic algorithms to adaptive subcarrier-and-bit allocation in multiclass multiuser OFDM systems, in *IEEE Vehicular Technology Conference*, vol. 3, May 2006, pp. 1416–1420.
11. R. B. Myerson, *Game Theory: Analysis of Conflict*. Cambridge: Harvard University Press, 1991.
12. D. Fudenberg and J. Tirole, *Game Theory*. Cambridge, MA: The MIT Press, 1991.
13. R. D. Luce and H. Raiffa, *Games and Decisions: Introduction and Critical Survey*. New York: Dover Publications, 1989.
14. V. Shah, N. B. Mandayam, and D. J. Goodman, Power control for wireless data based on utility and pricing, in *International Symposium on Personal, Indoor and Mobile Radio Communication*, vol. 3, 1998, pp. 1427–1432.

15. N. Nie and C. Comaniciu, Adaptive channel allocation spectrum etiquette for cognitive radio networks, in *IEEE International Symposium on New Frontiers in Dynamic Spectrum Access Networks*, November 2005, pp. 269–278.

16. B. MacKenzie and S. B. Wicker, Game Theory in Communications: Motivation, Explanation, and Application to Power Control, in *IEEE Globecom*, vol. 2, November 2001, pp. 821–826.

17. Z. Han, Z. Ji, and K. J. R. Liu, Power minimization for multi-cell OFDM networks using distributed non-cooperative game approach, in *IEEE Globecom*, vol. 6, Nov-Dec 2004, pp. 3742–3747.

18. Z. Han, Z. Ji, and K. J. R. Liu, Non-Cooperative Resource Competition Game by Virtual Referee in Multi-Cell OFDMA Networks, *IEEE Journal on Selected Areas in Communications*, 25, (6), 2007, pp. 1079–1090.

19. H. Kwon and B. G. Lee, Distributed resource allocation through noncooperative game approach in multicell OFDMA systems, in *IEEE International Conference on Communications*, vol. 9, June 2006, pp. 4345–4350.

20. L. Wang, Y. Xue, and E. Schulz, Resource Allocation in Multicell OFDM Systems Based on Noncooperative Game in *International Symposium for Personal, Indoor and Mobile Radio Communications*, vol. 17, September 2006, pp. 1–5.

21. X. Yu, T. Wu, J. Huang, and Y. Wang, A Non-Cooperative Game Approach for Distributed Power Allocation in Multi-Cell OFDMA-Relay Networks, in *IEEE Vehicular Technology Conference*, May 2008, pp. 1920–1924.

22. F. Chen, L. Xu, S. Mei, T. Zhenhui, and L. Huan, OFDM bit and power allocation based on game theory, in *International Symposium on Microwave, Antenna, Propagation and EMC Technologies for Wireless Communications*, August 2007, pp. 1147–1150.

23. Z. Han, Z. Ji, and K. J. R. Liu, Low-complexity OFDMA channel allocation with Nash bargaining solution fairness, in *IEEE Globecom*, vol. 6, Nov-Dec. 2004, pp. 3726–3731.

24. Z. Han, Z. Ji, and K. J. R. Liu, Fair multiuser channel allocation for OFDMA networks using nash bargaining solutions and coalitions, *IEEE Transactions on Communications*, 53, 8, 2005, pp. 1366–1376.

25. T. K. Chee, C.-C. Lim, and J. Choi, A cooperative game theoretic framework for resource allocation in OFDMA systems, in *IEEE Singapore International Conference on Communication Systems*, vol. 10, October 2006, pp. 1–5.

26. A. Ibing and H. Boche, Fairness vs. efficiency: comparison of game theoretic criteria for OFDMA scheduling, in *Conference Record of the Forty-First Asilomar Conference on Signals, Systems and Computers*, November 2007, pp. 275–279.

27. W. Thomson, Cooperative models of bargaining. In: *Handbook of Game Theory*, R.J. Aumann and S. Hart, (eds) vol. 2, Amsterdam: North-Holland, pp. 1237–1284, 1994.

28. S. Boyd and L. Vandenberghe, *Convex Optimization*, Cambridge University Press, Cambridge, UK; New York, 2004.

29. H. W. Kuhn, The Hungarian method for the assignment problem, *Naval Research Logistic Quarterly*, 2, 1955, 83–97.

30. W. Yu and J. M. Cioffi, FDMA capacity of Gaussian multiple-access channels with ISI, *IEEE Transactions on Communications*, 50, (1), 2002, 102–111.

31. IEEE 802.22 WRAN WG Website http://www.ieee802.org/22/ (accessed September 25, 2008).

32. J. O. Neel, J. H. Reed, and R. P. Gilles, Convergence of Cognitive Radio Networks, in *IEEE Wireless Communications and Networking Conference*, 2004, pp. 2250–2255.

33. Q. D. La, Y. H. Chew, W. H. Chin, and B. H. Soong, A Game Theoretic Distributed Dynamic Channel Allocation Scheme with Transmission Option, in *IEEE Military Communications Conference*, November 2008, pp. 1–7.

34. J. Huang, R. Berry, and M. L. Honig, Auction-based Spectrum Sharing, *ACM/Springer Mobile Networks and Apps.*, 2006, pp. 405–418.

Chapter 5

Resource Management Techniques for OFDMA

Didier Le Ruyet and Berna Özbek

Contents

5.1	Introduction	102
5.2	Resource Allocation	103
	5.2.1 System Model	103
	5.2.2 MA Optimization	103
	5.2.3 RA Optimization	105
5.3	Resource Allocation Algorithms	106
	5.3.1 Dynamic Subchannel Assignment	107
	5.3.1.1 Task 1: Bandwidth Assignment	108
	5.3.1.2 Task 2: Subchannel Assignment	109
	5.3.1.3 Combined Tasks 1 and 2	111
	5.3.2 APA Algorithms	111
	5.3.2.1 The Modified Levin–Campello Algorithm	113
	5.3.3 Simulations	113
	5.3.4 Extension to the Multiple Antennas Case	115
5.4	Scheduling	116
	5.4.1 Introduction	116
	5.4.2 Quality of Service	116
	5.4.3 Scheduler Properties	117
5.5	Scheduling Algorithms	118
	5.5.1 Classical Scheduling Algorithm	118
	5.5.1.1 Generalized Processor Sharing	118
	5.5.1.2 Weight Round Robin	118

            5.5.1.3   Weight Fair Queueing... 119
            5.5.1.4   Early Deadline First... 119
            5.5.1.5   Proportional Fair Scheduler.. 120
            5.5.1.6   Modified Largest Weighted Delay First Algorithm.................... 120
            5.5.1.7   Exponential Rule.. 121
      5.5.2   Schedulers for Heterogeneous Traffic... 121
      5.5.3   Utility-Based Function... 122
            5.5.3.1   Introduction.. 122
            5.5.3.2   Rate-Based Utility Function.. 123
            5.5.3.3   Delay-Based Utility Function.. 125
            5.5.3.4   Utility-Based Scheduling for Heterogeneous Traffic................... 126
5.6   Open Problems.. 126
References... 127

5.1 Introduction

Resource management is crucial for OFDMA wireless broadband networks where scarce spectral resources are shared by multiple users. Resource management is usually separated into two parts: scheduling and resource allocation [1,2]. The scheduler determines the quantity of packets that should be scheduled in the current frame and the resource controller assigns the subcarriers to each selected user in order to maximize the throughput. The resource allocation can also be divided into dynamic subchannel assignment and adaptive power allocation. However, recently, the resource management techniques have been changed to improve the overall system performance. The principles of multiuser downlink and medium access control (MAC) designs have moved from the traditional point of view to a multiuser network view with integrated adaptive design in a cross-layer fashion. For instance, channel-aware scheduling strategies have been proposed to adaptively transmit data and dynamically assign wireless resources based on physical layer considerations such as the channel state information (CSI) but also the intercell interference, and so on.

For cross-layer optimization, the scheduler in the MAC layer should effectively enhance the spectral efficiency and also handle multiple types of traffic (non-real-time and real-time) with different quality of service (QoS) requirements. Concerning the QoS, it is important to differentiate two types of criteria: the user data rate, which refers to the average amount of traffic generated by the user, and the utility function based on subjective measures depending on the applications [video streaming, Voice over IP (VoIP), etc.]. By exploiting the knowledge of the CSI and the traffic characteristics, the resource allocation tries to maximize the total utility, which is used to capture the satisfaction levels of users.

In this chapter, we will consider the resource management for OFDMA that offers performance gains needed for end-to-end QoS. We will focus on resource allocation and scheduling algorithms and explore the fundamental mechanisms such as throughput, fairness, and stability. The organization of the rest of the chapter is as follows: in Section 5.2, we will introduce the margin and rate adaptive optimization. The different algorithms for dynamic subchannel assignment (DSA) and adaptive power allocation (APA) will be reviewed in Section 5.3. Then, in Section 5.4, the properties of the scheduler will be given. In Section 5.5 we will study the classical scheduling algorithm, the scheduling techniques for heterogeneous traffic, and the utility-based schedulers. Finally, open research issues related to resource management are discussed.

5.2 Resource Allocation

5.2.1 System Model

We will consider multiuser OFDM systems with single transmit antennas. The base station will serve K users with only one receive antenna. The data are formed into OFDM symbols with N subcarriers and then transmitted through frequency selective channels. These channels are assumed to be constant over one OFDM frame and varying between the frames considering the Doppler frequency. Moreover, it is assumed that the channel taps are equal or smaller than the length of the cyclic prefix.

Assuming that the CSI about all the subcarriers for all the users is known at the transmitter, the adaptive resource allocation algorithm is used to optimize the system parameters in a way that maximizes the total number of bits received by all the users for a given total power or minimizes the required total power for given user rate constraints.

In this system, the channel that is between the base station and the kth user is described in the frequency domain as

$$H_k = \begin{bmatrix} H_{k,1} & H_{k,2} & \cdots & H_{k,N} \end{bmatrix}^{\mathrm{T}}, \tag{5.1}$$

where $H_{k,n}$ is the channel gain from the transmitter to the kth user for the nth subcarrier. In this scheme, the received signal is written as

$$Y_{k,n} = H_{k,n}S_{k,n} + N_{k,n}, \tag{5.2}$$

where $S_{k,n}$ is the transmitted symbol with the power of P_{T}/N, and $N_{k,n}$ is the additive white Gaussian noise, with zero mean and variance of $N_0/2N$. In this system, the total bandwidth, B, is equally divided into N orthogonal subcarriers and the bandwidth of a subcarrier is equal to B/N.

Once the subcarriers have been determined for each user, the base station has to inform each user which subcarriers have been allocated to it. Therefore, it is assumed that the subcarrier/bit allocation information is transmitted to each user through a separate control channel. The resource allocation must be performed on the order of the channel coherence time. However, it may be performed more frequently if many users are competing for resources. The resource allocation is usually formulated as a constrained optimization problem. Two different approaches are possible:

■ *Margin adaptive (MA) optimization*: Minimize the total transmit power with a constraint on the user data rate [3–5].
■ *Rate adaptive (RA) optimization*: Maximize the total data rate with a constraint on total transmit power [6–8].

5.2.2 MA Optimization

We define $p_{k,n}$ as the power allocated to the nth subcarrier for the kth user and denote $\rho_{k,n}$ as a binary variable for the nth subcarrier and the kth user:

$$\rho_{k,n} = \begin{cases} 1 & \text{if the } n\text{th subcarrier is used for the } k\text{th user,} \\ 0 & \text{else.} \end{cases} \tag{5.3}$$

The MA optimization problem can be described as follows:

$$\min_{c;\rho} P_T = \sum_{k=1}^{K} \sum_{n=1}^{N} p_{k,n} \rho_{k,n}$$

$$= \sum_{k=1}^{K} \sum_{n=1}^{N} \left(\frac{f(c_{k,n}) \rho_{k,n}}{|H_{k,n}|^2} \right) \tag{5.4}$$

subject to

$$r_k = \frac{B}{N} \sum_{n=1}^{N} c_{k,n} \rho_{k,n} \geq R_k \quad \forall k \in \{1, \ldots, K\} \quad (C_1), \tag{5.5}$$

$$p_{k,n} \geq 0 \quad \forall k, n \quad (C_2) \tag{5.6}$$

and

$$\sum_{k=1}^{K} \rho_{k,n} = 1 \quad \forall n \quad (C_3), \tag{5.7}$$

where $f(c_{k,n})$ is the required received power for a reliable reception of $c_{k,n}$ bits when the channel gain is equal to unity, and $c_{k,n}$ is the number of bits that are assigned to the nth subcarrier for the kth user.

In condition C_1, R_k is the minimum data rate required for the kth user.

Condition C_2 denotes the power allocation for each subcarrier, n.

Condition C_3 means that no more than one user can be allowed to transmit on the same subcarrier: if $\rho_{k,n} = 1$ then $\rho_{k',n} = 0$ for all $k' \neq k$.

Depending on the application, different relations between $c_{k,n}$ and $f(c_{k,n})$ have been proposed. One relation can be calculated using the Shannon capacity expression as

$$f(c_{k,n}) = \frac{BN_0}{N} (2^{c_{k,n}} - 1). \tag{5.8}$$

Another relation can be obtained taking into account the modulation and coding schemes (MCSs) available in the considered system. Depending on the application, the set of rates $c_{k,n}$ is selected from the finite set $[0, c_1, c_2, \ldots, c_{max}]$ where c_{max} is the maximum number of bits per symbol that can be transmitted by each subcarrier. A continuous approximation can be obtained from the set of rates $\{c_{k,n}\}$ and the associated transmitted power $\{f(c_{k,n})\}$ for a given target Bit Error Rate (BER) after demodulation and channel decoding [4].

A simpler relation can be given for M-ary quadrature amplitude modulation (M-QAM) schemes without considering the channel code. The bit error probability can be upper bounded by the symbol error probability, which is tightly approximated by $2\text{erfc}(d/2\sqrt{N_0})$ [9] where d is the minimum distance between the points in the signal constellation. Since the average energy of a M-QAM symbol is equal to $(M-1)d^2/6$, the required power $f(c_{k,n})$ for supporting $c_{k,n}$ bits per symbol at a required BER can be represented by

$$f(c_{k,n}) = \frac{B}{N}\frac{2N_0}{3}\left[\text{erfc}^{-1}\left(\frac{\text{BER}}{2}\right)\right]^2 (2^{c_{k,n}} - 1), \qquad (5.9)$$

where $\text{erfc}(x) = 2/\sqrt{\pi}\int_x^\infty e^{-t^2}\,dt$.

This relation can be simplified using the following inequality [10]:

$$\text{BER} \le 0.2\exp\left(\frac{-1.5\text{SNR}}{2^{c_{k,n}} - 1}\right). \qquad (5.10)$$

Then, we have

$$f(c_{k,n}) \le -\frac{B}{N}\frac{2N_0}{3}\log(5\text{BER})(2^{c_{k,n}} - 1). \qquad (5.11)$$

The term $\Gamma = -\frac{2}{3}\log(5\text{BER})$, which is a function of the required BER, is higher than 1 when $\text{BER} < 0.2\exp(-1.5) \approx 0.05$.

5.2.3 RA Optimization

Assuming that the available total transmit power is limited by P_T, in order to maximize the sum data rate, the RA optimization problem can be expressed as

$$\max_{c;\rho} r = \sum_{k=1}^{K} r_k = \frac{B}{N}\sum_{k=1}^{K}\sum_{n=1}^{N} c_{k,n}\rho_{k,n} \qquad (5.12)$$

subject to

$$r_k = \frac{B}{N}\sum_{n=1}^{N} c_{k,n}\rho_{k,n} \ge R_k \quad \forall k \in \{1,\dots,K\} \quad (C_1), \qquad (5.13)$$

$$p_{k,n} \ge 0 \quad \forall k, n \quad (C_2), \qquad (5.14)$$

$$\sum_{k=1}^{K} \rho_{k,n} = 1 \quad \forall n \quad (C_3), \qquad (5.15)$$

and

$$\sum_{k=1}^{K}\sum_{n=1}^{N}\left(\frac{f(c_{k,n})\rho_{k,n}}{|H_{k,n}|^2}\right) \le P_T \quad (C_4), \qquad (5.16)$$

In the multiuser OFDM system, the sum data rate r may be represented by

$$r = \sum_{k=1}^{K} r_k = \frac{B}{N} \sum_{k=1}^{K} \sum_{n=1}^{N} c_{k,n} \rho_{k,n}$$

$$= \frac{B}{N} \sum_{k=1}^{K} \sum_{n=1}^{N} \log_2 \left(1 + \frac{\text{SNR}_{k,n}}{\Gamma} \right) \rho_{k,n}, \tag{5.17}$$

where

$$\text{SNR}_{k,n} = \frac{p_{k,n} |H_{k,n}|^2}{N_0(B/N)}, \tag{5.18}$$

where the power per subcarrier $p_{k,n}$ can be chosen as equally shared and thus it is equal to P_{T}/N and $\Gamma = 1$ when using Equation 5.8 or $\Gamma = 2\log(5\text{BER})/3$ when considering Equation 5.11.

Without simplification, both classes of optimization belong to the class of discrete optimization and are known to be non-deterministic polynomial time (NP) complex problems. Consequently, several suboptimal algorithms have been proposed in the literature.

For OFDMA systems, we can perform the resource allocation algorithm at the cluster level. A cluster is a group of adjacent subcarriers and only the information about the strongest clusters of each user is used to apply allocation algorithms. In this structure, N_q adjacent subcarriers are grouped for each cluster and it is assumed that the correlation is high between the subcarriers within this cluster. The channel cluster vector $\bar{\mathbf{H}}_k$ associated with the kth user is given by

$$\bar{\mathbf{H}}_k = \begin{bmatrix} \bar{H}_{k,1} & \bar{H}_{k,2} & \cdots & \bar{H}_{k,Q} \end{bmatrix}^{\text{T}}, \tag{5.19}$$

where Q is the number of clusters with $N = Q \times N_q$, and $\bar{H}_{k,q}$ is the channel gain for the kth user and the qth cluster with $q = 1, 2, \ldots, Q$. The representative channel value for each cluster can be chosen as follows:

$$\eta = \arg \ \min\{|H_{k,qN_q}|, |H_{k,qN_q+1}|, \ldots, |H_{k,(q+1)N_q-1}|\}, \tag{5.20}$$

$$\bar{H}_{k,q} = H_{k,\eta}. \tag{5.21}$$

Below, we will focus on RA optimization, which is more relevant to the next generation of OFDMA systems.

5.3 Resource Allocation Algorithms

The rate maximization optimization problem is an NP-hard combinatorial problem. Some algorithms have been proposed to solve this problem after relaxing some constraints, but the complexity of the algorithm becomes prohibitive for systems with a large number of subcarriers. Consequently, the most common way of solving the problem is to split it into two different phases:

■ *DSA*: The aim of this first phase is to assign the different subchannels to the different users assuming the available total transmit power, P_{T}.
■ *APA optimization algorithms*: Once the subchannel assignment is fixed, the APA must maximize the sum data rate given the different constraints.

5.3.1 Dynamic Subchannel Assignment

DSA for RA optimization depends on the user rate constraints.

In case there is no minimum user rate constraint (no condition C_4), Jang and Lee [8] proved that the sum data rate is maximized when each subchannel is assigned to the user with the best subchannel gain assuming the power is shared equally between the subchannels.

Description of the Jang and Lee algorithm:

- *Initialization*: For each user k, initialize the associated set of subcarriers allocated to the users, $C_k = \emptyset$.
- For $n = 1$ to N:
  $k' = \arg \max_k |H_{k,n}|$, the subcarrier n is allocated to the user k', $C_{k'} = C_{k'} \bigcup \{n\}$.

In this case there is no fairness between the users and when the users have large path loss differences, the users with low average channel gains will be unable to receive data. Fairness requires a fair share of bandwidth among competing users. One of the representative types for the fairness is proportional fairness that provides each connection with a priority inversely proportional to its data rate.

The fairness index (FI) is calculated by using the Jain index [11] given as

$$\mathrm{FI} = \frac{\left(\sum_{k=1}^{K} x_k \right)^2}{K \sum_{k=1}^{K} x_k^2}, \tag{5.22}$$

where x_k can be equal to the allocated rate, r_k, or the difference between the allocated rate and minimum required rate, $r_k - R_k$. (Note that if R_k is higher than r_k, x_k will be equal to zero.)

The FI ranges between 0 (no fairness) and 1 (perfect fairness) in which all users would achieve the same data rate.

When a complete fairness is required between the users in the absence of the C_4 condition, the DSA is performed to maximize the minimum user data rate under the power constraint assuming that the power is shared equally between the subchannels. In Ref. [7], the authors proposed a reduced complexity suboptimal adaptive subchannel allocation algorithm to solve this problem.

Description of the Rhee and Cioffi max–min algorithm:

- *Initialization*: For each user k, initialize $r_k = 0$. Set $\mathcal{A} = \{1, 2, \ldots, N\}$.
- For $k = 1$ to K:
  (a) $n' = \arg \max_{n \in \mathcal{A}} |H_{k,n}|$, the subcarrier n' is allocated to the user k.
  (b) $r_k = r_k + (B/N) \log_2(1 + SNR_{k,n'})$.
  (c) $\mathcal{A} = \mathcal{A} - \{n'\}$.
- While $\mathcal{A} \neq \emptyset$
  (a) $k' = \arg \min_k r_k$.
  (b) $n' = \arg \max_{n \in \mathcal{A}} |H_{k',n}|$, the subcarrier n' is allocated to the user k'.
  (c) $r_{k'} = r_{k'} + (B/N) \log_2(1 + SNR_{k',n'})$.
  (d) $\mathcal{A} = \mathcal{A} - \{n'\}$.

Channel swapping can be performed to maximize the max–min capacity but the initial algorithm already achieves a good performance.

When there is a minimum user data rate constraint (condition C_4), the dynamic subchannel assignment can be solved in one step or divided into two tasks: bandwidth assignment and subchannel assignment.

5.3.1.1 Task 1: Bandwidth Assignment

In this task, the number N_u of subchannels per user is assigned.

A greedy bandwidth assignment based on SNR (BABS) algorithm has been proposed for MA optimization in Ref. [4] and for RA optimization in Ref. [12]. In this task, we assume that all the subchannels of a given user have the same gain. Let $|\bar{H}_k|^2$ be the average user gain:

$$|\bar{H}_k|^2 = \frac{1}{N} \sum_{n=1}^{N} |H_{k,n}|^2. \tag{5.23}$$

Description of BABS algorithm:

Initialization: Let $N_k = 1$ for each user k and $N_a = \sum_{k=1}^{K}$ where N_k is the number of allocated subcarriers. $P_k(N_k)$ is the transmit power required by user k to achieve the data rate R_k using N_k subcarriers.

$$P_k(N_k) = \frac{N_k}{|\bar{H}_k|^2} f\left(\frac{R_k}{N_k(B/N)}\right). \tag{5.24}$$

Iteration:

- If $\dfrac{\sum_{k=1}^{K} P_k(N_k)}{N_a} \leq \dfrac{P_T}{N}$, stop; otherwise continue.
- While $\dfrac{\sum_{k=1}^{K} P_k(N_k)}{N_a} > \dfrac{P_T}{N}$
  (a) Let $\Delta P_k = P_k(N_k) - P_k(N_k + 1)$ for $k = 1, 2, \ldots, K$.
  (b) $k' = \arg \max_k \Delta P_k$.
  (c) $N_{k'} = N_{k'} + 1$ and update N_a.

This algorithm gradually increases the number of subcarriers assigned to the users as N_k and gives the power value for assigned subcarriers as P_k/N_k. For BABS algorithm, it should be noticed that all the subcarriers are not necessarily allocated.

In order to determine the number of subcarriers for a given QoS criterion, the bandwidth allocation on rate estimation (BARE) algorithm [13] is described as follows:

Initialization: Let $N_k = \lfloor N/K \rfloor$ for each user k. We assume equal power allocation on all subcarriers, $p_{k,n} = P_T/N$. Then, compute the differences between the estimated rate of user k and the required minimum rate R_k:

$$G_k(N_k) = N_k C_k - R_k,$$

where $C_k = \frac{B}{N} \log_2(1 + \frac{P_T}{N} \frac{|\bar{H}_k|^2}{N_0(B/N)})$.
Iteration:

- While $\sum_{k=1}^{K} N_k < N$, find the user with minimum gap, $\kappa = \arg \min(G_k < 0)$. Then, the user κ receives one extra subcarrier.
- When $\sum_{k=1}^{K} N_k = N$ and at least one predicted rate is less than the required minimum rate:
  (a) $k'' = \arg \min_k(G_k(N_k) < 0)$ and provided that $k' = \arg \max_k(G_k(N_k - 1) > 0)$

(b) $N_{k'} = N_{k'} - 1$ for user k' and $N_{k''} = N_{k''} + 1$ for user k''.

■ Continue until $G_k(N_k) > 0$ for all k.

When the power is too low to meet the common user rate guaranty, a fairness mechanism that decreases the users' rate constraints is applied before restarting BARE.

5.3.1.2 Task 2: Subchannel Assignment

Once we determine the number of subcarriers allocated to each user, we perform the subchannel assignment.

The subchannel assignment is optimally solved using the Hungarian algorithm introduced by H. W. Kuhn in 1955. This problem is equivalent to the search of the optimum matching of a bipartite graph.

Since the cost matrix **R** must be square, we duplicate each user N_k times in order to have an $N \times N$ cost matrix (assuming that the number of users K is less than N).

Mathematically, the problem can be described as follows:

Given the $N \times N$ cost matrix $\mathbf{R} = [r'_{m,n}]$, find the $N \times N$ permutation matrix $\Psi = [\psi_{m,n}]$ such that

$$r' = \sum_{m=1}^{N} \sum_{n=1}^{N} r'_{m,n} \psi_{m,n}$$

is minimized.

The algorithm is briefly explained:

1. Find the minimum value of each row in the cost matrix **R** and subtract it from the corresponding row.
2. For the columns without zero, find the minimum value of the column and subtract it from the corresponding column.
3. Cover the zeros with the minimum number of horizontal and vertical lines in the updated cost matrix. If the minimum number of lines equals the dimension of the matrix, then stop. Else go to step 4.
4. Find the minimum value in the uncovered part of the cost matrix, and subtract it from the uncovered elements. Add it to the twice covered elements (elements at the intersection of a horizontal and a vertical line). Return to step 3.

Since we must minimize the cost r', we will use the following cost function:

$$r'_{m,n} = -\log_2 \left(1 + \frac{SNR_{m,n}}{\Gamma}\right), \tag{5.25}$$

where $SNR_{m,n}$ is defined in Equation 5.18 with $p_{k,n} = P_k/N_k$.

We will now give an example of the Hungarian algorithm. We consider a system with two users and four subcarriers. One subcarrier is assigned to the first user and three subcarriers are assigned to the second user. The data rate vector for the first and second user is [2 4 3 4.5] and [3.5 3 1 4], respectively. After forming the cost matrix **R**, we have

$$\mathbf{R} = \begin{pmatrix} -2 & -4 & -3 & -4.5 \\ -3.5 & -3 & -1 & -4 \\ -3.5 & -3 & -1 & -4 \\ -3.5 & -3 & -1 & -4 \end{pmatrix}.$$

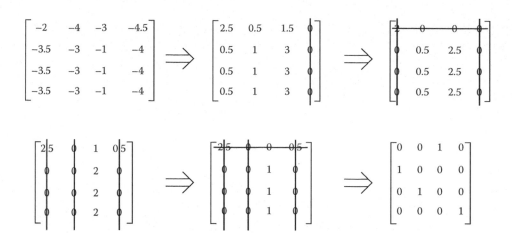

Figure 5.1 An example of the Hungarian algorithm.

The evolution of the Hungarian algorithm is illustrated in Figure 5.1.

In this example, according to the resulting permutation matrix, the optimal subcarrier allocation is obtained when the first user is allocated to subcarrier 3 and the second user is allocated to subcarriers 1, 2, and 4.

The Hungarian algorithm is computationally intensive and different suboptimal algorithms can be found to reduce the complexity. One of the most popular algorithms is the rate craving greedy (RCG) algorithm. In this algorithm, each subcarrier is first allocated to the user with the best rate. Then, some subcarriers are reassigned when users have more than N_k allocated subcarriers.

The description of the RCG algorithm [4]:

■ *Initialization*: For each user k, initialize the associated set of subcarriers allocated the users $C_k = \emptyset$.
■ *Initial allocation*: For $n = 1$ to N:
  $k' = \arg \max_k r_{k,n}$ the subcarrier n is allocated to the user $k' : C_{k'} = C_{k'} \bigcup \{n\}$.
■ *Reallocation*: For users k' such as $card(C_{k'}) > N_{k'}$ while $card(C_{k'}) > N_{k'}$

$$l' = \arg \min_{l:card(C_l)<N_l} \min_{n:C_{k'}} -r_{k',n} + r_{l,n},$$

$$n' = \arg \min_{n:C_{k'}} -r_{k',n} + r_{l',n},$$

$$C_k = C_k \backslash \{n'\} C_{l'} = C_{l'} \bigcup \{n'\}.$$

The ACG algorithm was initially proposed by Kivanc et al. [4] and an improved version has been proposed in Ref. [14]. Description of the improved ACG is given below:

■ *Initialization*: For each user k, initialize the associated set of subcarriers allocated $C_k = \emptyset$. Set $\mathcal{A} = \{1, 2, \ldots, N\}$.

■ *Allocation*: For $i = 1$ to N:

$$k', n' = \arg \max_{\substack{n \in \mathcal{A} \\ k : card(C_k) < N_k}} r_{k,n},$$

the subcarrier n' is allocated to the user k': $\mathcal{A} = \mathcal{A} \setminus \{n'\}$ and $\mathcal{C}_{k'} = \mathcal{C}_{k'} \bigcup \{n'\}$.

It should be noticed that the improved ACG algorithm can perform subcarrier assignment when the total number of allocated subcarriers is less than N.

5.3.1.3 Combined Tasks 1 and 2

In order to combine Tasks 1 and 2, in Ref. [5], a reduced complexity subcarrier and bit allocation algorithm has been proposed assuming equally shared power between subcarriers. The complexity of this algorithm has been significantly reduced by selecting the initial solution as an unconstrained optimal one. Besides that, only one constraint needs to be considered during each searching stage. The Zhang and Letaif algorithm is described briefly as follows.

Step 1: Optimization without inequality constraints: Firstly, the bit and subcarrier allocation is done without considering the rate constraint for each user as described in Jang and Lee algorithm.

Step 2: Subcarrier reallocation: The subcarrier allocation solution from Step 1 does not guarantee the fulfillment of every user's rate constraint. This subcarrier reallocation process is repeated until all the user's data rate requirements are satisfied. During the reallocation process, the following conditions must be satisfied:

■ A subcarrier that was originally assigned to user k_n^* cannot be reallocated to another user if the reallocation will cause the violation of user k_n^*s data rate requirement
■ Each subcarrier reallocation should cause the least possible reduction in the overall throughput
■ The number of reallocation operations should be kept as low as possible

The Zhang and Letaif algorithm is described in Figure 5.2 in detail.

5.3.2 APA Algorithms

Once the subchannel assignment has been performed, the APA is performed to maximize the sum data rate r according to the given total power constraint. In the absence of condition C_4, APA algorithm can be treated as a virtual single user OFDM system. Otherwise, it should be applied separately for each user.

In the absence of condition C_4, we have

$$\max_c \sum_{n=1}^{N} c_n \tag{5.26}$$

subject to

$$\sum_{n=1}^{N} p_n \leq P_T \tag{5.27}$$

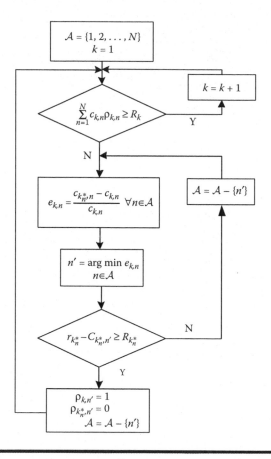

Figure 5.2 Subcarrier reallocation in Zhang and Letaif algorithm.

and

$$p_n \geq 0 \quad \forall n. \tag{5.28}$$

When continuous rate adaptation is used, the optimal power allocation for a fixed subcarrier assignment is waterfilling [15]. We have the following solution:

$$p_n^* = \frac{N_0 B \Gamma}{N} \left[\frac{1}{\lambda} - \frac{1}{|H_n|^2} \right]^+, \tag{5.29}$$

where $H_n = \mathbf{H}_n \rho_n^{\mathrm{T}}$ with $\mathbf{H}_n = [H_{1,n} \, H_{2,n}, \dots, H_{K,n}]$ and $\rho_n = [\rho_{1,n} \, \rho_{2,n}, \dots, \rho_{K,n}]$, $[a]^+ = \max\{a, 0\}$, and λ is a threshold associated with the total power constraint.

In practice, continuous rate adaptation is not feasible, and there are only several modulation levels. Consequently, the waterfilling algorithm cannot achieve the optimal power allocation.

For discrete modulation levels, a greedy power allocation algorithm has been proposed. The key idea of the greedy algorithm is to allocate bits and the corresponding power successively and to maximize the partial sum data rate in each step of bit loading [3]. In initialization, zero bits are assigned to all subcarriers. During each bit loading iteration, the subcarrier that needs the minimum additional power is assigned one more bit and the total partial power is updated. The iteration process will stop when the total transmission power constraint is reached. To apply the greedy power allocation for discrete rate adaptation, the modified Levin–Campello algorithm [16] can be used and it is described as follows:

5.3.2.1 The Modified Levin–Campello Algorithm

Let $\Delta P_n(c) = (f(c+1) - f(c))/|H_n|^2)$ denote the incremental power needed for the transmission of one additional bit at subcarrier n, and c is the number of loaded bits for the nth subcarrier.

- *Initialization*: For each subcarrier n, initialize $c_n = 0$ and evaluate $\Delta P_n(c = 0)$ with tentative transmit power $P_T^* = 0$.
- *Bit assignment iteration*: Repeat the following iterations until $P_T^* \geq P_T$:
  $n^* = \arg \min_n \Delta P_n(c_n)$
  $P_T^* = P_T^* + \Delta P_{n^*}(c_{n^*})$
  $c_{n^*} = c_{n^*} + 1$
  if $c_{n^*} = c_{\max}$, set $\Delta P_{n^*}(c_{n^*}) = \infty$, else evaluate $\Delta P_{n^*}(c_{n^*})$.
- *Finish*: The allocation result $\{c_n\}_{n=1}^N$ is the obtained optimal bit allocation solution.

5.3.3 Simulations

We evaluate the different resource allocation algorithms using the parameters listed in Table 5.1.

The sum capacity comparison results are obtained for different allocation algorithms of OFDMA systems in Figure 5.3. According to the results, the best sum capacity performance is obtained using the Jang–Lee algorithm and the Zhang–Letaif algorithm. However, the FI is also an important parameter to observe the distribution of the users' data rate. In Figure 5.4, the Jain FI is drawn and

Table 5.1   System Parameters for the Simulation

Parameter	Value
Cell radius	1.6 km
BS transmit power	43.10 dBm
Noise power	−174 dBm
Path loss L_p	$128.1 + 37.6 \log_{10}(d)$ dB
Channel model	3GPP TU
Number of clusters	48
Bandwidth	10 MHz
Carrier frequency	2.4 GHz
Velocity	3 km/h
Simulation time	5 s
User distribution	[0.3 0.4 0.5 0.8 1.0]km, equal probability

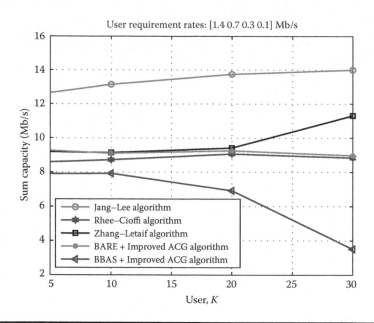

Figure 5.3   Sum capacity versus the number of users.

we can see that the Jang–Lee algorithm does not provide fairness. The Zhang–Letaif algorithm can bring limited fairness without sacrificing the sum capacity performance. By construction, the Rhee–Cioffi algorithm maximizes the fairness between the users when equal rate constraints are considered for all users. The BARE+Improved ACG algorithm gives a slightly better sum capacity performance than the Rhee–Cioffi algorithm and provides fairness compared to the Zhang–Letaif algorithm. The

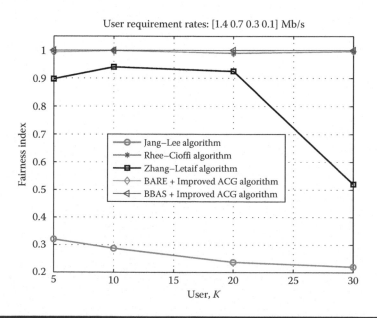

Figure 5.4   FI versus the number of users.

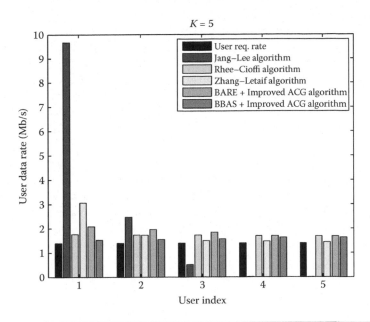

Figure 5.5 Capacity distribution among the users for *K* = 5.

BABS algorithm is based on subcarrier assignment that includes power loading; compared to the BARE algorithm, the BABS algorithm also provides fairness. It should be noticed that the BABS algorithm allocates the minimum number of subcarriers to satisfy the users' requirements and the total number of allocated subcarriers is not always equal to the number of total subcarriers in OFDMA, while the BARE algorithm fulfills all the subcarriers to increase the sum data rate performance.

In Figure 5.5, the distribution of the users' data rate is shown for *K* = 5 users in the systems. It is shown that the Zhang–Letaif and Rhee–Cioffi algorithms distribute the available resources and the capacity between the users and satisfy some users' QoS constraints, while the Jang–Lee algorithm does not consider the users' rate constraints. It is observed that the resource allocation based on subcarrier assignment using the BABS and BARE algorithms achieves all the users' rate requirement.

5.3.4 Extension to the Multiple Antennas Case

Multiple antennas at the base station enables a capacity increase by multiplexing spatially separable users in the same channel. A well-suited criterion to separate the spatial separation at the base station is zero forcing (ZF). It provides a reasonable performance loss compared to Dirty Paper precoding [17] or optimal downlink beamforming [18,19].

Resource allocation in OFDMA and exploiting multiple antennas have been mainly studied independently. Only some papers have considered the problem of resource allocation in multiple antennas and OFDM [20–22]. In Ref. [20], the authors have proposed a heuristic algorithm to allocate the subchannel to the users based on their spatial separability properties while adjusting beamforming weights for each user. Similar to the Zhang–Letaief algorithm, this is a two-step algorithm: the users are first allocated without considering the minimum rate requirement and then subcarrier exchange is performed between satisfied and unsatisfied users.

In the previous works, it was assumed that the CSI is available at the base station. In time division duplexing (TDD) systems, CSI may be estimated at the transmitter using channel reciprocity after a calibration phase. Frequency division duplexing (FDD) systems lack channel reciprocity and the CSI must be fed back to the base station over a limited rate feedback channel. In this case, it is possible to use quantized CSI using a codebook that is known to both the transmitter and the receiver. The codebook design for a narrowband single user communication system is a well-studied problem [23]. The extension to the OFDM case with multiple antennas has been considered in Refs. [24,25].

In OFDMA systems, the rate of the feedback link is high when all the users send their broadband CSI. In Ref. [26], the authors have proposed to reduce the feedback rate using the so-called clustered S-best criterion where adjacent subcarriers are grouped into clusters and only information about the strongest clusters of each user are fed back to the base station. This scheme reduces the feedback load per user; however, all the users send partial CSI to the BS in order to satisfy the minimum rate requirements. In Ref. [27], the authors have combined the clustered S-best and semi-orthogonal user selection criteria for spatial division multiple access (SDMA)–OFDMA systems with ZF precoding. A specific codebook design has been proposed by quantizing CSI for nonhomogenous user distribution and the proposed criterion considering both S-best cluster and semi-orthogonal user selection properties.

5.4 Scheduling

5.4.1 Introduction

In traditional networks, the resources are assigned to the users according to a predefined scheme (channel unaware scheduling). Only some feedback information bits are sent from the users to the base station to inform whether the QoS of the link is satisfied.

When the CSI is available at the base station, it is important to perform cross-layer scheduling by adding information from the physical layer such as the CSI and exploit the multiuser diversity to improve the spectral efficiency significantly (channel aware scheduling).

The aim of scheduling is to achieve objectives for spectral efficiency, fairness, and QoS. Usually, it is impossible to achieve the optimality of all these objectives. For instance, scheduling schemes aiming to maximize the total throughput are unfair to those users far away from a base station or with bad channel conditions. On the other hand, fairness may lead to low bandwidth efficiency. Therefore, an effective tradeoff between spectral efficiency, fairness, and QoS is desired in wireless resource allocation. The scheduling algorithms should dynamically exploit the channel variations and allocate different rates to different users. The main application of the scheduling is for downlink multiuser communication since all the data are available at the base station. However, the described schedulers can be also used for uplink multiuser communication.

Figure 5.6 shows the MAC-PHY block diagram of a downlink system. The scheduler and resource controller can work separately or with a cross-layer protocol.

5.4.2 Quality of Service

For many applications, a certain QoS must be obtained. Some applications are delay sensitive like voice, while some others are rate sensitive or error sensitive like video. The scheduler classifies

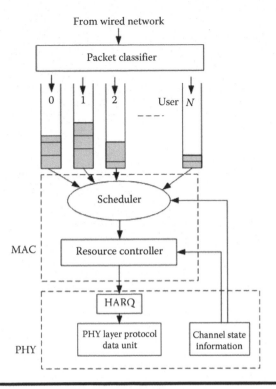

Figure 5.6 MAC-PHY block diagram of a downlink system.

connections into different numbers of QoS classes. As an example, we present below the four main different QoS classes in the IEEE 802.16 standard [2,28]:

■ *Unsolicited grant service (UGS)*: UGS support real-time data streams that generate fixed-rate data such as T1/E1 and VoIP without silence suppression. The UGS is granted periodically to reduce the delay. This class is both delay and rate sensitive and delayed packets are useless and will be dropped. The main QoS metric is the packet error rate.

■ *Real-time polling service (rtPS)*: rtPS is designed to support real-time data streams consisting of variable-sized data packets that are issued at periodic intervals, such as MPEG video. rtPS is rate and delay sensitive but with a higher tolerance than UGS. The main QoS metric is also the packet error rate.

■ *Non-real-time polling service (nrtPS)*: The nrtPS is designed to support delay-tolerant data streams consisting of variable-sized data packets for which a minimum data rate is required, such as the file transfer protocol (FTP). nrtPS is just data rate sensitive.

■ *Best effort (BE)*: BE is designed to support data streams for which no minimum service level is required. BE can be used for the hypertext transport protocol (HTTP).

5.4.3 Scheduler Properties

The wireless channels have different properties for different users, and vary in time and frequency. In addition, the wireless channel is usually a scarce resource that needs to be used efficiently. On the other hand, providing QoS like the data rate and packet delay constraints of real-time data users (e.g., live audio/video streams) over wireless channels (i.e., supporting as many users as possible with the desired QoS) is another requirement of the scheduling. In order to provide efficient operation in the wireless environment, the scheduler should have the following characteristics:

■ The scheduler must exploit efficiently the channel variations in both time and frequency. When the CSI is available, the scheduler should assign a transmission when the channel condition is good enough.
■ Since for delay-sensitive applications the delayed packets are useless, the scheduler must provide delay bound guarantees.
■ The scheduler should guarantee short-term or long-term throughput.
■ The scheduler must maintain fairness between users with the same priorities.

The overall performance is usually evaluated by increasing the number of users in the system and by measuring the associated percentage of satisfied users.

Since these QoS classes are associated with certain predefined sets, the MAC scheduler must support the appropriate data handling mechanisms for data transport according to each QoS. After classification of the users according to their QoS, we apply the scheduling algorithm using the CSI reported by the users on the uplink control channels in FDD or exploiting reciprocity in TDD.

5.5 Scheduling Algorithms

5.5.1 Classical Scheduling Algorithm

In this section, we introduce some of the most popular scheduling algorithms. We will first describe channel unaware schedulers.

5.5.1.1 Generalized Processor Sharing

The generalized processor sharing (GPS) [29,30] is a flexible and fair scheduler proposed for wired networks in an error-free environment. The GPS shares the users that have nonempty queues at time t proportionally to their rate weight. The allocated rate $r_i(t)$ for user i at time t is given by

$$r_i(t) = \frac{W_i c(t)}{\sum_{j \in \mathcal{U}(t)} W_j}, \tag{5.30}$$

where $\mathcal{U}(t)$ is the set of users with nonempty queues at time t, $c(t)$ is the total instantaneous channel capacity at time t, and W_i is the weight rate or the portion of $c(t)$ that should be allocated to user i.

GPS is an ideal scheduler that exactly achieves max–min fairness. However, this scheduler requires a fluid model for the traffic that is served, meaning that the traffic must be infinitely divisible. Furthermore, wireless schedulers must be able to exploit the time varying wireless channels that impose varying rates.

5.5.1.2 Weight Round Robin

The weight round robin (WRR) scheduler is the simplest emulation of the GPS scheduler. While GPS serves an infinitesimal amount of data from each nonempty queue, WRR serves a number of packets. Every user i has an integer weight rate W_i associated with it. The scheduler selects the users according to a precomputed sequence. The WRR tries to serve the user i at a rate $W_i / \sum_j W_j$.

When a packet of the ith user just misses its slot in a frame, it cannot be transmitted before the next ith user slot. If the system is heavily loaded, the packet may have to wait almost N slot times to be served. This is the main drawback of the WRR scheduler.

5.5.1.3 Weight Fair Queueing

The weight fair queueing (WFQ) is the packet-by-packet approximation of the GPS scheduler [30,31]. WFQ approximates GPS to within one packet transmission time regardless of the arrival patterns (so it is more efficient than WRR). Since time and packet are not infinitely divisible like for the GPS, the WFQ scheduler will keep track of a simulated GPS system virtual time. The virtual time $v(t)$ is given by

$$\frac{\mathrm{d}v(t)}{\mathrm{d}t} = \frac{1}{\sum_{j \in \mathcal{U}(t)} W_j} c(t), \tag{5.31}$$

where $\mathcal{U}(t)$ is the set of users with nonempty queues at time t, $c(t)$ is the total channel capacity at time t, and W_i is the weight rate or the portion of $c(t)$ that should be allocated to user i.

The start time S_i^k is calculated as follows:

$$S_i^k = \max(v(a_i^k), F_i^{k-1}), \tag{5.32}$$

where a_i^k is the arrival time of the kth packet and F_i^{k+1} is the finish time that is calculated as

$$F_i^k = S_i^k + \frac{L_i^k}{c_i}, \tag{5.33}$$

where L_i^k is the length of the kth packet for user i and c_i is the user i data rate.

Different improvements of the WFQ scheduler have been proposed in the literature. The worst case weighted fair queuing (WF^{2}Q) selects a packet for transmission at time t if it has the lowest virtual finish time. The WF^{2}Q provides a better emulation of the GPS than the WFQ [31].

5.5.1.4 Early Deadline First

The early deadline first (EDF) scheduler is a delay-based scheduler for real-time services. It allocates the user i with the minimum remaining time:

$$i = \arg \min_j \left(T_j - w_j(t) \right), \tag{5.34}$$

where $w_j(t)$ is the delay of the first packet in the user's queue and T_j is the maximum delay requirement.

The above schedulers are channel unaware schedulers since the CSI is not required to perform the scheduling. These schedulers are widely used in wired networks. In the wireless system, this implies that the MCS must be fixed in order to limit the number of packets retransmitted. On the other hand, channel aware schedulers exploits the available CSI to improve the throughput. We will now focus on the channel aware schedulers.

5.5.1.5 Proportional Fair Scheduler

The proportional fair (PF) scheduler [32,33] is one of the most popular schedulers to balance fairness and performance. It allocates the resource to the user i that can transmit at the highest rate, relative to its average throughput:

$$i = \arg \max_{j} \frac{c_j(t)}{\bar{r}_j(t)}, \tag{5.35}$$

where $c_j(t)$ is the instantaneous capacity of user j at time t and $\bar{r}_j(t)$ is the achieved data throughput for user j in a limited time window of size t_c. $\bar{r}_j(t)$ can be calculated as follows:

$$\bar{r}_j(t) = \left(1 - \frac{1}{t_c}\right)\bar{r}_j(t-1) + \frac{1}{t_c}c'_j(t), \tag{5.36}$$

$c'_j(t)$ is the effectively allocated rate to user j at time t.

The PF algorithm described above is a viable option for scheduling the BE data traffic: it utilizes the channel variation to improve the sum data rate, while giving each user a fair share of the throughput. However, this algorithm is less efficient for real-time users. An extension of the PF scheduler called modified proportional fair (MPF) scheduler has been proposed to support delay sensitive users [34].

The weighted multicarrier proportional fair algorithm (WMPF) is an extension of the PF algorithm for the multicarrier systems [35].

At each time index, the user i chosen to transmit on subcarrier n is given by

$$i = \arg \max_{j} \left(\frac{R_j c_{j,n}(t)}{\bar{r}_j(t)}\right), \tag{5.37}$$

where $c_{j,n}(t)$ is the achievable data rate of user j in subcarrier n at time t and R_j is the average required data rate for user j.

5.5.1.6 Modified Largest Weighted Delay First Algorithm

The QoS of a user can be defined in different ways. If the data user is a real-time user (e.g., it receives live audio or video streams), the delays of the data packet need to be kept below a certain threshold. The modified largest weighted delay first (M-LWDF) [1,36] is an extension of the largest weighted delay first [37]. More formally, the QoS requirement of user j is

$$Pr(w_j(t) > T_j) < \delta_j, \tag{5.38}$$

where $w_j(t)$ is the delay of the first packet in the user's queue, T_j is the maximum delay requirement, and δ_j is the desired probability to fulfill the delay requirement T_j:

The scheduling of M-LWDF assigns the user i as follows:

$$i = \arg \max_{j} p_j w_j(t) c_j(t), \tag{5.39}$$

where $c_j(t)$ is the channel capacity relative to the users j, and p_j is a QoS factor with the value

$$p_j = \frac{a_j}{\bar{c}_j}, \tag{5.40}$$

where $\bar{c}_j$ is the time average of $c_j(t)$.

The choice of the constants a_j is suggested to have the value

$$a_j = -\frac{\log(\delta_j)}{T_j}. \tag{5.41}$$

It has been shown that for delay-sensitive packet scheduling the M-LWDF is asymptotically optimal when the delay requirements T_j values are large and δ_j values are small.

While the M-LWDF scheduler was initially designed for delay constraints it is possible to modify the scheme to guarantee a minimum throughput R_i for each user [1].

For multiuser OFDMA systems, the M-LWDF scheduler can be modified further to exploit the multichannel allocation. In that case, the scheduling of M-LWDF assigns the user i and the subcarrier n as follows:

$$i = \arg\max_j c_{j,n}(t) p_j w_j(t), \tag{5.42}$$

where $c_{j,n}(t)$ is the achievable data rate of user j in subcarrier n at time t.

5.5.1.7 Exponential Rule

Exponential (EXP) scheduling rules have been proposed for Code Division Multiple Access (CDMA) downlink transmission in Ref. [38]. The user i is selected as follows:

$$i = \arg\max_j p_j c_j(t) \exp\left(\frac{a_j w_j(t)}{1 + \sqrt{\bar{w}}}\right), \tag{5.43}$$

where $c_j(t)$ is the channel capacity relative to the user j and $\bar{w} = (1/N) \sum_j a_j w_j(t)$.

It has been shown that the EXP scheduler is throughput optimal since the queue lengths will be bounded as long as the set of minimum rates required are feasible with any other rule. Further, no other scheduling algorithm can lead to a smaller value of the quantity $\max_i a_i w_i(t)$. Thus, this scheduler is optimal with respect to this criterion in the heavy traffic limit. Like previously, the EXP scheduler can be modified for the multichannel allocation.

5.5.2 Schedulers for Heterogeneous Traffic

The schedulers introduced above support only a specific QoS parameter. For example, PF provides a proportional fairness among users by only considering data rate, while M-LWDF considers maximum allowable delays. These scheduling algorithms have been designed to only support nrtPS service classes such as FTP. However, schedulers for next generation wireless networks based on the OFDMA system must be designed to support different applications with different QoSs such as VoIP, non-real-time, and video streaming services [39–43].

A first approach is to extend the classical scheduling rules (PF or M-LWDF) to the case of heterogeneous traffic. In Ref. [42], the authors have proposed a scheduler algorithm for QoS and

BE traffic. The scheduling metric of the RT traffic scheduler is waiting time, whereas the BE traffic scheduler uses the queue length information. The QoS traffic scheduler runs as long as there is any rtPS class packet in the system. A second scheduler, called joint scheduler, is proposed where the scheduler is more flexible since the QoS and BE packets treat together if the QoS packets do not approach the maximum allowable delay. This scheduling algorithm can be seen as a scheduler for multiple traffic classes where each connection is assigned a priority function (PRF), which is updated based on its channel and service status [41,42]. In Refs. [41,43], the authors have proposed a cross-layer scheduling algorithm for UGS, rtPS, nrtPS, and BE connections.

This scheduler is summarized as follows: At each time index, the user i chosen to transmit in subcarrier n is given by

$$i = \arg\max_j \phi_{j,n}(t), \tag{5.44}$$

where $\phi_{j,n}(t)$ is the PRF for connection i and subcarrier n at time t. For the rtPS, nrtPS, and BE, the PRF is defined as

$$\phi_{j,n}(t) = \begin{cases} \beta_{\text{class}} \dfrac{c_{j,n}(t)}{c_{\max}} \dfrac{1}{F_j(t)} & \text{if } F_j(t) \geq 1, c_{j,n}(t) \neq 0, \\ \beta_{\text{class}} & \text{if } F_j(t) < 1, c_{j,n}(t) \neq 0, \\ 0 & \text{if } c_{j,n} = 0, \end{cases} \tag{5.45}$$

where $c_{j,n}(t)$ is the achievable data rate of user j in subcarrier n at time t, $c_{j,n}(t)/c_{\max}$ is the normalized data rate since $c_{\max} = \max_n c_{j,n}(t)$ and $F_j(t) = T_j - w_j(t)$ is the delay satisfaction indicator for rtPS class, $F_j(t) = \bar{r}_j(t)/c_{\min}$ is the rate satisfaction indicator for nrtPS class and $F_j(t) = 1$ for BE class.

The role of β_{class} is to provide different priorities for the different QoS classes. In Refs. [41,43], the priority coefficients of each type of service class were fixed to $\beta_{\text{rtPS}} = 1$, $\beta_{\text{nrtPS}} = 0.8$, and $\beta_{\text{BE}} = 0.6$. Consequently, the rtPS user class can be satisfied prior to those of nrtPS and BE classes since the values of $\phi_{j,n}(t)$ are, respectively, upper bounded by β_{rtPS}, β_{nrtPS}, and β_{BE} for rtPS, nrtPS, and BE connections.

5.5.3 Utility-Based Function

5.5.3.1 Introduction

Efficient resource allocations have been well studied in economics, where utility functions are used to quantify the benefit of usage of certain resources. Shenker [44] proposes to use the utility theory in communication networks to evaluate the degree to which a network satisfies the service requirements of user applications, rather than in terms of throughput, outage probability, or packet loss. The basic idea of utility-pricing structures is to map the resource use (bandwidth, power) or performance criteria (rate or delay) into the corresponding utility or price values and optimize the established utility-pricing system. In wired networks, utility and pricing mechanisms have been used to describe QoS for flow control, congestion control, and routing. In wireless networks, the utility-pricing structure has been investigated for downlink power control in CDMA for voice and data applications in Ref. [45] and references therein.

Utility-function-based scheduling can exploit not only PHY information but also application layer information for cross-layer application [46,47].

5.5.3.2 Rate-Based Utility Function

We first focus on system performance during a short period of time when channel conditions for individual users remain the same, whereas they may vary for different users. The goal of the max utility resource assignment is to find a scheduling scheme that maximizes the total utility of the system, which is given by

$$
\max_{c;\rho} \sum_{k=1}^{K} U_k(r_k) = \sum_{k=1}^{K} U_k\left(\frac{B}{N}\sum_{n=1}^{N} c_{k,n}\rho_{k,n}\right),
\tag{5.46}
$$

where $U_k(r_k)$ is the utility function.

Equation 5.46 replaces Equation 5.4 in the rate maximization problem. Thus, the utility function $U_k(r_k)$ should be a nondecreasing function of the data rate r_k. Clearly, when $U_k(r_k) = r_k$, the utility is just the throughput, which is the objective of the rate maximization problem (maximum spectral efficiency). The utility function $U_k(r_k)$ maps the network resources that a user utilizes into a real number. In almost all wireless applications, a reliable data transmission rate is the most important factor to determine the satisfaction of users.

The utility function can be regarded as an extension of the traditional network optimizations. Utility functions serve as an optimization objective for the adaptive physical and MAC layer techniques. Consequently, they can be used to optimize radio resource allocation for different applications and to build a bridge between the physical, MAC, and upper layers. When a utility function is used to capture the level of satisfaction for assigned resources, it can be estimated from subjective surveys. By adjusting the shape of $U_k(r_k)$, a flexible tradeoff can be achieved between the overall total system data rate and fairness.

Figure 5.7 shows the utility functions for elastic, RA, and hard real-time applications.

When there is no minimum rate constraint (no condition C_4), the above optimization problem described by Equation 5.46 belongs to the family of nonlinear combinatorial optimization problems where there is no general approach to reach the optimal solution. However, when the utility function $U_k(r_k)$ is concave and differentiable and continuous rate adaptation is possible, the optimal subcarrier assignment can be described as follows [48]:

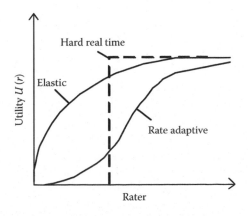

Figure 5.7 Utility function examples.

First, the set of subcarrier is $\mathcal{D}_k^*$ is globally optimal if and only if

$$U_k'(r_k^*)c_{k,n} \geq U_l'(r_l^*)c_{l,n} \qquad \forall n \in \mathcal{D}_k^*, \qquad \forall k, l \in \{1, 2, \ldots, K\}, \tag{5.47}$$

where

$$r_k^* = \frac{B}{N} \sum_{n \in \mathcal{D}_k^*} c_{k,n}. \tag{5.48}$$

Then, the subcarrier n should be allocated to the user k' according to the following closed-form expression:

$$k' = \arg \max_{k \in \{1,2,\ldots,K\}} U_k'(r_k^*)c_{k,n}, \tag{5.49}$$

where

$$U_k'(r_k) = \frac{\mathrm{d}U_k(r_k)}{\mathrm{d}r_k} \tag{5.50}$$

In Ref. [48] a sorting–search suboptimal algorithm with average complexity less than $(K - 1)^2(N + 1)\log_2(N)$ has been proposed to solve this problem.

We can define $U_k(r_k)$ based on scheduling criteria, which can rely not only on MAC and PHY requirements, but also on application layer requirements. A solution to obtain utility curves is through sophisticated subjective surveys, in which end users judge the application performance under a wide range of network conditions. Some studies have been performed to obtain utility functions for wireless systems in which many users share a very limited amount of resources. For example, for BE traffic, the utility function $U_k(r_k) = 0.16 + 0.8\ln(r_k - 0.3)$ has been proposed in Ref. [49], where r_k is given in kbps. It should be noted that to prevent assigning too much resource to the user with good channel conditions, the slope of the utility curves decreases when the data rate increases. A more general utility function is

$$U_k(r_k) = \begin{cases} a + b\ln(r - c) & \text{if } r_k \geq r_{th}, \\ 0 & \text{if } 0 \leq r_k < r_{th}, \end{cases} \tag{5.51}$$

where $U'(r) = b(r - c)^{-1}$. While this utility function $U_k(r_k)$ is not concave, it is strictly concave and differentiable when $r_k \geq r_{th}$. For a large class of utility functions, the nonconcavity does not affect significantly the above solution of the optimization problem.

For the high-priority (hard real-time) user, the utility function can be a step function [44]:

$$U(r_k) = u(r_k - R_k) = \begin{cases} 1 & \text{if } r_k \geq R_k, \\ 0 & \text{if } 0 \leq r_k < R_k. \end{cases} \tag{5.52}$$

While the above scheduler maximizes the aggregate utility given in Equation 5.46 based on instantaneous CSI, in practice the users care more about the average date rate than the instantaneous one. In that case, the optimization problem can be expressed as

$$\max_{c;\rho} \sum_{k=1}^{K} U_k(\bar{r}_k(t)). \tag{5.53}$$

Since the fairness is then relaxed over a given time window, the spectral efficiency will be increased. For this class of optimization problems, the subcarrier n must be allocated to the user k' according to the following closed-form expression:

$$k'(t) = \arg \max_{k \in \{1,2,...,K\}} U_k'(\bar{r}_k(t-1))c_{k,n}(t), \qquad (5.54)$$

where $\bar{r}_k(t)$ is the achieved data throughput for user k in a limited time window of size t_c. We can calculate $r_k(t)$ as follows:

$$\bar{r}_k(t) = \left(1 - \frac{1}{t_c}\right)\bar{r}_k(t-1) + \frac{1}{t_c}c_k'(t), \qquad (5.55)$$

where $c_k'(t)$ is the effectively allocated rate to user k at time t.

In a single carrier context, if we $U_k(\bar{r}_k) = \log \bar{r}_k$ where $\bar{r}_k$ is the achieved data rate of user k over a certain time assure that then the right term of Equation 5.55 can be rewritten as $\arg \max_{k \in \{1,...,K\}}, (c_k/r_k)$ [32]. Consequently, for this class of utility functions, the max utility resource assignment is obtained with the PF scheduler. Furthermore, if $U_k(r_k) = -\beta r_k^2$, where $\beta > 0$, then the optimal scheduling rule is the M-LWDF rule [46].

5.5.3.3 Delay-Based Utility Function

Since the scheduling rule using a rate-based utility function can be unfair for users that have bad channels for a long period of time, it can be important to design a scheduler using time or time average utility function for delay-sensitive traffic. The aim of this class of schedulers is to maximize the aggregate utility, where the utility function can be a decreasing function of the delay or average delay, incurred when serving a user. Like previously, this scheduler will also take into account the CSI [46].

Associated with each user k is an average waiting time $\bar{W}_k(t)$ and a corresponding utility function $U_k(\bar{W}_k(t))$. When the delay is high, the user has a low level of satisfaction. Consequently, we will assume that $U_k(\bar{W}_k(t))$ is decreasing with $\bar{W}_k(t)$.

The goal of max delay utility resource assignment is to find a scheduling scheme that maximizes the aggregate utility of the system, which is given by

$$\max_{c;\rho} \sum_{k=1}^{K} U(\bar{W}_k(t)). \qquad (5.56)$$

Then the scheduler allocates the subcarrier n to the user k' according to the following rule:

$$k'(t) = \arg \max_{k \in \{1,2,...,K\}} U_k'(\bar{W}_k(t))c_{k,n}(t), \qquad (5.57)$$

where $U_k'(\bar{W}_k(t)) = dU_k(\bar{W}_k(t))/d\bar{W}_k(t)$.

The stability region of the queueing system, which is the largest region on the arrival rates for which the queueing system can be stabilized by the scheduling policy, has been investigated using the Foster–Lyapunov method in Ref. [50]. It has been shown that under very loose conditions, the delay-based scheduling rule has the maximum stability region.

5.5.3.4 Utility-Based Scheduling for Heterogeneous Traffic

Scheduling algorithms must often allocate resources to different applications with different QoSs. Like the PF and M-LWDF scheduling rule, it is possible to extend the utility-based scheduling to the case of heterogeneous traffic. As seen previously, different utility functions will be used for different QoSs. For example, in Ref. [51], the authors propose a suboptimal solution for a resource allocation problem considering both real-time and BE user classes.

◼ *Initialization*: The users are sorted according to their rate R_k in ascending order. For each user k, initialize $C_k = \emptyset$ the set of subcarriers allocated to user k, $\mathcal{A} = \{1, 2, \ldots, N\}$, and $\mathcal{U} = \emptyset$ the set of satisfied users.
Set $r_k = 0$ and $t_k = 1 \quad \forall k$.

◼ *Subcarrier assignment*:

    while $\sum_k t_k > 0$ & $\mathcal{A} \neq \emptyset$
      for all users k
        if $\frac{B}{N} \sum_{n \in \mathcal{A} \cup C_k} c_{k,n} \leq R_k$ (we cannot satisfy the user's required rate)
          $t_k = 0$ and $C_k = \emptyset$
        end if
        if $r_k \geq R_k$ & $t_k = 1$
          $\mathcal{U} = \mathcal{U} \cup \{k\}$ and $t_k = 0$
        end if
        if $t_k = 1$
          $n' = \arg \max_{n \in \mathcal{A}} SNR_{k,n}$
          $C_k = C_k \cup \{n'\}$ $\mathcal{A} = \mathcal{A} \cap \{n'\}$
          $r_k = r_k + (B/N) \log_2(1 + SNR_{k,n'})$
        end if
      end for
    end while

This algorithm stops when all the users' rates are higher than the required user rates. The other subcarriers can be allocated to BE users. Then, it is possible to apply the greedy algorithm introduced in Section 5.3.2 to minimize the power allocation. In Ref. [52], the Satisfaction Oriented Resource Allocation (SORA) algorithm has also been proposed to increase the number of satisfied users using the same utility function.

5.6 Open Problems

◼ OFDMA with multiple transmit antennas is a promising technology for the flexible high-rate services in future mobile radio networks since a large number of degrees of freedom (space, frequency, and time) can be provided using CSI on the transmitter side. For SDMA–OFDMA systems, the sum data rate can be maximized when the optimal set of cochannel users is selected for each subcarrier. Then, the resource allocation, problem is more complex since it must include precoding vector selection, subcarrier allocation, and bit loading. This problem is far from been solved [53].

◼ The impact of nonperfect CSI on the performance of resource management techniques is another important issue that should be studied for OFDMA systems with and without multiple transmit antennas.

- The resource management strategies for OFDMA have been developed without considering the other cell interferences. In order to improve the overall capacity, the usage of cooperation between the base stations to manage the radio resources among the cells in the networks is another open research topic.
- While scheduling for the same class of traffic is now theoretically well understood, a unified theory for scheduling of heterogeneous traffic is still missing.

References

1. M. Andrews, K. Kumaran, K. Ramanan, A. Stolyar, and P. Whiting, Providing quality of service over a shared wireless link, *IEEE Commun. Mag.*, 39, 150–154, 2001.
2. K. Taesoo, L. Howon, C. Sik, K. Juyeop, and C. Dong-Ho, Design and implementation of a simulator based on a cross-layer protocol between MAC and PHY layers in a WiBro compatible IEEE 802.16e OFDMA system, *IEEE Commun. Mag.*, 43, 12, 136–146, 2005.
3. C. Y. Wong, R. S. Cheng, K. B. Letaief, and R. D. Murch, Multiuser OFDM with adaptive subcarrier, bit and power allocation, *IEEE J. Select. Areas Commun.*, 17, 1747–1758, 1999.
4. D. Kivanc, G. Li, and H. Liu, Computationally efficient bandwidth allocation and power control for OFDMA, *IEEE Trans. Wirel. Commun.*, 2, 6, 1150–1158, 2003.
5. Y. J. Zhang and K. B. Letaief, Multiuser adaptive subcarrier-and-bit allocation with adaptive cell selection for OFDM systems, *IEEE Trans. Wire. Commun.*, 3, 4, 1566–1575, 2004.
6. Z. Shen, J. G. Andrews, and B. L. Evans, Adaptive resource allocation in multiuser OFDM systems with proportional fairness, *IEEE Trans. Wirel. Commun.*, 4, 6, 2726–2737, 2005.
7. W. Rhee and J. M. Cioffi, Increase in capacity of multiuser OFDM system using dynamic subchannel allocation, In Proc. of IEEE Vehicular Technology Conference, Tokyo, May 2000, pp. 1085–1089.
8. J. Jang and K. B. Lee, Transmit power adaptation for multiuser OFDM Systems, *IEEE J. Sel. Areas Commun.*, 21, 2, 171–178, 2003.
9. J. G. Proakis, *Digital Communications*, McGraw-Hill, 4th Edition, Boston, 2000.
10. A. J. Goldsmith and S. G. Chua, Variable rate variable power MQAM for fading channels, *IEEE Trans. Commun.*, 45, 1218–1230, 1997.
11. R. Jain, D. M. Chiu, and W. R. Hawe, A Quantitative Measure of Fairness and Discrimination for Resource Allocation Shared Computer Systems, Digital Equipment Corporation technical report TR-301, 1984.
12. H. Yin and H. Liu, An efficient multiuser loading algorithm for OFDM-based broadband wireless systems, In Proc. of IEEE Global Telecommunications Conference (Globecom), no. 1, November 2000, pp. 103–107.
13. C. Lengoumbi, P. Godlewski, and P. Martins, An efficient subcarrier assignment algorithm for downlink OFDMA, In Proc. of VTC Fall'06, Montreal, Canada, September 2006, 25–28.
14. L. Zhen, Z. Geqing, W. Weihua, and S. Junde, Improved algorithm of multiuser dynamic subcarrier allocation in OFDM system, In Proc. of International Conference on Communication Technology Proceedings (ICCT), April 2003, pp. 1144–1147.
15. T. M. Cover and J. A. Thomas, Elements of Information Theory, New York: Wiley, 1991.
16. J. Campello, Pactical bit loading for DMT, In Proc. of IEEE International Conference on Communications, vol. 2, June 1999, pp. 6–10.
17. G. Caire and S. Shamai, On the achievable throughput of a multiantenna gaussian broadcast channel, *IEEE Trans. Inf. Theory*, 49, 7, 1691–1706, 2003.
18. E. Visotsky and U. Madhhow, Optimum beamforming using transmit antenna arrays, In Proc. of IEEE Vehicular Technology Conference (VTC), pp. 523–527, 1999.
19. F. Rashid Farrokhi, K. J. R. Liu, and L. Tassiulas, Transmit beamforming and power control for cellular wireless systems, *IEEE J. Sel. Areas Commun.*, 16, 8, 1437–1450, 1998.
20. I. Koutsopoulos and L. Tassiulas, Adaptive resource allocation in SDMA based wireless broadband networks with OFDM signaling, In Proc. of IEEE Infocom, pp. 1376–1385, 2002.

21. D. Bartholomé, A. I. Pérez-Neira, and C. Ibars, Practical bit loading schemes for multi-antenna multi-user wireless OFDM systems, 38th Asilomar Conference on Signals, Systems and Computers, California, 2004.

22. M. Petermann, C. Bockelmann, and K. D. Kammeyer, On allocation strategies for dynamic MIMO-OFDMA with multi-user beamforming, In Proc. of 12th International OFDM-Workshop (InOWo07), Germany, 2007.

23. D. J. Love, R. W. Heath, Jr., V. K. N. Lau, D. Gesbert, B. D. Rao, and M. Andrews, An Overview of Limited Feedback in Wireless Communication Systems, *IEEE J. Sel. Areas Commun.*, 26, 1341–1365, 2008.

24. J. Choi and R. W. Heath, Jr., Interpolation based transmit beamforming for MIMO-OFDM with limited feedback, *IEEE Trans. Sig. Proc.*, 53, 4125–4135, 2005.

25. T. Pande, D. J. Love, and J. V. Krogmeier, Reduced feedback MIMO-OFDM precoding and antenna selection, *IEEE Trans. Signal Process.*, 55, 5-2, 2284–2293, 2007.

26. P. Svedman, S. K. Wilson, L. J. Cimini, and B. Ottersten, Opportunistic beamforming and scheduling for OFDMA systems, *IEEE Trans. Commun.*, 55, 941–952, 2007.

27. B. Ozbek and D. Le Ruyet, Reduced feedback designs for SDMA-OFDMA systems, In Proc. of IEEE International Conference on Communications, Dresden, June 2009.

28. C. Y. Huang, H. Juan, M. Lin, and C. Chang, Radio resource management of heterogeneous services in mobile WiMAX systems, *IEEE Wirel. Commun.*, 14, 1, 20–26, 2007.

29. Parekh, A. and Gallager, R., A generalized processor sharing appproach to flow control in integrated services networks: the single node case, *IEEE/ACM Trans. Network.*, 1, 344–357, 1993.

30. V. Barghavan, S. Lun, and N. Nandagopal, Fair queueing in wireless networks: issues and approaches, *IEEE Pers. Commun.*, 6, 44–53, 1999.

31. J. C. Bennett and H. Zhang, WF2Q: worst-case fair weighted fair queueing, In Proc. of IEEE INFOCOM Conference on Computer Communications, pp. 120–128, 1996.

32. F. Kelly, Charging and rate control for elastic traffic, *Eur. Trans. Telecomm.*, 33–37, 1997.

33. J. M. Holtzman, Asymptotic analysis of proportional fair algorithm, In Proc. of IEEE PIMRC, San Diego, CA, pp. 33–37, 2001.

34. G. Barriac and J. Holtzman, Introducing delay sensitivity into the proportional fair algorithm for CDMA downlink scheduling, In Proc. of IEEE 7th International Symposium on Spread Spectrum Techniques and Applications, 3, 652–656, 2002.

35. H. Kim and Y. Han, A proportional fair scheduling of multicarrier transmission systems, *IEEE Commun. Lett.*, 9, 3, 210–213, 2005.

36. E. H. Choi, W. Choi, and J. Andrews, Throughput of the 1x EV-DO system with various scheduling algorithms, In Proc. of 8th IEEE International Symposium on Spread Spectrum Techniques and Applications, pp. 359–363, September 2004.

37. A. L. Stoylar and K. Ramanan, Largest weighted delay first scheduling: large deviations and optimality, *Ann. Appl. Probab.*, 11, 1, 1–48, 2001.

38. S. Shakkottai and A. L. Stolyar, Scheduling for multiple flows sharing a time-varying channel: the exponential rule, *Anal. Meth. Appl. Probab.*, 207, 185–202, 2002.

39. C. F. Tsai, C. J. Chang, F. C. Ren, and C. M. Yen, Adaptive radio resource allocation for downlink OFDMA/SDMA systems, In Proc. of IEEE International Conference on Communications, Glasgow, pp. 5683–5688, June 2007.

40. S. Ryu, B. Ryu, H. Seo, and M. Shin, Urgency and efficiency based packet scheduling algorithm for OFDMQ wireless system, In Proc. of IEEE VTC'05-Spring, pp. 1456–1462 , May 2005.

41. Q. Liu, X. Wang, and G. B. Giannakis, A cross layer scheduling algorithm with QoS support in wireless networks, *IEEE Trans. Veh. Technol.*, 55, 3, 839–847, 2006.

42. W. H. Park, S. Cho, and Q. Bahk, Scheduler design of multiple traffic classes in OFDMA networks, In Proc. of IEEE International Conference on Communications, pp. 790–795, June 2006.

43. L. Wan, W. Ma, and Z. Guo, A cross-layer Pachet scheduling and subchannel allocation scheme in 802.16e OFDMA system, In Proc. of IEEE WCNC 2007, pp. 1867–1872, Hong Kong, People's Republic of China, March 2007.

44. S. Shenker, Fundamental design issues for the future Internet, *IEEE J. Select. Areas Commun.*, 13, 1176–1188, 1995.
45. L. Song and N. B. Mandayam, Hierarchical sir and rate control on the forward link for CDMA data users under delay and error constraints, *IEEE J. Select. Areas Commun.*, 19, 1871–1882, 2001.
46. P. Liu, R. Berry, and M. L. Honig, Delay-Sensitive Packet Scheduling in Wireless Networks, In Proc. of IEEE WCNC 2003, 3, 1627–1632, March 2003.
47. G. Song and Y. (G.) Li, Cross-layer optimization for OFDM wireless networks. Part I: theoretical framework, *IEEE Trans. Wirel. Commun.*, 4, 2, 614–624, 2005.
48. G. Song and Y. (G.) Li, Cross-layer optimization for OFDM wireless networks. Part II: Algorithm development, *IEEE Trans. Wirel. Commun.*, 4, 2, 625–634, 2005.
49. Z. Jiang, Y. Ge, and Y. Li, Max-utility wireless resource management for best effort traffic, *IEEE Trans. Wirel. Commun.*, 4, 100–111, 2005.
50. G. Song and Y. (G.) Li, Utility-based joint physical-MAC layer optimization in OFDM, In Proc. of IEEE Global Communications Conf., vol. 1, pp. 671–675, November 2002.
51. M. S. Al Bashar and Z. Ding, QoS aware resource allocation for heterogeneous multiuser ofdm wireless networks, In Proc. of IEEE 9th International Workshop on Signal Processing Advances in Wireless Communications (SPAWC), pp. 535–539, Recifes, Brazil, 2008.
52. R. B. Santos, F. R. M. Lima, W. C. Freitas Jr., and F. R. P. Cavalcanti, QoS based radio resource allocation and scheduling with different user data requirements for OFDMA systems, In Proc. of IEEE International Symposium on Personal, Indoor and Mobile Radio Communications, vol. 18, pp. 1–5, 2007.
53. Y. J. Zhang and K. B. Letaif, An efficient resource allocation scheme for spatial multiuser access in MIMO/OFDM systems, *IEEE Trans. Commun.*, 53, 107–116, 2005.

Chapter 6

Scheduling and Resource Allocation in OFDMA Wireless Systems

Jianwei Huang, Vijay Subramanian,
Randall Berry, and Rajeev Agrawal

Contents

6.1 Introduction... 132
6.2 Related Work on OFDMA Resource Allocation.. 134
6.3 OFDMA Scheduling and Resource Allocation 135
  6.3.1 Gradient-Based Wireless Scheduling and Resource Allocation Problem
     Formulation... 135
  6.3.2 General OFDMA Rate Regions.. 136
     6.3.2.1 Self-Noise.. 137
     6.3.2.2 General Power Constraint—Single Cell Downlink, Uplink
        and Multicell Downlink with Frequency Sharing...................... 139
     6.3.2.3 Capacity Region—Max SNR and Min/Max Rate Constraints 139
  6.3.3 Optimal Algorithms... 140
     6.3.3.1 Dual of Problem... 141
     6.3.3.2 Optimizing the Dual Function over μ 145
     6.3.3.3 Optimizing the Dual Function over (α, λ)........................... 145
     6.3.3.4 Optimizing the Dual Function over α.............................. 147
  6.3.4 Primal Optimal Solution.. 148
  6.3.5 OFDMA Feasibility.. 149
  6.3.6 Power Allocation Given Subchannel Allocation 150

        6.3.6.1   Feasibility Check.. 151
6.4   Low Complexity Suboptimal Algorithms with Integer Channel Allocation.............. 151
     6.4.1   CA in SOA1: Progressive Subchannel Allocation Based on Metric Sorting...... 152
     6.4.2   CA in SOA2: Tone Number Assignment and Tone User Matching.............. 154
         6.4.2.1   SubChannel Number Assignment (CNA)............................ 154
         6.4.2.2   SubChannel User Matching (CUM) Step............................ 156
     6.4.3   Power Allocation (PA) Phase.. 158
     6.4.4   Complexity and Performance of Suboptimal Algorithms
         for the Uplink Scenario...158
6.5   Conclusions and Open Problems 160
Acknowledgments.. 161
References.. 161

6.1 Introduction

Scheduling and resource allocation are essential components of wireless data systems. Here by scheduling we refer to the problem of determining which users will be active in a given time-slot; resource allocation refers to the problem of allocating physical-layer (PHY) resources such as bandwidth and power among these active users. In modern wireless data systems, frequent channel quality feedback is available enabling both the scheduled users and the allocation of PHY resources to be dynamically adapted based on the users' channel conditions and quality of service (QoS) requirements. This has led to a great deal of interest both in practice and in the research community on various "channel aware" scheduling and resource allocation algorithms. Many of these algorithms can be viewed as "gradient-based" algorithms, which select the transmission rate vector that maximizes the projection onto the gradient of the system's total utility [1–9]. One example is the "proportionally fair rule" [3,4] first proposed for code division multiple access (CDMA) 1xEVDO based on a logarithmic utility function of each user's throughput. A larger class of throughput-based utilities is considered in Ref. [2] where efficiency and fairness are allowed to be traded off. The "Max Weight" policy (e.g., [5,10,11]) can also be viewed as a gradient-based policy, where the utility is now a function of a user's queue size or delay.

Compared to time division multiple access (TDMA) and CDMA technologies, Orthogonal frequency division multiple access (OFDMA) divides the wireless resource into nonoverlapping frequency–time chunks and offers more flexibility for resource allocation. It has many advantages such as robustness against intersymbol interference and multipath fading as well as lower complexity of receiver equalization. Owing to these, OFDMA has been adopted as the core technology for most recent broadband wireless data systems, such as IEEE 802.16 (WiMAX), IEEE 802.11a/g (Wireless LANs), and LTE for 3GPP.

This chapter discusses gradient-based scheduling and resource allocation in OFDMA systems. This builds on previous work specific to the single cell downlink [8] and uplink [7] settings (e.g., Figure 6.1). The key goal of this chapter is to provide a general framework that includes these as special cases and also applies to multiple cell/sector downlink transmissions (e.g., Figure 6.2). In particular, several important practical constraints are included in this framework, namely, (1) integer constraints on the tone allocation, that is, a tone can be allocated to at most one user; (2) constraints on the maximum signal-to-noise ratio (SNR) (i.e., rate) per tone, which models a limitation on the available modulation and coding schemes; (3) "self-noise" on tones due to channel estimation errors (e.g., Ref. [12]) or phase noise [13]; and (4) user-specific minimum and maximum rate constraints. We not only provide the optimal algorithm for solving the optimization problem corresponding to

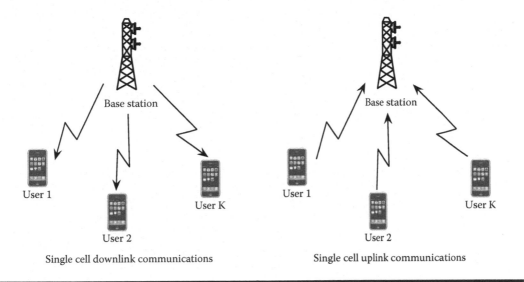

Figure 6.1 Example of single cell downlink (left) and uplink (right) scenerios.

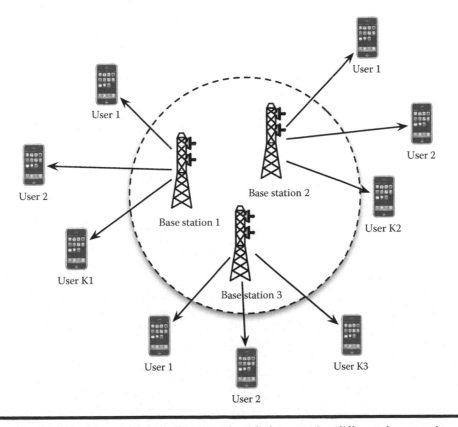

Figure 6.2 Example of a multiple cell/sector downlink scenerio (different base stations could represent different sectors of the same base station shown by the circle).

the generalized model, but also provide low complexity heuristic algorithms that achieve close to optimal performance.

Most previous work on OFDMA systems focused on solving the resource allocation problem without jointly considering the problem of user scheduling. We will briefly survey this work in the next section. Then we describe our general formulation together with the optimal and heuristic algorithms to solve the problem. Finally, we will summarize the chapter and outline some future research directions.

6.2 Related Work on OFDMA Resource Allocation

A number of formulations for single cell downlink OFDMA resource allocation have been studied (e.g., Refs. [14–23]). In Refs. [15,16], the goal is to minimize the total transmit power given target bit rates for each user. In Ref. [16], the target bit rates are determined by a fair queueing algorithm, which does not take into account the users' channel conditions. A number of papers including [17–20,22,23] have studied various sum-rate maximization problems, given a total power constraint. In Refs. [18–20] there is also a minimum bit rate per user that must be met. Jiao et al. [23] consider both minimum and maximum rate targets for each user and also take into account several constraints that arise in Mobile WiMax. In Ref. [22], certain "delay sensitive" users are modeled as having fixed target bit rate (i.e., their maximum and minimum rates are the same), while other "best effort" users have no bit rate constraints. Thus the scheduler attempts to maximize the sum rate of the best effort users while meeting the rate targets of the delay-sensitive ones. In Refs. [14,21], weighted sum-rate maximization is considered. This is a special case of the resource allocation problem we study here for a given time slot but does not account for constraints on the SNR per carrier, rate constraints, or self-noise. In Ref. [14], a suboptimal algorithm with constant power per tone was shown in simulations to have little performance loss. Other heuristics that use a constant power per tone are given in Refs. [17–19]; we will briefly discuss a related approach in Section 6.4. In Ref. [21], a dual-based algorithm similar to ours is considered, and simulations are given, which show that the duality gap of this problem quickly goes to zero as the number of tones increases. In Ref. [23], the information theoretic capacity region of a single cell downlink broadcast channel with frequency-selective fading using a time-division multiplexing (TDM) scheme is given; the feasible rate region we consider, without any maximum SNR and rate constraints, can be viewed as a special case of this region. None of these papers consider self-noise, rate constraints, or per user SNR constraints. Moreover, most of these papers optimize a static objective function, while we are interested in a dynamic setting where the objective changes over time according to a gradient-based algorithm. It is not *a priori* clear if a good heuristic for a static problem applied to each time step will be a good heuristic for the dynamic case, since the optimality result in Refs. [1–3,5,9–11] is predicated on solving the weighted-rate optimization problem exactly in each time slot. Simulation results in Ref. [8] show that this does hold for the heuristics presented in Section 6.4.

Resource allocation for a single cell OFDMA uplink has been presented in Refs. [24–31]. In Ref. [24], a resource allocation problem was formulated in the framework of Nash Bargaining, and an iterative algorithm was proposed with relatively high complexity. Pfletschinger et al. [25] proposed a heuristic algorithm that tries to minimize each user's transmission power while satisfying the individual rate constraints. In Ref. [26], the author considered the sum-rate maximization problem, which is a special case of the problem considered here with equal weights. The algorithm derived in Ref. [26] assumes Rayleigh fading on each subchannel; we do not make such an assumption here. In Ref. [27], an uplink problem with multiple antennas at the base station was considered; this enables spatial multiplexing of subchannels among multiple users. Here, we focus on single antenna systems where at most one user can be assigned per subchannel. The work in Refs. [28–31] is closer to our model. Kim et al. [28] also considered a weighted rate maximization problem in the uplink case, but assumed static weights. They proposed two algorithms, which are similar to one of the algorithms described in this chapter. We propose several other algorithms that outperform those in Ref. [28] with similar or slightly higher complexity. Ng and Sung [29] generalized the results in Ref. [28] by considering utility maximization in one time slot, where the utility is a function of the instantaneous rate in each time slot. Another work that focused on per time slot fairness is by Gao and Cui [31]. Finally, Kwon et al. [30] proposed a heuristic algorithm based on Lagrangian relaxation, which has high complexity due to a subgradient search of the dual variables.

Resource allocation and interference management of multicell downlink OFDMA systems were presented in Refs. [32–39]. A key focus of these works is on interference management among multiple cells. Our general formulation includes the case where resource coordination leads to no interference among different cells/sectors/sites. In our model, this is achieved by dynamically partitioning the subchannels across the different cells/sectors/sites. In addition to being easier to implement, the interference-free operation assumed in our model allows us to optimize over a large class of achievable rate regions for this problem. If the interference strength is of the order of the signal strength, as would be typical in the broadband wireless setting, then this partitioning approach could also be the better option in an information theoretic sense [40].*

6.3 OFDMA Scheduling and Resource Allocation

6.3.1 Gradient-Based Wireless Scheduling and Resource Allocation Problem Formulation

Let us consider a network with a total of K users. In each time slot t, the scheduling and resource allocation decision can be viewed as selecting a rate vector $r_t = (r_{1,t}, \ldots, r_{K,t})$ from the current feasible rate region $\mathcal{R}(e_t) \subseteq \mathbb{R}_+^K$. If a user is not scheduled his rate is simply zero. Here e_t indicates the time-varying channel state information of all users available at the scheduler at time t. The decision on the rate vector is made according to the gradient-based scheduling framework in Refs. [1–3,9] that is basically a stochastic version of the conditional gradient/Frank–Wolfe algorithm [41]. Namely, an $r_t \in \mathcal{R}(e_t)$ is selected that has the maximum projection onto the gradient of the system's total utility function

$$U(W_t) := \sum_{i=1}^{K} U_i(W_{i,t}),\tag{6.1}$$

where $U_i(\cdot)$ is an increasing concave utility function that measures user i's satisfaction for different values of throughput, and $W_{i,t}$ is user i's average throughput up to time t. In other words, the

---

* We note that our discussions do not directly apply to the case of frequency reuse, where different nonadjacent cells may use the same frequency bands. In practice, frequency reuse is typically considered together with fixed frequency allocations, while here we consider dynamic frequency allocations across different cells.

scheduling and resource allocation decision is the solution to

$$
\max_{r_t \in \mathcal{R}(e_t)} \nabla U(W_t)^{\mathrm{T}} \cdot r_t = \max_{r_t \in \mathcal{R}(e_t)} \sum_{i=1}^{K} U_i'(W_{i,t}) r_{i,t}, \tag{6.2}
$$

where $U_i'(\cdot)$ is the derivative of $U_i(\cdot)$. As a concrete example, it is useful to consider the class of commonly used iso-elastic utility functions given in Refs. [2,42],

$$
U_i(W_{i,t}) =
\begin{cases}
\dfrac{c_i}{\alpha}(W_{i,t})^{\alpha}, & \alpha \leq 1,\ \alpha \neq 0, \\[2mm]
c_i \log(W_{i,t}), & \alpha = 0,
\end{cases} \tag{6.3}
$$

where $\alpha \leq 1$ is a fairness parameter and c_i is a QoS weight. In this case, after taking derivatives, (6.2) becomes

$$
\max_{r_t \in \mathcal{R}(e_t)} \sum_{i} c_i(W_{i,t})^{\alpha-1} r_{i,t}. \tag{6.4}
$$

With equal class weights ($c_i = c$ for all i), setting $\alpha = 1$ results in a scheduling rule that maximizes the total throughput during each slot. For $\alpha = 0$, this results in the proportionally fair rule, and as α increases without bound, we get closer to a max–min fair solution. Thus, this family of utility functions yields a flexible class of policies: the α parameter allows for the choice of an appropriate fairness objective, while the c_i parameter allows one to distinguish relative priorities within each fairness class.

However, more generally, we consider the problem of

$$
\max_{r_t \in \mathcal{R}(e_t)} \sum_{i} w_{i,t} r_{i,t}, \tag{6.5}
$$

where $w_{i,t} \geq 0$ is a time-varying weight assigned to the ith user at time t. In the case of (6.4), we let $w_{i,t} = c_i(W_{i,t})^{\alpha-1}$. In Equation 6.4 these weights are given by the gradients of throughput-based utilities; however, other methods for generating the weights (possibly depending upon queue-lengths and/or delays [5,10,11]) are also possible. We note that Equation 6.5 must be resolved at each scheduling instance because of changes in both the channel state and the weights (e.g., the gradients of the utilities). While the former changes are due to the time-varying nature of wireless channels, the latter changes are due to new arrivals and past service decisions.

6.3.2 General OFDMA Rate Regions

The solution to Equation 6.5 depends on the channel state-dependent rate region $\mathcal{R}(e)$, where we suppress the dependence on time for simplicity. We consider a model appropriate for general OFDMA systems including single cell downlink and uplink as well as multiple cell/sector/site downlink with frequency sharing; related single cell downlink and uplink models have been considered in Refs. [7,8,14,43]. In this model, $\mathcal{R}(e)$ is parameterized by the allocation of tones to users and the allocation of power across tones. In a traditional OFDMA system at most one user may be assigned to any tone. Initially, as in Refs. [15,16], we make the simplifying assumption that multiple users can share one tone using some orthogonalization technique (e.g., TDM).* In practice, if a scheduling interval contains multiple OFDMA symbols, we can implement such sharing by giving a fraction of the symbols to each user; of course, each user will be constrained to use an integer number of symbols. Also, with a large number of tones, adjacent tones will have nearly identical gains, in which case this time sharing can also be approximated by frequency sharing. The two approximations become tight as the number of symbols or tones increases, respectively. We discuss the case where only one user can use a tone in Section 6.4.

Let $\mathcal{N} = \{1, \ldots, N\}$ denote the set of tones[†] and $\mathcal{K} = \{1, 2, \ldots, K\}$ denote the set of users. For each $j \in \mathcal{N}$ and user $i \in \mathcal{K}$, let e_{ij} be the received signal-to-noise ratio (SNR) per unit transmit power. We denote the transmit power allocated to user i on tone j by p_{ij}, and the fraction of that tone allocated to user i by x_{ij}. As tones are shared resources, the total allocation for each tone j must satisfy $\sum_i x_{ij} \leq 1$. For a given allocation, with perfect channel estimation, user i's feasible rate on tone j is

$$r_{ij} = x_{ij} B \log \left(1 + \frac{p_{ij} e_{ij}}{x_{ij}} \right),$$

which corresponds to the Shannon capacity of a Gaussian noise channel with bandwidth $x_{ij} B$ and received SNR $p_{ij} e_{ij}/x_{ij}$.[‡] This SNR arises from viewing p_{ij} as the energy per time slot that user i uses on tone j; the corresponding transmission power becomes p_{ij}/x_{ij} when only a fraction x_{ij} of the tone bandwidth is allocated. Similarly, this can also be explained by time sharing as follows: a channel of bandwidth B is used only a fraction x_{ij} of the time with average power p_{ij}, which causes the power during channel usage to be p_{ij}/x_{ij}. Without loss of generality, we set $B = 1$ in the following.

6.3.2.1 Self-Noise

In a realistic OFDMA system, imperfect carrier synchronization and channel estimation may result in "self-noise" (e.g., [12,13]). We follow a similar approach as in Ref. [12] to model self-noise. Let the received signal on the jth tone of user i be given by $y_{ij} = h_{ij} s_{ij} + n_{ij}$, where h_{ij}, s_{ij} and n_{ij} are the (complex) channel gain, transmitted signal, and additive noise, respectively, with $n_{ij} \sim \mathcal{CN}(0, \sigma^2)$.[§] Assume that $h_{ij} = \tilde{h}_{ij} + h_{ij,\delta}$, where $\tilde{h}_{ij}$ is receiver i's estimate of h_{ij} and $h_{ij,\delta} \sim \mathcal{CN}(0, \delta_{ij}^2)$. After matched-filtering, the received signal will be $z_{ij} = \tilde{h}_{ij}^* y_{ij}$ resulting in an effective SNR of

$$\text{Eff-SNR} = \frac{\|\tilde{h}_{ij}\|^4 p_{ij}}{\sigma_{ij}^2 \|\tilde{h}_{ij}\|^2 + \delta_{ij}^2 p_{ij} \|\tilde{h}_{ij}\|^2} = \frac{p_{ij} e_{ij}}{1 + \beta_{ij} p_{ij} e_{ij}}, \tag{6.6}$$

---

* We focus on systems that do not use superposition coding and successive interference cancellation within a tone, as such techniques are generally considered too complex for practical systems.
† In practice, tones may be grouped into subchannels and allocated at the granularity of subchannels. As discussed in Ref. [8], our model can be applied to such settings as well by appropriately redefining the subchannel gains $\{e_{ij}\}$ and interpreting $\mathcal{N}$ as the set of subchannels.
‡ To better model the achievable rates in a practical system we can renormalize e_{ij} by γe_{ij}, where $\gamma \in [0, 1]$ represents the system's "gap" from capacity.
§ We use the notation $x \sim \mathcal{CN}(0, b)$ to denote that x is a 0 mean, complex, circularly-symmetric Gaussian random variable with variance $b := \mathrm{E}(\|x\|^2)$.

where $p_{ij} = \mathrm{E}(\|s_{ij}\|^2)$, $\beta_{ij} = \delta_{ij}^2/\|\tilde{h}_{ij}\|^2$, and $e_{ij} = \|\tilde{h}_{ij}\|^2/\sigma_{ij}^2.$* Here, $\beta_{ij} p_{ij} e_{ij}$ is the self-noise term. As in the case without self-noise ($\beta_{ij} = 0$), the effective SNR is still increasing in p_{ij}. However, it now has a maximum of $1/\beta_{ij}$.

In general, β_{ij} may depend on the channel quality e_{ij}. For example, this happens when self-noise arises primarily from estimation errors. The exact dependence will depend on the details of channel estimation. As an example, using the model in Ref. [44, Section IV] it can be shown that when the pilot power is either constant or inversely proportional to channel quality subject to maximum and minimum power constraints (modeling power control), β is inversely proportional to the channel condition for large e. On the other hand, $\beta_{ij} = \beta$ is a constant when self-noise is due to phase noise as in Ref. [13]. For simplicity of presentation, we assume constant $\beta_{ij} = \beta$ in the remainder of the paper (except in Figure 6.4 where we allow $\beta(e) \propto 1/e$ to illustrate the impact of self-noise on the optimal power allocation). The analysis is almost identical if users have different β_{ij}'s.

We assume that e_{ij} is known by the scheduler for all i and j as is β. For example, in a frequency division duplex (FDD) downlink system, this knowledge can be acquired by having the base station transmit pilot signals, from which the users can estimate their channel gains and feedback to the base station. In a time division duplex (TDD) system, these gains can also be acquired by having the users transmit uplink pilots; for the downlink case, the base station can then exploit reciprocity to measure the channel gains. In both cases, this feedback information would need to be provided within the channel's coherence time.

With self-noise, user i's feasible rate on tone j becomes

$$r_{ij} = x_{ij} \log\left(1 + \frac{p_{ij} e_{ij}}{x_{ij} + \beta p_{ij} e_{ij}}\right) =: x_{ij} f\left(\frac{p_{ij} e_{ij}}{x_{ij}}\right), \qquad (6.7)$$

where again x_{ij} models the time sharing of a tone and the function $f(\cdot)$ is given by

$$f(s) = \log\left(1 + \frac{1}{\beta + 1/s}\right), \qquad \beta \geq 0. \qquad (6.8)$$

More generally, we assume that a user i's rate on channel j is given by

$$r_{ij} = x_{ij} f\left(\frac{p_{ij} e_{ij}}{x_{ij}}\right), \qquad (6.9)$$

for some function $f : \mathbb{R}_+ \to \mathbb{R}_+$ that is nondecreasing, twice continuously differentiable and concave with $f(0) = 0$, (without loss of generality)† $f'(0) := (\mathrm{d}f/\mathrm{d}s)(0) = \lim_{s \downarrow 0}(f(s)/s) =$

---

* This is slightly different from the Eff-SNR in Ref. [12] in which the signal power is instead given by $\|h_{ij}\|^4 p_{ij}$; the following analysis works for such a model as well by a simple change of variables. For the problem at hand, Equation 6.6 seems more reasonable in that the resource allocation will depend only on $\tilde{h}_{ij}$ and not on h_{ij}. We also note that Equation 6.6 is shown in Ref. [44] to give an achievable lower bound on the capacity of this channel.

† Using the idea that Shannon capacity $\log(1+s)$ is a natural upper bound for $f(s)$, it follows that $0 < (\mathrm{d}f/\mathrm{d}s)(0) \leq 1$. Therefore, if $f'(0) \neq 1$, then we can solve the problem using a scaled version of function, that is, $\tilde{f}(s) = f(s)/(\mathrm{d}f/\mathrm{d}s)(0)$, after scaling the rate constraints by the same amount; the power and subchannel allocations will be the same in the two cases. The Shannon capacity upper bound also yields that $0 \leq \lim_{t \to +\infty}(\mathrm{d}f/\mathrm{d}s)(t) \leq \lim_{s \to +\infty}(f(s)/s) \leq \lim_{s \to +\infty}(\log(1+s)/s) = 0$, as concavity of $f(\cdot)$ and $f(0) = 0$ imply that $(\mathrm{d}f/\mathrm{d}s)(t) \leq (f(t)/t)$ for all $t > 0$.

$\sup_{s>0}(f(s)/s) = 1$, and $\lim_{t \to +\infty}(df/ds)(t) = 0$. We also assume by continuity* that $xf(p/x)$ is 0 at $x = 0$ for every $p \geq 0$. From the assumptions on the function $f(\cdot)$ it follows that $xf(p/x)$ is jointly concave in x, p; this can be easily proved by showing that the Hessian is negative semidefinite [41,45]. It is easy to verify that f given by Equation 6.8 satisfies the above properties. We should, however, point out that using the theory of subgradients [41,45], our mathematical results easily extend to a general $f(\cdot)$ that is only nondecreasing and concave. For instance, it can be easily proved from first principles that $xf(p/x)$ is jointly concave in (x, p) if $f(\cdot)$ is merely concave. We consciously choose the simpler setting of twice continuously differentiable functions to keep the level of discussion simple, but to aid a more interested reader, we will strive to point out the loosest conditions needed for each of our results. Before proceeding we should point out that, operationally, $f(\cdot)$ is a function of the received signal-to-noise ratio, and thus, abstracts the usage of all possible single-user decoders, including the optimal decoder that yields Shannon capacity.

6.3.2.2 General Power Constraint—Single Cell Downlink, Uplink and Multicell Downlink with Frequency Sharing

Let $\{\mathcal{K}_m\}_{m=1}^M$ be nonempty subsets of the set of users $\mathcal{K}$ that form a covering, that is, $\cup_{m=1}^M \mathcal{K}_m = \mathcal{K}$. We assume that there is a vector of non-negative power budgets $\{P_m\}_{m=1}^M$ associated with these subsets, so that $\sum_{i \in \mathcal{K}_m} \sum_j p_{ij} \leq P_m$ for each m. This condition ensures that there is no user who is unconstrained in its power usage. This provides a common formulation of the single cell downlink and uplink scheduling problems as described in Refs. [8] and [7], respectively. For the single cell downlink problem $M = 1$ and $\mathcal{K}_1 = \mathcal{K}$, and for the single cell uplink problem $M = K$ and $\mathcal{K}_i = \{i\}$ for $i \in \mathcal{K}$. More generally, if $\{\mathcal{K}_m\}_{m=1}^M$ is a partition, that is, mutually disjoint, then we can view the "transmitters" for users $i \in \mathcal{K}_m$ as colocated with a single power amplifier. For example, such a model may arise in the downlink case where $\mathcal{M} := \{1, 2, \ldots, M\}$ represents sectors or sites across which we need to allocate common frequency/channel resources, but which have independent power budgets. A key assumption, however, is that we can make the transmissions from the different sectors/sites noninterfering by time sharing or by some other suitable orthogonalization technique.

6.3.2.3 Capacity Region—Max SNR and Min/Max Rate Constraints

Under these assumptions, the rate region can be written as

$$\mathcal{R}(e) = \left\{ r : r_i = \sum_j x_{ij} f\left(\frac{p_{ij} e_{ij}}{x_{ij}}\right) \text{ and } R_i^{\min} \leq r_i \leq R_i^{\max}, \forall i, \right.$$
$$\left. \sum_{i \in \mathcal{K}_m} \sum_j p_{ij} \leq P_m, \forall m, \sum_i x_{ij} \leq 1, \forall j, (x, p) \in \mathcal{X} \right\}, \tag{6.10}$$

where

$$\mathcal{X} := \left\{ (x, p) \geq 0 : x_{ij} \leq 1, p_{ij} \leq \frac{x_{ij} s_{ij}}{e_{ij}} \forall i, j \right\}. \tag{6.11}$$

---

* Using the Shannon capacity function, $\log(1 + s)$, upper bound, we have for $p > 0$, that $\lim_{x \downarrow 0} xf(p/x) = p \lim_{t \uparrow +\infty}(f(t)/t) \leq p \lim_{t \uparrow +\infty}(\log(1 + t)/t) = 0$. For $p = 0$, we directly get the property from $f(0) = 0$.

Here and in the following, a boldfaced symbol will indicate the vector of the corresponding scalar quantities, for example, $x := (x_{ij})$ and $p := (p_{ij})$. Also, any inequality such as $x \geq 0$ should be interpreted componentwise. The linear constraint on (x_{ij}, p_{ij}) in Equation (6.11) using s_{ij} models a constraint on the maximum rate per subchannel due to a limitation on the available modulation and coding schemes; if user i can send at a maximum rate of $\tilde{r}_{ij}$ on tone j, then $s_{ij} = f^{-1}(\tilde{r}_{ij})$. We have also assumed that each user $i \in \mathcal{K}$ has maximum and minimum rate constraints $R_i^{\max}$ and $R_i^{\min}$, respectively. In order to have a solution we assume that the vector of minimum rates $\{R_i^{\min}\}_{i \in \mathcal{K}}$ is feasible. For the vector of maximum rates, it is more convenient to assume that $\{R_i^{\max}\}_{i \in \mathcal{K}}$ is infeasible. Otherwise the optimization problem associated with feasibility (see Section 6.3.5) will yield an optimal solution. Typically we will set $R_i^{\min} = 0$ and $R_i^{\max}$ to be the (time-varying) buffer occupancy. However, with tight minimum throughput demands one can imagine using a nonzero $R_i^{\min}$ to guarantee this.

6.3.3 Optimal Algorithms

From Equations 6.5 and 6.10, the optimal scheduling and resource allocation problem can be stated as

$$\max_{(x,p) \in \mathcal{X}} V(x,p) := \sum_i w_i \sum_j x_{ij} f\left(\frac{p_{ij} e_{ij}}{x_{ij}}\right), \tag{P2}$$

$$\text{subject to: } \sum_j x_{ij} f\left(\frac{p_{ij} e_{ij}}{x_{ij}}\right) \geq R_i^{\min} \quad \forall i \in \mathcal{K}, \tag{η_i}$$

$$\sum_j x_{ij} f\left(\frac{p_{ij} e_{ij}}{x_{ij}}\right) \leq R_i^{\max} \quad \forall i \in \mathcal{K}, \tag{γ_i}$$

$$\sum_i x_{ij} \leq 1 \quad \forall j \in \mathcal{N}, \tag{μ_j}$$

$$\sum_{i \in \mathcal{K}_m} \sum_j p_{ij} \leq P_m \quad \forall m = 1, 2, \ldots, M, \tag{λ_m}$$

where set $\mathcal{X}$ is given in Equation 6.11. As a rule, variables at the right of constraints will indicate the dual variables that we will use to relax those constraints while constructing the dual problem later.

One important point to note is that as described above, the optimization problem (P2) is not convex and so we cannot appeal to standard results, such as Slater's conditions, to guarantee that it has zero duality gap [41,45]. In particular, note that the maximum rate constraints have a concave function on the left side. To show that we still have no duality gap, we will consider a related convex problem in higher dimensions that has the same primal and dual variables. The new optimization problem (P1) is given by

$$\max \sum_i w_i r_i, \tag{P1}$$

$$\text{subject to: } r_i \leq \sum_j x_{ij} f\left(\frac{p_{ij} e_{ij}}{x_{ij}}\right), \quad \forall i \in \mathcal{K}, \tag{α_i}$$

$$\sum_i x_{ij} \leq 1, \quad \forall j \in \mathcal{N}, \tag{μ_j}$$

$$\sum_{i \in \mathcal{K}_m} \sum_j p_{ij} \le P_m, \qquad \forall m = 1, 2, \ldots, M, \qquad\qquad (\lambda_m)$$

$$R_i^{\min} \le r_i \le R_i^{\max}, \qquad \forall i \in \mathcal{K},$$

$$(x, p) \in \mathcal{X}.$$

This problem is easily seen to be convex due to the joint concavity of $xf(p/x)$ as a function of (x, p) and also will satisfy Slater's condition.* Hence, it will have zero duality gap [41,45]. The problem (P1) can be practically motivated as follows: the (PHY) gives the scheduler (at the MAC layer) a maximum rate that it can serve per user based upon power and subchannel allocations, and the scheduler then drains from the queue an amount that obeys the minimum and maximum rate constraints (imposed by the network layer) and the maximum rate constraint from the PHY layer output. If the scheduler chooses not to use the complete allocation given by the PHY layer, then the final packet sent by the MAC layer is assumed to be constructed using an appropriate number of padded bits. However, we will now show that at the optimal, there is no loss of optimality in assuming that the scheduler never sends less than what the PHY layer allocates, that is, the first constraint in problem (P1) is always made tight at an optimal solution. This point of view is exemplified in the schematic shown in Figure 6.3.

Assume that there is an optimizer of (P1) at which for some user i, $r_i < \sum_j x_{ij} f(p_{ij} e_{ij}/x_{ij})$. We will now construct another feasible solution that will satisfy the above relationship with equality. Let $\gamma \in [0, 1]$ and set $\tilde{p}_{ij} := \gamma p_{ij}$. Note that by convexity, both the power and subchannel constraints are satisfied for every value of γ. Now $\sum_j x_{ij} f(\gamma(p_{ij} e_{ij}/x_{ij}))$ is a nondecreasing and continuous function of γ taking values 0 at $\gamma = 0$ and $\sum_j x_{ij} f(p_{ij} e_{ij}/x_{ij})$ at $\gamma = 1$. Therefore, there exists a $\gamma^* \in (0, 1)$ such that $r_i = \sum_j x_{ij} f(\gamma^* p_{ij} e_{ij}/x_{ij})$ as desired. This procedure can be followed for every user i for whom $r_i < \sum_j x_{ij} f(p_{ij} e_{ij}/x_{ij})$, so that at the end we satisfy $r_i = \sum_j x_{ij} f(\tilde{p}_{ij} e_{ij}/x_{ij})$ for a feasible $(x, \tilde{p})$. Therefore both the optimal value and an optimizer of problem (P1) coincides with those for problem (P2). The loosest condition needed for the above to hold is $f(\cdot)$ being nondecreasing and concave with $f'(0) = 0$. Henceforth, we will only work with problem (P1).

Before proceeding to solve the problem by dual methods, we first define some key notation. For two numbers, $x, y \in \mathbb{R}$ we set $x \wedge y := \min(x, y)$, $x \vee y := \max(x, y)$, and $(x)_+ = [x]_+ := x \vee 0$.

6.3.3.1 Dual of Problem

We now proceed to derive a closed-form expression for the dual function for problem (P1). The Lagrangian obtained by relaxing the marked constraints of (P1) using the corresponding dual variables is given by

$$L(r, x, p, \alpha, \mu, \lambda) = \sum_i (w_i - \alpha_i) r_i + \sum_j \mu_j + \sum_{m=1}^M \lambda_m P_m + \sum_{i,j} \alpha_i x_{ij} f\left(\frac{p_{ij} e_{ij}}{x_{ij}}\right)$$
$$- \sum_j \mu_j \sum_i x_{ij} - \sum_m \lambda_m \sum_{i \in \mathcal{K}_m} \sum_j p_{ij}. \qquad\qquad (6.12)$$

---

* More precisely, Slater's condition will be satisfied provided that the minimum rate $(R_i^{\min})$ are strictly in the interior of rate region $\mathcal{R}(e)$. If $R_i^{\min} = 0$ for all i, this will trivially be true.

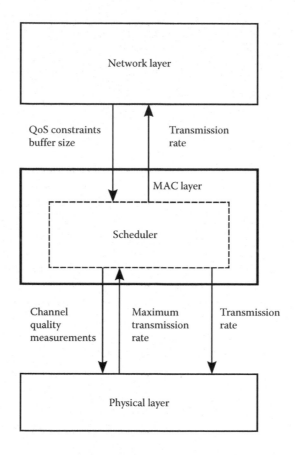

Figure 6.3 Schematic of a scheduler that has cross-layer visibility.

The corresponding dual function is then given by maximizing this Lagrangian over r, x, and p. First, optimizing over rate $r_i \in [R_i^{\min}, R_i^{\max}]$ and noting that the Lagrangian is linear in r_i we get

$$L(x, p, \alpha, \mu, \lambda) = \sum_i (w_i - \alpha_i)_+ R_i^{\max} - \sum_i (\alpha_i - w_i)_+ R_i^{\min} + \sum_j \mu_j + \sum_{m=1}^{M} \lambda_m P_m$$

$$+ \sum_{i,j} \alpha_i x_{ij} f\left(\frac{p_{ij} e_{ij}}{x_{ij}}\right) - \sum_j \mu_j \sum_i x_{ij} - \sum_m \lambda_m \sum_{i \in \mathcal{K}_m} \sum_j p_{ij}.$$

The optimizing r^* is given by the following:

$$\forall i \in \mathcal{K}, \; r_i^* \in \begin{cases} \{R_i^{\max}\} & \text{if } \alpha_i < w_i; \\ \{R_i^{\min}\} & \text{if } \alpha_i > w_i; \text{ and} \\ [R_i^{\min}, R_i^{\max}] & \text{if } \alpha_i = w_i. \end{cases} \tag{6.13}$$

Note that the last term of Equation 6.12 can be rewritten as

$$\sum_m \lambda_m \sum_{i \in K_m} \sum_j p_{ij} = \sum_{i,j} p_{ij} \sum_{m:i \in K_m} \lambda_m = \sum_{i,j} p_{ij} \hat{\lambda}_i, \tag{6.14}$$

where $\hat{\lambda}_i := \sum_{m:i \in K_m} \lambda_m$.

Now maximizing the Lagrangian over power p requires us to maximize

$$\alpha_i x_{ij} \left[f \left(\frac{p_{ij} e_{ij}}{x_{ij}} \right) - \frac{\hat{\lambda}_i}{\alpha_i e_{ij}} \frac{p_{ij} e_{ij}}{x_{ij}} \right] \tag{6.15}$$

over p_{ij} for each i, j. From the assumptions on the function f, it is easy to check that the maximizing p_{ij}^* will be of the form

$$\frac{p_{ij}^* e_{ij}}{x_{ij}} = g \left(\frac{\hat{\lambda}_i}{\alpha_i e_{ij}} \right) \wedge s_{ij}, \tag{6.16}$$

for some function $g : \mathbb{R}_+ \rightarrow [0, \infty]$ with $g(x) = 0$ for $x \geq f'(0)$. Specifically if df/ds is monotonically decreasing, we may show that $g(\cdot) = (df/ds)^{-1}(\cdot)$, that is, the inverse of the derivative of $f(\cdot)$. Otherwise, since df/ds is still a nonincreasing function, we can set $g(x) = \inf\{t : df/ds(t) = x\}$. Using the nonincreasing property of df/ds we can see that $g(x) \wedge y = g(x \vee (df/ds)(y))$. Note that we have assumed $df/ds(0) = 1$ and $\lim_{t \to +\infty} df/ds(t) = 0$, but we do not assume that $\lim_{s \to +\infty} f(s) = +\infty$ (e.g., see the self-noise example). In case $f(\cdot)$ is not differentiable, we would define function $g(\cdot)$ using the subgradients of $f(\cdot)$. In all cases, the key conclusion from Equation 6.16 is that the optimal value of p_{ij}^* is always a linear function of x_{ij}.

Note that when $f = \log(1 + 1/(\beta + 1/s))$, with $\beta \geq 0$, as given by Equation 6.8, then

$$g(x) = q((1/x - 1)_+),$$

where

$$q(z) = \begin{cases} z, & \text{if } \beta = 0, \\ \left(\dfrac{2\beta + 1}{2\beta(\beta + 1)} \right) \left(\sqrt{1 + \dfrac{4\beta(\beta + 1)}{(2\beta + 1)^2} z} - 1 \right) & \text{if } \beta > 0. \end{cases}$$

Figure 6.4 shows p_{ij}^* in Equation 6.16 as a function of e_{ij} for the specific choice of f from Equation 6.8 with three different values of $\beta = 0, 0.01, 0.1$. When $\beta = 0$, Equation 6.16 becomes a "water-filling" type of solution in which p_{ij}^* is nondecreasing in e_{ij}. For a fixed $\beta > 0$, this is not necessarily true, that is, due to self-noise, less power may be allocated to "better" subchannels. We also consider the case where $\beta = 10/e$ to model the case where self-noise is due to channel estimation error.

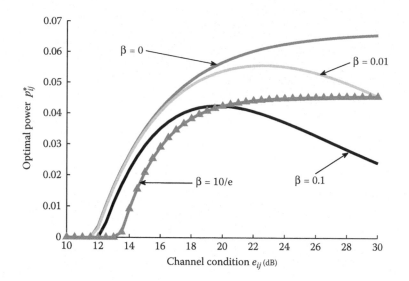

Figure 6.4 Optimal power p_{ij}^* as a function of the channel condition e_{ij}. Here $x_{ij} = 1$, $\alpha_i = 1$, $s_{ij} = +\infty$, and $\hat{\lambda}_i = 15$. (From J. Huang et al. *IEEE Transactions on Wireless Communications*, 8 (1), 288–296, 2009. With permission.)

Inserting the expression for p_{ij}^* into the Lagrangian yields

$$
L(x, \alpha, \mu, \lambda) = \sum_i (w_i - \alpha_i)_+ R_i^{\max} - \sum_i (\alpha_i - w_i)_+ R_i^{\min} + \sum_j \mu_j + \sum_{m=1}^M \lambda_m P_m
$$
$$
+ \sum_{i,j} x_{ij} \left[\alpha_i f \left(g \left(\frac{\hat{\lambda}_i}{\alpha_i e_{ij}} \right) \wedge s_{ij} \right) - \frac{\hat{\lambda}_i}{e_{ij}} \left(g \left(\frac{\hat{\lambda}_i}{\alpha_i e_{ij}} \right) \wedge s_{ij} \right) - \mu_j \right], \quad (6.17)
$$

which is now a linear function of $\{x_{ij}\}$. Thus, optimizing over x_{ij} yields the dual function for (P1),

$$
L(\alpha, \lambda, \mu) = \sum_i (w_i - \alpha_i)_+ R_i^{\max} - \sum_i (\alpha_i - w_i)_+ R_i^{\min} + \sum_j \mu_j + \sum_m \lambda_m P_m
$$
$$
+ \sum_{i,j} \left[\alpha_i f \left(g \left(\frac{\hat{\lambda}_i}{\alpha_i e_{ij}} \right) \wedge s_{ij} \right) - \frac{\hat{\lambda}_i}{e_{ij}} \left(g \left(\frac{\hat{\lambda}_i}{\alpha_i e_{ij}} \right) \wedge s_{ij} \right) - \mu_j \right]_+
$$
$$
= \sum_i \left((w_i - \alpha_i)_+ R_i^{\max} - (\alpha_i - w_i)_+ R_i^{\min} \right) + \sum_m \lambda_m P_m
$$
$$
+ \sum_j \left(\sum_i \left[\mu_{ij} \left(\alpha_i, \frac{\hat{\lambda}_i}{\alpha_i e_{ij}} \right) - \mu_j \right]_+ + \mu_j \right), \quad (6.18)
$$

where

$$\mu_{ij}(a, b) := a \left(f(g(b) \wedge s_{ij}) - b(g(b) \wedge s_{ij}) \right).$$

Note that any choice such that

$$x_{ij}^* \in \begin{cases} \{1\} & \text{if } \mu_{ij} \left(\alpha_i, \dfrac{\hat{\lambda}_i}{\alpha_i e_{ij}} \right) > \mu_j, \\[2em] [0, 1] & \text{if } \mu_{ij} \left(\alpha_i, \dfrac{\hat{\lambda}_i}{\alpha_i e_{ij}} \right) = \mu_j, \\[2em] \{0\} & \text{if } \mu_{ij} \left(\alpha_i, \dfrac{\hat{\lambda}_i}{\alpha_i e_{ij}} \right) < \mu_j \end{cases} \qquad (6.19)$$

will optimize the Lagrangian in Equation 6.17.

6.3.3.2 Optimizing the Dual Function over μ

From the duality theory of convex optimization [41,45] the optimal solution to problem (P1) is given by minimizing the dual function in Equation 6.18 over all $(\alpha, \lambda, \mu) \geq 0$. We do this coordinate-wise starting with the μ variables. The following lemma characterizes this optimization.

LEMMA 6.1

For all $\alpha, \lambda \geq 0$,

$$L(\alpha, \lambda) := \min_{\mu \geq 0} L(\alpha, \lambda, \mu)$$
$$= \sum_i \left((w_i - \alpha_i)_+ R_i^{\max} - (\alpha_i - w_i)_+ R_i^{\min} \right) + \sum_m \lambda_m P_m + \sum_j \mu_j^*(\alpha, \lambda), \quad (6.20)$$

where for every tone j, *the minimizing value of* μ_j^* *is achieved by*

$$\mu_j^*(\alpha, \lambda) := \max_i \mu_{ij} \left(\alpha_i, \dfrac{\hat{\lambda}_i}{\alpha_i e_{ij}} \right). \qquad (6.21)$$

The proof of Lemma 6.1 follows from a similar argument as in Ref. [6]. Note that Equation 6.21 requires searching for the maximum value of the metrics μ_{ij} across all users for each tone j. Since $L(\alpha, \lambda)$ is the minimum of a convex function over a convex set, it is a convex function of (α, λ).

6.3.3.3 Optimizing the Dual Function over (α, λ)

Now we are ready to optimize the remaining variables in the dual functions, namely, (α, λ). In the single cell downlink case with no rate constraints (and thus no α variables), this reduces to a one-dimensional problem in λ and hence, it can be minimized using an iterated one-dimensional search (e.g., the Golden Section method [41]). Since there is no duality gap, at $\lambda^* = \arg\min_{\lambda \geq 0} L(\lambda)$, $L(\lambda^*)$ gives the optimal objective value of problem (P1). Similarly, in the absence of rate constraints, the multiple sites/sectors problem with a partition of the users $\{\mathcal{K}_m\}_{m=1}^M$ also leads to a one-dimensional problem within each partition.

In general, however, one would need to use subgradient methods [41,45] to numerically solve for the optimal (α, λ). The following lemma characterizes the set of subgradients of $L(\alpha, \lambda)$ with respect to (α, λ).

LEMMA 6.2

About any $(\alpha^0, \lambda^0) \geq 0$,

$$L(\alpha, \lambda) \geq \sum_i d(\alpha_i^0)(\alpha_i - \alpha_i^0) + \sum_m d(\lambda_m^0)(\lambda_m - \lambda_m^0), \qquad (6.22)$$

with

$$d(\lambda_m) = P_m - \sum_{i \in \mathcal{K}_m} p_{ij}^* = P_m - \sum_{i \in \mathcal{K}_m} \frac{x_{ij}^*}{e_{ij}} g\left(\frac{\hat{\lambda}_i}{\alpha_i e_{ij}}\right) \wedge s_{ij}, \qquad (6.23)$$

$$d(\alpha_m) = \sum_j x_{ij}^* f\left(g\left(\frac{\hat{\lambda}_i}{\alpha_i e_{ij}}\right) \wedge s_{ij}\right) - r_i^*, \qquad (6.24)$$

where x_{ij}^*s satisfy

$$\sum_i x_{ij}^* \leq 1 \text{ and } \mu_j(\alpha, \lambda)\left(1 - \sum_i x_{ij}^*\right) = 0; \qquad \forall j,$$

and satisfy the Equation 6.19 with $\mu_j = \mu_j^*(\alpha, \lambda)$ as given in Equation 6.21, and r_i^* satisfy Equation 6.13. Thus the subgradients $d(\lambda_m)$ and $d(\alpha_i)$ are parameterized by (r^*, x^*) and are linear in these variables. Moreover, the permissible values of r^* lie in a hypercube and those of x^* in a simplex.

Observe that the dual function at any point (α, λ) is obtained by taking the maximum of the Lagrangian over (r^*, p^*, x^*) satisfying $\sum_i x_{ij} \leq 1, \forall j \in \mathcal{N}, (x, p) \in \mathcal{X}$. In case (r^*, p^*, x^*) is unique, then the resulting Lagrangian is a gradient to the dual function at (α, λ). In case there are multiple optimizers, the resulting Lagrangians are each a subgradient, and every subgradient can be obtained by a convex combination of these subgradients so that the set of subgradients is convex. The lemma follows easily by substituting for the optimal (r^*, p^*, x^*).

Having characterized the set of subgradients, a method similar to that used in Ref. [7] for the single cell uplink problem can be used to solve for the optimal dual variables (α^*, λ^*) numerically. In each step of this method we change the dual variables along the direction given by a subgradient subject to non-negativity of the dual variables. The convergence of this procedure (for a proper step-size choice) is once again guaranteed by the convexity of $L(\alpha, \lambda)$ (see Refs. [41] [Exer. 6.3.2], [7]).

6.3.3.4 Optimizing the Dual Function over α

Since the dimension of $\boldsymbol{\alpha}$ equals the number of users and the dimension of $\boldsymbol{\mu}$ equals the number of tones, it may be computationally better to optimize over $\boldsymbol{\alpha}$ instead of $\boldsymbol{\mu}$ if the number of users is greater, and then use numerical methods to solve the problem. Next we detail the means to optimize over $\boldsymbol{\alpha}$ before $\boldsymbol{\mu}$. The dual function contains many terms that have definitions with $(\cdot)_+$, and therefore we would need to identify exactly when these terms are nonzero. For this we need to solve a nonlinear equation that is guaranteed to have a unique solution. We first discuss this and then apply it to optimizing the dual function over $\boldsymbol{\alpha}$.

Given $y, z \geq 0$, define by $v(y, z)$, the unique solution with $1 \leq x < +\infty$ to

$$xf\left(g\left(\frac{1}{x} \vee \frac{df}{ds}(z)\right)\right) - g\left(\frac{1}{x} \vee \frac{df}{ds}(z)\right) = y,$$

where it is easy to show that $xf\left(g\left((1/x) \vee (df/ds)(z)\right)\right) - g\left((1/x) \vee ((df/ds)(z))\right)$ is a monotonically increasing function taking value 0 at $x = 1$ and increasing without bound as $x \to +\infty$. If $y > f(z)/(df/ds(z)) - z \ (\geq 0)$, then $v(y, z) = (y + z)/f(z)$ where it is easy to verify that $v(y, z) \geq z/f(z) \geq 1/(df/ds(z)) \geq 1/(df/ds(0)) = 1$ from the concavity of $f(\cdot)$ and from $f(0) = 0$. Otherwise, we need to solve for the unique $1 \leq x \leq 1/(df/ds(z))$ such that

$$xf\left(g\left(\frac{1}{x}\right)\right) - g\left(\frac{1}{x}\right) = y.$$

For our results we will be interested in $v\left(\mu_j e_{ij}/\hat{\lambda}_i, s_{ij}\right)$, using which we also define

$$v_{ij} := \frac{\hat{\lambda}_i v\left(\mu_j e_{ij}/\hat{\lambda}_i, s_{ij}\right)}{e_{ij}} \quad \text{and} \quad \zeta_{ij} := \frac{\mu_j + \left(s_{ij}\hat{\lambda}_i/e_{ij}\right)}{f(s_{ij})},$$

where $v_{ij} = \zeta_{ij}$ if $\mu_j e_{ij}/\hat{\lambda}_i \geq f(s_{ij})/(df(s_{ij})/ds) - s_{ij}$.

First, note that we can rewrite the function in Equation 6.18 as

$$L(\boldsymbol{\alpha}, \boldsymbol{\mu}, \boldsymbol{\lambda}) = \sum_j \mu_j + \sum_m \lambda_m P_m + \sum_i \tilde{L}_i,$$

where

$$\tilde{L}_i = (w_i - \alpha_i)_+ R_i^{\max} - (\alpha_i - w_i)_+ R_i^{\min}$$
$$+ \sum_j \frac{\hat{\lambda}_i}{e_{ij}}\left[\frac{\alpha_i e_{ij}}{\hat{\lambda}_i} f\left(g\left(\frac{\hat{\lambda}_i}{\alpha_i e_{ij}}\right) \wedge s_{ij}\right) - \left(g\left(\frac{\hat{\lambda}_i}{\alpha_i e_{ij}}\right) \wedge s_{ij}\right) - \frac{\mu_j e_{ij}}{\hat{\lambda}_i}\right]_+.$$

Now using the quantities defined earlier in this section, one can write $\tilde{L}_i$ as

$$\tilde{L}_i = \sum_j \frac{\hat{\lambda}_i}{e_{ij}} \left[1_{\{0 \le \alpha_i \le \zeta_{ij}\}} \left(\frac{\alpha_i e_{ij}}{\hat{\lambda}_i} f(s_{ij}) - s_{ij} - \frac{\mu_j e_{ij}}{\hat{\lambda}_i} \right) \right.$$

$$\left. + 1_{\{\zeta_{ij} < \alpha_i \le \upsilon_{ij}\}} \left(\frac{\alpha_i e_{ij}}{\hat{\lambda}_i} f\left(g\left(\frac{\hat{\lambda}_i}{\alpha_i e_{ij}} \right) \right) - g\left(\frac{\hat{\lambda}_i}{\alpha_i e_{ij}} \right) - \frac{\mu_j e_{ij}}{\hat{\lambda}_i} \right) \right]$$

$$+ (w_i - \alpha_i)_+ R_i^{\max} - (\alpha_i - w_i)_+ R_i^{\min}.$$

Minimizing $\tilde{L}_i$ over $\alpha_i \ge 0$ can now be accomplished by a simple one-dimensional search; we define the optimal vector of α_i's to be $\boldsymbol{\alpha}^*(\boldsymbol{\lambda}, \boldsymbol{\mu})$. Thereafter one would need to use a subgradient method [7,41] to numerically minimize over $(\boldsymbol{\mu}, \boldsymbol{\lambda})$. A subgradient of $\tilde{L}$ with respect to λ_m is given by $P_m - \sum_{i \in \mathcal{K}_m} p_{ij}^*$ where p_{ij}^* is taken from Equation 6.16 where one substitutes x_{ij}^* from Equation 6.19. A subgradient of $\tilde{L}$ with respect to μ_j is given by $1 - \sum_i x_{ij}^*$ where we substitute for x_{ij}^* from Equation 6.19. Note, however, that it is important that we also meet the following constraints for all i, namely

$$R_i^{\min} \le \sum_i x_{ij}^* f\left(\frac{p_{ij}^*}{x_{ij}^*} \right) \le R_i^{\max};$$

$$\text{if } \alpha_i^* < w_i, \text{ then } \sum_j x_{ij}^* f\left(\frac{p_{ij}^*}{x_{ij}^*} \right) = R_i^{\max}$$

and

$$\text{if } \alpha_i^* > w_i, \text{ then } \sum_j x_{ij}^* f\left(\frac{p_{ij}^*}{x_{ij}^*} \right) = R_i^{\min}.$$

The proof of this follows by retracing the steps of the proof of Lemma 6.2 with the roles of $\boldsymbol{\alpha}$ and $\boldsymbol{\mu}$ being switched.

6.3.4 Primal Optimal Solution

For the general OFDMA problem we presented two methods to solve for V^*: in the first method we showed how to characterize the dual variables $\boldsymbol{\mu}(\boldsymbol{\alpha}, \boldsymbol{\lambda})$ and then we proposed numerically solving for the optimal $(\boldsymbol{\alpha}^*, \boldsymbol{\lambda}^*)$ using subgradient methods, while in the second method we followed the same strategy after switching the roles of $\boldsymbol{\mu}$ and $\boldsymbol{\alpha}$. However, we still need to solve for the values of the corresponding optimal primal variables. Concentrating on the first method, we know by duality theory [41] that given $(\boldsymbol{\alpha}^*, \boldsymbol{\lambda}^*)$ we need to find one vector from the set of (r^*, x^*, p^*) that also satisfies primal feasibility and complementary slackness. These constraints can easily be seen to translate to the following:

$$d(\lambda_m^*) \ge 0, \qquad d(\lambda_m^*)\lambda_m^* = 0, \qquad \forall m; \qquad (6.25)$$
$$d(\alpha_i^*) \ge 0, \qquad d(\alpha_i^*)\alpha_i^* = 0, \qquad \forall i. \qquad (6.26)$$

From the linearity of $d(\lambda_m^*), d(\alpha_i^*)$ in (r^*, x^*), it follows that the primal optimal (r, x, p) are the solution of a linear program in (r^*, x^*).

For the single cell downlink case with no rate constraints, as we have previously noted searching for the dual optimal is a one-dimensional numerical search in λ. In that case, the search for primal optimal solution turns out to have additional structure as shown in Ref. [8].

6.3.5 OFDMA Feasibility

Next we turn to the corresponding feasibility problem, which can be stated as

$$V^* = \min \sigma \tag{6.27}$$

$$\text{subject to } R_i \le \sum_j x_{ij} f\left(\frac{p_{ij} e_{ij}}{x_{ij}}\right), \quad \forall i, \qquad (\alpha_i)$$

$$\sum_i x_{ij} \le 1 \quad \forall j, \qquad (\mu_j)$$

$$\sum_{i \in \mathcal{K}_m} \sum_j \frac{p_{ij}}{P_m} \le \sigma \quad \forall m, \qquad (\lambda_m)$$

$$(x, p) \in \mathcal{X}.$$

The vector of rates (R_i) is feasible if $V^* \le 1$, that is all the power constraints will also be satisfied by a vector (x^*, p^*). As mentioned earlier, we need to check that $(R_i) = (R_i^{\min})$ is indeed feasible; otherwise problems (P1) and (P2) are both infeasible as well. Moreover, if $(R_i) = (R_i^{\max})$ is also feasible, then $r = (R_i^{\max})$ is the optimizer for problems (P1) and (P2). In which case, the optimal solution to the problem above with $(R_i) = (R_i^{\max})$ will also yield an optimal solution to the scheduling problem. Observe that problem (6.27) is convex and satisfies Slater's conditions. Finally, we also note that other alternate formulations of the feasibility problem are possible where one could either apply the σ constraint also on the subchannel utilization or switch the roles of subchannel and power utilization. All of these will yield the same conclusion about feasibility although the actual solutions, in terms of (x^*, p^*), would possibly be different.

The Lagrangian considering the marked constraints is

$$L(\sigma, x, p, \alpha, \mu, \lambda) = \sigma\left(1 - \sum_m \lambda_m\right) - \sum_j \mu_j + \sum_i \alpha_i R_i$$

$$+ \sum_{ij} \mu_j x_{ij} - \sum_{ij} \left(\alpha_i x_{ij} f\left(\frac{p_{ij} e_{ij}}{x_{ij}}\right) + p_{ij} \tilde{\lambda}_i\right),$$

where $\tilde{\lambda}_i := \sum_{m: i \in \mathcal{K}_m} \lambda_m / P_m$. As before, minimizing over p_{ij} yields $p_{ij}^* e_{ij} / x_{ij} = g\left(\tilde{\lambda}_i / \alpha_i e_{ij}\right) \wedge s_{ij}$. Substituting this in the Lagrangian, we get

$$L(\sigma, x, \alpha, \mu, \lambda) = \sum_i \alpha_i R_i - \sum_j \mu_j + \sigma\left(1 - \sum_m \lambda_m\right)$$

$$- \sum_{i,j} x_{ij} \left[\alpha_i f\left(g\left(\frac{\hat{\lambda}_i}{\alpha_i e_{ij}}\right) \wedge s_{ij}\right) - \frac{\hat{\lambda}_i}{e_{ij}}\left(g\left(\frac{\hat{\lambda}_i}{\alpha_i e_{ij}}\right) \wedge s_{ij}\right) - \mu_j\right].$$

Minimizing over $0 \leq x_{ij} \leq 1$ yields

$$L(\sigma, \boldsymbol{\alpha}, \boldsymbol{\mu}, \boldsymbol{\lambda}) = \sum_i \tilde{L}_i - \sum_j \mu_j + \sigma \left(1 - \sum_m \lambda_m \right),$$

where

$$\tilde{L}_i = \alpha_i R_i - \sum_j \left[\alpha_i f \left(g \left(\frac{\hat{\lambda}_i}{\alpha_i e_{ij}} \right) \wedge s_{ij} \right) - \frac{\hat{\lambda}_i}{e_{ij}} \left(g \left(\frac{\hat{\lambda}_i}{\alpha_i e_{ij}} \right) \wedge s_{ij} \right) - \mu_j \right]_+.$$

Next we minimize L over all values of σ. Since there are no constraints on σ, it follows that the resulting L is finite only when $\sum_m \lambda_m = 1$; for all other values, we would get $L = -\infty$. Hereafter we will assume that $\sum_m \lambda_m = 1$. Thus

$$L(\sigma, x, \boldsymbol{\alpha}, \boldsymbol{\mu}, \boldsymbol{\lambda}) = \sum_i \tilde{L}_i - \sum_j \mu_j.$$

Note that as before, as a function of α_i the problem is now separable. Therefore we only need to maximize $\tilde{L}_i$ over $\alpha_i \geq 0$.

Similarly, we can write L as

$$L(\sigma, x, \boldsymbol{\alpha}, \boldsymbol{\mu}, \boldsymbol{\lambda}) = \sum_j \hat{L}_j + \sum_i \alpha_i R_i,$$

where we have

$$\hat{L}_j = -\left(\mu_j + \sum_i \left[\alpha_i f \left(g \left(\frac{\hat{\lambda}_i}{\alpha_i e_{ij}} \right) \wedge s_{ij} \right) - \frac{\hat{\lambda}_i}{e_{ij}} \left(g \left(\frac{\hat{\lambda}_i}{\alpha_i e_{ij}} \right) \wedge s_{ij} \right) - \mu_j \right]_+ \right).$$

As a function of μ_j the problem is now separable, and we only need to maximize $\hat{L}_j$ over $\mu_i \geq 0$.

Thus, we could optimize first either $\boldsymbol{\mu}$ or $\boldsymbol{\alpha}$, once again based on whether the number of users or subchannels is smaller. In either case, the methodology and the functions that appear are very similar to the corresponding problem in the scheduling problem (P1), and due to space constraints we do not elaborate on this. Care must be take, however, while evaluating subgradients with respect to $\boldsymbol{\lambda}$. Here we propose using a projected gradient method [41,45] based on the constraint $\sum_m \lambda_m = 1$ to numerically solve for the optimal $\boldsymbol{\lambda}$.

6.3.6 Power Allocation Given Subchannel Allocation

In many of the suboptimal scheduling algorithms that we will discuss, a central feature will be a computationally simpler (but still close to optimal) method to provide a subchannel allocation. Once the subchannel allocation has been made, all that will remain is the power allocation problem, subject to the various constraints that we discussed earlier. Here we discuss how this can be solved in an optimal manner. A similar question can also be asked about the feasibility problem, hence we also discuss this here. In all cases, we assume that we are given a feasible subchannel allocation.

Since we are given a feasible subchannel allocation x, the Lagrangian of the new scheduling problem (power allocation only) can be easily derived by setting $\mu = 0$. For this we once again use the formulation based on problem (P1). The optimal power allocation is then given by $p_{ij}^* = (x_{ij}/e_{ij})\left(g(\hat{\lambda}_i/\alpha_i e_{ij}) \wedge s_{ij}\right)$. The Lagrangian that results from substituting this formula is

$$
L(x, \alpha, \lambda) = \sum_m \lambda_m P_m + \sum_i (w_i - \alpha_i)_+ R_i^{\max} - \sum_i (\alpha_i - w_i)_+ R_i^{\min}
$$
$$
+ \sum_i \sum_j \alpha_i x_{ij} f\left(g\left(\frac{\hat{\lambda}_i}{e_{ij}\alpha_i}\right) \wedge s_{ij}\right) - \frac{\hat{\lambda}_i x_{ij}}{e_{ij}}\left(g\left(\frac{\hat{\lambda}_i}{e_{ij}\alpha_i}\right) \wedge s_{ij}\right).
$$

Now it is easy to argue that if $R_i^{\min} = 0$ and $R_i^{\max} = +\infty$ and if the $\mathcal{K}_m$'s form a partition, then within each partition the λ_m's can be solved for as in Section 6.3.3. In any case, in this setting, solving for the optimal $\alpha_i \geq 0$ is easier, but uses some of the functions described at the end of Section 6.3.3. However, after this step we would still need to solve for λ numerically; if the partitions assumption holds, then it would only need a single-dimensional search within each partition. A finite-time algorithm for achieving the optimal λ has been given in Refs. [7,8] under the assumption that $f(\cdot)$ represents the Shannon capacity as in Equation 6.8 with $\beta = 0$.

6.3.6.1 Feasibility Check

Under the assumption that a feasible subchannel allocation has already been provided, even the feasibility check problem becomes a lot easier. As before, we can assume $\sum_m \lambda_m = 1$, and that the optimal power allocation is given by $p_{ij}^* = (x_{ij}/e_{ij})\left(g\left(\tilde{\lambda}_i/e_{ij}\alpha_i\right) \wedge s_{ij}\right)$, and substituting this we get

$$
L(x, \alpha, \lambda) = \sum_i \alpha_i \hat{R}_i - \sum_j x_{ij}\left[\alpha_i f\left(g\left(\frac{\tilde{\lambda}_i}{e_{ij}\alpha_i}\right) \wedge s_{ij}\right) - \frac{\tilde{\lambda}_i}{e_{ij}}\left(g\left(\frac{\tilde{\lambda}_i}{e_{ij}\alpha_i}\right) \wedge s_{ij}\right)\right].
$$

Again solving for the optimal α_i is simpler. Once again the λ vector would need to be computed numerically, subject to it being a probability vector, that is, $\sum_m \lambda_m = 1$ and $\lambda_m \geq 0$ for each m.

6.4 Low Complexity Suboptimal Algorithms with Integer Channel Allocation

There are two shortcomings with using the optimal algorithm outlined in the previous section for scheduling and resource allocation: (i) the complexity of the algorithm in general is not computationally feasible for even moderate sized systems; (ii) the solution found may require a time-sharing channel allocation, while practical implementations typically require a single user per subchannel. One way to address the second point is to first find the optimal primal solution as in the previous section and then project this onto a "nearby" integer solution. Such an approach is presented in Ref. [8] for the case of a single cell downlink system ($M = 1$) without any rate constraints. In that setting, after minimizing the dual function over μ, one optimizes the function $L(\lambda)$, which only depends on a single variable. This function will have scalar subgradients which can then be used to develop rules for implementing such an integer projection. Moreover, in this case since $L(\lambda)$ is a one-dimensional function, the search for the optimal dual values is greatly simplified. However, in the general setting, this type of approach does not appear to be promising.*

In this section we discuss a family of suboptimal algorithms (SOAs) for the general setting that try to reduce the complexity of the optimal algorithm, while sacrificing little in performance. These algorithms seek to exploit the problem structure revealed by the optimal algorithm. Furthermore, all of these SOA enforce an integer tone allocation during each scheduling interval. In the following, we consider the general model from Section 6.3.1 with the restriction that $\{\mathcal{K}_m\}$ forms a partition of the user groups (i.e. each user is in only one of these sets) and that $R_i^{\min} = 0$ for all i. In a typical setting both of these assumptions will be true.

In the optimal algorithm, given the optimal λ and α, the optimal tone allocation up to any ties is determined by sorting the users on each tone according to the metric $\mu_{ij}(\alpha_i, \hat{\lambda}_i/\alpha_i e_{ij})$ (cf. Equation 6.19). Given an optimal tone allocation, the optimal power allocation is given by Equation 6.16. In each SOA, we use the same two phases with some modifications to reduce the complexity of computing (λ, α) and the optimal tone allocation. Specifically, we begin with a *subChannel Allocation (CA)* phase in which we assign each tone to at most one user. We consider two different SOAs that implement the CA phase differently. In SOA1, instead of using the metric given by the optimal λ and α, we consider metrics based on a constant power allocation over all tones assigned to a partition. In SOA2, we find the tone allocation, once again through a dual-based approach, but here we first determine the number of tones assigned to each user and then match specific tones and users. In all cases we assign the tones to distinct partitions that will, in turn, yield an interference-free operation. After the tone allocation is done in both SOAs, we execute the *Power Allocation (PA)* phase in which each user's power is allocated across the assigned tones using the optimal power allocation in Equation 6.16.

6.4.1 CA in SOA1: Progressive Subchannel Allocation Based on Metric Sorting

In this family of SOAs, tones are assigned sequentially in one pass based on a per user metric for each tone, that is, we iterate N times, where each iteration corresponds to the assignment of one tone. Let $\mathcal{N}_i(n)$ denote the set of tones assigned to user i after the nth iteration. Let $g_i(n)$ denote user i's metric during the nth iteration and let $l_i(n)$ be the tone index that user i would like to be assigned if he/she is assigned the nth tone. The resulting CA algorithm is given in Algorithm 1. Note that all the user metrics are updated after each tone is assigned.

We consider several variations of Algorithm 1, which correspond to different choices for steps 4 and 5. The choices for step 4 are:

(4A): Sort the tones based on the best channel condition among *all* users. This involves two steps. First, for each tone j, find the best channel condition among all users and denote it by $\tilde{\mu}_j := \max_i e_{ij}$. Second, find a tone permutation $\{\alpha_j\}_{j \in \mathcal{N}}$ such that $\tilde{\mu}_{\alpha_1} \geq \tilde{\mu}_{\alpha_2} \geq \cdots \geq \tilde{\mu}_{\alpha_N}$, and set $l_i(n) = \alpha_n$ for each user i at the nth iteration. Each max operation has a complexity of $O(K)$, and the sorting operation has a complexity of $O(N \log(N))$. The total complexity is $O(NK + N \log N)$. We note that this is a one-time "preprocessing" that needs to be done before

---

* See Ref. [7] for a more detailed discussion of this in the context of the uplink scenario.

ALGORITHM 1   CA Phase for SOA1

1: Initialization: set $n = 0$ and $\mathcal{N}_i(n) = \emptyset$ for each user i.
2: **while** $n < N$ **do**
3:     $n + 1$.
4:     Update tone index $l_i(n)$ for each user i.
5:     Update metric $g_i(n)$ for each user i.
6:     Find $i^*(n) = \arg\max_i g_i(n)$ (break ties arbitrarily).
7:     **if** $g_{i^*(n)}(n) \geq 0$ **then**
8:         Assign the nth tone to user $i^*(n)$:

$$\mathcal{N}_i(n) = \begin{cases} \mathcal{N}_i(n-1) \cup \{l_i(n)\}, & \text{if } i = i_n^*; \\ \mathcal{N}_i(n-1), & \text{otherwise.} \end{cases}$$

9:     **else**
10:         Do not assign the nth tone.
11:     **end if**
12: **end while**

the CA phase starts. During the tone allocation iterations, the users just choose the tone index from the sorted list.

(4B): Sort the tones based on the channel conditions for each *individual* user. For each user i at the nth iteration, set $l_i(n)$ to be the tone index with the largest gain among all unassigned tones, that is, $l_i(n) = \arg\max_{j \in \mathcal{N} \setminus \cup_i \mathcal{N}_i(n-1)} e_{ij}$. This requires K sorts (one per user); these also need to be performed only once (since each tone assignment does not change a user's ordering of the remaining tones) and can be done in parallel. The total complexity of the K sorting operations is $O(KN \log N)$, which is higher than that in (4A).

During the nth iteration, let $k_i(n) = |\cup_{j \in \mathcal{K}_{m(i)}} \mathcal{N}_j(n)|$ denote the number of tones assigned to users in the group to which user i belongs, that is, $m(i)$. The choices for Line 5 are:

(5A): Set $g_i(n)$ to be the total increase in user i's utility if assigned tone $l_i(n)$, assuming that the power for each user group is allocated uniformly over the tones assigned to that group, that is,

$$g_i(n) = \begin{cases} w_i \left[\left(\sum_{j \in \mathcal{N}_i(n-1) \cup \{l_i(n)\}} f\left(\frac{P_i e_{ij}}{k_i(n-1)+1} \wedge s_{ij} \right) \right) \wedge R_i^{\max} \right. \\ \quad \left. - \left(\sum_{j \in \mathcal{N}_i(n-1)} f\left(\frac{P_i e_{ij}}{k_i(n-1)} \wedge s_{ij} \right) \right) \wedge R_i^{\max} \right] & \text{if } k_i(n-1) > 0; \\ w_i \left[\left(\sum_{j \in \mathcal{N}_i(n-1) \cup \{l_i(n)\}} f\left(\frac{P_i e_{ij}}{k_i(n-1)+1} \wedge s_{ij} \right) \right) \wedge R_i^{\max} \right] & \text{otherwise.} \end{cases}$$

(6.28)

(5B): Set $g_i(n)$ to be user i's gain from only tone $l_i(n)$, again assuming that constant power allocation within each group, that is

$$g_i(n) = w_i \left[f\left(\frac{P_i e_{i,l_i(n)}}{k_i(n-1)+1} \wedge s_{ij} \right) \wedge R_i^{\max} \right].$$

Compared with (5A), this metric is simpler to calculate but ignores the change in user i's utility due to the decrease in power allocated to any tones in $\mathcal{N}_i(n-1)$. It also does not accurately enforce the maximum rate constraint, since it only considers one tone at a time.

The complexity of either of these choices over N iterations is $O(NK)$, and so the total complexity for the CA phase is $O\left(NK + N\log N\right)$ (if (4A) is chosen) or $O\left(KN\log N\right)$ (if (4B) is chosen). Algorithms similar to SOA1 with (4B) and (5B) have been proposed in the literature for both the single cell downlink setting [14]* and the uplink Ref. [28] without rate or SNR constraints. In the single cell downlink case, the algorithm in Ref. [14] is shown via numerical examples to have near optimal performance. In the uplink case, this also performs reasonably well in simulations [28], but [7] shows that better performance can be obtained using (4B) and (5A) instead.

6.4.2 CA in SOA2: Tone Number Assignment and Tone User Matching

SOA2 implements the CA phase through two steps: tone number assignment (CNA) and tone user matching (CUM). The algorithm is summarized in Algorithm 2.

ALGORITHM 2   CA Phase of SOA2

---

1. subChannel Number Assignment (CNA) step: determine the number of tones n_i allocated to each user i such that $\sum_{i \in \mathcal{K}} n_i \leq N$.
2. subChannel User Matching (CUM) step: determine the tone assignment $x_{ij} \in \{0, 1\}$ for all users i and tones j, such that $\sum_{j \in \mathcal{N}} x_{ij} = n_i$.

6.4.2.1 SubChannel Number Assignment (CNA)

In the CNA step, we determine the number of tones n_i assigned to each user $i \in \mathcal{K}$. The assignment is calculated based on the approximation that each user sees a flat wide-band fading tone. Notice that here we do not specify which tone is allocated to which user; such a mapping will be determined in the CUM step. The CNA step is further divided into two stages: a basic assignment stage and an assignment improvement stage.

Stage 1, Basic Assignment: Here, the assignment is based on the normalized SNR averaged over all tones. Specifically, we model each user i as having a normalized SNR $\bar{e}_i = (1/N) \sum_{j \in \mathcal{N}} e_{ij}$, and then determine a tone number assignment n_i for all i by solving

$$
\max_{\{n_i \geq 0, i \in \mathcal{K}\}} \sum_{i \in \mathcal{K}} w_i n_i f\left(\frac{P_{m(i)}\bar{e}_i}{\sum_{j \in \mathcal{K}_{m(i)}} n_j} \wedge s_i\right)
$$

$$
\text{subject to} \quad \sum_{i \in \mathcal{K}} n_i \leq N \qquad \text{(SOA2-CNA)}
$$

$$
n_i f\left(\frac{P_{m(i)}\bar{e}_i}{\sum_{j \in \mathcal{K}_{m(i)}} n_j} \wedge s_i\right) \leq R_i^{\max}.
$$

---

* The main difference with the algorithm in Ref. [14] is that after each iteration n, it then checks to see if $\sum_i w_i r_i$ is increasing and if not it stops at iteration $n-1$. Such a step can be added to Algorithm 1; however, unless the system is lightly loaded it is unlikely to have a large impact on the performance.

Here, we are again assuming that power is allocated uniformly over all the channels assigned to a given user group.

Unfortunately, in general, the objective in Problem SOA2-CNA is not concave. However, in the special case of the uplink ($\mathcal{K}_{m(i)} = \{i\}$) it will be.* In the case of the single cell downlink, if $nf(a/n)$ is increasing for all $a > 0$ (as in our general formulation), then the problem can be reformulated to have a concave objective by noting that in this case it must be that $\sum_{i \in \mathcal{K}} n_i = N$ at any optimal solution. Additionally, due to the maximum rate constraint, the constraint set may not be convex; this can be accommodated by considering a higher dimensional problem as in Section 6.3.3.

Next, we focus on solving Problem SOA2-CNA in the uplink setting without maximum rate constraints. In this case, the problem will have a unique and possibly noninteger solution, which we can again use a dual relaxation to find it. Consider the Lagrangian

$$L(\mathbf{n}, \lambda) := \sum_{i \in \mathcal{K}} w_i n_i f \left(\frac{P_i \bar{e}_i}{n_i} \wedge s_i \right) - \lambda \left(\sum_{i \in \mathcal{K}} n_i - N \right).$$

Optimizing $L(\mathbf{n}, \lambda)$ over $n \geq 0$ for a given λ is equivalent to solving the following K subproblems,

$$n_i^*(\lambda) = \arg \max_{n_i \geq 0} w_i n_i f \left(\frac{P_i \bar{e}_i}{n_i} \wedge s_i \right) - \lambda n_i, \quad \forall i. \tag{6.29}$$

Problem (6.29) can be solved by a simple line search over the range of $(0, N]$. Substituting the corresponding results into the Lagrangian yields

$$L(\lambda) := \sum_{i \in \mathcal{K}} w_i n_i^*(\lambda) f \left(\frac{P_i \bar{e}_i}{n_i^*(\lambda)} \wedge s_i \right) - \lambda \left(\sum_{i \in \mathcal{K}} n_i^*(\lambda) - N \right),$$

which is a convex function of λ [41]. The optimal value

$$\lambda^* = \arg \min_{\lambda \geq 0} L(\lambda) \tag{6.30}$$

can be found by a line section search over: $[0, \max_i w_i f(P_i \bar{e}_i / (N/K))]$.[†] For a given search precision, the maximum number of iterations needed to solve either Equation 6.29 or Equation 6.30 is fixed.[‡] Hence, the worst-case complexity of solving each subproblem is independent of K or N. Since there are K subproblems in Equation 6.29, it follows that the complexity of the basic assignment step

---

* Some care is required at the point where the SNR constraint becomes active as the objective is not differentiable there; nevertheless, by evaluating left and right derivatives the concavity can be shown.

† The upper-bound of the search interval can be obtained by examining the first-order optimality condition of Equation 6.29.

‡ For example, if we use bisection search to solve Equation 6.29 and stop when the relative error of the solution is less than $N/2^{10}$, then we only need a maximum of 10 search iterations.

is $O(K)$. If the resultant channel allocations contain noninteger values, we will approximate with an integer solution that satisfies $\sum_{i \in \mathcal{K}} n_i = N$.* Since each user is allocated only a subset of the tones, the normalized SNR $\bar{e}_i = (1/N) \sum_{j \in \mathcal{N}} e_{ij}$ is typically a pessimistic estimate of the averaged tone conditions over the allocated subset. This motivates us to consider the following assignment improvement stage of CNA.

Stage 2, Assignment Improvement: Here, assignment is performed by means of iterative calculations using the normalized SNR averaged over the best tone subset. Specifically, we iteratively solve the following variation of Problem SOA2-CNA (stated here for the uplink without maximum rate constraints):

$$\max_{\mathbf{n}(t) \geq 0} \sum_{i \in \mathcal{K}} w_i n_i(t) f\left(\frac{P_i \bar{e}_i(t)}{n_i(t)} \wedge s_i\right)$$

$$\text{subject to: } \sum_{i \in \mathcal{K}} n_i(t) \leq N \qquad \text{(SOA2-CNA-t)}$$

$$n_i f\left(\frac{P_{m(i)} \bar{e}_i(t)}{\sum_{j \in \mathcal{K}_{m(i)}} n_j} \wedge s_i\right) \leq R_i^{\max},$$

for $t = 1, 2, \ldots$. During the tth iteration, $\bar{e}_i(t)$ is a refined estimate of the normalized SNR based on the best $\lfloor n_i(t-1) \rfloor$ (or $\lceil n_i(t-1) \rceil$) tones of user i; additionally, $n_i(0) := N$ for all i. The iteration stops when the tone allocation converges or the maximum number of iterations allowed is reached. An integer approximation will be performed if needed.

The complete algorithm for the CNA phase of SOA2 is given in Algorithm 3. In order to perform the assignment improvement, we need to perform K sorting operations, with a total complexity $O(KN \log(N))$. Note that this only needs to be done once. Step 4 of each iteration has a complexity of $O(K)$ due to solving K subproblems for a fixed dual variable. The maximum number of iterations is fixed and thus is independent of N or K. The integer approximation stage requires a sorting with the complexity of $O(K \log(K))$. So the total complexity for the CNA phase of SOA2 is $O(KN \log(N) + K \log(K))$.

6.4.2.2 SubChannel User Matching (CUM) Step

After the CNA step, we know how many tones are to be allocated to each user. However, we still need to determine which specific tones are assigned to which user. This is accomplished in the CUM step by finding a tone assignment that maximizes the weighted sum rate assuming that each user

---

* One possible integer approximation is the following. Assume that n_i^* is the unique optimal solution of Problem SOA2-CNA. First, sort users in the descending order of the mantissa of n_i^*, $fr\left(n_i^*\right) = n_i^* - \lfloor n_i^* \rfloor$. That is, find a user permutation subset $\{\alpha_k, 1 \leq k \leq N\}$ such that $fr\left(n_{\alpha_1}^*\right) \geq fr\left(n_{\alpha_2}^*\right) \geq \cdots \geq fr\left(n_{\alpha_M}^*\right)$. Second, for each user i, let $\tilde{n}_i^* = \lfloor n_i^* \rfloor$. Third, calculate the number of unallocated tones, $N^A = N - \sum_i \tilde{n}_i^*$. Finally, adjust users with large mantissas such that all the tones are allocated, that is, $\tilde{n}_{\alpha_i}^* = \tilde{n}_{\alpha_i}^* + 1$ for all $1 \leq i \leq N^A$. The resulting $\{\tilde{n}_i^*\}_{i \in \mathcal{K}}$ give the integer approximation.

ALGORITHM 3   CNA Phase of SOA2

---

1. Initialization: integer MaxIte> 0, $t = 0$, $n_i(0) = N$ and $n_i(1) = N/2$ for each user i.
2. **while** $(n_i(t+1) \neq n_i(t)$ for some $i) \,\&\, (t <$ MaxIte) **do**
3. $t = t + 1$.
4.   For each user i, $\bar{e}_i(t) =$ average gain of user i's best $n_i(t-1)$ tones.
5.   Solve Problem (SOA2-CNA-t) to determine the optimal $n_i(t)$ for each user i.
6. **end while**
7. let $n_i^* = n_i(t)$ for each user i.

---

employs a flat power allocation, that is we solve the problem:

$$\max_{x_{ij} \in \{0,1\}} \sum_{i \in \mathcal{K}} \sum_{j \in \mathcal{N}} x_{ij} w_i f\left(\frac{P_i e_{ij}}{n_i^*} \wedge s_i\right)$$

$$\text{subject to: } \sum_{j \in \mathcal{N}} x_{ij} = n_i^*, \forall i \in \mathcal{K}, \qquad\qquad \text{(SOA2-CUM)}$$

$$\sum_{i \in \mathcal{K}} x_{ij} = 1, \forall j \in \mathcal{N},$$

where $n^* = (n_i^*, i \in \mathcal{K})$ is the integer tone allocation obtained in the CNA step. Since we solved Problem (SOA2-CNA-t) using the average of the best n^*, concavity of $f(\cdot)$ ensures that any feasible tone allocation for Problem (SOA2-CUM) will satisfy the maximum rate constraint.

Problem (SOA2-CUM) is an integer *Assignment Problem* whose *optimal* solution can be found by using the *Hungarian Algorithm* [46].* To use the Hungarian algorithm here, we need to perform "virtual user splitting" as explained next. For user i, let $r_{ij} = w_i f\left((P_i e_{ij}/n_i^*) \wedge s_{ij}\right)$, and let

$$r_i = [r_{i1}, r_{i2}, \ldots, r_{iN}]$$

be user i's achievable rates over all possible tones. We can then form a $K \times N$ matrix $R = \left[r_1^{\mathrm{T}}, r_2^{\mathrm{T}}, \ldots, r_M^{\mathrm{T}}\right]^{\mathrm{T}}$. Next, we split each user i into n_i^* virtual users by adding $n_i^* - 1$ copies of the row vector r_i to the matrix R. This expands R into a $N \times N$ square matrix. Solving Problem (SOA2-CUM) is then equivalent to finding a *permutation matrix* $C^* = \left[c_{ij}\right]_{N \times N}$ such that

$$C^* = \arg\min_{C \in \mathcal{C}} - C \cdot R := \arg\min_{C \in \mathcal{C}} - \sum_{i=1}^{N} \sum_{j=1}^{N} c_{ij} r_{ij}. \qquad\qquad (6.31)$$

Here $\mathcal{C}$ is the set of permutation matrices, that is, for any $C \in \mathcal{C}$, we have $c_{ij} \in \{0, 1\}$, $\sum_i c_{ij} = 1$, and $\sum_j c_{ij} = 1$ for all i and j. This problem can be solved by the standard Hungarian algorithm which has a computational complexity of $O(N^3)$, where N is the total number of tones. The detailed

---

* A similar idea has been used to solve various single cell downlink OFDMA resource allocation problems (e.g., Ref. [20]) as well as to find user coalitions for Nash Bargaining in an uplink OFDMA system in Ref. [24].

algorithm can be found in Ref. [46]. After obtaining C^*, we can calculate the corresponding tone allocation x^*. For example, if $c_{kj}^* = 1$ and virtual user k corresponds to the actual user i, then we know $x_{ij}^* = 1$, that is, tone j is allocated only to user i.

6.4.3 Power Allocation (PA) Phase

We can follow the tone allocation (CA) phase in either SOA1 or SOA2 with a power allocation phase in which power is optimally allocated among the tones assigned to the users in each partition.* After this optimization, it is possible that some tone is allocated zero power due to its poor tone gain. Alternatively, one can simply use a uniform power allocation as was assumed in the CA phase. For certain single cell downlink scenarios, such a uniform allocation has been shown to be nearly optimal in Refs. [8,14].

Since the tone allocation is given, optimizing the power allocation for each group is equivalent to the problem considered in Section 6.3.6 and can be addressed in a similar way, that is, by considering the dual formulation and numerically searching for the optimal dual variables. We note that in the uplink scenario without any maximum rate constraint, we need to solve one such problem for each user and for each problem only a single dual variable needs to be introduced (corresponding to the user's power constraint). Hence, the optimal dual value can be found through a simple line search, with a constant worst-case complexity given a fixed search precision as in our discussion of Equation 6.29.

6.4.4 Complexity and Performance of Suboptimal Algorithms for the Uplink Scenario

In this section, we discuss the complexity and performance of the suboptimal algorithms in an uplink scenario without any maximum rate constraints.† The worst-case computational complexities of the variations of SOA1 and SOA2 for this setting are summarized in Table 6.1.

Next, we briefly discuss the performance of this algorithms with a realistic OFDMA simulator assuming parameters and assumptions commonly found in the IEEE 802.16 standards [47]. These results are for a single cell with 40 users. All users are infinitely backlogged and assigned a throughput-based utility as in Equation 6.3 with parameter $c_i = 1$ and $\alpha = 0.5$. Each user i has a total transmission power constraint $P_i = 2$ W. We calculate the achievable rate of user i on tone j as

$$r_{ij} = Bx_{ij} \log \left(1 + \frac{p_{ij}e_{ij}}{x_{ij}} \right),$$

where B is the tone bandwidth and e_{ij} is generated according to a product of a fixed location-based term and a frequency-selective fast fading term. A detailed description of the simulation setup can be found in Ref. [7] with further results. Scheduling decisions are made at every 20 OFDM symbols, which corresponds to one fading block.

---

* In this section, we again consider the case where $\{\mathcal{K}_m\}$ forms a partition of the users and allow for maximum rate constraints.

† It can be argued that this will also be the worst-case setting for the general problem assuming partitions and no rate constraints.

Table 6.1 Worst-Case Computational Complexity of Suboptimal Algorithms

Suboptimal Algorithm			*Worst-Case Complexity*
SOA1	subChannel Allocation (CA)	4A and 5A	$O(NK + N \log N)$
		4A and 5B	$O(NK + N \log N)$
		4B and 5A	$O(KN \log N)$
		4B and 5B	$O(KN \log N)$
	Power Allocation (PA)		$O(KN)$
	Total (CA + PA)		$O(KN \log N)$
SOA2	subChannel Allocation (CA)	CNA	$O(KN \log N + K \log K)$
		CUM	$O(N^3)$
	Power Allocation (PA)		$O(KN)$
	Total (CA+PA)		$O(N^3 + KN \log N + K \log K)$

Source: J. Huang et al. *IEEE Journal on Selected Areas in Communications*, 27 (2), 226–234, 2009. With permission.

Table 6.2 shows simulation results for the following four algorithms:

1. *Integer-dual:* Integer tone allocation (with tie breaking) based on optimal dual-based algorithm and optimal power control. To reduce computational complexity in the case of too many ties, we randomly inspect up to 128 ways of breaking the ties with an integer allocation and select the allocation among these with the largest weighted sum rate (before reallocating the power).
2. *SOA1:* Tone allocation as in Section 6.4.1 and power control as in Section 6.4.3. There are four versions of SOA1, depending on how steps 4 and 5 in Algorithm 1 are implemented; we present results for each.
3. *SOA2:* Tone allocation as in Section 6.4.2 (with up to 10 iterations) and power control as in Section 6.4.3.
4. *Baseline:* Each tone j is allocated to the user i with the highest e_{ij}, without considering the weights w_i's and the power constraints. Each user's power is then allocated as in Section 6.4.3.

Table 6.2 Example Uplink Resource Allocation Performance

Algorithms		*Utility*	*log U*	*Rate*	*Scheduled Users*
Integer-Dual		53922	514.0	21.56	37.5
SOA1	4A and 5A	52494	510.7	22.86	34.6
	4A and 5B	51697	509.2	20.22	28.1
	4B and 5A	54165	513.3	22.25	35.0
	4B and 5B	53156	511.4	21.43	28.6
SOA2		54316	513.6	22.33	35.1
Base Line		21406	−1960.5	16.13	2.66

Source: J. Huang et al. *IEEE Journal on Selected Areas in Communications*, 27 (2), 226–234, 2009. With permission.

In this table it can be seen that SOA1 (with 4B and 5A) and SOA2 achieve the best performance in terms of total utility. Their performance is even better than the integer-dual approach, which was obtained based on the optimal value of the relaxed problem. This is likely because only 128 ways to break ties are considered, which is typically not sufficient. Since the integer-dual algorithm achieves an optimality ratio of 0.9412, this suggests that SOA1 and SOA2 achieve very close to optimal performance as well. The baseline algorithm always has poor performance.

Here, and in other uplink simulation reported in Ref. [7], all of the SOAs have good performance with SOA1 (with 4B and 5A) and SOA2 consistently achieving the best performance in terms of total utility. From Table 6.1, we note that these have slightly higher complexity than some of the other SOAs. Hence if lower complexity is desired, this can be provided with only a slight loss in performance. We also note that in each case the SOAs and the integer-dual algorithm schedule a large number of users on average in each time slot. A potential cost from this is that it may increase the needed signaling overhead. One way to reduce this cost is to add a penalty term to our objective which increases with the number of users scheduled.

6.5 Conclusions and Open Problems

In this chapter, we have considered a general model of gradient-based scheduling and resource allocation for OFDMA systems. This model includes single cell downlink, uplink, and multicell downlink with frequency sharing, and incorporates various practical constraints such as per carrier SNR constraints, self-noise due to imperfect channel estimates or phase noise, and minimum and maximum per user rate constraints. Essentially, the problem can be reduced to solving a weighted rate maximization problem in each time slot. We address this problem with a Lagrangian dual relaxation method. By exploiting the structure of the OFDMA rate region, we can express the dual function in terms of a small subset of dual variables. The optimal values of these variables can be found through standard numerical search methods. An interesting observation is that recovering the optimal primal solutions given optimal dual variables is rather straightforward in most cases, since the optimal channel allocations often turn out to be integer "automatically." In the case when this is not true, we need to calculate the channel allocation by either allowing time sharing or picking a good integer solution, and optimize the power allocation accordingly. Based on the intuition derived from the optimal algorithms, we demonstrate that it is possible to design a class of heuristic algorithms that are low in complexity but perform very well in simulation studies.

All algorithms presented in this chapter are centralized. This is not an issue for the single cell downlink case or even for a multisectored site, where the resource allocation decisions are made by the base station. In the uplink and multicell downlink cases, however, a distributed algorithm is more desirable since the decisions are made by the multiple network entities (either multiple mobile users or multiple base stations). Some preliminary results toward a fully distributed algorithm have been reported in Refs. [48,49] and more work is needed along this line. Another open issue regarding the multicell downlink case is to consider models that allow dynamic frequency reuse. A challenge in such settings is that the resulting optimization problem may no longer be convex even when the integer constraints are relaxed.

The algorithms presented here require that the scheduler has accurate channel quality information (although some inaccuracy may be accounted for via the self-noise terms). In OFDMA systems with many users and tones, the resulting feedback overhead can become significant. This overhead can be partially reduced by proper subchannelization methods (e.g., Ref. [8]) or by not reporting the channel quality on every subchannel as in Ref. [23]. However, there is little understanding of the interplay between these or other limited feedback schemes and the resulting scheduling performance. This is another area in which additional work is warranted.

Acknowledgments

J. Huang was supported by the Competitive Earmarked Research Grants (Project Number 412308) established under the University Grant Committee of the Hong Kong Special Administrative Region, People's Republic of China, the Direct Grant (Project Number C001-2050398) of The Chinese University of Hong Kong, and the National Key Technology R&D Program (Project Number 2007BAH17B04) established by the Ministry of Science and Technology of the People's Republic of China. V. Subramanian was supported by Science Foundation Ireland grant 07/IN.1/I901, associated with the Hamilton Institute, National University of Ireland, Maynooth. R. Berry was supported in part by the Motorola-Northwestern Center for Seamless Communications and NSF CAREER award CCR-0238382.

References

1. R. Agrawal and V. Subramanian, Optimality of certain channel aware scheduling policies, *Proc. of 2002 Allerton Conference on Communication, Control and Computing*, 2002.
2. R. Agrawal, A. Bedekar, R. La, and V. Subramanian, A class and channel-condition based weighted proportionally fair scheduler, *Proc. of ITC 2001*, Salvador, Brazil, September 2001.
3. H. Kushner and P. Whiting, Asymptotic properties of proportional-fair sharing algorithms, *40th Annual Allerton Conference on Communication, Control, and Computing*, 2002.
4. A. Jalali, R. Padovani, and R. Pankaj, Data throughput of CDMA-HDR a high efficiency-high data rate personal communication wireless system, *Proc. of IEEE Vehicular Technology Conference*, Spring, 2000.
5. M. Andrews, K. Kumaran, K. Ramanan, A. L. Stolyar, R. Vijayakumar, and P. Whiting, Providing quality of service over a shared wireless link, *IEEE Communications Magazine*, 39 (2), 150–154, 2001.
6. R. Agrawal, V. Subramanian, and R. Berry, Joint scheduling and resource allocation in CDMA systems, in *Proc. WiOpt*, Cambridge, UK, March 2004. Journal version under submission.
7. J. Huang, V. Subramanian, R. Berry, and R. Agrawal, Joint scheduling and resource allocation in uplink OFDM systems for broadband wireless access networks, *IEEE Journal on Selected Areas in Communications*, 27 (2), 226–234, 2009
8. J. Huang, V. G. Subramanian, R. Agrawal, and R. Berry, Downlink scheduling and resource allocation for OFDM systems, *IEEE Transactions on Wireless Communications*, 8 (1), 288–296, 2009.
9. A. L. Stolyar, Maximizing queueing network utility subject to stability: greedy primal-dual algorithm, *Queueing systems*, 50, 401–457, 2005.
10. A. L. Stolyar, MaxWeight scheduling in a generalized switch: State space collapse and workload minimization in heavy traffic, *Annals of Applied Probability*, 14 (1), 1–53, 2004.
11. L. Tassiulas and A. Ephremides, Dynamic server allocation to parallel queue with randomly varying connectivity, *IEEE Transactions on Information Theory*, 39, 466–478, 1993.
12. H. Jin, R. Laroia, and T. Richardson, Superposition by position, *2006 IEEE Information Theory Workshop*, March 2006.
13. J. Lee, H. Lou, and D. Toumpakaris, Analysis of phase noise effects on time-direction differential OFDM receivers, *IEEE GLOBECOM*, 2005
14. L. Hoo, B. Halder, J. Tellado, and J. Cioffi, Multiuser transmit optimization for multicarrier broadcast channels: asymptotic FDMA capacity region and algorithms, *IEEE Transactions on Communications*, 52 (6), 922–930, 2004.

15. C. Y. Wong, R. S. Cheng, K. B. Letaief, and R. D. Murch, Multiuser OFDM with adaptive subcarrier, bit, and power allocation, *IEEE Journal on Selected Areas in Communication*, 17 (10), 1747–1758, 1999.

16. Y. Zhang and K. Letaief, Adaptive resource allocation and scheduling for multiuser packet-based OFDM networks, *2004 IEEE ICC*, 2949–2953, 2004.

17. J. Jang and K. Lee, Transmit power adaptation for multiuser OFDM systems, *IEEE Journal on Selected Areas in Communications*, 21 (2), 171–178, 2003.

18. Y. Zhang and K. Letaief, Multiuser adaptive subcarrier-and-bit allocation with adaptive cell selection for OFDM systems, *IEEE Transactions on Wireless Communications*, 3 (5), 1566–1575, 2004.

19. T. Chee, C. Lim, and J. Choi, Adaptive power allocation with user prioritization for downlink orthogonal frequency division multiple access systems, *ICCS 2004*, 210–214, 2004.

20. H. Yin and H. Liu, An efficient multiuser loading algorithm for OFDM-based broadband wireless systems, *IEEE Globecom*, 2000.

21. K. Seong, M. Mohseni, and J. M. Cioffi, Optimal resource allocation for OFDMA Downlink Systems, in *IEEE ISIT*, 1394–1398, 2006.

22. M. Tao, Y. C. Liang and F. Zhang, Resource allocation for delay differentiated traffic in multiuser OFDM systems, *IEEE Transactions on Wireless Communications*, 7 (6), 2190–2201, 2008.

23. W. Jiao, L. Cai, and M. Tao, Competitive scheduling for OFDMA systems with guaranteed transmission rate, *Elsevier Computer Communications*, special issue on Adaptive Multicarrier Communications and Networks. Avaliable online 29 August 2008.

24. Z. Han, Z. Ji, and K. Liu, Fair multiuser channel allocation for OFDMA networks using nash bargaining solutions and coalitions, *IEEE Transactions on Communications*, 53 (8), 1366–1376, 2005.

25. S. Pfletschinger, G. Muenz, and J. Speidel, *Efficient subcarrier allocation for multiple access in OFDM systems,* in *7th International OFDM-Workshop 2002 (InOWo'02)*, 2002.

26. Y. Ma, Constrained rate-maximization scheduling for uplink OFDMA, in *Proc. IEEE MILCOM*, pp. 1–7, October 2007.

27. B. Da and C. Ko, Dynamic subcarrier sharing algorithms for uplink OFDMA resource allocation, in *Proc. 6th ICICS*, pp. 1–5, 2007.

28. K. Kim, Y. Han, and S. Kim, Joint subcarrier and power allocation in uplink OFDMA systems, *IEEE Comms. Letters*, 9 (6), 526–528, 2005.

29. C. Ng and C. Sung, Low complexity subcarrier and power allocation for utility maximization in uplink OFDMA systems, *IEEE Transactions on Wireless Communications*, 7 (5) Part 1, 1667–1675, 2008.

30. K. Kwon, Y. Han, and S. Kim, Efficient subcarrier and power allocation algorithm in OFDMA uplink system, *IEICE Transactions on Communication*, 90 (2), 368–371, 2007.

31. L. Gao and S. Cui, Efficient subcarrier, power, and rate allocation with fairness consideration for OFDMA uplink, *IEEE Transactions on Wireless Communications*, 7 (5) Part 1, 1507–1511, 2008.

32. C.-S. Chiu and C.-C. Huang, Combined partial reuse and soft handover in OFDMA downlink transmission, *IEEE VTC*, pp. 1707–1711, April 2008.

33. N. Damji and T. Le-Ngoc, Dynamic resource allocation for delay-tolerant services in downlink OFDM wireless cellular systems, *IEEE ICC*, 5, 3095–3099, 2005.

34. G. Li and H. Liu, Downlink dynamic resource allocation for multi-cell OFDMA system, *IEEE VTC*, 3, 1698–1702, 2003.

35. H. Lei, X. Zhang, and Y. Wang, Real-time traffic scheduling algorithm for MIMO-OFDMA systems, *IEEE ICC*, pp. 4511–4515, April 2008.

36. H. Xiaoben and K. Valkealahti, On distributed and self-organized inter-cell interference mitigation for OFDMA downlink in imt-advanced radio systems, *IEEE VTC*, pp. 1736–1740, January 2007.

37. I.-K. Fu and W.-H. Sheen, An analysis on downlink capacity of multi-cell OFDMA systems under randomized inter-cell/sector interference, *IEEE VTC*, pp. 2736–2740, March 2007.

38. L. Shao and S. Roy, Downlink multicell MIMO-OFDM: an architecture for next generation wireless networks, *IEEE WCNC*, 2, 1120–1125, 2005.

39. T. Thanabalasingham, S. Hanly, L. Andrew, and J. Papandriopoulos, Joint allocation of subcarriers and transmit powers in a multiuser ofdm cellular network, *IEEE ICC*, 1, 269–274, 2006.

40. T. M. Cover and J. A. Thomas, Elements of information theory," Second edition, Hoboken, NJ. Wiley-Interscience [John Wiley & Sons], 2006.

41. D. Bertsekas, *Nonlinear Programming*, Second edition, Belmont, MA: Athena Scientific, 1999.
42. J. Mo and J. Walrand, Fair end-to-end window-based congestion control, *IEEE/ACM Transactions on Networking*, 8 (5), 556–567, 2000.
43. L. Li and A. Goldsmith, Capacity and optimal resource allocation for fading broadcast channels. I. Ergodic capacity, *IEEE Transactions on Information Theory*, 47 (3), 1083–1102, 2001.
44. M. Medard, The Effect Upon Channel Capacity in Wireless Communications of Perfect and Imperfect Knowledge of the Channel, *IEEE Transactions on Information Theory*, 46 (3), 935–946, May 2000.
45. S. Boyd and L. Vanderberghe, *Convex Optimization*. Cambridge University Press, Cambridge, UK, 2004.
46. E. D. Nering and A. W. Tucker, *Linear Programs and Related Problems*. Academic Press Inc., 1993.
47. IEEE 802.16e-2005 and IEEE Std 802.16-2004/Cor1-2005, http://www.ieee802.org/16/
48. M. Chen and J. Huang, Optimal resource allocation for OFDM uplink communication: A primal-dual approach, in *Conference on Information Sciences and Systems (CISS)*, Princeton University, March 2008.
49. X. Zhang, L. Chen, J. Huang, M. Chen, and Y. Zhao, Distributed and optimal reduced primal-dual algorithm for uplink OFDM resource allocation, *IEEE Conference on Decision and Control*, Shanghai, People's Republic of China, December 2009.

Chapter 7

Spectrum Efficiency in OFDMA

Thomas Magesacher

Contents

7.1   Introduction... 165
7.2   System Model... 169
7.3   Transmit Spectra.. 171
7.4   Egress Reduction Techniques... 175
      7.4.1   Filtering.. 175
      7.4.2   Power Loading.. 175
      7.4.3   Transmit Windowing... 176
              7.4.3.1   Intersymbol Windowing....................................... 176
              7.4.3.2   Intrasymbol Windowing....................................... 177
              7.4.3.3   Nyquist Windowing... 180
      7.4.4   Spectral Compensation.. 181
7.5   Comparison 184
7.6   Conclusion and Open Issues... 186
Symbols.. 186
Abbreviations.. 187
References... 187

7.1 Introduction

Spectral efficiency or bandwidth efficiency is an important property of modern communication systems. In general, bandwidth can be a very costly resource. For example, the 3G mobile communication licenses in Germany and in the United Kingdom were auctioned off for an average price of roughly €850/Hz bandwidth. Furthermore, spectral efficiency can be the key prerequisite to applying orthogonal frequency-division multiple access (OFDMA) in certain setups. OFDM-based overlay systems for aviation communication, for example, share a frequency band with the already existing systems. High spectral efficiency is required to allow communication without degrading the performance of in-place systems by temporarily exploiting those parts of the common frequency band that are currently not used [1].

Spectral leakage, that is, leakage of spectral energy into undesired parts of the spectrum, can significantly lower spectral efficiency in practical systems. Figure 7.1 illustrates the importance of spectral leakage in OFDMA systems. Like all communication systems, also OFDMA-based systems have to maintain spectral compatibility with systems occupying neighboring bands—both in downlink (DL) and in uplink (UL) directions (cf. Figure 7.1a). Spectral energy outside the band covered by the subcarriers is eliminated by continuous-time postfiltering, which constructs the analog transmit signal by suppressing the mirror images of the digital-to-analog converter's output. A primary design goal in practical systems is to keep the requirements for the continuous-time postfilter fairly relaxed for the sake of hardware complexity and power consumption. Therefore, often up to a fifth of the available subcarriers are nulled to form guard bands with reduced spectral energy close to the two band edges.

Besides out-of-band emission, spectral leakage causes interference among users. In an ideal OFDMA system, subcarriers and thus users are orthogonal at the receiver(s). In practice, however, timing offsets, frequency offsets, phase noise, and so on disturb the perfect orthogonality, which causes spectral leakage and thus reduces spectral efficiency. Even when a *synchronous UL timing* scheme is employed, users are initially not synchronized, which, in general, results in severe intercarrier interference among adjacent subcarriers belonging to different users (or equivalently, to "out-of-band emission" caused by one user in another user's band) and has a detrimental effect on the performance of synchronization techniques. Typically, this problem is approached via a subband allocation scheme with frequency-domain guard bands (cf. Figure 7.1b), which allows the base station to separate adjacent users via filterbanks. However, guard bands reduce the exploitable bandwidth and thus lower the system's spectral efficiency.

When general subcarrier allocation and *asynchronous UL timing* are applied, the orthogonality among users is lost, which causes significant interference and thus reduces the achievable throughput in the UL direction. Although an asynchronous timing scheme in the UL direction annihilates the inherent orthogonality among users, asynchronous systems may be of interest in certain applications. For example, asynchronous timing allows very simple user terminals since all the complexity is shifted to the base station (which performs parameter detection and user alignment in time and frequency). Furthermore, there is no need to feed timing information back to the users. Another motivation is the desire to exploit the resource-scheduling capabilities of OFDMA to the full, which may lead to growing interest in general subcarrier assignment together with asynchronous transmission in future and may thus also emphasize the challenge of keeping out-of-band emissions low.

The out-of-band emission of a rectangularly windowed fast Fourier transform (FFT)-based OFDM subcarrier or a band of subcarriers is significant. Figure 7.2 depicts the power spectral densities (PSDs) caused by a spectral neighbor using OFDM versus distance in subcarriers. The smaller the distance to the neighbor and the wider the neighbor's band, the more significant the spectral leakage becomes. Assuming that the neighbor occupies more than just a few subcarriers, roughly 15 subcarriers closest to the neighboring band experience interfering PSD levels that are only 10–20 dB below the nominal PSD level.

An elegant aspect of OFDM as well as the basis for OFDMA is the fact that, in a perfectly synchronized scenario, the spectra of adjacent subcarriers overlap while orthogonality in time-dispersive

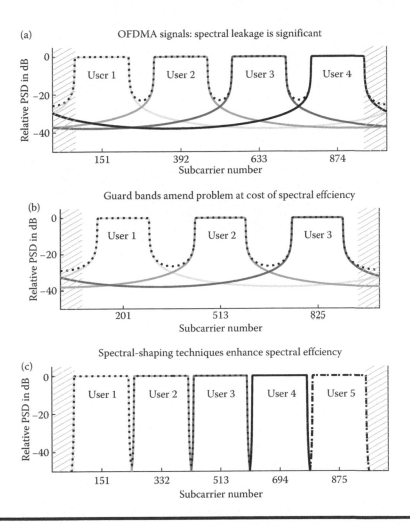

Figure 7.1 Impact of spectral leakage on spectral efficiency in OFDMA: (a) Spectral leakage of standard FFT-based OFDMA signals is significant and may cause two problems: First, spectral compatibility with coexisting systems needs to be assured, which requires power-backoff on subcarriers close to band edges (critical frequency regions are marked by pattern) both in OFDMA-UL and OFDMA-DL; second, spectral leakage causes interference among users both in the DL direction as well as in the UL direction of practical schemes, where imperfect synchronization destroys orthogonality. (b) The problem can be amended to a certain extent via frequency-domain guard bands at the cost of spectral efficiency (i.e., accommodating only three users instead of four in the example at hand). (c) Spectral-shaping techniques enhance spectral efficiency of FFT-based OFDMA (accommodating five users instead of four in the example at hand). ($N = 1024$ subcarriers in total. Each subband occupies 165 subcarriers. Dotted line: aggregate power spectral density (PSD) of all users.)

channels is preserved. However, in the presence of timing errors and frequency errors, orthogonality is compromised and intercarrier interference occurs. Clearly, the better the alignment in time and frequency (i.e., the better the synchronization of the interfering and the impaired system), the lower the interference level. Figure 7.3 depicts the worst-case interference power level (relative to the

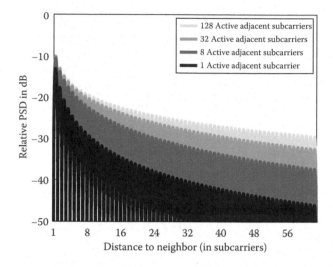

Figure 7.2 Spectral leakage PSD caused by an asynchronous neighbor versus distance (N = 1024).

power of the disturbing signal) caused by a *timing offset* versus distance to the disturbing neighbor's band edge for different band widths of the neighboring OFDM system [2]. A derivation similar to the one presented in Ref. [2] yields the worst-case interference levels caused by a *frequency offset* shown in Figure 7.4.

To conclude, the worst-case interference levels can lie above the interference level suggested by the PSD. However, it should be noted that the actual values of timing offset and frequency offset, which lead to the worst-case interference levels, vary with distance from the neighbor. Thus, it is

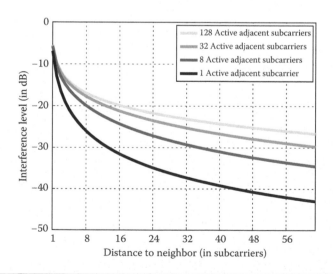

Figure 7.3 Worst-case intercarrier interference caused by a neighbor (occupying a band of 1, 8, 32, or 128 adjacent subcarriers) with *timing offset* versus distance according to Ref. [2] (N = 1024).

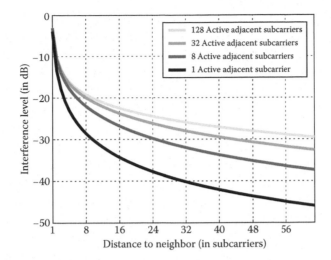

Figure 7.4   Worst-case intercarrier interference caused by a neighbor (occupying a band of 1, 8, 32, or 128 adjacent subcarriers) with *frequency offset* versus distance ($N = 1024$).

very unlikely to experience the worst-case interference levels over a wide range of subcarriers at the same time. The PSD is thus a reasonable measure for both out-of-band emission and intercarrier interference levels experienced by the impaired system.

Clearly, there is a trade-off between throughput and out-of-band emission: even the most stringent spectral constraints can be met by reducing the transmit power on subcarriers, which in turn reduces the achievable throughput. A viable approach, and the central topic of this chapter, are physical-layer remedies to reduce spectral leakage and thus increase the exploitable bandwidth (cf. Figure 7.1c). An optimization-based framework for emission-reduction techniques is introduced. The goal is to exploit the available bandwidth in the best possible way (in the sense of maximizing the throughput) while obeying a PSD limit.

The chapter's focus is on methods that preserve orthogonality among subcarriers in time-dispersive (frequency selective) channels. An alternative approach that has received a lot of attention is the design of pulse shapes (or equivalently, of basis functions) aiming at better spectral confinement in both time-domain and frequency-domain [3–8]. Most of these designs aim at basis functions that are orthogonal at the transmitter. However, once they pass the time-dispersive channel, orthogonality is lost, which results in both intercarrier interference and intersymbol interference.

7.2 System Model

In order to formalize the ideas, a matrix-based model of a standard FFT-based multicarrier system with N subchannels is introduced in the following. The ith time-domain OFDM transmit symbol $t_i \in \mathbb{C}^{N' \times 1}$ can be written as vector

$$t_i = \underbrace{Z_{\text{add}} W^{\text{H}} \text{diag}\{\sqrt{p}\}}_{A} x_i, \qquad (7.1)$$

where the vector $x_i \in \mathbb{C}^{N \times 1}$ contains the transmit data, diag$\{\cdot\}$ denotes a diagonal matrix with the argument vector on the main diagonal, and $p \in \mathbb{R}_+^{N \times 1}$ contains the power values assigned to individual subcarriers. The matrix $W \in \mathbb{C}^{N \times N}$ defined as

$$W[n,k] = \frac{1}{\sqrt{N}} e^{-j2\pi(n-1)(k-1)/N}, \quad n,k = 1,\ldots,N$$

is the normalized discrete Fourier transform (DFT) matrix. The matrix $Z_{\text{add}} \in \mathbb{R}^{N' \times N}$ defined by

$$Z_{\text{add}} = \begin{bmatrix} 0_{L \times (N-L)} & I_L \\ & I_N \\ I_{L'} & 0_{L' \times (N-L')} \end{bmatrix} \tag{7.2}$$

introduces cyclic extensions of length L and L' at the beginning and at the end of each symbol, respectively. For the standard transmitter, $L = L_{\text{cp}}$ and $L' = L_{\text{cs}}$ holds, that is, the lengths of the extensions are equal to the lengths L_{cp} and L_{cs} of the cyclic prefix (CP) and the cyclic suffix (CS), respectively. The total length of each such transmit symbol is $N' = N + L + L'$. The matrix A describes the transmit information processing.

In general, consecutive OFDM symbols may be transmitted with an overlap of L_w samples (cf. intersymbol windowing), which yields the transmit sequence

$$t(n) = \begin{cases} t_{\lfloor n/N'' \rfloor}[(n \bmod N'') + 1] \\ \quad + t_{\lfloor n/N'' \rfloor - 1}[(n \bmod N'') + 1 + N''], & (n \bmod N'') = 0,\ldots,L_w - 1, \\ t_{\lfloor n/N'' \rfloor}[(n \bmod N'') + 1], & \text{otherwise,} \end{cases}$$

where $N'' = N' - L_w$ is the effective symbol length accounting for all extensions and overlaps. For the standard transmitter, which transmits symbols without overlap, $N'' = N'$ holds.

Hereinafter, two assumptions are made that greatly simplify further analysis while introducing only a mild loss of generality. First, the data are assumed to be uncorrelated both over time and over subcarriers, that is,

$$\mathsf{E}\{x_k[m]x_\ell[n]\} = 0, \quad k \neq \ell \text{ or } m \neq n, \tag{7.3}$$

which is a reasonable assumption for coded (and interleaved) symbols. Second, the data are assumed to be proper complex Gaussian distributed with zero mean and unit variance:

$$x_k[m] \sim \mathcal{CN}(0,1). \tag{7.4}$$

The Gaussian assumption is only an approximation when finite alphabets are employed. However, it greatly simplifies the computation of the achievable throughput yielding upper bounds and is thus commonly used.

The time-dispersive Gaussian channel performs linear convolution of the transmit signal with the impulse response $h_n, n = 0,\ldots,M$ of length $M+1$ ($h_0 \neq 0$, $h_M \neq 0$, $h_n = 0$ for $n < 0$ and for $n > M$). Hereinafter, $L_{\text{cp}} \geq M$ is assumed, which eliminates intersymbol interference and intercarrier interference after removal of the CP so that the received time-domain symbol can be written as $r_i = Ht_i + n_i$ (assuming perfect synchronization). The noise vector $n_i \sim \mathcal{CN}(0, C_n)$ is a random vector of length N' with zero-mean proper complex Gaussian entries and covariance matrix $C_n = \mathrm{E}\{n_i n_i^{\mathrm{H}}\} \in \mathbb{C}^{N' \times N'}$. The channel convolution matrix $H \in \mathbb{C}^{N' \times N'}$ is defined as

$$H[n, k] = h_{n-k}, \qquad n, k = 1, \ldots, N'.$$

The frequency-domain receive signal at the output of the equalizer is given by

$$y_i = \underbrace{GWZ_{\mathrm{rem}}}_{B} r_i.$$

The matrix $Z_{\mathrm{rem}} \in \mathbb{R}^{N \times (L+N+L'')}$, defined as

$$Z_{\mathrm{rem}} = \begin{bmatrix} 0_{N \times L} & I_N & 0_{N \times L''} \end{bmatrix},$$

purges leading and trailing cyclic extensions. The diagonal matrix $G \in \mathbb{C}^{N \times N}$ performs linear per-subchannel equalization and the matrix B describes the receive information processing.

A performance measure that allows a fair comparison of different system variants is the normalized information rate

$$R = \frac{N}{N''} \sum_{k \in \mathcal{S}_i} \log_2 \left(1 + \frac{\left(BHA(BHA)^{\mathrm{H}} \right)[k, k]}{\left(BC_n B^{\mathrm{H}} \right)[k, k]} \right) \tag{7.5}$$

in bits per multicarrier symbol, where $\mathcal{S}_i$ denotes the set of all information-carrying subcarriers. Note that R/N is a measure of the number of bits per subchannel (of width $1/T_{\mathrm{sym}}$) per multicarrier symbol (of length T_{sym}), which corresponds to the number of bits per second per Hertz.

7.3 Transmit Spectra

Open literature on spectrum-shaping techniques includes several measures for spectral energy, which are often just loosely referred to as "spectra." This section presents a more careful view on these measures and their suitability. Furthermore, PSD is postulated as the actual measure of interest.

In the following, different spectral measures of a single-frequency finite-length complex exponential, which is the basis function (subcarrier) of standard rectangularly windowed OFDM, are discussed. In discrete time, this signal can be written as vector $w_k \in \mathbb{C}^{N \times 1}$ given by

$$w_k[n] = \frac{1}{\sqrt{N}} e^{j2\pi(n-1)(k-1)/N}, \quad n = 1, \ldots, N,$$

where $k \in \{1, \ldots, N\}$ denotes the subcarrier number and $(k-1)/N$ is the subcarrier's frequency. Note that the inverse DFT (IDFT) matrix W^{H} contains the basis functions: $W^{\mathrm{H}} = \begin{bmatrix} w_1 & w_2 & \cdots & w_N \end{bmatrix}$. The corresponding continuous-time version is given by

$$w_k(t) = \frac{1}{T} e^{j2\pi F_k t}, \qquad 0 \le t < T,$$

where $F_k = (k-1)/T$ is the frequency in Hertz and T is the symbol length in seconds.

The Fourier transform $W_k(f)$ of $\boldsymbol{w}_k$ is given by

$$W_k(f) = \sum_n w_k[n]\,e^{-j2\pi fn} = \frac{1}{\sqrt{N}}\frac{\sin(\pi N(f - k/N))}{\sin(\pi(f - k/N))}e^{-j\pi(f-k/N)(N-1)} \tag{7.6}$$

and features in all spectral measures in one way or another. A frequently used measure is $(1/\sqrt{N})|W_k(f)|$ depicted in Figure 7.5. Scaling by $1/\sqrt{N}$ yields $(1/\sqrt{N})|W_k(k/N)| = 1$ [note that $\lim_{f \to k/N}(\sin(\pi N(f - k/N))/\sin(\pi(f - k/N))) = N$].

Another measure is the Fourier transform $W_k(F)$ of $w_k(t)$, given by

$$W_k(F) = \int_t w_k(t)e^{-j2\pi Ft}\,dt = \frac{\sin(\pi(F - F_k)T)}{\pi(F - F_k)T}e^{-j\pi(F-F_k)T},$$

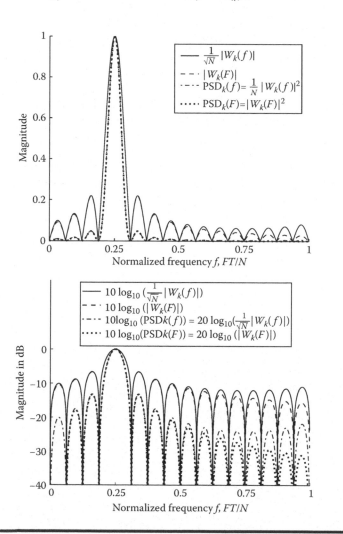

Figure 7.5 "Spectra" of finite-length complex exponential (OFDM basis function): Fourier transforms and PSDs. $N = 16$, $L_{cp} = 0$, $k = 5$, $p[5] = 1$.

whose absolute yields the classical "sinc-F" function*

$$|W_k(F)| = \text{sinc}((F - F_k)T).$$

Although both $|W_k(f)|$ and $|W_k(F)|$ are proportional to the energy distribution over frequency, the measure actually describing it is PSD, which is discussed next.

The blockwise processing of OFDM symbols results in a transmit sequence $t(n)$, which can be modeled as a cyclostationary random process [9]. The autocorrelation sequence

$$r(n, m) = \text{E}\{t(n)t^*(n - m)\}, \quad n, m = -\infty, \ldots, \infty \tag{7.7}$$

of $t(n)$ thus depends both on the time instant n and on the lag m and is periodic in n with period N'': $r(n, m) = r(n + N'', m)$. The averaged autocorrelation sequence [10] is given by

$$r(m) = \frac{1}{N''} \sum_{n=n_0}^{n_0+N''-1} r(n, m), \tag{7.8}$$

where the average is taken over a period of N'' samples starting at an arbitrary time instant n_0. The transmit PSD is the Fourier transform of $r(m)$:

$$\text{PSD}(f) = \sum_m r(m)\, \text{e}^{-\text{j}2\pi fm}. \tag{7.9}$$

Let us now look at the PSD of an OFDM signal when only a single subcarrier, say subcarrier number k, is in use. The transmit symbols are then given by $t_i = w_k\sqrt{\text{p}[k]}x_i[k]$. Assuming $L_{\text{cp}} = L_{\text{cs}} = 0$ and no overlap of consecutive blocks, $N'' = N$ holds and the averaged autocorrelation function is given by

$$r_k(m) \overset{(7.7),(7.8)}{=} \frac{1}{N} \sum_n w_k[n]\sqrt{\text{p}[k]}\, \underbrace{\text{E}\{x_i[k]x_i^*[k]\}}_{\overset{(7.4)}{=}\,1} \sqrt{\text{p}[k]}w_k^*[n - m],$$

which yields the transmit PSD

$$\text{PSD}_k(f) \overset{(7.9)}{=} \sum_m r_k(m)\, \text{e}^{-\text{j}2\pi fm} = \frac{\text{p}[k]}{N} \underbrace{\sum_n w_k[n]\text{e}^{-\text{j}2\pi fn}}_{\overset{(7.6)}{=}\, W(f)} \underbrace{\sum_\ell w_k^*[\ell]\text{e}^{\text{j}2\pi f\ell}}_{\overset{(7.6)}{=}\, W^*(f)} = \frac{\text{p}[k]}{N}|W_k(f)|^2$$

Analogously, the PSD of $w_k(t)$ is

$$\text{PSD}_k(F) = |W_k(F)|^2.$$

---

* $\text{sinc}(x) \overset{\wedge}{=} \sin(\pi x)/(\pi x)$

Figure 7.5 depicts Fourier transforms and PSDs both in linear and in logarithmic scales. The Fourier transforms are only proportional to the square root of energy distributed over frequency and thus not fully adequate as approximations for actual PSD levels. Still, they may be preferred in some settings since they are linear functions of the data modulating the basis functions. The sidelobes of $W_k(F)$ and $\text{PSD}_k(F)$ decay steadily with distance from the mainlobe's center frequency k/T. The periodicity of $W_k(f)$ and $\text{PSD}_k(f)$ causes the sidelobes to rise again for frequencies more than 0.5 away from k/N.

The larger the number of subcarriers in a system, the faster the decay of the sidelobes' magnitudes, as the basis functions depicted in the band center in Figure 7.6 show. At a first glance, spectral sidelobes do not seem to be a problem for large systems (e.g., $N = 2048$). However, the larger the system, the greater the number of subcarriers that contribute to the aggregate out-of-band PSD. In Figure 7.6, subcarriers within the normalized frequency range $k/N \in [1/8, 7/8]$ are in-band, while the rest of the carriers are out-of-band. Hereinafter, the out-of-band region is referred to as the union of all subbands within the normalized frequency range $[0, 1)$ that are not used by active subcarriers. Typically, subcarriers in the vicinity of the band edges (i.e., subcarriers with normalized frequencies close to 0 and close to 1) cause spectral energy that is critical in terms of spectral compatibility. Even in very large systems, the aggregate out-of-band PSD decays only slowly with distance from the in-band region.

Note that in the course of discrete-time to continuous-time conversion, the spectra and PSDs of discrete-time signals depicted in the previous figures are repeated in frequency and weighted by a function depending on the actual interpolation (e.g., a simple hold-element producing a step-like waveform corresponds to weighting with a sinc function). "True" out-of-band emission, that is, spectral energy outside the band corresponding to frequencies $[0, 1)$ caused by these replica, has to be taken care of by continuous-time filtering (or, equivalently, higher-quality interpolation in the course of discrete-time to continuous-time conversion). The critical frequency part is the band edges where the continuous-time filter is faced with the challenge of providing a brickwall-like magnitude

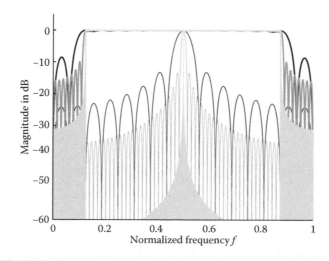

Figure 7.6 PSDs PSD(f) of basis functions (thin lines) and aggregate PSDs (thick lines) for different numbers of subcarriers: $N = 16$ (black), $N = 64$ (gray), and $N = 2048$ (light gray). Basis functions are depicted for frequency $k/N = 1/2$. Aggregate PSDs include all subcarriers with frequencies in the in-band range $k/N \in [1/8, 7/8]$.

response. Hereinafter, the focus is on techniques to shape the spectrum in the above-defined out-of-band part of the frequency range $[0, 1)$, which can be influenced by discrete-time processing and greatly relaxes the requirements on the continuous-time filter.

In case the data streams transmitted on individual subcarriers are mutually uncorrelated, the aggregate PSD, as for example depicted in Figure 7.6, is the sum of the PSDs of all in-band subcarriers. In case the data are correlated over subcarriers, the PSD is computed according to Equation 7.9. In order to cast optimization problems, a matrix-based formulation for the transmit PSD $q \in \mathbb{R}_+^{Q \times 1}$, given at Q points in the normalized frequency range $[0, 1)$, as a function of the transmit signal's autocorrelation matrix

$$C_t \triangleq \mathrm{E}\{t_i t_i^{\mathrm{H}}\} \overset{(7.1),(7.3),(7.4)}{=} AA^{\mathrm{H}}$$

proves useful and can be written as

$$q = Q \operatorname{vec}(C_t). \tag{7.10}$$

The transform matrix Q computes the averaged autocorrelation function and subsequently takes a length-Q DFT according to Equations 7.8 and 7.9, respectively, and vec $(\cdot)$ stacks the columns of its matrix argument into one column vector.

7.4 Egress Reduction Techniques

7.4.1 Filtering

Filtering may be the most straightforward approach to reducing out-of-band emission. Filtering of the transmit sequence $t(n)$, or an oversampled version of it, results in an effective impulse response of transmit filter (length $L_f + 1$) and channel (length $M + 1$) of length $M + L_f + 1$. In order to preserve orthogonality (avoid ISI and ICI) in time-dispersive channels, the CP has to be extended to length $L_{cp} = M + L_f$. Hence, one goal of the filter design for this purpose is clearly to keep L_f low. Apart from the desired suppression in the stopband, the filter may introduce an undesired ripple in the passband, which has a negative impact on the bit error rate performance. A predistortion in frequency-domain (before the IDFT) can be employed to mitigate this ripple.

7.4.2 Power Loading

Instead of best-effort emission reduction by simply nulling ("turning off") subcarriers close to band edges, a more advanced approach is to maximize the throughput under emission limits. Finding the power values $\mathbf{p}^{(\mathrm{opt})}$ that maximize R given by Equation 7.5 under a PSD constraint can be formulated as

$$\mathbf{p}^{(\mathrm{opt})} = \arg \max_p R \\ \text{subject to } q \leq m, \tag{7.11}$$

where $m \in \mathbb{R}_+^{Q \times 1}$ denotes the PSD mask and q is given by Equation 7.10. Problem 7.11 can be cast in convex form [11] and is, in essence, a constrained waterfilling problem. For the power-loading transmitter, all subcarriers potentially carry information ($\mathcal{S}_i = \{1, \ldots, N\}$). Each symbol is extended by CP ($L = L_{cp}$) and CS ($L' = L_{cs}$) before transmission without overlap ($L_w = 0$). The receiver purges the leading $L - L_{cp}$ and the trailing $L'' = L_{cs}$ samples. Note that through the choice $S_i = \{1, \ldots, N\}$ no subcarriers are *a priori* nulled—power loading will pick the best trade-off between a subcarrier's contribution to R given by Equation 7.5 and the spectral emission it causes.

7.4.3 Transmit Windowing

In general, the term *windowing* is used to denote multiplication of a signal by a finite-length weighting function—the so-called window (or window function). Based on the observation that the sharp transitions in time-domain at the symbol borders due to changing data cause high-frequency components in the frequency-domain, it is intuitively clear that a good window function should mitigate these transitions by scaling down the power at symbol borders. In fact, it can be shown that the quality of a continuous-time window function with respect to its out-of-band emission is proportional to its number of continuous derivatives. It is not surprising that the rectangular window (that is, no windowing) used in standard OFDM, which is not even continuous itself, has rather high out-of-band components.

It is worth noting that transmit windowing is equivalent to pulse-shape design. Hereinafter, the focus is on window designs that preserve orthogonality among subcarriers in time-dispersive channels, which may require window-related receive processing. Dedicated design of pulse shapes (or equivalently, basis functions) usually aims at improving the confinement of the pulse in time-domain and frequency-domain in order to enhance its immunity to time offsets and frequency offsets. Hereinafter, the goal is to maximize the achievable information rate under a PSD constraint.

7.4.3.1 Intersymbol Windowing

Intersymbol transmit windowing [11,12] introduces an additional cyclic extension of L_w at both the beginning and the end of a multicarrier symbol before shaping these extended parts with a window function, as depicted in Figure 7.7. The overall cyclic extensions of the beginning and end of each symbol are thus $L = L_{cp} + L_w$ and $L' = L_{cs} + L_w$, respectively. This kind of windowing appears in several communication standards, for example, in wireline communications [13]. A windowed

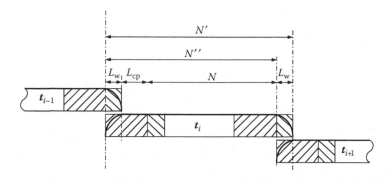

Figure 7.7 *Intersymbol windowing*: **Consecutive blocks are cyclically extended and only these extended parts are windowed. Consecutive blocks overlap and the sum of overlapping blocks is transmitted. Orthogonality among subcarriers in time-dispersive channels is preserved. (For simplicity, $L_{cs} = 0$.)**

transmit symbol of length N' can be written as

$$t = \underbrace{\mathrm{diag}\{u\}Z_{\mathrm{add}}W^{\mathrm{H}}\mathrm{diag}\{\sqrt{p}\}}_{A}x,$$

where the vector $u \in \mathbb{R}^{N' \times 1}$ is the time-domain transmit window which obeys

$$u[k] = \begin{cases} u[N' - k + 1], & k = 1, \ldots, L_{\mathrm{w}}, \\ 1, & k = L_{\mathrm{w}} + 1, \ldots, N' - L_{\mathrm{w}}. \end{cases} \tag{7.12}$$

The excess intervals of consecutive OFDM symbols may overlap (cf. Figure 7.7), which justifies the name "intersymbol windowing." The effective receive-block length is thus $N'' = L_{\mathrm{w}} + L_{\mathrm{cp}} + N + L_{\mathrm{cs}}$ samples and the length of the suffix removed by the receiver is $L'' = L_{\mathrm{cs}}$. Compared to standard OFDM, the symbols are thus L_{w} samples longer.

As an example, consider two practically relevant window shapes: the linear-slope window

$$u[k] = k/(L_{\mathrm{w}} + 1), \quad k = 1, \ldots, L_{\mathrm{w}} \tag{7.13}$$

and the root raised-cosine window

$$u[k] = \left(\frac{1}{2} \cos\left(\frac{\pi k}{L_{\mathrm{w}} + 1} - \pi \right) + \frac{1}{2} \right)^{1/2}, \quad k = 1, \ldots, L_{\mathrm{w}}, \tag{7.14}$$

for which all derivatives are continuous. While the root raised-cosine window keeps the average per-sample power of $t(n)$ constant, the linear-slope window might allow a more efficient implementation for certain values of L_{w}. Figure 7.8 compares the resulting PSDs of a single subcarrier with and without intersymbol windowing. The two exemplary window shapes perform similarly in terms of spectral leakage. Clearly, there is a trade-off between spectral leakage and window length.

Intersymbol windowing requires joint optimization of both window function u and power values p with R as an objective function, that is,

$$\{u^{(\mathrm{opt})}, p^{(\mathrm{opt})}\} = \arg\max_{u,p} R$$
$$\text{subject to } q \leq m,$$

which is a nonconvex problem. All subcarriers are used for the transmission of information ($\mathcal{S}_{\mathrm{i}} = \{1, \ldots, N\}$). Note that L_{w} is an optimization parameter concealed in u. It turns out that simple window shapes, such as the linear-slope window or the root raised-cosine window, often yield results that are close to those achieved with the optimal window shape [11]. In the sequel, the linear-slope window of optimal length is used together with optimal power loading (in the sense of maximizing R/N) according to Equation 7.11 for comparison with other methods.

7.4.3.2 Intrasymbol Windowing

In contrast to intersymbol windowing, an intrasymbol window does not extend the symbol but weights the length-N symbol before cyclic extension [14]. A transmit symbol can be written as

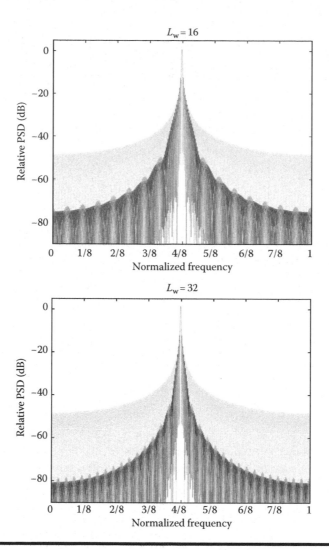

Figure 7.8 **PSDs of single subcarriers without windowing (light gray) and intersymbol windowing: linear-slope window given by Equation 7.13 (gray), root raised cosine window given by Equation 7.14 (black). $N = 256$, $L_w = 16$ (top), $L_w = 32$ (bottom).**

vector

$$t = Z_{\text{add}} \ \text{diag}\{u\} \, W^{\text{H}} \text{diag}\{\sqrt{p}\}x, \tag{7.15}$$

where $u \in \mathbb{C}^{N \times 1}$ is the intrasymbol window. After cyclic extension, the extended window $Z_{\text{add}}u$ itself has the cyclic-prefixed property (cf. Figure 7.9). This type of window is referred to as intrasymbol window—its extent is limited to a single symbol and no overlap occurs ($L_w = 0$) and all subcarriers are used for the transmission of information ($S_i = \{1, \ldots, N\}$).

Equivalently to Equation 7.15, a transmit block can also be written as

$$t = \underbrace{Z_{\text{add}} W^{\text{H}} U \ \text{diag}\{\sqrt{p}\}}_{A} x,$$

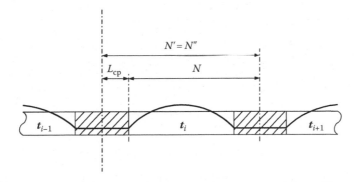

Figure 7.9 *Intrasymbol windowing:* **The entire block is windowed and consecutive blocks do not overlap. Windowing-related receive processing according to Equation 7.16 is required to restore orthogonality among subcarriers. (For simplicity, $L_{cs} = 0$.)**

where the transmit windowing is performed in DFT-domain by the circulant matrix $U \in \mathbb{C}^{N \times N}$ defined as

$$U[n, k] = \frac{1}{\sqrt{N}}(\boldsymbol{W}^{\mathrm{H}}\boldsymbol{u})[((k - n) \bmod N) + 1], \quad n, k = 1, \ldots, N.$$

Intrasymbol windowing destroys the orthogonality of the received basis functions and thus necessitates shaping-related receive processing, which in turn amplifies the noise and thus reduces the signal-to-noise power ratio (SNR). In order to restore the orthogonality among subcarriers, the receiver needs to invert U, which yields the data estimates

$$\widehat{x} = \underbrace{U^{-1}GWZ_{\mathrm{rem}}}_{\widehat{=}B} r. \tag{7.16}$$

Note that the shaping-related receive processing is channel independent (U^{-1} depends only on the window).

Joint optimization of both window function $\boldsymbol{u}$ and power values $\boldsymbol{p}$ with the normalized information rate as an objective function can be formulated as

$$\{\boldsymbol{u}^{(\mathrm{opt})}, \boldsymbol{p}^{(\mathrm{opt})}\} = \operatorname*{arg\,max}_{u,p} R$$
$$\text{subject to } \boldsymbol{q} \leq \boldsymbol{m},$$

where all subcarriers are used for transmission of information ($\mathcal{S}_i = \{1, \ldots, N\}$). The dependence of U^{-1} (and thus of B in R) on $\boldsymbol{u}$ renders the problem nonconvex [15]. An alternative design method that is optimal in the sense of maximizing the mainlobe energy of the transmit basis functions' spectra was suggested in Ref. [16] and a performance comparison was presented in Ref. [17]. Hereinafter, the maximum mainlobe-energy window of optimal length is employed with optimal power loading (in the sense of maximizing R/N) according to Equation 7.11 for comparison. Finally, it should be noted that intrasymbol windowing can be regarded as a special form of pulse shaping yielding non-orthogonal pulses that can, however, be restored at the receiver with comparably low complexity.

7.4.3.3 Nyquist Windowing

Nyquist windowing at the transmitter [18] is a variation of intrasymbol windowing applied together with zero padding. It allows intrasymbol shaping while preserving subcarrier orthogonality at the cost of very simple receive processing. A transmit symbol of length $N' = N + 2L_{nyq} + L_{cp}$ can be written as

$$t = \underbrace{\mathrm{diag}\{u\} Z_{add} W^H \, \mathrm{diag}\{\sqrt{p}\}}_{\hat{=}A} x,$$

where Z_{add} given by Equation 7.2 introduces cyclic extensions of length $L = L_{nyq}$ and $L' = L_{nyq} + L_{cp}$. The window $u \in \mathbb{C}^{N' \times 1}$ fulfills

$$u[k] = \begin{cases} 1 - u[k+N], & k = 1, \dots, 2L_{nyq}, \\ 1, & k = 2L_{nyq} + 1, \dots, N, \\ 0, & k = N + 2L_{nyq} + 1, \dots, N'. \end{cases}$$

Note that the last L_{cp} samples are nulled by the window, which corresponds to zero padding with L_{cp} samples. No overlap of adjacent symbols occurs ($L_w = 0$), as illustrated in Figure 7.10, and all subcarriers are used for the transmission of information ($\mathcal{S}_i = \{1, \dots, N\}$).

Instead of purging samples, the receiver sums leading and trailing transients according to

$$\hat{x} = GW \underbrace{\left[\begin{bmatrix} 0_{N-L_{nyq}, L_{nyq}} \\ I_{L_{nyq}} \end{bmatrix} \; I_N \; \begin{bmatrix} I_{L_{nyq}+L_{cp}} \\ 0_{N-L_{nyq}-L_{cp}, L_{nyq}+L_{cp}} \end{bmatrix} \right]}_{\hat{=}B} r. \tag{7.17}$$

It can be shown that $BHA = I$, that is, the orthogonality among subcarriers is preserved in time-dispersive channels.

Nyquist windowing allows weighting of the entire symbol, which should result in low spectral leakage. On the other hand, each symbol needs to be extended cyclically by $2L_{nyq}$ samples. Furthermore, zero padding makes each symbol L_{cp} samples longer. The transient-adding procedure at the receiver described by Equation 7.17 collects more noise compared to CP removal. On the other hand, zero padding allows for a higher average transmit power since, in contrast to cyclic

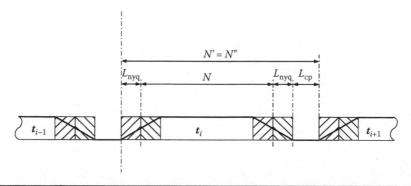

Figure 7.10 *Nyquist windowing*: **A block is cyclically extended and shaped by a Nyquist window. Zero padding together with simple receive processing (adding transients according to Equation 7.17) maintains orthogonality among subcarriers.**

prefixing, no power is wasted when padding zeros. To summarize, it is *a priori* not clear whether Nyquist windowing has an advantage over other windowing techniques, which motivates casting all schemes into a framework for comparison.

7.4.4 Spectral Compensation

An advanced approach, hereinafter referred to as spectral compensation, divides the N subcarriers into two sets: a set $\mathcal{S}_c$ of compensation subcarriers and a set $\mathcal{S}_i = \{1, \dots, N\} \backslash \mathcal{S}_c$ of information-carrying subcarriers. The transmitter modulates each compensation subcarrier $n \in \mathcal{S}_c$ with a linear combination of the data transmitted over a set $\mathcal{S}_r(n) \subseteq \mathcal{S}_i$ of reference subcarriers. Using properly chosen subcarriers as compensation subcarriers leads to a better exploitation of the spectral mask in the sense that the power on some of the information subcarriers can be increased. Spectral compensation does not require any shaping-related processing at the receiver, that is, simple subcarrier-wise equalization is possible since the orthogonality of the received basis functions is preserved. Consequently, the technique conforms with any standard.

A transmit block of N' samples can be written as vector

$$t - \underbrace{Z_{\text{add}} W^H S}_{A} x,$$

where the shaping matrix $S \in \mathbb{C}^{N \times N}$ defined as

$$S[n, k] = \begin{cases} \sqrt{p[n]}, & n = k \in \mathcal{S}_i, \\ s_{n,k}, & n \in \mathcal{S}_c, k \in \mathcal{S}_r(n), \\ 0, & \text{otherwise}, \end{cases} \qquad (7.18)$$

for $n, k = 1, \dots, N$ performs both the power loading for the subcarriers in $\mathcal{S}_i$ and the linear combination of the data for the subcarriers in $\mathcal{S}_c$. Figure 7.11 illustrates the structure of an exemplary shaping matrix. Apart from CP and CS, no additional extension is required, that is, $L = L_{\text{cp}}$ and $L' = L'' = L_{\text{cs}}$. Generally, the receiver processes only the subcarriers in $\mathcal{S}_i$ (a more complex receiver

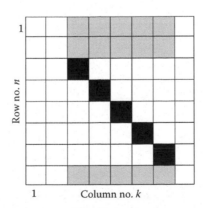

Figure 7.11 Structure of spectral compensation matrix S given by Equation 7.18. Black: information-subcarrier power-scaling coefficients; gray: spectral compensation coefficients; white: zero. Example for $N = 8$; $\mathcal{S}_c = \{1, 2, 8\}$; $\mathcal{S}_r(n) = \mathcal{S}_i, n \in \mathcal{S}_c$.

could exploit the redundancy available through the subcarriers in $\mathcal{S}_c$; however, this would imply shaping-related receive processing, which is not considered here).

To the best of the author's knowledge, the idea of spectral compensation was first mentioned in Refs. [19,20] followed by a considerable research effort to gain performance and insight [21–27]. Spectral compensation techniques can be classified as *active* or *passive*. Active methods re-compute the coefficients in S_i for each block based on the transmit data x_i [21–25]. Thus, active spectral compensation is very similar to tone reservation used for reducing the transmit signal's peak-to-average power ratio [28]. All active methods are based on criteria of more or less *ad hoc* nature, which makes it difficult to include them in any kind of structured framework. In Refs. [21–23], measures proportional to the spectral energy in specified out-of-band regions are minimized using essentially the least squares method. Cosovic et al. [23] additionally introduce some feasibility constraints on the magnitude of the compensation coefficients. Berthold et al. [1] suggest a combination of windowing and spectral compensation. In Ref. [25], an interesting time-domain criterion is proposed, which aims at improving the transmit signal's continuity at symbol borders. Passive methods use the same coefficients S once computed based on a given PSD mask and on second-order statistics of the data for many consecutive blocks and are updated only rarely, for example, to react to severely changing channel conditions [26,27].

Before focusing on the spectral compensation technique, two frequency-domain interpretations are presented that aim at supporting the intuitive understanding.

Interpretation 1: The first interpretation has in fact coined the term spectral compensation. The subcarriers in $\mathcal{S}_i$ remain untouched and function like in an ordinary multicarrier system. Assume for a moment that we only transmit information over the subcarriers in $\mathcal{S}_i$ with the optimal power values for a given PSD mask. Figure 7.12 illustrates the resulting transmit PSD for an example scenario

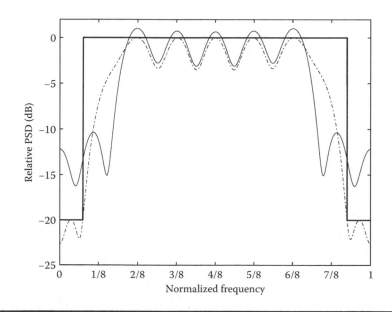

Figure 7.12 *Interpretation 1:* **The uncompensated PSD (solid line), which is the sum of the PSDs of the subcarriers in $\mathcal{S}_i$, exhibits violations of the mask (bold solid line). Proper modulation of the subcarriers in $\mathcal{S}_c$ yields the compensated PSD (dashed-dotted line). Example for $N = 8$, $L_{cp} = 2$, $\mathcal{S}_c = \{1, 2, 8\}$.**

(thin solid line). This uncompensated PSD exhibits considerable violations of the mask (bold solid line). Next, additionally the set $\mathcal{S}_c$ of compensation subcarriers is modulated with data correlated with the information sent over the subcarriers in $\mathcal{S}_i$, such that spectral violations are compensated for (dashed-dotted line). Since the data transmitted over the subcarriers are correlated, the spectral components of individual subcarriers may interfere in a constructive or destructive way and thus remove the mask violations and yield a better exploitation of the spectrum. In mathematical terms, data characterized by the covariance matrix SS^H modulate the N basis functions, which are complex exponentials defined by the columns of $Z_{add}W^H$.

Interpretation 2: The second interpretation is not based on the concept of subcarriers but uses a set of $|\mathcal{S}_i|$ basis functions, which are linear combinations of complex exponentials, specified by the kth columns of $Z_{add}W^H S$, where $k \in \mathcal{S}_i$. This set of basis functions is modulated with uncorrelated data such that the individual spectral components can be added. Figure 7.13 illustrates the individual PSDs corresponding to the information sent via the individual basis functions and the corresponding aggregate PSD (dashed-dotted line). The PSDs of the individual basis functions are neither of identical shape nor symmetric with respect to their center of mass. The particular shape of an individual basis function is the result of the optimization described in the following.

Hereinafter, the focus is on passive spectral compensation, which is a less complex technique than active compensation since the weights are determined only once in a while. The parameters of the optimal spectral-compensation transmitter—optimal in the sense of maximizing Equation 7.5—are given by

$$\{\mathcal{S}_c^{(opt)}, S^{(opt)}\} = \arg\max_{\mathcal{S}_c, S} R, \tag{7.19}$$
$$\text{subject to } q \leq m,$$

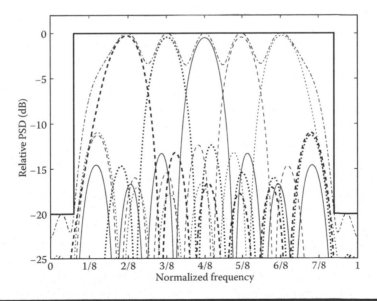

Figure 7.13 *Interpretation 2:* The column $i, i \in \mathcal{S}_i$ numbers of $Z_{add}W^H S$ are the $|\mathcal{S}_i|$ transmit basis functions. Their individual PSDs ($i = 3$: bold-dashed line; $i = 4$: bold-dotted line; $i = 5$: solid line; $i = 6$: dashed line; $i = 7$: dotted line) sum up to the aggregate PSD (dashed-dotted line), which obeys the mask (bold solid line). Example for $N = 8$; $L_{cp} = 2$; $\mathcal{S}_c = \{1, 2, 8\}$.

which is a nonconvex problem. In general, Problem 7.19 can be split into two subproblems: Finding the spectral compensation matrix S for a given subcarrier-set split and finding the subcarrier-set split itself. The latter is a problem general to all techniques using subcarriers for some purpose other than straightforward information transmission (including active spectral compensation and peak-to-average power ratio reduction).

In Ref. [26], a problem formulation using the data's autocorrelation matrix SS^H instead of S is proposed, which leads to a tractable solution for a given set S_c via solving a semidefinite program. Furthermore, a heuristics to find the set S_c of compensation subcarriers is suggested. These results are applied for the comparison with other techniques presented in the next section.

7.5 Comparison

This section presents a brief comparison of the techniques discussed in this chapter. The case study assumes a multicarrier system with $N = 64$ subcarriers, no CS ($L_{cs} = 0$), and different CP lengths. In order to allow for a straightforward interpretation of the results, an additive white Gaussian noise channel (i.e., the channel has no memory and its transfer function is constant) is chosen. The ratio between the receive signal PSD and the noise PSD is thus constant. A moderate SNR of 10 dB is assumed. As PSD limit m, the exemplary mask shown in Figures 7.12 and 7.13 is employed (bold solid line). The convex optimization problems are solved numerically, using dedicated software described in Refs. [29–32].

A simple upper bound for the achievable spectral efficiency R/N is given by

$$R/N \leq \frac{1}{N + L_{cp}} \sum_k \log_2(1 + \text{SNR } m[kQ/N]), \tag{7.20}$$

which assumes "sidelobe-free" basis functions. Consequently, there is no out-of-band emission and all subcarriers can be loaded with power values proportional to the PSD mask.

Figure 7.14 depicts the spectral efficiency R/N in bits per second per Hertz of the different techniques versus L_{cp}. With increasing L_{cp}, the oscillations of the in-band PSD become more pronounced and cause a decay in spectral efficiency R/N. The SNR-penalty of intrasymbol windowing caused by the receive processing (Equation 7.16) is severe and renders intrasymbol windowing inferior compared to the other techniques. Nyquist windowing has an advantage over other schemes for large L_{cp} since longer zero-padded portions of the transmit sequence allow higher power values in a PSD-constrained channel. Spectral compensation outperforms all other methods for small L_{cp}.

In general, it should be noted that the difference in performance of various methods is more pronounced for a low number of subcarriers. As the number of subcarriers increases while the sampling frequency of the system remains constant, both mainlobe width and sidelobe width of the basis functions' spectra decrease, which improves the inherent spectral confinement of the basis functions.

Table 7.1 summarizes the approximate* run-time complexity of different methods in terms of complex-valued multiply-and-add operations per multicarrier symbol. The matrix multiplication

---

* For simplicity, no explicit distinction between additions and multiplications is made. Plain additions are thus sometimes counted as multiply-and-add operations. Furthermore, certain parameters and parameter combinations allow a reduction of the nominal complexity stated in Table 7.1.

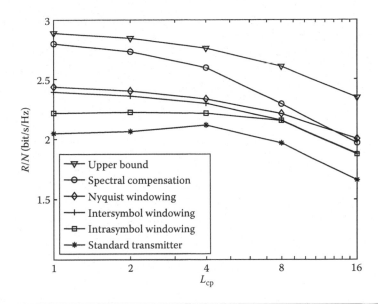

Figure 7.14 *Case study:* **Spectral efficiency R/N versus CP length L_{cp} of different egress-reduction techniques for the PSD mask shown in Figure 7.12 and a frequency-flat channel. Optimal power loading according to Equation 7.11 (star), maximum mainlobe-energy intrasymbol windowing [16] (square), linear-slope intersymbol windowing (plus), linear-slope Nyquist windowing [18] (diamond), spectral compensation [26] (circle), and upper bound given by Equation 7.20 (triangle). All windows have optimal length and are applied together with optimal power loading (optimal in the sense of maximizing R/N). ($N = 64$, SNR $= 10$ dB.)**

required at the receiver to restore orthogonality, as described by Equation 7.16, renders intrasymbol windowing most complex among all methods. Furthermore, the complexity scales with the number of subcarriers N. Spectral compensation allows a trade-off between performance and complexity—both $\mathcal{S}_c$ and $\mathcal{S}_r(n), n \in \mathcal{S}_c$, can be reduced at the cost of performance. The complexity thus does not scale with N but rather with the number of sharp edges in the PSD mask since each edge requires a couple of compensation subcarriers in order to achieve a shaping effect. In general, spectral

Table 7.1   Total Run-Time Complexity (of Transmitter and Receiver Processing) of Different Methods in Terms of Complex-Valued Multiply-and-Add Operations per Multicarrier Symbol

	Complex Multiply-and-Add Operations per Symbol	
Method	*General*	*Case Study, $L_{cp} = 16$*
Spectral compensation	$\sum_{n \in \mathcal{S}_c} \lvert \mathcal{S}_r(n) \rvert$	212
Nyquist windowing	$6L_{nyq} + L_{cp}$	34
Intersymbol windowing	$2L_w$	4
Intrasymbol windowing	$2N + N \log_2 N$	512

compensation is less complex than intrasymbol windowing but more costly than the other two windowing techniques. The complexity of Nyquist windowing scales not only with the window-function parameter L_{nyq} but also with L_{cp}. Note that Nyquist windowing requires dedicated processing at the receiver ($2L_w + L_{cp}$ additions). Intersymbol windowing is clearly the cheapest technique and its complexity scales exclusively with the window length L_w.

7.6 Conclusion and Open Issues

In any setup where bandwidth is costly, high spectral efficiency is desirable. Most practical OFDMA systems are FFT based and use rectangular pulse shapes, which emphasizes the trade-off between transmit power and out-of-band emissions on subcarriers close to band edges. The chapter has tackled the trade-off from an optimization perspective with the PSD as constraint and the throughput as an objective function—an approach that allows a fair comparison of techniques and can also serve as a basis for system design.

In the following, some open issues in the context of spectrum efficiency are summarized.

■ The framework presented in this chapter is based on maximizing the throughput under a PSD constraint while preserving orthogonality in time-dispersive channels. It may be worthwhile to explicitly include the effect of spectral leakage due to frequency offsets in the optimization process in order to enhance the robustness against frequency dispersion.

■ A rather general problem that has also applications outside the scope of spectrum efficiency is finding "good" subcarrier sets $\mathcal{S}_i$ and $\mathcal{S}_c$. If the objective function does not introduce structure, the problem requires testing all possible splits and is thus NP hard. Finding the optimal split may not be of vital importance for practical applications in communication; however, low-complexity solutions for quickly finding an acceptable split are clearly of interest. Good ideas and deeper insights may also be used for other applications, such as the design of tone-reservation schemes or precoders aiming at mitigating intercarrier interference.

■ Spectral compensation appears to be a promising technique. A conclusive comparison of active and passive methods including both performance and complexity is yet to be done.

Symbols

$f \in [0, 1)$	Discrete-time frequency
$F \in \mathbb{R}_+$	Frequency in Hertz
$L \in \mathbb{Z}_+$	Total leading cyclic extension in samples
$L' \in \mathbb{Z}_+$	Total trailing cyclic extension in samples
$L'' \in \mathbb{Z}_+$	Trailing cyclic extension purged by receiver
$L_{cp} \in \mathbb{Z}_+$	Cyclic prefix length in samples
$L_{cs} \in \mathbb{Z}_+$	Cyclic suffix length in samples
$L_f \in \mathbb{Z}_+$	Emission reducing transmit the filter's length in samples
$L_{nyq} \in \mathbb{Z}_+$	Nyquist-window extension in samples
$L_w \in \mathbb{Z}_+$	Overlap of consecutive symbols in samples
$M \in \mathbb{Z}_+$	Channel dispersion (length of channel impulse response minus one) in samples

$m \in \mathbb{R}_+^{Q \times 1}$	Transmit PSD mask at Q points in the frequency interval $[0, 1)$
$N \in \mathbb{Z}_+$	Symbol length in samples (or equivalently, FFT/IFFT size)
$N' = N + L + L'$	Symbol length in samples including cyclic extensions
$N'' = N' - L_\mathrm{w}$	Effective symbol length (N plus extensions minus overlap L_w)
$p \in \mathbb{R}_+^{N \times 1}$	Vector containing power values
$q \in \mathbb{R}_+^{Q \times 1}$	Transmit PSD at Q points in the frequency interval $[0, 1)$
$\mathcal{S}_\mathrm{i} \in \{1, \ldots, N\}$	Set of information-carrying subcarriers
$\mathcal{S}_\mathrm{c} \in \{1, \ldots, N\}$	Set of compensation subcarriers
$W \in \mathbb{C}^{N \times N}$	DFT matrix scaled by $1/\sqrt{N}$ (W^H is the IDFT matrix and $W W^\mathrm{H} = I$)
$Z_\mathrm{add} \in \{0, 1\}^{N' \times N}$	Matrix adding leading and trailing cyclic extensions
$Z_\mathrm{rem} \in \{0, 1\}^{N \times N''}$	Matrix purging leading and trailing cyclic extensions

Abbreviations

CP	Cyclic prefix
CS	Cyclic suffix
DFT	Discrete Fourier transform
DL	Downlink
FFT	Fast Fourier transform
IDFT	Inverse DFT
PSD	Power spectral density
SNR	Signal-to-noise power ratio
UL	Uplink

References

1. U. Berthold, F. Jondral, S. Brandes, and M. Schnell, OFDM-based overlay systems: A promising approach for enhancing spectral efficiency [topics in radio communications], *IEEE Communications Magazine*, 45(12), 52–58, 2007.
2. W. Yu, D. Toumpakaris, J. Cioffi, D. Gardan, and F. Gauthier, Performance of asymmetric digital subscriber lines in an impulse noise environment, *IEEE Transactions on Communications*, 51(10), 1653–1657, 2003.
3. A. Vahlin and N. Holte, Optimal finite duration pulses for OFDM, *IEEE Transactions on Communications*, 44(1), 10–14, 1996.
4. H. Bölcskei, P. Duhamel, and R. Hleiss, Design of pulse shaping OFDM/OQAM systems for high data-rate transmission over wireless channels, in *Proc. IEEE Intl. Conference on Communications (ICC '99)*, vol. 1, Vancouver, Canada, June 1999, pp. 559–564.
5. W. Kozek and A. Molisch, Nonorthogonal pulseshapes for multicarrier communications in doubly dispersive channels, *IEEE Journal on Selected Areas in Communication*, 16(8), 1579–1589, 1998.
6. R. Chang, Synthesis of band-limited orthogonal signals for multi-channel data transmission, *The Bell System Technical Journal*, 45, 1775–1796, 1966.
7. B. Maham and A. Hjorungnes, ICI reduction in OFDM by using maximally flat windowing, *Proc. IEEE International Conference on Signal Processing and Communications (ICSPC '07)*, pp. 1039–1042, November 2007.
8. P. Tan and N. Beaulieu, Reduced ICI in OFDM systems using the "better than" raised-cosine pulse, *IEEE Communications Letters*, 8, (3) 135–137, 2004.

9. V. P. Sathe and P. P. Vaidyanathan, Effects of multirate systems on the statistical properties of random signals, *IEEE Transactions on Signal Processing*, 41, 131–146, 1993.

10. W. A. Gardner, Exploitation of spectral redundancy in cyclostationary signals, *IEEE Signal Processing Magazine*, 8, 14–36, 1991.

11. T. Magesacher, P. Ödling, and P. O. Börjesson, Optimal intersymbol transmit windowing for multicarrier modulation, in *Proc. Nordic Signal Processing Symp. NORSIG 2006*, Reykjavik, Iceland, June 2006.

12. J. M. Cioffi, *Advanced Digital Communication*. Class reader EE379C, Stanford University, 2005, class web page http://www.stanford.edu/class/ee379c/

13. ETSI TM6, Transmission and multiplexing (TM); access transmission systems on metallic access cables; Very high speed Digital Subscriber Line (VDSL); Part 2: Transceiver specification, *TS 101 270-2, Version 1.1.5*, December 2000.

14. G. Cuypers, K. Vanbleu, G. Ysebaert, and M. Moonen, Intra-symbol windowing for egress reduction in DMT transmitters, *EURASIP Journal on Applied Signal Processing*, (1), 87–87, 2006.

15. T. Magesacher, P. Ödling, and P. O. Börjesson, Optimal intersymbol transmit windowing for multicarrier modulation, in *Proc. 7th Nordic Signal Processing Symp. (NORSIG '06)*, June 2006, pp. 70–73.

16. Y.-P. Lin and S.-M. Phoong, Window designs for DFT-based multicarrier systems, *IEEE Transactions on Signal Processing*, 53, 1015–1024, 2005.

17. T. Magesacher, Optimal intra-symbol transmit windowing for multicarrier modulation, in *Proc. Intl. Symp. on Communications, Control and Signal Processing (ISCCSP '06)*, Marrakech, Morocco, March 2006.

18. M. Sebeck and G. Bumiller, Effective configurable suppression of narrow frequency bands in multi-carrier modulation transmission, in *Proc. IEEE International Symp. on Power Line Communications and its Applications*, 2006, pp. 128–133.

19. J. Bingham and M. Mallory, RFI egress suppression for SDMT, *ANSI Contribution T1E1.4/96-085*, 1996.

20. J. Bingham, RFI suppression in multicarrier transmission systems, in *Proc. Global Telecommunications Conference (GLOBECOM '96)*, vol. 2, November 1996, pp. 1026–1030.

21. R. Baldemair, Suppression of narrow frequency bands in multicarrier transmission systems, in *Proc. European Signal Processing Conference (EUSIPCO '00)*, Tampere, Finland, September 2000, pp. 553–556.

22. H. Yamaguchi, Active interference cancellation technique for MB-OFDM cognitive radio, in *Proc. 34th European Microwave Conference*, vol. 2, October 2004, pp. 1105–1108.

23. I. Cosovic, S. Brandes, and M. Schnell, Subcarrier weighting: A method for sidelobe suppression in OFDM systems, *IEEE Communications Letters*, 10(6), 444–446, 2006.

24. S. Brandes, I. Cosovic, and M. Schnell, Reduction of out-of-band radiation in OFDM systems by insertion of cancellation carriers, *IEEE Communications Letters*, 10(6), 420–422, 2006.

25. J.-J. van de Beek and F. Berggren, Out-of-band power suppression in OFDM, *IEEE Communications Letters*, 12(9), 609–611, 2008.

26. T. Magesacher, P. Ödling, and P. O. Börjesson, Optimal intra-symbol spectral compensation for multicarrier modulation, in *Proc. Intl. Zurich Seminar on Broadband Communications (IZS '06)*, Zurich, Switzerland, February 2006, pp. 138–141.

27. T. Magesacher, Spectral compensation for multicarrier communication, *IEEE Transactions on Signal Processing*, 55(7), 3366–3379, 2007.

28. J. Tellado and J. Cioffi, Efficient algorithms for reducing PAR in multicarrier systems, in *Proc. IEEE International Symp. on Information Theory*, August 1998, p. 191.

29. B. Borchers, CSDP, a C library for semidefinite programming, *Optimization Methods & Software*, vol. 11-2, 1999, pp. 613–623, available from http://infohost.nmt.edu/~borchers/csdp.html.

30. J. Löfberg, YALMIP : A toolbox for modeling and optimization in MATLAB®, in *Proc. CACSD Conference*, Taipei, Taiwan, 2004, available from http://control.ee.ethz.ch/~joloef/yalmip.php.

31. J. F. Sturm, Using SeDuMi 1.02, a MATLAB® toolbox for optimization over symmetric cones, in *Optimization Methods and Software*, vol. 11–12, 1999, pp. 625–653, available from http://sedumi.mcmaster.ca

32. MathWorks, MATLAB® optimization toolbox (online documentation), 2005, http://www.mathworks.com

Chapter 8

Resource Allocation in IEEE 802.16e Mobile WiMAX*,†

Chakchai So-In, Raj Jain, and Abdel Karim Al-Tamimi

Contents

8.1 Introduction to Resource Allocation in Mobile WiMAX.................................. 190
    8.1.1 Key Features of WiMAX Networks.. 191
    8.1.2 IEEE 802.16 PHYs—SC, OFDM, and OFDMA 192
    8.1.3 WiMAX Frame Structure... 193
        8.1.3.1 Number of Bursts per Frame 195
        8.1.3.2 Two-Dimensional Rectangular Criterion.............................. 195
        8.1.3.3 Number and Size of MPDUs in a Burst.............................. 196
    8.1.4 WiMAX OFDMA MCSs.. 199
    8.1.5 WiMAX Configuration Parameters and Characteristics....................... 199
    8.1.6 Coding Schemes.. 203
    8.1.7 Automatic Repeat Request and Hybrid-ARQ 203
    8.1.8 MS Initialization Overview... 204
    8.1.9 Overview of WiMAX QoS Service Classes 204
    8.1.10 Section Summary.. 205
8.2 Triple Play Capacity Evaluation.. 205
    8.2.1 Traffic Models and Workload Characteristics................................. 206

---

* "WiMAX," "Mobile WiMAX," "Fixed WiMAX," "WiMAX Forum," "WiMAX Certified," "WiMAX Forum Certified," the WiMAX Forum logo, and the WiMAX Forum Certified logo are trademarks of the WiMAX Forum.

† Based on Chakchai So-In, Raj Jain, and Abdel-Karim Al Tamimi, Scheduling in IEEE 802.16e mobile WiMAX: Key issues and a survey, *IEEE JSAC*, February 2009.

      8.2.2   Overhead Analysis... 206
             8.2.2.1   Upper Layer Overhead... 206
             8.2.2.2   Lower Layer Overhead.. 209
      8.2.3   Pitfalls.. 211
      8.2.4   Section Summary.. 214
  8.3   Uplink Bandwidth Request/Grant Mechanisms.................................... 215
      8.3.1   Consider UGS... 215
      8.3.2   Consider the Delay Requirements.. 215
      8.3.3   Consider rtPS 216
      8.3.4   Consider ertPS 217
      8.3.5   Consider nrtPS.. 218
      8.3.6   Consider BE... 218
      8.3.7   Section Summary.. 218
  8.4   WiMAX Scheduler.. 218
      8.4.1   Design Factors... 219
      8.4.2   Classification of Schedulers 221
             8.4.2.1   Channel-Unaware Schedulers 221
             8.4.2.2   Channel-Aware Schedulers 225
      8.4.3   Section Summary.. 229
  8.5   Conclusion and Open Research Issues.. 229
  References... 230

8.1 Introduction to Resource Allocation in Mobile WiMAX

In this section, general concepts of resource allocation in orthogonal frequency division multiple access (OFDMA) are explained and compared to single carrier (SC) and orthogonal frequency division multiplexing (OFDM). Frame structure and mobile station (MS) initialization are introduced so that the reader can understand the constraints in designing resource allocation schemes.

IEEE 802.16 is a set of telecommunications technology standards aimed at providing wireless access over long distances in a variety of ways—from point-to-point links to full mobile cellular type access as shown in Figure 8.1. It covers a metropolitan area of several kilometers and is also called WirelessMAN. Theoretically, a worldwide interoperability for microwave access (WiMAX) base station (BS) can provide broadband wireless access (BWA) in the range of up to 30 miles (50 km) for fixed stations and 3–10 miles (5–15 km) for MSs with a maximum data rate of up to 70 Mbps [1,2], as compared to 802.11a with 54 Mbps to several hundreds of meters, enhanced data rates for global evolution (EDGE) with 384 kbps to a few kilometers, or code-division multiple access 2000 (CDMA2000) with 2 Mbps to a few kilometers.

The IEEE 802.16 standards group has been developing a set of standards for broadband (high-speed) wireless access (BWA) in a metropolitan area. Since 2001, a number of variants of these standards have been issued and are still being developed. Like any other standards, these specifications are also a compromise of several competing proposals and contain numerous optional features and mechanisms. The WiMAX Forum is a group of 400+ networking equipment vendors, service providers, and component manufacturers and users that decide which of the numerous options allowed in the IEEE 802.16 standards should be implemented so that equipment from different vendors will interoperate. Several features such as unlicensed band operation and 60 GHz operation, although specified in the IEEE 802.16, are not a part of WiMAX networks since they are not

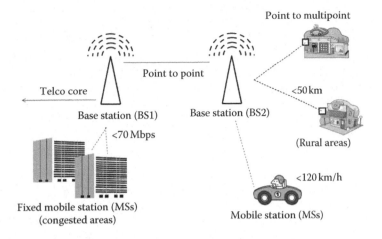

Figure 8.1 WiMAX deployment scenarios.

currently in the profiles agreed on at the WiMAX Forum. For an equipment to be certified as WiMAX compliant, the equipment has to pass the interoperability tests specified by the WiMAX Forum. For the rest of this chapter, the terms WiMAX and IEEE 802.16 are used interchangeably.

Note that although a mobile WiMAX standard allows several configurations such as mesh networks and relay networks, our focus is only on point-to-multipoint network configuration. Thus, the resource allocation problem is basically that the BS is the single resource controller for both the uplink (UL) and downlink (DL) directions.

8.1.1 Key Features of WiMAX Networks

The eight key features of WiMAX networks that differentiate them from other metropolitan area wireless access technologies are

- Use of OFDMA
- Scalable use of any spectrum width (varying from 1.25 to 28 MHz)
- Time and frequency division duplexing (TDD and FDD)
- Advanced antenna techniques such as beam forming and multiple input multiple output (MIMO)
- Per MS adaptive modulation
- Advanced coding techniques such as space-time coding and turbo coding
- Strong security
- Multiple quality of service (QoS) classes not only suitable for voice, but also designed for a combination of data, voice, and video services

Consider duplexing techniques. Unlike voice services, which make symmetric use of uplink (MS to BS) and downlink (BS to MS), data and video services make a very asymmetric use of link capacities and are, therefore, better served by TDD than FDD. This is because TDD allows the service provider to decide the ratio of uplink and downlink transmission times and match it to the expected usage. Thus, TDD will be the main focus of this chapter. However, the techniques mentioned here can be used for WiMAX networks using FDD as well.

In terms of guaranteed services, WiMAX includes several QoS mechanisms at the MAC layer. Typically, the QoS support in wireless networks is much more challenging than that in wired networks because the characteristics of the wireless link are highly variable and unpredictable both on a time-dependent basis and on a location-dependent basis. With a longer distance, multipath and fading effects are also put into consideration. The Request/Grant mechanism is used for MSs to access the media with a centralized control at BSs. WiMAX is a connection-oriented technology [with 16 bits Connection IDentifier (CID) shared for downlink and uplink]. Therefore, MSs are not allowed to access the wireless media unless they register and request the bandwidth allocations from the BS first, except for certain time slots reserved specifically for contention-based access.

To meet QoS requirements, especially for voice and video transmission with the delay and delay jitter constraints, the key issue is how to allocate resources among the users not only to achieve those constraints but also to maximize goodput and to minimize power consumption while keeping a feasible algorithm complexity and ensuring system scalability. IEEE 802.16 standard does not specify any resource allocation mechanisms or admission control mechanisms. Although a number of scheduling algorithms have been proposed in the literature, such as Fair Scheduling [3], Distributed Fair Scheduling [4], MaxMin Fair Scheduling [5], Channel State-dependent Round Robin (CSD-RR) [6], Feasible Earliest Due Date (FEDD) [7], and Energy Efficient Scheduling [8], these algorithms cannot be directly used for WiMAX due to specific features of the technology. Examples of these specific features are: the request/grant mechanism, OFDMA versus carrier sense multiple access/collision avoidance (CSMA-CA) for wireless LANs, the allocation unit being a slot with specific subchannel and time duration, the definition of fixed frame length, and the guaranteed QoS.

The purpose of this section is to provide a brief overview of WiMAX characteristics that need to be considered in developing a scheduler. Therefore, in Section 8.1.2, we provide a brief introduction to various WiMAX physical layers (PHYs) while we focus on the OFDMA-based PHY in the rest of the chapter. Section 8.1.3 gives an overview of the WiMAX frame structure, downlink map (DL-MAP) and uplink map (UL-MAP) for OFDMA. In this section, we also discuss the issues on the WiMAX frame structure, that is, number of bursts, number and size of MAC protocol data units (MPDUs), and two-dimensional rectangular mapping for downlink subframe. A description of different modulation and coding schemes (MCSs) and a brief overview of subchannelization are presented in Section 8.1.4. Section 8.1.5 discusses WiMAX configuration parameters and characteristics. We briefly described various coding schemes in Section 8.1.6. A brief introduction to Automatic Repeat Request (ARQ) and Hybrid ARQ (HARQ) is presented in Section 8.1.7. Then, the MS initialization process is briefly described in Section 8.1.8. Finally, WiMAX QoS classes are discussed.

8.1.2 IEEE 802.16 PHYs—SC, OFDM, and OFDMA

IEEE 802.16 supports a variety of physical layers. Each of these has its own distinct characteristics. First, WirelessMAN-SC (single carrier) PHY is designed for the 10–60 GHz spectrum. While IEEE has standardized this PHY, there are not many products implementing it because this PHY requires line of sight (LOS) communication. Rain attenuation and multipath also affect reliability of the network at these frequencies. To allow non-line of sight (NLOS) communication, IEEE 802.16 designed the OFDM PHY using the spectrum below 11 GHz. This PHY, popularly known as IEEE 802.16d, is designed for fixed MSs. The WiMAX Forum has approved several profiles using this PHY. Most of the current WiMAX products implement this PHY. In this PHY, also called OFDM-TDMA, multiple MSs use a time division multiple access (TDMA) to share the media. OFDM is a multicarrier transmission in which thousands of subcarriers are transmitted and each

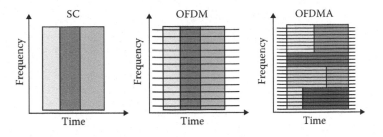

Figure 8.2   IEEE 802.16 PHYs: SC, OFDM, and OFDMA.

user is given a complete control of all subcarriers. The scheduling decision is simply to decide what time slots should be allocated to each MS. Like OFDM-TDMA, there is also OFDM-FDMA in which OFDM is used along with frequency division multiple access (FDMA). In this case, each user is allocated a subset of subcarriers. For mobile users, it is better to reduce the number of subcarriers and to have higher signal power per subcarrier. Therefore, multiple users are allowed to transmit using different subcarriers in the same time slot. The scheduling decision then is to decide which subcarriers and what time slots should be allocated to which user. This combination of TDMA and FDMA in conjunction with OFDM is called OFDMA. Figure 8.2 illustrates a schematic view of the three 802.16 PHYs discussed above. Details of these interfaces can be found in Ref. [1].

The scheduler for WirelessMAN-SC can be fairly simple because only time domain is considered. The entire frequency channel is given to the MS. For OFDM, it is more complex since each subchannel can be modulated differently, but it is still only in time domain. On the other hand, both time and frequency domains need to be considered for OFDMA. The OFDMA scheduler is the most complex one because each MS can receive some portions of the allocation for the combination of time and frequency so that the channel capacity is efficiently utilized. It can be shown that the OFDMA outperforms the OFDM [9]. The current direction of the WiMAX Forum, as well as most WiMAX equipment manufacturers, is to concentrate on mobile WiMAX, which requires OFDMA PHY. The authors of this chapter have been actively participating in the WiMAX Forum's activities. The Application Working Group (AWG) considers scheduling to be crucial for ensuring optimal performance for mobile WiMAX applications. Thus, the OFDMA will be the focus of the rest of this chapter.

8.1.3 WiMAX Frame Structure

IEEE 802.16 standard defines a frame structure as depicted logically in Figure 8.3 and a mapping from burst to MPDU in Figure 8.4. Each frame consists of DL and UL subframes. A preamble is used for time and frequency synchronization. The DL-MAP and UL-MAP define the burst start time and burst end time, modulation types, and forward error control (FEC) for each MS. Frame control header (FCH) defines these MAPs' lengths and usable subcarriers. The MS allocation is in terms of bursts. In the figure, we show one burst per MS; however, WiMAX supports multiple MSs in a single burst in order to reduce the burst overhead. In Figure 8.4, each burst can contain MPDUs—the smallest unit from MAC to physical layer. Basically, each MPDU is a MAC frame with MAC header (6 bytes), other subheaders such as fragmentation, packing, and grant management (GM) subheaders (2 bytes each) if needed; and, finally, a variable length of payload.

Owing to the nature of wireless media, the channel state condition keeps changing over time. Therefore, WiMAX supports adaptive modulation and coding, that is, the MCSs can be changed

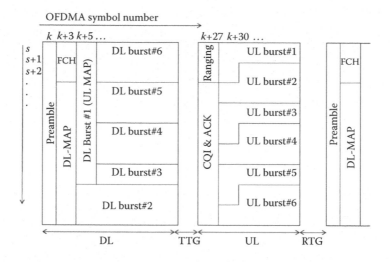

Figure 8.3 A sample OFDMA frame structure.

adaptively depending on the channel condition. Either MS or BS can do the estimation and then BS decides the most efficient MCS. Channel quality indicator (CQI) is used to pass the channel state condition information. Note that the standard does not define when the MCS should change. In general, changing MCS level depends on the signal-to-noise ratio (SNR) and the expected block error rate (BLER). Figure 8.5 shows the reference model of BLER curves from Ref. [2].

Figure 8.3 also shows TTG and RTG gaps. Transmit–receive transition gap (TTG) is when the BS switches from transmit to receive mode and receive–transmit transition gap (RTG) occurs when BS switches from receive to transmit mode. The MSs also use these gaps in the opposite way.

To design a WiMAX scheduler, some parameters and attributes need to be considered. The following five main issues related to the frame structure will be discussed below: number of bursts, two-dimensional rectangular mapping for downlink subframe, MPDU size, fragmentation, and packing considerations.

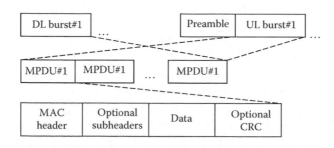

Figure 8.4 MPDU frame format.

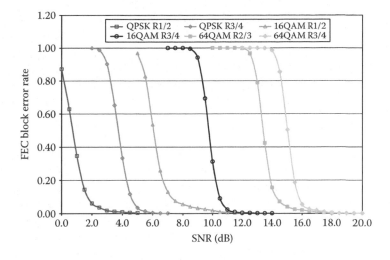

Figure 8.5   Reference BLER curves [2].

8.1.3.1 Number of Bursts per Frame

More bursts result in a larger burst overhead in the form of DL-MAP and UL-MAP information elements (IEs). For uplink, usually there is one burst per MS. Note that "burst" usually is defined when there is a different physical mode, such as one MS uses QPSK 1/4 and another may use 64-QAM 3/4. Moreover, all UL data bursts are allocated as horizontal stripes, that is, the transmission starts at a particular slot and continues until the end of UL subframe. Then it continues on the next subchannel horizontally. This minimizes the number of subcarriers used by the MS and thus maximizes the power per subcarrier and, hence, the SNR.

For downlink, although the standard allows more than one burst per MS, it increases DL-MAP overhead. The standard also allows more than one connection packed into one burst with the increased DL-MAP IE size. It is even possible to pack multiple MSs into one burst, particularly if they are parts of the same physical node. In this scenario, the unique CID helps separate the MSs. Packing multiple MSs into one burst reduces DL-MAP overhead. However, with an increase of burst size, there is a decoding delay at the receiving end. The DL and UL MAPs are modulated with reliable modulation and coding such as BPSK or QPSK. Also, these regions usually require two or four repetitions depending on the channel condition.

8.1.3.2 Two-Dimensional Rectangular Criterion

In the downlink direction, IEEE 802.16e standard requires that all DL data bursts be rectangular. In fact, the two-dimensional rectangular mapping problem is a variation of the bin packing problem, in which one is given bins to be filled with objects. The bins can be in two or more dimensions. If we restrict the bins to two dimensions, we have a "tiling" problem where the objective is to fill a given shape bin with tiles of another given shape. Note that this rectangular criterion is applied to IEEE 802.16e unless HARQ feature is implemented. We do not focus on HARQ/ARQ in this chapter since there are also issues related with HARQ/ARQ implementation such as the ARQ block boundary and ARQ retransmission.

The mapping problem in WiMAX is different from the original bin packing because, first, there are no fixed length and width limitations. Instead only bin sizes are given. Second, with an increase in the number of bursts (number of bins), the other end of the big bin (left side of the WiMAX frame) in which small bins are fitted also changes to allow an increase in size of the variable part of DL-MAP.

In Table 8.1, we compare and summarize several proposed mapping algorithms for WiMAX networks. Notice that each algorithm has its own pros and cons and complexity tradeoffs. Also, the performance tradeoff of increasing DL-MAP overhead versus number of bursts has not yet been studied in the literature.

With rectangular mapping, an MS is usually allocated more slots than its demand. Also, some left-over spaces are too small to allocate to any users. These two types of wasted slots are called over-allocation and unused slots, respectively. We present simulation results comparing an algorithm called "eOCSA" that we have developed with that proposed by Ohseki et al. [10]. The comparison is limited to these two algorithms for various reasons. For example, Ben-Shimol et al. [11] provide no details of how to map the resources to unused spaces if their sizes are over multiple rows. Bacioccola et al. [12] assume that it is possible to have more than one burst per MS. This violates our goal of minimizing burst overhead. The binary-tree full search can support only eight MSs [13] and so it is not of any practical use. Perez-Costa et al. [16] generalized the two-dimensional packing into one-dimensional packing. They categorized two types of bursts: urgent and unurgent. Their algorithm basically maximizes the sum of packed burst sizes with the condition that all urgent bursts are successfully mapped.

With partially used subchannelization (PUSC) mode, 10 MHz channel and a DL:UL ratio of 2:1, the DL frame consists of 14 columns of 30 slots each or 420 slots [17]. Assuming that we reserve the first two columns for DL/UL MAPs, we can allocate the remaining 12 columns, resulting in 360 slots per frame for the users. We also assume that each MS needs one burst. The number of MSs is randomly chosen from 1 to 49. The resource demand for each MS is also randomly generated so that the total demand is 360 slots. The over-allocations and unused slots are averaged over 100 trials. Note that using more realistic scenarios (using VoIP, Mobile TV, and Web traffic, for example) can change this conclusion.

The results for eOCSA are shown in Figure 8.6 in terms of the normalized over-allocation and unused slots versus the number of MSs. The normalization is done by dividing the total space required to map the demands. On an average, the normalized over-allocation and unused slots are 0.0088 and 0.0614, respectively.

Figure 8.7 shows the corresponding results for the algorithm by Ohseki et al. On an average, the normalized over-allocation slots and unused slots are 0.0029 and 0.5198, respectively. Notice that they have significantly higher unused slots than eOCSA because they do not allocate unused spaces below or above an allocated user's burst. On the other hand, eOCSA has a slightly higher over-allocation because we try to fit rectangles in these small unused spaces. More details on other tradeoffs in burst mapping are presented in [14].

8.1.3.3 Number and Size of MPDUs in a Burst

Each MPDU has a 6 bytes MAC header (see Figure 8.4). One can have large MPDU, but then the MPDU loss probability due to bit errors is higher. On the other hand, the MPDU header is significant if there are many small MPDUs. Note that in Ref. [18], the estimation of the optimal MPDU size was drawn. The equation is shown below.

$$\frac{O}{2} - \sqrt{\frac{(O \ln(1 - E))^2 - 4BO \ln(1 - E)}{2 \ln(1 - E)}}. \tag{8.1}$$

Table 8.1 Two-dimensional Rectangular Mapping for Downlink

	Algorithm Descriptions	*Pros*	*Cons*	*Complexity*
Ohseki et al. [10]	Allocate in time domain first and then the frequency domain (left to right and top to bottom)	Allows burst compaction if there are more than one bursts that belongs to the same physical node; Do not consider a variable part of DL-MAP	The algorithm does not consider the unused space	$O(N) + O$ (searching and compaction)
Ben-Shimol et al. [11] (Raster algorithm)	Assign the resource allocation row by row with the largest resource allocation first	Simple	There is no detailed explanation of how to map the resources to unused space in a frame when their sizes span over multiple rows; Do not consider a variable part of DL-MAP	N/A
Bacioccola et al. [12]	Allocate from right to left and bottom to top	Optimize frame utilization; Consider a variable part of DL-MAP	They map a single allocation into multiple rectangular areas that may result in increased DL MAP elements overhead	$O(N)$
Desset et al. [13]	Binary-tree full search algorithm	Optimize frame utilization	Only eight users at the most can be supported; Do not consider a variable part of DL-MAP	N/A
So-In et al. [14]	Allocate from right to left and bottom to top with the least width first vertically and the least height first horizontally for each particular burst	Optimize frame utilization; Consider a variable part of DL-MAP	Lack of detailed simulation	$O(N^2)$
Wand et al. [15]	Apply the less flexibility first (LFF) allocation (select the area with the least free space edge)	Consider all possible mapping pairs	Fixed resource reserved for DL-MAP	$O(N^2)$

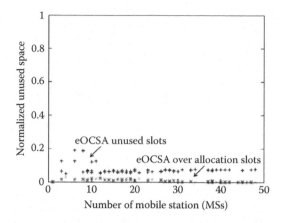

Figure 8.6 Normalized unused space versus number of MSs for eOCSA [14].

Here, O is the overhead measured by the number of bytes in headers, subheaders, and cyclic redundancy check (CRC). E is the BLER after FEC. B stands for FEC block size in bytes.

Depending on the number of retransmissions or losses, a dynamic change in MPDU size was introduced in Ref. [19] by typically adding more FEC for MPDU in the poor channel situation.

Notice that WiMAX also supports fragmentation and packing. Their overheads should also be taken into account. Consider fragmentation. Deficit round robin (DRR) with fragmentation was brought up in Ref. [20]. Without the fragmentation consideration, the WiMAX frame is underutilized since it may be possible that within a particular frame, all full packets cannot be transmitted. In Ref. [17], we have shown that with proper packing, especially for small packets such as voice packets, the number of users can be increased significantly; however, packet delays can also increase.

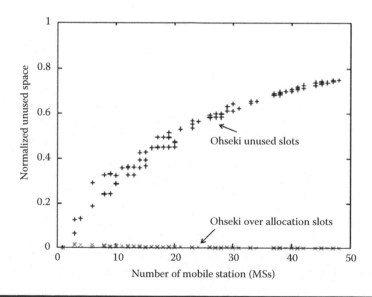

Figure 8.7 Normalized unused space versus number of MSs for Ohseki et al.'s algorithm.

8.1.4 WiMAX OFDMA MCSs

Unlike WiFi and many cellular technologies that use fixed width channels, WiMAX allows the use of almost any available spectrum width. Allowed channel bandwidths vary from 1.25 to 28 MHz. The channel is divided into many equally spaced subcarriers. For example, a 10 MHz channel is divided into 1024 subcarriers some of which are used for data transmission, while others are reserved for monitoring the quality of the channel (pilot subcarriers), for providing safety zone (guard subcarriers) between the channels, or for use as a reference frequency (DC subcarrier).

The data and pilot subcarriers are modulated using one of several available MCSs. Quadrature phase shift keying (QPSK) and quadrature amplitude modulation (QAM) are examples of modulation methods. Coding refers to the forward error correction (FEC) bits. Thus, QAM-64 1/3 indicates an MCS with 8-bit (64 combinations) QAM-modulated symbols and the error correction bits take up 2/3 of the bits, leaving only 1/3 for data.

In traditional cellular networks, the downlink (BS to MS) and uplink (MS to BS) use different frequencies. This is called FDD. WiMAX allows FDD but also allows TDD in which the downlink (DL) and uplink (UL) share the same frequency but alternate in time. The transmission consists of frames as shown in Figure 8.3. The DL subframe and UL subframe are separated by a TTG and an RTG. The frames are shown in two dimensions with frequency along the vertical axis and time along the horizontal axis.

In OFDMA, each MS is allocated only a subset of the subcarriers. The available subcarriers are grouped into a few subchannels and the MS is allocated one or more subchannels for a specified number of symbols. There are a number of ways to group subcarriers in subchannels; of these, PUSC is the most common. In PUSC, subcarriers forming a subchannel are selected randomly from all available subcarriers. Thus, the subcarriers forming a subchannel may not be adjacent in frequency.

Users are allocated a variable number of "slots" in the downlink and uplink. The exact definition of slots depends on the subchannelization method and on the direction of transmission (DL or UL). Figure 8.8 shows slot formation for PUSC. In uplink (Figure 8.8a), a slot consists of six "tiles" where each tile consists of four subcarriers over three symbol times. Of the 12 subcarrier–symbol combinations in a tile, four are used for pilot and eight are used for data. The slot, therefore, consists of 24 subcarriers over three symbol times. The 24 subcarriers form a subchannel and thus, at 10 MHz, 1024 subcarriers form 35 UL subchannels. The slot formation in downlink is different and is shown in Figure 8.8b. In the downlink, a slot consists of two clusters where each cluster consists of 14 subcarriers over two symbol times. Thus, a slot consists of 28 subcarriers over two symbol times. The group of 28 subcarriers is called a subchannel, resulting in 30 DL subchannels from 1024 subcarriers at 10 MHz.

Consider the downlink subframe (see Figure 8.3). Aside from one symbol-column of preamble, all other transmissions use slots as discussed above. The first field in the DL subframe after the preamble is a 24-bit FCH. For high reliability, FCH is transmitted with the most robust MCS (QPSK 1/2) and is repeated four times. The next field is DL-MAP, which specifies the burst profile of all user bursts in the DL subframe. DL-MAP has a fixed part that is always transmitted and a variable part that depends on the number of bursts in the DL subframe. This is followed by UL-MAP, which specifies the burst profile for all bursts in the UL subframe. It also consists of a fixed part and a variable part. Both DL and UL MAPs are transmitted using QPSK 1/2 MCS.

8.1.5 WiMAX Configuration Parameters and Characteristics

The key parameters of WiMAX PHY are summarized in Tables 8.2 through 8.4.

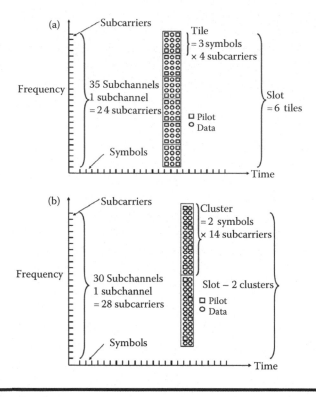

Figure 8.8 Symbols, tiles, clusters, and slots. (a) Uplink and (b) downlink.

Table 8.2   OFDMA Parameters for WiMAX

Parameters	Values						
System bandwidth (MHz)	1.25	5	10	20	3.5	7	8.75
Sampling factor	28/25				8/7		
Sampling frequency ($F_{s,MHz}$)	1.4	5.6	11.2	22.4	4	8	10
Sample time ($1/F_s$, ns)	714.3	178.6	89.3	44.6	250	125	100
FFT size (N_{FFT})	128	512	1024	2048	512	1024	1024
Subcarrier spacing (Δf, kHz)	10.93				7.81		9.76
Useful symbol time ($T_b = 1/\Delta f$, μs)	91.4				128		102.4
Guard time ($T_g = T_b/8$, μs)	11.4				16		12.8
OFDMA symbol time ($T_s = T_b + T_g$, μs)	102.8				144		115.2

Source: WiMAX Forum, WiMAX System Evaluation Methodology V2.1, July 2008, 230 pp.
Available: http://www.wimaxforum.org/technology/documents/

Table 8.3   Number of Subcarriers in PUSC

Parameters	Values				
(a) DL					
System bandwidth (MHz)	1.25	2.5	5	10	20
FFT size	128	N/A	512	1024	2084
# of guard subcarriers	43	N/A	91	183	367
# of used subcarriers	85	N/A	421	841	1681
# of pilot subcarriers	12	N/A	60	120	240
# of data subcarriers	72	N/A	360	720	140
(b) UL					
System bandwidth (MHz)	1.25	2.5	5	10	20
FFT size	128	N/A	512	1024	2084
# of guard subcarriers	31	N/A	103	183	367
# of used subcarriers	97	N/A	409	841	1681

Source: Data from H. Yaghoobi, *Intel Technol. J.*, 8, 201–212, 2004.

Table 8.4   Slot Capacity for Various MCSs

MCS	Receiver SNR Threshold (dB) [22]	Bits per Symbol	Coding Rate	DL Bytes per Slot	UL Bytes per Slot
QPSK 1/4	N/A	2	0.125	1.5	1.5
QPSK 1/4	N/A	2	0.250	3	3
QPSK 1/2	9.4	2	0.500	6	6
QPSK 3/4	11.2	2	0.750	9	9
QAM-16 1/2	16.4	4	0.500	12	12
QAM-16 2/3	N/A	4	0.667	16	16
QAM-16 3/4	18.2	4	0.750	18	16
QAM-64 1/2	22.7	6	0.500	18	16
QAM-64 2/3	N/A	6	0.667	24	16
QAM-64 3/4	24.4	6	0.750	27	N/A
QAM-64 5/6	N/A	6	0.833	30	N/A

Table 8.2 lists the OFDMA parameters for various channel widths. Note that the product of subcarrier spacing and FFT size is equal to the product of channel bandwidth and sampling factor. For example, for the 10 MHz channel, $10.93 \text{ kHz} \times 1024 = 10 \text{ MHz} \times (28/25)$. This table shows that at 10 MHz the OFDMA symbol time is 102.8 µs and so there are 48.6 symbols in a 5 ms frame. Of these, 1.6 symbols are used for TTG and RTG, leaving 47 symbols. If n of these are used for DL, then $47-n$ are available for uplink. Since DL slots occupy two symbols and UL slots occupy three symbols, it is best to divide these 47 symbols such that $47-n$ is a multiple of 3 and n is of the form $2k + 1$. For a DL:UL ratio of 2:1, these considerations would result in a DL subframe of 29 symbols and a UL subframe of 18 symbols. In this case, the DL subframe will consist of a total of 14×30 or 420 slots. The UL subframe will consist of 6×35 or 210 slots.

Table 8.3 lists the number data, pilot and guard subcarriers for various channel widths. A PUSC subchannelization is assumed, which is the most common subchannelization.

Table 8.4 lists the number of bytes per slot for various MCS values. For each MCS, the number of bytes is equal to (#bits per symbols × coding rate × 48 data subcarriers and symbols per slot/8 bits). Note that for UL, the maximum MCS level is QAM-16 2/3 [2].

This analysis method can be used for any allowed channel width, any frame duration, or any subchannelization. We assume a 10 MHz WiMAX TDD system with 5 ms frame duration, PUSC subchannelization mode, and a DL:UL ratio of 2:1. These are the default values recommended by the WiMAX Forum system evaluation methodology and are also common values used in practice.

The number of DL and UL slots for this configuration can be computed as shown in Table 8.5.

Table 8.5   WiMAX System Configuration

Configurations	Downlink	Uplink
DL/UL symbols excluding preamble	28	18
Ranging, CQI and ACK (column symbols)	N/A	3
# of symbol columns per cluster[a]/tile[b]	2	3
# of subcarriers per cluster[a]/tile[b]	14	4
Symbols×subcarriers per cluster[a]/tile[b]	28	12
Symbols×data subcarriers per cluster[a]/tile[b]	24	8
# of pilots per cluster[a]/tile[b]	4	4
# of clusters[a]/# of tiles[b] per slot	2	6
Subcarriers×symbols per slot	56	72
Data subcarriers×symbols per slot	48	48
Data subcarriers×symbols per DL/UL subframe	23,520	12,600
Number of slots	420	175

[a] Cluster for DL.

[b] Tile for UL.

8.1.6 Coding Schemes

This section gives a brief introduction to coding schemes used in mobile WiMAX. More detailed information can be found in Refs. [23–25].

The mobile WiMAX coding allows reducing and correcting transmission errors due to the unpredictable channel state condition. It is composed of four main steps: channel coding, interleaving, repetition, and modulation (Figure 8.9).

The channel coding consists of randomization and FEC. Randomization is used to avoid long sequences of ones or zeroes [22]. There are five methods for FEC coding: convolution coding (CC) with tail-biting, block turbo coding (BTC), low-density parity check coding (LDPC), and convolution turbo coding (CTC). Only tail-biting CC [22] and CTC [2] are mandatory. A performance comparison of CC, CTC, and LDPC is presented in Ref. [26]. It is shown that the performance of CTC and LDPC is very similar. LDPC decoding is less complex than the CTC decoding. Compared to LDPC and CTC, CC is the least complex and its performance is only slightly less than the other two, especially for rate 1/2 code.

Interleaving is used to minimize the impact of consecutive errors by distributing the code bits to nonadjacent subcarriers. Repetition is used to increase the decoding probability in the presence of channel errors. In general, repetition is by a factor of 2 or 4 and is applied to the slot boundary. Also, the number of allocated slots is in the range of RS and $RS + (R − 1)$ [25]. Here, S is the number of slots required for the data transmission and R is the repetition factor. For example, with QPSK 1/2, each slot can carry 6 bytes and so $\lceil 20/6 \rceil = 4$ slots are required to transmit 20 bytes. With a repetition of 4, the number of required slots is between 16 and 19 slots.

After the repetition, modulation is used to choose the proper MCS for each resource allocation. The MCS level is chosen based on the channel state condition. In general, SNR is used as an indication to change the level. Note that in Table 8.4, the SNR thresholds indicated are those required to achieve error rates (after FEC) of 10^{-6} or lower.

8.1.7 Automatic Repeat Request and Hybrid-ARQ

Automatic Repeat Request (ARQ) and Hybrid-ARQ (HARQ) are briefly described in this section. Again, more detailed information can be found in Refs. [23–25].

ARQ is a reliable control protocol at the data link layer. In general, a receiver asks the sender to retransmit the data if there is error or loss of data. The ARQ mechanism is based on the feedback of acknowledgement (ACK) or no ACK. As in transmission control protocol (TCP), a sliding window technique is used to limit the transmission. To reduce the number of ACKs, three types of feedbacks: cumulative ACK, selective ACK, and selective with cumulative ACK can be used in mobile WiMAX.

HARQ is another reliability mechanism that spans both MAC and PHY layers. In HARQ, subsequent retransmissions may be encoded differently at the sender and are combined with the previous transmissions of the same data at the receiver so as to improve the correct decoding rate. There are two types of HARQ: CC or type I HARQ and incremental redundancy (IR) or type II HARQ. The main difference lies in the fact that with CC, all retransmissions are identical, whereas with IR, which is especially used for convolutional coding and CTC, the redundancy of the encoded

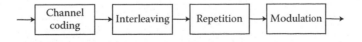

Figure 8.9 PHY channel coding.

bits is changed from one retransmission to the other. Typically, the code rate is effectively decreased at every retransmission, that is, additional parity bits are sent.

8.1.8 MS Initialization Overview

This section gives an overview of the MS initialization process. Details can be found in Refs. [1,2]. Since WiMAX is a connection-oriented technology, each MS must register and do the set up process such as the agreement on MCSs and QoS requirements. Basically, when the MS joins the network, it first scans for the downlink channel and obtains link parameters. Then, the MS goes through the ranging process that includes basic capability negotiation such as how much power needs to reach the BS. The MS uses a backoff mechanism if there is a contention during ranging. After the basic capabilities have been negotiated and QoS service class has been set up, the MS goes through the authorization and key exchange processes. Once these processes are complete, the MS registers with the BS and receives an IP address and is ready to transfer data. To transmit or receive data, the MS must request the bandwidth either explicitly or implicitly (we discuss the bandwidth request mechanism in detail in Section 8.3); the BS makes allocation decisions to grant the bandwidth for the MSs via DL-MAP entries for downlink (the MS receives the data) and via UL-MAP entries for uplink (the MS transmits the data).

8.1.9 Overview of WiMAX QoS Service Classes

IEEE 802.16e defines five QoS service classes: unsolicited grant scheme (UGS), extended real-time polling service (ertPS), real-time polling service (rtPS), non-real-time polling service (nrtPS), and best effort (BE) service. Each of these has its own QoS parameters such as minimum throughput requirement and delay/jitter constraints. Table 8.6 presents a comparison of these classes.

UGS: This service class provides a fixed periodic bandwidth allocation. Once the connection is set up, there is no need to send any other requests. This service is designed for constant bit rate

Table 8.6   Comparison of WiMAX QoS Service Classes

QoS	Pros	Cons
UGS	No overhead. Meet guaranteed latency for real-time service	Bandwidth may not be utilized fully since allocations are granted regardless of the current need
ertPS	Optimal latency and data overhead efficiency	Need to use the polling mechanism (to meet the delay guarantee) and a mechanism to let the BS know when the traffic starts during the silent period
rtPS	Optimal data transport efficiency	Require the overhead of bandwidth request and the polling latency (to meet the delay guarantee)
nrtPS	Provide efficient service for non-real-time traffic with minimum reserved rate	N/A
BE	Provide efficient service for BE traffic	No service guarantee; some connections may starve for a long period of time

(CBR) real-time traffic such as E1/T1 circuit emulation. The main QoS parameters are maximum sustained (MST) rate, maximum latency, and tolerated jitter (the maximum delay variation).

ertPS: This service is designed to support VoIP with silence suppression. No traffic is sent during silent periods. ertPS service is similar to UGS in that the BS allocates the MST rate in active mode, but no bandwidth is allocated during the silent period. There is a need to have the BS poll the MS during the silent period to determine if the silent period has ended. The QoS parameters are the same as those in UGS.

rtPS: This service class is for variable bit rate (VBR) real-time traffic such as moving picture experts group (MPEG) compressed video. Unlike UGS, rtPS bandwidth requirements vary and so the BS needs to regularly poll each MS to determine what allocations need to be made. The QoS parameters are similar to the UGS but minimum reserved traffic rate and MST traffic rate need to be specified separately. For UGS and ertPS services, these two parameters are the same, if present.

nrtPS: This service class is for non-real-time VBR traffic with no delay guarantee. Only the minimum rate is guaranteed. File transfer protocol (FTP) traffic is an example of applications using this service class.

BE: Most of the data traffic falls under this category. This service class guarantees neither delay nor throughput. The bandwidth will be granted to the MS if and only if there is a left-over bandwidth from other classes. In practice, most implementations allow specifying minimum reserved traffic rate and MST traffic rate even for this class.

Note that for non-real-time traffic, traffic priority is also one of the QoS parameters that can differentiate among different connections or MSs within the same service class.

Consider bandwidth request mechanisms for uplink. UGS, ertPS, and rtPS are real-time traffic. UGS has a static allocation. ertPS is a combination of UGS and rtPS. Both UGS and ertPS can reserve the bandwidth during set up. Unlike UGS, ertPS allows all kinds of bandwidth request including contention resolution. rtPS cannot participate in contention resolution. For the other traffic classes (non-real-time traffic), nrtPS and BE, several types of bandwidth requests are allowed such as piggybacking, bandwidth stealing, unicast polling, and contention resolution. These are further discussed in Section 8.3.

8.1.10 Section Summary

In this section, an overview of WiMAX was given to provide a better understanding of resource allocation management to be discussed in the rest of this chapter. Both PHY and MAC descriptions and characteristics were explained. We also briefly discussed the use of (H)ARQ along with the MS initialization process. A brief overview of WiMAX QoS was also presented.

8.2 Triple Play Capacity Evaluation

This section describes the capacity evaluation for triple play services: voice, video, and data. The purpose is to explain a simple model to estimate the number of supported users. This evaluation has been conducted for different MCSs assuming that the traffic model is constant over a long period. The results can be used not only for capacity planning but also for simulation validation. Note that this model is simple enough to be programmed in a spreadsheet program such as Microsoft Excel® [17].

We organize this section as follows: We present three sample workloads consisting of Mobile TV, VoIP, and data applications (Web traffic) in Section 8.2.1. Section 8.2.2 presents the analysis of overheads, namely, upper layer overheads and MAC and PHY overheads, and also ways to reduce these overheads. In Section 8.2.3, we describe the capacity evaluation procedure and also present the number of users supported for the three workloads.

8.2.1 Traffic Models and Workload Characteristics

The WiMAX Forum classifies applications into five categories as shown in Table 8.7. Each application class has its own characteristics such as the bandwidth, latency, and jitter constraints in order to ensure a good quality of user experience. The traffic models for these applications can be found in Ref. [2].

However, in this section, a capacity modeling study, to simplify the model three sample workloads, namely VoIP, Mobile TV, and Web traffic, are used. Note that to simplify the capacity evaluation model, only average packet size is used in the model. Second-order statistics (e.g., standard deviation) are not modeled.

First, the VoIP workload is symmetric in that DL data rate is equal to the UL data rate. It consists of very small packets that are generated periodically. The packet size and the period depend upon the Vocoder used. G723.1 is used in our analysis and results in a data rate of 5.3 kbps, 20 bytes voice packet every 30 ms.

Second, the Mobile TV workload depends upon the quality and size of the display. In our analysis, a sample measurement on a small screen Mobile TV device produced an average packet size of 984 bytes every 30 ms, resulting in an average data rate of 350.4 kbps [27,28]. Note that Mobile TV workload is highly asymmetric with almost all of the traffic going downlink.

Finally, for data workload, we selected the hypertext transfer protocol (HTTP) workload recommended by the Third Generation Partnership Project (3GPP) [29].

A characteristic summary of the three workloads is presented in Table 8.8.

8.2.2 Overhead Analysis

In this section, we consider both upper (network, transport, etc.) and lower (MAC and PHY) layer overheads. We consider only either real-time transport protocol (RTP) or TCP, and IP for the upper layer and these overheads can apply for both downlink and uplink. For lower layer protocol, MAC overhead basically consists of MAC header and other subheaders. Finally, the PHY overhead can be divided into DL overhead and UL overhead. Each of these overheads is discussed next.

8.2.2.1 Upper Layer Overhead

Table 8.8, which lists the characteristics of our Mobile TV, VoIP, and data workloads, includes the type of transport layer used: RTP or TCP. This affects the upper layer protocol overhead. Both RTP over UDP over IP (12 + 8 + 20) or TCP over IP (20 + 20) can result in a per packet header overhead of 40 bytes. This is significant and can severely reduce the capacity of any wireless system.

There are two ways to reduce upper layer overheads and to improve the number of supported users: payload header suppression (PHS) and robust header compression (ROHC). PHS is a WiMAX feature. It allows the sender to not send fixed portions of the headers and can reduce the 40 byte header overhead down to 3 bytes. ROHC, specified by the Internet Engineering Task Force (IETF), is another higher layer compression scheme. It can reduce the higher layer overhead to 1–3 bytes. In our analysis, we use ROHC-RTP packet type 0 with R-0 mode. In this mode, all RTP sequence number functions are known to the decompressor. This results in a net higher layer overhead of just 1 byte [30,31].

Table 8.7   WiMAX Application Classes

Classes	Applications	Bandwidth Guideline		Latency Guideline		Jitter Guideline		QoS Classes
1	Multiplayer interactive gaming	Low	50 kbps	Low	<25 ms	N/A		rtPS and UGS
2	VoIP and video conference	Low	32–64 kbps	Low	<160 ms	Low	<50 ms	UGS and ertPS
3	Streaming media	Low to high	5 kbps to 2 Mbps	N/A		Low	<100 ms	rtPS
4	Web browsing and instant messaging	Moderate	10 kbps to 2 Mbps	N/A		N/A		nrtPS and BE
5	Media content downloads	High	>2 Mbps	N/A		N/A		nrtPS and BE

Table 8.8   Workload Characteristics

Parameters	Mobile TV	VoIP	Data
Type of transport layer	RTP	RTP	TCP
Average packet size (bytes)	983.5	20.0	1200.2
Average data rate (kbps) w/o headers	350.0	5.3	14.5
UL:DL traffic ratio	0	1	0.006
Silence suppression (VoIP only)	N/A	Yes	N/A
Fraction of time user is active		0.5	
ROHC packet type	1	1	TCP
Overhead with ROHC (bytes)	1	1	8
PHS	No	No	No
MAC SDU size with header	984.5	21.0	1208.2
Data rate (kbps) after headers	350.4	5.6	14.6
Bytes/frame per user (DL)	219.0	3.5	9.1
Bytes/frame per user (UL)	0.0	3.5	0.1

For small packet size workloads, such as VoIP, header suppression and compression can make a significant impact on the capacity. We have seen several published studies that use uncompressed headers, resulting in significantly reduced performance, which would not be the case in practice.

Table 8.9   Downlink Fragmentation and Packing Subheaders

Parameters	Mobile TV	VoIP	Data
Average packet size with higher level header (bytes)	984.5	21.0	1208.2
Simple scheduler (every frame scheduling)			
Bytes/5 ms frame per user	219.0	3.5	9.1
Number of fragmentation subheaders	1	1	1
Number of packing subheaders	0	0	0
Enhanced scheduler (scheduling within deadline)			
Deadline (ms)	10	60	250
Bytes/5 ms frame per user	437.9	42.0	454.9
Number of fragmentation subheaders	1	0	1
Number of packing subheaders	0	2	0

PHS or ROHC can significantly improve the capacity and should be used in any capacity planning or estimation.

One option with VoIP traffic is that of silence suppression, which if implemented can increase the VoIP capacity by the inverse of the fraction of time the user is active (not silent). As a result, in this analysis, given the silence suppression option, a number of supported users are twice that without this option.

8.2.2.2 Lower Layer Overhead

In this section, we analyze the overheads at MAC and PHY layers. Basically, there is a 6-byte MAC header and several 2-byte subheaders. For PHY overheads, both downlink and uplink overheads are discussed in detail.

8.2.2.2.1 MAC Overhead

At the MAC layer, the smallest unit is MPDU. As shown in Figure 8.10, each MPDU has at least 6 bytes of MAC header and a variable length payload consisting of a number of optional subheaders, data, and an optional 4-byte CRC. The optional subheaders include fragmentation, packing, mesh, and general subheaders. Each of these is 2 bytes long.

In addition to generic MPDUs, there are bandwidth request protocol data units (PDUs). These are 6 bytes in length. Bandwidth requests can also be piggybacked on data PDUs as a 2-byte subheader.

Consider fragmentation and packing subheaders. As shown in Table 8.9, the user bytes per frame in the downlink are 219, 3.5, and 9.1 bytes for Mobile TV, VoIP, and Web, respectively. In each frame, a 2-byte fragmentation subheader is needed for all types of traffic. Packing is not used for the simple scheduler used here.

However, in the enhanced scheduler, given a variation of the deadline, packing multiple SDUs is possible. Table 8.9 also shows an example when the deadline is put into consideration. In this analysis, the deadlines of Mobile TV, VoIP, and Web traffic are set to 10, 60, and 250 ms. As a result, 437.9, 42, and 454.9 bytes are allocated per user. These configurations result in one 2-byte fragmentation overhead for Mobile TV and Web traffic but two 2-byte packing overheads with no fragmentation for VoIP. Table 8.10 also shows a detailed explanation of fragmentation and packing overheads in the downlink. Note that the calculation for uplink is very similar.

8.2.2.2.2 Downlink Overhead

In the DL subframe, overhead consists of preamble, FCH, DL-MAP, and UL-MAP. The MAP entries can result in a significant amount of overhead since they are repeated four times. The WiMAX Forum recommends using compressed MAP [2], which reduces the DL-MAP entry overhead to 11 bytes including 4 bytes for CRC [1,2]. The fixed UL-MAP is 6 bytes long with an optional 4-byte CRC. With a repetition code of 4 and QPSK 1/2, both fixed DL-MAP and UL-MAP take up 16 slots.

UL preamble	MAC/BW-REQ header	Other subheaders	Data	CRC (optional)

Figure 8.10 UL burst preamble and MPDU.

Table 8.10 Capacity Evaluation using a Simple and Enhanced Scheduler

Parameters	Simple Scheduler			Enhanced Scheduler		
	Mobile TV	VoIP	Data	Mobile TV	VoIP	Data
MAC SDU size with header (bytes)	984.5	21.0	1208.2	984.5	21.0	1208.2
Data rate (kbps) with upper layer headers	350.4	5.6	14.6	350.4	5.6	14.6
Deadline (ms)	N/A	N/A	N/A	10	60	250
(a) DL						
Bytes/5 ms frame per user (DL)	219.0	3.5	9.1	437.9	42.0	454.9
Number of fragmentation subheaders	1	1	1	1	0	1
Number of packing subheaders	0	0	0	0	2	0
DL data slots per user with MAC header + packing and fragmentation subheaders	38	2	3	75	9	78
Total slots per user (data + DL-MAP IE + UL-MAP IE)	46	18	19	83	25	94
Number of users (DL) slot-based	8	22	21	8	192	200
Number of users (DL) upper bound (with rounding error)	9	35	33	10	269	233
(b) UL						
Bytes/5 ms frame per user (UL)	0.0	3.5	0.1	0.0	42.0	2.9
# of fragmentation subheaders	0	1	1	1	0	1
# of packing subheaders	0	0	0	0	2	0
UL data slots per user with MAC header+packing and fragmentation subheaders	0	2	2	0	9	2
Number of users (UL)	∞	87	87	∞	228	4350
Number of users (min of UL and DL)	9	35	33	10	228	233
Number of users with silence suppression	9	70	33	10	456	233

The variable part of DL-MAP consists of one entry per burst and requires 60 bits per entry. Similarly, the variable part of UL-MAP consists of one entry per burst and requires 52 bits per entry. These are all repeated four times and use only QPSK 1/2 MCS. It should be pointed out that repetition consists of repeating slots (and not bytes). Thus, both DL and UL MAP entries also take up 16 slots each per burst.

Equations 8.2 through 8.5 show the details of UL and DL MAPs overhead computation.

$$UL\_MAP(bytes) = \frac{(48 + 52 \times \#UL\_users)}{8}, \tag{8.2}$$

$$DL\_MAP(bytes) = \frac{(88 + 60 \times \#DL\_users)}{8}, \tag{8.3}$$

$$UL\_MAP(slots) = \lceil UL\_MAP/S_i \rceil \times r, \tag{8.4}$$

$$DL\_MAP(slots) = \lceil DL\_MAP/S_i \rceil \times r. \tag{8.5}$$

Here, r is the repetition factor and S_i is the slot size (bytes) given the ith MCS. Note that basically QPSK 1/2 is used for the computation of UL and DL MAPs.

8.2.2.2.3 Uplink Overhead

The UL subframe also has fixed and variable parts (see Figure 8.3). Ranging and contention are in the fixed portion. Their size is defined by the network administrator. These regions are allocated not in units of slots but in units of "transmission opportunities." For example, in CDMA initial ranging, one opportunity is six subchannels and two symbol times.

The other fixed portion is CQI and ACK. These regions are also defined by the network administrator. Obviously, more fixed portions are allocated; less number of slots is available for the user workloads. In our analysis, we allocated three OFDM symbol columns for all fixed regions.

Each UL burst begins with a UL preamble. One OFDM symbol is used for a short preamble and two for a long preamble. In this analysis, we do not consider one short symbol (a fraction of one slot); however, users can add an appropriate size of this symbol to the analysis.

8.2.3 Pitfalls

Many WiMAX analyses ignore the overheads described in Section 8.2.2, namely, UL-MAP, DL-MAP, and MAC overheads. In this section, we show that these overheads have a significant impact on the number of users supported. Since some of these overheads depend upon the number of users, the scheduler needs to be aware of this additional need while admitting and scheduling the users.

Given the user workload characteristics and the overheads discussed so far, it is straightforward to compute the system capacity for any given workload. Using the slot capacity indicated in Table 8.4, for various MCSs, we can compute the number of users supported.

One way to compute the number of users is simply to divide the channel capacity by the bytes required by the user payload and overhead [1]. This is shown in Table 8.10. The table assumes QPSK 1/2 MCS for all users. This can be repeated for other MCSs. The final results are as shown in Figure 8.11. The number of users supported varies from 2 to 46 depending upon the workload and the MCS.

The computation to calculate the number of supported users is illustrated below in Equations 8.6 through 8.8. However, the number of supported users calculated in this analysis is an approximate from Equations 8.6 through 8.8 by considering the ceiling function of #users and #the size of information element as one single term.

$$\#DL\_slots = \left\lceil \frac{(DL\_MAP + CRC + \#DL\_users \times DIE)}{S_i} \right\rceil \times r$$

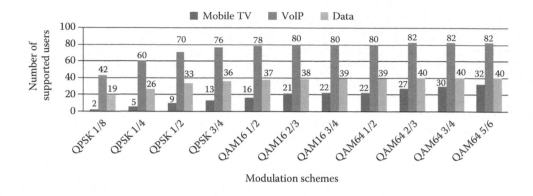

Figure 8.11 Number of users supported in lossless channel (simple scheduler).

$$+ \left\lceil \frac{(\text{UL\_MAP} + \text{CRC} + \#\text{UL\_users} \times \text{UIE})}{S_i} \right\rceil \times r$$
$$+ \#\text{DL\_users} \times \left\lceil \frac{D}{S_k} \right\rceil, \tag{8.6}$$

$$\#\text{UL\_slots} = \#\text{UL\_users} \times \left\lceil \frac{D}{S_k} \right\rceil, \tag{8.7}$$

$$D = B + \text{MAC\_header} + \text{subheaders}. \tag{8.8}$$

Here, D is the data size (per frame) including overheads, B is the bytes per frame, *MAC_header* is 6 bytes. *Subheaders* are fragmentation and packing subheaders, 2 bytes each if present. DIE and UIE are the size of DL-MAP and UL-MAP information elements (IEs). Note that DL and UL MAPs are fixed MAP parts and also in terms of bytes. Again, r is the repetition factor and S_i is the slot size (bytes) given the ith MCS. #DL_slots is the total number of DL slots without preamble and #UL_slots is the total number of UL slots without ranging, ACK, and CQICH.

For example, consider VoIP with QPSK 1/2 (slot size=6 bytes) and repetition of four, for downlink from Equation 8.6, the number of DL users is 35 users. The derivation is shown below:

$$\#\text{DL\_slots} = 420 = \left\lceil \frac{(11 + 4 + \# \text{DL\_users} \times (60/8))}{6} \right\rceil \times 4$$
$$+ \left\lceil \frac{(6 + \# \text{DL\_users} \times (52/8))}{6} \right\rceil \times 4$$
$$+ N\# \text{DL\_users} \times \left\lceil \frac{11.5}{6} \right\rceil.$$

For uplink, from Equations 8.7 and 8.8, the number of UL users is 87.

$$\#\text{UL\_slots} = 175 = \#\text{UL\_users} \times \lceil (3.5 + 6 + 2)/6 \rceil.$$

Finally, after calculating the number of supported users for both DL and UL, the total number of supported users is the minimum number of those two. In this example, the total number of supported users is 35, (minimum of 35 for downlink and 87 for uplink). The downlink is the bottleneck mostly due to a large MAP overhead. Together with silence suppression, the absolute number of supported users can be up to $2 \times 35 = 70$ users. Figure 8.10 shows an approximate number of supported users for various MCSs.

The main problem with the analysis presented above is that it assumes that every user is scheduled in every frame. Since there is a significant per burst overhead, this type of allocation will result in too much overhead and too little capacity. Also, since every packet (SDU) is fragmented, a 2-byte fragmentation subheader is added to each MPDU.

What we discussed above is a common pitfall. The analysis assumes a dumb scheduler. A smarter scheduler will try to aggregate payloads for each user thus minimizing the number of bursts. We call this an enhanced scheduler. It works as follows. Given n users with any particular workload, we divide the users in k groups of n/k users each. The first group is scheduled in the first frame, the second group is scheduled in the second frame, and so on. The cycle is repeated every k frames. Of course, k should be selected to match the delay requirements of the workload. For example, with VoIP users, a VoIP packet is generated every 30 ms, but assuming that 60 ms is an acceptable delay, we can schedule a VoIP user every 12th WiMAX frame (recall that each WiMAX frame is 5 ms) and send two VoIP packets in one frame as compared to the previous scheduler, which would send 1/6 of the VoIP packet in every frame thereby aggravating the problem of small payloads. A 2-byte×2 packing overhead has to be added to the MAC payload along with the two SDUs.

Table 8.10 also shows the capacity analysis for the three workloads with QPSK 1/2 MCS and the enhanced scheduler. The results for other MCSs can be similarly computed. These results are plotted in Figure 8.12. Note that the number of users supported has gone up from 2 to 800. Compared to Figure 8.10, there is a capacity improvement by a factor of 1–20 depending upon the workload and MCS.

> Proper scheduling can change the capacity by an order of magnitude. Making less frequent but bigger allocations can reduce the overhead significantly.

The number of supported users for this scheduler is derived from the equations described in when we calculated these with a simple scheduler; however, again the scheduler allocates as large a size as possible given the deadlines; for example, for Mobile TV with 10 ms deadline instead of 219 bytes, the scheduler allocates 437.9 bytes within a single frame, and for VoIP with 60 ms deadline,

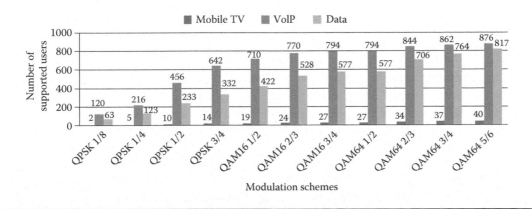

Figure 8.12   Number of users supported in lossless channel (enhanced scheduler).

instead of 3.5 bytes per frame, it allocates 42 bytes and this results in two packing overheads instead of one fragmentation overhead.

Moreover, in Table 8.10, the number of supported users for VoIP is 228. This number is calculated over 1 s since 42 bytes are allocated for one user in every 60 ms. As a result, over 1 s, the number of supported users is shown as

$$\left\lceil \frac{\#slots\_subframe}{\#slots\_aggregated\_users} \right\rceil \times \left(\frac{deadline}{5\ ms} \right). \tag{8.9}$$

With the configuration in Table 8.10, the number of supported users is $\lceil 175/9 \rceil \times 60/5 = 228$. Moreover with silence suppression, the absolute number of supported users is $2 \times 228 = 456$. Note that the computation of DL users is derived from Equations 8.6 through 8.8 and then Equation 8.9 can be applied. A calculation of Mobile TV and data are similar to that for VoIP.

Note that the per user overheads impact the downlink capacity more than the uplink capacity. The downlink subframe has DL-MAP and UL-MAP entries for all DL and UL bursts and these entries can take up a significant part of the capacity; hence, minimizing the number of bursts increases the capacity.

Note that there is a limit to aggregation of payloads and minimization of bursts. First, the delay requirements for the payload should be met and, hence, a burst may have to be scheduled even if the payload size is small. In these cases, multiuser bursts in which the payload for multiple users is aggregated in one DL burst can help reduce the number of bursts. This is allowed by the IEEE 802.16e standards and applies only to the downlink bursts.

The second consideration is that the payload cannot be aggregated beyond the frame size. For example, with QPSK 1/2, a Mobile TV application will generate enough load to fill the entire DL subframe every 10 ms or every two frames. This is much smaller than the required delay of 30 ms between the frames.

8.2.4 Section Summary

In this section, we explained how to compute the capacity of a WiMAX system and account for various overheads. We illustrated the methodology using three sample workloads consisting of Mobile TV, VoIP, and data users.

In order to maximize the number of supported users, especially voice users, the overheads, namely, DL-MAP and UL-MAP, should be reduced. One technique is that the scheduler must aggregate the data and packets. Note that in Ref. [32], a persistent resource allocation for voice users was investigated. With the persistent technique, the burst allocation for voice traffic is fixed in a particular location for each WiMAX frame. The IEs are not required for the following frame. However, this reduces the flexibility of downlink frame mapping.

Analysis such as the one presented in this section can be easily programmed in a simple program or a spreadsheet and the effects of various parameters can be analyzed instantaneously. This can be used to study the sensitivity to various parameters so that parameters that have a significant impact can be analyzed in detail by simulation. This analysis can also be used to validate simulations.

Note that this capacity evaluation is based on several assumptions. For example, we did not include the overhead of bandwidth request mechanisms or the effect of two-dimensional downlink mapping. We assumed an error-free channel and, therefore, we did not model (H)ARQ. The error-prone channel analysis is presented in Ref. [17]. We also assumed that fixed UL-MAP is always present in the DL subframe even if there is no UL traffic such as in Mobile TV. Dynamic properties of the different traffic types, time and frequency channel, noise, cell interference, and automatic request strategies are also important in the evaluation of the numbers of supported users.

8.3 Uplink Bandwidth Request/Grant Mechanisms

Consider the BS scheduler. This scheduler has to decide slot allocation for traffic going to various MSs. It also has to grant slots to various MSs to be able to send the traffic upward. For downlink, the BS has complete knowledge of the traffic such as queue lengths and packet sizes to help make the scheduling decisions.

For uplink traffic, the MSs need to send bandwidth request (BWR) packets to the BS, which then decides how many slots are granted to each MS in the subsequent uplink subframes. Although originally the standard allowed BS to allocate the bandwidth per connection, namely grant per connection (GPC), or per station, namely grant per subscriber station (GPSS), the latest version of the standard recommends only GPSS and leaves the allocation for each connection to the MS scheduler.

Basically, there are two types of BWRs: incremental and aggregate. When the BS receives an incremental BWR, it adds the requested bytes to the previous outstanding amount. On the other hand, an aggregate BWR indicates total demand overwriting any previous unsatisfied demands. There are also a number of ways to request bandwidth. These methods can be categorized as implicit or explicit based on the need for polling as shown in Tables 8.11 and 8.12. Explicit bandwidth request means the BS needs to explicitly allocate an opportunity for the MS to transmit a BWR. However, for implicit bandwidth requests, the MS can send the BWR to the BS without needing additional allocation for the request itself. The MS does not need to wait the transmission opportunity from the BS. As indicated in these two tables, the BWR mechanisms are: unsolicited request, poll-me bit, piggybacking, bandwidth stealing, codeword over channel quality indicator channel (CQICH), CDMA code-based BWR, unicast polling, multicast polling, broadcast polling, and group polling. Table 8.13 provides a comparison of these mechanisms. The optimal way to request the bandwidth for a given QoS requirement is still an open research area [33–42].

In addition, we briefly discuss the issue of bandwidth request mechanisms for each QoS class. Obviously, there is a tradeoff between the flexibility of resource utilization and QoS requirements. For example, unicast polling can guarantee the delay; however, resources can be wasted if there are no enqueued packets at the MS. On the other hand, multicast or broadcast polling may utilize the resource but the delay cannot be guaranteed.

8.3.1 Consider UGS

There is no polling (static allocation) but the scheduler needs to be aware of the resource requirements and should be able to schedule the flows so that the resources can be optimized. For example, given 10 UGS flows, each flow requiring 500 bytes every five frames, if only 2500 bytes are allowed in one frame, all 10 flows cannot start in the same frame. The scheduler needs to rearrange (phase) these flows in order to meet the delay jitter while maximizing frame utilization. The problem gets more difficult when the UGS flows dynamically join and leave.

8.3.2 Consider the Delay Requirements

Polling in every frame is the best way to ensure the delay bound; however, this results in a significant polling overhead as mentioned earlier. Some research papers recommend polling in every video

Table 8.11 Implicit Bandwidth Request Mechanisms

Types	Mechanisms	Overhead	QoS Classes
Unsolicited request	Periodically allocates bandwidth at setup stage	N/A	UGS and ertPS
Poll-me bit (PM)	Asks BS to poll non-UGS connections	N/A (implicitly in MAC header)	UGS
Piggybacking	Piggyback BWR over any other MAC packets being sent to the BS	Grant management (GM) subheader (2 bytes)	ertPS, rtPS, nrtPS, and BE
Bandwidth stealing	Sends BWR instead of general MAC packet	BWR (6 bytes = MAC header)	nrtPS and BE
Contention region (WiMAX)	MSs use contention regions to send BWR.	Adjustable	ertPS, nrtPS, and BE
Codeword over CQICH	Specifies codeword over CQICH to indicate the request to change the grant size	N/A	ertPS
CDMA code-based BWR (Mobile WiMAX)	MS chooses one of the CDMA request codes from those set aside for bandwidth requests.	Six subchannels over 1 OFDM symbol for up to 256 codes	nrtPS and BE

frame such as one every 33 ms [43] because video frames are generated every 30–40 ms. Without the arrival of information of packets, it is difficult for BS to guarantee the delay requirements. As a result, the polling optimization is still in an open research topic.

8.3.3 Consider rtPS

There is a strict or loose requirement of delay. If any packets are over the deadline, those packets will be dropped. Video applications also have their own characteristics such as the size and the duration of

Table 8.12 Explicit Bandwidth Request Mechanisms

Types	Mechanisms	Overhead	QoS Classes
Unicast polling	BS polls each MS individually and periodically	BWR (6 bytes) per user	ertPS, rtPS, nrtPS, and BE
Multicast polling	BS polls a multicast group of MSs	BWR (6 bytes) per multicast	ertPS, nrtPS, and BE
Broadcast polling	BS polls all MSs	Adjustable	ertPS, nrtPS, and BE
Group polling	BS polls a group of MSs periodically	BWR (6 bytes) per group	ertPS, rtPS, nrtPS, and BE

Table 8.13   Comparisons of Bandwidth Request Mechanisms

Types	Pros	Cons
Unsolicited request	No overhead and meet guaranteed latency of MS for real-time service	Wasted bandwidth if bandwidth is granted and the flow has no packets to send
Poll me bit	No overhead	Still needs the unicast polling
Piggybacking	Do not need to wait for poll, less overhead; 2 bytes versus 6 bytes	N/A
Bandwidth stealing	Do not need to wait for poll	6 bytes overhead
Contention region	Reduced polling overhead	Need the backoff mechanism
Codeword over CQICH	Makes use of CQI channel	Limit number of bandwidth on CQICH
CDMA code-based BWR	Reduced polling overhead compared to contention region	Results in one more frame delay compared to contention region
Unicast polling	Guarantees that MS has a chance to ask for bandwidth	More overhead (6 bytes per MS) periodically
Multicast, broadcast and group polling	Reduced polling overhead	Some MSs may not get a chance to request bandwidth; need contention resolution technique

intra-coded pictures (I-frame), bi-directionally predicted pictures (B-frame), and predicted pictures (P-frame) frames for MPEG video. Basically, I-frames are very large and occur periodically. Therefore, the scheduler can use this information to avoid overlapping among connections. The BS can phase new connections so that the new connection's I-frames do not overlap with the existing connections' I-frames [44].

8.3.4 Consider ertPS

This service is used for VoIP traffic, which has active and silent periods. As an example, if adaptive multirate (AMR) coding is used, only 33 bytes are sent every 20 ms during the active periods and 7 bytes during silent periods. The silent period can be up to 60% [45–47]. Schedulers for voice users need to be aware of these silent periods. Bandwidth is wasted if an allocation is made when there are no packets (which happens with UGS). With rtPS or ertPS in the uplink direction, although the throughput can be optimized, the deadline is the main factor to be considered. The key issue is how to let the BS know whether there is a packet to transmit or not. The polling mechanism should be smart enough so that once there is traffic, the BS allocates a grant for the MS in order to send the bandwidth request and then transmit the packet within the maximum allowable delay. Moreover, BS does not need to allocate the bandwidth during the silent period. To indicate the end of a silent period, an MS can piggyback a zero bandwidth request, make use of a reserved bit in the MAC header to indicate their on/off states [45], or send a management message directly to the BS.

During the active period, the MS can use piggybacking or bandwidth stealing mechanisms in order to reduce the polling overhead and delay, otherwise use contention region (WiMAX) or CDMA bandwidth request (Mobile WiMAX). The scheduler should be aware of this and should make predictions accordingly.

There is also a provision for a contention region and for CDMA bandwidth requests. The number of contention slots should be close to the number of connections enqueued, so there is no extra delay in contention resolution. Obviously, this region should be adaptively changed over time. Therefore, BS needs to make a prediction on how many MSs and/or connections are going to send the bandwidth request.

In addition, recent research shows how to optimize the backoff algorithm including backoff start and stop timer [41]. In fact, the efficiency is just 33% with the random binary exponential backoff [48].

8.3.5 Consider nrtPS

The only constraint for nrtPS is the minimum guaranteed throughput. Polling is allowed for this service. Some proposed schemes recommend polling intervals of over 1 s [43]. The polling should be issued if and only if the average rate that is calculated from proportional fairness (PF) is less than the minimum reserved rate [49]. We will describe PF in Section 8.4.

8.3.6 Consider BE

All bandwidth request mechanisms are allowed for BE but contention resolution is most commonly used. The main issue for BE is fairness. The problem is whether the scheduler should be fair in a short term or a long term. For example, over 1 s, a flow can transmit 1 byte every 5 ms or 200 bytes every 1 s. Also, the scheduler should prevent starvation.

As can be seen from this discussion, with the combination of different types of traffic and many types of bandwidth request mechanisms, WiMAX scheduler design is complicated.

8.3.7 Section Summary

In this section, we described different kinds of uplink bandwidth request/grant mechanisms. The issues of these mechanisms are categorized and discussed in each QoS class.

8.4 WiMAX Scheduler

In this section, we focus on Mobile WiMAX scheduling architecture: downlink/uplink at the BS and uplink at the MS. Several factors for scheduler design are discussed such as QoS assurance, throughput optimization, fairness, energy consumption, and implementation complexity. Finally, a brief survey of proposed WiMAX scheduling algorithms is presented. The algorithms are classified according to their channel awareness/unawareness.

A connection admission control (CAC) also plays an important role in assuring the QoS requirements and needs to be designed along with the scheduler. Before joining the network, the MSs need to have a permission from the BS to transmit data with a QoS agreement. The CAC basically maintains the current system load and QoS parameters for each existing connection. Then, it can make a decision if a new connection should be admitted and, if admitted, what

QoS the BS can provide. It should be obvious that if the CAC cannot support at least the minimum reserved rate for a new flow, that connection should be rejected. Otherwise, the QoS requirements of the existing flows can be broken. For example, instead of admitting another UGS flow, a BE flow is accepted if there is no way to guarantee the maximum allowable delay. We do not include the survey of CAC in this chapter; however, further information can be found in Refs. [50–55].

Scheduling is the main component of the MAC layer that helps ensure QoS for various service classes. The scheduler works as a distributor to allocate the resources among MSs. The allocated resource can be defined as the number of slots and then these slots are mapped into a number of subchannels and time duration. In OFDMA, the smallest logical unit for bandwidth allocation is a slot. The definition of a slot depends upon the direction of traffic and subchannelization modes.

The mapping process from a logical subchannel to multiple physical subcarriers is called a permutation. PUSC, discussed above, is one of the permutation modes. Others include fully used subchannelization (FUSC) and adaptive modulation and coding (band-AMC). The term "band-AMC" distinguishes the permutation from AMC MCS selection procedure. Basically, there are two types of permutations: distributed and adjacent. The distributed subcarrier permutation is suitable for mobile users, while the adjacent permutation is for fixed (stationary) users. Detailed information again can be found in Ref. [1].

Scheduler designers need to consider the allocations logically and physically. Logically, the scheduler should calculate the number of slots based on QoS service classes. Physically, the scheduler needs to select which subchannels and time intervals are suitable for each user. The goal is to minimize power consumption, to minimize bit error rate, and to maximize the total throughput. Notice that in PUSC mode, the scheduler designer has less flexibility to choose a set of subcarriers for the resource allocation since randomness is introduced during the subchannelization process.

There are three distinct scheduling processes: two at the BS (one for downlink and the other for uplink) and one at the MS (for uplink) as shown in Figure 8.13. At the BS, packets from the upper layer are put into different queues, which ideally is per-CID queue in order to prevent head of line (HOL) blocking. However, the optimization of queue can be done and the number of required queues can be reduced. Then, based on the QoS parameters and some extra information, such as the channel state condition, the DL-BS scheduler decides which queue to service and how many service data units (SDUs) should be transmitted to the MSs.

Since the BS controls the access to the medium, the second scheduler—the UL-BS scheduler—makes the allocation decision based on the bandwidth requests from the MSs and the associated QoS parameters. Several ways to send bandwidth requests were described earlier in Section 8.3. Finally, the third scheduler is at the MS. Once the UL-BS grants the bandwidth for the MS, the MS scheduler decides which queues should use that allocation. Recall that while the requests are per connections, the grants are per MS and the MS is free to choose the appropriate queue to service. The MS scheduler needs a mechanism to allocate the bandwidth in an efficient way.

8.4.1 Design Factors

To decide which queue to service and how much data to transmit, one can use a very simple scheduling technique such as first in first out (FIFO). This technique is very simple but unfair. A little more complicated scheduling technique is round robin (RR). This technique provides the fairness among the users but it may not meet the QoS requirements. Also, the definition of fairness is questionable if the packet size is variable. In this section, we describe the factors that the scheduler designers need to consider. Then, we present a survey of recent scheduling proposals in Section 8.4.2.

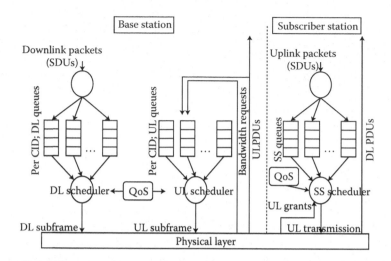

Figure 8.13 Component schedulers at BS and MSs.

QoS Parameters: The first factor is whether the scheduler can assure the QoS requirements for various service classes. The main parameters are the minimum reserved traffic, the minimum sustained traffic, the maximum allowable delay (latency), the tolerated jitters, the traffic priority, the unsolicited grant interval, and the unsolicited polling interval. Earliest deadline first (EDF) [56] is an example of a technique used to guarantee the delay requirement. Similarly, largest weighted delay first (LWDF) has been used to guarantee the minimum throughput [57].

Throughput Optimization: Since the resources in wireless networks are limited, another important consideration is how to maximize the total system throughput. The metrics here could be the maximum number of supported MSs or whether the link is fully utilized. One of the best ways to represent throughput is using the goodput. The overheads include MAC overhead, fragmentation and packing overheads, and burst overhead. This leads to the discussion of how to optimize the number of bursts per frame and how to pack or fragment the SDUs into MPDUs (see Section 8.1.2).

The bandwidth request is indicated in number of bytes. This does not translate straightforwardly to number of slots since one slot can contain different numbers of bytes depending upon the modulation technique used. For example, with QPSK 1/2, the number of bits per symbol is 1. Together with PUSC at 10 MHz system bandwidth and 1024 FFT, that leads to 6 bytes per slot. If the MS asks for 7 bytes, the BS needs to give two slots thereby consuming 12 bytes. Moreover, the percentage of packets lost is also important. The scheduler needs to use the channel state condition information and the resulting bit error rate in deciding the MCS for each user.

Fairness: Aside from assuring the QoS requirements, the left-over resources should be allocated fairly. The time to converge to fairness is important since the fairness can be defined as short term or long term. The short-term fairness implies long-term fairness but not vice versa [58].

Energy Consumption and Power Control: The scheduler needs to consider the maximum power allowable particularly in uplink direction. Given the BER and SNR that the BS can accept for transmitted data, the scheduler can calculate the suitable power to use for each MS depending upon their location. For mobile users, the power is very limited, so the energy consumption and power control are additional uplink issues. Therefore, UL scheduler needs to optimize the transmission power.

Implementation Complexity: Since the BS has to handle many simultaneous connections and decisions have to be made within 5 ms WiMAX frame duration [2], the scheduling algorithms have to be simple, fast, and use minimum resources such as memory. The same applies to the scheduler at the MS.

Scalability: The algorithm should efficiently operate as the number of connections increases.

8.4.2 Classification of Schedulers

In this section, we present a survey of recent scheduler proposals for WiMAX. Most of these proposals focus on the scheduler at BS, especially DL-BS scheduler. For this scheduler, the queue length and packet size information are easily available. To guarantee the QoS for MS at UL-BS scheduler, the polling mechanism is involved. Once the QoS can be assured, how to split the allocated bandwidth among the connections depends on the MS scheduler.

Recently published scheduling techniques for WiMAX can be classified into two main categories: channel-unaware schedulers and channel-aware schedulers as shown in Figure 8.14. Basically, the channel-unaware schedulers use no information of the channel state condition in making the scheduling decision. In the discussion that follows, we apply the metrics discussed to schedulers in each of these two categories.

Channel-unaware schedulers generally assume error-free channel since it makes it easier to prove assurance of QoS. However, in wireless environment where there is a high variability of radio link such as signal attenuation, fading, interference, and noise, the channel-awareness is important. Ideally, scheduler designers should take into account the channel condition in order to optimally and efficiently make the allocation decision.

8.4.2.1 Channel-Unaware Schedulers

This type of schedulers makes no use of channel state conditions such as the power level, channel state condition, and loss rates. These basically assure the QoS requirements among five classes—mainly the delay and throughput constraints. Although jitter is also one of the QoS parameters, so far none of the published algorithms can guarantee jitter. A comparison of the scheduling disciplines is presented in Table 8.14 and also the mappings between the scheduling algorithms and the QoS classes are shown in Table 8.15.

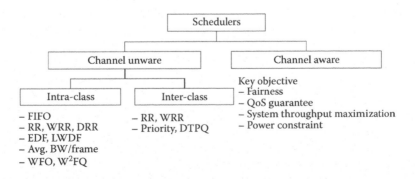

Figure 8.14 Classifications of WiMAX schedulers.

Table 8.14 Comparison of Channel-Unaware Schedulers

Scheduling	Pros	Cons
FIFO	Fast and simple	Unfair and cannot meet QoS requirements
RR	Very simple	Unfair (variable packet size), cannot meet QoS requirements
WRR	Simple; meets the throughput guarantee	Unfair (variable packet size)
DRR/DWRR	Simple, supports variable packet sizes	Not fair on a short time scale
Priority	Simple, meets the delay guarantee	Some flows may starve, lower throughput
DTPQ	Trades-off the packet loss rate of rtPS and average data throughput of nrtPS	Lower throughput
EDF	Meets the delay guarantee	Nonwork conservative
LWDF	Guarantees the minimum throughput	N/A
WFQ	With proper and dynamic weight, guarantees throughput and delay, fairness	Complex
WF^{2}Q	WFQ with worst-case fairness property	Complex

8.4.2.1.1 Intra-Class Scheduling

Intra-class scheduling is used to allocate the resource within the same class given the QoS requirements.

Round Robin (RR) algorithm: Aside from FIFO, round-robin allocation can be considered the very first simple scheduling algorithm. RR fairly assigns the allocation one by one to all connections. The fairness considerations need to include whether allocation is for a given number of packets or a given number of bytes. With packet-based allocation, stations with larger packets have an unfair advantage.

Moreover, RR may be nonwork conserving in the sense that the allocation is still made for connections that may have nothing to transmit. Therefore, some modifications need to be made to skip the idle connections and allocate only to active connections. However, now the issues become how to calculate average data rate or minimum reserved traffic at any given time and how to allow for the possibility that an idle connection later has more traffic than average? Another issue is what should be the duration of fairness? For example, to achieve the same average data rate, the scheduler can allocate 100 bytes every frame for 10 frames or 1000 bytes every 10th frame.

Since RR cannot assure QoS for different service classes, RR with weight, weighted round robin (WRR), has been applied for WiMAX scheduling [59–61]. The weights can be used to adjust for

Table 8.15 Comparison of Channel-Aware Schedulers

Category	Scheduling Algorithms	Pros/Cons	Class of Service
Fairness	Variation of PFS [49, 80–83]	Achieves long-term fairness but cannot guarantee the delay constraint	BE
QoS Guaranteed (minimum through-put and delay)	Variation of M-LWDF [85–88]	Meet the throughput and delay guarantee with threshold probability	ertPS, rtPS, and nrtPS
System throughput maximization	Variation of maximum C/R [89–91]	Maximizes the total system throughput but cannot meet QoS requirement especially delay as well as unfairness	BE
Power constraint	LWT [92], linear programming [92,93]	Minimizes the power consumption but cannot meet QoS requirement especially delay as well as unfairness	BE

the throughput and delay requirements. Basically, the weights are in terms of queue length, packet delay, and number of slots. The weights are dynamically changed over time. In order to avoid the issue of missed opportunities, variants of RR such as DRR or deficit weighted round robin (DWRR) can be used for the variable size packets [59]. The main advantage of these variations of RR is their simplicity. The complexity is $O(1)$ compared to $O(log(N))$ and $O(N)$ for other fair queuing algorithms. Here, N is the number of queues.

Weighted Fair Queuing algorithm (WFQ): WFQ is an approximation of general processor sharing (GPS). WFQ does not make the assumption of infinitesimal packet size. Basically, each connection has its own FIFO queue and the weight can be dynamically assigned for each queue. The resources are shared in proportion to the weight. For data packets in wired networks with leaky bucket, an end-to-end delay bound can be provably guaranteed. With the dynamic change of weight, WFQ can be also used to guarantee the data rate. The main disadvantage of WFQ is the complexity, which could be $O(N)$.

To keep the delay bound and to achieve worst-case fairness property, a slight modification of the WFQ, worst-case fair weighted fair queueing (WF^{2}Q) was introduced. Similar to WFQ, WF^{2}Q uses a virtual time concept. The virtual finish time is the time GPS would have finished sending the packet. WF^{2}Q looks for the packet with the smallest virtual finishing time and whose virtual start time has already occurred instead of searching for the smallest virtual finishing time of all packets in the queue. The virtual start time is the time GPS starts to send the packet [62]. Note that Iera et al. [62] also introduced the concept of flow compensation with leading and lagging flows.

In achieving the QoS assurance, the procedure to calculate the weight plays an important role. The weights can be based on several parameters. Aside from queue length and packet delay we mentioned above, for uplink the size of bandwidth request can be used to determine the weight of queue (the larger the size, the more the bandwidth) [63]. The ratio of a connection's average data rate to the total average data rate can be used to determine the weight of the connection [64]. The minimum reserved rate can be used as the weight [48]. The pricing can be also used as a weight [65]. Here, the goal is to maximize service provider revenue.

Delay-based algorithms: This set of schemes is specifically designed for real-time traffic such as UGS, ertPS, and rtPS service classes, for which the delay bound is the primary QoS parameter and basically the packets with unacceptable delays are discarded. EDF is the basic algorithm for the scheduler to serve the connection based on the deadline. LWDF [57] chooses the packet with the largest delay to avoid missing its deadline.

Note that variants of RRs, WFQs, and delay-based algorithms can resolve some of the QoS requirements. However, there are no published papers considering the tolerated delay jitter in the context of WiMAX networks. Especially for UGS and ertPS, the simple idea is to introduce a zero delay jitter by the fragmentation mechanism. Basically, BS transfers the last fragmented packet at the end of period. However, this fragmentation increases the overhead and also requires fixed buffer size for two periods. Compared to EDF, this simple technique may require more bursts. This needs to be investigated further.

8.4.2.1.2 Interclass Scheduling

As shown in Figure 8.14, RR, WRR, and priority-based mechanism have been applied for interclass scheduling in the context of WiMAX networks. The main issue for interclass is whether each traffic class should be considered separately, that is, have its own queue. For example, in Ref. [66] rtPS and nrtPS are put into a single queue and moved to the UGS (highest priority) queue once the packets approach their deadline. Similarly, in Ref. [67], UGS, rtPS, and ertPS queues are combined to reduce the complexity. Another issue here is how to define the weights and/or how much resources each class should be served. There is a loose bound on service guarantees without a proper set of weight values.

Priority-based algorithm (PR): In order to guarantee the QoS for different classes of service, priority-based schemes can be used in a WiMAX scheduler [67–69]. For example, the priority order can be: UGS, ertPS, rtPS, nrtPS, and BE, respectively. Or packets with the largest delay can be considered at the highest priority. Queue length can be also used to set the priority level, for example more bandwidth is allocated to connections with longer queues [70].

Delay Threshold Priority Queuing (DTPQ) [71] was proposed for use when both real-time and non-real-time traffic are present. A simple solution would be to assign higher priority to real-time traffic but that could harm the non-real-time traffic. Therefore, urgency of the real-time traffic is taken into account only when the head-of-line packet delay exceeds a given delay threshold. This scheme is based on the tradeoff of the packet loss rate performance of rtPS with average data throughput of nrtPS with a fixed data rate. Rather than fixing the delay, the author also introduced an adaptive delay threshold-based priority queuing scheme that takes both the urgency and channel state condition for real-time users adaptively into consideration [72].

The direct negative effect of priority is that it may starve some connections of lower priority service classes. The throughput can be lower due to increased number of missed deadlines for the lower service classes' traffic. To mitigate this problem, deficit fair priority queuing (DFPQ) with a counter was introduced to maintain the maximum allowable bandwidth for each service class [73,74]. The counter decreases according to the size of the packets. The scheduler moves to another class once the counter falls to zero.

To sum up, since the primary goal of a WiMAX scheduler is to assure the QoS requirements, the scheduler needs to support at least the five basic classes of services with QoS assurance. To ensure this, some proposed algorithms have indirectly applied or modified existing scheduling disciplines for each WiMAX QoS class of services. Each class has its own distinct characteristics such as the hard-bound delay for rtPS and ertPS. Most proposed algorithms have applied some basic algorithms proposed in wired/wireless networks to WiMAX networks such as variations of RR and WFQ. For example, to schedule within a class, RR and WFQ are common approaches for nrtPS and BE and EDF for UGS and rtPS [69,75]. The priority-based algorithm is commonly used for scheduling between the classes. For example, UGS and rtPS are given the same priority, which is also the highest priority [63].

Moreover, the "two-step scheduler" [76] is a generic name for schedulers that try first to allocate the bandwidth to meet the minimum QoS requirements—basically the throughput in terms of the number of slots or subcarrier and time duration and delay constraints. Then, especially in WiMAX networks (OFDMA-based) in the second step, they consider how to allocate the slots for each connection. This second step of allocating slots and subcarriers is still an open research area. The goal should be to optimize the total goodput, to maintain the fairness, to minimize the power, and to optimize delay and jitter. Note that the second step usually may not be directly applied to the scheduler in PUSC mode because the randomization is made after the first step in the mobile environment.

8.4.2.2 Channel-Aware Schedulers

The scheduling disciplines we discussed so far make no use of the channel state condition. In other words, they assume perfect channel condition, no loss, and a unlimited power source. Channel unaware schedulers are widely used in wired networks. However, due to the nature of the wireless medium and the user mobility, channel awareness is important in wireless networks. For example, an MS may receive allocation but may not be able to transmit successfully due to a high loss rate and, as a result, the number of retransmission attempts may increase resulting in a reduced goodput. In this section, we discuss the use of channel state conditions in scheduling decisions.

The channel aware schemes can be classified into four classes based on the primary objective: fairness, QoS guarantee, system throughput maximization, or power optimization. A comparison of the scheduling disciplines is presented in Table 8.15. In this table, we also mapped the scheduling techniques in each category to various classes of service. In general, the scheduling techniques in QoS guarantee category can apply for service classes with minimum rate and maximum latency guarantee such as ertPS, rtPS, and nrtPS. For other categories, the scheduling techniques are mostly applied to the BE traffic. For example, with the throughput maximization objective, the scheduler may favor the users with better channel conditions, which can introduce unfairness. Similarly, optimization of power consumption may result in unfairness. Fair scheduling techniques may not be able to guarantee the maximum latency, which is required for real-time traffic classes.

Basically, the BS downlink scheduler can use the carrier to interference and noise ratio (CINR), which is reported back from the MS via the CQI channel. For UL scheduling, the CINR is measured directly on previous transmissions from the same MS. Based on the CINR, the scheduler basically determines the proper MCS for the MS. Note that most of the purposed algorithms have the common assumption that the channel condition does not change within the frame period. Also, it is assumed that the channel information is known at both the transmitter and the receiver.

In general, schedulers favor the users with better channel quality since to exploit the multiuser diversity and channel fading, the optimal resource allocation is to schedule the user with the best channel or perhaps the scheduler does not allocate any resources for the MS with high error rate because the packets would be dropped anyway and, consequently, more packets are retransmitted.

However, the schedulers also need to consider other users' QoS requirements such as the minimum reserved rate and may need to introduce some compensation mechanisms. The schedulers basically use the property of multiuser diversity in order to increase the system throughput and to support more users.

Consider the compensation issue. Unlike the wireless LANs, WiMAX users pay for their QoS assurance. Thus, in Ref. [101], the argument of what is the level of QoS was brought up because of the question whether the service provider should provide a fixed number of slots. If the user happens to choose a bad location (such as the basement of a building on the edge of the cell), the provider will have to allocate a significant number of slots to provide the same QoS as a user who is outside and near the BS. Since the providers have no control over the locations of users, they can argue that they will provide the same resources to all users and the throughput observed by the user will depend upon their location. A generalized weighted fairness (GWF) concept, which equalizes a weighted sum of the slots and the bytes, was introduced in Ref. [101]. WiMAX equipment manufacturers can implement generalized fairness. The service providers can then set a weight parameter to any desired value and achieve either slot fairness or throughput fairness or some combination of the two. The GWF can be illustrated with the following equation:

$$\text{total\_slots} = \sum_{i=1}^{N} S_i,$$

$$wS_i + (1-w)B_i/M = wS_j + (1-w)B_j/M$$

for all MSs *i* and *j* in *N*

$$B_i = b_i S_i.$$

Here, S_i and B_i are the total numbers of slots and bytes for MS *i*; b_i is the number of bytes per slot for MS *i*; *N* is the number of active MSs; *M* is the highest level MCS size in bytes; and *w* is a general weight parameter.

It has been observed that allowing unlimited compensation to meet the QoS requirements may lead to bogus channel information to gain resource allocations [77]. The compensation needs to be taken into account with leading/lagging mechanisms [78]. The scheduler can reallocate the bandwidth left-over either due to a low channel error rate or due to a flow not needing its allocation. It should not take the bandwidth from other well-behaved flows. In case there is still some left-over bandwidth, the leading flow can also gain the advantage of that left-over. However, another approach can be by taking some portion of the bandwidth from the leading flows to the lagging flows. When the error rate is high, a credit history can be built based on the lagging flows and the scheduler can allocate the bandwidth based on the ratio of their credits to their minimum reserved rates when the error rate is acceptable [79]. In either case, if and how the compensation mechanism should be put into consideration are still open questions.

8.4.2.2.1 Fairness

This metric mainly applies for the BE service. One of the commonly used baseline schedulers in published research is the proportional fairness scheme (PFS) [80,81]. The objective of PFS is to maximize the long-term fairness. PFS uses the ratio of channel capacity [denoted as $W_i(t)$] to the long-term throughput [denoted as $R_i(t)$] in a given time window T_i of queue *i* as the preference metric instead of the current achievable data rate. $R_i(t)$ can be calculated by exponentially averaging the ith queue's throughput in terms of T_i. Then, the user with the highest ratio of $W_i(t)/R_i(t)$ receives the transmission from the BS. Note that defining T_i affects the fluctuation of the throughput. There are several proposals that have applied and modified the PFS. For example, T_i derivation with delay considerations is described in [82]. In [49], given a 5 ms frame duration, setting T_i to 50 ms is shown to result in an average rate over 1 s instead of 10 s with $T_i=1000$ ms. In Ref. [83], the moving average was modified to not update when a user queue is empty. A starvation timer was introduced in Ref. [84] to prevent users from starving longer than a predefined threshold.

8.4.2.2.2 QoS Guarantee

Modified LWDF (M-LWDF) [85] can provide QoS guarantee by ensuring a minimum throughput guarantee and by maintaining delays lower than a predefined threshold value with a given probability for each user (rtPS and nrtPS). And, it is provable that the throughput is optimal for LWDF [57]. The algorithm can achieve the optimal whenever there is a feasible set of minimal rates. The algorithm explicitly takes both the current channel condition and the state of the queue into account. The scheme serves the queue j for which "$\rho_i W_j(t) r_j(t)$" is maximal, where ρ_i is a constant that could be different for different service classes (the difficulty is how to find the optimal value of ρ_i). $W_i(t)$ can be either the delay of the HOL packet or the queue length. $r_i(t)$ is the channel capacity for traffic class i.

There are several proposals that have used or modified M-LWDF. For example, in Ref. [86], the scheduler selects the users on each subcarrier during every time slot. For each subcarrier k, the user selection for the subcarrier is expressed by

$$i = \max\left[\text{channel\_gain}(i, k) \times \text{HOL\_delay}(i) \times \{a(i)/d(i)\}\right].$$

In this equation, a is the mean windowed arrival and d is the mean windowed throughput. "a" and "d" are averaged over a sliding window. HOL_delay is the head of line delay. The channel state information is indirectly derived from the normalized channel gain. Then, the channel gain and the buffer state information are both used to decide which subcarriers should be assigned to each user. The buffers state information consists of HOL_delay, a and d.

Similar to M-LWDF, urgency and efficiency-based packet scheduling (UEPS) [87] was introduced to make use of the efficiency of radio resource usage and the urgency (time-utility as a function of the delay) as the two factors for making the scheduling decision. The scheduler first calculates the priority value for each user based on the urgency factor expressed by the time-utility function [denoted as $U_i'(t)$]× the ratio of the current channel state to the average [denoted as $R_i(t)/R_i'(t)$]. After that, the subchannel is allocated to each selected user i where

$$i = \max|U_i'(t)| \times \left[\frac{R_i(t)}{R_i'(t)}\right].$$

Another modification of M-LWDF has been proposed to support multiple traffic classes [88]. The UEPS is not always efficient when the scheduler provides higher priority to nrtPS and BE traffic than rtPS, which may be near their deadlines. This modification handles QoS traffic and BE traffic separately. The HOL packet's waiting time is used for QoS traffic and the queue length for BE traffic.

8.4.2.2.3 System Throughput Maximization

A few schemes, for example [89–91], focus on maximizing the total system throughput. In these, Max C/I (Carrier to Interference) is used to opportunistically assign resources to the user with the highest channel gain.

Another maximum system throughput approach is the exponential rule [90] in that it is possible to allocate the minimum number of slots derived from the minimum modulation scheme to each connection and then adjust the weight according to the exponent (p) of the instant modulation scheme over the minimum modulation scheme. This scheme obviously favors the connections with better modulation scheme (higher p). Users with better channel conditions receive exponentially higher bandwidth. Two issues with this scheme are that additional mechanisms are required if the total slots are less than the total minimum required slots. And, under perfect channel conditions, connections with zero minimum bandwidth can gain higher bandwidth than those with nonzero minimum bandwidth.

Another modification for maximum throughput was proposed in Ref. [91] using a heuristic approach of allocating a subchannel to the MS so that it can transmit the maximum amount of data on the subchannel. Suppose a BS has n users and m subchannels, let λ_i be the total uplink demand (bytes in a given frame) for its UGS connections, R_{ij} be the rate for MS_i on channel j (bytes/slot in the frame), and N_{ij} be the number of slots allocated to MS_i on subchannel j; the goal of scheduling is to minimize the unsatisfied demand, that is,

$$\text{Minimize} \sum_{1 \leq i \leq n} \left[\lambda_i - \left(\sum_{1 \leq j \leq m} R_{ij} N_{ij} \right) \right]$$

subject to the following constraints:

$$\sum_{1 \leq i \leq n} N_{ij} < N_j' \quad \text{and} \quad \sum_{1 \leq j \leq m} R_{ij} N_{ij} \leq \lambda_i.$$

Here, N_j' is the total number of slots available for data transmission in the jth subchannel. A linear programming approach was introduced to solve this problem, but the main issue is the complexity, which is $O(n^3 m^3 N)$. Therefore, a heuristic approach with a complexity of only $O(nmN)$ was also introduced by assigning channels to MSs that can transmit the maximum amount of data.

8.4.2.2.4 Power Constraint

The purpose of this class of algorithms is not only to optimize the throughput but also to meet the power constraint. In general, the transmitted power at an MS is limited. As a result, the maximum power allowable is introduced as one of the constraints. Minimum amount of transmission power is preferred for mobile users due to their limited battery capacities and also to reduce the radio interference.

Link-adaptive largest-weighted-throughput (LWT) algorithm has been proposed for OFDM systems [93]. LWT takes the power consumption into consideration. If assigning the nth subcarrier to the kth user at power $p_{k,n}$ results in a slot throughput of $b_{k,n}$, the algorithm first determines the best assignment that maximizes the link throughput (max $\sum b_{k,n}$). The bit allocation is derived from the approximation function of received SNR, transmission power and instantaneous channel coefficient. Then, the urgency is introduced in terms of the difference between the delay constraint and the waiting time of HOL packets. After that, the scheduler selects the HOL packet with the minimum value of the transmission time and the urgency. The main assumption here is that the packets are of equal length.

The integer programming (IP) approach has also been used to assign subcarriers [93]. However, IP complexity increases exponentially with the number of constraints. Therefore a suboptimal approach was introduced with fixed subcarrier allocation and bit loading algorithm. The suboptimal Hungarian or Linear Programming [92] algorithm with adaptive modulation is used to find the subcarriers for each user and then the rate of the user is iteratively incremented by a bit loading algorithm, which assigns one bit at a time with a greedy approach to the subcarrier. Since this suboptimal and iterative solution is greedy in nature, the user with worse channel condition will mostly suffer.

A better and fairer approach could be to start the allocation with the highest level of modulation scheme. The scheduler has to try to find the best subcarriers for the users with the highest number of bits. This is also a greedy algorithm in a sense as the algorithm is likely to fill the un-allocated subcarriers to gain the power reduction. To minimize the transmit power, a horizontal and vertical swapping technique can also be used. The bits can be shifted horizontally among subcarriers of the same user if the power reduction is needed. Or, the swapping can be done vertically (swap subcarriers between users) to achieve the power reduction.

IEEE 802.16e standard [1] defines power saving class (PSC) type I, II, and III. Basically, PSC I increases the sleep window size by a power of 2 every time there is no packet (similar to binary backoff). Sleep window size for PSC type II is constant. PSC III defines a pre-determined long sleep interval without the existence of the listen period.

Most of the proposals on this topic concentrate on constructing the analytical models for the sleep time, to figure out the optimal sleep time with guaranteed service, especially delay (the more the sleep time, the more the packet delay and the more the buffer length). The models are basically based on the arrival process; for example, in [94], the Poisson distribution is used for the arrival process. Hyper-Erlang distribution is used for self-similarity of Web traffic in Ref. [95].

In order to reduce waking period for each MS, Burst scheduling was proposed in Ref. [96]. A rearrangement technique for unicast and multicast traffic is used so that an MS can wake up and receive both types of traffic at once if possible [97].

In Ref. [98], a hybrid energy-saving scheme was proposed by using a truncated binary exponential algorithm to decide sleep cycle length for VoIP with silence suppression (voice packets are generated periodically during talk-spurt but not generated at all during the silent period).

8.4.3 Section Summary

In this section, we discussed the scheduling mechanism primarily at the BS. The design factors are also discussed. A brief survey of recent scheduling techniques applied to WiMAX networks is categorized into channel-unaware and channel-aware scheduling.

8.5 Conclusion and Open Research Issues

In this chapter, we provided an extensive survey of recent scheduling proposals for WiMAX and discussed key issues and design factors. The scheduler designers need to be thoroughly familiar with WiMAX characteristics such as the physical layer, frame format, registration process, and so on as described in Section 8.1. The goals of the schedulers are basically to meet QoS guarantees for all service classes, to maximize the system goodput, to maintain the fairness, to minimize power consumption, to have as less a complexity as possible, and finally to ensure the system scalability. To meet all these goals is quite challenging since achieving one may require that we have to sacrifice the others.

We classified recent scheduling disciplines based on the channel awareness in making the decision. Well-known scheduling disciplines can be applied for each class such as EDF for rtPS and WFQ for nrtPS and WRR for interclass. With the awareness of channel condition and with a knowledge of applications, schedulers can maximize the system throughput or support more users.

Optimization for the WiMAX scheduler is still an ongoing research topic. There are several holes to fill in, for example, the polling mechanism, backoff optimization, overhead optimization, and so on. WiMAX can support a reliable transmission with ARQ and HARQ [99,100]. Future research on scheduling should consider the use of these characteristics. The use of MIMO with multiple antennas to increase the bandwidth makes the scheduling problem even more sophisticated. Also, the multihops scenario also needs to be investigated for end-to-end service guarantees. With user mobility, future schedulers need to handle BS selection and the subsequent hand off. All these issues are still open for research and new discoveries.

This work was sponsored in part by a grant from Application Working Group of WiMAX Forum.

References

1. IEEE P802.16Rev2/D2, DRAFT Standard for Local and Metropolitan area Networks, Part 16: Air Interface for Broadband Wireless Access Systems, December 2007, 2094 pp.
2. WiMAX Forum, WiMAX System Evaluation Methodology V2.1, July 2008, 230 pp. Available: http://www.wimaxforum.org/technology/documents/
3. S. Lu, V. Bharghavan, and R. Srikant, Fair scheduling in wireless packet networks, *IEEE/ACM Trans. Netw.*, 7, 473–489, 1999.
4. N. H. Vaidya, P. Bahl, and S. Gupta, Distributed fair scheduling in a Wireless LAN, *IEEE Trans. Mobile Comput.*, 4, 616–629, 2005.
5. L. Tassiulas and S. Sarkar, Maxmin fair scheduling in wireless networks, in *Proc. IEEE Computer Communication Conf.*, New York, NY, 2002, vol. 2, pp. 763–772.
6. P. Bhagwat, P. Bhattacharya, A. Krishna, and S. K. Tripathi, Enhancing throughput over Wireless LANs using channel state dependent packet scheduling, in *Proc. IEEE Computer Communication Conf.*, San Francisco, CA, 1996, vol. 3, pp. 1133–1140.
7. S. Shakkottai and R. Srikant, Scheduling real-time traffic with deadlines over a wireless channel, *ACM/Baltzer Wirel. Netw.*, 8, 13–26, 2002.
8. E. Jung and N. H. Vaidya, An energy efficient MAC protocol for Wireless LANs, in *Proc. IEEE Computer Communication Conf.*, New York, NY, 2002, vol. 3, pp. 1756–1764.
9. X. Zhang, Y. Wang, and W. Wang, Capacity analysis of adaptive multiuser frequency-time domain radio resource allocation in OFDMA systems, in *Proc. IEEE Int. Symp. Circuits and Systems*, Island of Kos, Greece, 2006, pp. 4–7.
10. T. Ohseki, M. Morita, and T. Inoue, Burst construction and packet mapping scheme for OFDMA downlinks in IEEE 802.16 systems, in *Proc. IEEE Global Telecomunications Conf.*, Washington, DC, 2007, pp. 4307–4311.
11. Y. Ben-Shimol, I. Kitroser, and Y. Dinitz, Two-dimensional mapping for wireless OFDMA systems, *IEEE Trans. Broadcast.*, 52, 388–396, 2006.
12. A. Bacioccola, C. Cicconetti, L. Lenzini, E. A. M. E. Mingozzi, and A. A. E. A. Erta, A downlink data region allocation algorithm for IEEE 802.16e OFDMA, in *Proc. Sixth Int. Conf. Information, Communications & Signal Processing*, Singapore, 2007, pp. 1–5.
13. C. Desset, E. B. de Lima Filho, and G. Lenoir, WiMAX downlink OFDMA burst placement for optimized receiver duty-cycling, in *Proc. IEEE Int. Conf. Communications*, Glasgow, Scotland, 2007, pp. 5149–5154.
14. C. So-In, R. Jain, and A. Al-Tamimi, eOCSA: An algorithm for burst mapping with strict QoS Requirements in IEEE 802.16e mobile WiMAX networks, Accepted to appear in *Proc. IFIP Wireless Days*, Paris, French, 2009. Available: http://www.cse.wustl.edu/~jain/papers/eocsa.htm.
15. T. Wand, H. Feng, and B. Hu, Two-dimensional resource allocation for OFDMA system, in *Proc. IEEE Int. Conf. Communications Workshop*, Beijing, People's Republic of China, 2008, pp. 1–5.

16. X. Perez-Costa, P. Favaro, A. Zubow, D. Camps, and J. Arauz, On the challenges for the maximization of radio resources usage in WiMAX networks, in *Proc. IEEE Consumer Communications and Networking Conf.*, Las Vegas, NV, 2008, pp. 890–896.

17. C. So-In, R. Jain, and A. Al-Tamimi, Capacity evaluation in IEEE 802.16e mobile WiMAX networks, Accepted to appear in *J. of Comp. Systems, Networks, and Comm. (Special Issue on WIMAX, LTE, and WiFi Interworking)*, 2010. Available: http://www.cse.wustl.edu/~jain/papers/capmodel.htm.

18. H. Martikainen, A. Sayenko, O. Alanen, and V. Tykhomyrov, Optimal MAC PDU size in IEEE 802.16, in *Proc. Telecommunication Networking Workshop on QoS in Multiservice IP Networks*, Venice, Italy, 2008, pp. 66–71.

19. S. Sengupta, M. Chatterjee, and S. Ganguly, Improving quality of VoIP streams over WiMAX, *IEEE Trans. Comput.*, 57, 145–156, 2008.

20. C. So-In, R. Jain, and A. Al-Tamimi, A deficit round robin with fragmentation scheduler for IEEE 802.16e mobile WiMAX, in *Proc. IEEE Sarnoff Symp.*, Princeton, NJ, 2009, pp. 1–7. Available: http://www.cse.wustl.edu/~jain/papers/drrf.htm

21. H. Yaghoobi, Scalable OFDMA physical layer in IEEE 802.16 WirelessMAN, *Intel Technol. J.*, 8, 201–212, 2004.

22. IEEE P802.16–2004, IEEE standard for local and metropolitan area networks–Part 16: Air interface for fixed broadband wireles access systems, October 2004, 893 pp.

23. C. Eklund, R-B. Marks, S. Ponnuswamy, K-L. Stanwood, and N-V. Waes, *WirelessMAN Inside the IEEE 802.16 Standard for Wireless Metropolitan Networks*, IEEE Standards Information Network/IEEE Press, NY, USA, 2006, 400 pp.

24. G. Jeffrey, J. Andrews, A. Arunabha Ghosh, and R. Muhamed, *Fundamentals of WiMAX Understanding Broadband Wireless Networking*, Prentice Hall PTR, NY, USA, 2007, 496 pp.

25. L. Nuaymi, *WiMAX: Technology for Broadband Wireless Access*, Wiley, West Sussex, England, 2007, 310 pp.

26. B. Baumgartner, M. Reinhardt, G. Richter, and M. Bossert, Performance of forward error correction for IEEE 802.16e, in *Proc. the 10th Int. OFDM Workshop*, Hamburg, Germany, 2005. Available: http://tait.e-technik.uni-ulm.de/~richter/papers/ofdm2005.pdf

27. D. Ozdemir and F. Retnasothie, WiMAX capacity estimation for triple play services including Mobile TV, VoIP and Internet, WiMAX Forum, June 2007.

28. R. Srinivasan, T. Papathanassiou, and S. Timiri, Mobile WiMAX VoIP capacity system level simulations, WiMAX Forum, March 2007.

29. 3GPP2-TSGC5, HTTP and FTP Traffic Model for 1xEV-DV Simulations, 3GPP2-C50-EVAL-2001022–0xx, 2001.

30. L-E. Jonsson, F. Pelletier, and K. Sandlund, Framework and four profiles: RTP, UDP, ESP and uncompressed, RFC 3095, July 2001.

31. G. Pelletier, K. Sandlund, L-E. Jonsson, and M. West, RObust header compression (ROHC): A profile for TCP/IP (ROHC-TCP), RFC 4996, January 2006.

32. M. Fong, R. Novak, S. Mcbeath, and R. Srinivasan, Improved VoIP capacity in mobile WIMAX systems using persistent resource allocation, *IEEE Commun. Mag.*, 46, 50–57, 2008.

33. M. Hawa and D. W. Petr, Quality of service scheduling in cable and broadband wireless access systems, in *Proc. IEEE Int. Workshop Quality of Service*, Miami Beach, MI, 2002, pp. 247–255.

34. Q. Ni, A. Vinel, Y. Xiao, A. Turlikov, and T. Jiang, Wireless broadband access: WiMAX and beyond – investigation of bandwidth request mechanisms under point-to-multipoint mode of WiMAX networks, *IEEE Commun. Mag.*, 45, 132–138, 2007.

35. L. Lin, W. Jia, and W. Lu, Performance analysis of IEEE 802.16 multicast and broadcast polling based bandwidth request, in *Proc. IEEE Wireless Communication and Networking Conf.*, Hong Kong, 2007, pp. 1854–1859.

36. B. Chang and C. Chou, Analytical modeling of contention-based bandwidth request mechanism in IEEE 802.16 wireless network, *IEEE Trans. Veh. Technol.*, 57, 3094–3107, 2008.

37. P. Rastin, S. Dirk, and M. Daniel, Performance evaluation of piggyback requests in IEEE 802.16, in *Proc. IEEE Vehicular Technology Conf.*, Baltimore, MD, 2007, pp. 1892–1896.

38. V. Alexey, Z. Ying, N. Qiang, and L. Andrey, Efficient request mechanism usage in IEEE 802.16, in *Proc. IEEE Global Telecomunications Conf.*, San Francisco, CA, 2006, pp. 1–5.

39. O. Alanen, Multicast polling and efficient VoIP connections in IEEE 802.16 networks, in *Proc. Int. Workshop Modeling Analysis and Simulation Wireless and Mobile Systems*, Crete Island, Greece, 2007, pp. 289–295.

40. A. Doha, H. Hassanein, and G. Takahara, Performance evaluation of reservation medium access control in IEEE 802.16 Networks, in *Proc. ACS/IEEE Int. Cont. Computer Systems and Applications.*, Dubai, UAE, 2006, pp. 369–374.

41. A. Sayenko, O. Alanen, and T. Hamalainen, On contention resolution parameters for the IEEE 802.16 base station, in *Proc. IEEE Global Telecomunications Conf.*, Washington, DC, 2007, pp. 4957–4962.

42. J. Yan and G. Kuo, Cross-layer design of optimal contention period for IEEE 802.16 BWA systems, in *Proc. IEEE Int. Conf. Communications*, Istanbul, Turkey, 2006, vol. 4, pp. 1807–1812.

43. C. Cicconetti, A. Erta, L. Lenzini, and E. A. M. E. Mingozzi, Performance evaluation of the IEEE 802.16 MAC for QoS support, *IEEE Trans. Mobile Comput.*, 6, 26–38, 2006.

44. O. Yang and J. Lu, New scheduling and CAC scheme for real-time video application in fixed wireless networks, in *Proc. IEEE Consumer Communications and Networking Conf.*, Las Vegas, NV, 2006, vol. 1, pp. 303–307.

45. H. Lee, T. Kwon, and D. Cho, An enhanced uplink scheduling algorithm based on voice activity for VoIP services in IEEE 802.16d/e system, *IEEE Commun. Lett.*, 9, 691–693, 2005.

46. H. Lee, T. Kwon, and D. Cho, Extended-rtPS algorithm for VoIP services in IEEE 802.16 systems, in *Proc. IEEE Int. Conf. Communications*, Istanbul, Turkey, 2006, vol. 5, pp. 2060–2065.

47. P. T. Brady, Model for generating on-off speech patterns in two-way conversation, *Bell Syst. Tech. J.*, 2445–2472, 1969.

48. M. Hawa and D. W. Petr, Quality of service scheduling in cable and broadband wireless access systems, in *Proc. IEEE Int. Workshop Quality of Service*, Miami Beach, FL, 2002, pp. 247–255.

49. J. Wu, J. Mo, and T. Wang, A method for non-real-time polling service in IEEE 802.16 wireless access networks, in *Proc. IEEE Vehicular Technology Conf.*, Baltimore, MD, 2007, pp. 1518–1522.

50. C. Jiang and T. Tsai, Token bucket based CAC and packet scheduling for IEEE 802.16 broadband wireless access networks, in *Proc. IEEE Consumer Communications and Networking Conf.*, Las Vegas, NV, 2006, pp. 183–187.

51. H. Wang, B. He, and D. P. Agrawal, Admission control and bandwidth allocation above packet level for IEEE 802.16 wireless MAN, in *Proc. 12th Int. Conf. Parallel and Distributed Systems*, Minneapolis, MN, 2006, pp. 6–13.

52. D. Niyato and E. Hossain, Joint bandwidth allocation and connection admission control for polling services in IEEE 802.16 broadband wireless networks, in *Proc. IEEE Int. Conf. Communications*, Istanbul, Turkey, 2006, vol. 12, pp. 5540–5545.

53. B. Chang, Y. Chen, and C. Chou, Adaptive hierarchical polling and cost-based call admission control in IEEE 802.16 WiMAX networks, in *Proc. IEEE Wireless Communication and Networking Conf.*, Hong Kong, 2007, pp. 1954–1958.

54. H. Wang, W. Li, and D. P. Agrawal, Dynamic admission control and QoS for 802.16 wireless MAN, in *Proc. Wireless Telecommunications Symp.*, Pomona, CA, 2005, pp. 60–66.

55. B. Rong, Y. Qian, and K. Lu, Downlink call admission control in multiservice WiMAX networks, in *Proc. IEEE Int. Conf. Communications*, Glasgow, Scotland, 2007, pp. 5082–5087.

56. M. Andrews, Probabilistic end-to-end delay bounds for earliest deadline first scheduling, in *Proc. IEEE Computer Communication Conf.*, Israel, 2000, vol. 2, pp. 603–612.

57. A. L. Stolyar and K. Ramanan, Largest weighted delay first scheduling: Large deviations and optimality, *Ann. Appl. Probab.*, 11, 1–48, 2001.

58. C. E. Koksal, H. I. Kassab, and H. Balakrishnan, An analysis of short-term fairness in wireless media access protocols, in *Proc. ACM SIGMETRICS Performance Evaluation Review*, Santa Clara, CA, 2000, vol. 28, pp. 118–119.

59. C. Cicconetti, L. Lenzini, E. Mingozzi, and C. Eklund, Quality of service support in IEEE 802.16 networks, *IEEE Netw.*, 20, 50–55, 2006.

60. A. Sayenko, O. Alanen, J. Karhula, and T. Hamaainen, Ensuring the QoS requirements in 802.16 scheduling, in *Proc. Int. Workshop Modeling Analysis and Simulation Wireless and Mobile Systems*, Terromolinos, Spain, 2006, pp. 108–117.

61. A. Sayenko, O. Alanen, and T. Hamaainen, Scheduling solution for the IEEE 802.16 base station, *Int. J. Comput. Telecomm. Netw.*, 52, 96–115, 2008.
62. A. Iera, A. Molinaro, S. Pizzi, and R. Calabria, Channel-aware scheduling for QoS and fairness provisioning in IEEE 802.16/WiMAX broadband wireless access systems, *IEEE Netw.*, 21, 34–41, 2007.
63. N. Liu, X. Li, C. Pei, and B. Yang, Delay character of a novel architecture for IEEE 802.16 systems, in *Proc. Int. Conf. Parallel and Distributed Computing, Applications and Technologies*, Dalian, People's Republic of China, 2005, pp. 293–296.
64. K. Wongthavarawat and A. Ganz, Packet scheduling for QoS support in IEEE 802.16 broadband wireless access systems, *Int. J. Commun. Syst.*, 16, 81–96, 2003.
65. A. Sayenko, T. Hamalainen, J. Joutsensalo, and J. Siltanen, An adaptive approach to WFQ with the revenue criterion, in *Proc. IEEE Int. Symp. Computers and Communication.*, 2003, vol. 1, pp. 181–186.
66. J. Borin and N. Fonseca, Scheduler for IEEE 802.16 networks, *IEEE Commun. Lett.*, 12, 274–276, 2008.
67. Y. Wang, S. Chan, M. Zukerman, and R.J. Harris, Priority-based fair scheduling for multimedia WiMAX uplink traffic, in *Proc. IEEE Int. Conf. Communications*, Beijing, People's Republic of China, 2008, pp. 301–305.
68. L. F. M. de Moraes and P. D. Jr. Maciel, Analysis and evaluation of a new MAC protocol for broadband wireless access, in *Proc. Int. Conf. Wireless Networks, Communications and Mobile Computing*, Kaanapali Beach Maui, Hawaii, 2005, vol. 1, pp. 107–112.
69. W. Lilei and X. Huimin, A new management strategy of service flow in IEEE 802.16 systems, in *Proc. IEEE Conf. Industrial Electronics and Applicaitions*, Harbin, People's Republic of China, 2008, pp. 1716–1719.
70. D. Niyato and E. Hossain, Queue-aware uplink bandwidth allocation for polling services in 802.16 broadband wireless networks, in *Proc. IEEE Global Telecomunications Conf.*, St. Louis, MO, 2005, vol. 6, pp. 5–9.
71. D. H. Kim and C. G. Kang, Delay threshold-based priority queueing packet scheduling for integrated services in mobile broadband wireless access system, in *Proc. IEEE Int. Conf. High Performance Computing and Communications*, Kemer-Antalya, Turkey, 2005, pp. 305–314.
72. J. M. Ku, S. K. Kim, S. H. Kim, S. Shin, J. H. Kim, and C. G. Kang, Adaptive delay threshold-based priority queueing scheme for packet scheduling in mobile broadband wireless access system, in *Proc. IEEE Wireless Communication and Networking Conf.*, Las Vegas, NV, 2006, vol. 2, pp. 1142–1147.
73. J. Chen, W. Jiao, and H. Wang, A service flow management strategy for IEEE 802.16 broadband wireless access systems in TDD mode, in *Proc. IEEE Int. Conf. Communications*, Seoul, Korea, 2005, vol. 5, pp. 3422–3426.
74. J. Chen, W. Jiao, and Q. Quo, An integrated QoS control architecture for IEEE 802.16 broadband wireless access systems, in *Proc. Global Telecommunications Conf.*, St. Louis, MO, 2005, pp. 6–11.
75. K. Wongthavarawat and A. Ganz, IEEE 802.16 based last mile broadband wireless military networks with quality of service support, in *Proc. IEEE Military Communications Conf.*, Boston, MA, 2003, vol. 2, pp. 779–784.
76. A. K. F. Khattab and K. M. F. Elsayed, Opportunistic scheduling of delay sensitive traffic in OFDMA-based wireless networks, in *Proc. Int. Symp. World of Wirless Mobile and Multimedia Networks*, Buffalo, NY, 2006, pp. 10–19.
77. Z. Kong, Y. Kwok, and J. Wang, On the impact of selfish behaviors in wireless packet scheduling, in *Proc. IEEE Int. Conf. Communications*, Beijing, People's Republic of China, 2008, pp. 3253–3257.
78. S. A. Filin, S. N. Moiseev, M. S. Kondakov, A. V. Garmonov, D. H. Yim, J. Lee, S. Chang, and Y. S. Park, QoS-guaranteed cross-layer transmission algorithms with adaptive frequency subchannels allocation in the IEEE 802.16 OFDMA system, in *Proc. IEEE Int. Conf. Communications*, Istanbul, Turkey, 2006, vol. 11, pp. 5103–5110.
79. W. K. Wong, H. Tang, S. Guo, and V. C. M. Leung, Scheduling algorithm in a point-to-multipoint broadband wireless access network, in *Proc. IEEE Vehicular Technology Conf.*, Orlando, FL, 2003, vol. 3, pp. 1593–1597.
80. P. Bender, P. Black, M. Grob, R. Padovani, N. Sindhushayana, and A. Viterbi, CDMA/HDR: A bandwidth-efficient high-speed wireless data service for nomadic users, *IEEE Commun. Mag.*, 38, 70–77, 2000.

81. H. Kim and Y. Han, A proportional fair scheduling for multicarrier transmission systems, *IEEE Commun. Lett.*, 9, 210–212, 2005.

82. F. Hou, P. Ho, X. Shen, and A. Chen, A novel QoS scheduling scheme in IEEE 802.16 networks, in *Proc. IEEE Wireless Communication and Networking Conf.*, Hong Kong, 2007, pp. 2457–2462.

83. N. Ruangchaijatupon and Y. Ji, Simple proportional fairness scheduling for OFDMA frame-based wireless systems, in *Proc. IEEE Wireless Communication and Networking Conf.*, Las Vegas, NV, 2008, pp. 1593–1597.

84. J. Qiu and T. Huang, Packet scheduling scheme in the next generation high-speed wireless packet networks, in *Proc. IEEE Int. Wireless and Mobile Computing, Networking and Communications*, Montreal, Canada, 2005, pp. 224–227.

85. M. Andrews, K. Kumaran, K. Ramanan, A. Stolyar, P. Whiting, and R. Vijayakumar, Providing quality of service over a shared wireless link, *IEEE Commun. Mag.*, 39, 150–154, 2001.

86. P. Parag, S. Bhashyam, and R. Aravind, A subcarrier allocation algorithm for OFDMA using buffer and channel state information, in *Proc. IEEE Vehicular Technology Conf.*, Dallas, TX, 2005, vol. 1, pp. 622–625.

87. S. Ryu, B. Ryu, H. Seo, and M. Shi, Urgency and efficiency based wireless downlink packet scheduling algorithm in OFDMA system, in *Proc. IEEE Vehicular Technology Conf.*, Stockholm, Sweden, 2005, vol. 3, pp. 1456–1462.

88. W. Park, S. Cho, and S. Bahk, Scheduler design for multiple traffic classes in OFDMA networks, in *Proc. IEEE Int. Conf. Communications*, Istanbul, Turkey, 2006, vol. 2, pp. 790–795.

89. P. Viswanath, D. Tse, and R. Laroia, Opportunistic beamforming using dumb antennas, *IEEE Trans. Inf. Theory*, 48, 1277–1294, 2002.

90. S. Shakkottai, R. Srikant, and A. Stolyar, Pathwise optimality and state space collapse for the exponential rule, in *Proc. IEEE Int. Symp. Information Theory.*, 2002, pp. 379.

91. V. Singh and V. Sharma, Efficient and fair scheduling of uplink and downlink in IEEE 802.16 OFDMA Networks, in *Proc. IEEE Wireless Communication and Networking Conf.*, Las Vegas, NV, 2006, vol. 2, pp. 984–990.

92. Z. Liang, Y. Huat Chew, and C. Chung Ko, A linear programming solution to subcarrier, bit and power allocation for multicell OFDMA systems, in *Proc. IEEE Wireless Communication and Networking Conf.*, Las Vegas, NV, 2008, pp. 1273–1278.

93. Y. J. Zhang and S. C. Liew, Link-adaptive largest-weighted-throughput packet scheduling for real-time traffics in wireless OFDM networks, in *Proc. IEEE Global Telecomunications Conf.*, St. Louis, MO, 2005, vol. 5, pp. 5–9.

94. Z. Yan, Performance modeling of energy management mechanism in IEEE 802.16e mobile WiMAX, in *Proc. IEEE Wireless Communication and Networking Conf.*, Hong Kong, 2007, pp. 3205–3209.

95. X. Yang, Performance analysis of an energy saving mechanism in the IEEE 802.16e wireless MAN, in *Proc. IEEE Consumer Communications and Networking Conf.*, Las Vegas, NV, 2006, pp. 406–410.

96. J. Shi, G. Fang, Y. Sun, J. Zhou, Z. Li, and E. Dutkiewicz, Improving mobile station energy efficiency in IEEE 802.16e WMAN by burst scheduling, in *Proc. IEEE Global Telecomunications Conf.*, San Francisco, CA, 2006, pp. 1–5.

97. L. Tian, Y. Yang, J. Shi, E. Dutkiewicz, and G. Fang, Energy efficient integrated scheduling of unicast and multicast traffic in 802.16e WMANs, in *Proc. IEEE Global Telecomunications Conf.*, Washington, DC, 2007, pp. 3478–3482.

98. H. Choi and D. Cho, Hybrid energy-saving algorithm considering silent periods of VoIP traffic for mobile WiMAX, in *Proc. IEEE Int. Conf. Communications*, Glasgow, Scotland, 2007, pp. 5951–5956.

99. A. Sayenko, O. Alanen, and T. Hamalainen, ARQ aware scheduling for the IEEE 802.16 base station, in *Proc. IEEE Int. Conf. Communications*, Beijing, People's Republic of China, 2008, pp. 2667–2673.

100. F. Hou, J. She and P. Ho, and X. Shen, Performance analysis of ARQ with opportunistic scheduling in IEEE 802.16 networks, in *Proc. IEEE Global Telecomunications Conf.*, Washington, DC, 2007, pp. 4759–4763.

101. C. So-In, R. Jain, and A. Al-Tamimi, Generalized weighted fairness and its application for resource allocation in IEEE 802.16e mobile WiMAX, Accepted to appear in *Proc. the 2nd Int. Conf. on Comp. and Automation Engineering*, Singapore, 2010. Available: http://www.cse.wustl.edu/~jain/papers/gwf.htm

Chapter 9

Resource Management in MIMO-OFDM Systems

Malte Schellmann, Lars Thiele, Thomas Wirth, Volker Jungnickel, and Thomas Haustein

Contents

9.1	Introduction	236
9.2	System Concept	237
	9.2.1 Fixed Unitary Beams—GoB Concept	239
9.3	Resource Allocation Algorithm	240
	9.3.1 Channel Evaluation at UT, Step 1	241
	9.3.2 Determination of the Scores	242
	9.3.3 Resource Scheduling at BS, Step 2	243
	9.3.3.1 CQI-Based Feedback	243
9.4	Investigations on Link Level (Isolated Cell)	244
	9.4.1 System Setup	244
	9.4.2 Throughput Performance	245
	9.4.3 Capacity Scaling	249
9.5	Investigations on System Level (Multicell Environment)	251
	9.5.1 System Model—Extensions to Multicell Application	251
	9.5.2 Multicell System Setup	252
	9.5.3 Capacity Scaling with Number of Receive Antennas and Users	254
9.6	Experimental Results with the LTE Test Bed	255
	9.6.1 Experimental Real-Time Test Bed	256
	9.6.2 Measurement Scenario	257
	9.6.3 Results	258

9.7   Conclusion and Future Work 261
Acknowledgments 262
Abbreviations 262
References 263

9.1 Introduction

In future wireless systems, transmitting and receiving ends may be equipped with multiple antennas, forming a so-called multiple-input multiple-output (MIMO) link. Such an antenna configuration extends the radio channel by a spatial dimension. This spatial dimension can be utilized for simultaneous transmission of parallel data streams [spatial multiplexing (SMUX) mode] or for transmission of a single data stream with an improved signal quality (spatial diversity mode). By accessing the spatial channel according to these two modes, substantial increase in the system capacity can be achieved. For an isolated point-to-point link, these capacity gains have been shown to scale with the minimum of the number of transmit and receive antennas [1]. For an introduction into the basics of transmission via MIMO channels, refer to Ref. [2].

Recently, a fundamental tradeoff between the two spatial transmission modes mentioned above has been pointed out in Ref. [3], revealing that the mode maximizing the capacity depends on the actual channel state. This fact motivated the development of an adaptive transmission system, which selects the transmission mode depending on the actual channel quality in order to improve the error rate performance for fixed data rate transmission [4,5] or to increase the spectral efficiency [6–8].

If multiple users are present in the system, the spatial dimension of the MIMO channel can be used to grant multiple users simultaneous access to the same time/frequency resource. This enables a significant increase in system capacity if the multiuser diversity is properly exploited. The multiple access mode in the spatial domain, which is in fact a special case of the SMUX mode, is denoted by the term multiuser MIMO (MU-MIMO). In contrast to that, single-user MIMO (SU-MIMO) denotes the case where the simultaneously transmitted streams in SMUX mode are assigned to a single user only. The capacity of the MIMO broadcast channel (BC) was shown to be achievable with dirty paper coding (DPC) [9,10]. Based on that work, solutions were proposed that enable one to approach this capacity by applying less complex algorithms [11,12]. However, the major drawback of these approaches is that they require full knowledge of channel state information (CSI) at the transmitter, which cannot be provided easily. Especially in frequency division duplex (FDD) systems, obtaining CSI at the transmitter is problematic, as this information would have to be conveyed via a high-rate feedback link. To alleviate the problem and thus come to practically applicable solutions, several approaches have been proposed that rely on partial CSI at the transmitter only (see [13,14] and references therein). By these approaches, the demands on the feedback rate are significantly relaxed. A common solution here is to quantize channel vectors according to a given codebook and to characterize the channel quality by so-called channel quality identifiers (CQIs), yielding a limited amount of information constituting the partial CSI.

When assigning the available resources to distinct users, accounting for fairness is of crucial importance to ensure that each user is granted access to the shared medium, even if it experiences relatively poor channel conditions. Fairness can be provided by applying adequate scheduling policies like the *proportional fair scheduler* [15], which enables each user to realize a constant fraction of its total achievable rate. The target to achieve proportional fairness was also adopted for the score-based scheduler [16], which aims at assigning each user its best resources according to current signal and interference conditions while offering a simple structure and high efficiency. These properties make it a suitable solution for practical application under the full buffer assumption.

In this chapter, we combine the idea of spatial transmission mode selection with score-based scheduling to realize a practical concept for resource allocation in the downlink (DL) of a MU-MIMO-OFDM system, which aims at high overall system throughput while meeting proportional fairness constraints. The base station (BS) provides fixed unitary beam sets that will be used for the simultaneous transmission of spatial data streams. The beam sets are evaluated per resource block (RB)* by the terminals, which then report to the BS the beams they want to be served on as well as the corresponding CQI values. Based on this reported information, the BS carries out the resource allocation process, where it determines for each RB the number of beams that are simultaneously active and assigns the beams to the users. Consequently, transmission mode selection is performed inherently in this process. With the proposed concept, we establish a system that adapts to the spatial properties of the MU-MIMO-OFDM channel in a frequency-selective fashion, enabling one to realize a considerable proportion of the capacity of the MIMO BC.

Evaluation is carried out for an isolated cell as well as for a multicell environment based on hexagonal cells with a frequency reuse factor equal to one. In the latter system, the same frequencies are used in all available cells, resulting in high inter-cell interference that affects the signal conditions of all users. The multiple antennas at the user terminals (UTs) are used to actively suppress cochannel interference (CCI). For the isolated cell, we examine the influence of mean signal-to-noise ratio (SNR) as well as the number of users on the achievable throughput performance for a 2×2 MIMO system configuration. Herein, we show that the MU-MIMO mode clearly promotes the selection of SMUX mode even at low SNR conditions. In the multicell context, application of the fixed beams translates to a spatial partitioning of the transmit space. This allows the resource allocation process to coordinate the users so that they can be served on resources where they experience low CCI. Within our investigations of the achievable system and user throughput, it turns out that the concept of spatial adaptation based on fixed beams together with the interference suppression at the UTs are the key drivers for the efficient application of MU-MIMO mode on system level. Furthermore, we show that the MIMO multiplexing gains known for isolated links can also be achieved in the interference-limited multicell environment. Thus, the capacity scaling law for MIMO systems [1] is also found to hold true in the multicell context. In the last section, findings from analysis are substantiated by results from practical experiments.

The proposed resource allocation concept may be applied in future wireless networks supporting MIMO technology as well as fixed precoding, where a signaling link for feedback and feed-forward information is also provided. In particular, this holds for the current specification of 3G long term evolution (LTE) as well as worldwide interoperability for microwave access (WiMAX).

9.2 System Concept[†]

We consider the DL of a broadband MU-MIMO-OFDM system, where a BS with N_t antennas communicates with K UTs equipped with N_r antennas each (refer to Figure 9.1). The BS provides $B \geq N_t$ fixed beams $\mathbf{b}_u$, which are used for spatial precoding of the transmission signals [grid of beams (GoB) concept, presented in Ref. [17]]. Up to N_t beams may be simultaneously served, each one being used to transmit an independent data stream. Consequently, we denote the supported spatial transmission modes as single stream (ss) and multistream (ms), which directly relate to the diversity and SMUX mode, respectively. Based on the GoB, the UTs can evaluate the channel and

---

* A resource block is defined as a set of contiguous subcarriers, thus forming a frequency subband.
† Reprinted from M. Schellmann et al., *IEEE Transaction on Vehicular Technology*, Jan. 2010. With permission.

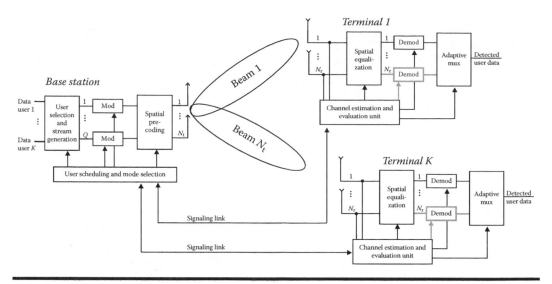

Figure 9.1 System concept for channel-adaptive transmission in MU-MIMO DL.

determine the rates they are able to achieve in the different spatial modes with each beam. Beams achieving the highest rate are selected for potential transmission. Each user conveys information on the beam selection per spatial mode and the achievable rate per beam to the BS via a signaling link. At the BS, user scheduling is performed, and a decision in favor of the most suitable spatial transmission mode is made. The scheduling decision is then signaled forward to the UTs, and transmitter and terminal receivers are configured correspondingly. Note that in ms mode, a user can receive up to N_r simultaneously transmitted spatial streams. Therefore, Figure 9.1 exhibits a gray-shaded signal path at the terminal side, which is used to decode any additional stream dedicated to that user.

Transmission is based on a radio frame structure, where each frame is constituted of several consecutive OFDM symbols. The total transmission resources of a radio frame are given by the subcarriers available for signal transmission. By subdividing the signal bandwidth into single subbands confined to a fixed number of consecutive subcarriers, we partition these transmission resources into blocks, which are denoted as RBs in the following (refer to Figure 9.2). RBs form the basic scheduling resources that can be individually assigned to distinct users. Each RB is processed separately and thus may support an individual spatial transmission mode.

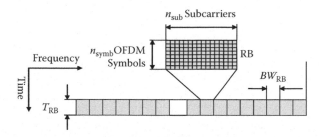

Figure 9.2 Partitioning of the resources in time and frequency into RBs. (Reprinted from M. Schellmann et al., *IEEE 40th Asilomar Conference on Signals, Systems and Computers*, pp. 1635–1639, 2006. With permission.)

To facilitate system design, we assume a uniform transmit power allocation over all subcarriers; hence the transmission equation for each subcarrier signal is given by

$$\mathbf{r} = \mathbf{HCs} + \mathbf{n}, \tag{9.1}$$

where $\mathbf{H}$ is the $N_r \times N_t$-dimensional MIMO channel and $\mathbf{C}$ is the $N_t \times N_t$ precoding matrix comprising N_t of the B beams $\mathbf{b}_u$ with unitary property, which may be simultaneously served. $\mathbf{s}$ is the transmit vector containing up to N_t nonzero transmit symbols with constant transmit power $E[\mathbf{s}^H \mathbf{s}] = P_s$, that is, the power P_s is uniformly distributed over all nonzero transmit symbols in $\mathbf{s}$. Here, $E[\cdot]$ denotes the expectation operator and $(\cdot)^H$ the conjugate transpose operator. Finally, $\mathbf{n}$ is the noise vector with N_r circularly symmetric complex Gaussian entries, and its covariance is given by $E[\mathbf{nn}^H] = N_0 \cdot \mathbf{I}$, with $\mathbf{I}$ representing the identity matrix and N_0 being the power of the additive white Gaussian noise (AWGN).

We assume that linear equalization techniques are being applied at multiantenna receivers. Hence, to recover the uth symbol s_u within vector $\mathbf{s}$, we multiply $\mathbf{r}$ with the equalization vector $\mathbf{w}_u$ according to

$$\hat{s}_u = \mathbf{w}_u^H \mathbf{r}. \tag{9.2}$$

In this work, we consider maximum ratio combining (MRC) for ss mode and minimum mean square error (MMSE) equalization for ms mode. For these techniques, closed-form expressions for the signal to interference and noise ratios (SINRs) of the spatial streams after equalization exist, facilitating the evaluation and proper comparison of the rates that can be achieved with the different spatial modes.

The applied scheduling process is based on the score-based scheduling strategy (SB) [16], which is a simple heuristic process aiming to assign each user its best resources from a set of resources defined over arbitrary dimensions. The resources of each user within the set are ranked by their quality, and corresponding scores are assigned. The BS then assigns a resource to the user providing the best score. Averaged over many channel realizations, this scheduling strategy assigns an equal amount of resources to each user, enabling them to realize a constant fraction of their total achievable rate and thus yielding (asymptotically) a performance similar to that of the proportional fair scheduler [15]. In our research, we confine the set of resources to those comprised in a transmission frame; hence all users will be scheduled within the same time-slot and fairness is achieved on a short time scale.

9.2.1 Fixed Unitary Beams—GoB Concept

In this section, we give some details on the proper choice of the beam vectors constituting the GoB. The GoB concept itself can be understood as a quantization scheme for the precoding vectors to be used for signal transmission. It is well known that the optimum choice of precoding vectors in SU-MIMO links is the eigenvectors of the matrix product $\mathbf{H}^H \mathbf{H}$. The optimum quantization of these vectors in uncorrelated MIMO Rayleigh-fading channels has been elaborated in Ref. [18] for diversity transmission and in Ref. [19] for SMUX transmission. For SMUX transmission, the optimum precoding matrices were shown to be $N_r \times N_t$ matrices with unitary property.

However, due to per-antenna power constraints encountered in practice, it is desirable to use quantized beams for precoding that distribute the power uniformly over all transmit antennas. In the literature, this technique is referred to as equal gain transmission (EGT). Quantization of precoding vectors for EGT has been studied in Ref. [20], where it was shown for uncorrelated Rayleigh fading that a set of N_t unitary vectors is sucient to guarantee achieving the full diversity gain of channel $\mathbf{H}$. To obtain $B = \Omega N_t$ suitable precoding vectors for EGT, with Ω being an integer specifying the number unitary beam sets with N_t beams, Love and Heath [20] propose taking the first N_t rows of the ΩN_t-dimensional discrete Fourier transform (DFT) matrix. These precoding vectors are commonly referred to as DFT beams.

In practice, MIMO channels are often spatially correlated to some extent, so that the uncorrelated Rayleigh-fading assumption, which was the basis for the investigations mentioned above, does not hold in general. One of the major differences is that in correlated channels, the distribution of the channel's eigenmodes is biased. In particular, the dominant eigenmodes are concentrated around the dominant eigenvector of the channel's covariance matrix $\mathbf{R}_H = E[\mathbf{H}^H\mathbf{H}]$. If the channels between different transmit/receive antenna pairs have the same statistical properties—which is a common assumption also for realistic MIMO channels—$\mathbf{R}_H$ is a Toeplitz matrix with hermitian property, that is $\mathbf{R}_H^H = \mathbf{R}_H$, with a real-valued diagonal and, in general, with all entries being nonzero. The dominant eigenvector of a matrix with these properties is close to the N_t-dimensional vector with equal-weighted entries. This vector fulfils the properties of an EGT vector, suggesting that precoding based on DFT beams as described above may be a suitable choice especially for the case of correlated MIMO channels.

In a multiuser scenario, a solution achieving near-optimal performance is to serve multiple users simultaneously on their dominant channel eigenmodes [21]. This result suggests that the same quantization techniques from above can also be deemed suitable in the multiuser context.

The above remarks suggest that DFT beams are a convenient choice for GoB-based precoding in practical systems, and hence we have also adopted them for our research. For SMUX transmission, we allow only sets of unitary beams to be active simultaneously, since precoding matrices of this kind were shown to optimize various performance measures in single user links (see Ref. [5] and references therein). Hence, the beams provided by the BS are constituted from Ω independent sets of N_t unitary DFT beams, that is $B = \Omega N_t$. For $N_t = 2$, the $\Omega = 2$ unitary DFT beam sets $\mathbf{C}_1$ and $\mathbf{C}_2$ that will be used as precoding matrices in Equation 9.1 are given as

$$\mathbf{C}_1 = \frac{1}{\sqrt{2}} \cdot \begin{bmatrix} 1 & 1 \\ i & -i \end{bmatrix}, \quad \mathbf{C}_2 = \frac{1}{\sqrt{2}} \cdot \begin{bmatrix} 1 & 1 \\ 1 & -1 \end{bmatrix}, \tag{9.3}$$

where the $B = 4$ beams $\mathbf{b}_u$, $u \in \{1, \ldots, B\}$, are given as columns of the two matrices.

9.3 Resource Allocation Algorithm*

We will now describe the adaptive transmission concept, which is based on a two-step algorithm consisting of an evaluation of the transmission channel at the side of the UT (step 1) and following the resource scheduling and transmission mode selection carried out at the BS (step 2). In step 1, a user carries out an RB-wise evaluation of the different transmission modes and determines the achievable rates per beam. The single per-beam rates from all modes over all RBs are then ranked by their quality, and corresponding scores are assigned. This also yields a ranking of the single RBs of that user. The scores are used by the BS in step 2 to assign the beams in an RB individually to the users and to make a final selection on the transmission mode. The objective for this scheduling

---

* Reprinted from M. Schellmann et al., *IEEE Transaction on Vehicular Technology*, Jan. 2010. With permission.

process is to assign each user his best resources, and the decision on the spatial mode is taken under the premise of guaranteeing a high throughput for each user.

The structure of the two steps of the algorithm is depicted in Figure 9.3; they will be described in detail in the following subsections.

9.3.1 Channel Evaluation at UT, Step 1

Based on the actual channel $\mathbf{H}$, a UT determines for each transmission mode the beams it can achieve the highest data rate with. Evaluation is carried out for each RB separately, which is represented by the different layers in Figure 9.3a.

In ss GoB mode (upper branch of Figure 9.3a), a single beam is assumed to be powered with full transmit power P_s. At the receiver, MRC is used, where the equalization vector for beam $\mathbf{b}_u$ is defined as $\mathbf{w}_u = \mathbf{H}\mathbf{b}_u$. Based on MRC, the postequalization SINR for each beam $\mathbf{b}_u$ can be determined according to

$$\text{SINR}_{ss,u} = \frac{P_s}{N_0}\|\mathbf{H}\mathbf{b}_u\|^2, \quad u \in \{1, \ldots, B\}, \tag{9.4}$$

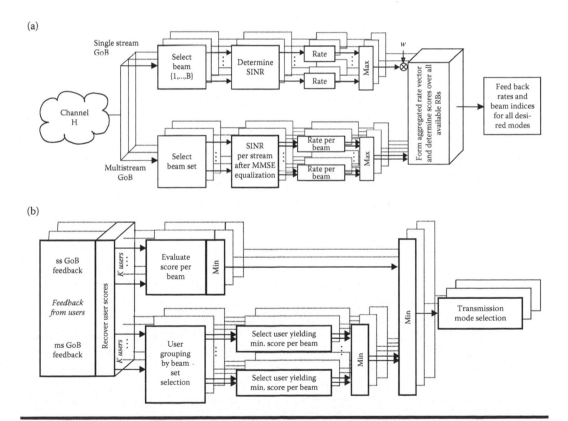

Figure 9.3 Structure of the two-step algorithm constituting the adaptive transmission concept. (a) Step 1: Channel evaluation and determination of the scores at the UT. (b) Step 2: Resource scheduling with transmission mode selection at the BS. (Reprinted from M. Schellmann et al., _IEEE Transaction on Vehicular Technology_, Jan. 2010. With permission.)

where $\|\mathbf{x}\|^2 = \mathbf{x}^H\mathbf{x}$ is the square of the Euclidean norm of vector $\mathbf{x}$. The equation yields the SINR for a single subcarrier signal; the SINR for the entire RB can be obtained by averaging over all subcarriers of that RB.* Once the SINR values for all B beams are obtained, the achievable rate $r_{ss,u}$ per beam $\mathbf{b}_u$ can be determined by using a suitable mapping function. The maximum rate $R_{ss} = \max_u r_{ss,u}$ determines the beam favored for ss mode.

In ms GoB mode (lower branch of Figure 9.3a), up to N_t unitary beams may be simultaneously active. As a general case, we thus assume that $Q \leq N_t$ unitary beams are served in parallel with equal transmit power P_s/Q per beam. As Ω unitary beam sets of dimension N_t are provided, there exists a total of $n = \Omega\binom{N_t}{Q}$ sets of unitary beams of dimension Q. The following evaluation is performed for each of the n sets of beams separately. To recover the data stream transmitted on beam $\mathbf{b}_u$, $u \in \{1, \ldots, Q\}$, we use the MMSE equalizer. The corresponding MMSE equalization vector is defined as

$$\mathbf{w}_u = \mathbf{R}_{rr}^{-1}\mathbf{H}\mathbf{b}_u, \quad \mathbf{R}_{rr} = E[\mathbf{r}\mathbf{r}^H] = \frac{QN_0}{P_s} \cdot \mathbf{I} + \sum_{k=1}^{Q} \mathbf{H}\mathbf{b}_k\mathbf{b}_k^H\mathbf{H}^H. \tag{9.5}$$

The postequalization SINR for each beam $\mathbf{b}_u$ is then given by

$$\text{SINR}_{ms,u} = \frac{\|\mathbf{w}_u^H\mathbf{H}\mathbf{b}_u\|^2}{\mathbf{w}_u^H\mathbf{R}_{rr}\mathbf{w}_u - \|\mathbf{w}_u^H\mathbf{H}\mathbf{b}_u\|^2}. \tag{9.6}$$

From the SINR values obtained from Equation 9.6, one can determine the achievable rate per beam by applying the mapping function again. The beam set preferred for ms transmission is the one comprising the beam achieving highest overall rate. Once the beam set is selected, the Q per-beam rates $R_{ms,u}$ belonging to that set are stored, while all others may be discarded.

9.3.2 Determination of the Scores

In the following, the score concept is introduced. The scores are used to rank the per-beam rates of each user over all RBs according to their quality. We use a single score set for the ranking of the user rates from all transmission modes to enable an implicit selection of the transmission mode within the SB process following in step 2. However, this requires a direct comparison of the single per-beam rates from different spatial modes, whereby it must be taken into account that each mode supports a different number of simultaneously active beams. A practical solution to enable the desired comparison with simple means is the introduction of a penalty factor w, which is used to weight the rates of the ss mode in order to have some compensation for its lacking SMUX capabilities. For a proper choice of w, we will take into account some basic considerations.

As we aim for a high user throughput, spatial mode selection should follow the rationale to favor ss mode whenever the user rate can be expected to be larger than the rate expected in ms mode. Consider that if a user decides globally for ms mode, the available spatial streams compared to ss mode are increased by factor Q. As a general result from that, we can assume that the user will be assigned also Q times the streams it would get if it globally selected ss mode. Hence, we can

---

* If the channel conditions do not vary considerably over the frequency width of an RB, it may be sufficient to determine the SINR for the center subcarrier within the RB only.

conclude that decision in favor of ss mode should be taken if the rates for the different modes in an RB fulfil

$$Q^{-1}R_{ss} > \max_{u} R_{ms,u}, \tag{9.7}$$

suggesting $w = Q^{-1}$ as a suitable choice for the penalty factor of per-beam rates in ss mode.

We now return to the generation of the scores. The per-beam rates from ms mode as well as the weighted rates from ss mode from all RBs are aggregated into one vector, which is sorted by magnitude in descending order.* The index within the sorted vector represents the score of each beam.

Optionally, the user may use the scores to make a selection of its best RBs, which could be a convenient measure to further reduce the amount of feedback.[†] For all selected RBs, the user finally feeds back the achievable rates for the beams supported by the desired transmission modes as well as the corresponding beam indices.

9.3.3 Resource Scheduling at BS, Step 2

The second step of the process comprises resource scheduling with implicit transmission mode selection and is carried out at the BS, which collects feedback information from the K users, see Figure 9.3b. As a first step, it recovers the user scores from the provided rates for the selected RBs. Resource allocation with transmission mode selection is then carried out for all RBs successively, whereby we track the number of beams assigned to each user.

For each RB, the available user scores are first partitioned according to the transmission mode they refer to. The user selection process is then carried out for each transmission mode separately. For ss mode (upper branch of Figure 9.3b), the favored user is the one providing the minimum score for that mode.[‡] The lower branch of Figure 9.3b illustrates user selection for ms mode. Here, users that chose the same beam set are possible candidates for MU-MIMO access and are thus put into one group. In each group, each of the Q available beams is assigned to the user providing the minimum score for that beam. Obviously, this user selection implicitly includes the SU-MIMO mode, as all spatial streams will be assigned to a single user if it provides the highest scores for all Q available beams. After user selection has been carried out for all groups, we pick the group containing the user with minimum score. Finally, we compare the score of that ms user with the score of the user favored for ss mode and select the transmission mode yielding the total minimum. In essence, the transmission mode and beam set selection is thus dictated by the user providing minimum overall score within each RB. The decision on the mode and the user allocation per RB is then signaled forward to the UTs, who configure their receivers accordingly.

9.3.3.1 CQI-Based Feedback

For our evaluations in the following sections, we assume full feedback from the UTs on the achievable rates per transmission mode for each RB. For practical applications, however, the concept offers potential for further feedback reduction, as the score-based ranking allows each user to preselect its best RBs as well as its preferred mode to be served with. A score-based preselection of the

---

* In the case of multiple resources yielding identical rates, these are ordered in a random fashion.
† This kind of user-driven resource selection has also been suggested in Ref. [22].
‡ In the case of multiple users providing identical scores, the stream is given to the one who has been assigned the least beams so far.

transmission mode may result in severe performance loss though which is illustrated by the following two examples. If there are not sufficient users in the system (sparse network), it may occur that no MU-MIMO partner can be found for a user who provided the best score for the ms mode. In this case, this user could be served via multiple beams in SU-MIMO mode or, alternatively, in diversity mode. To achieve highest possible throughput for that user, the final mode selection should be based on his achievable rate—which requires also the availability of the ss rate at the BS. On the other hand, if a user prefers ss mode, but an ms user is selected, the former could still be assigned resources if it provided appropriate rates for ms mode.

However, as the scheduling process aims at assigning each user its best resources only, it is certainly not economical to let the users report on all available resources, but on their best RBs only. These can easily be selected after the scores have been determined by the UTs, which is related to the suggestion in Ref. [22] denoted as *Top-M feedback*. Furthermore, a similar selection can also be done for rates referring to the transmission modes in an RB. A practical solution could be to let the users report two rates: the best ms rate enabling MU-MIMO access, and additionally the ss rate (enabling diversity mode) or the next ms rate (enabling dual-stream SMUX for that user). The adequate choice of the second rate to report could be based on the rate achievable with diversity mode and dual-stream SU-MIMO mode, respectively.

Moreover, the frequency-selective feedback information for the utilized transmission band will be highly correlated, so that proper compression techniques may conveniently be applied, yielding a further reduction of the required feedback per user. Elaborating on the adequate amount of total required feedback is an interesting field for further studies, but is beyond the scope of this work.

9.4 Investigations on Link Level (Isolated Cell)*

The properties of the scheduler with transmission mode switching are investigated in a single-cell link-level simulation environment. These kinds of link-level investigations enable an isolated examination of system behavior depending on fixed SNR conditions, which are common for all involved users. Thus, we are able to gain insights into the basic relationships that influence the performance of MU-MIMO communication systems.

9.4.1 System Setup

As an example, we focus on a system setup with K UTs being equipped with $N_r = 2$ antennas each and a BS with $N_t = 2$ antennas, which provides the two unitary DFT beam sets C_1 and C_2 introduced in Equation 9.3. Possible ms modes are thus two-user MU-MIMO or dual-stream SU-MIMO. We use the channel model provided by the European Wireless World Initiative New Radio (WINNER) project, WINNER channel model (WIM), in its configuration for a wide area urban macroscenario. This model assumes a uniform linear array of copolarized antennas; antenna spacing is set to 4λ at the BS and to 0.5λ at the UTs, yielding a low degree of correlation between the paths emanating from the antennas. The mean channel energy is normalized to unity; channels for different users are modeled independently. The mean reception SNR is thus P_s/N_0 for any user. An OFDM system with 1024 subcarriers spanning a bandwidth of 40 MHz is assumed, accommodating 128 RBs of eight subcarriers width. To obtain the per-stream SINR γ for the different transmission

---

* Reprinted from M. Schellmann et al., *IEEE Transaction on Vehicular Technology*, Jan. 2010. With permission.

modes per RB, Equations 9.4 and 9.6 are calculated for the subcarrier in the center of the RB*
based on ideal knowledge of channel **H**.

From the SINR values, the corresponding achievable rates are determined via the Shannon
information rate $\log_2(1 + \gamma)$. To obtain rates that are closer to those achievable in practice, we
instead use a quantized rate mapping function, which was introduced as a component of the
WINNER link to system interface, presented in Ref. [23]. This rate mapping function is based on
a puncturable low-density parity check (LDPC) code with a constant block length of 1152 bits and
supports fixed symmetric modulation formats up to 64QAM. The Discrete steps of the mapping
function are derived from the SINR values required to meet a block error rate performance of 10^{-2}
in an equivalent AWGN channel. As both the block length of the code as well as the modulation
formats are limited, the rates supported by the mapping function are confined to a minimum rate
of 0.5 and a maximum rate of 5.538 bit/s/Hz. The former value corresponds to BPSK modulation
with code rate 0.5, while the latter is achieved with 64QAM modulation with code rate 24/26. All
simulation results are obtained from a total of 10,000 independent channel realizations.

9.4.2 Throughput Performance

First we examine the system performance of the adaptive system when only beam set C_1 is available.
We focus on the low SNR regime, $P_s/N_0 = 0$ dB, which is relevant for cell-edge users, where we
expect the benefits from switching to ss mode to become prominent. First results are based on
Shannon information rates. Figure 9.4a presents cumulative distribution functions (CDFs) of the
achievable user throughput divided by the signal bandwidth (left) and the spectral efficiency in
the cell (right) for $P_s/N_0 = 0$ dB for $K = 10$ users. Focussing on user throughput (left), we first
point out that the scheduler conveniently meets the fairness constraint, as all users achieve nonzero

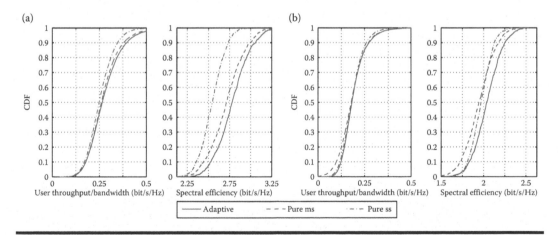

Figure 9.4 **CDFs of achievable user throughput and spectral efficiency in the cell. 1 beam set,
$K = 10$ users, SNR $= 0$ dB. (a) Shannon information rates and (b) quantized rate mapping.
(Reprinted from M. Schellmann et al., *IEEE Transaction on Vehicular Technology*, Jan. 2010.
With permission.)**

---

* Note that with the chosen set of system parameters, the channel variations over the frequency width of an RB can
be considered negligible.

rates. We compare the adaptive system described above to a system supporting either ss or ms mode exclusively. For user throughput (left), we observe that the performance of the adaptive system benefits slightly from switching in the region where the CDF is above 0.5. Further, for the CDF region below 0.2, the ss curve is nearly identical to the ms curve, and hence no gains from switching can be realized here. This observation can be explained as follows: Recall that a beam in ss mode is served with double the power used for an ms beam. In the low SNR regime, where noise dominates the interference from simultaneously active beams, we can thus expect that the SINR of the selected beam for ss mode is about twice as large as the SINR γ for the corresponding beam in ms mode. Moreover, in ms mode the amount of beams assigned to each user is twice as large as in ss mode. As

$$\log_2(1 + 2\gamma) \approx 2\log_2(1 + \gamma) \quad \text{if } \gamma \ll 1$$

holds, the rates achievable with the two different modes are nearly identical. Considering the spectral efficiency within the cell (right subfigure in Figure 9.4a), we observe that the adaptive system benefits significantly from the mode switching over the entire CDF region.

Figure 9.4b depicts performance curves for the same setting, but this time the quantized rate mapping function is used. For user throughput (left), we observe here that the CDF of the adaptive system represents a hull curve of the two other single-mode schemes. As the minimum supported rate to be assigned is bound here to 0.5 bit/s/Hz, the adaptive system now significantly gains from switching to ss mode if the SINR conditions are low (left region of the CDF curve). The CDF of the adaptive system is quite close to the one supporting ss mode only, suggesting that this mode is predominantly chosen at low SNR. Considering the spectral efficiency in the cell (right subfigure in Figure 9.4b), we observe that only the left tail of the adaptive system's CDF approaches the curve of the pure ss mode. In the remaining region, substantial gains from mode switching become visible.

In the next step, we will examine the system behavior for varying SNR. Therefore, we focus on the median spectral efficiency in the cell based on quantized rate mapping, that is, the value determined from the CDF for a probability of 0.5. Furthermore, we draw our attention to the probabilities of mode selection, which reveal the dominantly chosen mode depending on the SNR conditions. Figure 9.5 depicts the probabilities of mode selection for two users (a)* as well as for 10 users (b), while Figure 9.6 shows the corresponding median spectral efficiency in the cell. Three different configurations of the adaptive mode switching system are considered here:

1. *Adaptive MU-MIMO:* The adaptive system as described in Section 9.3 with beam set C_1 from Equation 9.3 being available. Simultaneously active beams can be assigned independently to different users. The mode per user is selected per RB, that is, a user may be served in different modes simultaneously.
2. *Adaptive SU-MIMO:* The MU-MIMO option is switched off, that is, ms mode reduces to SU-MIMO. Now only one user is served per RB either in ss or in SU-MIMO mode.
3. *2 beam sets:* The adaptive MU-MIMO system with the two beam sets C_1 and C_2 from Equation 9.3 being available.

The crossing point of the curves for ss and ms modes in Figure 9.5 highlights the point in the SNR region where the ms mode becomes the dominantly selected one. From both figures, we observe

---

* If none of the users can support a rate $r > 0$ for a given resource, the scheduler does not assign this resource. For that reason, the selection probability of ss mode drops down to 75% at $P_s/N_0 = -5$ dB in Figure 9.5a.

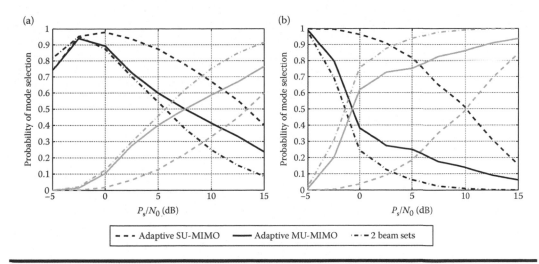

Figure 9.5 Probability of mode selection versus SNR. Black: ss mode; gray: ms mode. (a) $K = 2$ users and (b) $K = 10$ users. (Reprinted from M. Schellmann et al., *IEEE Transaction on Vehicular Technology*, Jan. 2010. With permission.)

that going from SU-MIMO to MU-MIMO promotes selection of the ms mode substantially, as the crossing point is shifted by 5 dB in the case of two users and by more than 10 dB in the case of 10 users down toward the low SNR regime. For 10 users, the crossing point falls below an SNR of 0 dB. These results strongly emphasize that MU-MIMO is the key for the efficient use of SMUX transmission even at low SNR conditions.

MU-MIMO also allows one to benefit from multiuser diversity in the assignment of spatial beams that are active in one RB, and correspondingly the higher probability of ms mode translates

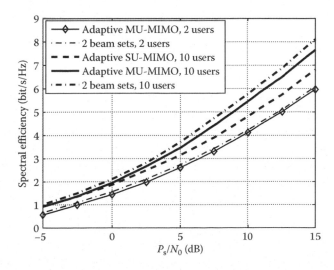

Figure 9.6 Median spectral efficiency in the cell for the different system configurations. (Reprinted from M. Schellmann et al., *IEEE Transaction on Vehicular Technology*, Jan. 2010. With permission.)

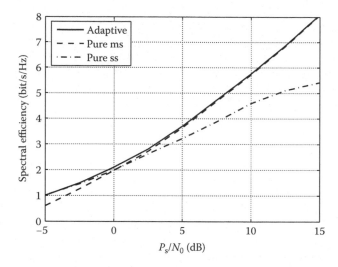

Figure 9.7 **Median spectral efficiency in the cell achievable with two beam sets and comparison to single-mode systems.** $K = 10$ **users. (Reprinted from M. Schellmann et al.,** *IEEE Transaction on Vehicular Technology***, Jan. 2010. With permission.)**

into the performance gains seen in Figure 9.6. Providing an additional beam set shifts the crossing point further down and also delivers additional performance gains, which can be attributed to the finer granularity in the quantization of the transmit vector space. For 10 users, this additional gain in throughput amounts to about 5% compared to MU-MIMO with one beam set. Moreover, the crossing point in Figure 9.5b can be shifted down to about -1.5 dB now. In total, we can point out here that by extending the degrees of freedom in the spatial domain, utilization of the SMUX mode is substantially promoted within the resource allocation process.

Considering the probability curves for two beam sets in Figure 9.5b gives us some more interesting insights: We observe that at 5 dB SNR, the probability of selecting the ss mode is below 0.1, suggesting that mode switching loses its significance here. This conjecture is confirmed by Figure 9.7, where we compare the cell throughput for configuration 3 from the list above (adaptive MU-MIMO, two beam sets) with a similar system supporting either ss mode or ms mode exclusively, as done at the beginning of this section. Furthermore, the figure suggests that in the low SNR regime, the adaptive system achieves the same performance as the system supporting ss mode only. Correspondingly, the curve of the adaptive system represents a hull curve of the performance of the two single-mode systems. This observation thus suggests that by providing more than one beam set, the adaptive system tends to behave like a system that exclusively uses the ss mode in the low and the ms mode in the high SNR regime, with a switching point given at a fixed SNR level.

We conclude this subsection with the important observation that proper application of the MU-MIMO mode enables one to conveniently serve even users in the ms mode who experience relatively poor SNR conditions.* Thus, the MU-MIMO mode establishes a win–win situation for low- and high-rate users competing for a frequency or time resource, as a low-rate user can now be

---

* In general, these are the users at the cell edge.

served without blocking this resource for any high-rate user, which can support a rate on any of the available beams.

9.4.3 Capacity Scaling

Finally, we examine the capacity achievable with the adaptive system and compare it to the upper bound, which is the capacity of the 2×2 BC when full CSI is available at the receivers as well as the transmitter. As mentioned in the introduction, the capacity of the BC was shown to be achievable with the DPC, and in Ref. [24] an algorithm was presented to compute it in an iterative fashion for any given set of flat-fading user channels. While maintaining equal power distribution over all RBs, we use this algorithm to compute optimal user allocation and the corresponding precoding matrices per RB to obtain an upper bound for the (flat-fading) capacity of the BC, which is depicted in Figure 9.8 versus the SNR for $K = 10$ users. The achievable capacity of our adaptive system is obtained by applying Shannon's information rates and carrying out a maximum throughput scheduling (MT) based on the reported rates at the BS, which selects for each RB the user (ss) or user constellation (ms) that achieves the highest throughput. In Figure 9.8 we observe that for an SNR above 0 dB, the capacity of our adaptive system utilizing partial CSI achieves a constant fraction of the capacity of the BC, which amounts to about 80% if one beam set is available. Utilization of two beam sets provides an extra gain in capacity of about 5%. Additionally, we included the capacity of the single-input single-output (SISO) channel achieved in an equivalent scenario. While we observe here that the capacity of the BC scales with factor $\alpha = 2$ [corresponding to $\min(N_r, N_t)$] compared to the capacity of the SISO channel in the high SNR range, the capacity of our adaptive system (with one beam set) achieves a factor of $\alpha = 1.6$. For comparison, we also added the spectral efficiency achievable with the fair SB technique. It can be seen that the price we

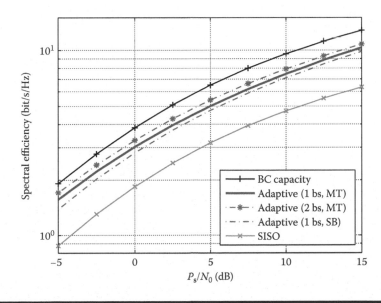

Figure 9.8 Capacity scaling with SNR for $K = 10$ users. Shannon information rates. bs—beam sets; scheduling: MT—max. throughput; SB—score based. (Reprinted from M. Schellmann et al., *IEEE Transaction on Vehicular Technology*, Jan. 2010. With permission.)

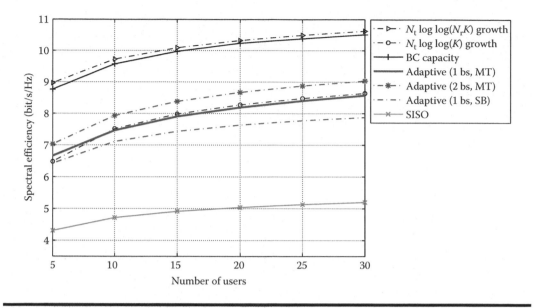

Figure 9.9   Capacity scaling with number of users at SNR = 10 dB. (Reprinted from M. Schellmann et al., *IEEE Transaction on Vehicular Technology*, Jan. 2010. With permission.)

have to pay to obtain user fairness within one radio frame is only marginal, as the loss in spectral eciency is only about 5%.*

In Figure 9.9 we examine the system capacity for a constant high SNR = 10 dB for a variable number of users. First we compare the capacity of the adaptive system (MT) with the achievable spectral efficiency of the fair scheduling approach (SB). Although the loss in throughput to provide the desired fairness is again not exceptionally high, we observe that the gap between MT and SB scheduling increases with increasing number of users. This is not very surprising, as with a growing number of users, the probability of a user experiencing poor channel conditions increases, who imperatively has to be served by the system based on fairness. The support of these users by the fair scheduler thus costs a growing proportion of the maximum sum capacity.

Next we focus on the scaling of system capacity versus the number of users and compare it to the BC and the SISO systems. At first glance, the capacity of the adaptive system seems to scale similarly as the flat-fading BC capacity achieved with DPC. However, a closer look reveals slight differences, as the BC capacity grows with a slope equivalent to $N_t \log \log(N_r K)$, which actually represents the scaling law for the BC capacity for a large number of users as shown in Ref. [26], while the capacity of the adaptive system grows with a slope equivalent to $N_t \log \log(K)$. This observation suggests that, compared to the BC capacity scaling term, the scaling term for our adaptive system is reduced by N_r degrees of freedom. This is also reasonable, as the $(N_r - 1)$ additional receive antennas are used for suppressing the interference from simultaneously active beams rather than exploiting the additional receive diversity for capacity enhancements.

We also plot the capacity of the SISO system, which grows roughly with a slope of $\log \log(K)$ and thus is less steep than the adaptive system. We observe that the capacity scaling factor relative

---

* Note that this loss may increase substantially if users with different mean SNRs are considered, as shown in Ref. [25].

to the SISO system of $\alpha = 1.6$ is achieved by the adaptive system for $K = 10$, remaining about constant for further increasing K.

9.5 Investigations on System Level (Multicell Environment)

In the following section we will extend the SB concept for its application in a multicell simulation environment. These investigations enable an evaluation of system behavior in a multicell simulation environment, where SINRs are significantly varying for distinct users in the cell. Our simulations explicitly consider the spatial correlation of CCI caused by MIMO channels seen from different BSs. Thus, we are able to obtain insights into the performance of MU-MIMO in interference-limited multicell scenarios, enabling one to assess the spectral eciencies achievable in cellular systems.

9.5.1 System Model—Extensions to Multicell Application

For multicell application we need to extend the system model, which was introduced for single-cell applications in Section 9.2. The DL MIMO-OFDM transmission per subcarrier via N_t transmit and N_r receive antennas in cellular deployment is described as follows. Assume that all BSs provide Ω fixed unitary beam sets $\mathbf{C}_\omega$, $\omega \in \{1, \ldots, \Omega\}$. Each beam set contains N_t fixed beams $\mathbf{b}_{\omega,u}$ with $u \in \{1, \ldots, N_t\}$. Each BS i independently selects one of these sets. In the following, we denote $\mathbf{b}_{i,u}$ as the uth precoding vector from the beam set selected by BS i. The received DL signal $\mathbf{r}^k$ at the UT k in the cellular environment is given by

$$\mathbf{r}^k = \underbrace{\mathbf{H}_i^k \mathbf{b}_{i,u}}_{\bar{\mathbf{h}}_{i,u}} s_{i,u} + \underbrace{\sum_{\substack{j=1 \\ j \neq u}}^{N_t} \mathbf{H}_i^k \mathbf{b}_{i,j} s_{i,j}}_{\substack{\zeta_{i,u} \\ \text{intracell interference}}} + \underbrace{\sum_{\substack{\forall l \\ l \neq i}} \sum_{j=1}^{N_t} \mathbf{H}_l^k \mathbf{b}_{l,j} s_{l,j}}_{\substack{\mathbf{z}_{i,u} \\ \text{intercell interference}}} + \mathbf{n}. \tag{9.8}$$

The desired data stream $s_{i,u}$ transmitted on the uth beam from the ith BS is distorted by intracell and intercell interference aggregated in $\zeta_{i,u}$ and $\mathbf{z}_{i,u}$, respectively. Note that in case of single-cell evaluation, the term for intercell interference reduces to $\mathbf{z}_{i,u} = \mathbf{n}$.

Each BS i may select a limited number $Q_i \leq N_t$ of active beams from the set to serve the users simultaneously. Therefore, the transmit power per beam is uniformly distributed over all nonzero transmit symbols $s_{i,j}$ with p_i/Q_i, where $p_i = \sum_{j=1}^{N_t} E\{|s_{i,j}|^2\}$ is the total available power for BS i. As defined in Section 9.2, ss mode refers to $Q_i = 1$, while ms mode refers to $Q_i > 1$. For $N_t > 2$, we have to distinguish between several additional transmission modes employing multiple beams. As already indicated in Section 9.3.1, there exists a total of $n = \Omega\binom{N_t}{Q_i}$ subsets of unitary beams of dimension Q_i. For $N_t = 4$ transmit antennas, all possible cases $1 < Q_i \leq N_t$ are considered in the evaluation process depicted in the lower branch of Figure 9.3a. Thus, the BS can choose from ss transmission and MU-MIMO with 2, 3, and 4 active data streams. For $N_r = N_t = 4$, the MU-MIMO mode implicitly includes SU-MIMO transmission with up to four data streams to serve a single user and several hybrid MU-MIMO modes. In these hybrid modes, at least one user is assigned more than one beam, but not all of the active beams. To enable comparison of the rates from the ms modes based on different Q_i, which is needed to determine the scores, we introduce a slightly modified penalty factor (Section 9.3.2) of $w = Q_i/N_t$.

Assuming a linear equalizer at the UT, we yield the postequalization SINR for beam $\mathbf{b}_{i,u}$ according to

$$\text{SINR}_u = p_i \frac{\mathbf{w}_u^H \overline{\mathbf{h}}_{i,u} \overline{\mathbf{h}}_{i,u}^H \mathbf{w}_u}{\mathbf{w}_u^H \mathbf{Z}_u \mathbf{w}_u}, \tag{9.9}$$

where $\mathbf{w}_u$ is the equalization vector for beam $\mathbf{b}_{i,u}$ and $\mathbf{Z}_u$ is the covariance matrix of the interference signals $\zeta_{i,u} + \mathbf{z}_{i,u}$, that is, $\mathbf{Z}_u = E\left[(\zeta_{i,u} + \mathbf{z}_{i,u})(\zeta_{i,u} + \mathbf{z}_{i,u})^H\right]$. We use the interference aware MMSE receiver for both ss and ms transmissions [25,27].

$$\mathbf{w}_u^{\text{MMSE}} = \frac{p_i \mathbf{R}_{rr}^{-1} \overline{\mathbf{h}}_{i,u}}{Q_i} \quad \Rightarrow \quad \text{SINR}_u = \frac{\frac{p_i}{Q_i} \overline{\mathbf{h}}_{i,u}^H \mathbf{R}_{rr}^{-1} \overline{\mathbf{h}}_{i,u}}{1 - \frac{p_i}{Q_i} \overline{\mathbf{h}}_{i,u}^H \mathbf{R}_{rr}^{-1} \overline{\mathbf{h}}_{i,u}}, \tag{9.10}$$

where $\mathbf{R}_{rr}$ denotes the covariance matrix of the received signal $\mathbf{r}^k$, that is, $\mathbf{R}_{rr} = E\left[\mathbf{r}^k (\mathbf{r}^k)^H\right]$.

9.5.2 Multicell System Setup

Performance is investigated in a triple-sectored hexagonal cellular network with 19 BSs in total as indicated in Figure 9.10 (left). The BSs are depicted as black dots; BS sectors are indicated as hexagonal areas, which are continuously numbered. The widely used spatial channel model extended

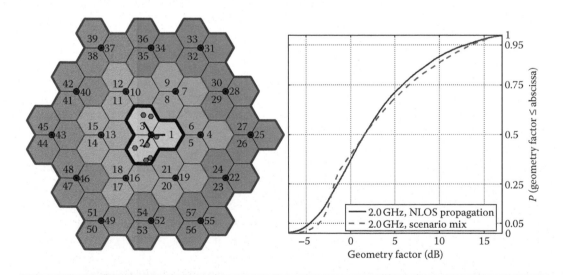

Figure 9.10 Triple-sectored hexagonal cell layout (left) used for system level simulations and its corresponding user geometries (right) for different propagation conditions. (Reprinted from L. Thiele et al., *IEEE 42nd Asilomar Conference on Signals, Systems and Computers*, October 2008. With permission.)

Table 9.1   Simulation Assumptions

Parameter	Value
Channel model	3GPP SCME
Scenario	Urban-macro
Additional modifications	Scenario mix
f_C	2 GHz
System bandwidth	31.72 MHz
Signal bandwidth	18 MHz, 100 RBs
Frequency reuse	1
Intersite distance	500 m
Number of BSs	19 having three sectors each
Antenna elements; spacing	2; 4λ
Transmit power	46 dBm
Sectorization	Triple, with FWHM[a] of 68°
BS height	32 m
Antenna elements; spacing	2; λ/2
UT height	2 m

[a]   Full width at half maximum.

(SCME)* with urban macroscenario parameters is used, yielding an equivalent user's geometry as reported in Ref. [28]. Simulation assumptions for these Monte Carlo drops are given in Table 9.1.

Based on channel measurements in an urban macrocellular deployment [29], a so-called scenario mix has been introduced. For distinct UT positions, different channel conditions may be experienced, for example, line of sight (LOS) or NLOS propagation to different BSs. This is more realistic than assuming the same conditions for all channels from all BSs, which is the general simulation assumption made in the SCME. We model the state of switching between these two main propagation conditions as changing for different channel realizations in each simulation run, following a distance-dependent stochastic process derived from experimental results. This scenario-mix slightly changes the interference statistics, as shown in Figure 9.10 (right). Due to the possibility of LOS propagation, MIMO channels of the UTs may experience rank deficiency. In principle, this results in the fact that a single-user system may not allow supporting more than one spatial data stream. However, in multiuser systems this does not necessarily need to result in a restricted choice of ms mode. Due to the unitary precoding and direction constancy in the propagation paths, each user may be able to receive one beam $\mathbf{b}_{i,u}$, while another may be invisible to this user. With a moderate number of UTs in the system, the BS may find a second user having conditions contrary to the first one, which enables to serve both users in MU-MIMO mode.

For the sectorization, the simulation scenario is initialized cellwise, that is, independently for each BS. The large-scale parameters are kept fixed for all three sectors belonging to the same BS, while the small-scale parameters are randomized as indicated in Ref. [30]. The UTs are always served by the BS whose signal is received with highest average power over the entire frequency band and in each independent channel realization. For capacity evaluation, only UTs being placed inside

---

* Note that the WIM employed for single-cell evaluation is based on the SCME and thus broadband statistics are similar for the urban macroscenario in case of non-line of sight (NLOS) propagation.

the center cell, encircled by the black solid line in Figure 9.10, are considered, so that BS signals transmitted from the first and second tiers of cells model the intercell interference properly [25].

9.5.3 Capacity Scaling with Number of Receive Antennas and Users

Throughput performance is evaluated for both sum throughput in a sector and throughput for individual users. Both values are normalized by signal bandwidth, yielding a sector's overall spectral efficiency and normalized user throughput, respectively. Achievable rates are determined from the SINRs by the use of the WINNER link to system interface, as already done in Section 9.4.1. From these results, CDF plots are obtained.

In the following, we concentrate on the case where all BSs provide a single fixed unitary beam set, that is, $\Omega = 1$. Note that $\Omega > 1$ would render the system noncausal as independent selection of C_ω for all BSs would result in a system concept where intercell interference caused by surrounding BSs is no longer predictable. This would lead to a decrease in achievable spectral efficiencies, since predictable interference is the fundamental requirement of the concept described in this chapter, as interference directly affects R_{rr} in Equation 9.10.

Our results in Figure 9.11 (solid lines) indicate a capacity increase in the median sector's spectral efficiency by a factor of $\alpha = 1.95$ and $\alpha = 3.43$ for the MIMO 2×2 ($N_t \times N_r$) and 4×4 system with respect to the SISO reference case for $K = 20$ UTs. Comparing these results with those obtained for an isolated cell from Section 9.4.3, we observe that the gains here are higher. For an explanation, note that in the multicell environment performance is mainly limited due to spatially correlated interference in contrast to additive white noise, which is the general assumption for single-cell evaluation. By taking the spatial color of the interference into account, the MMSE equalizer can suppress the interference and thus achieve better performance as compared to AWGN, where the noise can only be averaged. For both the system as well as the user spectral efficiencies, we demonstrate substantial gains from MIMO transmission in multicell environments, which are close to those known for point-to-point links [1,25]. We emphasize that the spectral efficiency

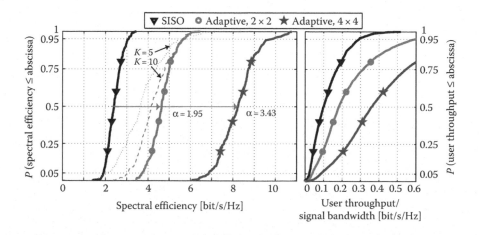

Figure 9.11 System performance for the SISO, MIMO 2×2 ($N_t \times N_r$) and 4×4 system for $K = 20$ users assigned to the BS. Dashed lines indicate the performance achievable with $K = \{5, 10\}$ users. (Reprinted from L. Thiele et al. *IEEE 69th Vehicular Technology Conference, VTC2009-Spring*, April 2009. With permission.)

grows proportionally with the minimum number of receive and transmit antennas, even in the interference-limited multicell environment.

For a decreasing number of users K sharing all resources, the system shows inferior throughput performance for all transceiver setups, indicated by the dashed lines in Figure 9.11 (left) for the MIMO 2×2 system. This may be explained by considering that a reduced number of active users in the system results in an increasing amount of resources being assigned to each user. It is obvious that this also includes resources with degraded quality. In addition, the multiuser diversity available in the user selection process is decreased. Interestingly, throughput performance in the high SINR regime, represented by the upper CDF region, is hardly influenced by a reduced degree of multiuser diversity, suggesting that the scheduler most likely decides to serve a single user in the SU-MIMO mode here. In contrast to that, in the lower CDF region, where users experience degraded channel quality, the system needs to group multiple users to be served in the MU-MIMO mode or selects a single user for ss transmission. A reduced number of active users clearly results in a higher variance of the distribution of achievable spectral efficiency, which can be seen in the figure as the slope of the CDF loses its steepness.

Figure 9.12 indicates the selection probabilities of the different spatial modes supported by the 2×2 and the 4×4 configurations, respectively. It is worth emphasizing the high selection probability of the MU-MIMO mode, which is chosen in more than 90% of all cases. For the 4×4 configuration, the system subdivides the MU-MIMO mode into several submodes, resulting in two additional modes being selected here: MU-MIMO hybrid, where at least one user is assigned more than one of the $Q_j = 4$ active beams, and MU-MIMO with $Q_j = 3$ active beams assigned to three different users. For the configuration considered here ($N_t = N_r = 4$), it turns out that the selection probability for both the SU-MIMO and the ss mode tends to zero. The results from this section thus clearly show that the MU-MIMO mode can be efficiently applied also in the interference-limited multicell environment.

9.6 Experimental Results with the LTE Test Bed

Implementing the MU-MIMO concept in a real-time test bed and realizing the gains discussed in the previous chapters is challenging. The performance evaluation is therefore computed offline for a real-world scenario, which is based on CSI and CQI recorded in DL measurements with a real-time

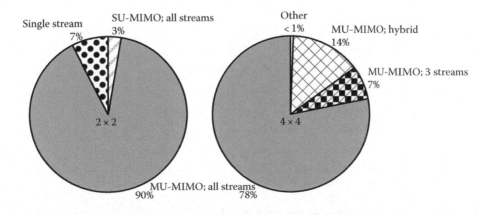

Figure 9.12 Selection probabilities for different transceiver modes.

LTE test bed. The data were recorded during measurement campaigns in Berlin in 2008. During measurements with the LTE test bed, channel coefficients and SNRs were recorded. The recorded data were used for offline simulation using the same algorithms previously introduced. The results were finally compared to results obtained with WIM and SCME.

9.6.1 Experimental Real-Time Test Bed

The real-time LTE test bed realizes MU-MIMO orthogonal frequency division multiple access (OFDMA) scheduling and has previously been used in test trials, refer to Refs. [31] and [32]. The parameter set for the physical layer (PHY layer) implementation was chosen according to a tentative working assumption issued around November 2005, which is shown in Table 9.2. The system bandwidth was set to 20 MHz in the digital domain. The highlights of the system are multiuser scheduling, spatial mode switching, and link adaptation, all performed in a frequency-selective manner.

The PHY layer and the medium access layer (MAC layer) for BS and UT were implemented on a signal processing platform previously used for 1 Gbit/s MIMO-OFDM transmission experiments [33]. The new MAC layer connects via 1 Gbit/s Ethernet to the IP layer. The implemented PHY layer is similar to the one currently discussed, LTE R8. Differences concern only synchronization, reference signals, and the size of RBs. DL and uplink (UL) signals are transmitted simultaneously in FDD mode. RF front-ends include duplex filters, mixers, local oscillators, and automatic gain control. In-phase and quadrature component (IQ) imbalance correction and analog filtering support MIMO-OFDMA transmission with up to 64QAM modulation in both directions. Front-end and base band processing are digitally coupled via a parallel link at the UT side and via a serial link at the BS. The serial link protocol used at the BS is common public radio interface (CPRI) 1.2 Gbit/s. The

Table 9.2   DL MIMO-OFDM Test-Bed Parameter Set

System Parameter	Value
DL UMTS extension band	2.68 GHz
Antennas BS/UT	2 Tx/Rx
Bandwidth used	Up to 18 MHz
Symbol period	71.4 μs
Cyclic prefix	4.7 μs
Total no. of subcarriers	2048
No. of used subcarriers	1200
Radio frame duration	10 ms
Transmit-time interval (TTI)	0.5 ms
TTIs per radio frame	20
Symbols per TTI	7
Resource block size	25 subcarries
No. of resource blocks	48
Modulation	4, 16, 64QAM
Channel coding rates	Conventional with 1/2, 3/4
Feedback update rate	10 ms
Feedback granularity	3 RBs
BS transmit power per antenna	10 W
UT transmit power	200 mW

PHY chain includes coarse and fine synchronization, channel estimation, MIMO equalization, and a forward error correction (FEC) based on convolutional coding. The resource map is transmitted as broadcast over the control channel in the DL. It is decoded by each UT. The receiver demaps signals on its granted resources and feeds them into the FEC.

This MU-MIMO concept is implemented in hard- and software as shown in Figure 9.13. The scheduling algorithm is implemented on a standard floating point DSP platform. The DSP calculates a radio resource map (RRM) from the received terminal feedback, taking into account the states of the queueing system. The queueing system, FEC, resource block pre-processor (RBPP), and adaptive modulator are implemented in hardware on an field programmable gate array (FPGA) platform. The RBPP processes the RBs according to the RRM. The modulated data are then transmitted via the radio front-end.

9.6.2 Measurement Scenario

The DL measurement was carried out with 1 BS and 1 UT, which was synchronized to the BSs. There was no interference from other BSs. The scenario was a typical macrourban scenario with a mixture of LOS and NLOS and multipath propagation between buildings (see Figure 9.14). The measurement area consists of a mix of office buildings, general residential area, and is to the east adjoined by the Tiergarten park. The BS antenna was placed on top of the building of the TU-Berlin main building, above rooftop, at a height of 37 m. The BS sector used was the 30° north-east sector. For the antenna setup, $N_t = 2$ transmit and $N_r = 2$ receive antennas were used at the BS and UT, respectively. Both antenna sets were cross-polarized antennas($\pm 45°$). The BS antennas had a downtilt of 10° and an antenna gain of 18 dBi. The UT was placed in a car that moved at speeds of approximately 15–30 km/h on the measurement track. The measurement track is depicted in Figure 9.14. The gray scale code corresponds to the measured SNR along the track. The outdoor test route covers a BS to UT distance ranging from 60 to 1200 m.

For the analysis, we consider two special areas, marked on the map by concentric circles. The first area is within the 300 m radius around the BS, which represents the cell center and will be denoted as scenario 1 in the following. The second area covers distances from 300 to 1200 m, which is the outer cell annulus and will be referred to as scenario 2.

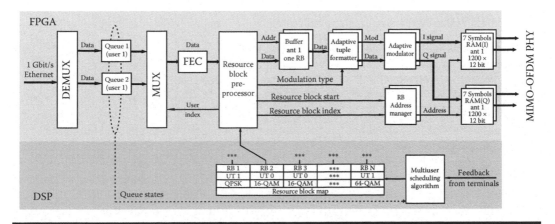

Figure 9.13 Concept: multiuser MIMO MAC hardware/software implementation. (Reprinted T. Wirth et al., *Wireless Communications and Networking Conference, WCNC 2008*. IEEE, pp. 1328–1333, April 2008. With permission.)

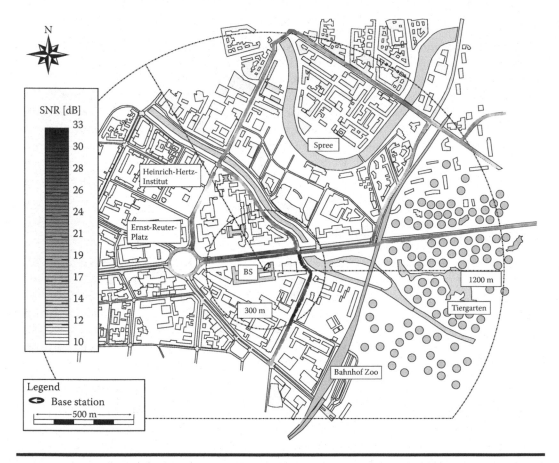

Figure 9.14 Measurement scenario in a macrourban environment in Berlin showing measured reception SNR (dB), averaged over receive antennas and transmission bandwidth, on the measurement track.

9.6.3 Results

In the following, the results obtained from measurements will be compared with those from simulations presented in the preceding sections. Figure 9.15a shows the statistics of the reception SNR, averaged over the receive antennas, which is determined per RB over all channel realizations. CDFs from the measurements are compared to those from the WIM link level and the SCME system level channel model, respectively. Measurements were carried out at high SNR conditions; average measured SNRs were 28.9 and 21.8 dB for scenario 1 and scenario 2, respectively. Hence, the channels could be recorded with high accuracy. For the evaluations here, artificial noise was introduced in order to achieve SNR conditions similar to those experienced in interference-limited multicell systems, which allows comparison to the simulation results from the preceding sections. This noise addition operation can be seen as an equivalent reduction of transmit power at the BS. In particular, SNR conditions were lowered by approximately 14.2 dB, yielding average SNRs of 14.7 and 7.6 dB for scenario 1 and scenario 2, respectively. Hence, we may compare the results for the two scenarios with the link-level results based on WIM for 15 and 7.5 dB, respectively. From Figure 9.15a we observe that the statistics of WIM are very similar to those from the measured channels. For scenario

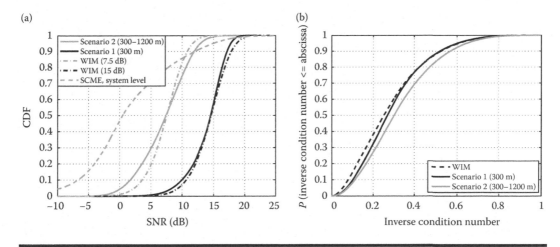

Figure 9.15 (a) CDFs of the SNR statistics for the different system setups and (b) CDF of the inverse channel condition number for WIM and measured channels.

1, the corresponding CDFs are very close, while for scenario 2 larger deviations can be pointed out. This larger deviation for scenario 2 can be attributed to the wide range of distances considered (300–1200 m), resulting in the fact that the path loss will have a non-negligible influence on the SNR statistics.* The CDF of SNRs from the SCME system level model shows a broad statistic with a mean SNR of 0.5 dB crossing the SNR curve of scenario 1 at around 11 dB. The broadening and the significant reduction of the mean SNR can clearly be attributed to the presence of interference from adjacent cells.

To point out further differences between the conditions for measured and simulated channels, we also compare the statistics of the channel condition numbers. Figure 9.15b depicts the CDFs of the inverse channel condition number for the measured scenarios and simulated WIM channels. Note that a larger inverse channel condition number denotes a better conditioned channel. We observe that the measured channels exhibit slightly better channel conditions than those generated by the WIM. This can be attributed to the cross-polarized antennas used in the real-world setup, which enable the support of SMUX transmission even under strong LOS conditions (polarization multiplexing). We point out further that the inverse channel condition number increases with increasing distance to the BS, as can be seen by comparing the CDF of scenario 1 and scenario 2. This can be attributed to the fact that the scattering resulting from object reflections becomes richer for increasing distance between BS and UT.

Next the achievable spectral efficiency based on the score-based scheduler will be determined for the two scenarios and compared to the corresponding cases based on WIM. Evaluations will be carried out similar to the results presented in Figure 9.4. We assume two users to be randomly distributed in the cell service area, that is, we randomly pick two measured channels from the scenario and carry out the channel evaluation and scheduling as presented in Section 9.3. The results are shown in Figures 9.16a and b for scenario1 and scenario 2, respectively. Figure 9.16a reveals that the CDFs for the three different transceiver configurations (mode fixed to ss, mode fixed to ms,

---

* Note that the link-level investigations based on WIM consider only the small-scale fading effects of the channel; hence the path loss is neglected totally.

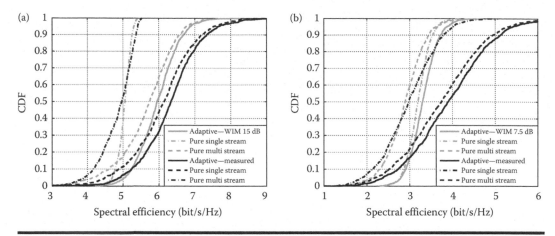

Figure 9.16 Comparison of achievable spectral efficiency in the cell for measurements and WIM link-level simulations, $K = 2$ users. Rates based on quantized rate mapping. (a) Scenario 1: Cell center and (b) Scenario 2: Outer cell annulus.

and adaptive mode switching) are quite similar for measured and modeled channels, yielding also similar gains from applying spatial transmission mode switching. These results could be expected from similar SNR distribution for measured and modeled channels shown in Figure 9.15a. However, the spectral efficiency for the ms mode is slightly higher for the measured channels, which can be related to the better channel conditioning (compare with Figure 9.15b). A better channel condition will clearly yield a smaller norm of the equalization vector $\mathbf{w}_u$ in Equation 9.5, and correspondingly the SINR achievable in ms mode will be higher. As a result, application of ms mode will be promoted in this case. In contrast to that, for the configuration where the mode is fixed to ss, we observe that the spectral efficiency for the measured channels is smaller than that for WIM channels. This can be related to the cross-polarized antennas being employed in the measurements, as these do not allow one to achieve beamforming gains realizable with co-polarized antennas, which have been assumed in WIM.

In scenario 2, relations are more different. First of all, we note that the CDF for the measured channels, especially for the adaptive and the pure ms mode system, is much broader. This can be related to the path loss, which influences the SNR and SINR conditions additionally. However, most interestingly, we observe that the CDF of the adaptive system is closest to the CDF of pure ms mode for the measured channels, while for WIM channels the corresponding CDF is closest to the CDF of pure ss mode. Again, this can be attributed to the further improved conditioning of the measured channels (see Figure 9.15b), promoting the application of ms mode.

The results from resource scheduling based on measured channels as well as channels from link- and system-level simulations are finally compared in Table 9.3, which comprises the probabilities of mode selection for ss and the two options of ms mode: MU-MIMO or SU-MIMO. Comparison of the results from measured scenario1 with simulations based on WIM at 15 dB SNR (first row of the Table 9.3) reveals that the selection probabilities are very similar; however, the probability of MU-MIMO mode is slightly higher for measured scenario 1. Achievable spectral efficiencies are also quite close. In the second row, where results for scenario 2 are opposed to those from WIM at 7.5 dB SNR, we observe that now there are large differences in the selection probabilities. For evaluations based on WIM, selection of SU-MIMO is significantly decreased in favor of ss mode, while for the measured channels selection of SU-MIMO decreases only slightly. As already mentioned above, this

Table 9.3   Comparison of Results from Link Level, Measurements, and System Level

Scenario	ss	ms (MU-MIMO)	ms (SU-MIMO)	Spectral eff. (bit/s/Hz)
Scenario 1 WIM 15 dB	15.7% 23.5%	45.4% 37.6%	38.9% 38.9%	6.4 6.0
Scenario 2 WIM 7.5 dB	20.2% 50.2%	44.9% 34.1%	34.8% 15.7%	3.9 3.3
System-level SCME	27.1%	43.7%	29.2%	2.7

can be attributed to the significantly better channel conditions in the measured scenario, which also yields significantly improved spectral efficiency. It is interesting to note further that the probability to select MU-MIMO mode barely changes for scenarios 1 and 2 as well as the corresponding cases based on WIM. This suggests that the decreased SNR conditions mainly affect the application of SU-MIMO mode, which complies with the famous finding concerning the tradeoff between spatial diversity and multiplexing in Ref. [3].

Finally, we also compare the results from system-level simulations based on the SCME channel model. We observe that the selection probabilities are very similar to those for scenario 2, where users are distributed in the outer cell annulus. As the mean SNR of the system-level channel statistics is further decreased compared to scenario 2 (compare with Figure 9.15a), SU-MIMO mode is selected less frequently. However, note that the selection probability of MU-MIMO mode is nearly identical to that of scenario 1 and scenario 2. This observation substantiates the promotion of the MU-MIMO mode within our resource allocation concept again, as it suggests that the selection probability for this mode becomes independent of the actual SNR conditions for the case of $K = 2$ users. The broader distribution of SNRs and its decreased mean value also result in a significantly decreased spectral efficiency of the multicell system, which amounts to 2.7 bit/s/Hz, emphasizing that intercell interference is the dominant performance limiting factor when going to a realistic multicell scenario.

9.7 Conclusion and Future Work

We have presented a practical concept for resource allocation in the DL of a multiuser MIMO-OFDM system, which combines spatial transmission mode switching with a fair scheduling approach. Based on unitary fixed beamforming, users report information on the rates they can achieve for distinct beams in the supported transmission modes via a low-rate feedback channel. According to the feedback given, the BS then selects the transmission mode per RB and assigns each user the beams yielding highest rates. Performance evaluation has been carried out for an isolated cell for a 2×2 MIMO configuration as well as in a multicell environment for 2×2 and 4×4 MIMO systems. It turned out that utilization of the MU-MIMO mode, which allows the assignation of spatial beams in an RB to distinct users, is the key to enabling SMUX transmission even at low SNR conditions. For the isolated cell, we have shown that the resource allocation concept allows the achievement of a considerable proportion of the MIMO BC capacity. Performance evaluation in the interference-limited multicell environment has revealed that the concept of assigning users to resources where they experience low interference conditions promotes application of the MU-MIMO mode also in this scenario. Hence, the MU-MIMO mode dominates the overall selection and yields a convenient throughput performance even for cell-edge users. Furthermore, we have demonstrated that the concept achieves a scaling of system capacity that is similar to that known for isolated point-to-point links. In the last section, we compared the results from link- and system-level investigations to those obtained from measurements in a real-world scenario in Berlin. Close agreement could be shown, confirming the validity of the findings from the preceding investigations.

In this work, we considered ideal knowledge of CSI at the UTs. In a practical system, however, the CSI has to be estimated, and correspondingly errors will be introduced that may degrade system performance. Channel estimation as well as the effect of estimation errors on system behavior have been investigated in continuative work presented in Refs. [34–36] and will remain in the focus of our future studies. Future work will also address adequate selection and compression of user feedback, so that the required capacity of the feedback channel can be kept as limited as possible. Another issue is the application of the channel-adaptive transmission concept in time-varying environments, which are encountered under user mobility. If the channels vary over time, the SINR conditions serving as the basis for scheduling at the BS may be outdated when the scheduling decision is finally being applied. Therefore, it is a challenge to search for suitable extensions of the channel- adaptive transmission concept to cope with channel aging effects. Initial results for such a possible extension have been published in Ref. [37], and further studies on that topic are still ongoing. Finally, the transmission concept is currently being implemented in our real-time test-bed, where its full functionality is supposed to be shown in a real world application.

Acknowledgments

The authors thank the German Ministry for Education and Research (BMBF) and Nokia Siemens Networks for financial support in the project 3GeT and ScaleNet. Part of this work was performed in the framework of the IST project IST-4-027756 WINNER II, which is partly funded by the European Union. The authors would like to acknowledge the contributions of their colleagues.

Abbreviations

AWGN	Additive white Gaussian noise
BC	Broadcast channel
BPSK	Binary phase-shift keying
BS	Base station
CCI	Cochannel interference
CDF	Cumulative distribution function
CPRI	Common public radio interface
CQI	Channel quality identifier
CSI	Channel state information
DFT	Discrete Fourier transform
DL	Downlink
DPC	Dirty paper coding
EGT	Equal gain transmission
FDD	Frequency division duplex
FEC	Forward error correction
GoB	Grid of beams
LDPC	Low density parity check
LOS	Line of sight

LTE	3G Long Term Evolution
MAC layer	Medium access layer
MIMO	Multiple-input multiple-output
MRC	Maximum ratio combining
MMSE	Minimum mean square error
ms	Multistream
MT	Maximum throughput scheduling
MU-MIMO	Multiuser MIMO
NLOS	Non-line of sight
OFDM	Orthogonal frequency division multiplexing
OFDMA	Orthogonal frequency division multiple access
PHY layer	Physical layer
QAM	Quadrature amplitude modulation
RB	Resource block
RBPP	Resource block preprocessor
RRM	Radio resource map
SB	Score-based scheduling
SCME	Spatial channel model extended
SINR	Signal to interference and noise ratio
SISO	Single-input single-output
SMUX	Spatial multiplexing
SU-MIMO	Single-user MIMO
ss	Single stream
UL	Uplink
UT	User terminal
WIM	WINNER channel model
WiMAX	Worldwide Interoperability for Microwave Access
WINNER	Wireless World Initiative New Radio

References

1. G. Foschini and M. Gans, On limits of wireless communications in a fading environment when using multiple antennas, *Wirel. Pers. Commun.*, (3), 311–335, 1998.
2. A. Paulraj, D. Gore, R. Nabar, and H. Bölcskei, An overview of MIMO communications—a key to gigabit wireless, *Proc. IEEE*, 92(2), 198–218, 2004.
3. L. Zheng and D. Tse, Diversity and multiplexing: A fundamental tradeoff in multiple-antenna channels, *IEEE Trans. Inf. Theory*, 49(5), 1073–1096, 2003.
4. R. Heath and A. J. Paulraj, Switching between diversity and multiplexing in MIMO systems, *IEEE Trans. Commun.*, 53(6), 2005.
5. D. Love and R. Heath, Multi-mode precoding using linear receivers for limited feedback MIMO systems, *Communications, 2004 IEEE International Conference on*, pp. 448–452, 2004.
6. S. Catreux, V. Erceg, D. Gesbert, and J. Heath, R.W., Adaptive modulation and MIMO coding for broadband wireless data networks, *IEEE Commun. Mag.*, 40(6), 108–115, 2002.
7. S. Chung, A. Lozano, H. Huang, A. Sutivong, and J. Cioffi, Approaching the MIMO capacity with a low-rate feedback channel in V-BLAST, *EURASIP JASP*, (5), 762–771, 2004.
8. M. Schellmann, V. Jungnickel, A. Sezgin, and E. Costa, Rate-maximized switching between spatial transmission modes, *IEEE 40th Asilomar Conference on Signals, Systems and Computers*, pp. 1635–1639, 2006.
9. G. Caire and S. Shamai, On the achievable throughput of a multiantenna Gaussian broadcast channel, *IEEE Trans. Inf. Theory*, 49(7), 1691–1706, 2003.

10. P. Viswanath and D. Tse, Sum capacity of the vector Gaussian broadcast channel and uplink-downlink duality, *IEEE Trans. Inf. Theory*, 49(8), 1912–1921, 2003.

11. Z. Shen, R. Chen, J. Andrews, R. Heath, and B. Evans, Sum capacity of multiuser MIMO broadcast channels with block diagonalization, *IEEE Trans. Wirel. Commun.*, 6(6), 2040–2045, 2007.

12. Q. H. Spencer, A. Swindlehurst, and M. Haardt, Zero-forcing methods for downlink spatial multiplexing in multiuser MIMO channels, *IEEE Trans. Signal Process.*, 52(2), 461–471, 2004.

13. N. Jindal, Finite rate feedback MIMO broadcast channels, *Workshop on Information Theory and its Applications*, Feb. 2006.

14. D. Love, J. Heath, R. W., W. Santipach, and M. Honig, What is the value of limited feedback for MIMO channels? *Commun. Mag., IEEE*, 42(10), 54–59, 2004.

15. P. Viswanath, D. Tse, and R. Laroia, Opportunistic beamforming using dumb antennas, *IEEE Trans. Inf. Theory*, 48(6), 1277–1294, 2002.

16. T. Bonald, A score-based opportunistic scheduler for fading radio channels, *Proc. of European Wireless*, 2004.

17. IST-2003-507581 WINNER –D2.7, Assessment of advanced beam forming and MIMO technologies, Feb. 2005.

18. D. Love, J. Heath, R. W., and T. Strohmer, Grassmannian beamforming for multiple-input multiple-output wireless systems, *IEEE Trans. Inf. Theory*, 49(10), 2735–2747, 2003.

19. D. Love and J. Heath, R. W., Limited feedback unitary precoding for spatial multiplexing systems, *IEEE Trans. Inf. Theory*, 51(8), 2967–2976, 2005.

20. D. Love and J. Heath, R. W., Equal gain transmission in multiple-input multiple-output wireless systems, *IEEE Trans. Commun.*, 51(7), 1102–1110, 2003.

21. F. Boccardi and H. Huang, A near-optimum technique using linear precoding for the MIMO broadcast channel, *Acoustics, Speech and Signal Processing. ICASSP 2007. IEEE International Conference on*, (3), III–17–III–20, 2007.

22. J. van de Beek, Channel quality feedback schemes for 3GPP's Evolved-UTRA downlink, *Global Telecommunications Conference. GLOBECOM '06. IEEE*, pp. 1–5, Nov. 2006.

23. IST-4-027756 WINNER II – D2.2.3, Modulation and coding schemes for the WINNER II system, Nov. 2007.

24. N. Jindal, W. Rhee, S. Vishwanath, S. Jafar, and A. Goldsmith, Sum power iterative water-filling for multi-antenna Gaussian broadcast channels, *IEEE Trans. Inf. Theory*, 51(4), 1570–1580, 2005.

25. L. Thiele, M. Schellmann, W. Zirwas, and V. Jungnickel, Capacity scaling of multi-user MIMO with limited feedback in a multi-cell environment, *41st Asilomar Conference on Signals, Systems and Computers*, Nov. 2007, invited.

26. M. Sharif and B. Hassibi, On the capacity of MIMO broadcast channels with partial side information, *IEEE Trans. Inf. Theory*, 51(2), 506–522, 2005.

27. J. Winters, Optimum combining in digital mobile radio with cochannel interference, *IEEE J. Sel. Areas Commun.*, 2(4), 528–539, 1984.

28. H. Huang, S. Venkatesan, A. Kogiantis, and N. Sharma, Increasing the peak data rate of 3G downlink packet data systems using multiple antennas, *Vehicular Technology Conference, 2003. VTC 2003-Spring. The 57th IEEE Semiannual*, vol. 1, pp. 311–315, April 2003.

29. L. Thiele, M. Peter, and V. Jungnickel, Statistics of the Ricean k-factor at 5.2 GHz in an urban macro-cell scenario, *17th IEEE Intern. Symp. on Personal, Indoor and Mobile Radio Communications (PIMRC)*, Sep. 2006.

30. J. Salo, G. D. Galdo, J. Salmi, P. Kyösti, M. Milojevic, D. Laselva, and C. Schneider, MATLAB® implementation of the 3GPP Spatial Channel Model (3GPP TR 25.996). Implementation Documentation, Technical Report, 2005.

31. V. Venkatkumar, T. Haustein, H. Wu, E. Schulz, T. Wirth et al., Field trial results on multi-user MIMO downlink OFDMA in typical outdoor scenario using proportional fair scheduling, *Proc. International ITG/IEEE Workshop on Smart Antennas (WSA)*, Feb. 2008.

32. T. Wirth, V. Jungnickel, A. Forck, S. Wahls, T. Haustein et al., Realtime multiuser multi-antenna downlink measurements, *Wireless Communications and Networking Conference, WCNC 2008. IEEE*, pp. 1328–1333, April 2008.

33. V. Jungnickel, A. Forck, T. Haustein, S. Schiffermüller, C. von Helmolt et al., 1 Gbit/s MIMO-OFDM Transmission Experiments, *Vehicular Technology Conference, VTC Fall 2005, IEEE*, Sep. 2005.
34. L. Thiele, M. Schellmann, S. Schiffermüller, V. Jungnickel, and W. Zirwas, Multi-cell channel estimation using virtual pilots, *IEEE 67th Vehicular Technology Conference VTC2008-Spring*, May 2008.
35. L. Thiele, M. Schellmann, T. Wirth, and V. Jungnickel, On the value of synchronous downlink MIMO-OFDMA systems with linear equalizers, *IEEE International Symposium on Wireless Communication Systems 2008 (ISWCS08)*, Oct. 2008.
36. L. Thiele, M. Schellmann, T. Wirth, and V. Jungnickel, Interference-aware scheduling in the synchronous cellular multi-antenna downlink, *IEEE 69th Vehicular Technology Conference, VTC2009-Spring*, April 2009, invited.
37. M. Schellmann, L. Thiele, and V. Jungnickel, Predicting SINR conditions in mobile MIMO-OFDM systems by interpolation techniques, *42nd Asilomar Conference on Signals, Systems and Computers*, Oct. 2008.
38. M. Schellmann, L. Thiele, T. Haustein, and V. Jungnickel, Spatial transmission mode switching in multi-user MIMO-OFDM systems with user fairness, *IEEE Transactions on Vehicular Technology*, Jan. 2010.
39. L. Thiele, M. Schellmann, T. Wirth, and V. Jungnickel, Cooperative multi-user MIMO based on reduced feedback in downlink OFDM systems, *IEEE 42nd Asilomar Conference on Signals, Systems and Computers*, Oct. 2008.

Chapter 10

Differential Space Time Block Codes for MIMO–OFDM

Benigno Rodríguez Díaz

Contents

10.1   Introduction..268
10.2   MIMO–OFDM Systems..268
      10.2.1   MIMO Channel Models....................................269
      10.2.2   MIMO–OFDM Implementation...........................271
10.3   Differential Space Time Block Codes..................................272
      10.3.1   Origin and Evolution of DSTBC Schemes...............272
      10.3.2   Description of DSTBC Schemes...........................273
10.4   Differential Space Frequency Block Codes...........................274
      10.4.1   Differential Modulation in Time Direction for an OFDM System..........274
      10.4.2   Differential Modulation in Frequency Direction for an OFDM System.....275
10.5   Differential Space Time Frequency Block Codes.....................276
10.6   A New Class of DSTBCs..277
      10.6.1   New Modulation Schemes for DSTBCs...................278
           10.6.1.1   "4A16PSK"......................................278
           10.6.1.2   "2A32PSK"......................................279
           10.6.1.3   "APSK1"...281
           10.6.1.4   "APSK2"...282
           10.6.1.5   "APSK3"...283
      10.6.2   Power Control Mechanism in this New Class of DSTBCs...................284
           10.6.2.1   PCM1...284
      10.6.3   System Performance in AWGN Channels 285
      10.6.4   Influence of the PCM in New Class of DSTBCs...........285

10.6.4.1 Analysis of a New PCM Scheme (PCM2) 286

10.6.4.2 Conclusion about the Influence of Different PCMs..............287

10.6.5 System Performance in WSSUS Channels...................................288

10.6.6 New Technique with Receive Diversity 293

10.6.6.1 Equal Gain Combining...295

10.6.6.2 Maximum Ratio Combining......................................296

10.6.7 Conclusion 298

10.6.8 Open Problems and Possible Extensions.....................................298

References..299

10.1 Introduction

Broadband wireless mobile data networks are an extremely important technical topic nowadays. These networks are probably the most convenient alternative for achieving the objective of providing and increasing connectivity and Internet access. For underdeveloped countries, these networks are an excellent alternative to diminish the so-called digital gap. For developed countries, they represent an opportunity to increase services, security, and comfort.

Orthogonal frequency division multiplexing (OFDM) transmission technique plays an important role in this area. OFDM is a very convenient transmission technique present in the most successful air interface standards of WiFi and WiMAX. Multiple input multiple output (MIMO) systems have an increasing importance to guarantee high data rate transmission schemes. MIMO techniques are well-known techniques to increase the data rate maintaining a given signal to noise ratio (SNR) and bit error rate (BER). The combination of the OFDM transmission technique and multiple antenna MIMO systems is the key focus of this chapter. Robustness and bandwidth efficiency (BE) of new differential transmission techniques, improved by the use of diversities, in particular spatial diversity and the related advantages, are discussed. Differential modulation schemes have the advantage that any radio channel estimation procedure can be avoided which reduces the computation complexity dramatically. In this chapter a differential MIMO–OFDM technique called differential space time block code (DSTBC) is described.

The combination of MIMO–OFDM is a very powerful and flexible one that presents high performance. When coherent systems are considered, this high performance pays an important price in system complexity due to the channel estimation process. In MIMO systems when N_T transmit antennas and N_R receive antennas are considered, $N_T \times N_R$ channels have to be estimated and a valid estimation for each channel has to be maintained during the data transmission. This important increase in system complexity can be avoided by using differential transmission techniques with incoherent demodulation. DSTBC is a kind of differential transmission technique with incoherent demodulation applicable to MIMO–OFDM systems.

In the following sections, DSTBCs are explained and also a new kind of DSTBCs [1] that has an excellent performance is introduced, keeping a low level of system complexity.

10.2 MIMO–OFDM Systems

MIMO is one of the most active areas in wireless systems. The reason is that MIMO offers an alternative to improve the performance of a wireless system without needing more spectrum: only

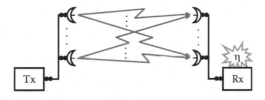

Figure 10.1   Scheme of a MIMO system. (From Cuvillier Verlag Göttingen, Göttingen, Germany, http://www.cuvillier.de/flycms/en/html/1/-/Home.html?SID=ekrseVJbeed5. With permission.)

by increasing the number of antennas (as shown in Figure 10.1), related components, and the complexity of the system. Considering the extremely high cost of the radio-electric spectrum and the always falling prices of electronic systems, this is a very attractive option for increasing the channel capacity.

Usually, multipath fading is a cause of degradation for wireless systems, but in the case of MIMO, this random fading is used to improve the performance [2–4]. MIMO can also exploit the multipath delay spread [5,6]. These are the key features of MIMO systems.

MIMO techniques can be classified according to the *channel state information* (CSI) required in the transmitter and in the receiver. Table 10.1 shows a classification according to this criterion. For a description of the techniques mentioned in Table 10.1 see Refs. [7,8]. This table shows the simplicity of DSTBC techniques from the point of view that these techniques do not require CSI; this is a good reason to try to improve the performance of them.

10.2.1 MIMO Channel Models

MIMO is a technique that achieves spatial diversity by using more than one antenna in the transmitter and in the receiver; there are several MIMO techniques proposed to take advantage of this spatial diversity. In this section, an introduction to this topic will be presented; for a deeper description see Ref. [9]. In general, these techniques provide more reliable (less BER for the same SNR) or higher data rate (for the same BER) communications without increasing the used bandwidth, which generally is a limited and very expensive resource.

Considering a system with N_T transmit antennas and N_R receive antennas, the MIMO channel is composed of a set of individual channels between each transmit antenna and each receive antenna, as it is indicated in Figure 10.2.

Table 10.1   Classification of MIMO Techniques Considering the Required CSI

Rx \ Tx	Full	Partial	None
Full	SVD + bit loading, Eigenbeamforming	Tx selection diversity	STBC, SFBC, spatial multiplexing
None	Joint transmission		DSTBC

Source: Cuvillier Verlag Göttingen, Göttingen, Germany, http://www.cuvillier.de/flycms/en/html/1/-/Home.html?SID=ekrseVJbeed5. With permission.

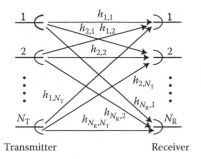

Figure 10.2 MIMO channel for N_T transmit antennas and N_R receive antennas. (From Cuvillier Verlag Göttingen, Göttingen, Germany, http://www.cuvillier.de/flycms/en/html/1/-/Home.html?SID=ekrseVJbeed5. With permission.)

If the channel is frequency-nonselective, as it is for each subcarrier in an OFDM system, then the MIMO channel can be represented by an $N_R \times N_T$ matrix H

$$H = \begin{bmatrix} h_{1,1} & h_{1,2} & .. & .. & .. & h_{1,N_T} \\ h_{2,1} & h_{2,2} & .. & .. & .. & h_{2,N_T} \\ .. & .. & .. & .. & .. & .. \\ h_{i,1} & h_{i,2} & .. & h_{i,j} & .. & h_{i,N_T} \\ .. & .. & .. & .. & .. & .. \\ h_{N_R,1} & h_{N_R,2} & .. & .. & .. & h_{N_R,N_T} \end{bmatrix}.$$

In this matrix, the element $h_{i,j}$ represents the channel transfer factor between the transmit antenna j and the receive antenna i (arriving to i coming from j). This means for example, that if only the transmit antenna j transmits a signal c_j and all the other transmit antennas do not transmit, then the received signal in the receive antenna i (r_i) will be $r_i = h_{i,j} \cdot c_j + n_i$, where n_i is the channel noise added in the receiver. In matrix notation we can represent the received signal $\vec{r}$, for each time slot as

$$\vec{r} = H \cdot \vec{c} + \vec{n}, \tag{10.1}$$

where $\vec{r}$ is the received signal (a $N_R \times 1$ vector with elements r_i), $\vec{c}$ is the transmitted signal (an $N_T \times 1$ vector whose elements are c_j), and $\vec{n}$ is the channel noise added in the receiver (an $N_R \times 1$ vector whose elements are n_i).

In a MIMO system, it is necessary to specify how the data are sent to the transmit antennas and how they are collected from the receive antennas; these tasks are performed by the *MIMO encoder* and the *MIMO decoder*, respectively, as shown in Figure 10.3.

Here $\vec{s}$ is the information signal (also a vector) and $\hat{\vec{s}}$ is the estimator of the information signal ($\vec{s}$) obtained in the receiver. The relation between $\hat{\vec{s}}$ and $\vec{s}$ is given by

$$\hat{\vec{s}} = D \cdot (H \cdot E \cdot \vec{s} + \vec{n}). \tag{10.2}$$

In Section 10.3.2, a particular example of a *MIMO encoder* is described. In Section 10.6.6, two particular examples of *MIMO decoders* are explained. These examples are based on a particular

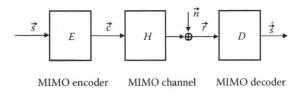

Figure 10.3 Representation of a MIMO system. (From Cuvillier Verlag Göttingen, Göttingen, Germany, http://www.cuvillier.de/flycms/en/html/1/-/Home.html?SID=ekrseVJbeed5. With permission.)

MIMO system $2 \times n$ with $n = 2$ or 3 (2 transmit antennas and 2 or 3 receive antennas) called DSTBC, which will be explained in the following sections.

10.2.2 MIMO–OFDM Implementation

Starting from Figure 10.3 it is clear that a MIMO–OFDM system can be defined just by repeating the structure in this figure for each subcarrier in the OFDM system. This is a very simple way of implementing a MIMO–OFDM system but also a flexible and effective one. The system would be as represented in Figure 10.4, where H_i is the channel matrix corresponding to the subcarrier i, and E_i and D_i are the matrices corresponding to the MIMO encoder i and the MIMO decoder i, respectively. N_S is the number of subcarriers in the OFDM system.

From Figure 10.4 it is clear that this structure allows the use of different MIMO techniques (encoders–decoders) for each subcarrier, which represents an interesting grade of flexibility.

As was explained, MIMO techniques increase the capacity of a telecommunication system. Figure 10.5 shows an example of the difference in performance between a single input single output (SISO) system using a 64DPSK and a MIMO system using a 64-PSK modulation scheme in DSTBCs.

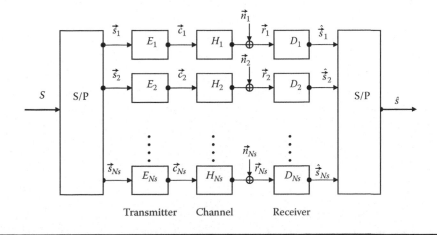

Figure 10.4 Representation of a MIMO–OFDM system. (From Cuvillier Verlag Göttingen, Göttingen, Germany, http://www.cuvillier.de/flycms/en/html/1/-/Home.html?SID=ekrseVJbeed5. With permission.)

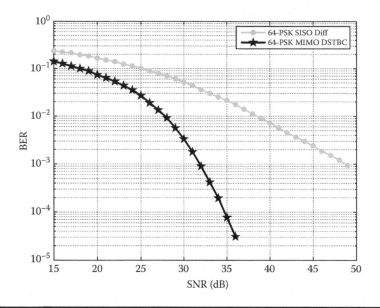

Figure 10.5 **Comparison of differential 64-PSK modulation in a SISO case and in a 2 × 3 MIMO case in an uncorrelated Rayleigh fading channel. (From Cuvillier Verlag Göttingen, Göttingen, Germany, http://www.cuvillier.de/flycms/en/html/1/-/Home.html?SID=ekrseVJbeed5. With permission.)**

10.3 Differential Space Time Block Codes

As it was already mentioned, the spatial diversity given by MIMO systems is very promising and successful. This is the reason why so many different MIMO techniques have been proposed. In this chapter an improvement for one of these techniques will be discussed.

When we consider coherent transmission systems, the radio channel *impulse response* or *transfer function* must be estimated precisely. In MIMO systems, the radio channel must be estimated between all transmit and all receive antennas. In this case, the test signal overhead and the computation complexity are increased. This is the reason why differential modulation schemes are especially interesting for MIMO systems, because with these systems the radio channel does not need to be estimated. The test signal overhead can be avoided and the computation complexity is significatively reduced.

DSTBCs are a very interesting kind of differential MIMO systems. The performance of these systems is quite good and the complexity is not too high. For this reason, these systems are discussed in this chapter. To get high channel capacity, higher-level modulation schemes are needed. Reason why a new high level modulation technique, based in the use of new amplitude and phase shift keying (APSK) modulation schemes, will be introduced in this chapter.

10.3.1 Origin and Evolution of DSTBC Schemes

The origin of DSTBCs is in the STBCs proposed in Refs. [10] and [11]. After that, *differential unitary space time modulations* were proposed [12,13]. In this case, the data bits to be transmitted are mapped in a unitary matrix that is differentially encoded with the previous transmitted matrix for obtaining the present matrix to transmit.

Later, some extensions that consider a simultaneous differential phase and amplitude modulation have been proposed [14–21]. Previous to the development of STBCs some differential amplitude and phase shift keying (DAPSK) schemes [22] have been proposed for the SISO system environment.

In this chapter, a simultaneous APSK, modulation technique is discussed for DSTBC schemes [23]. This differential encoding technique is combined with a new receiver structure.

10.3.2 Description of DSTBC Schemes

A classical DSTBC procedure is described here for a 64-PSK modulation scheme over a 2×1 multiple input single output (MISO) system as the one in Figure 10.6.

The first step in the encoding procedure is to construct the 2×2 block code (BC) described by the 2×2 information matrix S_k, which contains the two complex-valued modulation symbols $s1_k$ and $s2_k$

$$S_k = \begin{bmatrix} s1_k & s2_k \\ -s2_k^* & s1_k^* \end{bmatrix}. \tag{10.3}$$

By taking the amplitude of the 64-PSK modulation equal to $\sqrt{0.5}$, the S_k is a unitary matrix.

For the assumed 64-PSK modulation scheme, each matrix S_k transmits 12 bits by the two constellation points $s1_k$ and $s2_k$.

The differential modulation scheme is described by the 2×2 transmit matrix C_k, which is recursively calculated by the product of the information matrix S_k and the previous transmitted matrix (C_{k-1}). The identity matrix can be used as the first transmitted matrix (C_0). The differential modulation scheme can be described mathematically by the following matrix product

$$C_k = S_k \cdot C_{k-1} = \begin{bmatrix} c1_k & c2_k \\ -c2_k^* & c1_k^* \end{bmatrix}. \tag{10.4}$$

These complex-valued symbols described in matrix C_k will be transmitted via two adjacent antennas at the time slot k (ts_k). Time slot k is composed of two time slots ts_{k1} and ts_{k2} as shown in Figure 10.7. This is the way in which the information is differentially encoded and transmitted.

Concerning the radio channel influence, usually a Rayleigh fading model is used. The noise added in the receiver is considered to be additive white gaussian noise (AWGN). In this way, the received signals $r1_k$ and $r2_k$ in adjacent time samples of a single receive antenna are corresponding to the two adjacent time slots ts_{k1} and ts_{k2}. The received signals can be written in matrix notation as

$$\begin{bmatrix} r1_k & -r2_k^* \\ r2_k & r1_k^* \end{bmatrix} = \begin{bmatrix} c1_k & c2_k \\ -c2_k^* & c1_k^* \end{bmatrix} \cdot \begin{bmatrix} h1_k & -h2_k^* \\ h2_k & h1_k^* \end{bmatrix} + \begin{bmatrix} n1_k & -n2_k^* \\ n2_k & n1_k^* \end{bmatrix} \tag{10.5}$$

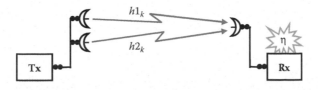

Figure 10.6 General scheme of the considered MISO system. (From Cuvillier Verlag Göttingen, Göttingen, Germany, http://www.cuvillier.de/flycms/en/html/1/-/Home.html?SID= ekrseVJbeed5. With permission.)

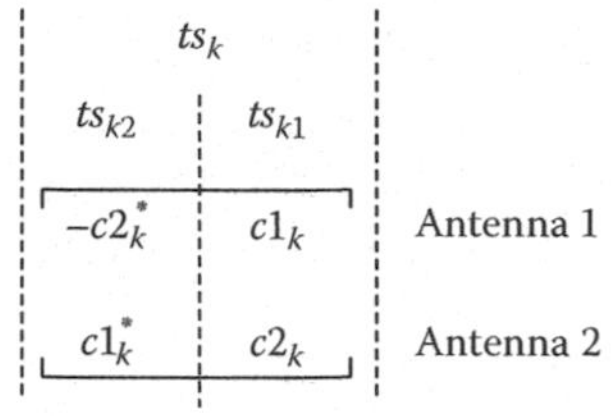

Figure 10.7 C_k **matrix transmitted at time slot k. (From Cuvillier Verlag Göttingen, Göttingen, Germany, http://www.cuvillier.de/flycms/en/html/1/-/Home.html?SID=ekrseVJbeed5. With permission.)**

and in a summarized form as

$$R_k = C_k \cdot H_k + N_k \tag{10.6}$$

Concerning the classical receiver structure, the received signal matrix R_k is conventionally processed by a matrix multiplication

$$D_k = R_k \cdot R_{k-1}^{\mathrm{H}}, \tag{10.7}$$

where R_{k-1}^{H} is the Hermitian matrix of R_{k-1} (the transpose conjugate of R_{k-1}).

10.4 Differential Space Frequency Block Codes

In a multicarrier differential OFDM system, the differential encoding can be performed in the direction of time or in the direction of frequency. Differential block codes (BCs) encoded in the direction of frequency are referred as differential space frequency block codes (DSFBCs). As was discussed in the previous section, when the differential encoding is performed in the direction of the time, the encoding and decoding processes are performed as follows.

10.4.1 Differential Modulation in Time Direction for an OFDM System

Considering an OFDM system with a differential modulation in the direction of time, the transmitted value in time symbol k and subcarrier l will be $C_{k,l}$

$$C_{k,l} = S_{k,l} \cdot C_{k-1,l}, \tag{10.8}$$

where $S_{k,l}$ is the constellation point containing the user information after the modulation and $C_{k-1,l}$ is the previous transmitted value in time symbol $k-1$ and subcarrier l. $C_{k,l}$, $S_{k,l}$, and $C_{k-1,l}$ are complex values; the use of capital letters here denotes that the equations are written in the space of frequency.

At the receiver it will be received

$$R_{k,l} = C_{k,l} \cdot H_{k,l} + N_{k,l}. \tag{10.9}$$

The previous received value is

$$R_{k-1,l} = C_{k-1,l} \cdot II_{k-1,l} + N_{k-1,l}. \tag{10.10}$$

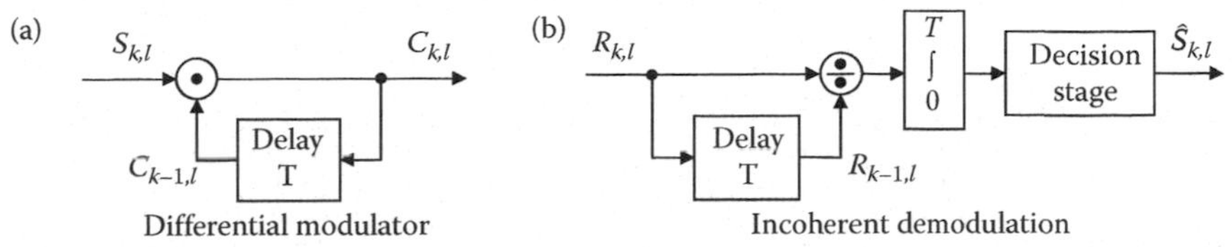

Figure 10.8 Block diagram of (a) differential modulation and (b) incoherent demodulation in time direction. (From Cuvillier Verlag Göttingen, Göttingen, Germany, http://www.cuvillier.de/flycms/en/html/1/-/Home.html?SID=ekrseVJbeed5. With permission.)

Assuming

$$H_{k,l} = H_{k-1,l} \tag{10.11}$$

by neglecting the noise, the following can be obtained:

$$D_{k,l} = \frac{R_{k,l}}{R_{k-1,l}} \simeq \frac{C_{k,l}}{C_{k-1,l}} = S_{k,l}. \tag{10.12}$$

Observe that Equation 10.12 is also valid when $H_{k,l} \simeq H_{k-1,l}$, which is valid unless the channel is a very time-variant channel.

From Equation 10.12, it is clear that with this differential procedure, no channel estimation is needed to recover the transmitted information.

Figure 10.8 shows a block diagram of the differential modulator and incoherent demodulator described previously.

When the differential encoding process is performed in the direction of frequency, the encoding and decoding processes are performed as follows.

10.4.2 Differential Modulation in Frequency Direction for an OFDM System

For an OFDM system with a differential modulation in the direction of frequency, the transmitted value in time symbol k and subcarrier l will be $C_{k,l}$

$$C_{k,l} = S_{k,l} \cdot C_{k,l-1}, \tag{10.13}$$

where $C_{k,l-1}$ is the transmitted value in subcarrier $l - 1$ (for the same time symbol k).

The received signal in subcarrier l will be

$$R_{k,l} = C_{k,l} \cdot H_{k,l} + N_{k,l}. \tag{10.14}$$

At the same time the received signal in subcarrier $l - 1$ will be

$$R_{k,l-1} = C_{k,l-1} \cdot H_{k,l-1} + N_{k,l-1}. \tag{10.15}$$

Assuming

$$H_{k,l} = H_{k,l-1} \tag{10.16}$$

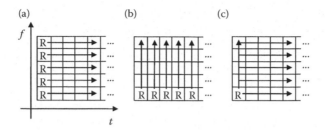

Figure 10.9 Alternatives for the use of reference values in differential modulations for OFDM systems. (From Cuvillier Verlag Göttingen, Göttingen, Germany, http://www.cuvillier.de/flycms/en/html/1/-/Home.html?SID=ekrseVJbeed5. With permission.)

by neglecting the noise, we obtain

$$D_{k,l} = \frac{R_{k,l}}{R_{k,l-1}} \simeq \frac{C_{k,l}}{C_{k,l-1}} = S_{k,l}. \tag{10.17}$$

Observe that Equation 10.17 is also valid when $H_{k,l} \simeq H_{k,l-1}$, which is valid unless the channel was a very frequency-selective channel.

Equation 10.17 shows that it is also possible to transmit information with a differential modulation in the direction of frequency.

Figure 10.9 shows different alternatives to select the reference values (first transmitted values) in differential modulations. In (a), the modulation is carried out in the direction of time, starting from one reference value (R) for each subcarrier. In (b), it is carried out in the direction of frequency, starting from one reference value for each time slot. For avoiding the performance degradation due to the transmission of so many reference values, strategy (c) can be used where only one reference value is transmitted.

10.5 Differential Space Time Frequency Block Codes

DSTBC schemes are considered because of their simplicity and good performance. But in order to exploit space diversity in a differential modulation scheme, there are other alternatives, for example, differential space time frequency block codes (DSFBCs or DSTFBCs).

In DSFBCs, instead of using two time slots to transmit the two rows of the C_k matrix, two neighbor subcarriers to transmit these two rows are used. In terms of an OFDM system, it means that two subcarriers of the same OFDM symbol are used instead of the same subcarrier in two consecutive OFDM symbols.

One obvious advantage of these systems is to reduce the transmission delay. DSFBCs present also advantages for channels where the time variability is high, because one C_k matrix is fully transmitted in one time slot (by using different subcarriers), which means that the condition $H_k = H_{k-1}$ can be fulfilled just by having an unchanged channel in two consecutive time slots (considering a 2×2 C_k) instead of in four as for DSTBCs.

DSFBCs have disadvantages in channels with high frequency selectivity, because to have a good performance in this case it is necessary that the channel remains unchanged in two consecutive subcarriers (for a 2×2 C_k). This means that a similar restriction than the one in the direction of time, which has been mentioned previously, appears now in the direction of frequency. Usually, one alternative to fulfill this requirement in the direction of frequency is to consider a high number of subcarriers, which results in a narrower band for each subcarrier and then less variation in the channel coefficients for neighbor subcarriers. But finally it could result in an excessive high number of subcarriers, if the channel is quite frequency selective.

The problems of this technique in frequency-selective channels are still increased when four transmit antennas are used, where the channel must remain unchanged along eight consecutive subcarriers. Some simulation results performed in Ref. [24] show that the restriction of constant channel coefficients over an orthogonal design (block code) is much more critical for SFBCs than for STBCs, usually needing a high number of subcarriers for the first one in order to cope with the channel frequency selectivity. While this restriction is usually easily fulfilled by STBCs.

As was discussed previously, DSTBCs have problems in channels with fast variations in time, while DSFBCs have problems in channels with high frequency selectivity. This is due to the restriction of constant channel coefficients over two successive transmit matrices (C_{k-1} and C_k). For providing more flexibility in order to handle with this restriction, the DSTFBCs were proposed.

In the case of DSTFBCs the elements of the C_k matrix can be distributed in different ways considering several subcarriers and time slots for each antenna. This option is particularly attractive when more than two transmit antennas are used. Considering the particularities of the channel (time variations and frequency selectivity) can be decided to use more time slots and less subcarriers to map the C_k elements in a mainly frequency-selective channel or vice versa. In this way, the degradation of the transmission can be prevented.

One advantage of DSTFBCs is to diminish the transmission delay compared with DSTBCs (advantage shared with DSFBCs). Provided that STBCs have a good performance at normal high speed for terrestrial vehicles (even using a reasonable large number of subcarriers (up to 2048) [24]), the improvement in transmission delay would be probably the most useful advantage of DSTFBCs with respect to DSTBCs. DSTFBCs being a little bit more complex than DSTBCs and needing some assumptions over the channel behavior in order to optimize the distribution of the C_k matrix's elements in time slots and subcarriers, the second ones are still a very attractive alternative for future wireless systems.

10.6 A New Class of DSTBCs

This new class of DSTBCs allows the use of modulations whose constellation points have different amplitudes. Using modulation schemes that include variations in amplitude (not only in phase) allows one to obtain an improved BER. The proposed DSTBC structure is quite similar to the one described previously but with the following differences:

■ Given that the modulations to be used have constellation points with different amplitudes, the S_k matrices will not be unitary matrices. Instead of it, the following normalized matrix U is a unitary matrix in this case

$$U = \frac{1}{\sqrt{|s1_k|^2 + |s2_k|^2}} \cdot S_k,$$

that is, $U^{\mathrm{H}} = U^{-1}$, where U^{H} is the Hermitian matrix of U. This observation is important because DSTBC theory is generally described for unitary matrices but in this case they are not strictly unitary.

■ There will be a power control mechanism (PCM) in order to control the power of the transmitted symbols. In previous propositions, this PCM is usually implicit in the definition of the modulation scheme. Using an explicit PCM allows the use of a greater and more varied range of modulation schemes.

■ The new decoding procedure is mathematically described by the inverse matrix R_{k-1}^{-1} as

$$
\begin{aligned}
D_k = R_k \cdot R_{k-1}^{-1} &\simeq C_k \cdot H_k \cdot (C_{k-1} \cdot H_{k-1})^{-1} \\
&= C_k \cdot H_k \cdot H_{k-1}^{-1} \cdot C_{k-1}^{-1}.
\end{aligned} \tag{10.18}
$$

Here the decoding procedure is different compared to Equation 10.7 because in this case it is necessary to decode amplitude and phase, not only phase as for 64-PSK. The noise influence is neglected in the received matrices R_k and R_{k-1}^{-1}. If furthermore the two adjacent radio channel matrices do not vary in this short time interval, which means

$$
H_k = H_{k-1},
$$

and if Equation 10.4 is applied, the following relation is valid:

$$
D_k \simeq C_k \cdot C_{k-1}^{-1} = S_k, \tag{10.19}
$$

where S_k is the information matrix containing the transmitted symbols from the constellation diagram. If the added noise was small, the demodulation matrix D_k is strongly related to the information matrix S_k transmitted in the 2×1 MISO system.

10.6.1 New Modulation Schemes for DSTBCs

Here several amplitude and phase modulation schemes are presented and compared with a 64-PSK modulation (fixed amplitude modulation) used in DSTBCs.

64-PSK modulation used in DSTBCs is considered here as the reference modulation scheme, because this modulation does not have amplitude variations and the objective is to show the improvement in performance obtained by using this new technique which makes it feasible to use amplitude and phase modulations in DSTBCs with quite freedom.

The problem when using amplitude variable modulations is that the transmitted power can be increased or decreased up to ∞ or 0, respectively. For avoiding such a situation, the amplitude of the constellation points in the 64-PSK modulation is normalized to $a1 = \sqrt{0.5}$.

10.6.1.1 "4A16PSK"

The first kind of 64-APSK modulation that we are going to consider is "4A16PSK." This is a modulation with 4 amplitude values and 16 phase values for each subconstellation. Each constellation has two subconstellations with one amplitude value in common ($a1 = \sqrt{0.5}$); seven different amplitudes for the whole constellation (see Figure 10.10).

In the "4A16PSK" modulation scheme, the set of possible amplitudes A is determined by a parameter a as

$$
A \in \{(\sqrt{0.5}) \cdot [1/a^3, 1/a^2, 1/a, 1, a, a^2, a^3]\}
$$

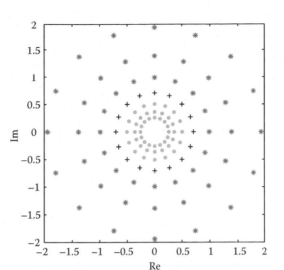

Figure 10.10 "4A16PSK" modulation scheme used in DSTBCs. (From Cuvillier Verlag Göttingen, Göttingen, Germany, http://www.cuvillier.de/flycms/en/html/1/-/Home.html?SID= ekrseVJbeed5. With permission.)

(coding $[01, 11, 10, 00, 01, 11, 10]$, respectively) and the phases are 16 equal-spaced phase states starting at $0°$ (see Figure 10.10). These phases map 4 bits onto one modulation symbol in a Gray coding way. "4A16PSK" has two subconstellations: the small one composed of those constellation points with amplitude smaller or equal to $a1 = \sqrt{0.5}$ ("·" and "+" in Figure 10.10) and the big one whose constellation points have bigger or equal to $a1 = \sqrt{0.5}$ amplitudes ("+" and "∗" in the same figure). The responsible unit for using the small or big subconstellation to modulate the data bits is the transmitter, depending on the *spectral norm* of the previous C_k matrix (C_{k-1}) (definition in Section 10.6.2.1); if it is smaller than a predefined value *maxl* the big subconstellation will be used, else the small one will be used. In this way, the control of transmitted power is performed. This mechanism to control the power (described in Section 10.6.2.1) will be referred as power control mechanism 1 (PCM1).

Due to the fact that all the constellation points of "4A16PSK" are different between them (no matter if they are in the small or big subconstellation), no extra information is needed in the receiver to demodulate these constellation points.

The value of a that produces acceptable transmitted powers and optimizes the performance of the modulation was estimated by simulations. This value is $a = 1.4$ and it can be seen in Figure 10.11 that the performance is approximately 5.5 dB better at a BER $= 10^{-2}$ than for the corresponding classical PSK modulation (64-PSK, BE 6 bit/s/Hz).

All the APSK modulation schemes presented here were tested under the PCM named PCM1, described later in this section. This PCM demonstrated to be completely robust and quite effective for all the analyzed modulation schemes.

10.6.1.2 "2A32PSK"

An other 64-APSK tested modulation was "2A32PSK," it consists of three circles as shown in Figure 10.12. Here, as in the previous case, the small constellation is composed by the points "·"

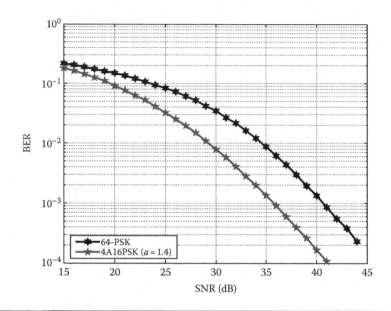

Figure 10.11 Comparison of "4A16PSK PCM1" and 64-PSK for DSTBCs in an uncorrelated Rayleigh fading channel. (From Cuvillier Verlag Göttingen, Göttingen, Germany, http://www.cuvillier.de/flycms/en/html/1/-/Home.html?SID=ekrseVJbeed5. With permission.)

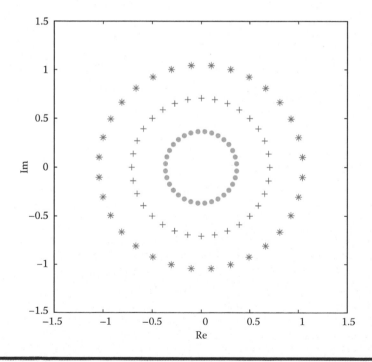

Figure 10.12 "2A32PSK" modulation applied to DSTBC scheme. (From Cuvillier Verlag Göttingen, Göttingen, Germany, http://www.cuvillier.de/flycms/en/html/1/-/Home.html?SID=ekrseVJbeed5. With permission.)

and "+" and the big one by the points "+" and "*" (see Figure 10.12). For this modulation, as in the previous cases, the central circle has a radius $a1 = \sqrt{0.5}$. An a parameter that specifies the difference between the circles radii was optimized. The optimum value obtained for the parameter a was $a = 0.34$.

As can be seen in Figure 10.16, "2A32PSK" is approximately 4.2 dB better at a BER $= 10^{-2}$ than 64-PSK when they are used in DSTBCs. This result is reasonable considering that when 32-PSK is used instead of 64-PSK, the BER is improved to 6 dB and the BE is decreased in 1 bit/s/Hz. In this case, the BE is maintained by adding a second circle (for each subconstellation) with other 32-PSK set of constellation points. By this, the 6 dB of improvement are reduced to 4.2 dB.

10.6.1.3 "APSK1"

This modulation scheme is a variation of "4A16PSK" where an additive rule is used instead of an exponential rule for the radii variations of the circles in the constellation diagram.

For "APSK1" modulation scheme, the set of possible amplitudes A is determined by the parameter a as

$$A \in \{(\sqrt{0.5}) \cdot [1 - 3a, 1 - 2a, 1 - a, 1, 1 + a, 1 + 2a, 1 + 3a]\}$$

(coding $[01, 11, 10, 00, 01, 11, 10]$, respectively) and the phases are 16 equally spaced, starting at 0° (see Figure 10.13).

The parameter a was optimized in order to obtain the best possible performance for this modulation scheme and the optimum value for a was $a = 0.25$. As can be seen in Figure 10.16, the performance of this modulation scheme is not as good as the one corresponding to "4A16PSK."

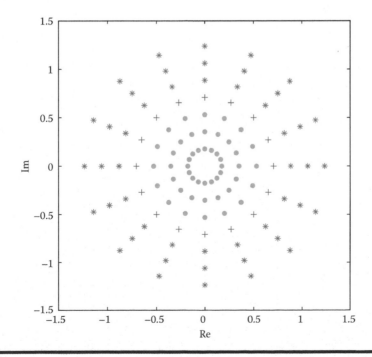

Figure 10.13 "APSK1" modulation scheme used in DSTBCs. (From Cuvillier Verlag Göttingen, Göttingen, Germany, http://www.cuvillier.de/flycms/en/html/1/-/Home.html?SID= ekrseVJbeed5. With permission.)

10.6.1.4 "APSK2"

"APSK2" tests the performance of a modulation scheme with equal-spaced constellation points. Here the parameter that determine the distance between points is a. In this case, a was taken as $a = \sqrt{1/41}$; for this value, the points in the common ring has an amplitude of $\sqrt{0.5}$. This value ensures that even without considering the sequence of bits in the input, the modulation scheme will be always (for any sequence of bits in the input) able to control the transmitted power. "APSK2" is shown in Figure 10.14.

If for example, $a > \sqrt{1/41}$ is selected, it could happen that while the transmitter is trying to reduce the transmit power by using the small subconstellation, it continues to grow. This situation would happen for example, for a given sequence of bits, whose modulation always repeats one constellation point over the common ring. In this case being the amplitude of this point greater than $\sqrt{0.5}$, the transmit power would continue growing. For this reason, for ensuring the robustness of the system, a was selected as $a = \sqrt{1/41}$.

It could be considered that such a situation has very low probability to appear, specially due to scrambling process (pseudo randomizing) in the transmitter and that is true. But in any case the proposed value for a makes the proposed modulation scheme absolutely robust.

The performance of this modulation scheme for the optimum value of a (without considering the robustness of the system) was also evaluated. The optimum value obtained was $a = 0.20$. The difference in performance for $a = \sqrt{1/41}$ and $a = 0.20$ is meaningless; hence in Figure 10.16, only the result for $a = \sqrt{1/41}$ is shown.

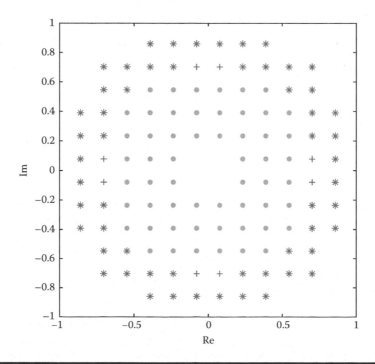

Figure 10.14 "APSK2" modulation scheme used in DSTBCs. (From Cuvillier Verlag Göttingen, Göttingen, Germany, http://www.cuvillier.de/flycms/en/html/1/-/Home.html?SID= ekrseVJbeed5. With permission.)

10.6.1.5 "APSK3"

Later more flexible modulation schemes were studied; the number of circles, their radii, the number of constellation points in each circle, and also one phase shift for each circle were considered as variables to optimize. One of these modulation schemes is presented in Figure 10.15 under the name of "APSK3."

For this specific modulation scheme, the set of radii was $(0.34, 0.52, \sqrt{0.5}, 0.88, 1.07)$, the number of points in each circle was respectively $(16, 23, 25, 16, 23)$ and the phase shift for each circle was respectively $(0, 0.15, 0.07, 0.2, 0)$. Although the flexibility for placing the constellation points was quite high for this kind of modulation schemes, a better result than for "4A16PSK" was not obtained (see Figure 10.16).

Several different modulation schemes were tested and some of them were presented in this section. By observing the different modulation schemes presented here and the results for them, it seems clear that it is not a simple task to identify which are the factors that make a modulation scheme more effective in this particular case of DSTBCs. There are some well-known factors that have influence over the performance of a normal modulation scheme (e.g., distance between constellation points, power ratio between smaller and bigger constellation points in the scheme, etc.), but it is not a simple task to determine how all of them affect the performance of the system in our case. Then the simulation process is a very good alternative to look for more effective modulation schemes.

At this point it is clear that "4A16PSK" is the best of the modulation schemes discussed here. In future sections mainly this modulation scheme will be considered.

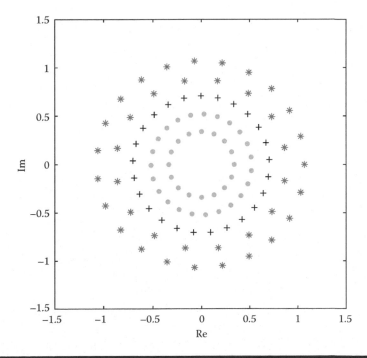

Figure 10.15 "APSK3" modulation scheme used in DSTBCs. (From Cuvillier Verlag Göttingen, Göttingen, Germany, http://www.cuvillier.de/flycms/en/html/1/-/Home.html?SID =ekrseVJbeed5. With permission.)

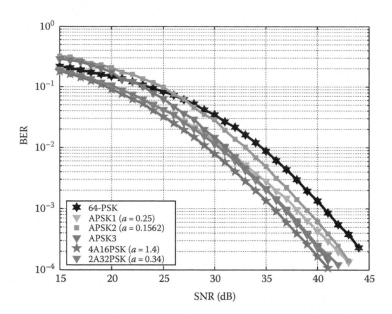

Figure 10.16 Results for all the tested APSK modulation schemes with PCM1 in an uncorrelated Rayleigh fading channel. (From Cuvillier Verlag Göttingen, Göttingen, Germany, http://www.cuvillier.de/flycms/en/html/1/-/Home.html?SID=ekrseVJbeed5. With permission.)

10.6.2 Power Control Mechanism in this New Class of DSTBCs

When an amplitude modulation is used in DSTBCs, a PCM must be used in order to control the transmitted power; in other case, the transmitted power can go to infinity or null depending on the data to transmit. Here the first PCM considered for this new class of DSTBCs is described. In Section 10.6.4, the influence of the PCM over the general performance of the system will be discussed.

10.6.2.1 PCM1

The first PCM considered to be used with this new class of DSTBCs was PCM1. In this case, the transmitter observes the *spectral norm* (*norm2*) of the previous transmitted block ($\left\| C_{k-1} \right\|_2$).

$$\left\| C_{k-1} \right\|_2 = (\text{maximum eigenvalue } (C_{k-1}^H \cdot C_{k-1}))^{1/2}.$$

Considering this value, the transmitter takes the decision of using the big or small subconstellation. It can be easily demonstrated that in our case

$$\left\| C_{k-1} \right\|_2 = (|c1_{k-1}|^2 + |c2_{k-1}|^2)^{1/2}.$$

If the *norm2* of the previous transmitted block is smaller than a certain limit (*maxl*), then the big subconstellation is used to modulate the bits, else the small one is used.

No redundancy is transmitted; the receiver considers both subconstellations in the demodulation process and depending on the received values, elements of D_k in Equation 10.19, they will be recognized as points of the small or big subconstellation and demodulated.

The results obtained for the modulation schemes discussed in this section (where PCM1 was used) are summarized in Figure 10.16.

For the understanding of these results, it is necessary to have in mind that they correspond to a system without any channel coding or any other data treatment in order to improve the BER. This is the reason why these results need a high SNR to obtain a good BER value. This model is very effective to compare directly (without factors to cover the differences) the performance of different modulation schemes.

10.6.3 System Performance in AWGN Channels

Initially the technique was evaluated in uncorrelated Rayleigh fading channels, because this is the channel model usually used for comparing different techniques in DSTBC research. In Figure 10.17, the results for "4A16PSK" and "2A32PSK" with PCM1 in AWGN channels are shown.

The discussed new class of DSTBCs is very flexible; several APSK modulation schemes could be used with PCM1 and also with different PCMs. In the next section, the influence of the PCMs over the general performance of the system is studied.

10.6.4 Influence of the PCM in New Class of DSTBCs

Initially in Section 10.6.2.1 PCM1 was described, the first PCM used with this technique. This PCM is based on observing $\|C_{k-1}\|_2$ and then taking the decision on which subconstellation is to be used, the small (in order to decrease or maintain the transmit power) or the big one (for increasing or maintaining the transmit power). In this section, new PCM strategies are considered.

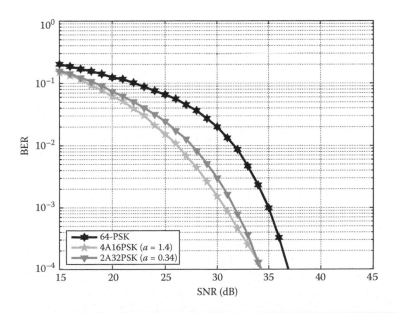

Figure 10.17 Results for 64-PSK, "4A16PSK," and "2A32PSK" with PCM1 in an AWGN channel. (From Cuvillier Verlag Göttingen, Göttingen, Germany, http://www.cuvillier.de/flycms/en/html/1/-/Home.html?SID=ekrseVJbeed5. With permission.)

Probably the most important difference between the PCMs considered in this section and PCM1 is that now, the possible C_k that can be obtained by using the small and the big subconstellations are going to be calculated and then, the decision of transmitting the C_k matrix that better complies the defined power control strategy will be taken.

For studying the influence of the PCM in the performance of this new class of DSTBCs, two *criteria* were defined. Criterion 1 (C1) is a PCM based on controlling $\|C_k\|_2$. In this case, the C_k matrix which has the smallest value of $d = (\|C_k\|_2 - maxl)$ will be used, where $maxl$ is a predefined value that does not play any perceptible role in the performance of the technique, at least while it varies in the range $0.4 \leq maxl \leq 1$. Criterion 2 (C2) is based on controlling $|c1_k|^2$ and $|c2_k|^2$. For C2, the C_k matrix that has the smallest value of $d = (\|c1_k\|^2 - maxl1| + \|c2_k\|^2 - maxl1|)$ will be used. $maxl1$ was taken as $maxl1 = maxl2/2$ in order to work with similar levels of transmit power. In both cases, the idea is to maintain the observed variables as close to a given value as possible.

Also, two *selection procedures* were defined: selection procedure 1 (SP1) uses for $s1_k$ and $s2_k$ the same subconstellation (the small or the big one); while selection procedure 2 (SP2) can use different subconstellations for $s1_k$ and $s2_k$.

This gives place to four new PCMs; considering PCM1 as the first used PCM, we could define PCM2 as the one obtained by the combination of C1 and SP1, PCM3 by using C1 and SP2, PCM4 by using C2 and SP1, and PCM5 by using C2 and SP2 (summarized in Table 10.2). C1 demonstrated to be more effective than C2. SP1 demonstrated to be more effective than SP2, which can sound curious at the beginning because SP2 is more flexible than SP1. SP2 achieves a narrower fluctuation of the transmitted power than SP1, but it was observed that this does not result in a better performance from a BER point of view. Considering that C1 and SP1 (PCM2) was the most successful combination, only this case will be compared with PCM1 in the following subsections.

10.6.4.1 Analysis of a New PCM Scheme (PCM2)

PCM1 makes a decision about the two possible modulation symbol constellation diagrams based on the matrix norm $\|C_{k-1}\|_2^2$. In this subsection, it is shown that the performance can be slightly improved by directly controlling the complex-valued symbols in the current and instantaneous transmit matrix C_k. The general objective is to keep the transmit power of C_k as close as possible to the average transmit power value. Large fluctuations of the transmit power are avoided in this case.

For PCM2, it is also decided in the modulator from which subconstellation the two modulation symbols of the information matrix S_k should be taken. In this case, there are two different possibilities to map the user bits to modulation symbols (by using the small subconstellation or the big one). This results in two different transmit matrices C_k. The final decision, for PCM2, will be taken by choosing the C_k matrix which has the smallest value of $d = (\|C_k\|_2 - maxl)$.

Table 10.2 New PCMs Evaluated

C \ SP	SP1	SP2
C1	PCM2	PCM3
C2	PCM4	PCM5

Source: Cuvillier Verlag Göttingen, Göttingen, Germany, http://www.cuvillier.de/flycms/en/html/1/-/Home.html?SID=ekrseVJbeed5. With permission.

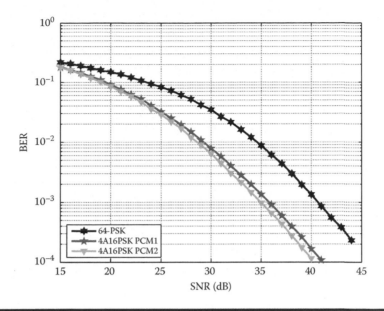

Figure 10.18 Comparison of "64-PSK" and "4A16PSK" used with different PCM techniques. (From Cuvillier Verlag Göttingen, Göttingen, Germany, http://www.cuvillier.de/flycms/en/html/1/-/Home.html?SID=ekrseVJbeed5. With permission.)

This new PCM scheme (PCM2) has been applied to the two APSK modulation schemes "4A16PSK" and "2A32PSK" in a DSTBC system.

10.6.4.1.1 Results of PCM2 Applied to "4A16PSK"

Figure 10.18 shows the system performance of a pure PSK-based modulation scheme without any amplitude variations ("64-PSK") as a reference. As an example, all the modulation schemes used in this work have a BE of 6 bit/s/Hz. It is well known that in the comparison between PSK and APSK modulation techniques, the difference in performance is increased for higher BE while it is decreased for lower BE.

Other BER performance curve that can be appreciated in Figure 10.18 is the one that corresponds to the "4A16PSK" modulation scheme using PCM2. This BER curve shows an advantage of 6.1 dB at BER $= 10^{-2}$ compared to the corresponding "64-PSK" modulation scheme used in DSTBC systems. It was improved by 0.6 dB at BER $= 10^{-2}$ compared to the previous PCM (PCM1).

10.6.4.1.2 Results of PCM2 Applied to "2A32PSK"

Figure 10.19 shows the BER performance for the "2A32PSK" modulation scheme, under two different PCM techniques. For this particular modulation scheme, the difference in performance for PCM1 and PCM2 is imperceptible.

10.6.4.2 Conclusion about the Influence of Different PCMs

The different PCMs influence BER performance. Better BER performance can be achieved in this new class of DSTBC systems by selecting an adequate PCM. It has been shown that an improvement

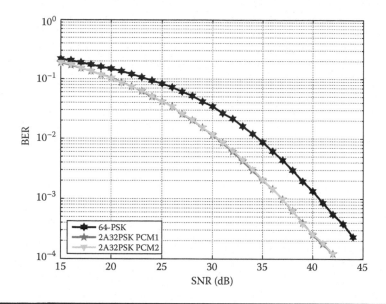

Figure 10.19 Comparison of "64-PSK" and "2A32PSK" used with different PCM techniques. (From Cuvillier Verlag Göttingen, Göttingen, Germany, http://www.cuvillier.de/flycms/en/html/1/-/Home.html?SID=ekrseVJbeed5. With permission.)

in BER performance of 6.1 dB at BER $= 10^{-2}$ can be gained by using an APSK modulation scheme instead of a pure PSK one in DSTBC systems.

It was also shown that it is better to control the transmit power based on the instantaneous matrix C_k instead of C_{k-1}. In this case, an improvement in the BER performance of 0.6 dB at BER $= 10^{-2}$ is feasible (difference between PCM1 and PCM2 for "4A16PSK").

10.6.5 System Performance in WSSUS Channels

The evaluation of systems in WSSUS channels is quite appreciated because it is a more realistic model than a pure AWGN model or a Rayleigh fading model. For this reason, in this section, the performance of the introduced technique under wide-sense stationary with uncorrelated scattering (WSSUS) channel condition is evaluated.

Considering that the simulations are being performed for a single subcarrier (f_0), the time variance is described as

$$h(\tau, t) = \frac{1}{\sqrt{P}} \cdot \sum_{p=1}^{P} \delta(\tau - \tau_p) \cdot e^{j(2\pi f_{D,p} t + \theta_p)} \tag{10.20}$$

by making a Fourier transform of Equation 10.20 in the direction of τ, and using the time as discrete ($t = nT$), we obtain

$$H(f, nT) = \frac{1}{\sqrt{P}} \cdot \sum_{p=1}^{P} e^{j2\pi f_{D,p} nT} \cdot e^{j\theta_p} \cdot e^{-j2\pi f \tau_p}, \tag{10.21}$$

then by evaluating Equation 10.21 in $f = f_0$, the used equation is obtained:

$$H(f_0, nT) = \frac{1}{\sqrt{P}} \cdot \sum_{p=1}^{P} e^{j2\pi f_{D,p} nT} \cdot e^{j\theta_p} \cdot e^{-j2\pi f_0 \tau_p}, \tag{10.22}$$

where θ_p, τ_p, and $f_{D,p}$ are obtained by using their respective *probability density functions*.

The set of parameters used in order to perform the simulations was inspired in WiMAX standard and are contained in Table 10.3.

In the first place, it was decided to maintain the assumption $H_k = H_{k-1}$ and evaluate the performance of the system for a mobile velocity of 60 km/h. In this case, a result very similar to the one obtained for Rayleigh fading channels was obtained.

In Figure 10.20, it can be observed that the improvement by using "4A16PSK PCM2" instead of 64-PSK in DSTBC is approximately 6.1 dB at a BER = 10^{-2}; approximately the same value obtained in Section 10.6.4 for a Rayleigh fading channel.

Then it was decided to work with a more realistic assumption, now for successive transmitted matrices (C_k and C_{k-1}), successive samples of the WSSUS channel (H_k and H_{k-1}) will be used. Then ($H_k \approx H_{k-1}$) will be valid instead of ($H_k = H_{k-1}$), which is a much more realistic assumption.

Table 10.3 Simulation Parameters

Parameter	Value
Carrier frequency	$f_c = 5\,\text{GHz}$
Bandwidth	$B = 10\,\text{MHz}$
Number of subcarriers	$N_{FFT} = 128$
Subcarrier spacing	$\Delta f = \frac{B}{N_{FFT}} = 78,125\,\text{Hz}$
Symbol duration	$T_s = 12.8\,\mu\text{s}$
Guard interval	$T_G = \frac{T_s}{8} = 1.6\,\mu\text{s}$
Symbol interval	$T_{S+G} = T_s + T_G = 14.4\,\mu\text{s}$
Number of paths	$P = 30$
Number of clusters (groups of paths)	$N_c = 1$
Maximum time delay	$\tau_{max} = 1\,\mu\text{s}$
Mobile velocity	$v = 3, 60, 120\,\text{km/h}$
Maximum Doppler shift	$f_{D\,max} = f_0 \cdot \frac{v}{c} \approx 14, 278, 556\,\text{Hz}$ using $f_0 = f_c$
Time delay distribution	$b = \frac{\tau_{max}}{\ln(1000)} = 0.1448\,\mu\text{s}$

Source: Cuvillier Verlag Göttingen, Göttingen, Germany, http://www.cuvillier.de/flycms/en/html/1/-/Home.html?SID=ekrseVJbeed5. With permission.

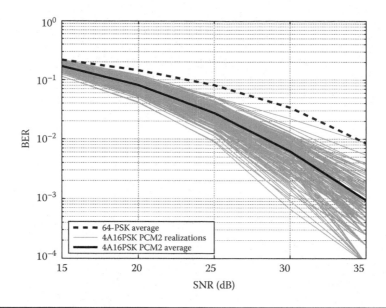

Figure 10.20 Performance of "4A16PSK PCM2" used in DSTBCs under WSSUS channels ($H_k = H_{k-1}$, $v = 60$ km/h). (From Cuvillier Verlag Göttingen, Göttingen, Germany, http://www. cuvillier.de/flycms/en/html/1/-/Home.html?SID=ekrseVJbeed5. With permission.)

By using this approach ($H_k \approx H_{k-1}$), the expected result is obtained, the performances of both systems—the one based 64-PSK and the one based "4A16PSK" with PCM2 are degraded. It can be observed in Figure 10.21.

In Figure 10.21, the performance of the average of 64-PSK in DSTBC with the one corresponding to "4A16PSK PCM2" in DSTBCs is compared. There it can be observed that the degradation is not equal for both systems, being higher for the first one. This increases the improvement obtained by using the proposed technique instead of 64-PSK in DSTBCs; now it is approximately 6.8 dB at a BER $= 4 \times 10^{-2}$ (see Figure 10.21) instead of approximately 5.8 dB at a BER $= 4 \times 10^{-2}$ (see Figure 10.20). This is an interesting result for the new technique; by changing to a more realistic channel model, the original improvement calculated for the Rayleigh model (and still valid for WSSUS channels under the assumption of $H_k = H_{k-1}$) is increased.

Considering the question of how the performance of the proposed technique varies when the velocity of the mobile terminal is increased, the following simulations were performed. In Figure 10.22, the performance for 64-PSK in DSTBC and "4A16PSK PCM2" in DSTBC were evaluated for a mobile terminal velocity of 120 km/h. By comparing the average results for both techniques in Figure 10.22 it can be observed that an improvement of 9.3 dB at BER $= 4 \times 10^{-2}$ is obtained for the new technique. This means that the improvement is significantly increased with the increment of the mobile terminal velocity. To verify this result, new simulations at a mobile velocity of 3 km/h were performed.

In Figure 10.23, the results obtained for the reference system and for the new system, when the mobile terminal velocity is 3 km/h, are shown. There, it can be observed that the improvement for the proposed technique is reduced (approx. 6.0 dB at BER $= 4 \times 10^{-2}$). This confirms that the improvement produced for the proposed technique is higher when the velocity of the mobile terminal is higher.

In Table 10.4, in a summarized way, the variation of the improvement with the velocity of the mobile terminal is shown.

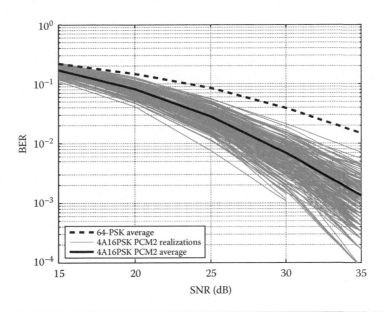

Figure 10.21 **Performance of "4A16PSK PCM2" used in DSTBCs under WSSUS channels** ($H_k \approx H_{k-1}$, $v = 60$ km/h). (From Cuvillier Verlag Göttingen, Göttingen, Germany, http://www. cuvillier.de/flycms/en/html/1/-/Home.html?SID=ekrseVJbeed5. With permission.)

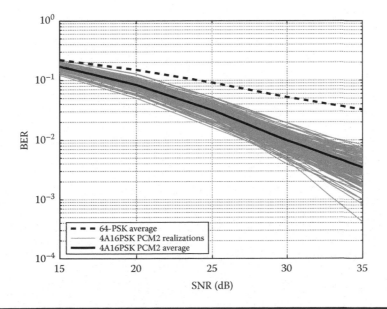

Figure 10.22 **Performance of "4A16PSK PCM2" used in DSTBCs under WSSUS channels** ($H_k \approx H_{k-1}$, $v = 120$ km/h). (From Cuvillier Verlag Göttingen, Göttingen, Germany, http://www. cuvillier.de/flycms/en/html/1/-/Home.html?SID=ekrseVJbeed5. With permission.)

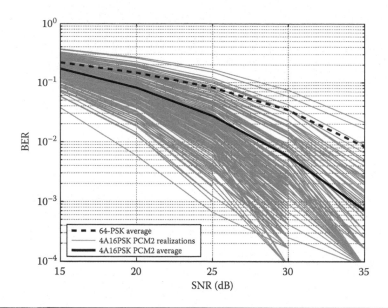

Figure 10.23 Performance of "4A16PSK PCM2" used in DSTBCs under WSSUS channels ($H_k \approx H_{k-1}$, $v = 3$ km/h). (From Cuvillier Verlag Göttingen, Göttingen, Germany, http://www. cuvillier.de/flycms/en/html/1/-/Home.html?SID=ekrseVJbeed5. With permission.)

Then, in this section, two very important results were obtained for the new technique. The first one is that the original improvement is increased when a more realistic channel model is used. The second one is that the improvement is increased when the velocity of the mobile terminal is increased, which makes the new technique particularly useful for mobile systems, specially in high mobility scenarios. Also, it can be said that the improvement is very good for low mobility scenarios, an improvement of approximately 6.4 dB can be obtained at BER $= 10^{-2}$ for a mobile velocity of only 3 km/h (pedestrian mobility), see Figure 10.23. Observe that the previous discussion is about the improvement in the performance of "4A16PSK PCM2" with respect to 64-PSK in DSTBCs, and not about the absolute performance of "4A16PSK PCM2." For sure it diminishes when the velocity of the mobile terminal is increased; but it diminishes less than for 64-PSK in DSTBCs.

To verify whether this behavior with respect to the velocity of the mobile terminal is a characteristic of the proposed technique shared with other APSK modulation schemes for DSTBCs,

Table 10.4   Improvement of the New Technique (with Respect to 64-PSK) for Different Mobile Terminal Velocities in WSSUS Channels with ($H_k \approx H_{k-1}$)

Velocity (km/h)	Improvement (dB at BER $= 4 \times 10^{-2}$)
3	≈ 6.0
60	≈ 6.8
120	≈ 9.3

Source: Cuvillier Verlag Göttingen, Göttingen, Germany, http://www.cuvillier.de/flycms/en/html/1/-/Home.html? SID=ekrseVJbeed5. With permission.

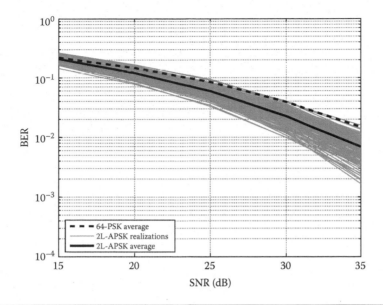

Figure 10.24 **Performance of "2L-APSK" used in DSTBCs under WSSUS channels** ($H_k \approx H_{k-1}, v = 60$ **km/h). (From Cuvillier Verlag Göttingen, Göttingen, Germany, http://www. cuvillier.de/flycms/en/html/1/-/Home.html?SID=ekrseVJbeed5. With permission.)**

or not, the study carried out for the new technique was also carried out for another APSK modulation technique for DSTBCs called "2L-APSK" or "$2^{2L} - $ APSK" reported in Refs. [25] and [26].

In Figure 10.24, the results for "2L-APSK" in a WSSUS channel as the one defined in Table 10.3, with a mobile terminal velocity of 60 km/h, are shown. It can be appreciated that there exists an improvement of approximately 2.8 dB at BER $= 4 \times 10^{-2}$ with respect to 64-PSK in DSTBCs.

In Figure 10.25, it is shown that the improvement for "2L-APSK" with respect to 64-PSK in DSTBCs is approximately 4.4 dB at BER $= 4 \times 10^{-2}$ when the velocity of the mobile terminal is 120 km/h.

Finally, in Figure 10.26, it can be observed that the improvement for "2L-APSK" (with respect to 64-PSK in DSTBC) is approximately 2.5 dB at BER $= 4 \times 10^{-2}$ when the velocity of the mobile terminal is 3 km/h.

The results in Figures 10.25 and 10.26 show that the increment of the relative improvement (relative to 64-PSK in DSTBCs) when the mobile terminal velocity is increased is also a characteristic shared for "2L-APSK" (not exclusive of "4A16PSK PCM2"). This means that the performance degradation that the techniques suffer when the mobile terminal velocity is increased is worse for 64-PSK in DSTBCs than for APSK techniques as "2L-APSK" and "4A16PSK PCM2." This is something important in favor of the use of APSK techniques. The relative improvements of "2L-APSK" and "4A16PSK PCM2" with respect to 64-PSK in DSTBCs as a function of the mobile terminal velocity are graphically summarized in Figure 10.27.

10.6.6 New Technique with Receive Diversity

The advantages of using *Spatial Diversity*, that is, MIMO systems was already discussed in Section 10.2.2. For the defined technique there is a very simple alternative to increase the *Spatial Diversity* order by using several antennas at the receiver side (*Receive Diversity*).

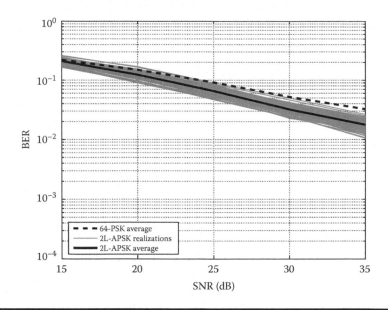

Figure 10.25 Performance of "2L-APSK" used in DSTBCs under WSSUS channels ($H_k \approx H_{k-1}$, $v = 120$ km/h). (From Cuvillier Verlag Göttingen, Göttingen, Germany, http://www.cuvillier.de/flycms/en/html/1/-/Home.html?SID=ekrseVJbeed5. With permission.)

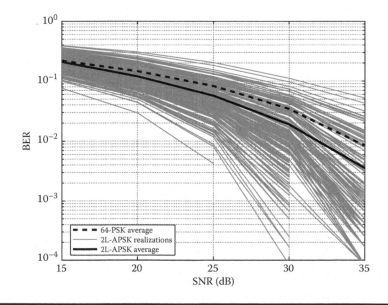

Figure 10.26 Performance of "2L-APSK" used in DSTBCs under WSSUS channels ($H_k \approx H_{k-1}$, $v = 3$ km/h). (From Cuvillier Verlag Göttingen, Göttingen, Germany, http://www.cuvillier.de/flycms/en/html/1/-/Home.html?SID=ekrseVJbeed5. With permission.)

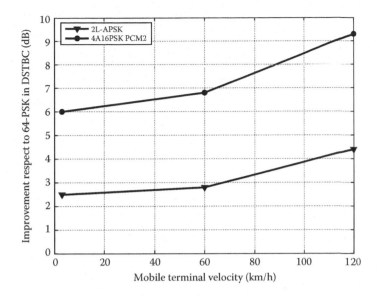

Figure 10.27 Relative improvements of "2L-APSK" and "4A16PSK PCM2" with respect to 64-PSK used in DSTBCs versus mobile terminal velocity, under WSSUS channels ($H_k \approx H_{k-1}$). (From Cuvillier Verlag Göttingen, Göttingen, Germany, http://www.cuvillier.de/flycms/en/html/1/-/Home.html?SID=ekrseVJbeed5. With permission.)

When several antennas are used in the receiver a combining technique has to be applied in order to combine the information received by each antenna. In this section, we explain how to apply the most well-known combining techniques for the new system and also show some results.

The discussion starts by analyzing two well known combination techniques, *Equal Gain Combining* and *Maximum Ratio Combining* (MRC).

10.6.6.1 Equal Gain Combining

This technique consists only of making an average of the D_k matrices (see Equations 10.18 and 10.19) obtained through each antenna, before making the demodulation. If we refer to the D_k matrix obtained through the receive antenna i $D_{k,i}$, the combination rule for *Equal Gain Combining* will be

$$D_k = \frac{1}{M} \cdot \sum_{i=1}^{M} D_{k,i}, \tag{10.23}$$

where M is the number of antennas used in the receiver. Figure 10.28 clarifies how the combination is made for the new technique. The "Combination Module" for *Equal Gain Combining* applies (Equation 10.23).

In Figure 10.29, the results of using two and three receive antennas are shown and are also compared with the reception with a single antenna. In all the three cases "4A16PSK" modulation scheme under PCM2 was used as defined in Section 10.6.4. There, it can be appreciated that an improvement of approximately 5 dB (at BER = 10^{-2}) can be obtained when three receive antennas are used instead of one.

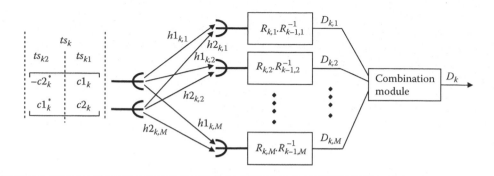

Figure 10.28 Combination system for the proposed technique with multiple receive antennas. (From Cuvillier Verlag Göttingen, Göttingen, Germany, http://www.cuvillier. de/flycms/en/html/1/-/Home.html?SID=ekrseVJbeed5. With permission.)

10.6.6.2 Maximum Ratio Combining

In MRC, the criterion to make the combination is to make a weighted average of the $D_{k,i}$ matrices obtained by the different receive antennas. It is clear how to apply MRC for a coherent system, where the channel gain coefficients can be estimated. For DSTBCs, the channel gain coefficients are not available, because it is a differential technique with incoherent demodulation. In this section, a way of applying a similar concept corresponding to MRC, for DSTBCs, without adding any redundancy is discussed [27].

In this case, the weighting factor ($Wf_{k,i}$) for each $D_{k,i}$ matrix will be related with the sum of the square absolute values of the corresponding channel gain coefficients (see Equation 10.30). For

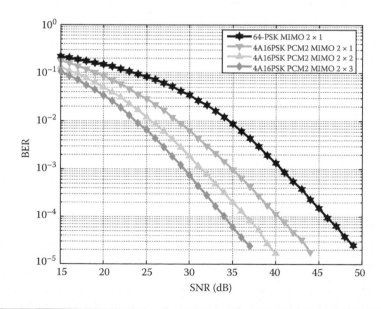

Figure 10.29 Results for the *Equal Gain Combining* technique in a multiple receive antenna system. (From Cuvillier Verlag Göttingen, Göttingen, Germany, http://www.cuvillier.de/ flycms/en/html/1/-/Home.html?SID=ekrseVJbeed5. With permission.)

each receive antenna i, the channel gain coefficients of the channels through which the information arrives to antenna i coming from the transmit antennas will be considered.

The weighting factor for antenna i ($Wf_{k,i}$) will be

$$Wf_{k,i} = \frac{\det(R_{k,i})}{\sum_{i=1}^{M} \det(R_{k,i})}, \tag{10.24}$$

where $\det(R_{k,i})$ is the *Determinant* of the matrix $R_{k,i}$ and M is the number of receive antennas, and then the combination rule will be

$$D_k = \frac{1}{\sum_{i=1}^{M} \det(R_{k,i})} \cdot \sum_{i=1}^{M} D_{k,i} \cdot \det(R_{k,i}). \tag{10.25}$$

The "Combination Module" in Figure 10.28 will apply (Equation 10.25).

For an arbitrary transmit matrix C_k, the receive matrix at antenna i is

$$R_{k,i} \simeq C_k \cdot H_{k,i}. \tag{10.26}$$

Using now that the *Determinant* of a product of matrices is the product of the *Determinants* of those matrices, we can write

$$\det(R_{k,i}) \simeq \det(C_k) \cdot \det(H_{k,i}). \tag{10.27}$$

By combining Equations 10.25 and 10.27 and eliminating the common factor $\det(C_k)$

$$D_k \simeq \frac{1}{\sum_{i=1}^{M} \det(H_{k,i})} \cdot \sum_{i=1}^{M} D_{k,i} \cdot \det(H_{k,i}). \tag{10.28}$$

By observing the matrix H in Equation 10.5, it is clear that

$$\det(H_{k,i}) = |h1_{k,i}|^2 + |h2_{k,i}|^2. \tag{10.29}$$

Then the weighting factors are

$$Wf_{k,i} \simeq \frac{|h1_{k,i}|^2 + |h2_{k,i}|^2}{\sum_{i=1}^{M}(|h1_{k,i}|^2 + |h2_{k,i}|^2)}. \tag{10.30}$$

Finally, it is clear that by using Equation 10.24 as weighting factors, the $D_{k,i}$ matrices are being weighted approximately by the channel gain coefficients of the channels through which the information arrives to antenna i coming from the transmit antennas (as shown in Equation 10.30).

In Figure 10.30, the results of using two and three receive antennas with this combining technique are shown and are also compared with the reception with a single antenna. For all the three cases the "4A16PSK" modulation scheme under PCM2 as defined in Section 10.6.4 was used. There, an improvement of approximately 7 dB (at BER = 10^{-2}) can be observed when three receive antennas are used instead of one.

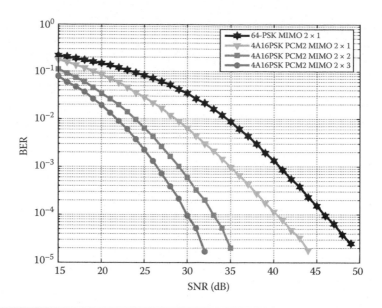

Figure 10.30 Results for the MRC technique in a multiple receive antenna system. (From Cuvillier Verlag Göttingen, Göttingen, Germany, http://www.cuvillier.de/flycms/en/html/1/-/ Home.html?SID=ekrseVJbeed5. With permission.)

Other different methods were considered by using different weighting factors in the "Combination Module," also making the combination after the demodulation. But they were not more successful than the previously explained technique.

10.6.7 Conclusion

In this chapter, a new alternative to use APSK modulation schemes in DSTBCs was discussed. By using APSK modulation schemes instead of pure PSK ones, the bit error rate is diminished for the same throughput. In order to have more freedom in the selection of APSK modulation schemes, than in previously proposed techniques, a PCM was introduced. This freedom makes it feasible to look for new highly efficient modulation schemes. Then several new APSK modulation schemes were considered and evaluated. In this way, "4A16PSK" was found, a new highly efficient modulation scheme for DSTBCs.

Later the optimization of the PCM was considered. The goal was to find a more efficient PCM (PCM2) for "4A16PSK." The combination "4A16PSK PCM2" was the most successful among the tested ones in Ref. [1].

Finally, "4A16PSK PCM2" was tested in different scenarios and with different number of receive antennas, showing in all these cases an excellent performance, among the best reported up to now, as far as the author knows. The relative improvement (with respect to the use of 64-PSK in DSTBCs) is in some cases twice as large as that in some previously proposed techniques.

10.6.8 Open Problems and Possible Extensions

One open problem in this area is to obtain an analytical expression to characterize the performance of a modulation scheme; up to now, in most of the cases it is done by using simulations.

In multiuser scenarios orthogonal frequency division multiple access (OFDMA) is a very promising alternative allowing an optimized use of broadband frequency-selective fading channels [28]. Also the DSTBC–OFDMA [29] combination is an attractive option for mobile communication systems. In general, this is a research area with high potential to improve the efficiency of future mobile communication technologies.

Considering Ref. [1], a motivation to look for more efficient modulation schemes or PCM, is the simplicity and good performance of the proposed technique.

One possible extension of this work is to verify the performance improvement (with respect to 64-PSK in DSTBCs) obtained for an adapted version of the introduced technique when more than two transmit antennas are used.

References

1. B. Rodríguez, Differential STBC for OFDM based wireless systems, PhD Thesis, Department of Telecommunications, Technische Universität Hamburg-Harburg (TUHH), Germany, December 2007.
2. G. J. Foschini and M. J. Gans, On limits of wireless communications in a fading environment when using multiple antennas, *Wirel. Pers. Commun.*, 6, 311–335, 1998.
3. G. J. Foschini, Layered space-time architecture for wireless communication in a fading environment when using multielement antennas, *Bell Labs Tech. J.*, 41–59, 1996, http://www.ece.mtu.edu/faculty/ztian/ee5535s08/4-1-BLAST.pdf.
4. E. Telatar, Capacity of multi-antenna gaussian channels, AT&T Bell Laboratories, Technical Memo, June 1995.
5. G. G. Raleigh and J. M. Cioffi, Spatio-temporal coding for wireless communication, *IEEE Trans. Commun.*, 46 (3), 357–366, March 1998.
6. D. Gesbert, M. Shafi, D. Shiu, P. J. Smith, and A. Naguib, From theory to practice: An overview of MIMO space-time coded wireless systems, *IEEE J. Sel. Areas Commun.*, 21 (3), 281–302, 2003.
7. L. Correia, *Mobile Broadband Multimedia Networks*, Elsevier Ltd., Oxford, UK, 2006.
8. D. Tse and P. Viswanath, *Fundamentals of Wireless Communication*, Cambridge University Press, New York, 2005.
9. A. Goldsmith, *Wireless Communications*, Cambridge University Press, New York, 2005.
10. S. M. Alamouti, A simple transmit diversity technique for wireless communication, *IEEE J. Sel. Areas Commun.*, 16 (8), 1451–1458, 1998.
11. V. Tarokh, H. Jafarkhani, and A. Calderbank, Space-time block codes from orthogonal designs, *IEEE Trans. Inform. Theory* 45, 1456–1467, 1999.
12. B. Hochwald and W. Swelden, Differential unitary space-time modulation, *IEEE Trans. Commun.*, 48, 2041–2052, 2000.
13. B. L. Huges, Differential space-time modulation, *IEEE Trans. Inform. Theory*, 46, 2567–2578, 2000.
14. V. Tarokh and H. Jafarkhani, A differential detection scheme for transmit diversity. *IEEE J. Sel. Areas Commun.*, 18, 1169–1174, 2000.
15. M. Tao and R. S. Cheng, Differential space-time block codes, in *Proceedings of the IEEE Globecom*, November 2001.
16. X.-G. Xia, Differentially en/decoded orthogonal space-time block codes with APSK signals, *IEEE Commun. Lett.*, 6, 150–152, 2002.
17. Z. Chen, G. Zhu, J. Shen, and Y. Liu, Differential space-time block codes from amicable orthogonal designs, in *Proceedings of the IEEE Wireless Communications and Networking Conference*, March 2003.
18. C. S. Hwang, S. H. Nam, J. Chung, and V. Tarokh, Differential space-time block codes using nonconstant modulus constellations, *IEEE Trans. Signal Process* 51 (11), 2955–2964, 2003.
19. G. Bauch, A bandwidth-efficient scheme for non-coherent transmit diversity, *IEEE Globecom*, December 2003.

20. G. Bauch, Differential amplitude and unitary space-time modulation, in *Proceedings of the 5th International ITG Conference on Source and Channel Coding*, Erlangen, Germany, January 14th/16th 2004.

21. G. Bauch and A. Mengi, Non-unitary orthogonal differential space-time modulation with non-coherent soft-output detection, in *Proceedings of the IEEE 62nd Vehicular Technology Conference (VTC'05-Fall)*, Dallas, USA, September 2005.

22. H. Rohling and V. Engels, Differential Amplitude Phase Shift Keying (DAPSK) – a New Modulation Method for DTVB, International Broadcasting Convention, 1995.

23. B. Rodríguez and H. Rohling, A new class of differential space time block codes, in *Proceedings of the 11th International OFDM-Workshop*, Hamburg, Germany, August 30th/31st 2006.

24. G. Bauch, Space-time block codes versus space-frequency block codes, in *Proceedings of the IEEE Vehicular Technology Conference*, volume 1, April 22nd–25th 2003.

25. A. Vanaev and H. Rohling, Design of amplitude and phase modulated signals for differential space-time block codes, *Wirel. Pers. Commun.*, 39 (4), 401–413, 2006.

26. A. Vanaev and H. Rohling, Design of differential space-time block codes for high bandwidth efficiency, in *Proceedings of the 10th International OFDM-Workshop*, Hamburg, Germany, August 31st/ September 1st 2005.

27. B. Rodríguez and H. Rohling, Receive diversity in DSTBC using APSK modulation schemes, in *Proceedings of the 12nd International OFDM-Workshop*, Hamburg, Germany, August 29th/ 30th 2007.

28. M. Stemick, S. Olonbayar, and H. Rohling, PHY-mode selection and multiuser diversity in OFDM based transmission systems, in *Proceedings of the 11th International OFDM-Workshop*, Hamburg, Germany, August 30th/ 31st 2006.

29. Z. Chen, G. Zhu, W. Cai, and Q. Ni, Differential space-time block coded OFDMA for frequency-selective fading channels, in *Proceedings of the IEEE 14th Personal, Indoor and Mobile Radio Communications (PIMRC 2003)*, volume 3, September 2003.

30. Cuvillier Verlag Göttingen, Göttingen, Germany, http://www.cuvillier.de/flycms/en/html/1/-/Home.html?SID=ekrseVJbeed5.

Chapter 11

Adaptive Modulation

Víctor P. Gil Jiménez and Ana García Armada

Contents

11.1	Introduction	302
	11.1.1 Approaching Capacity	303
	11.1.2 Multiuser Diversity	305
11.2	Organizing the Time–Frequency Grid	305
	11.2.1 Time–Frequency Grid Description	306
	11.2.2 Some Examples	307
	11.2.2.1 Long-Term Evolution	307
	11.2.2.2 WINNER Project Proposal	307
	11.2.2.3 IEEE 802.16e (WiMaX Mobile)	307
11.3	Algorithms for Adaptive Modulation	309
	11.3.1 Optimization Problem	310
	11.3.2 SNR Gap Approximation and Constant Power Allocation	311
	11.3.3 Algorithms for OFDMA	311
	11.3.3.1 Two-Step Allocation	312
	11.3.3.2 Joint Subcarrier, Bit, and Power Allocation	313
	11.3.3.3 Multicell Considerations	313
	11.3.4 Complexity and Performance	313
11.4	Channel Feedback and Compression	314
	11.4.1 Motivation	314
	11.4.2 Algorithms in the Literature	315
	11.4.2.1 Data-Limited	315
	11.4.2.2 Quantized	317
	11.4.2.3 Compressed	318
	11.4.3 Impact of Errors on Feedback Data or Delay	319
	11.4.4 Fairness and QoS Remarks	319

11.4.5 Comparison of Different Algorithms 320
11.5 Summary, Open Issues, and Challenges 324
References. 324

11.1 Introduction

Adaptive modulation (sometimes referred to as link adaptation [1] or adaptive transmission [2]) implies matching the transmitted signal to the channel state. The parameters to be changed may be the modulation and coding scheme (MCS) and/or transmitted power. The time scale of the adaptation may vary from instantaneously following the channel state to a long-term adjustment, depending on the availability of channel state information (CSI) at the transmitter side and the variability of the channel.

The effectiveness of adaptive modulation depends on the number and possible values of the parameters to be tuned. The time–frequency structure of orthogonal frequency division multiple access (OFDMA) signals lends itself very well to adaptation because modulation, coding, and power for each user may not only be changed at the orthogonal frequency division multiplexing (OFDM) symbol level but also at the subcarrier level, that is, independently for each subcarrier or group of subcarriers, provided that proper channel information is made available [3].

The use of adaptive modulation in OFDMA is very advantageous in multiuser environments. Each user will be assigned one or a group of subcarriers and its signal adapted to the characteristics of its particular communication channel. The fact that different users will likely perceive different channels will bring about a new degree of diversity, namely multiuser diversity. In general, CSI is obtained at the receiver side by an estimation algorithm performed over the received values of pilot subcarriers. This information must be conveyed to the transmitter through a feedback channel. The amount of channel knowledge and the degree of adaptation that may be practically achieved depend on the scenario where the communication takes place. Figure 11.1 shows the block diagram of an OFDMA system with adaptive modulation.

OFDMA may be used in centralized scenarios where there is a coordinator (e.g., base station, access point, or central office) that may have access to CSI from every user so that jointly optimizing

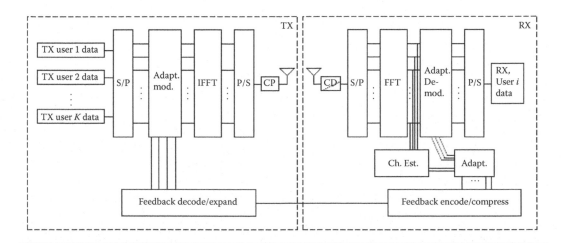

Figure 11.1 Block diagram of an OFDMA system with adaptive modulation. Base station with *K* users and receiver user *i*.

the transmission of all of them is feasible. Even in this case, uplink (return channel) and downlink (forward channel) directions must be distinguished since they do not offer the same degree of complexity for the adaptation process. In the downlink, the coordinator, which will be called base station without loss of generality, transmits toward users; knowledge of the forward channel response to each user may be made available to the base station through the uplink. Sending a synchronized OFDMA signal tailored to the channel of each user is thus feasible. Moreover, OFDMA is optimum for the downlink with any adaptive modulation scheme whose transmission rate can be approximated as a convex function in terms of signal-to-noise ratio/signal-to-interference-noise ratio (SNR/SINR) with independent decoding [4]. In the uplink, each user will transmit information toward the base station. Rather than transmitting back to each user the CSI relative to all of them, it is more feasible to perform the multiuser adaptation algorithm at the base station, cognizant of all users' uplink channel state, and inform each user in the downlink about the modulation, coding, and power appropriate for its transmission. In general, either CSI or the modulation, coding, and/or power to be used in one direction must be transmitted in the other direction of the communication. Examples of centralized scenarios where OFDMA may be used with adaptive modulation are digital subscriber lines (DSLs), cellular systems, infrastructure-based wireless local area networks (WLANs), or wireless metropolitan area networks (WMANs). The use of adaptive modulation in distributed environments is much more involved. In these scenarios, the availability of a node having knowledge of all channels is not possible any more and then suboptimal algorithms with reduced channel knowledge must be looked for. Examples of distributed scenarios where OFDMA may be used with adaptive modulation are *ad hoc* WLANs and wireless personal area networks (WPANs).

Adaptive modulation in OFDM or OFDMA is often referred to as bit and power loading because the power and the number of bits per symbol (that is, modulation order) to be "loaded" in each subcarrier are determined with the goal of optimizing the transmission: either minimizing the required power or maximizing the information rate. Several bit and power loading algorithms may be found in the literature. Often they are inspired in the capacity-achieving distributions that information theory states.

In this chapter we will explain the organization of the OFDMA signal in time and frequency and how adaptive modulation algorithms are able to determine the modulation, coding, and power to be used at each element of the time–frequency grid. We will examine feasible ways of conveying CSI in the feedback channel and explain compression mechanisms useful to minimize the transmission rate of this channel. We will conclude the chapter with a summary and will highlight open issues and challenges in this topic.

First, before finishing this introduction, we will recall some concepts related to capacity and multiuser diversity.

11.1.1 Approaching Capacity

Transmission of an OFDM signal with N subcarriers through a frequency-selective channel will result in no intersymbol interference (ISI) as far as a cyclic prefix longer than the channel impulse response is used [5]. In this case the transmission of each OFDM symbol through the frequency-selective channel and the addition of additive white gaussian noise (AWGN) at the receiver are equivalent to a bank of N frequency-flat parallel channels since the information carried by one subcarrier is not affected by the channel seen by the remaining $N - 1$ subcarriers. This equivalent model is depicted in Figure 11.2.

In Figure 11.2, X_i, $i = 0 \dots N - 1$, are the transmitted information symbols with power P_i, H_i denotes the channel frequency response at subcarrier i (that is samples of the discrete Fourier

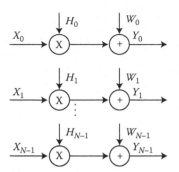

Figure 11.2 OFDM transmission is equivalent to a bank of frequency-flat parallel AWGN channels.

transform of the channel impulse response), W_i are samples of AWGN with power N_i, and Y_i are the received information symbols. An interesting problem approached by information theory is the maximization of mutual information between the input and the output of a set of parallel channels with a common power restriction P_{max}. The mutual information is maximized when the inputs are Gaussian and the common power is distributed among the inputs following the so-called waterfilling distribution [6]:

$$P_i = 0, \qquad \text{if } \frac{N_i}{|H_i|^2} \geq Q, \tag{11.1}$$

$$P_i = Q - \frac{N_i}{|H_i|^2}, \quad \text{if } \frac{N_i}{|H_i|^2} < Q. \tag{11.2}$$

The constant Q, called water level, is chosen to satisfy the total power restriction:

$$\sum_{i=0}^{N-1} P_i \leq P_{max}. \tag{11.3}$$

The capacity achieved by the waterfilling distribution is

$$C = \sum_{i=0}^{N-1} \log_2 (1 + G_i P_i), \quad G_i = \frac{N_i}{|H_i|^2}. \tag{11.4}$$

Interestingly, the structure of the OFDM signal, as shown in Figure 11.2, lends itself to optimization of the spectrum inspired in the waterfilling distribution. The power assigned to each subcarrier can be chosen in order to maximize the bit rate (Figure 11.3). In a realistic case of using discrete constellations [e.g., M-ary quadrature amplitude modulation (M-QAM) or M-ary phase shift keying (M-PSK)] rather than Gaussian inputs, capacity is not achieved. The relationship between the transmitted symbol rate on each subcarrier and the used power will depend on the bit error rate (BER) that is required.

When multiple users are present, there is no single channel frequency response but a multiplicity of them. In Ref. [7] multiuser waterfilling is described to achieve capacity in Gaussian multiple access channels with ISI.

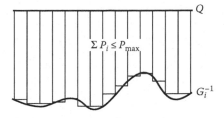

Figure 11.3 Power loading of single-user OFDM.

11.1.2 Multiuser Diversity

The random nature of the propagation channels suggests that when one user experiences a poor channel, some other user will likely perceive a good channel. If we are able to adapt each user's transmission to the instantaneous channel characteristics, we will end up with each user transmitting under the most favorable conditions and the transmission rate of the system as a whole will increase compared to what could be achieved by each user individually transmitting all the time. This strategy of selecting each time the best users to transmit introduces the so-called multiuser diversity [8]. The idea is illustrated in Figure 11.4 for a two-user system. The multiuser diversity increases with the number of users. Channel fluctuations are essential in order to obtain multiuser diversity. However, when the channel varies too rapidly channel estimation becomes complicated. Furthermore, if the channel state cannot be estimated accurately, the performance of the adaptive system will be impaired. Therefore there is a tradeoff between multiuser diversity and mobility as pointed out in Ref. [9]. The advantage of OFDMA here is that the channel fluctuations do not have to happen necessarily in time; they can take place in the frequency domain. Corroborating this idea, it is shown in Ref. [7] that when the channel is highly frequency selective the capacity region of a two-user multiple access channel is greater in the case of users having very dissimilar channels.

The strategy of selecting the user with the best channel characteristics for transmission is also called opportunistic user scheduling (see e.g., Ref. [10]).

11.2 Organizing the Time–Frequency Grid

In this section, the structure and organization of the time–frequency grid will be described. Depending on the way it is organized, the granularity, efficiency, and performance of the

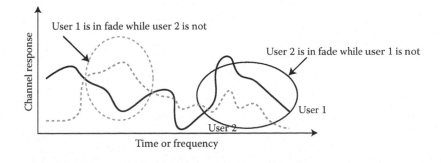

Figure 11.4   Illustration of multiuser diversity.

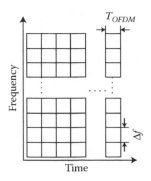

Figure 11.5 Time–frequency grid.

adaptive modulation will be different. Several examples from standards or project proposals will also be presented to illustrate these concepts.

11.2.1 Time–Frequency Grid Description

As has been shown in this book, the OFDM signal has two dimensions: *time* and *frequency*, which can be described as a two-dimensional (2D) grid where users' data can be allocated (see Figure 11.5, where each square represents a subcarrier). Besides, in the case of using OFDMA, the grid is shared by different users. Therefore, the way in which this grid is organized has an impact on the performance of the adaptive modulation scheme.

These schemes can be classified into three main categories, depending on granularity of resource allocation:

- *Adaptation at frame level:* The minimum resource unit that may be used to adapt the transmission is a frame (i.e., several complete OFDM symbols) such as in WiFi [11].* It is the simplest alternative and the one that requires less feedback data; however, the adaptation is poor, especially for fast time-varying scenarios where it obtains limited performance. Besides, the multiuser diversity is not fully leveraged.
- *Adaptation at bin level:* The subcarriers and symbols can be grouped into sets denoted as *bins*, *chunks*, *resource blocks*, *slots*, or *clusters* and this group is used as the minimum resource unit. Throughout this chapter we will use the term *bin*. Actual standards such as IEEE 802.16e [12] or LTE (long-term evolution) [13] and most proposals such as IST WINNER project [14,15] or Wireless IP project [16] belong to this category. This option offers a tradeoff between flexibility, performance, complexity, and amount of feedback data. A careful design of the size of the *bin* is important. If it is too small, the feedback data rate will increase unnecessarily whereas a large *bin* will probably exhibit high BER. In Ref. [17] for example, the *cluster* is designed keeping the *cluster*'s edges above 3 dB of the *cluster*'s average.
- *Adaptation at subcarrier level:* Each subcarrier can be allocated to a user and independently encoded. The granularity is minimum, the flexibility is maximum here, and the performance is the highest one. However, the amount of feedback data can be extremely high, depending

---

* IEEE 802.11a [11] could be taken as an OFDMA system in the limit: different users share the whole OFDM symbol in different frames.

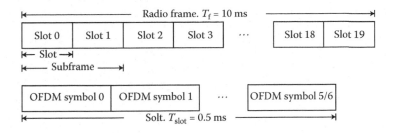

Figure 11.6   LTE frame structure (type I).

on updating frequency and channel conditions. A proposal belonging to this group may be found in the IST PACWOMAN Project [18].

11.2.2 Some Examples

In the following, some examples of the time–frequency grid organization will be described in order to give to the reader a picture of real systems (or proposals) that use adaptive modulation for OFDMA.

11.2.2.1 Long-Term Evolution

The FDD (frequency division duplexing) downlink OFDMA signal in the LTE [13] consists of radio frames of 10 ms. Each frame is divided into 10 subframes and each subframe has two slots, each one containing six or seven OFDM symbols (depending on the type of frame) as can be seen in Figure 11.6. A *resource block* (the minimum structure for adaptive modulation to be allocated to one user, which is denoted in this chapter as *bin*) is defined as 12 consecutive subcarriers out of 2048 during one slot, that is 180 kHz ×0.5 ms. In Ref. [13] it is denoted as a physical resource block (PRB), which can be independently assigned to the different terminals (users) [19].

This structure relies on short frame times that can be accommodated to a time- and frequency-selective channel. It also allows the use of time–frequency correlation properties in the design of the algorithms.

11.2.2.2 WINNER Project Proposal

Another proposal that organizes the resources into *bins* is the IST WINNER project [14,15]. The downlink/uplink OFDMA signal is divided into *chunks* consisting of eight subcarriers (out of 1024) over six OFDM symbols, that is 156.248 kHz ×0.3372 ms, as depicted in Figure 11.7.

Since the structure is similar to the one in LTE, the former comments apply here too.

11.2.2.3 IEEE 802.16e (WiMaX Mobile)

The WirelessMAN OFDMA PHY layer of the IEEE 802.16e standard [12] is another example where the resources are organized into *bins*, denoted in the standard as *slots*. A *slot* is thus the minimum amount of allocable subcarriers, and it is a 2D set of logical *subchannels* and OFDMA symbols (see Figure 11.8). Each *subchannel* is composed of 48 logical adjacent subcarriers.* However,

---

* Depending on the permutation mode, the mapping between logical subcarriers and physical subcarriers will vary.

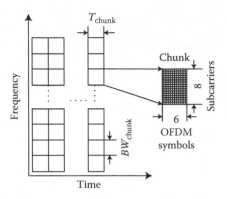

Figure 11.7 Chunk description for the FDD mode. Every chunk is eight subcarriers ×6 OFDM symbols.

the structure is much more complex than for example in the LTE or WINNER projects. The size of the slot depends on the mode of operation. Data are organized in the *data region*, which is a set of adjacent *slots* as can be seen in Figure 11.9 taken from the standard [12].

Here two examples will be briefly described, namely, *downlink FUSC (full usage of subchannels) with distributed permutation* and *Downlink PUSC (partial usage of subchannels) with adjacent permutation*.

Downlink FUSC with distributed permutation: In this mode, all the subcarriers are used and a *slot* is composed of one *subchannel* by one OFDMA symbol. Although logically all the subcarriers are adjacent, physically, depending on the permutation algorithm (in this example the distributed one), logical subcarriers are mapped into physical subcarriers in the following way. First, physical subcarriers are classified into 32 groups (in the case of 2048 subcarriers mode)* of 48 subcarriers. Then each logical *subchannel* takes one subcarrier from each group and then, according to a permutation sequence, these subcarriers are permuted and mapped into physical ones. This is carried out to avoid having a complete slot within deep fade in a mobile environment by spreading the

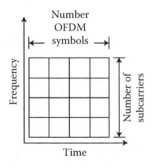

Figure 11.8 Slots in IEEE 802.16e [12]. Each square represents a logical subchannel.

---

* In the WirelessMAN OFDMA PHY [12] there exist four possible number of subcarriers: 2048, 1024, 512, and 128.

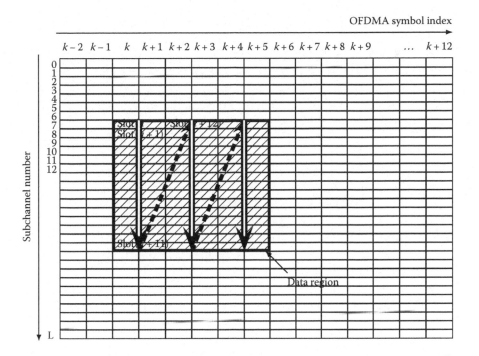

Figure 11.9 Data region for the IEEE 802.16e WirelessMAN OFDMA PHY layer. (Adapted from IEEE. IEEE 802.16e: Air interface for fixed broadband wireless access systems. Technical Report, IEEE, 2005.)

subcarriers of the slot across the whole bandwidth. Besides, the permutation scheme changes on each slot and OFDMA symbol. This will be important for the algorithms that allocate information, perform channel prediction, or compress the feedback data.

Downlink PUSC with adjacent permutation: In this mode, not all the subcarriers need to be used, and a slot is composed of one *subchannel* during two, three, or six OFDMA symbols. Here, the permutation is fixed during these OFDMA symbols and it maps logical adjacent subcarriers into adjacent physical subcarriers. However, each *subchannel* is still formed the same way as in the downlink FUSC mode explained above. This configuration is desirable for fixed communications where the channel exhibits large coherence time and bandwidth. Algorithms that allocate resources can leverage on time and frequency correlation properties for this operation mode.

11.3 Algorithms for Adaptive Modulation

When matching the OFDMA signal to the channel state perceived by each user, there are several resources that may be subject to optimization:

■ Subcarriers (either individually or grouped) that may be assigned to each user. In order to reduce the complexity of the resource assignment problem, the subcarriers are often grouped, as explained in Section 11.2, and the same resources are assigned to a group of subcarriers for the duration of several OFDM symbols depending on the channel perceived for the group as a whole.

- MCSs that will result in the choice of the number of bits/symbol to be transmitted at each subcarrier. Even though some schemes adapt only modulation per subcarrier (or group), the use of coding will result in better performance and finer granularity in the obtained data rates.
- Power to be allocated at each subcarrier for the successful transmission of a given MCS choice.

The allocation of these resources is a coupled problem: the optimum solution should perform an exhaustive search over users, subcarriers, modulation-coding schemes, and power levels. Since this exhaustive search is not feasible, several suboptimum approaches have emerged and will be described in this section.

Adaptation to large-scale channel variations requires only that the values of these resources are changed at a very slow pace: several OFDM symbols are transmitted with the same number of bits/symbol (in all subcarriers) and power. This scheme is called nonfrequency adaptive [2], and corresponds to adaptation at the frame-level scheme (see Section 11.2). Adaptation to small-scale channel variations (i.e., frequency-adaptive [2]) offers higher performance at the expense of complexity. In this chapter we will concentrate on frequency-adaptive schemes.

In the following, we formulate the optimization problem and describe some approaches that may be used to simplify adaptation algorithms: the so-called SNR gap and the use of constant power allocation. Then we describe algorithms that make use of these concepts to perform the adaptation of the OFDMA signal and we compare their complexity and performance.

11.3.1 Optimization Problem

In a multiuser OFDMA system with K users and N subcarriers, the problem of assigning subcarriers, bits/symbol, and power to the users may be solved with several possible goals in mind. We describe here the goal of maximizing the sum of the users' bit rates and that of minimizing the total required power. Other criteria such as fairness among users are discussed in Section 11.3.3.

We denote by H_{ik} the channel frequency response seen by user k ($k = 0 \ldots K - 1$) in subcarrier i ($i = 0 \ldots N - 1$), by R_{ik} the bit rate achieved with the MCS allocated to user k in subcarrier i, and by P_{ik} the power used by user k in subcarrier i.

Maximizing rate: The goal is achieving the maximum sum rate under a total power constraint $P_{\max}$.

$$\max_{R_{ik}, P_{ik}} \sum_{k=0}^{K-1} \sum_{i=0}^{N-1} R_{ik} \tag{11.5}$$

$$\text{s.t.} \sum_{k=0}^{K-1} \sum_{i=0}^{N-1} P_{ik} \leq P_{\max}. \tag{11.6}$$

Minimizing power: The goal is achieving the minimum transmission power while satisfying a required total number of bits/symbol $R_{\min}$.

$$\min_{R_{ik}, P_{ik}} \sum_{k=0}^{K-1} \sum_{i=0}^{N-1} P_{ik} \tag{11.7}$$

$$\text{s.t.} \sum_{k=0}^{K-1} \sum_{i=0}^{N-1} R_{ik} \geq R_{\min}. \tag{11.8}$$

Each subcarrier can only be assigned to one user, that is, for each k if $R_{ik} \neq 0$ for a given i, then $R_{jk} = 0 \forall j \neq i$.

The required power to be used at a given subcarrier in order to transmit a number of bits/symbol depends on the error probability demand of the transmission system. A useful way of relating them is obtained with the SNR gap approximation.

11.3.2 SNR Gap Approximation and Constant Power Allocation

The number of bits/symbol that may be transmitted over subchannel i with power P_i and channel-to-noise ratio G_i (as defined in Equation 11.4) with a given MCS may be approximated by

$$R_i = \log_2\left(1 + \frac{P_i G_i}{\Gamma_i}\right), \tag{11.9}$$

where Γ_i is the SNR gap that represents the reduction in SNR relative to capacity that is due to the fact that a given MCS is used rather than the theoretical capacity-achieving input [20].

For M-QAM modulation, when error probability (P_e) is specified by symbol error rate (SER), the SNR gap is given by

$$\Gamma = \tfrac{1}{3}\big[Q^{-1}(P_e/4)\big]^2,$$

where the Q function is defined as

$$Q(x) = \int\limits_{x}^{\infty} \frac{e^{-u^2/2}}{\sqrt{2\pi}}\,\mathrm{d}u.$$

The SNR gap expression is accurate for rectangular M-QAM constellations with size equal to a power of 2 and $R_i > 1$. In general it constitutes a good approximation that simplifies bit loading algorithms. An equivalent SNR formulation for M-PSK modulation is obtained in Ref. [21]. The introduction of coding together with adaptive modulation may be accounted for by decreasing the SNR gap by the amount of coding gain.

Waterfilling, which gives the power distribution to achieve capacity, allocates more power to the channels with better SNRs. This means that channels with different SNRs are allocated different powers. In Ref. [22] it was reported from empirical observation that as long as the correct subchannels are used, a constant power allocation has a negligible performance loss compared to waterfilling. This fact has been proved recently by theoretical analysis [23]. Many bit loading algorithms inspired in waterfilling will use constant power in order to reduce the computational complexity at a very low performance cost.

11.3.3 Algorithms for OFDMA

Algorithms that perform adaptive modulation for OFDM and OFDMA often make use of single-user bit and power loading as a basic tool in the more complex optimization problem with multiple users and multiple channels.

Waterfilling solves the optimization problem of Section 11.3.1. However, it gives as a result noninteger numbers of bits/symbol that may be impractically small for some subchannels. Discrete loading algorithms optimize power and bit allocations under the restriction of using an integer (or noninteger but quantized, that is, discrete, in case of applying coding) number of bits/symbol at each subcarrier. Among them the Hughes-Hartogs [24] algorithm provides the optimum solution for the integer loading problem. However, it has slow convergence and high computational cost. Chow's algorithm [25] starts with a solution close to waterfilling and then the bit distribution is rounded to account for the requirement of using discrete information units. It makes use of the fact that constant power allocation is close to waterfilling. However, its complexity is also high. The Levin-Campello [26] algorithm provides the optimal solution of Ref. [24] with a lower complexity. It is a greedy algorithm that places every incremental data unit in the subcarrier that will require the lowest power increment to transmit it.

Recently, Dardari [27] proposed ordering and selection of the more reliable subcarriers while maintaining uniform bit and power allocation among them. This uniform bit allocation reduces its complexity compared to other algorithms. By using only half of the more reliable subchannels and by quadrupling the constellation size, a gain greater than 5 dB in terms of SNR can be achieved in a 12 Mb/s system, with a little loss compared to the optimum case [26].

In the multiuser scenario it is known that in order to achieve multiuser capacity, superposition coding and successive interference cancelation are needed [7]. However, FDMA approaches (i.e., each subcarrier is assigned to at least one user) are advisable since they will lead to easier implementation [28]. In Ref. [28] it is also shown that constant power allocation, although suboptimal, introduces very little performance degradation in the multiuser case too. We will focus on FDMA approaches in this chapter.

Given the complexity of the optimization problem, many authors take a two-step approach for practical implementation, first assigning subcarriers to users and after that performing bit and power allocation. However, some authors tackle directly the joint optimization problem. We describe both types of algorithms in the following.

11.3.3.1 Two-Step Allocation

Assuming knowledge of the users' channels and a given set of required user data rates, in Ref. [3] the allocation of subcarriers, bits, and power in a multiuser OFDM system is analyzed with the goal of minimizing the total transmit power. After examining the joint optimization of subcarriers, bits, and power, they propose a suboptimal algorithm in which subcarriers are assigned to the users in a first step and then a greedy approach similar to that of Ref. [24] is used to assign bits and power to individual users in the allocated subcarriers. They compare the performance of this algorithm to static subcarrier allocation schemes when R_i can take the values $0, 2, 4,$ and 6: the algorithm saves 5–10 dB in SNR when compared to OFDM-TDMA and OFDM-FDMA (with predetermined time slots and subcarriers assigned to users independently of their channel responses) and adaptive modulation and 3–5 dB compared to these schemes without adaptive modulation.

In Ref. [29] the same two-step approach is followed. In this case the subcarriers are assigned assuming constant transmit power for each of them. Next, bits and power are loaded to the subcarriers with a modified Levin-Campello algorithm.

Mohanram and Bhashyam [30] also use a two-step approach; the procedure is different though. In the first step, the number of subcarriers that each user will get is determined based on the users' average SNR. In the second stage, it finds the best assignment of subcarriers to users. Several algorithms are proposed that minimize the transmitted power while ensuring that each user transmits at a minimum desired rate (or higher).

The algorithm in Ref. [31] also separates subcarrier allocation and power allocation. In the proposed algorithm, subcarrier allocation is first performed by assuming equal power distribution; then an optimal power allocation algorithm maximizes the sum capacity while maintaining proportional fairness. Simulation results show that this suboptimal algorithm can achieve above 95% of the optimal performance in a two-user system.

Another algorithm that divides the search into two steps, first optimizing the subcarriers and assigning bits and power afterwards described in Ref. [32].

11.3.3.2 Joint Subcarrier, Bit, and Power Allocation

Two-step algorithms are suboptimal approaches to overcome an exhaustive search. Some approaches consider jointly the optimization of subcarriers, bits/symbol, and power, such as Refs. [33–35]. They obtain better performance than that in Ref. [30].

In Ref. [33], knowing the channel characteristics of all the users at the base station, the subcarrier allocation algorithm assigns subcarriers to the users in the downlink in a way that the total transmit power is minimized. The proposed subcarrier and power allocation algorithm offers low complexity and is very stable with the number of users in the system.

YuMing et al. [34] try to minimize the total transmission power, while satisfying the BER constraint and rate requirement of each user. The proposed algorithm combines the subcarriers and power allocation simultaneously, by comparing the power increment to swap the subcarriers between users.

Mohanram and Bhashyam [35] perform joint subcarrier and power allocation in order to maximize the overall rate while achieving proportional fairness among users under a total power constraint. It avoids the two-step approach by alternating between subcarriers and power allocation. Power allocation to the subcarriers belonging to a particular user is performed by waterfilling. However, unlike the schemes of the previous section, this waterfilling of each user plays a role in deciding subcarrier allocation.

In Ref. [36], a heuristic noniterative method that is an extension of Ref. [27] for OFDMA is applied for minimization of the average BER performance under fixed transmit power and fairness between the users in terms of bit rates. The algorithm is refined by using channel power gain inversion rather than constant power as in Ref. [27].

11.3.3.3 Multicell Considerations

In the algorithms detailed above, a single cell is considered; hence frequency reuse is not addressed. Kim et al. [37] propose subchannel (subcarriers are grouped in subchannels) allocation algorithms considering frequency reuse. They maximize the total system throughput while guaranteeing the quality of service (QoS) of each user by equal power allocation to subchannels and the same modulation scheme for subcarriers in a subchannel. Zhang and Letaief [38] also tackle the multicell environment by introducing adaptive cell selection. Besides, they lower the complexity of the algorithm in Ref. [3] by the use of constant power allocation.

11.3.4 Complexity and Performance

The work in Ref. [3] is a reference that established the basis for OFDMA subcarrier, bit, and power allocation. The algorithm has been improved subsequently in terms of both performance and complexity. The complexity of Ref. [3] is $\mathcal{O}(INK)$, $I \gg N$, while Kivanc et al. [30] propose two algorithms of complexity $\mathcal{O}(KN + N \log N)$ and $\mathcal{O}(KN)$. The improved version of Ref. [3] that also takes into account the multicell environment [38] is $\mathcal{O}(3KN + 2N^2)$. Also, the outage probability of Ref. [3] is two to three times that of Ref. [30]. More recent proposals improve the performance maintaining the complexity. Reference [35] has, for example, 5–25% gain in spectral efficiency over Ref. [3].

Recently, the concept of fairness has attracted much interest. In case of rate maximization, purely maximizing the total bit rate may bring the unfair result that users experiencing weak channels are never transmitting. Hence algorithms are required to guarantee that a minimum bit rate is transmitted by each user. Algorithms [34] and [30] include constraints on the minimum bit rate per user. References [31], [35], and [36] include fairness restrictions in the optimization problem.

11.4 Channel Feedback and Compression

11.4.1 Motivation

As has been mentioned in the introduction to this chapter, most of the times channel or signal-to-noise and/or interference estimation is performed at the receiver side and needs to be fed back to the transmitter by some means, in order to be used by the adaptive modulation algorithms at that side. The amount of feedback data can be very large, especially in OFDMA systems designed for wireless high data rate terminals to be deployed in high-mobility scenarios. Besides the feedback, it is also worth mentioning the signaling data required to inform terminals about the resources that may be used. The amount of signaling information can also be high and it affects the performance as shown in Refs. [39,40]. The feedback and signaling data will depend on the time–frequency grid organization (Section 11.2). Another important design issue is the feedback period, that is how long the feedback data can be assumed to be valid or how often it must be fed back [41].

Another aspect to take into account is what kind of information is going to be fed back, namely, the values of the channel coefficients (in time or frequency), the MCS, or the SNR. Depending on that choice, the amount of information and the algorithms will be different. For example, transmitting the channel coefficients requires the maximum data rate while it offers the best performance since it allows one to perform algorithms such as waterfilling [6], beamforming in multiple-input multiple-output (MIMO) or multiple-input single-output (MISO) systems [42], or encoding/precoding such as dirty paper coding (DPC) [43] or Tomlinson–Harashima [44]. Thus, these schemes are used mainly for MIMO systems and will not be covered in this chapter. For a review of these algorithms, the reader is referred to Ref. [45].

To feed back the MCS requires that receivers perform the adaptive modulation algorithms themselves (or a coarse version of them), and thus the complexity increases at the receiver side. Besides, in Ref. [46] it is shown that the scheduler can do a better job (in terms of performance) with the SNR instead of the MCS because it can exploit the better accuracy of a real-valued SNR level.

Algorithms for reducing/compressing feedback data can thus be classified in the following three main categories. The first two correspond to a reduction of the amount of feedback data and the last one to a compression of the information:

- *Data-limited:* The SNR or MCS is only fed back when some condition is fulfilled. Thus, only a set of the terminals (usually small) will feed back their data. In Ref. [47] it is shown that in a wireless mobile channel the reduction can be more than 50% of the feedback need.
- *Quantized:* Instead of sending the real-valued data, it is quantized to reduce the amount of information. If quantization regions are designed carefully and asymmetrically, the feedback data can be considerably reduced. There exists, however, some degradation of the quality of feedback information.

■ *Compressed:* These algorithms compress the feedback data, removing redundancy from information (usually exploiting time-frequency correlations).

Within each category, the data to be fed back may be the channel coefficients, the MCS, or the SNR.

11.4.2 Algorithms in the Literature

The algorithms described here will be classified using the three above-mentioned categories. Most of the algorithms have been proposed for single carrier systems, but they can however be easily adapted to OFDMA, as we will address.

11.4.2.1 Data-Limited

The common factor of all the algorithms in this category is that feedback information is only transmitted if some quality measurement performed by the terminal is above or below a threshold that is usually broadcasted by the base station or coordinator. It has been shown in Refs. [47–49] that there exists a high probability that a user (terminal) experiences poor channel conditions, especially in wireless mobile scenarios with high speed. In Refs. [50,51] it is shown that the scheduling algorithm that maximizes the average spectral efficiency (ignoring feedback loss) is the one that schedules the terminal with highest SNR, which is called *max-SNR scheduling*. This scheduler, however, does not take into account fairness (see Section 11.4.4 for more details).

Gesbert and Alouini [48] proposed *Selective MultiUser Diversity (SMUD)*. In this algorithm, terminals estimate the SNR and if the value is above a threshold γ_{th} they feed back the SNR or the MCS to the BS. If no terminal has SNR above the threshold, the scheduler randomly selects a receiver to transmit data. This is actually one of the weaknesses of the algorithm, because when a random user is selected to transmit in this particular situation the multiuser diversity is reduced. For this reason, the authors in Ref. [52] propose an improvement of the SMUD algorithm, where in the case of no terminal with SNR above the threshold, the scheduler requests a full feedback and all the terminals send to the transmitter their SNR, and thus the scheduler selects the one with highest SNR. In this way, multiuser diversity is held (with an increment in feedback data rate). Also in that paper, a closed-form expression to calculate the optimum threshold γ_{th}^* is derived to be

$$\gamma_{\text{th}}^* = -\bar{\gamma} \ln\left(1 - \left(\frac{1}{K}\right)^{1/(K-1)}\right), \tag{11.10}$$

where K is the number of users and $\bar{\gamma}$ is the average SNR. As can be observed, the optimum threshold depends on the number of users in the system and the average SNR. This lack of generality is one of the main drawbacks of the proposal.

Although the SMUD algorithm was proposed for a single-carrier system, it can be easily adapted to OFDMA: performing the measurement of SNR on each subcarrier, *bin/chunk*, or symbol (depending on time–frequency grid organization) and sending some index jointly with feedback data to indicate which resources correspond to the measured SNR values. The same applies when changing the SNR values by the MCS in Ref. [53], where the authors refine the algorithm requesting feedback sequentially with lower thresholds, reducing in this way the feedback load since not all terminals will send information back as in Ref. [52].

This algorithm is adapted to MIMO and analyzed in Refs. [54] and [55], respectively.

In Ref. [17] a similar approach is proposed, but it is applied to an OFDMA system* where the subcarriers are grouped into *clusters* (which is denoted in this chapter as a *bin*). Instead of using a threshold, SNR values of the S strongest *clusters* are always fed back with their indexes. The obtained performance is close to a waterfilling solution. Although this scheme may waste feedback data rate (the strongest *cluster* for some users might not be good enough to be used for transmission), it obtains better performance than SMUD in terms of fairness since the scheduler has more information about the terminals (see Section 11.4.4). Then, in Refs. [56] and [57,58] the authors adapt this scheme to MIMO (with beamforming), improve it in terms of fairness (see Section 11.4.4), and refine the feedback data distribution while keeping the feedback load constant: when the number of users is reduced they are allowed to feed back information about more *clusters* than when there are many terminals sharing the link. This refinement makes sense. In a scenario where only a few users are active, if they only feed back a small set of *clusters* there will exist a high probability that some *clusters* will be unused (and therefore wasted). On the other hand, if there are many users, it is likely that using only the strongest *cluster* of each one will be enough to capture the whole bandwidth. Of course, the number of *clusters* to be fed back is a parameter that needs to be broadcasted; authors claim that the number of users is a slowly changing parameter and thus introduces little feedforward overhead. They even propose quantizing it to reduce the impact. However, the overhead might not be negligible in burst-traffic highly dense scenarios.

A similar idea is presented in Ref. [59] based on Refs. [60,61]. The allocation is split into two parts and so is the feedback. The subcarriers are grouped into *bins*. Then, each terminal selects its best S *bins* and feeds back the index of these selected *bins* and the average SNR over all of them. The BS allocates users into different *bins* and broadcasts this information to terminals. Finally, users feed back the SNR on each of the allocated *bins*. In this way, up to 83% of feedback reduction (with respect to full feedback) can be obtained and 23% reduction with respect to the original algorithm in Ref. [60]. Besides, a higher degree of Multiuser diversity is attained. Going a step further, in Ref. [62] the authors propose to adapt the number of strongest *bins* depending on the number of unfed-back *bins*. It is possible that if the number of strongest *bins* (S) is small and the number of active users (K') is reduced too, some *bins* do not have information feedback and, therefore, a random user is selected to transmit on them. In Ref. [62], when this situation appears, the value of S is increased to guarantee that the BS receives feedback for all the *bins* at least from one terminal. The reduction of feedback rate obtained is around 83%.

In Ref. [63] a distributed access protocol denoted as *channel-aware ALOHA* is proposed. It is included here, although no feedback is needed from users since decisions are taken locally at each terminal based on a threshold (broadcasted by the BS). If a terminal experiences a channel gain above this threshold, it chooses to transmit with probability p or not to transmit with probability $1 - p$. There can exist collisions if two or more terminals decide to transmit at the same time. In this case, an ALOHA protocol is used [64]. The authors extend this idea to multicarrier-like systems where waterfilling is performed on subchannels (or subcarriers). It can be easily applied to OFDMA selecting subcarriers for different users. Besides, the authors show that the algorithm can reach capacity. The users with best channel conditions are usually transmitting if the channel is invariant over the necessary time to manage collisions.

In Ref. [41] the authors determine the feedback period depending on terminal speed (which is known at the base station by estimation [65]). The scheduling period is thus, in order to be able to schedule all users, the minimum of the feedback period for every terminal. In this way, the feedback

---

* Although the paper is proposed for a 2×1 MIMO system, it can be straightforwardly adapted to SISO.

data are reduced since all equipment will only feedback according to feed back period. It should be noted that in all the schemes presented so far, all the terminals have the same feedback period and this must be adjusted to the minimum one to have accuracy on channel conditions. So, this idea can be applied jointly with the others presented here. Since BS must inform terminals about the feedback period (which may change during the transmission), they propose quantizing the period according to the number of available periods in the system. A different approach but similar concept, the reduction of feedback adaptively by each user, is proposed in Ref. [66], where the authors fix the packet length for all terminals and the users feed back an amount of data inversely proportional to the modulation scheme. To give an example, if a terminal uses 64QAM, for a fixed packet length, the number of *slots* needed to transmit the packet will be six times less than if transmitting with binary phase shift keying (BPSK). So the first user will send six times less feedback data because the next feedback will be sent once it has finished sending the packet. Of course, the modulation scheme may vary depending on channel conditions. They also apply an adaptive feedback period as in Ref. [41]. They obtain reductions around 20% of full feedback.

Regarding signaling, in Ref. [39] the authors propose two schemes, namely, *fixed* and *variable*. In the first one the base station always broadcasts the whole signaling information in a tupla [*index of bin, scheme (bit, power, coding), and terminal index*] while in the second one only different assignments are sent. They obtain some interesting conclusions from the point of view of the signaling:

- Increasing the number of subcarriers in an adaptive system increases the throughput until a maximum number of subcarriers (around 128 for a bandwidth of 16.25 MHz) is reached and then the net throughput decreases due to signaling overhead.
- Increasing the bandwidth causes the optimum number of subcarriers to be increased. So, from the point of view of signaling, if we want to use a large number of subcarriers the bandwidth should be increased.
- Increasing the frame length in general increases the throughput since it requires less signaling, although the BER may increase.

The authors reach the conclusion that for large bandwidth systems such as future 4G, signaling overhead can be neglected. On the other hand, in Ref. [40] the authors evaluate the impact of imposing constraints on the signaling to the throughput.

One of the main problems of these techniques is that BS does not have full information of all users and thus fairness policies at the scheduler cannot be applied.

11.4.2.2 Quantized

Since SNR is real valued, it will be necessary to quantize it somehow in order to feed it back to the transmitter. A straightforward way of reducing feedback is to reduce the number of bits (or quantization levels). However, the more the reduction, the more the degradation in performance (because of quantization errors) [67]. However, for the max-SNR scheduler, only one-bit quantization is enough to obtain 90% of throughput [67–69]. Indeed, the idea proposed in Ref. [68] is that each user feeds back one bit: 0 means that the SNR of the terminal is below some threshold and 1 means that the SNR is above. This technique could be applied jointly to the ones described in the previous section (11.4.2) and thus terminals only send back information when they are above the threshold. In this case, since we have one feedback bit for more refinement, we can quantize the SNR value above the threshold γ_{th} with one bit and extra performance can be achieved. The threshold must be broadcasted by the base station according to the load conditions, and as it was shown in Refs. [17] or [54] the threshold also depends on the number of active terminals in the system. However, it can be assumed in general that the frequency of this broadcasting information does not need to be very high. The adaptation to OFDMA system is trivial (one bit for each *bin*).* In order to exploit multiuser diversity and introduce some fairness, the same authors proposed a modification in Refs. [70,71]. Their proposal is to apply a Round Robin strategy at the scheduler and then sort the terminals by their maximum SNR. After that, each time one terminal is scheduled and it is removed from the list. Once all the active terminals have been scheduled, the process starts again, and thus all the terminals have access to the channel once every K' (number of active users) slot's time. Moreover, the multiuser diversity is still exploited every K' slots, that is in average a multiuser diversity of $K'/2$. The proposal assumes that all the terminals in the system know how many active terminals exist, and therefore the feedback is even reduced by avoiding to feed back once the terminal has been scheduled until the next round.

Since adaptive modulation depends on the feedback scheme, in Ref. [72] the authors propose designing a quantizer jointly with the power, rate, and subcarrier allocation algorithm. Power and rate allocation is performed continuously instead of discretely. Although the hardware complexity is higher [73], it offers more flexibility and less algorithmic complexity by applying the Karush–Kuhn–Tucker (KKT) conditions and Lagrange multipliers.

11.4.2.3 Compressed

It has been shown in Refs. [47,48] that there exists a correlation between channel conditions in time and frequency, that is the channel response in a concrete *bin* at a time is very similar to that in adjacent *bins* and near in time (depending on the Doppler and delay spreads). Therefore, since there exists inherent redundancy, information can be compressed using lossless methods or lossy techniques. The first ones are less powerful (in terms of compression) but do not distort data, that is information can be recovered exactly as it was transmitted (except for errors in the channel), and the second ones offer more compression but however introduce some distortion on the data that could reduce global performance.

In Ref. [75] the authors make use of time correlation to differentially encode the number of bits on each subcarrier. Although the original idea is applied to a single-user multicarrier system, it can be easily extended to a multiuser multicarrier system such as OFDMA. The authors propose to split the feedback into an integer part and a fractional part and they encode the difference between the previous time and the current one. They also propose using the same feedback information during a longer interval to reduce the feedback rate even more at the expense of increasing degradation due to changes in channel conditions.

Lestable et al. [76] proposed the use of Lempel–Ziv–Welch (LZW) coding (as in compression programs) for eliminating redundancy on the feedback data in multicarrier systems. They obtained a reduction of around 1/4 at best for slowly moving terminals. In Ref. [77] the authors applied image compression techniques to compress the feedback data since the number of bits on each subcarrier can be viewed as an image (frequency × time grid). Thus, on this *image* they first apply a decorrelation transform and then an entropy encoder based on a Run-Length coding, followed by a universal variable length code (UVLC) [78]. In Ref. [77] it is shown that only with two elements decorrelation is enough to obtain the maximum performance, and introducing more elements does not provide any larger gain but only more complexity. The decorrelation transform is performed

---

* It should be noted that no indexes are needed since data are sent sorted, that is first bit for first *bin* and so forth.

first in time (differentially encoded) and then in frequency with adjacent elements. Again, the best performance (10% of uncompressed data at best) is obtained for slow movement scenarios.

In Ref. [79] a lossy compression method is proposed to reduce feedback information. The proposed method is divided into two parts to leverage on the time and frequency correlation properties of mobile wireless channels. The authors use the *bin*'s structure (denoted as *chunk*) from the WINNER project [80]. First, at the terminals, SNR is estimated and a discrete cosine transform (DCT) is performed to leverage on frequency correlation properties, then, the coefficients are scalarly quantized.* Finally, a sub-sampled version is used as the feedback data to leverage on the time correlation properties. At the BS, the missed data are estimated by using the minimum mean square error (MMSE) estimator. Great compression gains can be obtained (0.5 bits/bin/user) with negligible throughput degradation due to lossy compression but with an increase in complexity at the BS.

In Ref. [47] that also uses the WINNER project for experiments, the time and frequency correlation properties are exploited too, but in a different way and using Huffman coding [82] instead of generalized Lloyd. The time redundancy is removed by differentially encoding the different *bins* since only two of them are enough to capture the time correlation properties [77]. The frequency redundancy is exploited by designing (off-line) the Huffman codes for groups of one *bin* and its adjacent *bins*. Two algorithms are proposed, *iterative* and *block-wise*, obtaining both similar performance and compression gains. At the on-line stage, only a lookup table is needed (known at both sides) to encode/decode the feedback data, so the complexity at both sides is low and a feedback of 1 bit/bin/user can be obtained. Since there exists a differentially encoding process that can accumulate errors, it is suggested that one refresh the feedback data every certain time.

11.4.3 Impact of Errors on Feedback Data or Delay

Although performance can be seriously affected if feedback data are corrupted, there is not much literature available that analyzes the effect of errors on these sensitive data. In Ref. [47] the effect of errors on feedback data is simulated, giving as a conclusion that data must be refreshed every 15–20 frames when there exist errors on the feedback channel. Besides, due to the sensitivity of these feedback data, a convolutional 1/2 encoder is suggested to protect that information.

In Ref. [52], a closed-form expression for the degradation of the SMUD algorithm is obtained when there exists a delay between the time when the SNR is estimated and when it is used. It is shown that for normalized delays (with respect to feedback period) up to 2×10^{-2} degradation is negligible whereas there is a sharp degradation for higher values.

11.4.4 Fairness and QoS Remarks

Important issues to take into account when talking about future wireless communications systems are QoS and fairness. Both are closely related. In a wired network, QoS and fairness can be managed with a reasonable complexity since all the terminals may have the same link quality, but when dealing with the mobile wireless link, the problem becomes harder. Maximization of the throughput usually assigns resources to the terminals with better channel conditions (max-SNR scheduler), which may

---

* In order to design the quantization tables off-line, the generalized Lloyd algorithm is used [81]. After that, for each component it is only needed to use on-line the quantizer table with the correct number of bits and rescale and translate it to fit the component variance and mean.

not be fair for all users and may not offer the required QoS. Thus, some strategies and modifications to the algorithms must be performed since most of them are based on the max-SNR scheduler, especially *opportunistic* ones, which are tailored to the max-SNR scheduler.

QoS and fairness are closely connected because if all users experience the required QoS (users receive what they have paid for), the system is fair. Recent studies on internet satisfaction indicate that the satisfaction of an already well-served user increases only marginally if the system offers to him even further quality [83,84], whereas if the service level is decreased below some threshold the satisfaction level drops significantly.

In Ref. [57] the authors introduce a modified *Proportional Fairness* (PF) algorithm [85] to offer fairness and QoS guarantees. Besides, this modified PF allows one to adjust the fairness* and QoS of each user.

Since the channel must be predicted at the receiver (to take into account the channel that the terminal will experience when feedback data are ready), in Ref. [86] the idea of using this extra information (channel prediction) to obtain a more fair system is investigated. It is based on the well-known Proportional Fairness algorithm but leveraging on this extra information. The conclusion is that the more restrictive the fairness and time requirement, the better the results obtained with this method. The increment on complexity at the receiver is not very large and the tradeoff with performance is good.

11.4.5 Comparison of Different Algorithms

A comparison of the different algorithms presented throughout this chapter is shown in Table 11.1, where many of the criteria discussed above have been taken into account. The table can be useful when designing the feedback algorithm depending on the desired criteria by the user. The following parameters used in the table are summarized below:

- N: Number of subcarriers
- N_{bins}: Number of *bins*, or *clusters* or *slots*
- K: Number of users in the system
- K': Number of active users in the system
- K_{th}: Number of active users with SNR above the threshold in the system
- M: Number of modulation schemes
- U: Number of bits needed to send the real-valued SNR
- N_r: Number of receive antennas
- N_t: Number of transmit antennas
- S: Number of *bins* to be fed back (strongest)

Besides, in Figure 11.10 the average number of total feedback bits has been drawn for different algorithms in order to evaluate how they behave with the number of users. The scenario was a system with $N_{bins} = 32$, with $K' = K$ active users, $M = 8$ modulation schemes and feedback $U = 4$ bits to represent the real-valued SNR and $S = 3$ strongest *bins* for the algorithms that use this parameter. Looking at this figure, the best algorithms in terms of feedback amount of information are given in Refs. [17,62,68,70,72,79] and [47], especially [72,79]. In Figure 11.11 where the complexity

---

* The weights for a specific *cluster* can be retained to increase fairness for those users. This adjustment can be done at a system level, that is at the scheduler's higher layers.

Table 11.1 Comparison of Feedback Algorithms

Algorithm's Name	Feedback Load (Bits)	Complexity at BS	Complexity at Users	Fairness	Type of Feedback	Throughput	Observations
SMUD-SNR [48]	$[K_{th} \times N_{bins} \times U, K \times N_{bins} \times U]$	Low	Low	Unfair	SNR	High	Multiuser diversity is not well exploited
SMUD-MCS [48]	$[K_{th} \times N_{bins} \times \log_2(M), K \times N_{bins} \times \log_2(M)]$	Low	Low	Unfair	MCS	High	Multiuser diversity is not well exploited
M-SMUD-SNR [52]	$[K_{th} \times N_{bins} \times U, K \times N_{bins} \times U]$	Low	Low	Unfair	SNR	Higher	Multiuser diversity can be better exploited
M-SMUD-MCS [53]	$[K_{th} \times N_{bins} \times \log_2(M), K \times N_{bins} \times \log_2(M)]$	Low	Low	Some fairness	MCS	Higher	Multiuser diversity can be better exploited. Fairness and QoS can be tackled
MIMO-SMUD [54]	$[N_t \times N_r \times K_{th} \times N_{bins} \times \log_2(M), K \times N_{bins} \times \log_2(M)]$	Medium	Medium	Some fairness	SNR	Higher	Multiuser diversity can be better exploited. Fairness and QoS can be tackled. MIMO
CL [17]	$K' \times S \times (U + \log_2(N_{bins}))$	Low	Low	Unfair	SNR	Higher	
CL-BF [57]	$N_t \times N_r \times K' \times S \times \log_2(S) \times \log_2(N_{bins}) \times U\times$	Low	Medium	Some fairness	SNR	Higher	It offers fairness and QoS
CA-ALOHA [64]	Collisions	Low	Low	Unfair	No	High	The number of users must be high
1BIT-JOHANSSON [68]	$K' \times N_{bins}$	Low	Low	Unfair	SNR	Medium	
1BIT-JOHANSSON-MOD [70]	$K'/2 \times N_{bins}$	Low	Low	Some fairness	SNR	Medium	

continued

Table 11.1 (continued) Comparison of Feedback Algorithms

Algorithm's Name	Feedback Load (Bits)	Complexity at BS	Complexity at Users	Fairness	Type of Feedback	Throughput	Observations
QCSI-OFDMA [72]	$N_{bins} \times \log_2 \times (K \times (2^U - 1) + 1)$	Medium	Medium	Unfair	SNR	High	The number of bits is $\sum_{n=1}^{N_{bins}} \log_2 \times \left(\sum_{k=1}^{K} (2^{U_{n,k}} - 1) + 1\right)$
DIFF-T [75]	$[K' \times N_{bins} \times 3, K' \times N_{bins} \times (2 \times \log_2(M) + 2)]^a$	Medium	Medium	Unfair	SNR	Higher	
UVLC [77]	$[0.2 \times K' \times N_{bins} \times \log_2(M), 0.85 \times K' \times N_{bins} \times \log_2(M)]$	Medium–high	Medium–high	Unfair	MCS	Higher	
ADAPT-CH [41]	$0.81 \times K' \times N_{bins} \times \log_2(M)$	Medium	Medium	Unfair	MCS	High	
BEST-M [60]	$(\lceil \log_2 N_{bins} \rceil \times S + (S+1) \times U) \times K \approx 0.21 \times K' \times N_{bins} \times U$	Medium	Medium	Unfair	SNR	High	
M-BEST-M [59]	$(\lceil \log_2 N_{bins} \rceil \times S + U) \times K' + S \times U \times K \approx 0.16 \times K' \times N_{bins} \times U$	Medium	Medium	Unfair	SNR	High	In the formula K and K' can be equal
M2-BEST-M [62]	$\approx 0.17 \times K' \times N_{bins} \times U$	Medium	Medium	Unfair	SNR	Higher	
LOSSY [79]	$\approx 0.5 \times K' \times N_{bins}$	Medium–high	Medium–high	Fair	SNR	Higher	Since feedbacks are sent by all the users, fairness algorithms can be applied
HUFFMAN [47]	$1.1 \times K' \times N_{bins}$	Low	Low	Fair	MCS	Higher	Since feedbacks are sent by all the users (removing from [47] the opportunistic part), fairness algorithms can be applied

a Assuming high time correlation (for the lower bound) and so the maximum of the integer part 1, and 1 bit for the fractional part (both limits).

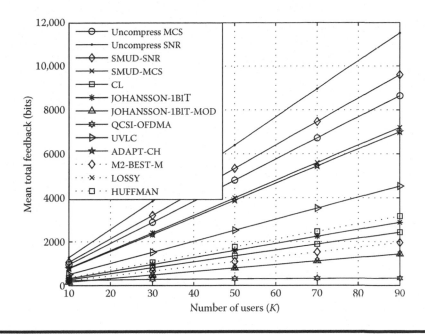

Figure 11.10 Average total number of feedback bits for different algorithms as a function of number of users.

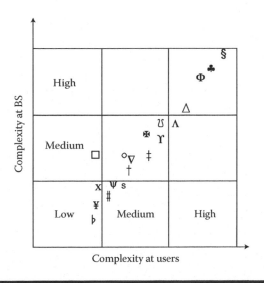

Figure 11.11 Complexity comparison of the feedback algorithms. x—SMUD-SNR [48], s—SMUD-MCS [48], □—M-SMUD-SNR [52], ◇—M-SMUD-MCS [53], §—MIMO-SMUD [54], †—CL [17], ‡—CL-BF [57], ♭—CA-ALOHA [64], ¥—1BIT-JOHANSSON [68], ♯—1BIT-JOHANSSON-MOD [70], ♣—QCSI-OFDMA [72], ∇—DIFF-T [75], △—UVLC [77], Λ—ADAPT-CH [41], ⊠—BEST-M ⊠ [60], Υ—M-Best-M [59], ℧—M2-BEST-M [62], Φ—Lossy [79], Ψ—HUFFMAN [47].

at the user terminal and at the base station is shown, it can be seen that the simplest algorithms at both sides (especially at the user terminal) are in Refs. [17,47,48,64,68,70].

Another important aspect is the possibility of implementing fairness policies or the maximum throughput attainable. At the end, a tradeoff between different criteria will configure the final decision on what algorithm to choose.

11.5 Summary, Open Issues, and Challenges

We have shown that the time–frequency structure of OFDMA signals lends itself very well to adaptation because modulation, coding, and power may be changed independently for each subcarrier (or group of subcarriers) provided that proper channel information is made available. This adaptation constitutes a way of approaching channel capacity, and besides in environments with several users it provides an additional gain known as multiuser diversity.

In general, CSI is obtained at the receiver side and must be conveyed to the transmitter through a feedback channel. We have seen that the amount of channel knowledge and the degree of adaptation that may be practically achieved depend on the scenario where communication takes place and we have examined several examples.

Adaptive modulation in OFDMA is often referred to as bit and power loading. We have explained the algorithms developed for OFDMA that may be found in the literature. Some algorithms perform the joint optimization of subcarriers, bits, and power. Some others proceed in two steps: first subcarrier allocation is performed and then the bits and power are loaded to subcarriers once they are assigned to users.

We have examined feasible ways of conveying CSI in the feedback channel and explained compression mechanisms that are useful in minimizing the transmission rate of this channel.

There are several open issues and challenges associated with this topic. Regarding bit and power loading algorithms for OFDMA, in general they are designed taking into account a single cell with a base station transmitting to the users. However, the whole system is composed of a multiplicity of cells and the algorithms in the literature very seldom consider the complete perspective. The work in Ref. [37] considers frequency reuse and Zhang and Letaief [38] introduce adaptive cell selection obtaining intercell as well as intracell multiuser diversity. In future research, probably the cooperation of the neighboring base stations will be exploited to a higher degree. Then channel knowledge must be shared among base stations, which looks feasible through the communication backbone.

Another important challenge is the appearance of standards and products that bring it all to reality. Nowadays the adaptation capabilities of existing standards and systems are rather limited. It seems that LTE, defining the direction of 3GPP (Third Generation Partnership Project) toward fourth-generation mobile communications, will embrace multiuser OFDM with a much higher degree of adaptability.

References

1. K.L. Baum, T.A. Kostas, P.J. Sartori, and B.K. Classon. Performance characteristics of cellular systems with different link adaptation strategies. *IEEE Transactions on Vehicular Technology*, 52(6):1497–1507, 2003.
2. M. Sternad, T. Svensson, T. Ottosson, A. Ahlen, A. Svensson, and A. Brunstrom. Towards systems beyond 3G based on adaptive OFDMA transmission. *Proceeding of IEEE*, 95(12):2432–2455, 2007.

3. C.Y. Wong, R.S. Cheng, K.B. Letaief, and R.D. Murch. Multiuser OFDM with adaptative subcarrier, bit, and power allocation. *IEEE Journal on Selected Areas in Communications (JSAC)*, 17(10):1747–1758, 1999.

4. G. Li and H. Liu. On the optimality of the OFDMA network. *IEEE Communications Letters*, 9(5):438–440, 2005.

5. L. Hanzo, W. Webb, and T. Keller. *Single- and Multi-carrier Quadrature Amplitude Modulation: Principles and Applications for Personal Communications, WLANs and Broadcasting*. Wiley, New York, 1 edition, June 2000.

6. R.G. Gallager. *Information Theory and Realiable Communication*. Wiley, New York, 1 edition, 1968.

7. R.S. Cheng and S. Verdú. Gaussian multiaccess channels with ISI: Capacity region and multiuser water-filling. *IEEE Transactions on Information Theory*, 39(3):773–785, 1993.

8. D.N. Tse and P. Viswanath. *Fundamentals of Wireless Communication*. Cambridge University Press, Cambridge, 2005.

9. D. Piazza and L.B. Milstein. Analysis of multiuser diversity in time-varying channels. *IEEE Transactions on Wireless Communications*, 6(12):4412–4419, 2007.

10. E. Yoon, D. Tujkovic, and A. Paulraj. Statistical opportunistic scheduling with tap correlation information for an ofdma system in uplink. *IEEE Transactions on Vehicular Technology*, 57(3):1708–1714, 2008.

11. IEEE. WLAN medium access control (MAC) and physical layer (PHY) specifications: High-speed physical layer in the 5 GHz Band. Technical Report 802.11a, IEEE, June 2001.

12. IEEE. IEEE 802.16e: Air interface for fixed broadband wireless access systems. Technical Report, IEEE, 2005.

13. 3rd Generation Partnership Project. Evolved universal terrestrial radio access (E-UTRA) and evolved universal terrestrial radio access network (E-UTRAN): Overall description. Stage 2 (Release 8). Technical Report V 8.5.0, 3GPP TS 36.300, May 2008.

14. D2.10: Final report on identified RI key technologies, system concept, and their assessment. Technical Report, IST-2003-507581 WINNER, December 2005.

15. D2.4: Assessment of adaptive transmission technologies. Technical Report, IST-2003-507581 WINNER, February 2005.

16. Wireless IP Project. [Online]. Available: www.signal.uu.se/Research/PCCwirelessIP.html

17. P. Svedman, S.K. Wilson, L. J. Cimini Jr, and B. Ottersten. A simplified opportunistic feedback and scheduling scheme for OFDM. In *Proceedings of IEEE Vehicular Technology Conference. VTC Spring*, Vol. 4, pp. 1878–1882, 17–19 2004.

18. PACWOMAN. PACWOMAN: Power Aware Communication for Wireless OptiMised personal Area Networks (IST-2001-34157). Proposal, 2001.

19. H. Ekström, A. Furuskär, J. Karlsson, M. Meyer, S. Parkvall, J. Torsner, and M. Wahlqvist. Technical solutions for the 3G long-term evolution. *IEEE Commun. Mag.*, 44(3):38–45, 2006.

20. J.M. Cioffi, G. D. Dudevoir, M. V. Eyubouglu, and G. D. Forney Jr. MMSE decision-feedback equalizers and coding – Part II: Coding results. *IEEE transactions on Communications*, 43(10):2595–2604, 1995.

21. A. G. Armada. SNR gap approximation for M-PSK-based bit loading. *IEEE Transactions on Wireless Communications*, 5(1):57–60, 2006.

22. P.S. Chow. Bandwidth Optimized Digital Transmission Techniques for Spectrally Shaped Channels. PhD thesis, Standford University, May 1993.

23. W. Yu and J.M. Cioffi. Constant-power waterfilling: Performance bound and low-complexity implementation. *IEEE Transactions on Communications*, 54(1):23–28, 2006.

24. D. Hughes-Hartogs. Ensemble modem structure for imperfect transmission media. U.S. Patents no. 4 679 227-4 731 816, March 1988.

25. P.S. Chow, J.M. Cioffi, and J.A.C. Bingham. A practical discrete multitone transceiver loading algorithm for data transmission over spectrally shaped channels. *IEEE Transactions on Communications*, 43(2/3/4):773–775, 1995.

26. J. Campello. Practical bit loading for DMT. In *Proceedings of IEEE International Conference on Communications (ICC)*, Vol. 2, pp. 801–805, 1999.

27. D. Dardari. Ordered subcarrier selection algorithm for OFDM-based high-speed wlans. *IEEE Transactions on Wireless Communications*, 3(5):1452–1458, 2004.

28. L.M.C. Hoo. Multiuser Transmit Optimization for Multicarrier Modulation Systems. PhD thesis, Standford University, December 2000.

29. C. Suh and J. Mo. Resource allocation for multicast services in multicarrier wireless communications. *IEEE Transactions on Wireless Communications*, 7(1):27–31, 2008.

30. D. Kivanc, G. Li, and H. Liu. Computationally efficient bandwidth allocation and power control for OFDMA. *IEEE Transactions on Wireless Communications*, 2(6):1150–1158, 2003.

31. Z. Shen, J.G. Andrews, and B.L. Evans. Adaptive resource allocation in multiuser OFDM systems with proportional rate constraints. *IEEE Transactions on Wireless Communications*, 4(6):2726–2737, 2005.

32. I. Kim, H.L. Lee, B. Kim, and Y.H. Lee. On the use of linear programming for dynamic subchannel and bit allocation in multiuser OFDM. In *Proceedings of the IEEE Global Telecommunications Conference (GLOBECOM)*, pp. 3648–3652, 2001.

33. K. el Baamrani, A.A. Ouahman, V.P.G. Jiménez, A.G. Armada, and S. el Allaki. Subcarrier and power allocation for the downlink of multiuser ofdm transmission. *Wireless Personal Communications*, 39(4):457–465, 2006.

34. Z. YuMing, G. JingBo, W. ZanJi, and X. Yi. Novel sub-carrier and power allocation algorithm for multiuser OFDM systems. In *Proceedings of the IEEE Global Telecommunications Conference (GLOBECOM)*, pp. 5639–5642, 2006.

35. C. Mohanram and S. Bhashyam. A sub-optimal joint subcarrier and power allocation algorithm for multiuser OFDM. *IEEE Communications Letters*, 9(8):685–687, 2005.

36. N.Y. Ermolova and B. Makarevitch. Low complexity adaptive power and subcarrier allocation for OFDMA. *IEEE Transactions on Wireless Communications*, 6(2):433–437, 2007.

37. H. Kim, Y. Han, and J. Koo. Optimal subchannel allocation scheme in multicell OFDMA systems. In *Proceedings of IEEE Vehicular Technology Conference (VTC) – Spring*, Vol. 3, pp. 1821–1825, May 2004.

38. Y.J. Zhang and K.B. Letaief. Multiuser adaptive subcarrier-and-bit allocation with adaptive cell selection for OFDM systems. *IEEE Transactions on Wireless Communications*, 3(5):1566–1575, 2004.

39. J. Gross, H-F. Geerdes, H. Karl, and A. Wolisz. Performance analysis of dynamic OFDMA systems with inband signaling. *IEEE Journal on Selected Areas in Communications (JSAC)*, 24(3):427–436, 2006.

40. M. Einhaus and M. Schinnenburg. Impact of signaling constraints on the MAC level performance of OFDMA systems. In *Proceedings of the International Conference on Wireless Communications, Networking and Mobile Computing (WiCom)*, pp. 747–750, September 2007.

41. H. Oh and H-M Kim. An adaptive determination of channel information feedback period in OFDMA systems. In *Proceedings of the IEEE 10th International Symposium on Consumer Electronics*, pp. 1–5, 2006.

42. M. Olfat, F.R. Farrokhi, and K.J.R. Liu. Power allocation for OFDM using adaptive beamforming over wireless networks. *IEEE Transactions on Communications*, 53(3):505–514, 2005.

43. M. Costa. Writing on dirty paper (corresp.). *IEEE Transactions on Information Theory*, 29(3):439–441, 1983.

44. R.D. Wesel and J.M. Cioffi. Achievable rates for tomlinson-harashima precoding. *IEEE Transactions on Information Theory*, 44(2):824–831, 1998.

45. D. Gesbert, M. Kountouris, R.W.Heath Jr., C.-B. Chae, and T. Sälzer. Shifting the MIMO paradigm. *IEEE Signal Processing Magazine*, 24(5):36–46, 2007.

46. T. Eriksson and T. Ottosson. Compression of feedback for adaptive transmission and scheduling. *Proceeding of IEEE*, 95(12):2314–2321, 2007.

47. V.P.G. Jiménez, T. Eriksson, A.G. Armada, M.J.F-G. García, T. Ottosson, and A. Svensson. Methods for compression of feedback in adaptive multi-carrier 4G schemes. *Wireless Personal Communications*, 47(1):101–112, 2008.

48. D. Gesbert and M.-S. Alouini. How much feedback is multiuser diversity really worth? In *Proceedings IEEE International Conference on Communications (ICC)*, Vol. 1, pp. 234–238, June 2004.

49. L. Yang, M.-S. Alouini, and D. Gesbert. Further results on selective multiuser diversity. In *Proceedings of 7th ACM International Symposium on Modeling, Analysis and Simulation of Wireless and Mobile Systems*, 2004.

50. R. Knopp and P.A. Humblet. Information capacity and power control in single-cell multiuser communications. In *Proceedings of the IEEE International Conference on Communications (ICC)*, Vol. 1, pp. 331–335, Seatle, June 1995.

51. M. Johansson and M. Sternad. Resource allocation under uncertainty using the maximum entropy principle. *IEEE Transactions on Information Theory*, 51(12):4103–4117, 2005.

52. V. Hassel, M.-S. Alouini, G.E. Øien, and D. Gesbert. Rate-optimal multiuser scheduling with reduced feedback load and analysis of delay effects. *EURASIP Journal on Wireless Communications and Networking*, 7 p, 2006. doi:10.1155/WCN/2006/36424.

53. V. Hassel, D. Gesbert, M.-S. Alouini, and G.E. Øien. A threshold-based channel state feedback algorithm for modern cellular systems. *IEEE Transactions on Wireless Communications*, 6(7):2422–2426, 2007.

54. J.L. Vicario and C. Antón-Haro. Robust explotation of spatial and Multiuser diversity in limited-feedback systems. In *Proceedings IEEE International Conference Audio, Speech and Signal Processing (ICASSP)*, Vol. 3, pp. 417–420, May 2005.

55. J.L Vicario, R. Bosisio, U. Spagnolini, and C. Anton-Haro. A throughtput analysis for opportunistic beamforming with quantized feedback. In *Proceedings of the 17th Annual IEEE International Symposium on Personal, Indoor and Mobile Radio Communications (PIMRC)*, 2006.

56. P. Svedman, L.J. Cimini Jr., M. Bengtsson, S.K. Wilson, and B. Ottersten. Exploiting temporal channel correlation in opportunistic SD-OFDMA. In *Proceedings of the IEEE International Conference on Communications*, Vol. 12, pp. 5307–5312, June 2006.

57. P. Svedman, S. Kate Wilson, L.J. Cimini Jr., and B. Ottersten. Opportunistic beamforming and scheduling for OFDMA systems. *IEEE Transactions on Communications*, 55(5):941–952, 2007.

58. P. Svedman, S.K. Wilson, L.J. Cimini Jr., and B. Ottersten. Corrections to opportunistic beamforming and scheduling for OFDMA systems. *IEEE Transactions on Communications*, 55(6):1266–1266, 2007.

59. J.H. Kwon, D. Rhee, I.M. Byun, K.S. Kim, and K.C. Whang. Efficient adaptive transmission technique for multiuser OFDMA systems with reduced feedback rate. In *Proceedings of the Wireless Telecommunications Symposium (WTS)*, pp. 1–5, April 2006.

60. Z.-H. Han and Y.-H. Lee. Opportunistic scheduling with partial channel information in OFDMA/FDD systems. In *Proceedings of the IEEE Vehicular Technology Conference (VTC)-Fall*, Vol. 1, pp. 511–514, September 2004.

61. 3GPP TR 25.813. Physical layer aspects for evolved Universal Terrestrial Radio Access (UTRA). Technical Report Release 7.1, 3GPP, September 2006.

62. X. Wu, J. Zhou, G. Li, and M. Wu. Low overhead CQI feedback in multi-carrier systems. In *Proceedings of the IEEE Global Telecommunications Conference (GLOBECOM)*, pp. 371–375, November 2007.

63. X. Qin and R.A. Berry. Distributed approaches for exploiting multiuser diversity in wireless networks. *IEEE Transactions on Information Theory*, 52(2):392–413, 2006.

64. X. Qin and R. Berry. Distributed power allocation and scheduling for parallel channel wireless networks. In *Proceedings of the International Symposium on Modeling and Optimization in Mobile, Ad Hoc, and Wireless Networks (WIOPT)*, pp. 77–85, April 2005.

65. C. Tepedelenliglu and G.B. Giannakis. On velocity estimation and correlation properties of narrow-band mobile communication channels. *IEEE Transactions on Vehicular Technology*, 49(12):344–353, 2003.

66. S. Iijima, R. Zhou, and I. Sasase. Reduction of feedback overload by exploiting adaptive channel in OFDMA systems. In *Proceedings of the IEEE Pacific Rim Conference on Communications, Computers and Signal Processing (PacRim)*, pp. 153–156, August 2007.

67. F. Floren, O. Edfors, and B.A. Molin. The effect of feedback quantization on the throughput of a multiuser diversity scheme. In *Proceedings of the IEEE Global Conference on Communications (GLOBECOM)*, 2003.

68. M. Johansson. Benefits of multiuser diversity with limited feedback. *IEEE 4th Workshop on Signal Processing Advanced in Wireless Communications*, pp. 155–159, 2003.

69. S. Sanayei and A. Nosratinia. Exploiting multiuser diversity with only 1-bit feedback. In *Proceedings of the IEEE Wireless Communications and Networking Conference (WCNC)*, Vol. 2, pp. 978–983, March 2005.

70. M. Johansson. Diversity-enhanced equal access – considerable throughput gains with 1-bit feedback. In *Proceedings of IEEE Workshop on Signal Processing Advances in Wireless Communications (WSPAWC)*, Lisboa, Portugal, 2004.

71. M. Johansson. On Scheduling and Adaptive Modulation with Limited Feedback. PhD thesis, Signal and Systems, Uppsala University, Sweden, 2004.

72. A.G. Marques, G.B. Giannakis, F.F. Digham, and F.J. Ramos. Power-efficient wireless ofdma using limited-rate feedback. *IEEE Transactions on Wireless Communications*, 7(2):685–696, 2008.

73. A.J. Goldsmith and S.-G. Chua. Variable-rate variable-power mqam for fading channels. *IEEE Transactions on Communications*, 45(10):1218–2130, 1997.

74. W. Wang, T. Ottosson, M. Sternad, A. Ahlen, and A. Svensson. Impact of multiuser diversity and channel variability on adaptive OFDM. In *Proceedings of IEEE Vehicular Technology Conference (VTC Fall)*, Vol. 1, pp 547–551, September 2003.

75. H. Cheon, B. Park, and D. Hong. Adaptive multicarrier system with reduced feedback information in wideband radio channels. *Proceedings of the IEEE 50th Vehicular Technology Conference. VTC Fall*, 5, pp. 2880–2884, September 1999.

76. T. Lestable and M. Bartelli. LZW adaptive bit loading. In *Proceedings of the International Symposium on Advances Wireless Communications (ISAWC)*, 2002.

77. H. Nguyen, J. Brouet, V. Kumar, and T. Lestable. Compression of associated signaling for adaptive multicarrier systems. In *Proceedings of the IEEE Vehicular Technology Conference (VTC). Spring*, pp. 1916–1919, 2004.

78. Y. ltoh and N.M. Cheung. Universal variable length code/or dct coding. In *Proceedings of the IEEE International Conference on Image Processing*, Vol. 1, pp. 940–943, 2000.

79. T. Eriksson and T. Ottosson. Compression of feedback in adaptive OFDM-based systems using scheduling. *IEEE Communications Letters*, 11(11):859–861, 2007.

80. M. Stenard, T. Svensson, and G. Klang. WINNER MAC for cellular transmission. In *Proceedings of IST Mobile Summit*, Mikonos, Greece, June 2006.

81. S.P. Lloyd. Least square quantization in pcm. *IEEE Transactions on Information Theory*, 28(2):129–136, 1982.

82. D.A. Huffman. A method for the construction of minimum-redundancy codes. *Proceedings of I.R.E*, pp. 1098–1102, September 1952.

83. Z. Jiang, H. Mason, B.J. Kim, N.K. Shankaranarayanan, and P. Henry. A subjective survey of user experience for data applications for future cellular wireless networks. In *Proceedings IEEE Symposium Applications and the Internet*, pp. 167–175, January 2001.

84. N. Enderlt and X. Lagrange. User satisfaction models and scheduling algorithms for packet-switched services in UMTS. In *Proceedings of the IEEE Vehicular Technology Conference (VTC – Spring)*, Vol. 3, pp. 1704–1709, April 2003.

85. F. Kelly. Charging and rate control for elastic traffic. *European Transactions on Telecommunications*, 8:33–37, 1997.

86. H.J. Bang, T. Ekman, and D. Gesbert. Channel predictive proportional fair scheduling. *IEEE Transactions on Wireless Communications*, 7(2):482–487, 2008.

Chapter 12

Training Sequence Design in Multiuser OFDM Systems

Jianwu Chen, Yik-Chung Wu, Tung-Sang Ng,
and Erchin Serpedin

Contents

12.1   Introduction..330
12.2   System Model..331
12.3   CRB and Asymptotic CRB..332
      12.3.1   Finite Sample CRB..333
      12.3.2   Asymptotic CRB 334
12.4   Optimal Condition for the Training Sequence 335
      12.4.1   Training Design for Channel Estimation Only 335
      12.4.2   Training Design for Frequency Offset Estimation Only...336
            12.4.2.1   Channel-Dependent Training..........................336
            12.4.2.2   Channel-Independent Training.......................337
            12.4.2.3   Summary of Training Design for CFO Estimation 338
      12.4.3   Training Design for Joint Frequency and Channel Estimation...............339
12.5   Realization of Optimal Training...340
      12.5.1   Time-Domain Realizations...340
      12.5.2   Frequency-Domain Realizations 341
            12.5.2.1   Frequency Division Multiplexing Sequences.....................342
            12.5.2.2   Code Division Multiplexing Sequences in the Frequency
                    Domain..343
      12.5.3   Performance Comparisons 344
12.6   Concluding Remarks and Open Issues 346
References..346

12.1 Introduction

Multiuser orthogonal frequency-division multiplexing (OFDM) systems have attracted much attention recently and are widely regarded as potential candidates for fourth-generation (4G) broadband wireless communication systems due to their ability to mitigate intersymbol interference (ISI) efficiently and enhance system capacity [1,2]. Depending on the mode of multiple access, multiuser OFDM systems can be implemented in two popular architectures: orthogonal frequency-division multiple access (OFDMA) [3,4] and OFDM with spatial multiplexing [5–7].

In OFDMA systems, several users share a spectrum simultaneously by modulating exclusive sets of orthogonal subcarriers; thus each user's signal can be easily separated in the frequency domain. OFDMA allows straightforward dynamic channel assignment, and, furthermore, multiuser and frequency diversities can be easily exploited. On the other hand, in OFDM spatial multiplexing systems, independent data from several users are transmitted on the same set of subcarriers simultaneously, thus promising higher spectral efficiency. The combination of multiple-input multiple-output (MIMO) architecture and OFDM basically turns a frequency-selective MIMO channel to a set of frequency-flat MIMO channels. This makes all the MIMO techniques in a flat-fading channel also applicable in a frequency-selective fading channel.

In general, to fully exploit the potentials of multiuser OFDM systems, two critical issues should be considered: frequency synchronization and channel estimation. Similar to other multicarrier-based techniques, multiuser OFDM systems are highly sensitive to carrier frequency offsets (CFOs) caused by oscillator mismatches and/or Doppler shifts [8]. Without accurate frequency synchronization, the orthogonality among subcarriers is lost, and intercarrier interference (ICI) as well as multiple-access interference (MAI) will be introduced. In addition, accurate channel estimation is indispensable for coherent data detection in multiuser OFDM systems. In OFDMA systems, channel estimates are used for per-tone equalization, while for OFDM with spatial multiplexing systems, channel estimates are used for separating multiple users' data.

For multiuser OFDM systems, CFO and channel estimation are relatively simple in the downlink, as each user only needs to estimate one CFO and one channel vector of its own. The real challenge exists in the uplink multiuser OFDM systems, where signals from multiple users pass through distinct channels and with different CFOs overlapped together. In the uplink, the CFO and channel estimation problem is a multiparameter and high-dimensional estimation problem [9–12].

In order to aid the estimation of CFOs and channel coefficients, pilot symbols are usually used. Many different kinds of pilot symbols have been proposed in the literature, including preamble-based training [13–16], superimposed training [17–19], and pilot symbol-assisted modulation (PSAM) [20–23]. While each category of training has its own advantages, in this chapter we focus on preamble-based training, since a lot of current OFDM-based standards employ this kind of training (e.g., various wireless local area networks (WLANs) [24–26] and worldwide interoperability for microwave access (WiMAX) [4]).

In this chapter, the optimal training design for channel and CFO estimations in multiuser OFDM systems will be discussed. First the Cramer–Rao bound (CRB) and asymptotic Cramer–Rao bound (asCRB) for channel and CFO estimations will be presented. Then, the condition for optimal training will be derived by minimizing the CRB or asCRB. Three cases will be considered in deriving the optimal condition: (1) channel estimation only, (2) CFO estimation only, and (3) joint channel and CFO estimation. It turns out that the optimal conditions for all three cases are basically the same. After that, several types of training sequences that satisfy the optimal condition will be discussed. Performance comparisons between different types of training sequences will also be presented.

The following notations are used throughout this chapter. Symbol $j \triangleq \sqrt{-1}$. The operator diag$\{\mathbf{x}\}$ denotes a diagonal matrix with the elements of $\mathbf{x}$ located on the main diagonal, while $\Re\{\cdot\}$ and $\Im\{\cdot\}$ take the real and imaginary part of the argument, respectively. Superscripts $(\cdot)^{-1}, (\cdot)^{\mathrm{T}}$, and $(\cdot)^{\mathrm{H}}$ denote the inverse, transpose, and conjugate transpose operations, respectively. Operators $\otimes$ and Tr$\{\cdot\}$ represent the Kronecker product and trace of a matrix, respectively. The $k \times k$ identity and zero matrix are denoted by $\mathbf{I}_k$ and $\mathbf{0}_k$, respectively, and $\|\mathbf{x}\|$ represents the norm operation for a vector $\mathbf{x}$. Notation $\mathbb{E}\{\cdot\}$ is the expectation of a random variable, and $\lambda_{\min}\{\cdot\}$ is the smallest eigenvalue of the matrix in the argument. The symbol $\delta(k)$ is the discrete unit impulse function.

12.2 System Model

In this chapter, a packet-based multiuser OFDM uplink system is considered. The packet structure is shown in Figure 12.1. In each of the packets, the first section is dedicated for training and the rest are data OFDM symbols. The preamble consists of two portions. The first part is for packet detection, automatic gain control (AGC), and frame synchronization. The second part is for channel and CFO estimations. This chapter focuses on designing the second portion of the preamble. Without loss of generality, it is assumed that the second portion of the preamble comprises one OFDM symbol.

In the considered multiuser OFDM system, K users transmit different data streams simultaneously using the same set of subcarriers to the base station (BS), which is equipped with M_{r} antennas and is responsible for decoding the symbols for each user. For each user, only one transmit antenna is assumed. It is reasonable to assume that the receive antennas at the BS share the same oscillator while all users are driven by different oscillators. Thus the data streams from different users experience different CFOs. Before initiating the transmission, the timing for each user is acquired by using the downlink synchronization channel from the BS. Consequently, transmissions from all users can be regarded as quasi-synchronous [16].

The data stream from user k is first segmented into blocks of length N (denoted as $\mathbf{d}_k = [d_k(0), \ldots, d_k(N-1)]^{\mathrm{T}}$) and then modulated onto different subcarriers by left multiplying the N-point inverse fast Fourier transform (FFT) matrix $\mathbf{F}^{\mathrm{H}}$, where $\mathbf{F}$ is the FFT matrix with $\mathbf{F}(p,q) = (1/\sqrt{N}) \exp(-(j2\pi pq/N))$. After inserting a cyclic prefix (CP) of length L_{cp} into each block of the time-domain signal (denoted as $\mathbf{F}^{\mathrm{H}}\mathbf{d}_k$), the augmented block is serially transmitted through the multipath channel. Let the channel impulse response (including all transmit/receive filtering effects) between the kth user and the jth receive antenna be denoted as $\boldsymbol{\xi}_{k,j} = [\xi_{k,j}(0), \ldots, \xi_{k,j}(L_{k,j}-1)]^{\mathrm{T}}$, where $L_{k,j}$ is the channel length. Denoting the timing offset caused by propagation delay as $\theta_{k,j}$, the compound channel response can be written as $\mathbf{h}_{k,j} \triangleq [\mathbf{0}_{\theta_{k,j} \times 1}^{\mathrm{T}} \quad \boldsymbol{\xi}_{k,j}^{\mathrm{T}} \quad \mathbf{0}_{(L-\theta_{k,j}-L_{k,j}) \times 1}^{\mathrm{T}}]^{\mathrm{T}}$, where L is the upper bound on the compound channel length. Since for coherent data detection, only the estimate of the compound channel $\mathbf{h}_{k,j}$ is required, the estimation of timing offset becomes

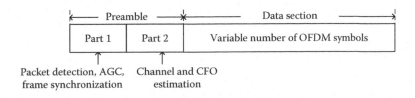

Figure 12.1 **Packet structure of the multiuser OFDM system.**

dispensable. Through combining the timing offset $\theta_{k,j}$ and the exact channel $\xi_{k,j}$, the system can be regarded as a perfectly synchronous system with the compound channel $\mathbf{h}_{k,j}$. For user k, the normalized CFO (between the oscillator of user k and that of the BS) is denoted as ε_k. At the BS, after removal of CP, the signal from user k to the jth receive antenna is given by

$$\mathbf{x}_{k,j} \triangleq [x_{k,j}(0), \ldots, x_{k,j}(N-1)]^{\mathrm{T}} = \mathbf{\Gamma}(\omega_k)\mathbf{A}_k\mathbf{h}_{k,j}, \tag{12.1}$$

where

$$\mathbf{\Gamma}(\omega_k) \triangleq \mathrm{diag}\{[1, \ \exp(j\omega_k), \ldots, \exp(j(N-1)\omega_k)]\}, \tag{12.2}$$

$$\mathbf{A}_k = \mathbf{F}^{\mathrm{H}}\mathbf{D}_k\mathbf{W}, \tag{12.3}$$

$$\mathbf{D_k} \triangleq \mathrm{diag}\{\mathbf{d_k}\}, \tag{12.4}$$

$$\omega_k \triangleq 2\pi\varepsilon_k/N, \tag{12.5}$$

$$\mathbf{W} = \sqrt{N}\mathbf{F}(:, 1:L) \tag{12.6}$$

with $\mathbf{F}(:, 1:L)$ denoting the first L columns of $\mathbf{F}$. Since the received signal at the jth receive antenna of the BS is the sum of signals from all users and noise, the received signal at this antenna is given by

$$\mathbf{x}_j = \sum_{k=1}^{K}\mathbf{x}_{k,j} + \mathbf{n}_j = \sum_{k=1}^{K}\mathbf{\Gamma}(\omega_k)\mathbf{A}_k\mathbf{h}_{k,j} + \mathbf{n}_j. \tag{12.7}$$

In the above, vector $\mathbf{n}_j$ represents complex white Gaussian noise with zero mean and covariance matrix $\sigma^2\mathbf{I}_N$.

Denoting $\mathbf{x} = [\mathbf{x}_1^{\mathrm{T}} \ \mathbf{x}_2^{\mathrm{T}} \ldots \mathbf{x}_{M_{\mathrm{r}}}^{\mathrm{T}}]^{\mathrm{T}}$, $\mathbf{\omega} = [\omega_1 \ \omega_2 \ldots \omega_K]$, $\mathbf{n} = [\mathbf{n}_1^{\mathrm{T}} \ \mathbf{n}_2^{\mathrm{T}} \ldots \mathbf{n}_{M_{\mathrm{r}}}^{\mathrm{T}}]^{\mathrm{T}}$, $\mathbf{h} = [\mathbf{h}_1^{T}\mathbf{h}_2^{T} \ldots \mathbf{h}_{M_{\mathrm{r}}}^{T}]^{\mathrm{T}}$ with $\mathbf{h}_j = [\mathbf{h}_{1,j}^{\mathrm{T}} \ \mathbf{h}_{2,j}^{\mathrm{T}} \ldots \mathbf{h}_{K,j}^{\mathrm{T}}]^{\mathrm{T}}$, the signal model from Equation 12.7 can be compactly rewritten as

$$\mathbf{x} = \mathbf{Q}(\mathbf{\omega})\mathbf{h} + \mathbf{n}, \tag{12.8}$$

where

$$\mathbf{Q}(\mathbf{\omega}) = \mathbf{I}_{M_{\mathrm{r}}} \otimes [\mathbf{\Gamma}(\omega_1)\mathbf{A}_1 \ldots \mathbf{\Gamma}(\omega_K)\mathbf{A}_K]. \tag{12.9}$$

Remark 1: The above system model is valid for OFDMA, OFDM spatial multiplexing, or a combination of both systems. In OFDMA systems, only one user transmits in each subcarrier. In OFDM spatial multiplexing systems, each subcarrier is shared by more than one user. If some subcarriers are dedicated to a single user, while some subcarriers are shared by several users, then the system is a combination of both OFDMA and OFDM spatial multiplexing systems.

12.3 CRB and Asymptotic CRB

The optimal training sequence design in this chapter is based on CRB analyses because the optimality of the resulting sequences would be independent of the estimation method. Furthermore, the maximum likelihood (ML) estimator can asymptotically approach the CRB [14,27,28], meaning that the performance limit given by the optimal sequence is reachable in practice.

12.3.1 Finite Sample CRB

Since the additive noise $\mathbf{n}$ in the signal model in Equation 12.8 is circular Gaussian and white, the received signal $\mathbf{x}$ is circular complex Gaussian with mean $\boldsymbol{\mu} \triangleq \mathbf{Q}(\boldsymbol{\omega})\mathbf{h}$ and covariance matrix $\sigma^2 \mathbf{I}_{M_r N}$. The Fisher information matrix (FIM) for $\boldsymbol{\eta} \triangleq [\boldsymbol{\omega}, \Re\{\mathbf{h}\}, \Im\{\mathbf{h}\}]$ is given by [29]

$$
\begin{aligned}
\boldsymbol{\Phi} &= \frac{2}{\sigma^2} \Re\left[\frac{\partial \boldsymbol{\mu}^H}{\partial \boldsymbol{\eta}} \frac{\partial \boldsymbol{\mu}}{\partial \boldsymbol{\eta}^T}\right] \\
&= \frac{2}{\sigma^2} \begin{bmatrix} \Re\{\mathbf{Z}^H\mathbf{Z}\} & \Im\{\mathbf{Z}^H\mathbf{Q}\} & \Re\{\mathbf{Z}^H\mathbf{Q}\} \\ -\Im\{\mathbf{Q}^H\mathbf{Z}\} & \Re\{\mathbf{Q}^H\mathbf{Q}\} & -\Im\{\mathbf{Q}^H\mathbf{Q}\} \\ \Re\{\mathbf{Q}^H\mathbf{Z}\} & \Im\{\mathbf{Q}^H\mathbf{Q}\} & \Re\{\mathbf{Q}^H\mathbf{Q}\} \end{bmatrix},
\end{aligned}
\tag{12.10}
$$

where $\mathbf{Q}$ denotes $\mathbf{Q}(\boldsymbol{\omega})$ for expression simplicity and

$$
\mathbf{Z} = (\mathbf{I}_{M_r} \otimes \mathbf{M})\mathbf{Q}\mathbf{H},
\tag{12.11}
$$

with

$$
\mathbf{M} = \text{diag}\{[0, 1, \ldots, N-1]\},
\tag{12.12}
$$

$$
\mathbf{H} = [\mathbf{H}_1^T \, \mathbf{H}_2^T \ldots \mathbf{H}_{M_r}^T]^T,
\tag{12.13}
$$

$$
\mathbf{H}_j = \text{diag}\{[\mathbf{h}_{1,j} \, \mathbf{h}_{2,j} \, \ldots \, \mathbf{h}_{K,j}]\}.
\tag{12.14}
$$

The CRB matrix can be obtained by inverting the above FIM $\boldsymbol{\Phi}$. Through some manipulations, the result is given by

$$
\text{CRB} = \frac{\sigma^2}{2} \cdot
$$

$$
\begin{bmatrix} \boldsymbol{\gamma}^{-1} & \boldsymbol{\gamma}^{-1}\Im\{\boldsymbol{\beta}^T\} & -\boldsymbol{\gamma}^{-1}\Re\{\boldsymbol{\beta}^T\} \\ \Im\{\boldsymbol{\beta}\}\boldsymbol{\gamma}^{-1} & \Re\{(\mathbf{Q}^H\mathbf{Q})^{-1}\} + \Im\{\boldsymbol{\beta}\}\boldsymbol{\gamma}^{-1}\Im\{\boldsymbol{\beta}^T\} & -\Im\{(\mathbf{Q}^H\mathbf{Q})^{-1}\} - \Im\{\boldsymbol{\beta}\}\boldsymbol{\gamma}^{-1}\Re\{\boldsymbol{\beta}^T\} \\ -\Re\{\boldsymbol{\beta}\}\boldsymbol{\gamma}^{-1} & \Im\{(\mathbf{Q}^H\mathbf{Q})^{-1}\} - \Re\{\boldsymbol{\beta}\}\boldsymbol{\gamma}^{-1}\Im\{\boldsymbol{\beta}^T\} & \Re\{(\mathbf{Q}^H\mathbf{Q})^{-1}\} + \Re\{\boldsymbol{\beta}\}\boldsymbol{\gamma}^{-1}\Re\{\boldsymbol{\beta}^T\} \end{bmatrix},
$$

$$
\tag{12.15}
$$

where

$$
\boldsymbol{\beta} = (\mathbf{Q}^H\mathbf{Q})^{-1}\mathbf{Q}^H\mathbf{Z},
\tag{12.16}
$$

$$
\boldsymbol{\gamma} = \Re\{\mathbf{Z}^H \boldsymbol{\Pi}_Q^\perp \mathbf{Z}\},
\tag{12.17}
$$

$$
\boldsymbol{\Pi}_Q^\perp = \mathbf{I}_{M_r N} - \mathbf{Q}(\boldsymbol{\omega})(\mathbf{Q}^H(\boldsymbol{\omega})\mathbf{Q}(\boldsymbol{\omega}))^{-1}\mathbf{Q}^H(\boldsymbol{\omega}).
\tag{12.18}
$$

Notice that when there is only one user, $\boldsymbol{\gamma}$ will be a scalar and Equation 12.15 reduces to Equation 12.9 of Ref. [28]. Using similar mathematical manipulations as those in Appendix I of

Ref. [28], the CRBs for $\boldsymbol{\omega}$ and $\mathbf{h}$ are obtained as

$$\mathrm{CRB}(\boldsymbol{\omega}) = \frac{\sigma^2}{2}\left(\Re\{\mathbf{Z}^H\boldsymbol{\Pi}_Q^{\perp}\mathbf{Z}\}\right)^{-1}, \tag{12.19}$$

$$\mathrm{CRB}(\mathbf{h}) = \frac{\sigma^2}{2}\Big(2(\mathbf{Q}^H(\boldsymbol{\omega})\mathbf{Q}(\boldsymbol{\omega}))^{-1}$$
$$+ (\mathbf{Q}^H(\boldsymbol{\omega})\mathbf{Q}(\boldsymbol{\omega}))^{-1}\mathbf{Q}^H(\boldsymbol{\omega})\mathbf{Z}\big(\Re\{\mathbf{Z}^H\boldsymbol{\Pi}_Q^{\perp}\mathbf{Z}\}\big)^{-1}\mathbf{Z}^H\mathbf{Q}(\boldsymbol{\omega})(\mathbf{Q}^H(\boldsymbol{\omega})\mathbf{Q}(\boldsymbol{\omega}))^{-H}\Big). \tag{12.20}$$

As can be seen from Equations 12.19 and 12.20, the dependencies of finite sample CRBs on the training sequences are very complicated. In order to get more insights into the relationship between estimation performance limit and training sequences, we turn to asymptotic CRBs (when the number of subcarriers is infinite).

12.3.2 Asymptotic CRB

In asymptotic analysis, it is assumed that the time-domain trainings are zero-mean random sequences that satisfy the mixing conditions [27]. Then, exploiting the results of Equation 51 in Ref. [27], we have the following equations:

$$\lim_{N\to\infty} \frac{\mathbf{Q}^H\mathbf{Q}}{N} = \mathbf{R}, \tag{12.21a}$$

$$\lim_{N\to\infty} \frac{\mathbf{Q}^H(\mathbf{I}_{M_\mathrm{r}} \otimes \mathbf{M})\mathbf{Q}}{N^2} = \frac{1}{2}\mathbf{R}, \tag{12.21b}$$

$$\lim_{N\to\infty} \frac{\mathbf{Q}^H(\mathbf{I}_{M_\mathrm{r}} \otimes \mathbf{M})^2\mathbf{Q}}{N^3} = \frac{1}{3}\mathbf{R}, \tag{12.21c}$$

where

$$\mathbf{R} = \lim_{N\to\infty} \frac{\mathbf{Q}^H(\boldsymbol{\omega})\mathbf{Q}(\boldsymbol{\omega})}{N} = \mathbf{I}_{M_\mathrm{r}} \otimes \begin{bmatrix} \mathbf{R}_{1,1} & \mathbf{R}_{1,2} & \cdots & \mathbf{R}_{1,K} \\ \mathbf{R}_{2,1} & \mathbf{R}_{2,2} & \cdots & \mathbf{R}_{2,K} \\ \vdots & \vdots & \ddots & \vdots \\ \mathbf{R}_{K,1} & \mathbf{R}_{K,2} & \cdots & \mathbf{R}_{K,K} \end{bmatrix}, \tag{12.22}$$

$$\mathbf{R}_{k_1,k_2} = \mathbb{E}\{\mathbf{A}_{k_1}^H\mathbf{A}_{k_2}\}\delta(\omega_{k_1} - \omega_{k_2}). \tag{12.23}$$

Following Ref. [28], the normalized FIM is defined as

$$\overline{\boldsymbol{\Phi}} = \mathbf{K}_N^{-T}\boldsymbol{\Phi}\mathbf{K}_N^{-1} \tag{12.24}$$

with the block-diagonal matrix $\mathbf{K}_N \triangleq \mathrm{diag}(N^{3/2}\mathbf{I}_K, N^{1/2}\mathbf{I}_{M_\mathrm{r}KL}, N^{1/2}\mathbf{I}_{M_\mathrm{r}KL})$ and the corresponding CRB is denoted as $\overline{\mathrm{CRB}} = \overline{\boldsymbol{\Phi}}^{-1}$. From the results of Equations 12.10 and 12.21, it is easy to obtain

$$\lim_{N\to\infty} \overline{\Phi} = \frac{2}{\sigma^2} \begin{bmatrix} \frac{1}{3}\Re\{H^H RH\} & \frac{1}{2}\Im\{H^H R\} & \frac{1}{2}\Re\{H^H R\} \\ -\frac{1}{2}\Im\{RH\} & \Re\{R\} & -\Im\{R\} \\ \frac{1}{2}\Re\{RH\} & \Im\{R\} & \Re\{R\} \end{bmatrix}. \tag{12.25}$$

Since $\lim_{N\to\infty} \overline{CRB} = [\lim_{N\to\infty} \overline{\Phi}]^{-1}$, using a similar approach in the above CRB derivation, the asymptotic CRBs for ω and h are obtained as

$$\mathbf{asCRB}(\omega) = \frac{6\sigma^2}{N^3}\left(\Re\{H^H RH\}\right)^{-1}, \tag{12.26}$$

$$\mathbf{asCRB}(h) = \frac{\sigma^2}{N}\left(R^{-1} + \frac{3}{2}H\left(\Re\{H^H RH\}\right)^{-1}H^H\right). \tag{12.27}$$

Notice that $\mathbf{R}$ is assumed to be a positive definite matrix such that the matrix inversions are meaningful in Equations 12.26 and 12.27.

12.4 Optimal Condition for the Training Sequence

12.4.1 Training Design for Channel Estimation Only

In the CFO-free case, the joint CFO and channel estimation reduces to a pure channel estimation problem. For this case, the CRB for channel estimation in Equation 12.20 is simplified to

$$\mathbf{CRB}(h)_{\text{CFO-free}} = \sigma^2 I_{M_r} \otimes (A^H A)^{-1}, \tag{12.28}$$

where

$$A \triangleq [A_1 \ A_2 \ \dots \ A_K] \tag{12.29}$$

is assumed to be of full column rank to guarantee that $A^H A$ is invertible. In order to have the best estimation performance, training is designed to minimize $\mathrm{Tr}\{\mathbf{CRB}(h)_{\text{CFO-free}}\}$. With the result that for an $L \times L$ positive-definite matrix $\mathbf{B}$, $\mathrm{Tr}\{B^{-1}\} \geq \sum_{i=1}^{L}([B]_{ii})^{-1}$ with equality holds if $\mathbf{B}$ is diagonal [30], the condition for optimal training is therefore that $A^H A$ is a diagonal matrix or equivalently

$$A_{k_1}^H A_{k_2} = \alpha_{k_1} I_L \delta(k_1 - k_2), \tag{12.30}$$

where $\alpha_k = \mathrm{Tr}\{A_k^H A_k\}$ is the power of training from the kth user.

Remark 2: Interestingly, the condition for the optimal training sequence in the CFO-free case can also be obtained by minimizing the MSE of the ML channel estimator [13]. In the CFO-free case, the system model Equation 12.8 becomes $x = (I_{M_r} \otimes A)h + n$. The ML channel estimator can be shown to be $h = (I_{M_r} \otimes (A^H A)^{-1}A^H)x$, with the corresponding MSE given by $M_r\sigma^2\mathrm{Tr}\{(A^H A)^{-1}\}$ [29]. Therefore, in order to minimize the MSE, the training should also satisfy $A_{k_1}^H A_{k_2} = \alpha_{k_1} I_L \delta(k_1 - k_2)$.

12.4.2 Training Design for Frequency Offset Estimation Only

As can be seen from Equation 12.19, the $\mathbf{CRB}(\boldsymbol{\omega})$ depends on the training $\mathbf{A}_k$ in a complicated way. Therefore, designing training by minimizing Equation 12.19 is very difficult. On the other hand, the asymptotic CRB Equation 12.26 has a much simpler relationship with the training than the finite sample CRB, hence making its minimization much more feasible. Therefore, in this section, training sequences are designed by minimizing the $\mathbf{asCRB}(\boldsymbol{\omega})$.

Based on the fact that $\Re\{\mathbf{H}^{\mathrm{H}}\mathbf{RH}\}$ is a positive-definite matrix, it is easy to prove that [31]

$$\left[\left(\Re\{\mathbf{H}^{\mathrm{H}}\mathbf{RH}\}\right)^{-1}\right]_{k,k} \geq \frac{1}{\left[\Re\{\mathbf{H}^{\mathrm{H}}\mathbf{RH}\}\right]_{k,k}} = \frac{1}{\sum_{j=1}^{M_{\mathrm{r}}} \mathbf{h}_{k,j}^{\mathrm{H}}\mathbf{R}_{k,k}\mathbf{h}_{k,j}}, \tag{12.31}$$

where the equality holds if and only if $\mathbf{R}_{k_1,k_2} = \mathbf{0}_L$, $1 \leq k_1, k_2 \leq K$ and $k_1 \neq k_2$. From the $\mathbf{asCRB}(\boldsymbol{\omega})$ in Equation 12.26 and using Equation 12.31, we can obtain the following inequality:

$$\mathrm{Tr}\{\mathbf{asCRB}(\boldsymbol{\omega})\} \geq \frac{6\sigma^2}{N^3} \sum_{k=1}^{K} \frac{1}{\sum_{j=1}^{M_{\mathrm{r}}} \mathbf{h}_{k,j}^{\mathrm{H}}\mathbf{R}_{k,k}\mathbf{h}_{k,j}}, \tag{12.32}$$

where the equality holds if and only if

$$\mathbf{R}_{k_1,k_2} = \mathbf{0}_L, \quad 1 \leq k_1, k_2 \leq K \text{ and } k_1 \neq k_2. \tag{12.33}$$

The above inequality in Equation 12.32 means that for any nonblock diagonal matrix $\mathbf{R}$, its block diagonal version defined as $\bar{\mathbf{R}} = \mathbf{I}_{M_{\mathrm{r}}} \otimes \mathrm{diag}\{\mathbf{R}_{1,1}, \ldots, \mathbf{R}_{K,K}\}$ always has smaller $\mathrm{Tr}\{\mathbf{asCRB}(\boldsymbol{\omega})\}$. This implies that the matrix $\mathbf{R}$ should be designed as a block diagonal matrix.

From the previous discussion, the first condition for optimal training is $\mathbf{R}_{k_1,k_2} = \mathbf{0}_L$ for $k_1 \neq k_2$. On top of this, training of an individual user should be designed such that the right-hand side of Equation 12.32 is minimized. Notice that minimizing the right-hand side of Equation 12.32 is equivalently minimizing $1/\sum_{j=1}^{M_{\mathrm{r}}} \mathbf{h}_{k,j}^{\mathrm{H}}\mathbf{R}_{k,k}\mathbf{h}_{k,j}$ independently for each user k. In the following, two training design methods are presented that further minimize $1/\sum_{j=1}^{M_{\mathrm{r}}} \mathbf{h}_{k,j}^{\mathrm{H}}\mathbf{R}_{k,k}\mathbf{h}_{k,j}$ for each user k.

12.4.2.1 Channel-Dependent Training

If channel realization is known or has been estimated, training is designed as follows. First notice that $\mathbf{R}_{k,k} = \mathbb{E}\{\mathbf{A}_k^{\mathrm{H}}\mathbf{A}_k\}$ and $\mathbf{A}_k = \mathbf{F}^{\mathrm{H}}\mathbf{D}_k\mathbf{W}$; then minimizing $1/\sum_{j=1}^{M_{\mathrm{r}}} \mathbf{h}_{k,j}^{\mathrm{H}}\mathbf{R}_{k,k}\mathbf{h}_{k,j}$ is equivalent to

$$\max_{\substack{[\mathbf{P}_k]_{\ell,\ell}, \\ \ell=0,\ldots,N-1}} \left\{ \sum_{j=1}^{M_{\mathrm{r}}} \mathbf{h}_{k,j}^{\mathrm{H}}\mathbf{W}^{\mathrm{H}}\mathbf{P}_k\mathbf{W}\mathbf{h}_{k,j} \right\}, \tag{12.34}$$

where $\mathbf{P}_k \triangleq \mathbb{E}\{\mathbf{D}_k^{\mathrm{H}}\mathbf{D}_k\}$ is a diagonal matrix with the (ℓ, ℓ)th element $[\mathbf{P}_k]_{\ell,\ell} = \mathbb{E}\{|d_k(\ell)|^2\}$ representing the power of training at the ℓth subcarrier. Denoting $\mathbf{V}_{k,j} = \mathbf{W}\mathbf{h}_{k,j}$ as the frequency-domain representation of channel $\mathbf{h}_{k,j}$, the above training design problem can be rewritten as

$$\max_{\substack{[\mathbf{P}_k]_{\ell,\ell}, \\ \ell=0,\ldots,N-1}} \left\{ \sum_{\ell=0}^{N-1} [\mathbf{P}_k]_{\ell,\ell} \beta_{k,\ell} \right\}, \tag{12.35}$$

where $\beta_{k,\ell} = \sum_{j=1}^{M_r} |[\mathbf{V}_{k,j}]_\ell|^2$ is the sum of powers of subchannel ℓ from the k user to all the receiver antennas.

Here, a power constraint on the training is needed for meaningful maximization. Without loss of generality, it is assumed that $\mathrm{Tr}\{\mathbf{R}_{k,k}\} \leq \alpha_k L$, which can be shown to be equivalent to $\sum_{\ell=0}^{N-1} [\mathbf{P}_k]_{\ell,\ell} \leq \alpha_k$. It is obvious that, for user k, in order to maximize $\sum_{\ell=0}^{N-1} [\mathbf{P}_k]_{\ell,\ell} \beta_{k,\ell}$ under the power constraint, all the available power α_k should be put into the subcarrier $\ell_k^{\mathrm{opt}} = \arg\{\max_\ell \{\beta_{k,\ell}\}\}$. This is intuitively appealing since it corresponds to transmitting a single tone at the strongest subchannel in the frequency domain, and CFO is estimated by measuring how much the transmitted tone has shifted. As a special case, if $M_r = 1$, then there is only one multipath channel from user k to the BS, and the above optimal training reduces to that of Ref. [32].

In summary, for the kth user, the optimal training for CFO estimation should satisfy

$$\mathbf{R}_{k,k} = \alpha_k \mathbf{W}^{\mathrm{H}} \mathbf{e}_{\mathrm{opt}} \mathbf{e}_{\mathrm{opt}}^{\mathrm{H}} \mathbf{W}, \tag{12.36}$$

where $\mathbf{e}_{\mathrm{opt}} = [0, 0, \ldots 1, \ldots 0]^{\mathrm{T}}$ with the single 1 appearing in the ℓ_k^{opt} position. However, if the training is selected according to Equation 12.36, the resulting $\mathbf{R}_{k,k}$ is a rank-one matrix, and $\mathbf{R}$ is not invertible in the asymptotic CRB Equation 12.26. This is not desirable, and it was proposed in Refs. [32] and [15] that a diagonal loading term is added to Equation 12.36, which gives

$$\mathbf{R}_{k,k} = (\alpha_k - \gamma) \mathbf{W}^{\mathrm{H}} \mathbf{e}_{\mathrm{opt}} \mathbf{e}_{\mathrm{opt}}^{\mathrm{H}} \mathbf{W} + \gamma \mathbf{I}_L, \tag{12.37}$$

where γ is a small number.

12.4.2.2 Channel-Independent Training

In general, knowledge of channel realization is needed in order to implement optimal training Equation 12.37. If there is no knowledge on the channel, the best we can do is to distribute the training power evenly across all subcarriers, that is, setting $\mathbf{R}_{k,k} \triangleq \mathbb{E}\{\mathbf{A}_k^{\mathrm{H}} \mathbf{A}_k\} = \alpha_k \mathbf{I}_L$.

In the following, another approach, which results in channel-independent training, will be presented. This training design method is based on the minimax idea, which seeks the training that minimizes the worst-case asCRB among all possible channel realizations. Since $\sum_{j=1}^{M_r} \mathbf{h}_{k,j}^{\mathrm{H}} \mathbf{R}_{k,k} \mathbf{h}_{k,j} = \tilde{\mathbf{h}}_k^{\mathrm{H}} (\mathbf{I}_{M_r} \otimes \mathbf{R}_{k,k}) \tilde{\mathbf{h}}_k$, where $\tilde{\mathbf{h}}_k \triangleq [\mathbf{h}_{k,1}^{\mathrm{T}} \ \mathbf{h}_{k,2}^{\mathrm{T}} \ldots \mathbf{h}_{k,M_r}^{\mathrm{T}}]^{\mathrm{T}}$, the minimax approach corresponds to [32]

$$\min_{\mathbf{R}_{k,k}} \ \max_{\tilde{\mathbf{h}}_k: \, ||\tilde{\mathbf{h}}_k||=1} \left\{ (\tilde{\mathbf{h}}_k^{\mathrm{H}} (\mathbf{I}_{M_r} \otimes \mathbf{R}_{k,k}) \tilde{\mathbf{h}}_k)^{-1} \right\}, \tag{12.38}$$

where the constraint $||\tilde{\mathbf{h}}_k|| = 1$ is added to avoid the trivial solution $\tilde{\mathbf{h}}_k = \mathbf{0}$. Solving inner maximization, the minimax problem becomes

$$\min_{\mathbf{R}_{k,k}} \left\{ [\lambda_{\min}\{\mathbf{I}_{M_r} \otimes \mathbf{R}_{k,k}\}]^{-1} \right\} \tag{12.39}$$

$$= \max_{\mathbf{R}_{k,k}} \left\{ 1 + \lambda_{\min}\{\mathbf{R}_{k,k}\} \right\} \tag{12.40}$$

Obviously, a proper constraint on $\mathbf{R}_{k,k}$ is needed in order to obtain meaningful maximization in Equation 12.40. Here, to be consistent with the power constraint for designing channel-dependent training, we also constrain the training power to be $\mathrm{Tr}\{\mathbf{R}_{k,k}\} \leq \alpha_k L$. Therefore, the minimax training design problem is stated as

$$\max_{\mathbf{R}_{k,k}} \left\{ \lambda_{\min}\{\mathbf{R}_{k,k}\} \right\} \quad \text{subject to} \quad \mathrm{Tr}\{\mathbf{R}_{k,k}\} \leq \alpha_k L . \tag{12.41}$$

It can be shown that [32] the solution to the above training design problem is given by

$$\mathbf{R}_{k,k} = \alpha_k \mathbf{I}_L. \tag{12.42}$$

12.4.2.3 Summary of Training Design for CFO Estimation

In the first part of this subsection, it was shown that the first condition for optimal training in CFO estimation is $\mathbf{R}_{k_1,k_2} \triangleq \mathbb{E}\{\mathbf{A}_{k_1}^H \mathbf{A}_{k_2}\}\delta(\omega_{k_1} - \omega_{k_2}) = \mathbf{0}_L$ for $k_1 \neq k_2$. Since the CFOs of users k_1 and k_2 may be equal (i.e., $\omega_{k_1} = \omega_{k_2}$), we need

$$\mathbb{E}\{\mathbf{A}_{k_1}^H \mathbf{A}_{k_2}\} = \mathbf{0}_L \quad \text{for } k_1 \neq k_2 \tag{12.43}$$

in order to make the design valid in all possible situations. This condition can be interpreted as uncorrelated time-domain training sequences among users.

Then, depending on whether channel realization is known or not, the second condition of optimal training for individual user k is

$$\mathbf{R}_{k,k} \triangleq \mathbb{E}\{\mathbf{A}_k^H \mathbf{A}_k\} = \begin{cases} (\alpha_k - \gamma)\mathbf{W}^H \mathbf{e}_{\mathrm{opt}} \mathbf{e}_{\mathrm{opt}}^H \mathbf{W} + \gamma \mathbf{I}_L & \text{if channel is known,} \\ \alpha_k \mathbf{I}_L & \text{if channel is unknown,} \end{cases} \tag{12.44}$$

where α_k is the power constraint on the training such that $\mathrm{Tr}\{\mathbf{R}_{k,k}\} \leq \alpha_k L$, γ is a small number, and $\mathbf{e}_{\mathrm{opt}} = [0, 0, \ldots 1, \ldots 0]^T$ with the single 1 appearing in the $\ell_k^{\mathrm{opt}} = \arg\{\max_\ell\{\beta_{k,\ell}\}\}$ position.

Notice that for channel-independent training, the first and second conditions can be combined to give

$$\mathbb{E}\{\mathbf{A}_{k_1}^H \mathbf{A}_{k_2}\} = \alpha_{k_1} \mathbf{I}_L \delta(k_1 - k_2), \tag{12.45}$$

which is analogous to the optimal condition for channel estimation only in Equation 12.30.

12.4.3 Training Design for Joint Frequency and Channel Estimation

In the previous two subsections, optimal conditions on training for channel and CFO estimation are derived separately. In this section, optimal conditions on training for joint CFO and channel estimations are derived. This problem is not trivial because

1. Even for the asCRBs, which are simpler than the CRBs, the effect of CFO and channel parameters are coupled in complicated ways as shown in the expressions of **asCRB(ω)** and **asCRB(h)** in Equations 12.26 and 12.27.
2. As discussed in the previous two subsections, in general, the forms of optimal training for CFO and channel estimation are different, meaning that no single training sequence allows one to simultaneously minimize both **asCRB(ω)** and **asCRB(h)** [15].

In the following, the condition of optimal training for joint CFO and channel estimation is derived using the minimax approach. The minimax approach is considered because it results in a single channel-independent training that provides good performances for both CFO and channel estimations. Moreover, as shown in Ref. [28], the "true" optimal training, even if it exists, cannot outperform the minimax optimal design by much.

First notice that since $\mathbf{R}$ is a positive-definite matrix, based on Theorem 7.7.8 in Ref. [33], we have

$$\mathrm{Tr}\{\mathbf{R}^{-1}\} \geq M_{\mathrm{r}} \sum_{k=1}^{K} \mathrm{Tr}\{\mathbf{R}_{k,k}^{-1}\} \tag{12.46}$$

with equality holding if and only if $\mathbf{R}_{k_1,k_2} = \mathbf{0}_L$, $1 \leq k_1, k_2 \leq K$ and $k_1 \neq k_2$. Based on Equations 12.31 and 12.46, we obtain the following two inequalities:

$$\mathrm{Tr}\{\mathbf{asCRB}(\boldsymbol{\omega})\} \geq \frac{6\sigma^2}{N^3} \sum_{k=1}^{K} \frac{1}{\sum_{j=1}^{M_{\mathrm{r}}} \mathbf{h}_{k,j}^{\mathrm{H}} \mathbf{R}_{k,k} \mathbf{h}_{k,j}}, \tag{12.47}$$

$$\mathrm{Tr}\{\mathbf{asCRB}(\mathbf{h})\} \geq \frac{\sigma^2}{N} \left(M_{\mathrm{r}} \sum_{k=1}^{K} \mathrm{Tr}\{\mathbf{R}_{k,k}^{-1}\} + \frac{3}{2} \sum_{k=1}^{K} \frac{\sum_{j=1}^{M_{\mathrm{r}}} \mathbf{h}_{k,j}^{\mathrm{H}} \mathbf{h}_{k,j}}{\sum_{j=1}^{M_{\mathrm{r}}} \mathbf{h}_{k,j}^{\mathrm{H}} \mathbf{R}_{k,k} \mathbf{h}_{k,j}} \right), \tag{12.48}$$

with the equalities holding if and only if

$$\mathbf{R}_{k_1,k_2} = \mathbf{0}_L, \quad 1 \leq k_1, k_2 \leq K \text{ and } k_1 \neq k_2. \tag{12.49}$$

With the same reason as in the training design for frequency estimation only, the above two inequalities imply that the matrix $\mathbf{R}$ should be designed as a block diagonal matrix, which means that the training sequences from different users should be uncorrelated.

In order to design channel-independent training, we further minimize the sum of the right-hand sides of Equations 12.47 and 12.48 for each individual user using the minimax approach, which can be stated as

$$\min_{\substack{\mathbf{R}_{k,k}: \\ \mathrm{Tr}\{\mathbf{R}_{k,k}\} \leq \alpha_k L}} \max_{\tilde{\mathbf{h}}_k: \|\tilde{\mathbf{h}}_k\|=1} \left\{ M_{\mathrm{r}} \mathrm{Tr}\{\mathbf{R}_{k,k}^{-1}\} + \frac{2\|\tilde{\mathbf{h}}_k\|^2/3 + 6/N^2}{(\tilde{\mathbf{h}}_k^{\mathrm{H}} (\mathbf{I}_{M_{\mathrm{r}}} \otimes \mathbf{R}_{k,k}) \tilde{\mathbf{h}}_k)} \right\}. \tag{12.50}$$

Performing inner maximization, the minimax problem becomes

$$\min_{\substack{\mathbf{R}_{k,k}: \\ \text{Tr}\{\mathbf{R}_{k,k}\} \leq \alpha_k L}} \left\{ M_{\text{r}} \text{Tr}\{\mathbf{R}_{k,k}^{-1}\} + \frac{(2/3) + (6/N^2)}{1 + \lambda_{\min}\{\mathbf{R}_{k,k}\}} \right\}. \tag{12.51}$$

Hence for an $L \times L$ positive-definite matrix $\mathbf{B}$, $\text{Tr}\{\mathbf{B}^{-1}\} \geq \sum_{i=1}^{L} ([\mathbf{B}]_{ii})^{-1}$ with equality holds if $\mathbf{B}$ is diagonal [30], together with Equations 12.41 and 12.42, and we have that minimax optimal training for user k should satisfy

$$\mathbf{R}_{k,k} = \alpha_k \mathbf{I}_L. \tag{12.52}$$

Notice that the same optimal condition Equation 12.52 can still be obtained even if we minimize a weighted sum of $\text{Tr}\{\mathbf{asCRB}(\boldsymbol{\omega})\}$ in Equation 12.47 and $\text{Tr}\{\mathbf{asCRB}(\mathbf{h})\}$ in Equation 12.48.

In summary, for joint multiuser CFO and channel estimation, optimal training should satisfy $\mathbf{R}_{k_1,k_2} \triangleq \mathbb{E}\{\mathbf{A}_{k_1}^{H}\mathbf{A}_{k_2}\} = \alpha_{k_1} \mathbf{I}_L \delta(k_1 - k_2)$, which coincides with that of channel-independent optimal training for CFOs estimation only Equation 12.45.

Remark 3: If there is only one user, we have $\mathbb{E}\{\mathbf{A}_1^{H}\mathbf{A}_1\} = \alpha_1 \mathbf{I}_L$, which represents the results in Ref. [28]. Moreover, when there are multiple users but the channels are flat fading, we can set $L = 1$, and this reduces to the results in Ref. [27].

12.5 Realization of Optimal Training

As can be seen from the previous discussion, all the optimal training designs (except the channel-dependent one) point to a single condition: $\mathbf{A}_{k_1}^{H}\mathbf{A}_{k_2} = \alpha_{k_1} \mathbf{I}_L \delta(k_1 - k_2)$. Notice that although the optimal condition from asymptotic analysis is $\mathbb{E}\{\mathbf{A}_{k_1}^{H}\mathbf{A}_{k_2}\} = \alpha_{k_1} \mathbf{I}_L \delta(k_1 - k_2)$, we simply consider $\mathbf{A}_{k_1}^{H}\mathbf{A}_{k_2} = \alpha_{k_1} \mathbf{I}_L \delta(k_1 - k_2)$ as the optimal condition for all the cases, since (1) $\mathbf{A}_{k_1}^{H}\mathbf{A}_{k_2} = \alpha_{k_1} \mathbf{I}_L \delta(k_1 - k_2)$ is a stronger condition than $\mathbb{E}\{\mathbf{A}_{k_1}^{H}\mathbf{A}_{k_2}\} = \alpha_{k_1} \mathbf{I}_L \delta(k_1 - k_2)$ and (2) asymptotically, $\mathbf{A}_{k_1}^{H}\mathbf{A}_{k_2}$ is very close to $\mathbb{E}\{\mathbf{A}_{k_1}^{H}\mathbf{A}_{k_2}\}$. There are many ways of designing training sequences satisfying $\mathbf{A}_{k_1}^{H}\mathbf{A}_{k_2} = \alpha_{k_1} \mathbf{I}_L \delta(k_1 - k_2)$. In this section, we discuss several representative realizations of training that satisfy this condition. Without loss of generality, $\alpha_k = 1$ for all k is assumed in the following discussion, and the optimal condition is represented by $\mathbf{A}^{H}\mathbf{A} = \mathbf{I}_{KL}$. Extension to the case where α_k's are different for different users is straightforward.

12.5.1 Time-Domain Realizations

In the time domain, the optimal condition $\mathbf{A}^{H}\mathbf{A} = \mathbf{I}_{KL}$ implies that optimal sequences have impulse-like auto-correlation and zero cross-correlation. One natural idea is to transmit white sequences from different users. Although this is a simple choice, for one realization of finite length white sequence, $\mathbf{A}^{H}\mathbf{A}$ can only be approximately an identity matrix.

Another class of sequences that satisfies $\mathbf{A}^{H}\mathbf{A} = \mathbf{I}_{KL}$ is the L-perfect sequences [34]. The idea of L-perfect sequences comes from the fact that $\mathbf{A}^{H}\mathbf{A} = \mathbf{I}_{KL}$ actually only requires L taps zero auto-correlation and cross-correlation. Therefore, a set of sequences satisfying $\mathbf{A}^{H}\mathbf{A} = \mathbf{I}_{KL}$ can be constructed from a single sequence with all of its out-of-phase periodic auto-correlation terms being zero. L-perfect sequences are constant-modulus sequences in the time domain, and the construction procedure is as follows.

1. Construct a sequence $\mathbf{s} = [s(0)\ s(1)\ldots s(N-1)]$ with length N such that all of its out-of-phase periodic auto-correlation terms are equal to zero. Such a sequence is called a *perfect* sequence and a unified method for constructing perfect sequence is presented in Ref. [35].

2. Construct another sequence $\mathbf{s}' = [s'(0)\ s'(1)\ldots s'(N+KL-1)]$ of length $N+KL$ as follows:

$$\mathbf{s}' = [\underbrace{s(0)\ s(1)\ldots s(N-1)}_{\mathbf{s}}\ s(0)\ s(1)\ldots s(KL-1)]. \tag{12.53}$$

Note that $N \geq KL$ must be satisfied. That is, if the number of user K is large, we cannot use training sequences with short length.

3. The time-domain training sequence of the kth user $(k = 1, \ldots, K)$ is given by

$$\mathbf{c}_k = [s'((k-1)L),\ s'((k-1)L+1),\ \ldots,\ s'((k-1)L+N-1)]/\sqrt{N}. \tag{12.54}$$

For example, let us consider $N = 32$, $L = 3$, and $K = 2$. First we construct a perfect sequence of length 32 according to Ref. [35]. Then we cyclically extend the sequence by copying the first $2 \times 3 = 6$ symbols and putting them at the back. Finally, $\mathbf{c}_1 = [s'(0)\ s'(1)\ \ldots\ s'(31)]/\sqrt{32}$ and $c_2 = [s'(3)\ s'(4)\ \ldots\ s'(34)]/\sqrt{32}$.

Similar time-domain constant-modulus sequences satisfying $\mathbf{A}^H\mathbf{A} = \mathbf{I}_{KL}$ were also proposed in Refs. [14] and [36].

12.5.2 Frequency-Domain Realizations

The optimal condition $\mathbf{A}_{k_1}^H\mathbf{A}_{k_2} = \mathbf{I}_L\delta(k_1 - k_2)$ can be translated into requirements in the frequency domain as follows. First notice that $\mathbf{A}_k = \mathbf{F}^H\mathbf{D}_k\mathbf{W}$; then the optimal condition can be rewritten as $\mathbf{W}^H\mathbf{D}_{k_2}^H\mathbf{D}_{k_2}\mathbf{W} = \mathbf{I}_L\delta(k_1 - k_2)$. Furthermore, the (p,q)th element of $\mathbf{W}^H\mathbf{D}_{k_1}^H\mathbf{D}_{k_2}\mathbf{W}$ is given by

$$[\mathbf{W}^H\mathbf{D}_{k_1}^H\mathbf{D}_{k_2}\mathbf{W}]_{p,q} = \sum_{n=0}^{N-1} d_{k_1}^*(n)d_{k_2}(n)\exp(\mathrm{j}2\pi n(p-q)/N), \tag{12.55}$$

where $0 \leq p, q \leq L-1$. Therefore, $\mathbf{W}^H\mathbf{D}_{k_1}^H\mathbf{D}_{k_2}\mathbf{W} = \mathbf{I}_L\delta(k_1 - k_2)$ can be equivalently expressed as the following three conditions:

$$\sum_{n=0}^{N-1} |d_k(n)|^2 = 1, \tag{12.56}$$

$$\sum_{n=0}^{N-1} |d_k(n)|^2\exp(\mathrm{j}2\pi np/N) = 0 \quad \text{for } p = \pm 1, \pm 2, \ldots, \pm(L-1), \tag{12.57}$$

$$\sum_{n=0}^{N-1} d_{k_1}^*(n)d_{k_2}(n)\exp(\mathrm{j}2\pi np/N) = 0$$

$$\text{for } p = 0, \pm 1, \pm 2, \ldots, \pm(L-1), \forall k_1 \neq k_2. \tag{12.58}$$

Condition 12.56 is the power constraint on the training of the kth user, which can be easily satisfied by proper normalization of the training. The more critical constraints on the training come from Equations 12.57 and 12.58.

Before we discuss how to choose $d_k(n)$ in order to satisfy Equations 12.56 through 12.58, three basic properties of discrete Fourier transform (DFT) are reviewed below. Let $X[n] = \sum_{k=0}^{N-1} x[k] e^{-j2\pi kn/N}$ and $x[k] = \frac{1}{N} \sum_{n=0}^{N-1} X[n] e^{j2\pi kn/N}$ be a Fourier transform pair (i.e., $x[k] \overset{DFT}{\longleftrightarrow} X[n]$); then we have the following:

Property 1: If $X[n] = a$ for all n, where a is a constant, then $x[k] = a\delta(k)$.

Property 2: Assume $N = M_1 M_2$, where M_1 and M_2 are integers. If $X[n] = a \sum_{i=0}^{M_2-1} \delta(n - iM_1)$, then $x[k] = aM_2/N \sum_{i=0}^{M_1-1} \delta(k - iM_2)$.

Property 3: $X[(n - l)_N] \overset{DFT}{\longleftrightarrow} e^{j2\pi lk/N} x[k]$, where $(\cdot)_N$ represents the modulo-N operation.

12.5.2.1 Frequency Division Multiplexing Sequences

Let $N = ML_o$, where both M and L_o are integers and $L_o \geq L$. By DFT properties 2 and 3, condition 12.57 can be satisfied if the training of the kth user is designed such that

$$|d_k(n)|^2 = a_k^{(\ell)} \sum_{i=0}^{L_o-1} \delta(n - (\ell + iM)), \tag{12.59}$$

where $\ell \in \{0, 1, \ldots, M - 1\}$ and $a_k^{(\ell)}$ is a real number depending on k and ℓ but independent of i. This means that the power of training for the kth user in the frequency domain should be in the form of a comb-shaped impulse train with equal impulse magnitude $a_k^{(\ell)}$ and separation between impulses being $M = N/L_o$. Notice that there are totally M sets of such impulse trains with different offsets determined by ℓ, and by property 3 of DFT, user k can use more than one set of impulse train simultaneously.

In addition to Equation 12.59, trainings from different users should also satisfy Equation 12.58. Since there are multiple sets of impulse trains defined in Equation 12.59, one simple way of making sure that Equation 12.58 is satisfied is to put the trainings of different users on nonoverlapping sets of impulse trains in the frequency domain. This results in the frequency division multiplexing (FDM) sequences proposed in Ref. [13]. Similar designs were also proposed in Ref. [14]. Notice that since there are totally M nonoverlapping sets of impulse trains, the maximum number of users supported by FDM sequences is M.

In summary, for FDM sequences,

1. Each user should transmit the training on nonoverlapping frequency sets with subcarrier indices of the ℓth set defined by $\{\ell, \ell + M, \ldots, \ell + (L_o - 1)M\}$.
2. The power transmitted on different subcarriers in each set should be the same, but the powers on different sets can be different.
3. The training of each user should be normalized such that Equation 12.56 is satisfied.

Mathematically, the FDM sequence for user k can be expressed as

$$d_k(n) = \sum_{i=0}^{L_o-1} \sum_{\ell \in \mathbb{L}_k} \gamma_\ell c_{k,i}^{(\ell)} \delta(n - (\ell + iM)) \tag{12.60}$$

with $\qquad L_o \sum_{\ell \in \mathbb{L}_k} |\gamma_\ell|^2 = 1,$

$$\bigcap_k \mathbb{L}_k = \{\} \quad \text{and} \quad \bigcup_k \mathbb{L}_k = \{0, 1, \dots, M\},$$

where γ_ℓ is used to control the power of training at the ℓth set of frequencies, $\mathbb{L}_k$ is a set containing the ℓ index assigned to user k, and $\{c_{k,i}^{(\ell)}\}$ are unit amplitude complex numbers.

As a simple example for FDM sequences, consider a system with $N = 16$, $L = 3$, and $K = 2$. Since $L_o \geq L$ and must be a factor of N, we choose $L_o = 4$ and therefore $M = N/L_o = 4$. There are totally $M = 4$ sets of nonoverlapping frequencies defined in point 1 of the summary above. These four sets of frequencies can be assigned to two users evenly, or one user with three sets of frequencies and another user with one set of frequencies. In this example, it is assumed that each user uses two sets of frequencies, which correspond to each user using $2L_o = 16$ pilot tones for training. One possible realization of the training sequences is

$$d_1(n) = \sum_{i=0}^{L_o-1} \left\{ \sqrt{\frac{1}{4L_o}} c_{1,i}^{(0)} \delta(n - iM) + \sqrt{\frac{3}{4L_o}} c_{1,i}^{(2)} \delta(n - (2 + iM)) \right\}, \tag{12.61}$$

$$d_2(n) = \sum_{i=0}^{L_o-1} \left\{ \sqrt{\frac{1}{2L_o}} c_{1,i}^{(1)} \delta(n - (1 + iM)) + \sqrt{\frac{1}{2L_o}} c_{1,i}^{(3)} \delta(n - (3 + iM)) \right\}, \tag{12.62}$$

where $\{c_{k,i}^{(\ell)}\}$ are unit amplitude complex numbers.

12.5.2.2 Code Division Multiplexing Sequences in the Frequency Domain

Notice that FDM is simple but it is a very strong condition that guarantees that Equation 12.58 is satisfied for all p, not only $p = 0, \pm 1, \dots, \pm(L - 1)$. In case all pilot tones are used by all users (i.e., FDM is not used), further analysis is needed to find the condition to make Equation 12.58 satisfied. Again, assuming $N = ML_o$ with M and $L_o \geq L$ being integers, by properties 2 and 3 of DFT, it can be shown that Equation 12.58 is satisfied if the following two conditions are met:

$$d_{k_1}^*(n)d_{k_2}(n) = \sum_{\ell=0}^{M-1} \sum_{i=0}^{L_o-1} a_{k_1,k_2}^{(\ell)} \delta(n - (\ell + iM)) \quad \forall k_1 \neq k_2, \tag{12.63}$$

$$\sum_{n=0}^{M-1} d_{k_1}^*(n)d_{k_2}(n) = 0 \quad \forall k_1 \neq k_2, \tag{12.64}$$

where $\{a_{k_1,k_2}^{(\ell)}\}$ are complex numbers independent of i. Condition 12.63 states that the products $d_{k_1}^*(n)d_{k_2}(n)$ should be periodic with period M, while condition 12.64 states that the sum of the first M products $d_{k_1}^*(n)d_{k_2}(n)$ is zero. One way to have these two requirements satisfied is by using the following constraints:

$$d_{k_2}(n) = d_{k_1}(n) \exp(-j2\pi n(k_2 - k_1)/M) \exp(j\mu_{k_1,k_2}), \tag{12.65}$$

$$|d_k(n)| = 1, \tag{12.66}$$

where $\mu_{k_1,k_2} \in [0, 2\pi]$ depends on k_1 and k_2 only (i.e., independent of n). Coincidentally, using property 1 of DFT, it can be seen that Equation 12.66 also makes Equations 12.56 and 12.57 satisfied; thus Equations 12.65 and 12.66 are conditions of optimal training when all the pilot tones are used by all users. This is known as code division multiplexing in frequency-domain (CDM(F)) sequences [13].

Summarizing Equations 12.65 and 12.66, the CDM(F) sequences can be obtained by first generating a constant-modulus sequence (in the frequency domain) for one of the user; then the training sequences for other users can be obtained from Equation 12.65. Mathematically, the CDM(F) sequence for the kth user is given by

$$d_k(n) = \exp(j\zeta_n) \exp(j\phi_k) \exp(-j2\pi(k-1)n/M),$$
$$k = 1, \ldots, K; \quad n = 0, 1, \ldots, N-1, \quad (12.67)$$

where ζ_n and ϕ_k are random variables in $[0, 2\pi]$ with respect to n and k, respectively.

For example, consider again a system with $N = 16$, $L = 3$, and $K = 2$. We also choose $L_o = 4$ and so $M = N/L_o = 4$. From Equation 12.67, the training sequences (in the frequency domain) for the two users are

$$\mathbf{d}_1^{\mathrm{T}} = [c_0, \quad c_1, \quad c_2, \quad c_3, \quad \ldots, \quad c_{15}], \quad (12.68)$$
$$\mathbf{d}_2^{\mathrm{T}} = [c_0, \ c_1 e^{-j2\pi/4} \ c_2 e^{-j2\pi(2)/4}, \ c_3 e^{-j2\pi(3)/4}, \ \ldots, \ c_{15} e^{-j2\pi(15)/4}] e^{j\phi_2}, \quad (12.69)$$

where $\{c_i\}$ are unit amplitude complex numbers and ϕ_2 is a random number generated from $[0, 2\pi]$.

Remark 4: By dividing the whole spectrum into several FDM groups and with CDM(F) sequences transmitted in each FDM group, a combination of FDM and CDM(F) sequences is also possible. More details can be found in Ref. [13].

12.5.3 Performance Comparisons

It is clear that for all the training sequences derived in this section, they are optimal for channel estimation in the absence of CFO. On the other hand, for joint CFO and channel estimation, they are only asymptotically optimal, meaning that for a practical system with a finite number of subcarriers, it is still unclear how these training sequences behave.

In this subsection, simulation results are presented to compare the effectiveness of different types of training under a finite number of subcarriers. For simplicity, it is assumed that $\theta_{k,j} = 0, \forall k, j$ in the simulations. The channel response for each user is generated according to the HIPERLAN/2 channel model with eight paths ($L = 8$) [26]. In detail, channel taps are modeled as independent and complex Gaussian random variables with zero mean, and the channel follows an exponential power delay profile

$$\mathbb{E}\{|\mathbf{h}_{k,j}(l)|^2\} = \lambda_{k,j} \cdot \exp\{-l\}, \quad l = 0, \ldots, L-1, \quad (12.70)$$

where parameter $\lambda_{k,j}$ denotes the power of user k observed at the receive-antenna j. Since the receive antennas at the BS are colocated, we have $\lambda_{k,1} = \cdots = \lambda_{k,M_r} \triangleq \lambda_k$. For simplicity, it is also assumed that the users' signals arrive at the BS with equal power, so $\lambda_1 = \cdots = \lambda_K$ holds. Other parameters of the simulations are $K = 4$ and SNR $= 10$ dB, and the normalized CFO ε_k for

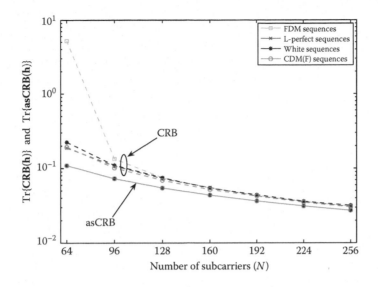

Figure 12.2   Tr{CRB(h)} and Tr{asCRB(h)} versus *N*.

all users are generated as random variables uniformly distributed in [−10,10]. This represents that the maximum CFO is 10 subcarriers spacing, which is considered to be a very large CFO range. Notice that since all the CRBs and asCRBs are directly proportional to the noise variance σ^2 (see Equations 12.19, 12.20, 12.26, and 12.27), the relative positions of the bounds will be the same for other SNRs; therefore only SNR = 10 dB is shown here. All points in the figures are averaged over 1000 Monte Carlo runs.

Figure 12.2 shows Tr{**CRB(h)**} and Tr{**asCRB(h)**} as a function of number of subcarriers *N*. From the figure, it can be seen that the asCRBs for all trainings are the same, while the CRBs are different. Notice that all the CRBs asymptotically converge to the asCRB, which justifies the training design from asymptotic analysis.

For a small number of subcarriers (e.g., at $N = 64$), the CRB of FDM sequences is higher than that of CDM(F) sequences (the CRB of FDM sequences is the highest among all four sequences considered). This is because for FDM sequences, the corresponding time-domain signals are repetitive due to the equally spaced pilot tones in the frequency domain. In particular, if the number of nonzero tones in the FDM training is small, the time-domain training signal will be highly correlated because of the small repetition length. As pointed out in Ref. [28], colored training sequences with high correlation coefficients could have their finite sample CRBs depart from the asCRBs. Compared with the FDM sequences, the CDM(F) sequences have long correlation lengths, so CDM(F) sequences are expected to have better performances than FDM-like sequences.

For white sequences, due to their good correlation properties, they consistently provide good performances, similar to the CDM(F) sequences. Notice that in the simulation, only one set of white sequences is generated for each *N*, and the same set of sequences is used for different channel and CFO realizations. For the L-perfect sequences, also because of their good correlation properties in the time domain, their performances are very close to that of white sequences and CDM(F) sequences for all *N*. Similar observations can be obtained from the CRBs and asCRBs of CFOs shown in Figure 12.3.

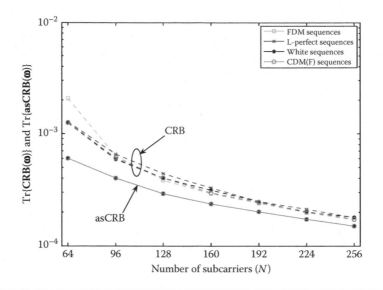

Figure 12.3 Tr{CRB(ω)} and Tr{asCRB(ω)} versus *N*.

12.6 Concluding Remarks and Open Issues

In this chapter, based on CRB analyses, we introduced the optimal conditions of training sequences in multiuser OFDM systems. Three cases were considered: channel estimation only, CFO estimation only, and joint CFO and channel estimation. Almost all the cases point to the single condition that trainings from different users should be uncorrelated, and the training of each user should exhibit impulse-like auto-correlation property. Then we reviewed several representative training sequences that satisfy the optimal conditions, and their performances were compared in systems with a finite number of subcarriers.

It is important to note that the training design problem can be approached from many different angles and with different criteria. For example, training sequence design has also been linked with a lower bound on training-based channel capacity [37], the MSE of the linear minimum mean square error (LMMSE) estimator of the channel [38], or both [22]. Recently, the MSE of data estimation at the output of the Wiener-type equalizer was used as a design criterion for training [39]. This work represents an important first step having a criterion that automatically balances the effect of CFO and channel estimation errors in training design. However, Ciblat et al. [39] only derived the training in single-carrier systems with frequency-selective channels, and extension to multicarrier systems is still an open problem.

Another trend of training design is to formulate the training design problem as a mathematical programming problem [40,41], so that results from optimization theory can be used to solve the problem. It is expected that with the powerful tools of mathematical programming, more constraints (e.g., peak to average power ratio, individual user estimation accuracy, etc.) can be incorporated into the problem so that additional desirable properties of the training sequences can be obtained.

References

1. H. Sampath, S. Talwar, J. Tellado, V. Erceg, and A. Paulraj, A fourth-generation MIMO-OFDM broadband wireless system: Design, performance, and field trial results, *IEEE Commun. Mag.*, 40, 143–149, 2002.

2. G. L. Stuber, J. R. Barry, S. W. Mclaughlin, Y. Li, M. A. Ingram, and T. G. Pratt, Broadband MIMO-OFDM wireless communications, *Proc. IEEE*, 92 (2), 271–294, 2004.
3. H. Yin and S. Alamouti, OFDMA: A broadband wireless access technology, in *Proc. IEEE Srnoff Symp. 2006*, pp. 1–4, March 2006.
4. IEEE Std. 802.16a Standard for Local and Metropolitan Area Networks, Part 16: Air interface for fixed broadband wireless access systems amendment 2, IEEE Std. 802.16a, 2003.
5. M. Jiang, J. Akhtman, and L. Hanzo, Soft-information assisted near-optimum nonlinear detection for BLAST-type space division multiplexing OFDM systems, *IEEE Trans. Wirel. Commun.*, 6 (4) 1230–1234, 2007.
6. H. Lee and I. Lee, New approach for error compensation in coded V-BLAST OFDM systems, *IEEE Trans. Commun.*, 55 (2), 345–355, 2007.
7. H. Lee, B. Lee, and I. Lee, Iterative detection and decoding with an improved V-BLAST for MIMO-OFDM systems, *IEEE J. Sel. Areas Commun.*, 24 (3), 504–513, 2006.
8. L. Weng, E. K. S. Au, P. W. C. Chan, R. D. Murch, R. S. Cheng, W. H. MOW, and V. K. N. Lau, Effect of carrier frequency offset on channel estimation for SISO/MIMO-OFDM systems, *IEEE Trans. Wire. Commun.*, 6 (5), 1854–1863, 2007.
9. S. Barbarossa, M. Pompili, and G. B. Giannakis, Channel-independent synchronization of orthogonal frequency division multiple access systems, *IEEE J. Sel. Areas Commun.*, 20 (2), 474–486, 2002.
10. Z. Cao, U. Tureli, and Y. D. Yao, Deterministic multiuser carrier-frequency offset estimation for interleaved OFDMA uplink, *IEEE Trans. Commun.*, 52 (9), 1585–1594, 2004.
11. M.-O. Pun, M. Morelli, and C.-C. Jay Kuo, Maximum-likelihood synchronization and channel estimation for OFDMA uplink transmissions, *IEEE Trans. Commun.*, 54 (4), 726–736, 2006.
12. M. Morelli, C.-C. J. Kuo, and M.-O. Pun, Synchronization techniques for orthogonal frequency division multiple access (OFDMA): A tutorial review, *Proc. IEEE*, 95 (7), 1394–1427, 2007.
13. H. Minn and N. Al-Dhahir, Optimal training signals for MIMO OFDM channel estimation, *IEEE Trans. Wirel. Commun.*, 5 (5), 1158–1168, 2006.
14. M. Ghogho and A. Swami, Training design for multipath channel and frequency-offset estimation in MIMO systems, *IEEE Trans. Signal Process.*, 54 (10), 3957–3965, 2006.
15. S. Sezginer, P. Bianchi, and W. Hachem, Asymptotic Cramér-Rao bounds and training design for uplink MIMO-OFDMA systems with frequency offsets, *IEEE Trans. Signal Process.*, 55 (7), 3606–3622, 2007.
16. J. Chen, Y.-C. Wu, S. Ma, and T.-S. Ng, Joint CFO and channel estimation for multiuser MIMO-OFDM systems with optimal training sequences, *IEEE Trans. Signal Process.*, 56 (8), 4008–4019, 2008.
17. H. Zhu, B. Farhang-Boroujeny, and C. Schlegel, Pilot embedding for joint channel estimation and data detection in MIMO communication systems, *IEEE Commun. Lett.*, 7 (1), 30–32, 2003.
18. S. He, J. K, Tugnait, and X., Meng, On superimposed training for MIMO channel estimation and symbol detection, *IEEE Trans. Signal Process.*, 55 (6), 3007–3021, 2007.
19. J. Wang and X. Wang, Superimposed training-based noncoherent MIMO systems, *IEEE Trans. Commun.*, 54 (7), 1267–1276, 2006.
20. M. Dong and L. Tong, Optimal design and placement of pilot symbols for channel estimation, *IEEE Trans. Signal Process.*, 50 (12), 3055–3069, 2002.
21. S. Ohno and G. B. Giannakis, Optimal training and redundant precoding for block transmissions with application to wireless OFDM, *IEEE Trans. Commun.*, 50 (12), 2113–2123, 2002.
22. X. Ma, L. Yang, and G. B. Giannakis, Optimal training for MIMO frequency-selective fading channels, *IEEE Trans. Wirel. Commun.*, 4 (2), 453–466, 2005.
23. A. Vosoughi and A. Scaglione, Everything you always wanted to know about training: Guidelines derived using the affine precoding framework and the CRB, *IEEE Trans. Signal Process.*, 54 (3), 940–954, 2006.
24. IEEE Std 802.11a, Supplement to Part 11: Wireless LAN medium access control (MAC) and physical layer (PHY) specification: High-speed physical layer in the 5GHz band IEEE Std. 802.11a-1999, 1999.
25. E. Perahia, IEEE 802.11n Development: History, Process, and Technology, *IEEE Commun. Mag.*, 46 (7), 48–55, 2008.
26. ETSI. 2001. BRAN; HIPERLAN Type 2; Physical (PHY) Layer Specification (2nd ed.) Technical Specification 101 475. [Online]. Available: http://www.etsi.org

27. O. Besson and P. Stoica, On parameter estimation of MIMO flat-fading channels with frequency offsets, *IEEE Trans. Signal Process.*, 51 (3), 602–613, 2003.

28. P. Stoica and O. Besson, Training sequence design for frequency offset and frequency-selective channel estimation, *IEEE Trans. Commun.*, 51 (11), 1910–1917, 2003.

29. S. M. Kay, *Fundamentals of Statistical Signal Processing*, Englewood Cliffs, NJ: Prentice-Hall, 1993.

30. M. Biguesh and A. B. Gershman, Training-based MIMO channel estimation: A study of estimator tradeoffs and optimal training signals, *IEEE Trans. Signal Process.*, 54 (3), 884–893, 2006.

31. J. Chen, Y.-C. Wu, S. Ma, and T.S. Ng, Training design for joint CFO and channel estimation in multiuser MIMO OFDM system, in *Proc. IEEE Globecom*, pp. 3008–3012, November 2007.

32. O. Besson and P. Stoica, Training sequence selection for frequency offset estimation in frequency-selective channels, *Digital Signal Process.: Rev. J.*, 13 (1), 106–127, 2003.

33. R. A. Horn and C. R. Johnson, *Matrix Analysis*, Cambridge, London: Cambridge University Press, 1985.

34. C. Fragouli, N. Al-Dhahir, and W. Turin, Training-based channel estimation for multiple-antenna broadband transmissions, *IEEE Trans. Wirel. Commun.*, 2 (2), 384–391, 2003.

35. W. H. Mow, A new unified construction of perfect root-of-unity sequences, in *Proc. Spread-Spectrum Techniques and Applications*, vol. 3, pp. 955–959, 1996.

36. J. Coon, M. Beach, and J. McGeehan, Optimal training sequences for channel estimation in cyclic-prefix-based single-carrier systems with transmit diversity, *IEEE Signal Proc. Lett.*, 11 (9), 729–732, 2004.

37. B. Hassibi and B. Hochwald, How much training is needed in multiple-antenna wireless links, *IEEE Trans. Inf. Theory.*, 49 (4), 951–963, 2003.

38. Y. Liu, T. F. Wong, and W. W. Hager, Training signal design for estimation of correlated MIMO channels with colored interference, *IEEE Trans. Signal Process.*, 55 (4), 1486–1497, 2007.

39. P. Ciblat, P. Bianchi, and M. Ghogho, Training sequence optimization for joint channel and frequency offset estimation, *IEEE Trans. Signal Process.*, 56 (8), 3424–3436, 2008.

40. F. Gao, T. Cui, and A. Nallanathan, Optimal training design for channel estimation in decode-and-forward relay networks with individual and total power constraints, *IEEE Trans. Signal Process.*, 56 (12), 5937–5949, 2008.

41. H. Minn and S. Xing, An optimal training signal structure for frequency-offset estimation, *IEEE Trans. Commun.*, 53 (2), 343–355, 2005.

## Chapter 13

Fundamentals of OFDMA Synchronization

Romain Couillet and Merouane Debbah

Contents

13.1   Introduction...350
13.2   Effects of Timing and Frequency Shifts...................................352
     13.2.1   Origins of Parameter Offsets....................................352
          13.2.1.1   Static Effects.......................................352
          13.2.1.2   Dynamic Effects..................................353
     13.2.2   Performance Impacts of Parameter Offsets...................355
          13.2.2.1   Effects of STO.....................................355
          13.2.2.2   Effects of CFO.....................................357
          13.2.2.3   The Uplink Case...................................358
13.3   Synchronization Recovery..359
     13.3.1   STO Estimation..360
          13.3.1.1   Downlink STO Estimation.......................360
          13.3.1.2   Uplink STO Estimation..........................361
     13.3.2   CFO Estimation..362
          13.3.2.1   Rough CFO Estimation.........................362
          13.3.2.2   Fine CFO Estimation............................363
          13.3.2.3   CFO Estimation in Uplink.......................364
     13.3.3   Advanced Methods..366
13.4   A Case Study: 3GPP-LTE...366
13.5   Discussion..369
     13.5.1   Bayesian Framework..369
     13.5.2   Case Study: Bayesian CFO Estimation 370

13.6 Conclusion...373
References...373

13.1 Introduction

Before the fundamental work of Shannon in 1948 [1] and the introduction of the channel capacity, no theoretical bound for data transmission rate had been proposed; therefore, at that time, communication-related questions were investigated without any objective comparison tool or any performance evaluation bound. In the realm of synchronization, some 60 years later, the time has not yet come for such a unification of the field. That is, there exist no theoretical bounds on the amount of energy (or time) required for a transmitter–receiver link to synchronize their system parameters, for example reference frequency, timing, clock speed, and so on. Besides, there does not exist any theoretical derivation of the capacity of a time-limited communication taking into account the need for synchronization and channel estimation. In spite of a few recent contributions [2,3], the amount of bits dedicated to synchronization that is needed to maximize the transmission capacity is as yet unknown; too little synchronization effort leads to numerous decoding errors, while too much synchronization effort leaves little room for bits dedicated to the actual communication. In both extreme scenarios, the impact on capacity is disastrous but still no satisfying *reliability* versus *spectrum efficiency* trade-off study has yet been performed. In fact, one might even say that the processes dedicated to synchronization should not be isolated from the effective data to be transmitted, as both synchronization parameters and useful data are equally unknown entities to the receiver; instead the whole *data plus synchronization parameters* should be encoded in such a way to achieve the optimal transmission rate in a finite time. Therefore the whole field of synchronization is not yet fully understood and the set of proposed synchronization parameter recovery processes is only based on many different solutions, which are not unified by strong theoretical bases. The latter attempt to tackle either individual parameter estimation problems or joint decoding, joint channel estimation and synchronization, and so on. In the coming sections, those solutions will be divided into *rough* (also referred to as *coarse*) or *fine* estimators and *data-aided* (DA) or *non-data-aided* (NDA) algorithms; however, this division is merely conventional and does not reflect any theoretical foundation for synchronization, as will be discussed in Section 13.5.

If most academic studies on orthogonal frequency-division multiplexins (OFDM) often consider ideal synchronization, the reader must understand that synchronization is an important task and, as such, should not be undermined. solutions such as wireless local area networks (WLAN) were already available in the early 90's, some 15 years were needed for the first mobile OFDM systems to appear. The difficulty in OFDM is to preserve the orthogonality between subcarriers when mobile terminals are in motion and thus are subject to Doppler frequency shifts. Besides, wireless networks encourage more and more *packet-switched* (e.g., IEEE 802.16 WiMax [4], Third Generation Partnership Project (3GPP)-Long Term Evolution (LTE) [5]) than *connected* (e.g., digital video broadcasting [6] and digital audio broadcasting [7]) transmission modes: the former has the strong advantage to be highly dynamic and has coped with its past latency problem. Future communication technologies will therefore rely on short data (i.e., packet) transmissions, compelling the synchronization recovery processes to operate very fast.* In most OFDM technologies, the synchronization phase consists first a *power detection* process, meant to roughly identify a power source. The next procedure is classically

---

* Note also that the progressive integration of mobile data transfers to the Internet requires one to adapt to packet-switched communications.

an *acquisition* phase aiming to give a rough estimate of the system synchronization parameters (e.g., slot start time, a reference frequency). First reliable data exchanges are in general possible at a low rate at this point. During the rest of the communication, especially in the connected mode, the synchronization processes enter the *tracking* phase to regularly update the parameter estimates. In the rest of this chapter, we concentrate on the following three main synchronization parameters:

■ The carrier frequency offset (CFO), which corresponds to a mismatch between the transmitter and the receiver frequency references. Even small values of CFO are detrimental to the system performance; as a consequence, frequency offsets must be efficiently corrected. Particularly, in a mobile system, CFO is fast varying due to Doppler shifts. It is thus very challenging for mobile OFDM designers to ensure a continuous quality of service in high mobility conditions.

■ The symbol timing offset (STO), which is defined as the time difference between the real and estimated beginnings of the received OFDM symbol. As detailed in the subsequent sections, small STOs are not critical for OFDM since the length of the cyclic prefix (CP), usually slightly longer than the maximum channel delay spread, can absorb negative time offsets: this avoids intersymbol interference; moreover, STO correction might not even be necessary, since the induced frequency rotation effect can be considered as part of the communication channel; therefore, channel estimation usually allows one to conceal the STO problem. As a consequence, STO estimation for OFDM is less tackled in the literature than CFO. However, when no channel estimation is performed (so typically during the initial synchronization procedure), the phase rotations introduced by timing offsets in the received frequency-domain signal might disturb the synchronization processes.

■ The sampling clock offset (SCO), which is in general a negligible effect of misalignment between the local oscillators of the two communicating ends. Typically, a shift in those oscillators is due to the physical sensitivity (e.g., temperature and pressure variations) of embedded crystal oscillators. Since symbol timing shifts due to SCO are classically harmful only after the reception of hundreds or thousands OFDM symbols, SCO synchronization does not need to be performed very often.

In a mobile distributed system, those parameters need to be estimated both at the mobile devices and at the fixed base station to ensure reliable communication in both downlink and uplink. The synchronization procedure at the mobile receiver is often treated as multiple *single-parameter estimation* problems or as a *joint parameter estimation* problem. In a single-user (SU) scenario in which the communication link to the base station is exclusive to a specific user, synchronization at the base station is similar to synchronization at the mobile terminal. However, in systems based on orthogonal frequency division multiple access (OFDMA), all terminals transmitting in the uplink have different parameter offsets, so that the allocated user bands overlap one another. For the base station, this means the users cannot be separated in the frequency domain. One major consequence is the introduction of strong intercarrier interference (ICI), which means, from a processing viewpoint, that discrete Fourier transform (DFT) operations are better avoided for synchronization purposes at the base station. Most of the classical frequency-domain synchronization algorithms are therefore unavailable to the base station; this constitutes a fundamental difference between OFDM and OFDMA systems. Note that the parameter offsets in the uplink are in general very small when downlink and uplink communications are scheduled along a time division multiplexing (TDD) strategy, for which time and frequency references are the same in uplink and downlink. For this data duplexing scheme, usually in practice, the transmitting terminals already have a good estimate of the time and frequency offsets based on the primary downlink transmissions [i.e., primary synchronization sequences (PSS)] that allow for an initial rough synchronization. As a consequence, in TDD, it will be in general acceptable to consider small STO and CFO for synchronization algorithms in the uplink. On the contrary, in frequency division duplex (FDD) mode, which is used more often in practice,* one can only share timing synchronization between downlink and uplink transmissions, if both links are time-synchronous. However, frequency references being different in downlink and uplink, in FDD, one has to assume potentially large CFO.

Notation: In the following, boldface lower-case symbols represent vectors, capital boldface characters denote matrices ($\mathbf{I}_N$ is the $N \times N$ identity matrix). The transpose and Hermitian transpose are denoted $(\cdot)^T$ and $(\cdot)^H$, respectively. The symbol E[·] denotes expectation.

13.2 Effects of Timing and Frequency Shifts

13.2.1 Origins of Parameter Offsets

The synchronization offsets originate from various physical phenomena. Some are due to static hardware defects (e.g., SCO and CFO) or to an imperfect initial synchronization process (e.g., STO, CFO) while others are mainly due to dynamic effects that depend on the channel conditions (e.g., Doppler shifts in CFO).

13.2.1.1 Static Effects

The central bandwidth frequency and the sampling frequency are always imposed by the technology standard. All communicating entities are then required to align to these frequencies. However, the precision of the hardware material is often impacted by environmental conditions. Typically, in a mobile phone device, both SCO and CFO are aligned on the embedded *crystal oscillator* frequency. Those crystals are sensible to external conditions such as pressure, temperature, and aging. A mismatch between the oscillator frequencies at the transmitter and the receiver causes frequency offsets. In practice, this mismatch is the main reason for SCO. Since both offsets are closely related, oscillator mismatch might also be the main explanation for CFO. However, this statement only stands when the communication channel is static. Indeed, when at least one communicating side is in motion, then dynamic Doppler effects come into play and usually become the critical reason to explain CFO. In Figure 13.1, the typical response to temperature of the cheap digital crystal oscillator (DXO) and the onerous voltage controlled temperature compensated crystal oscillator (VCTCXO) is depicted. Note that the effect of temperature is rather important: if the central frequency is set to 2 GHz, then a frequency drift of ±5 ppm corresponds to ±10 KHz, which is of the order of the typical subcarrier spacings.

STO is of a different nature. Indeed, while SCO and CFO might be ideally null before the beginning of any communication (assuming perfect crystal oscillators at both communication ends and no motion), STO appears when the communication begins, since too little prior information is available for both communicating entities before their first handshake. To align the timing references, the beginning of each OFDM symbol must be identified. However, as will be presented in Section 13.2.2, the symbol timing parameter is not required to be finely tracked since even a rough estimate

---

* The TDD mode has the strong disadvantage to require a thorough synchronization of transmissions in the time-domain. In particular, guard periods need be taken into account that absorb the (potentially long) propagation delay. The strong advantage of TDD, however, lies in an easy tuning of the ratio downlink rate/uplink rate, which is fixed in SU OFDM with FDD.

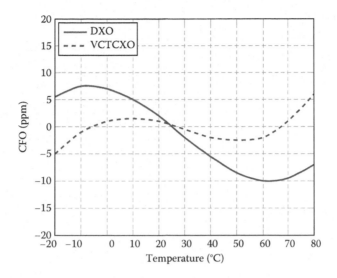

Figure 13.1 **CFO effect on crystal oscillators due to temperature.**

might not lead to any performance loss. By "rough," we mean here that the timing error is not larger than the CP duration.

Those parameters usually do not encounter practical synchronization issues. Once the reference timing and sampling rate are appropriately estimated, those parameters do not significantly change during the overall communication process. If the time for communication is rather long, for example long enough for the local temperature to change, then refinements on the STO and SCO are desirable but do not usually face any difficulty. In mobile multicell networks, when a terminal hands over neighboring unsynchronized base stations, this initial synchronization process will be triggered anew.

13.2.1.2 Dynamic Effects

By dynamic effects, we refer to the fast varying phenomena that impact the synchronization parameters of the system. In particular, in mobile OFDM systems, the relative distance $d(t)$ between the transmitting and receiving entities varies with time t. Consider the simple scenario of a fixed base station transmitting a sinusoidal waveform $x(t)$ of period T_0, and a mobile handset at initial distance $d_0 = d(0)$ from the base station moving at a constant speed $v \ll c$ (with c being the light speed) at an angle ϕ from the base station–handset direction. This scenario is presented in Figure 13.2. Using Al-Kashi's geometrical relations, at time $t = T_0$,

$$d(T_0) = \left(d_0^2 + v^2 T_0^2 - 2vd_0 T_0 \cos(\phi)\right)^{1/2}.$$

Therefore, the relative frequency of the sent signal, that is $f_{BS} = 1/T_0$ from the base station's viewpoint, is different from the handset's viewpoint and equals

$$f_H = \left(T_0 \sqrt{1 + \left(\frac{v}{c}\right)^2 - 2\frac{v}{c}\cos(\phi)}\right)^{-1}, \tag{13.1}$$

$$\simeq f_{BS}\left(1 - \frac{v}{c}\cos(\phi)\right)^{-1}, \tag{13.2}$$

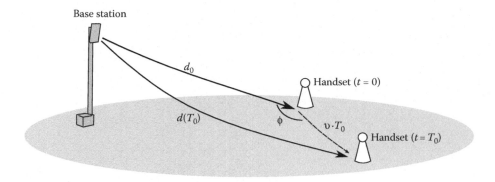

Figure 13.2 Doppler frequency shift effect.

$$\simeq f_{BS}\left(1 + \frac{v}{c}\cos(\phi)\right) \tag{13.3}$$

for ϕ not too close to $\pi/2$ (otherwise the Taylor coefficient of second order must be taken into account).

This relative shift in frequency is referred to as the *Doppler effect*. The received signal $y(t)$ then reads

$$y(t) = \rho x(t - \tau)e^{2\pi i\xi t} + w(t) \tag{13.4}$$

with $w(t)$ being the additive noise process, ρ being the channel attenuation, τ being the propagation delay and $\xi = f_H - f_{BS}$ being the Doppler shift. To understand the $e^{2\pi i\xi t}$ factor, one needs to observe that in the Fourier domain, due to the Doppler effect, the received signal originates from the transmitted signal convolved by a frequency shift, that is, a Dirac function, of amplitude ξ; back in the original domain, this Dirac convolution turns into a complex exponential product.

The model 13.4 is generalized to practical realistic situations where not only one but numerous scatterers are present in the medium. Those scatterers are gathered into subsets of common Doppler shift and propagation delay. This yields the model

$$y(t) = \int_\tau \int_\xi \rho(\xi, \tau)x(t - \tau)e^{2\pi i\xi t} \, d\xi \, d\tau, \tag{13.5}$$

where $\rho(\xi, \tau)$ accounts for the mean (complex) fading of the subset of scatterers that induce a propagation delay τ and a Doppler shift ξ.

The *Doppler spectrum* $D(\xi)$, which denotes the relative signal power received at Doppler shift ξ, is computed as

$$D(\xi) = \int_\tau E[|\rho(\xi, \tau)|^2] \, d\tau \tag{13.6}$$

from which one derives the *Doppler spread* B_d, defined as the standard deviation of the random variable ξ (whose density function is given by $D(\xi)/\{\int_\xi D(\xi)d\xi\}$),

$$B_d = \left(\frac{\int_\xi (\xi - \xi_0)^2 D(\xi) \, d\xi}{\int_\xi D(\xi) \, d\xi}\right)^{1/2} \tag{13.7}$$

with ξ_0 being the mean value of ξ.

CFO estimation consists of tracking the value of ξ_0 so as to minimize the effects of frequency offsets on the received signal. Note in particular that a large Doppler spread would be detrimental to the decoding of the received OFDM symbol. Indeed, as shall be discussed in Section 13.2.2, if much power is received outside the exact subcarrier frequency, then the decoding bit error rate (BER) dramatically increases. However, large Doppler spreads typically come along with very short *channel coherence time* (i.e., the time during which the consecutive channel realizations are strongly correlated), which does not satisfy the OFDM fundamentals that require the channel realization to be constant at least during an OFDM symbol.*

13.2.2 Performance Impacts of Parameter Offsets

Thanks to the time–frequency duality of the OFDM modulation, the effects of STO, SCO, and CFO are very similar. Basically, a constant offset in a representation domain translates into a phase rotation in the dual Fourier domain. However, for every particular offset, some fundamental differences arise that we develop in the following.

Consider an OFDM system of N subcarriers, N_{CP} CP samples and sampling period T_s. Therefore, the subcarrier spacing Δ_f equals $1/(NT_s)$. For notational ease, the entries of the discrete vectors $\mathbf{a} = [a_1, \ldots, a_N]^T$ sampled from a continuous waveform $a(t)$ are denoted by $a_k = a(t_0 + kT_s)$ with t_0 being the beginning of the OFDM symbol. The OFDM data symbol to transmit is denoted by $\mathbf{s} = [s_1, \ldots, s_N]^T$; its variance $E[\mathbf{s}^H\mathbf{s}]$ is denoted by P. The time-domain OFDM symbol vector $\mathbf{x} = [x_1, \ldots, x_N]^T = \mathbf{F}^H\mathbf{s}$, with $\mathbf{F}$ by the Fourier matrix of size N, is sent through the channel $\mathbf{h} = [h_1, \ldots, h_L]^T$ of delay spread L symbols (we assume that $L \leq N_{CP}$). The noisy time-domain received signal is called $\mathbf{y} = [y_1, \ldots, y_N]^T$ and its Fourier dual is denoted by $\mathbf{r} = [r_1, \ldots, r_N]^T = \mathbf{F}\mathbf{y}$. The OFDM system aims to decode $\mathbf{r}$ from the original data $\mathbf{s}$ with the smallest possible BER. Under perfect synchronization, we have the classical discrete channel convolution effect, for all $n \in \{1, \ldots, N\}$

$$y_n = \sum_{l=0}^{L-1} h_l x_{n-l} + w_n \tag{13.8}$$

with $\mathbf{w} = [w_1, \ldots, w_N]^T$ by the noise process of variance $E[\mathbf{w}^H\mathbf{w}] = \sigma_w^2$.

13.2.2.1 Effects of STO

Consider now that the system comprises a single transmitter and that the timing synchronization to the receiver under investigation is offset by θT_s. Equation 13.8 becomes

$$y_n = \sum_{l=0}^{L-1} h_l x_{n-l-\theta} + w_n. \tag{13.9}$$

Assuming an infinitely small energy acquisition time at the receiver and a perfect square pulse shape for the time-domain signal, θ can be taken as an integer without generality restriction.

If $0 \leq \theta \leq N_{CP} - L$, then the received OFDM symbol does not suffer from the channel leakage due to previous blocks. Then the CP property and hence the orthogonality property hold. The

---

* Indeed, if the channel varies during one OFDM symbol, the channel matrix in the time domain is no longer circulant and then no longer diagonalizable in the Fourier basis.

output of the DFT block at the receiver therefore outputs, for $k \in \{1, \ldots, N\}$,

$$r_k = e^{2\pi i(k\theta/N)} H_k s_k + W_k, \tag{13.10}$$

where $[H_1, \ldots, H_N]^T = \mathbf{Fh}$ and $[W_1, \ldots, W_N]^T = \mathbf{Fw}$, respectively, denote the Fourier transform of $\mathbf{h}$ and $\mathbf{w}$.

This results in a phase rotation of the received symbols. As shall be detailed in Section 13.3, this effect is easily corrected and might even be harmless. By incorporating the phase rotation into the channel frequency response: $e^{2\pi i(k\theta/N)} H_k$, a mere channel estimation process suffices to absorb the STO effect.

However, if $\theta \notin [0, N_{CP} - L]$, then the system orthogonality collapses, with the direct consequence of introducing intersymbol interference (ISI) from adjacent OFDM blocks. The system output is generally modeled [8] as the expected DFT weighted by an attenuation factor $\alpha(\theta)$ plus an additional interference process $I(\theta)$ of power σ_I^2 due to ISI,

$$r_k = e^{2\pi i(k\theta/N)} \alpha(\theta) H_k s_k + I(\theta) + W_k. \tag{13.11}$$

The relative performance loss is classically measured through the degradation $\gamma_{STO}(\theta)$ between the signal-to-noise ratio (SNR) in the synchronized (SNR$_{\text{sync}} = P/\sigma_w^2$) and the unsynchronized cases (SNR$_{\text{unsync}} = P\alpha(\theta)^2/\sigma_w^2 + \sigma_I(\theta)^2$) [8],

$$\gamma_{STO}(\theta) = \frac{SNR_{\text{sync}}}{SNR_{\text{unsync}}} = \frac{1}{\alpha(\theta)^2} \left(1 + \frac{\sigma_I(\theta)^2}{\sigma_w^2}\right), \tag{13.12}$$

the behavior of which is depicted in Figure 13.3 for an OFDM system with $N = 128$ DFT size, $N_{CP} = 9$ CP, communicating through an exponential decaying channel of length $L = 5$.

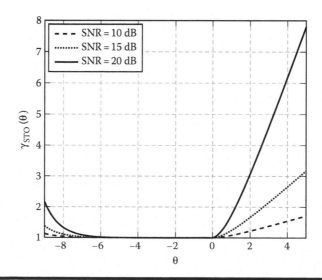

Figure 13.3 **Performance decay due to STO—$N = 128$, $N_{CP} = 9$, exponential decaying channel with $L = 8$.**

In multiple access uplink scenarios, the problem is more involved and might be very harmful to the system performances. Indeed, every user k faces a different STO θ_k so that, even when all θ_k belong to the ISI-free region (i.e., $0 \leq \theta_k \leq N_{CP} - L_k$ with L_k being the length of the channel seen from user k), the DFT output at the receiver introduces multiple access interference (MAI). In those situations, the performance limiting factor is linked to the largest STO gap ($\max_{k,k'} |\theta_k - \theta_{k'}|$) among all pairs of users. Therefore system performance is dictated by the ill-conditioned users; this is one of the reasons [the peak to average power ratio (PAPR) problem and the similar SCO and STO effects studied in the following sections are other reasons] why OFDMA is rarely used in uplink schemes in practice.

13.2.2.2 Effects of CFO

Suppose now perfect timing synchronization (i.e., $\theta = 0$) and introduce a frequency offset δ between the transmitter and the receiver. To observe the consequences of frequency offsets, the continuous frequency representation of the OFDM signals must be examined. The receive symbol of Equation 13.8 is there updated as

$$y_n = e^{2\pi i (n\delta/N)} \sum_{l=0}^{L-1} h_l x_{n-l} + w_n.$$ (13.13)

Assuming again perfect square pulse shaping in time, after some computation, the signal at the output of the receiver DFT is [9]

$$r_k = e^{2\pi i \delta ((N+N_{CP})/N)} \sum_{k'=1}^{N} H_{k'} s_{k'} \, \text{sinc} \left(\pi [\delta + k' - k] \right) e^{i\pi((\delta+k'-k)(N-1)/N)} + W_k$$ (13.14)

in which we remark that when δ is a multiple of the subcarrier spacing Δ_f, that is $\delta = p \in \mathbb{Z}$, the sum in Equation 13.14 reduces to a single nonnull term that corresponds, up to a constant phase, to the data symbol intended to the pth next subcarrier. Therefore, *integer* frequency offsets merely engender a phase rotation and a circular shift of all subcarriers. The adaptive decoding processes required to compensate for integer CFOs are therefore not a challenging task.

However, if δ is fractional, every received sample r_k suffers from ICI from all subcarriers (and not only from neighboring subcarriers). Following Speth's SNR degradation measure γ_{CFO} [10], the performance loss for small values of δ is approximated by

$$\gamma_{CFO}(\delta) = \frac{SNR_{sync}}{SNR_{unsync}} \simeq 1 + \frac{\pi^2 \delta^2}{3} SNR_{sync}.$$ (13.15)

As depicted in Figure 13.4, the performance is dramatically impacted even for small values of δ. For instance, at $SNR_{sync} = 20$ dB, a CFO of 4% $\times \Delta_f$ leads to $SNR_{unsync} \simeq 18$ dB, which might turn out to be a sufficient loss to prevent the transmission of a 64-QAM modulation for instance. Fast CFO estimators are then required to recover synchronization.

Identically to the STO increased complexity in OFDMA schemes, CFO in multiple access technologies is more involved to compensate. These topics are discussed in the subsequent sections.

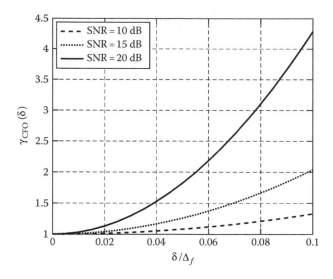

Figure 13.4   Performance decay due to CFO.

13.2.2.3 The Uplink Case

We dedicate a section to the OFDMA uplink, since the major difficulty in OFDMA synchronization lies in the uplink. For this reason and also because of the PAPR problem, OFDMA is not often used in the uplink of centralized mobile networks. For instance, in 3GPP-LTE, single-carrier frequency division multiple access (SC-FDMA) is used in the uplink in place of OFDMA. The uplink synchronization issue is twofold: (i) multiple users face different STO and CFO, turning the parameter estimation problem into a vectorial-parameter estimation problem, (ii) contrary to the SU scenario where STO and CFO effects can be counteracted at the receiver (e.g., counter-rotation of CFO shift and clock-adjustment to STO delay), the problem of general multiparameter offsets is only solved via maximum-likelihood NP-hard algorithms.

The model for the uplink of an OFDMA cell with M transmitting users indexed by $m \in \{1,\ldots,M\}$, with respective STO θ_m and CFO δ_m, reads

$$y_n = \sum_{m=1}^{M} e^{2\pi i \delta_m(n/N)} \sum_{l=0}^{L-1} h_l^{(m)} s_{n-l-\theta_m}^{(m)} + w_n, \qquad (13.16)$$

where the subscript (m) indicates that the considered channel and signal belong to the mth user.

In the frequency domain, assuming $\delta_m = 0$ for all m,[*] from Equation 13.14, the post-DFT receive signal reads

$$r_k = \sum_{m=1}^{M} e^{2\pi i \delta_m((N+N_{CP})/N)} \sum_{k' \in \mathcal{S}_m} H_{k'}^{(m)} s_{k'}^{(m)} \operatorname{sinc}\left(\pi[\delta_m + k' - k]\right) e^{i\pi((\delta_m+k'-k)(N-1)/N)} + W_k,$$

$$(13.17)$$

---

[*] Or, as will be seen later, assuming equivalently that δ_m is included into the channel $H^{(m)}$.

where $\mathcal{S}_m$ is the set of subcarriers allocated to user m (these sets are obviously mutually exclusive in this case).

As suggested above, while in Equation 13.14 it is clear that changing k to $k - \delta$ solves the CFO problem (i.e., by an appropriate shift of the radio interface frequency reference), it is impossible here to undo the ICI effect by a mere frequency shift at the receiver. Assuming large frequency offsets δ_m, the ICI effect on the general performance is dramatic and cannot be completely annihilated for orthogonality between users cannot be recovered.

In the following sections, we provide techniques and algorithms that allow us to recover STO and CFO. The scope of these sections is restricted to the main key methods used in practice. In the literature on OFDM synchronization, and synchronization at large, there exist a large number of other methods, so the authors do not claim that they have gathered in the following pages all the contributions in the synchronization field. Also, some recent work from the authors are presented in the last sections, which provide an information-theoretical Bayesian view on synchronization.

13.3 Synchronization Recovery

Synchronization recovery is an information theoretic dilemma. The ideal transmission scheme on a given channel, whose performance is assessed thanks to its ergodic capacity [1], contains no *excess bandwidth* (i.e., no *useful* information is transmitted to an extent more than necessary). However, synchronization parameters, which need to be shared or estimated by both communication ends, are considered *nonuseful* information for the data transmission purpose. As a consequence, two situations classically arise in practical OFDMA systems:

■ Specific pilot sequences are transmitted to allow fast synchronization at the receiver. These techniques, qualified as DA, have been used in most of the existing telecommunication systems for they are easier to implement and allow for fast synchronization. However, they imply transmitting nonuseful data at the expense of a reduction in the achievable useful data rate. This statement is even more verified for systems such as mobile communication handsets that require us to constantly track the synchronization parameters: in those scenarios (see e.g., 3GPP-LTE, Section 13.4), many pilot sequences might be used for parameter estimation purposes.

■ Parameter estimation is conducted by exploiting excess bandwidth inherent to the system. If the communication scheme shows good transmission rate performance relative to the channel capacity, then little excess bandwidth is available so that those estimators are usually very slowly converging. Moreover, the excess bandwidth might turn out very impractical to exploit, contrary to pilot sequences designed for synchronization purpose. Therefore, these processes, often referred to as NDA, are usually complemented by DA methods. Some other schemes similarly exploit excess bandwidth due to transmitted constellations, redundancy due to excessive channel coding, and so on. These are usually isolated from the NDA group into the special class of *blind techniques*. Following our excess bandwidth philosophy, we shall indifferently qualify them as NDA or *blind* in what follows.

Regarding inherent redundancy, the OFDM case is particularly simple. Thanks to the subcarrier orthogonality, the *spectral efficiency* (i.e., how much of the frequency spectrum is used) achieves the theoretic Shannon's ergodic capacity. However, in the time domain, the CP duration is completely lost for useful communications for it consists of a mere copy of transmitted symbols that are discarded at the receiver. This CP therefore constitutes the major part of the OFDM excess bandwidth. That is why, even 15 years ago, the pioneering work on OFDM synchronization [11,12] exploited symbol repetition either in dedicated pilots or in the CP.

As previously mentioned, the classical approach to synchronization is a multistep process: quick and rough estimators are firstly used before advanced tools perform refined estimates. We shall review in the following the main historical synchronization techniques found in the current literature.

13.3.1 STO Estimation

13.3.1.1 Downlink STO Estimation

A first very rough STO estimation is often handled as a first synchronization step in OFDM. Indeed, as long as the beginning of the OFDM sequence is not approximately found, pilot sequences cannot be read and in particular DFT operations cannot be performed without being impacted by a strong ISI from consecutive OFDM blocks. The very rough STO estimator often consists of a mere correlation process with a pilot sequence designed to enjoy desirable correlation properties. This is the case, in particular, with the popular Zadoff–Chu (ZC) sequences [13] with properties detailed in Section 13.4. At the end of this first STO process, the OFDM symbol timing error is expected to be less than the CP length.

When this very rough estimation is obtained, classical methods are used to perform the so-called rough, or coarse, STO acquisition. The latter consists of exploiting the time correlation properties of a repetitive structure *insensitive to CFO* so that CFO can be evaluated in a posterior phase. It is desirable to carry out the first estimates in the time domain since, as we already noted, even small mismatches in the synchronization parameters spawn dramatic signal distortion after DFT processing. For instance, in Ref. [14], a pilot sequence $\mathbf{x}$ made of the concatenation of two identical vectors $\{x_1, \ldots, x_{N/2}\} = \{x_{N/2+1}, \ldots, x_N\}$ of size $N/2$ is designed for STO estimation. The time-domain received sequence reads

$$\begin{cases} y_n & = e^{2\pi i \delta n/N} \sum_{l=0}^{L-1} h_l x_{n-l-\theta} + w_n, \\ y_{n+N/2} & = e^{2\pi i \delta n/N} e^{\pi i \delta} \sum_{l=0}^{L-1} h_l x_{n-l-\theta} + w_{n+N/2}. \end{cases} \qquad n < N/2 \qquad (13.18)$$

Thanks to a window of size $N/2$ sliding along hypothetic values for θ, the absolute value of the cross-correlation between the first and second part of $\mathbf{y}$ is computed. This allows us to remove the CFO rotation effect. The maximum value $\hat{\theta}$ of the correlations is then sought to generate the STO estimate,

$$\hat{\theta} = \arg\max_{\tilde{\theta}} \frac{\left| \sum_{n=1}^{N/2} y_{n+\tilde{\theta}} y^*_{n+N/2+\tilde{\theta}} \right|}{\sum_{n=1}^{N/2} \left| y_{n+\tilde{\theta}} \right|^2}. \qquad (13.19)$$

Note that this maximum is usually not unique. Indeed, as described in Section 13.2.2, when the CP is longer than the channel delay spread, then as long as $L - N_{CP} < \theta < 0$ the fundamental subcarrier orthogonality is preserved. This indicates that the solution of Equation 13.19 is not a unique value but a continuum of size $N_{CP} - L$. Note that this also allows for a rough estimation of the channel length. This is pictured in Figure 13.5 in which an exponential decaying channel of

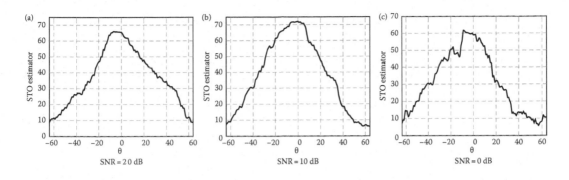

Figure 13.5 STO estimation for different SNR values, $N = 128$, $N_{CP} = 9$, exponential decaying channel with $L = 5$.

length $L = 5$ is used for an OFDM system with $N = 128$, $N_{CP} = 9$, under different SNR values. Some, seeing in this *plateau* a synchronization inconvenience, proposed refined algorithms [15] that result in a smaller continuum of solutions (containing the perfect synchronization value) at the expense of a higher false alarm rate in the detection of the maximum.

The same method can be used without reference signals thanks to the OFDM inner redundancy. Indeed, if the CP is larger than the channel length, then $N_{CP} - L$ symbols are duplicated in the signal and the STO can be therefore blindly estimated by cross-correlation of the CP symbols. However, this technique is rarely used in practice for its reliability depends on the channel conditions (e.g., N_{CP} might not be fairly larger than L and the correlation size might be very small). Note that all those techniques have the strong advantage to be independent of the channel realization, which is a feature typically sought when one does not have access to any channel estimation.

There does not exist a large literature for fine OFDM timing estimation, at least in the downlink case. Indeed, provided that the compensated STO after the estimation processes verifies $L - N_{CP} \leq \theta \leq 0$, the consequence of a synchronization mismatch is a mere symbol rotation in frequency. When performing channel estimation, this rotation might be seen as part of the channel, with an increased frequency selectivity. As a consequence, as long as the channel estimation procedure can cope with the increase of the channel frequency selectivity, the performance of OFDM decoding in downlink is not impaired.

13.3.1.2 Uplink STO Estimation

In uplink OFDMA, as already mentioned, the STO problem is slightly more involved due to the multiple STO values involved. A classical solution to cope with this multiple STO issue is for the whole system to align downlink and uplink timing. Indeed, from the downlink timing information, the user already has a synchronized uplink STO up to twice the propagation delay. If the system allows for a large enough CP length (i.e., large enough to over the channel delay spread and the double propagation delay), timing offsets coupled with channel estimation for every user's handset do not produce any harm to the system performance. However, spectral efficiency and overall throughput performance suffer from the CP extension and therefore only short spatial coverage is tolerable in such uplink OFDMA technologies. If the CP is limited to the maximum channel delay spread of all users, then the synchronization problem is heavily more critical and requires exhaustive multiparameter search is (e.g., joint decoding and timing acquisition) for all θ_k, $k \in \{1, \ldots, K\}$.

Another classical solution for uplink synchronization, which has benefits both in the time and frequency domains, is to allocate sets of contiguous subcarriers to every uplink user. To avoid frequency overlap due to additional CFO problems, frequency guard bands, that is nonallocated subcarriers, are placed between these sets. This allows the receiver to individually treat each user by filtering out the other users, with a minimal impact of ICI due to hypothetical CFO problems. Then the STO of every user can be estimated independently of the other users. The same technique as in the SU case can then be used. Equation 13.19 is still valid on a per-user basis, but here the noise term w_n in Equation 13.18 also contains interference contribution from residual ICI. However, using contiguous blocks for all users reduces the available frequency selectivity for every user, especially when a large number of users are present in the OFDMA cell. Indeed, such a subcarrier allocation makes every user very sensitive to deep channel fades. In practice, a simple workaround consists using high-level user scheduling, such as frequency-hopping techniques [16]. If for some reason, such as short packet transmissions,* interleaved carrier allocation is demanded, then practical computationally cheap STO estimations are yet unknown.

In order to cope with short-time transmissions issues in OFDMA when users are allocated sets of contiguous subcarriers, the authors propose in Ref. [17] an alternative solution to the OFDM modulation, referred to as α-OFDM, which provides additional frequency diversity at a minimal implementation cost. This novel modulation scheme allows users to dynamically exploit side frequency bands by sacrificing a few subcarriers on the edge of the total bandwidth. α-OFDM brings in a particularly significant outage capacity gain when users are allocated very small frequency bands, compared to the total bandwidth.

13.3.2 CFO Estimation

13.3.2.1 Rough CFO Estimation

Acquisition and tracking of the frequency offsets are the most critical synchronization tasks. The first reason was studied in Section 13.2.2: a small mismatch between local oscillators entails dramatic system performance losses. CFO estimation is also made difficult by the Doppler effect, introduced in Section 13.2.1; in short coherence time channels, every new data transmission is subject to a different frequency shift, which demands fast CFO tracking.

Similarly to the STO case, it is common to perform a very rough CFO estimation prior to any accurate CFO estimation so as to align the DC-equivalent frequencies from the base station and the terminals up to more or less one subcarrier spacing. This can be handled, like in the STO case, by correlating a training sequence with different frequency-shifted copies of this sequence. Since the channel is not known at this early step and this estimate can be impaired by different sources of interference, the process is not very reliable. Therefore, the estimation range sought for the CFO at this stage is typically of the order of the subcarrier spacing. From this point on, rough STO estimation is performed and then proper CFO evaluation can be processed.

Historically, Moose [11] was the first to provide a DA technique for CFO estimation, which is independent of the channel realization. Similarly to the STO estimation, Moose proposes a pilot OFDM symbol **x** composed of two identical vectors of size $N/2$. Assuming a prior STO estimation, the CFO effect in time (see Section 13.2.2) is a phase rotation of the transmitted symbols by an angle proportional to the time index. Therefore, the correlation of the first and second halves of the

---

* Short packet transmissions, moreover, lead us to consider performance in terms of outage capacity, instead of long-term ergodic capacity.

received data symbol results, for all $n \in \{1, \ldots, N\}$, in

$$y_n y^*_{n+N/2} = \left(e^{2\pi i \delta n/N} \sum_{l=0}^{L-1} h_l x_{n-l} + w_n \right) \left(e^{2\pi i \delta n/N} e^{\pi i \delta} \sum_{l=0}^{L-1} h_l x_{n-l} + w_{n+N/2} \right)^*, \quad (13.20)$$

$$= e^{-\pi i \delta} \left| \sum_{l=0}^{L-1} h_l x_{n-l} \right|^2 + \tilde{w}_n, \quad (13.21)$$

where $\tilde{w}_n$ includes the double products and the noise correlation, of null average.

Summing up coherently the $N/2$ correlations leads to the estimate $\hat{\delta}$ of δ,

$$\hat{\delta} = \frac{1}{\pi} \tan^{-1} \left(\frac{\sum_{n=1}^{N/2} \Im(y_n y^*_{n+N/2})}{\sum_{n=1}^{N/2} \Re(y_n y^*_{n+N/2})} \right), \quad (13.22)$$

The two main limitations in this approach are: (i) the effective acquisition range that is limited to $\delta \in [-\pi, \pi]$ (or equivalently to the length of the subcarrier spacing), and (ii) in the low SNR region, the noise $\tilde{w}_n$ is very strong since it contains components originating from cross-correlation to the pilot. As a consequence of (i), only the decimal part of the frequency offsets can be identified through this method. Moose proposes [11] solutions to enlarge the acquisition range at the expense of a reduction in the estimation resolution. Many schemes based on the latter were then successively proposed to enhance the performance trade-off between acquisition range and resolution, the most popular of these being the Schmidl and Cox [14] and the Morelli and Mengali algorithms [18]. All those schemes are particularly adapted to circuit-switched communications or low-speed mobile systems and show high accuracy in the CFO acquisition, especially for high SNR regimes. Indeed, they require specific pilot sequences that should not be made available at many time symbols (otherwise having a strong impact on the system spectral efficiency). Those are therefore not suitable for fast varying channels or short data transmissions.

To cope with this double issue, Van de Beek [12] considers a blind CFO estimator based on the CP redundancy. The principle is essentially identical to Moose's estimator but here the correlation is obtained between the last samples of each OFDM symbol and its CP copy. Unfortunately, this method ideally works in additive white Gaussian noise (AWGN) channels, which are mono-path channels. The ISI leakage due to multipath environments is detrimental to the performance of Van de Beek's method. To correct this deficiency, Ai et al. [19] suggest a reduction in the correlation window to result in an ISI-free estimate. Typical performances of these methods are depicted in Figure 13.6 for $N = 128$, $N_{CP} = 9$ and different channel configurations and SNR. These NDA estimators have the strong advantage not requiring any dedicated sequence, but pose the main issue that they are slowly converging, especially for high channel delay spread and short CP length.

13.3.2.2 Fine CFO Estimation

When channel estimation can be performed, the previous CFO estimation problem is less involved and advanced accurate algorithms can be performed, which can take into account all the available information (received data, pilots, source coding structure, etc.). While the previous algorithms

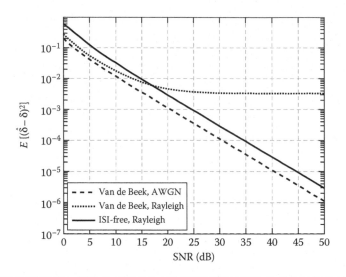

Figure 13.6 Rough CFO estimation versus SNR values, $N = 128$, $N_{CP} = 9$, AWGN and Rayleigh channel with $L = 5$.

were designed to acquire a rough CFO estimation, this other set of algorithms is meant to perform CFO tracking (i.e., constant refinement of the CFO estimation). In the acquisition phase the objective is to find a rough estimate $\hat{\delta}$ for the CFO δ, whose estimation variance was depicted for instance in Figure 13.6; in the tracking phase, this estimation is usually refined in a closed-loop operation to significantly reduce the error variance and to adapt to the hypothetical Doppler shift. Those closed-loop systems, which originate from Wiener feedback loops back in 1948 [20], are a useful tool to come up with parameter estimates in a system whose behavior is rather complex to model. In this specific case, the complexity lies in the anticipation of the Doppler shift dynamics.

The historical feedback loop example for CFO estimation was proposed by Daffara and Adami [21] and was followed by various derived contributions whose working structures are essentially the same. The typical Daffara and Adami's block diagram is depicted in Figure 13.7. In this loop, every symbol is first fed to the "error function f_e" block that evaluates some error due to frequency offsets. Most enhancements of Daffara and Adami's solution are based on alternative f_e functions. The next step consists of evaluating the residual term $\hat{\delta} - \delta$. This is performed through the estimation of the CFO for all symbol indexes $n \in \mathbb{N}$. This allows one to update $\hat{\delta}$ by successive refinement of the angle of symbol rotation due to CFO for every new incoming symbol. This symbol rotation update is denoted by the function $\phi(n)$ in Figure 13.7. The loop is then closed by feeding back the reconstructed (i.e., counter-rotated) symbols y_n, $n \in \mathbb{N}$.

13.3.2.3 CFO Estimation in Uplink

In uplink OFDMA, the frequency synchronization model stumbles on the same multiparameter issue as its STO counterpart. But here the problem is heavily more involved. Indeed, frequency offsets engender ICI that corrupts the data in the destination to the base station such that only complex techniques would help decode the overlapping data streams. For this reason, system-wide solutions are usually exploited. Consider the situation of *localized subcarrier allocation* (i.e., every

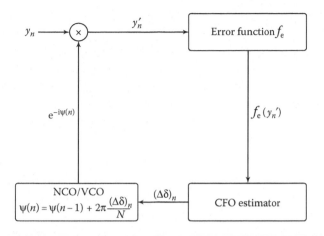

Figure 13.7 Closed-loop CFO tracking.

user is allocated a frequency subband of contiguous subcarriers). As mentioned previously for the STO case, a common approach is to insert frequency guard bands between adjacent users, so that the individual data can be easily filtered and the ICI minimized (since only remote subcarriers would leak on the individual user's data). The same techniques as in downlink can then be applied to individually estimate all the frequency offsets. Therefore, Equation 13.22 is still valid, on a per-user basis, in which again ICI is added to the noise w_n. When many users share the bandwidth, the number of available subcarriers per user (especially for NDA techniques) can, however, be so small that the CFO estimation performance is heavily impacted in the frequency domain; more computationally demanding time-domain processes are more desirable in such situations.

When the subcarrier allocation is *distributed*, as opposed to a localized allocation, it is practically impossible to separate users' frequency subbands and the ICI effect is even more detrimental. Some authors recently tackled the problem of CFO estimation in these scenarios; among those, a joint STO and CFO maximum-likelihood solution was proposed by Morelli [22], which comes along with a high complexity, since then an exhaustive search of a two-dimensional grid is demanded. However, Morelli uses the major assumption that at most one user in the OFDMA network is imperfectly aligned in time and frequency. For this particular user, who is assigned the set $\mathcal{S}$ of L nonconsecutive subcarriers $\mathcal{S} = \{k_1, \ldots, k_L\}$, $L < N$, whose channel fades are assumed from approximately known to the base station and whose STO and CFO are, respectively, denoted by θ and δ, the DA ML joint estimate, that is when pilot sequences are used, is given by

$$(\hat{\theta}, \hat{\delta}) = \arg \min_{(\theta, \delta)} \|\mathbf{r}^{\mathcal{S}} - \mathbf{z}(\delta, \theta)\|^2, \tag{13.23}$$

where $\mathbf{r}^{\mathcal{S}}$ is the restriction of $\mathbf{r}$ to the set of subcarriers in $\mathcal{S}$, and $\mathbf{z}(\delta, \theta)$ models the noiseless received pilot plus ICI due to STO and CFO, when all other users are perfectly aligned in time and frequency. In general, the joint STO–CFO ML solution requires a search of a 2D grid spanning of possible values θ and δ. In practice, the hypothesis that the base station knows the user's channel is not true and then the ML problem actually encompasses also the search over the channel $\mathbf{h}$ but Morelli manages still to turn the problem into a 2D search, independent of the channel realization. Practical solutions with lesser complexity, for example with decoupled 1D searches over δ and θ, are also proposed in Ref. [22], under the assumption of small CFO, which is often met in practice.

Most of the previously detailed algorithms make use of several *ad hoc* methods that do not minimize a given performance metric. The reason for *ad hoc* methods to be the major techniques used in practice is twofold: (i) they are usually simple to implement and very little computationally demanding, which is very important for synchronization processes that might be used very often in mobile networks; (ii) they cope with the absence of knowledge of major system parameters such as information about the communication channel. In the following, we briefly evoke advanced solutions that rely on optimal orthodox or Bayesian approaches, in the sense that they achieve Cramer–Rao bounds or use maximum entropy-based methods to deal with limited knowledge.

13.3.3 Advanced Methods

Few complete studies and optimal methods (with respect to some given performance metric) found in the literature on OFDMA synchronization for the mathematical derivations are usually not tractable; this is mainly due to the difficulty to model systems from which one does not know much *a priori* (i.e., when trying to estimate rough STO and CFO, channel state information is usually not known). Lately, the recursive expectation-maximization (EM) algorithm has grown into a handy solution to tackle such incomplete data problems and particularly suits multiple parameter estimation problems. Indeed, EM is a recursive technique that allows one to turn an *a priori* difficult problem with some missing parameters (from an *incomplete parameter set*) into a simple problem in which those parameters are known (forming then the *completed parameter set*). Under some adequate conditions [23], this method converges to a solution whose complete parameter set contains consistent parameter values. In our synchronization framework, such problems as channel estimation, decoding, and so on are simple problems when all system parameters are known, while the marginal problems when some parameters are not known *a priori* are often more involved. This gives birth to joint EM estimation techniques such as joint channel estimation and parameter offset estimation [24], joint decoding and CFO estimation [25], and so on. Other joint estimation studies are considered in Refs. [26,27] that give hints on the achievable performance to be expected in orthodox probability-based approaches. In particular, theoretical limits in terms of the Cramér–Rao bound of the joint SCO and channel estimation are given in Ref. [27]. Unfortunately these orthodox techniques do not take into consideration any prior knowledge of the unknown parameters and might lead to incongruous solutions, especially if the problem to maximize is not convex as a function of the unknown parameters. Optimal Bayesian maximum entropy approaches have been proposed by the authors [28,29] in place of EM-like solutions. This is further discussed in Section 13.5.

In Tables 13.1 and 13.2, the main synchronization algorithms for coarse and fine STO and CFO estimation, are recalled.

13.4 A Case Study: 3GPP-LTE

Due to the synchronization problems discussed in Section 13.2.2 and to the major PAPR problem in the uplink, 3GPP decided against an OFDMA uplink setup in the first releases of the Long Term Evolution (3GPP-LTE) standard. Therefore, only downlink OFDMA is considered in the following. Few synchronization sequences are utilized in LTE to minimize the system overhead. Therefore, at the receiver, synchronization is only performed:

■ either on the PSS, on the secondary synchronization sequence (SSS), and/or on any pilot sequence present in the LTE frame
■ or blindly, thanks to such methods as the NDA techniques described in Section 13.3

Table 13.1 Main OFDMA Coarse Synchronization Techniques

	DA	*NDA*
STO Estimation		
	Rough correlation to pilot	CP-based correlation
	Schmidl's double-half sequence [14]	
	Serpedin's finer sequence [15]	
CFO Estimation		
	Rough correlation to pilot	Van de Beek's CP-based algorithm [12]
	Moose's double-half sequence [11]	Ai's ISI-free CP-based method [19]
	Moose's redundant sequence [11]	
	Schmidl's refinement of Moose's sequence [14]	
	Morelli's algorithm [18]	

Note that, contrary to the uplink scenario, low-power-consuming methods are demanded at the receiving interface to minimize battery usage. The standard is then demanded to provide simple synchronization sequences while minimizing the system overhead. A typical LTE synchronization phase unfolds as follows:

1. When the user equipment is switched on, the first physical layer operation consists of detecting a power source along the licensed LTE bandwidth. This is referred to as a public land mobile network (PLMN) search. This is typically handled by a mere mean power measure on the receive antenna array. A threshold on this receive power is set to decide on the presence or absence of the OFDM source.

2. When a source of power is detected, it then has to be identified. This operation is done thanks to a set of three orthogonal time-domain ZC sequences of length 63 that enjoy the following properties:
   - Two ZC sequences of different indexes show very small cross-correlation
   - The cross-correlation of a ZC sequence with itself shifted by an integer number of samples is very small

Table 13.2 Main OFDMA Fine Synchronization Techniques

STO and CFO Estimation
Daffara's CFO closed-loop [21]
Joint ML CFO-STO [22]
EM-based algorithms [25]
Decision directed algorithms [26]
Maximum entropy-based algorithms [29]

■ The frequency response of a ZC sequence is flat
■ Third-order statistics of ZC sequences are small to mitigate nonlinearities in the analog front-end (e.g., analog amplifiers)

These three ZC sequences allow us to map the different transmitting base stations into three groups (those groups are organized such that, in the hexagonal cell planning, two cells of a given group are never adjacent). Those sequences are called PSS and are found every 5 ms on the central frequency band of size 1.4 MHz. Through the ZC sequence detection, a first rough STO estimation is performed, since the beginning of the ZC sequence is then identified. Depending on the detection technique used (e.g., the classical technique is a point-to-point correlation with the three ZC sequences on different hypothetical central frequencies), a first rough CFO estimation is also performed. Note however that a very low sampling frequency is used at cell detection step to match the 1.4 MHz central band. Therefore the STO estimates cannot be very accurate if the effective signal bandwidth is as large as 20 MHz (i.e., the maximum usable bandwidth), since one symbol sampled at 1.4 MHz then corresponds to a set of 16 symbols at a sampling rate of 20 MHz. It is also important to note that 5 ms of PSS detection over different frequency references represents a considerable amount of processing. Therefore, time and frequency acquisition cannot be made using a thin time–frequency grid. To cope with the constraint of large frequency grid steps, which does not allow for a good CFO estimate, the DA but pilot-independent technique proposed in Ref. [29] and presented briefly in Section 13.5 is of particular interest.

3. When the PSS sequence is discovered, the cell identification is completed thanks to the SSS in the frequency domain. The latter is scrambled by one of three possible codes given by the index of the identified ZC sequence. This sequence uniquely identifies the selected cell (in reality, as many as 504 identifiers are available, so that, with a correct cell planning, two cells with the same identifier should never interfere). Those PSS sequences are based on two interleaved binary *maximum length sequences* of size 32 whose main property is to have good cyclic cross-correlation properties. In terms of CFO estimation, they allow for a coarse evaluation of δ, reducing the search range to the subcarrier spacing Δ_f.

STO and CFO acquisitions can be performed thanks to the pilot sequences introduced in the LTE standard. Note nonetheless that PSS and SSS do not have the repetitive structure advised by Moose [11] due to the structure of ZC and maximum length sequences, respectively; alternative schemes must then be produced to adapt the standard. In particular, the time-domain PSS sequence **x** is *close to* a two-half mirrored sequence: $x_{N-k+2} = x_k$, $k \in [2, N/2]$. This symmetrical structure might allow us to design specific CFO and STO recovery techniques. This structure is, however, less adequate since the treatment of channel leakage on mirrored signals is more involved than for Moose's double-half sequences. In the next section, the authors provide a CFO estimation technique whose performance is independent of the pilot sequence (as long as all entries of the time-domain sequence have the same amplitude); this method particularly targets the current LTE standard.

As for the tracking phase, the problem is more difficult for no dedicated sequence allows for fine parameter estimation. PSS and SSS are length-64 sequences that pop up only every 5 ms. *Reference symbols* (RS) dedicated to channel estimation purposes are scattered along the whole frequency bandwidth and spaced every six subcarriers; as they were designed with no synchronization consideration, only advanced (and often computationally demanding) STO or CFO estimation techniques can be performed. As a consequence, NDA techniques are suitable to achieve accurate parameter estimates. Joint estimations based on the EM algorithm, despite their apparent complexity, turn out to be an interesting compromise since, practically speaking, they do not require additional software treatment (e.g., turbo decoders, channel estimators, etc. are already part of the terminal software) and they actually perform reusable tasks (e.g., soft decoding and channel estimation can be reused). Those advanced solutions are especially demanded for communications over the large 64-QAM constellation whose tolerable SNR (for a channel coding rate 1/3) is of the order of 30 dB in single-input single-output (SISO) channels and of the order of 25 dB in multi-input multi-output (MIMO) 2×2 channels. Simulations show that this constrains the CFO estimation to be constantly of the order of $\delta \simeq 200$ Hz for a subcarrier spacing $\Delta_f = 15$ KHz.

From formula 13.3, on a working LTE frequency of around 3 GHz and under a vehicular speed of 120 km/h, the typical Doppler shift is around 300 Hz, which is more than the maximum tolerable CFO. Therefore the CFO estimation task is highly critical and must be performed accurately in order to ensure a satisfying working SNR under high mobility conditions.

13.5 Discussion

13.5.1 Bayesian Framework

We already mentioned that the synchronization field does not rely on strong information theoretic grounds. This is mainly because synchronization is usually isolated from the rest of the communication chain and, as such, is considered an independent field. The authors suggest the modeling of synchronization parameters as simply *a priori* unknown parameters in the same way as the transmitted data **s**.

The objective of a communication scheme is to optimize the useful data decoding process given some information at the receiver. This information often summarizes as the exact knowledge of the transmit codebook, the noisy receive sequences, the exact synchronization pilots, and so on. The general soft decoding decision for a transmitted vector **x** of a known codebook $\mathcal{X}$ and received vector **y** is based on the Bayesian probabilities,

$$p(\mathbf{x} \mid \mathbf{y}, I) = \int p(\mathbf{x} \mid \mathbf{y}, \delta, \theta, I) p(\delta, \theta \mid I) \, \mathrm{d}\delta \, \mathrm{d}\theta \qquad (13.24)$$

where I stands for the prior information on the system.

When CFO and STO estimation is handled, the classical approach is to simplify Equation 13.24 by (erroneously) setting $p(\delta, \theta \mid I) = \delta_\delta^{\hat{\delta}} \cdot \delta_\theta^{\hat{\theta}}$, which differs from the updated probability $p(\mathbf{x} \mid \mathbf{y}, \hat{\theta}, \hat{\delta}, I)$ that should now be considered. When the parameter estimators are very poor (which can happen in many situations, such as short packet transmission, low SNR, scarce synchronization resources, etc.), soft decisions on the estimates will prove of key importance, instead of wrong hard decisions on $\hat{\theta}, \hat{\delta}$. Of course, it is usually difficult to perform the integration Equation 13.24. Still, the latter leads to a few relevant considerations:

■ As shown in Section 13.3, parameter estimations often come from the minimization of some error measure. The choice of this measure is often directed by computational simplicity or common usage (e.g., minimizing the quadratic error). Instead, the estimation error should be minimized in accordance with the operations performed in the data decoding step, that is, so as to end up with a satisfying approximation of $p(\mathbf{x} \mid \mathbf{y}, I)$.

■ As a first refinement of the hard decisions of the synchronization parameters, $p(\mathbf{x} \mid \mathbf{y}, I)$ can be better approximated by

$$p(\mathbf{x} \mid \mathbf{y}, I) \simeq \sum_{(\delta,\theta)\in\mathcal{D}} p(\mathbf{x} \mid \mathbf{y}, \delta, \theta, I)p(\delta, \theta \mid I), \qquad (13.25)$$

where $\mathcal{D}$ is some discrete space of high joint probability for the couple (δ, θ). The diameter of this space is chosen to meet the best computation load/decoding quality compromise. Couillet and Debbah [28,30] show that this approach shows significant results in the sense that even parameter estimates with low probability can be *resurrected* by a high joint data-parameter probability $p(\mathbf{x} \mid \mathbf{y}, \delta, \theta, I)$. This observation is often difficult to anticipate, mainly because of the complexity hidden in the probability $p(\mathbf{x} \mid \mathbf{y}, \delta, \theta, I)$.

In light of those considerations already envisioned in schemes such as joint decoding–synchronization, joint channel estimation–synchronization, and so on, one realizes that, with the ever-growing computation performance of embedded hardware, much progress can be achieved in the synchronization field. Through the recent interest in cognitive radio systems, novel adaptive synchronization schemes could appear and replace the already too old classical algorithms. A simple example of an optimal Bayesian CFO estimator is detailed in the following.

13.5.2 Case Study: Bayesian CFO Estimation

In this section, the authors propose a generalized study of optimal Bayesian parameter estimation from which important conclusions shall be drawn. For a more complete study, the reader is invited to refer to Ref. [29]. Say one wants to perform CFO estimation using all the provided *a priori* system information I. As recalled in Section 13.2.2, the time-domain effect of a CFO is a mere symbol phase rotation, proportional to the time index. From the transmitted data $\mathbf{x}$, received with perfect timing synchronization as $\mathbf{y}$, one has

$$\mathbf{y} = \mathbf{D}_\delta \mathbf{H} \mathbf{x} + \mathbf{w}, \qquad (13.26)$$

where $\mathbf{H}$ is the circulant matrix originating from the channel vector $\mathbf{h} = [h_0, \ldots, h_{L-1}]^{\mathrm{T}}$ of length L, $\mathbf{w}$ is the noise process with entries of power $1/\mathrm{SNR}$, and $\mathbf{D}_\delta$ is the diagonal matrix of the main diagonal $\{1, e^{i\delta}, \ldots, e^{(N-1)i\delta}\}$. For simplicity, in the upcoming derivations, let us rewrite the model as

$$\mathbf{y} = \mathbf{D}_\delta \mathbf{X} \mathbf{h} + \mathbf{w} \qquad (13.27)$$

with $\mathbf{X}$ being the pseudo-circulant matrix

$$\mathbf{X} = \begin{pmatrix} x_0 & x_{N-1} & \cdots & x_{N-L-1} \\ x_1 & x_0 & \cdots & x_{N-L-2} \\ \vdots & \vdots & \vdots & \vdots \\ x_{L-2} & x_{L-3} & \cdots & x_{N-1} \\ x_{L-1} & x_{L-2} & \cdots & x_0 \\ \vdots & \vdots & \vdots & \vdots \\ x_{N-1} & x_{N-2} & \cdots & x_{N-L} \end{pmatrix}. \qquad (13.28)$$

The estimation problem consists of evaluating, for all $\delta \in \mathbb{R}$,

$$p(\delta \mid \mathbf{y}, I) = p(\mathbf{y} \mid \delta, I) \frac{p(\delta \mid I)}{p(\mathbf{y})}. \tag{13.29}$$

The term $p(\mathbf{y} \mid \delta, I)$ can be further developed

$$p(\mathbf{y} \mid \delta, I) = \int p(\mathbf{y} \mid \delta, \mathbf{h}, \mathbf{x}, I) p(\mathbf{h}, \mathbf{x} \mid I) \, d\mathbf{h} \, d\mathbf{x} \tag{13.30}$$

in which $p(\mathbf{y} \mid \delta, \mathbf{h}, \mathbf{x}, I)$ is easily evaluated from the linear model 13.26. Equation 13.30 does not consider at all the selected synchronization method (i.e., DA or NDA). This property is actually hidden in the expression $(\mathbf{x} \mid I)$. Two cases can arise:

■ Either the information on $\mathbf{x}$ is completely included in the prior information I of the receiver. This makes of $\mathbf{x}$ a pilot sequence. In this scenario, Equation 13.30 simply becomes

$$p(\mathbf{y} \mid \delta, I) = \int p(\mathbf{y} \mid \delta, \mathbf{h}, I) p(\mathbf{h} \mid I) \, d\mathbf{h} \tag{13.31}$$

■ Either the information on $\mathbf{x}$ contained in I is limited to some statistical properties (e.g., mean and variance), constellation knowledge, and so on. This leads to a semiblind or fully blind analysis, whose solution can be found in the works on joint decoding–parameter estimation.

The *a priori* distribution $p(\mathbf{h} \mid I)$ can be evaluated through the available prior information contained in I; if the average power and the typical length L of the channel, that is the expected channel delay spread, are known, then, thanks to the *maximum entropy principle* [31], the most appropriate distribution to represent $\mathbf{h}$ is a multivariate Gaussian independent and identically distributed (i.i.d.) density function [32]. Ordinarily, L is not supposed to be known, but let us assume perfect knowledge on L in the following (for deeper analysis when L is unknown, refer to Ref. [30]). After further development, in the DA case (i.e., $\mathbf{x}$ is *a priori* known), one can show that

$$p(\delta \mid \mathbf{y}, I) = \alpha(\mathbf{y}) \cdot p(\delta \mid I) e^{-C(\mathbf{y}, \delta)} \tag{13.32}$$

for some function $\alpha(\mathbf{y})$ independent of δ and

$$C(\mathbf{y}, \delta) = \mathbf{y}^H \left[\mathbf{I}_N + \mathbf{D}_\delta^H \mathbf{X} \left(\mathbf{X}^H \mathbf{X} + \frac{1}{\text{SNR}} \mathbf{I}_L \right)^{-1} \cdot \text{SNR} \cdot \mathbf{X}^H \mathbf{D}_\delta \right] \mathbf{y}. \tag{13.33}$$

If no prior information is known about δ apart from its belonging to a finite range $\mathcal{D}$, $p(\delta|I)$ can be considered uniform over $\mathcal{D}$ [31] and a simple CFO estimator consists of the value $\delta \in \mathcal{D}$ for which C is minimized.* Unfortunately $C(\mathbf{y}, \delta)$ is not convex in δ in general, which makes the solution difficult to grasp and requires exhaustive search methods. However, from exhaustive

---

* The choice of this estimator is purely subjective since no mention is made here of any ultimate objective apart CFO estimation.

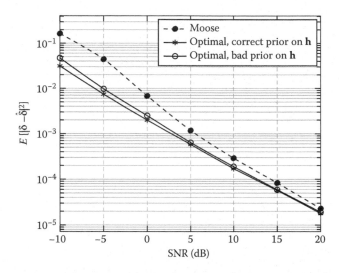

Figure 13.8 Optimal Bayesian CFO estimation—Comparison between Moose's methods [11] and the Bayesian method. *Conditions: N = 128, random double-half OFDM symbol, h is i.i.d. Gaussian with length L = 3, prior on h either perfect $L_{as} = L$ or imperfect $L_{as} = 9$.*

simulations, it appears that $C(\mathbf{y}, \delta)$ is convex on $[-\pi, \pi]$, that is, in the range of one subcarrier spacing. Therefore, iterative algorithms can be produced based on steepest descent methods to find the maximum likelihood estimate $\hat{\delta}$. Figure 13.8 provides a comparison between Moose's correlation method discussed in Section 13.3 and the proposed Bayesian method, whose performance plot is produced from the iterative algorithm provided in Ref. [29]. The results show an increase in performance in the CFO estimate. We also provided simulation results when the prior for $\mathbf{h}$ is initially wrong, that is, the assumed channel length L_{as} is not the true channel length L; for not-too-low SNR, the solution is close to optimal. Note, moreover, that this optimal solution in the Bayesian sense is better than Moose's solution while being applicable to any pilot sequence $\mathbf{x}$. Also, the complexity of the algorithm, which is obviously more than Moose's correlation method, can be dynamically controlled by the number of required iteration steps. In the low SNR regime in particular, an important synchronization time is gained, at the expense of a small increase in processing complexity.

Those Bayesian considerations lead to a new approach regarding problems of parameter estimation: the authors propose here an information theoretic solution based on the state of knowledge of the receiver (which provides upper bounds on estimation performance in this information theoretic framework) and envision simplifications of the optimal solution to better suit the computational complexity requirements. This allows one to keep a constant control on the performance.

Similar maximum entropy Bayesian studies are carried out in thorough detail for blind MIMO signal detection [28] and pilot-based channel estimation for OFDM [30]. In the latter, we particularly observe that, even when the channel length L is unknown to the receiver, channel estimation can be equally performed as if L were known (with almost as good performance) since the missing information, carried by the incoming signal, can be automatically recovered. The resulting algorithms are shown to be more complex than classical methods, making simplification algorithms only a matter of mathematical complexity reduction, and not a matter of finding an adequate *ad hoc* alternative.

13.6 Conclusion

To ensure reliable communication in an OFDMA system, a first timing and frequency synchronization step is necessary, for local oscillators in both the communication ends generally do not match. In addition, mobility in recent OFDM-based technologies engenders Doppler effects, which dynamically contribute to the frequency mismatch (CFO). For these reasons, STO and CFO need be estimated at device initialization and then tracked during the proper communication phase. We showed that STO estimation is not in general a critical task, while CFO can lead to dramatic performance impairment. Synchronization mainly consists of a multiparameter estimation problem. No optimal solution has ever been proposed since there exists no strong theoretical foundation for synchronization that aims at optimizing the useful data transmission capacity. As a consequence, STO and CFO recoveries in the literature consist of a multitude of various solutions aim at different objectives. From those solutions, we selected the historical and most used algorithms, either based on dedicated pilot sequences, designed for synchronization purposes, or based on the redundancy found in the system overhead, and in particular in the CP. However, we showed that in a concrete application such as 3GPP-LTE, most of those schemes are not adequate. This has led to recent proposals using information theoretic grounds on synchronization issues to conclude that DA and NDA methods are just particular cases of a more general Bayesian parameter estimation approach. In the near future, with the availability of high embedded computation rates, the synchronization field is expected to join the current trend for cognitive radios and to move from low complex solutions join more involved but information theoretically optimal processes.

References

1. C. E. Shannon, A mathematical theory of communication, *The Bell System Technical Journal*, 27, 379–423, 623–656, 1948.
2. L. Zheng and D. N. C. Tse, Communication on the Grassmann manifold: A geometric approach to the noncoherent multiple-antenna channel, *IEEE Transactions on Information Theory*, 48 (2), 359–383, 2002.
3. G. Caire, N. Jindal, M. Kobayashi, and N. Ravindran, Multiuser MIMO downlink made practical: Achievable rates with simple channel state estimation and feedback schemes, Arxiv preprint arXiv:0711.2642, 2007.
4. http://www.wimaxforum.org/technology/
5. S. Sesia, I. Toufik, and M. Baker, *LTE, the UMTS Long Term Evolution: From Theory to Practice*, Wiley, New York, 2009.
6. http://www.dvb.org/
7. http://www.worlddab.org/
8. M. Speth, S. Fechtel, G. Fock, and H. Meyr, Optimum receiver design for wireless broad-band systems using OFDM – Part I, *IEEE Transactions on Communications*, 47 (11), 1668–1677, 1999.
9. M. Morelli, C. Kuo, and M. Pun, Synchronization techniques for orthogonal frequency division multiple access (OFDMA): A tutorial review, *IEEE Proceedings*, 95 (7), 2007.
10. T. Pollet, P. Spruyt, and M. Moeneclaey, The BER performance of OFDM systems using nonsynchronized sampling, *IEEE Global Telecommunications Conf.*, pp. 253–257, 1994.
11. P. H. Moose, A technique for orthogonal frequency-division multiplexing frequency offset correction, *IEEE Transactions on Communications*, 42 (10), 2908–2914, 1994.
12. J. J. Van de M. Beek, Sandelland, and P. O. Börjesson, ML estimation of time and frequency offset correction in OFDM systems, *IEEE Transactions on Signal Processing*, 45 (7), 1800–1805, 1997.
13. D. C. Chu, Polyphase codes with good periodic correlation properties, *IEEE Transactions Information Theory*, vol. 18, pp. 531–532, July 1972.

14. T. M. Schmidl and D. C. Cox, Robust frequency and timing synchronization for OFDM, *IEEE Transactions on Communications*, 45 (12), 1613–1621, 1997.

15. K. Shi and E. Serpedin, Coarse frame and carrier synchronization of OFDM systems: A new metric and comparison, *IEEE Transactions on Wireless Communications*, 3 (4), 1271–1284, 2004.

16. M. Gudmunson, Generalized frequency hopping in modile radio systems, *Processing of IEEE Vehicular Technologies Conf.*, pp. 788–791, 1993.

17. R. Couillet and M. Debbah, Outage performance of flexible OFDM schemes in packet-switched transmissions, *EURASIP Journal on Advances in Signal Processing*, ID 698417, 2009.

18. M. Morelli and U. Mengali, An improved frequency offset estimator for OFDM applications, *IEEE Communications Letters*, 3 (3), 75–77, 1999.

19. B. Ai, J. Ge, Y, Wang, S. Yang, and P. Liu, Decimal frequency offset estimation in COFDM wireless communications, *IEEE Transactions on Broadcasting*, 50 (2), 2004.

20. N. Wiener, Cybernetics, or control and communication in the animal and the machine, Herman et Cie/The Technology Press, Hermann et Cie, Paris, 1948.

21. F. Daffara and O. Adami, A novel carrier recovery technique for orthogonal multicarrier systems, *European Transactions on Telecommunication*, vol. 7, pp. 323–334, July–Aug. 1996.

22. M. Morelli, Timing and frequency synchronization for the uplink of an OFDMA system, *IEEE Transactions on Communications*, 52 (2), 296–306, 2004.

23. T. Moon, The expectation-maximization algorithm, *IEEE Signal Processing Magazine*, 13 (6), 47–60, 1996.

24. X. Ma, H. Kobayashi, and S. Schwartz, Joint frequency offset and channel estimation for OFDM, *IEEE Global Telecommunications Conf.*, 1–5, December 2003.

25. N. Noels, C. Herzet, A. Dejonghe, V. Lottici, H. Steendam, M. Moeneclaey, M. Luise, and L. Vandendorpe, Turbo synchronization: an EM algorithm interpretation, *IEEE Int. Conf. on Communications*, 4, pp.11–15, 2003.

26. K. Shi, E. Serpedin, and P. Ciblat, Decision-directed fine synchronization in OFDM systems, *IEEE Transactions on Communications*, 53 (3), 408–412, 2005.

27. S. Gault, W. Hachem, and P. Ciblat, Joint sampling clock offset and channel estimation for OFDM signals: Cramer–Rao bound and algorithms, *IEEE Transactions on Signal Processing*, 54 (5), 1875–1885, 2006.

28. R. Couillet and M. Debbah, Bayesian inference for multiple antenna cognitive receivers, Proceedings IEEE WCNC'09 Conference, 2009.

29. R. Couillet and M. Debbah, Information theoretic approach to synchronization: the OFDM carrier frequency offset example, *IEEE Trans. on Signal Processing submitted*, Unpublished work.

30. R. Couillet and M. Debbah, A maximum entropy approach to OFDM channel estimation, *Proceedings of IEEE Signal Processing Advances in Wireless Communications Conference*, Perugia, Italy, 2009.

31. E. T. Jaynes, *Probability theory: The logic of science*, Cambridge University Press, New York, 2003.

32. M. Debbah and R. Müller, MIMO channel modelling and the principle of maximum entropy, *IEEE Transactions on Information Theory*, 51 (5), 2005.

Chapter 14

Synchronization for OFDM and OFDMA

Thomas Magesacher, Jungwon Lee,
Per Ödling, and Per Ola Börjesson

Contents

14.1   Introduction..376
14.2   Synchronization in OFDM and OFDMA DL.......................................378
    14.2.1   Symbol-Timing Acquisition..378
        14.2.1.1   Tradeoff: Symbol-Synchronization Precision versus Robustness
                    to Time Dispersion 378
        14.2.1.2   Elements of Pilot-Based Timing Synchronization.................380
        14.2.1.3   CP-Based Timing Acquisition....................................380
    14.2.2   Carrier-Frequency Acquisition...381
        14.2.2.1   Elements of Pilot-Based Frequency Acquisition..................381
        14.2.2.2   CP-Based Frequency Acquisition................................382
    14.2.3   Tracking 382
14.3   Synchronization in OFDMA UL...383
    14.3.1   Synchronous and Asynchronous Multiuser Systems......................383
    14.3.2   Offset Estimation in OFDMA UL.......................................385
    14.3.3   Offset Correction in OFDMA UL.......................................387
14.4   Example: Synchronization for WiMAX 389
14.5   Literature..392
14.6   Summary and Conclusion 393
Symbols...393
Abbreviations...393
References...394

14.1 Introduction

The whole synchronization issue arises from the fact that transmitter and receiver are spatially separated and not connected in any sense except for the fact that the receiver processes a signal that is a modified version of what the transmitter once sent. Synchronization-related tasks required in every multicarrier system are

■ *Timing acquisition*: After down-conversion and analog-to-digital conversion, the receiver sees a sequence of baseband samples. Before block processing (cyclic-prefix (CP) removal, fast Fourier transform (FFT), etc.) can begin, the symbol boundaries have to be determined. Timing offsets are caused by the inherent lack of a timing reference at the receiver.

Even with access to the transmitter's timing reference (which is not available in a practical system though), there are distance-dependent path delays of significant duration. For example, the time a block needs to travel from a base station (BS) to a user that is 5 miles (ca. eight kilometers) away, is roughly* 27 μ*s*, which corresponds to about a quarter of symbol duration (roughly 100 μ*s*) in a WiMAX (worldwide interoperability for microwave access) system (IEEE 802.16 Wireless MAN standard).

■ *Frequency offset*: After down-conversion in the receiver, the resulting baseband signal may exhibit an unwanted frequency offset of ΔF_c Hz for two reasons. First, down-conversion is based on a reference carrier generated by a local oscillator, which may have a slightly different frequency compared to the carrier generated in the transmitter. Second, the signal itself may experience a motion-incurred Doppler shift in frequency on its way through the channel.

Frequency offsets caused by carrier-oscillator mismatch can be significant. For example, in a system operating at $F_c = 5.2$ GHz with carrier oscillators of 1 ppm[†] precision, the frequency offset[‡] can be as large as 10.4 kHz, which is in the order of the subcarrier spacing (roughly 11.16 kHz) for a WiMAX system.

Compared to oscillator-incurred offsets, motion-incurred frequency offsets are rather harmless. For example, a user moving at 40 mph (ca. 64 km/h) experiences a motion-incurred Doppler shift[§] of roughly 308 Hz, which corresponds to roughly 2.7% of the subcarrier spacing. A simple motion-incurred shift occurs only in a single-path propagation scenario. In a multipath propagation scenario, signal components arrive from several different angles and thus with several different relative velocities, which yield several different Doppler shifts. Consequently, the receive signal experiences a dispersion in frequency rather than a shift. The practical approach to handle motion-incurred dispersion in frequency is to ensure a large enough subcarrier spacing.

■ *Clock offset*: At some point in the receive chain, the continuous-time receive signal is sampled in the time domain (and, in practice, quantized in amplitude at the same time by an analog-to-digital converter). If the difference $\Delta F_s = F_s^{(rx)} - F_s^{(tx)}$ between the sampling clocks of the transmitter and the receiver is smaller (larger) than zero, elements of the receive signal's discrete Fourier transform (DFT) have a frequency spacing that is smaller (larger) than the spacing of the transmit signal's DFT. Furthermore, the receiver's sampling period is larger (smaller) than the transmitter's sampling period.

---

* Assuming a propagation velocity of $c = 3 \times 10^8$ m/s, the path delay is given by $8047/c \approx 26.83$ μ*s*.
† ppm stands for parts per million, a common measure to specify the precision of an oscillator.
‡ Assuming that the offsets of transmitter and receiver have opposite signs, we have $\Delta F_c = 2 \times 10^{-6} \cdot F_c = 10.4$ kHz.
§ Assuming a propagation velocity of $c = 3 \times 10^8$ m/s, the Doppler shift is given by $\Delta F_c = 64/3.6/(F_c/c) \approx 308.15$ Hz.

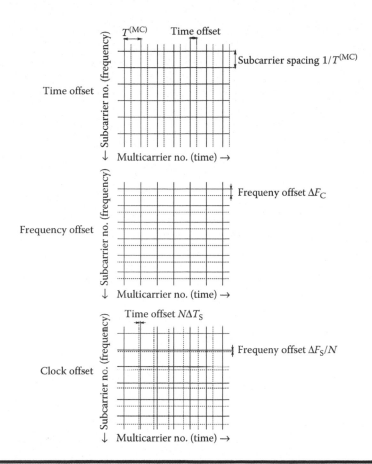

Figure 14.1 Illustration of timing offset, frequency offset, and clock offset in the time–frequency plane: vertical and horizontal lines indicate symbol boundaries and subcarriers' center frequencies, respectively (for simplicity, $T^{(CP)} = 0$). The receiver assumes the time/frequency grid illustrated by dashed lines while the true time/frequency grid is marked by solid lines.

While a carrier-oscillator mismatch causes a frequency offset, a clock offset results in a subcarrier-spacing mismatch. Consequently, subcarriers at the band edges experience stronger intercarrier interference (ICI) while subcarriers close to the band center experience almost no ICI. The resulting frequency mismatch is small compared to a carrier-oscillator in-curred offset. For example, in a WiMAX system with 2048 subcarriers (sampling frequency $B = 22.857$ MHz), clock oscillators with 1 ppm cause an offset* of up to ca. 23 Hz, which corresponds to roughly 0.2 % of the subcarrier spacing. The time error accumulates and corresponds to roughly 1.3 orthogonal frequency division multiplexing (OFDM) symbols per minute.

Figure 14.1 illustrates the offsets discussed above in the time–frequency plane. Time offsets and frequency offsets deserve special attention in OFDM(A) systems and are thus central to this

---

* The maximum frequency error is half the maximum sampling-clock difference, given by $\Delta F_s = 2 \times 10^{-6} B = 45.71$ Hz.

chapter. Clock offsets can be dealt with using standard techniques applied in digital transceivers. Furthermore, the sampling clock can be derived from the carrier clock both at the BS and at the user side. Thus, once the carrier frequency is corrected, the sampling clock offset is negligible.

In general, synchronization consists of two tasks: *estimation* of appropriate parameters (frequency offset, time offset) and *correction* of offsets based on these estimates. Estimation techniques are generally classified as *pilot-based methods*, which are supported by deliberately inserted synchronization-assisting signals (pilot signals, synchronization symbols) or *blind methods*, which operate without dedicated pilots. Synchronization is usually carried out in two steps: an initial synchronization procedure (often referred to as *acquisition*) and periodic update steps to cope with varying conditions (often referred to as *tracking*).

14.2 Synchronization in OFDM and OFDMA DL

The techniques discussed in the following are elementary in the sense that they can be applied directly to single-user (point-to-point) systems (traditional OFDM) and constitute the basic building blocks for multiuser (point-to-multipoint) setups, that is, OFDMA downlink (DL).

14.2.1 Symbol-Timing Acquisition

14.2.1.1 Tradeoff: Symbol-Synchronization Precision versus Robustness to Time Dispersion

Before dealing with dedicated synchronization techniques, it is helpful to realize that a CP longer than the channel delay spread can ease the symbol-timing problem. Figure 14.2 schematically depicts a sequence of time-domain blocks both before and after passing through the channel. An OFDM symbol of length $T^{(MC)} = NT_s$ extended by a CP of length $T^{(CP)}$ is sent through a time-dispersive channel of length $\tau^{(CIR)}$. Often, multicarrier symbols are extended by an additional CP and an additional cyclic suffix, each of length $T^{(W)}$, followed by time-domain windowing of these additional extensions in order to reduce out-of-band emissions.

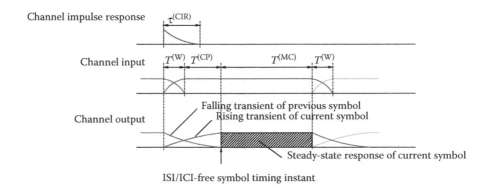

Figure 14.2 Symbol timing when CP length is equal to channel dispersion ($T^{(CP)} = \tau^{(CIR)}$): the ISI/ICI-free timing instant is unique.

Taking an FFT at the true symbol boundary captures exclusively the channel's steady-state response to the current symbol. When the symbol-boundary estimate is ahead of the true boundary (too early in time), the FFT captures a portion of the transient stemming from the CP of the current symbol [which causes intercarrier interference (ICI)] as well as a piece of the falling transient of the previous symbol [which causes intersymbol interference (ISI)]. Similarly, when the symbol-boundary estimate lags behind the true boundary (too late in time), the FFT captures a portion of the current symbol's falling transient (causing ICI) and a part of the succeeding symbol's rising transient (causing ISI). The performance decay caused by ISI and ICI induced by the deviation from the optimal symbol boundary is gradual, where late timing typically results in a larger performance decay than early timing, since most practical channels exhibit a decaying channel delay profile. Consequently, late timing results in capturing "early-response" components from the succeeding symbol that are stronger than "tail" components from the preceding symbol.

Clearly, there is a tradeoff between the delay spread that can be handled and the admissible timing error. A channel impulse response of length $\tau^{(\text{CIR})} < T^{(\text{CP})}$ increases the length of the received block's steady-state part by $T^{(\text{CP})} - \tau^{(\text{CIR})}$, as shown in Figure 14.3. Consequently, the symbol boundary can be up to $T^{(\text{CP})} - \tau^{(\text{CIR})}$ seconds ahead of the true timing instant without capturing the falling transient of the preceding block (i.e., there is no ISI) nor capturing a transient of the current symbol (i.e., there is no ICI). However, compared to the receive block captured at the true timing instant, the receive block captured n samples earlier corresponds to a version that is cyclically right-shifted by n samples. A cyclic time-domain right-shift of $r(n), n = 0, \dots, N - 1$, by s samples, resulting in the block $r'(n) = r((n - s) \bmod N)$, corresponds to a phase rotation of all subcarriers in the DFT domain. The kth subcarrier is rotated by $2\pi sk/N$, that is,

$$r'(n) = r((n - s) \bmod N) \longleftrightarrow R'(k) = R(k)e^{-j2\pi sk/N}, \quad k = 0, \dots, N - 1, \quad (14.1)$$

where $R(k)$ and $R'(k)$ denote the DFT of $r(n)$ and $r'(n)$, respectively. Differential modulation in time (in combination with noncoherent detection) is inherently immune to time-invariant phase rotations. For absolute modulation (in combination with coherent detection), these phase rotations can be viewed as part of the channel and are thus implicitly taken care of by frequency-domain equalization.

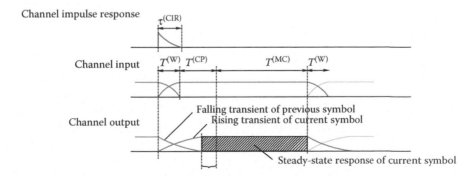

Figure 14.3 Symbol timing when CP is longer than channel dispersion ($T^{(\text{CP})} > \tau^{(\text{CIR})}$) yields an ISI/ICI-free timing window of length $T^{(\text{CP})} - \tau^{(\text{CIR})}$.

14.2.1.2 Elements of Pilot-Based Timing Synchronization

The basic idea of pilot-based synchronization is to transmit a piece of signal (a so-called synchronization symbol or pilot symbol) that is known to the receiver and can thus be detected to mark a reference timing instant. Most detectors are based on evaluating the similarity of a piece of receive signal and a piece of reference signal through computing their correlation. Unless the pilot symbol is simply a sequence of zeros, it is modified by the time-dispersive channel and thus correlation at the receiver with the sent pilot symbol will perform poorly.* Most practically relevant schemes thus revert to transmitting a periodic piece of signal, which yields (apart from the channel noise) a periodic response (assuming that a long-enough CP is applied). Rather than correlating with a known reference, the receive stream is correlated with a shifted version of itself (shifted exactly by one period)—an idea that has been used in single-carrier systems long before OFDM became popular. Under the (reasonable) assumption that the data do not exhibit periodic behavior, strong correlation will pinpoint the pilot symbol.

More concretely, the classical pilot-based synchronization method [1], tackling both time and frequency offsets, uses a time-domain pilot symbol with two identical halves, which can be generated easily by setting odd subcarriers to zero. The normalized correlation measure

$$M_{\text{time}}(d) = \frac{\sum_{m=d}^{d+N/2-1} r^*(m)r(m+N/2)}{\sum_{m=d}^{d+N-1} |r(m)|^2} \tag{14.2}$$

computed at the receiver based on the receive stream $r(n)$ measures the similarity between a length-$N/2$ block starting at time instant d and the consecutive length-$N/2$ block. The timing instant

$$\widehat{d} = \arg\max_d M_{\text{time}}(d) \tag{14.3}$$

yields the maximum value of $M_{\text{time}}(d)$ and pinpoints the first sample of the pilot symbol's steady-state part. For $\tau^{(\text{CIR})} = T^{(\text{CP})}$, the correlation metric $M_{\text{time}}(d)$ yields a unique peak and the desired symbol-boundary instant (first sample following the pilot symbol) is given by $\widehat{d} + N$. When the CP is longer than the channel delay spread ($\tau^{(\text{CIR})} < T^{(\text{CP})}$), $M_{\text{time}}(d)$ exhibits a nonunique plateau-like maximum. Clearly, choosing the last time instant of the plateau yields largest immunity to time dispersion—however, reliably pinpointing this instant in the presence of channel noise using a detection algorithm may be nontrivial.

14.2.1.3 CP-Based Timing Acquisition

In setups that do not use dedicated pilots for acquisition, like broadcast systems, the receiver can exploit the repetitive pattern of the receive stream for identifying the symbol boundaries. Similarly to the pilot-based approach, CP-based timing acquisition uses a correlation measure

$$\gamma(d) \widehat{=} \sum_{m=d}^{d+L-1} r^*(m)r(m+N) \tag{14.4}$$

---

* Note that one could, in principle, use the channel's response to the pilot symbol for correlation. However, computation of the latter is difficult since, during the synchronization phase, there is usually no channel knowledge available yet.

and a power measure

$$\Phi(d) \hat{=} \frac{1}{2} \sum_{m=d}^{d+L-1} |r(m)|^2 + |r(m+N)|^2.$$

Joint maximum-likelihood estimation of both timing offset and frequency offset yields the estimate pinpointing the first sample of the pilot symbol's steady-state part

$$\widehat{d} = \arg \max\{|\gamma(d)| - \rho\Phi(d)\} + L, \tag{14.5}$$

where ρ is the magnitude of the correlation coefficient between two N-spaced samples.

In an AWGN (additive white Gaussian noise) channel, $r(m)$ and $r(m+N)$ are identical up to the noise introduced by the channel. A time-dispersive channel, however, additionally introduces ISI and ICI, which reduces the estimator's performance. However, power-delay profiles found in real systems often decay rapidly enough (e.g., exponentially) in order to yield sufficient correlation between $r(m)$ and $r(m+N)$ to achieve an acceptable performance.

14.2.2 Carrier-Frequency Acquisition

14.2.2.1 Elements of Pilot-Based Frequency Acquisition

Like timing acquisition schemes, carrier-frequency synchronization methods used in practice almost exclusively rely on pilot symbols. Preferably, the timing pilots are also exploited for frequency synchronization reducing overhead and taking advantage of the channel-independence property of periodic pilot signals.

A carrier-frequency offset between transmitter and receiver causes an unwanted, steadily growing phase rotation: denoting $r'(n)$ as the baseband receive signal without carrier-frequency offset ($\Delta F_c = 0$), the receive stream with carrier-frequency offset can be written as $r(n) = r'(n) \, e^{j2\pi(n_0+n)\Delta F_c/F_s}$. Frequency-offset measures rely on the following observation: if $r'(n)$ is periodic with period P, any two samples $r(n)$ and $r(n+P)$ are identical up to a constant phase difference

$$\arg(r(n+P)) - \arg(r(n)) = \arg(r(n+P)r^*(n)) = 2\pi P \Delta F_c/F_s + 2\pi i \tag{14.6}$$

and noise. The integer multiple $i \in \mathbb{Z}$ of 2π accounts for the fact that $\arg(\cdot) \in [-\pi, \pi)$. Averaging over P samples starting at $\widehat{d}$, which captures the (periodic) pilot symbol and mitigates noise, yields the fractional part* of the frequency-offset estimate

$$\widehat{\Delta F_c} = \frac{F_s}{2\pi P} \arg \left(\sum_n r(\widehat{d} + n + P) r^*(\widehat{d} + n) \right).$$

If the frequency offset is known to be small enough so that $|\Delta F_c| \leq F_s/2P$ can be guaranteed, then $i = 0$ and the fractional part yields the correct estimate. Thus, a smaller value P yields a larger acquisition range. However, there is a tradeoff since smaller P results in less averaging and thus more susceptibility to noise. The scheme proposed in Ref. [1], for example, uses $P = N/2$, which corresponds to an offset range of $\pm F_s/N$ (i.e., $\pm$ one subcarrier spacing).

---

* Fraction of F_s/P.

The fractional frequency offset is easily corrected in the time domain before FFT processing through counter-rotation yielding $r'(n) = r(n) e^{-j2\pi n\widehat{\Delta F_c}/F_s}$. Estimating the remaining integer part of the frequency offset can be done with the help of a second training symbol that carries differentially modulated data with respect to the pilot symbol used up to now. The pilot symbols in the FFT domain, denoted by R_1 and R_2, exhibit no (or only residual) ICI. However, all subcarriers are shifted by iF_s/P (for the scheme [1], this corresponds to $i2F_s/N$—an integer multiple of twice the subcarrier spacing), which is the basis for estimating the integer part of the frequency offset. Differential modulation can be formulated as $R_2 = R_1 P$, where P is a pseudo-random number (PN) vector with points on the unit circle. In order to find the integer part of the offset, the similarity between $R_2(k)/R_1(k)$ and a shifted version $P(k - iN/P)$ of P is measured by*

$$M_{\text{freq}}(i) = \sum_k \left| \frac{R_2(k)}{R_1(k)} P(k - iN/P) \right| = \sum_k \frac{|R_2(k)R_1^*(k)P(k - iN/P)|}{|R_1(k)|^2}$$

and its maximum indicates the integer part

$$\widehat{i} = \arg \max_i M_{\text{freq}}(i)$$

of the frequency offset, which can then be included in the time-domain rotation.

Finally, a carrier-phase offset $e^{j2\pi n_0 \Delta F_c/F_s}$ remains, which either can be ignored (when differential modulation in time is used) or is taken care of through estimation and equalization of the channel.

14.2.2.2 CP-Based Frequency Acquisition

In the absence of dedicated acquisition pilots, the receiver can exploit the CP also for estimating the fractional frequency offset using the same principle as before: repetitive signal parts are identical up to a phase difference. Joint maximum-likelihood estimation of offsets in time and frequency (cf. Ref. [2]) yields the frequency-offset estimate

$$\widehat{\Delta F_c} = \frac{F_s}{2\pi N} \arg(\gamma(\widehat{d})), \tag{14.7}$$

where the correlation measure $\gamma(d)$ is given by Equation 14.4. Note that Equation 14.7 is identical to Equation 14.6 with $P = N$. As in the case of CP-based timing acquisition, the performance of CP-based frequency acquisition degrades with stronger time dispersion and with increasing channel noise.

14.2.3 Tracking

After acquisition, time offsets and frequency offsets are updated on a regular basis exploiting dedicated tracking pilots, which are arranged in time and frequency following a predefined pattern known to the receiver. A small timing offset manifests itself in the frequency domain as a phase offset that increases linearly with the subcarrier index, which is a consequence of the circular shift theorem Equation 14.1. Knowing both the pilot symbols and the channel, the time shift corresponding to

---

* Note that the sum index k includes only subcarriers for which $R_1(k) \neq 0$ and $k - iN/P \in [1, N]$.

this phase offset can be estimated. The frequency offset can be tracked using again the same principle as for acquisition, exploiting the linearly increasing phase offset between consecutive pilots in time (cf. Equation 14.6).

14.3 Synchronization in OFDMA UL

Synchronization in the uplink (UL) direction in an OFDMA system with K users bears the challenge that the receive stream is a superposition of K components, which pass through potentially different channels and thus experience different path delays (reflecting different distances of users from the BS) and different frequency offsets (reflecting different velocities of users). Furthermore, in an asynchronous setup, the transmission of each component may begin at different points in time.

Two aspects render synchronization in the UL direction more difficult than synchronization in the DL direction. First, the parameter space increases by a factor of K and, unless structured subcarrier assignment is applied, these parameters have to be estimated jointly. Second, straightforward correction of an individual user's offset applied to the UL stream affects the offsets of all other users. Even if all parameters are known, correction of offsets in the UL direction is thus not at all straightforward. Hence, as appealing as the principal idea of using OFDM as an access scheme may seem, the requirement of synchronized users in OFDMA UL introduces a "chicken–egg problem": separating users' streams through FFT processing is only possible after correcting offsets of individual users—which, in turn, requires access to individual users' streams.

Clearly, synchronization in the UL direction is closely connected to multiuser detection. A rigorous approach embracing the synchronization problem in OFDMA-UL is joint detection of all users' data sequences. Such a multiuser detector that is optimal in the maximum *a posteriori* probability (MAP) sense is formulated in Ref. [3]. Instead of separately estimating and correcting offsets, the multiuser detector jointly decides the coded transmitted sequences of all users in the presence of timing offsets and frequency offsets for arbitrary subcarrier allocation. However, the underlying trellis, which includes the three dimensions time, frequency, and users, quickly grows in complexity to a level that prohibits implementation for real-world parameters.

Practical schemes thus separate the tasks of parameter estimation and offset correction, treated in Sections 14.3.2 and 14.3.3, respectively. Structured subcarriers assignment allows dedicated low-complexity solutions. Moreover, as discussed next in Section 14.3.1, dedicated timing schemes exploiting the DL timing reference can greatly simplify the synchronization problem.

14.3.1 Synchronous and Asynchronous Multiuser Systems

In OFDMA UL, we distinguish the following elementary timing schemes:

- In an *asynchronous system*, the users transmit their UL blocks without considering a timing reference. Additionally, the UL path delay may vary greatly from user to user (as a consequence of different distances from the BS). Consequently, different users' UL components arrive with different time offsets (and different frequency offsets as a consequence of different velocities) at the BS, which renders the asynchronous case very challenging.
- In a *quasi-synchronous system*, each terminal exploits the timing reference established through DL synchronization. Together with using a long enough CP, timing acquisition can be entirely avoided. More concretely, the BS transmits a DL block at time instant t_0. User no. k receives

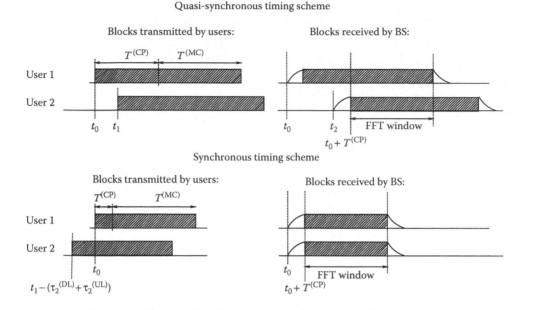

Figure 14.4   Quasi-synchronous (top) and synchronous (bottom) OFDMA UL timing scheme: UL block of user 1 (with path delay ≈ 0) and user 2 (with longest path delay among all users) at the user side (left) and at the BS (right).

the block $\tau_k^{(DL)}$ seconds later, implicitly identifies this time instant $t_0 + \tau_k^{(DL)}$ in the course of DL synchronization, and starts transmitting its UL block at $t_0 + \tau_k^{(DL)}$. Figure 14.4 depicts the UL blocks of two users: user 1 is very close to the BS ($\tau_1^{(DL)} \approx 0$) and thus transmits its UL block virtually at t_0; user 2 is farthest away for the BS among all users ($\tau_2^{(DL)} = \max_k \tau_k^{(DL)}$) and starts its UL transmission at $t_1 = t_0 + \tau_2^{(DL)}$. Under the assumption of quasi-reciprocal path delays ($\tau^{(UL)} \approx \tau^{(DL)}$) and negligible processing delays, UL components from user 1 and user 2 reach the BS at t_0 and $t_2 = t_1 + \tau_2^{(UL)}$, respectively. In a quasi-synchronous system, the CP is chosen long enough to accommodate the maximum roundtrip delay as well as the maximum channel delay spread:

$$T^{(CP)} \geq \max_k \tau_k^{(UL)} + \max_k \tau_k^{(DL)} + \max_k \tau_k^{(CIR)}. \tag{14.8}$$

The BS takes an FFT of the aggregate UL stream beginning at $t_0 + T^{(CP)}$. As illustrated in Figure 14.4, the choice of the CP length ensures that the FFT window extends exclusively over steady-state parts of both the component with the largest and the component with the shortest delay: $t_0 + T^{(CP)}$ is the earliest possible timing to avoid the rising transient of the component with the largest delay (user 2); at the same time, $t_0 + T^{(CP)}$ is the latest possible timing to avoid the falling transient of the component with the shortest delay (user 1). In this manner, multiple access interference (MAI) is entirely avoided without explicit estimation of path delays. Note, however, that the FFT window in general captures cyclically shifted versions of individual users' components. A cyclic shift in the time domain corresponds to phase rotations in the frequency domain (cf. Equation 14.1), which can be easily corrected by a per-tone phase shift carried out by the equalizer without explicitly estimating the phase factors or, equivalently, the number of shifted samples.

An alternative way of viewing a quasi-synchronous system is the following. Assume that all UL blocks are sent at t_0 and treat the roundtrip delays $\tau_k^{(DL)} + \tau_k^{(UL)}$ as part of the channel impulse responses. Consequently, there are no timing offsets to be estimated or corrected explicitly.

To summarize, in a quasi-synchronous OFDMA UL system, there is no need for estimating and/or correcting timing offsets. The price to pay is reduced efficiency caused by increasing the CP length by the largest roundtrip delay. Frequency offsets, however, remain and have to be estimated and corrected.

■ In a *synchronous system*, each user estimates its roundtrip delay $\tau_k^{(DL)} + \tau_k^{(UL)}$ and transmits its UL block $\tau_k^{(DL)} + \tau_k^{(UL)}$ seconds before the arrival of the DL block. Consequently, the UL blocks of all users arrive aligned at t_0 at the BS. Figure 14.4 again depicts the two users closest to and farthest from the BS. The CP is chosen long enough to accommodate the maximum channel delay spread:

$$T^{(CP)} \geq \max_i \tau_i^{(CIR)}. \tag{14.9}$$

The BS takes an FFT of the aggregate UL stream beginning at $t_0 + T^{(CP)}$, which ensures that the FFT window extends exclusively over steady-state parts of both the component with the largest and the component with the shortest delay.

To summarize, in a synchronous system, offsets are corrected by the users such that blocks arrive in an aligned manner at the BS. The CP only needs to enclose the maximum channel delay spread. The offsets are estimated either by the users themselves or by the BS and are subsequently fed back to the users.

14.3.2 Offset Estimation in OFDMA UL

Considering the complexity of this task, it comes as no surprise that the open literature does not seem to include a viable solution for jointly estimating time and frequency offsets in an asynchronous setup. There has been, however, considerable progress under certain simplifying assumptions:

■ *Consecutive estimation of parameters* can be applied assuming that users begin to access the UL channel one at a time, which eliminates the need for joint estimation [4]. In practice, however, it may be difficult to exclude the case when two users enter the network simultaneously.

■ *Structured subcarrier allocation* can greatly simplify the estimation task. Assigning groups of adjacent subcarriers forming subbands to individual users, for example, essentially allows the separation of users in the frequency domain applying filterbank-like techniques. Another popular allocation scheme is regular interleaving, which assigns subcarrier k and each consecutive K-spaced subcarrier to user no. $k \in \{1, \ldots, K\}$. Provided that $N/K \in \mathbb{N}$, using every Kth subcarrier yields a periodic time-domain signal with period N/K—a distinct signal structure to be exploited by estimation techniques such as multiple signal classification (MUSIC) [5]. However, subcarrier allocation through resource allocation schemes aiming at maximizing some performance metric, which in general yields an arbitrary, unstructured allocation, is one of the most appealing advantages of OFDMA. Structured subcarrier allocation may thus be a too stringent assumption.

■ *Quasi-synchronous timing* leaves only frequency offsets to be estimated. Dedicated pilot-based estimation schemes have been proposed in Refs. [6,7] using the SAGE (space-alternating generalized expectation-maximization) algorithm [8,9] described in the following. Under the quasi-synchronous timing assumption, a receive block arriving at the BS after purging the cyclic extension can be written as

$$r = \sum_{k=1}^{K} \text{diag}\left\{\begin{bmatrix} 1 & e^{j2\pi\epsilon_k/N} & \cdots & e^{j2\pi\epsilon_k(N-1)/N} \end{bmatrix}\right\} S_k h_k + n, \qquad (14.10)$$

where ϵ_k is the kth user's frequency offset, S_k is a Toeplitz matrix containing the kth user's pilot data known to the receiver, h_k is channel impulse response, and n is noise. A possibly existing phase offset is absorbed by h_k in Equation 14.10 and can be taken care of by the equalizer.

The SAGE algorithm applied to the signal model (Equation 14.10) iteratively improves estimates of ϵ_k and h_k in a two-step procedure and approaches the maximum likelihood solution. Given initial estimates $\widehat{\epsilon}_i^{(0)}$ and $\widehat{h}_i^{(0)}$, initial estimates of the users' receive signals can be calculated as

$$\widehat{r}_k^{(0)} = \text{diag}\left\{\begin{bmatrix} 1 & e^{j2\pi\widehat{\epsilon}_k^{(0)}/N} & \cdots & e^{j2\pi\widehat{\epsilon}_k^{(0)}(N-1)/N} \end{bmatrix}\right\} S_k \widehat{h}_k^{(0)}.$$

In each iteration $j = 1, 2, \ldots$, for each user $k = 1, 2, \ldots, K$, the algorithm performs the following steps:

– *E-step (compute expectation):* Compute an estimate of the kth user's receive signal based on the (best available) estimates of all other users' signals according to

$$\bar{r}_k^{(j)} = r - \sum_{i=1}^{k-1} \widehat{r}_i^{(j)} - \sum_{i=k+1}^{K} \widehat{r}_i^{(j-1)}.$$

– *M-step (maximize expectation):* Compute improved estimates of ϵ_k and h_k by minimizing

$$\left\| \bar{r}_k^{(j)} - \text{diag}\left\{\begin{bmatrix} 1 & e^{j2\pi\bar{\epsilon}_k/N} & \cdots & e^{j2\pi\bar{\epsilon}_k(N-1)/N} \end{bmatrix}\right\} S_k \bar{h}_k \right\|^2$$

with respect to $\bar{\epsilon}_k$ and $\bar{h}_k$. Since the problem allows one to decouple $\bar{\epsilon}_k$ and $\bar{h}_k$, we can first estimate the frequency offset according to

$$\widehat{\epsilon}_k^{(j)} = \arg\max_{\bar{\epsilon}_k} \left\{ \left\| S_k (S_k^H S_k)^{-1} S_k^H \text{diag}\left\{\begin{bmatrix} 1 & e^{-j2\pi\bar{\epsilon}_k/N} & \cdots & e^{-j2\pi\bar{\epsilon}_k(N-1)/N} \end{bmatrix}\right\} \bar{r}_k^{(j)} \right\|^2 \right\}$$

and subsequently use $\widehat{\epsilon}_k^{(j)}$ to calculate

$$\widehat{h}_k^{(j)} = (S_k^H S_k)^{-1} S_k^H \text{diag}\left\{\begin{bmatrix} 1 & e^{-j2\pi\widehat{\epsilon}_k^{(j)}/N} & \cdots & e^{-j2\pi\widehat{\epsilon}_k^{(j)}(N-1)/N} \end{bmatrix}\right\} \bar{r}_k^{(j)}.$$

Finally, using these estimates of frequency offset and channel, an update of the kth user's receive signal can be computed for the next E-step(s):

$$\widehat{r}_k^{(j)} = \operatorname{diag}\left\{\begin{bmatrix} 1 & e^{j2\pi\widehat{\epsilon}_k^{(j)}/N} & \cdots & e^{j2\pi\widehat{\epsilon}_k^{(j)}(N-1)/N} \end{bmatrix}\right\} S_k \widehat{b}_k^{(j)}.$$

Note that this scheme can be applied when unstructured subcarrier allocation is used. Although quasi-synchronous timing is appealing because of its simplicity, it can impose serious limits on the allowed roundtrip time and thus on the system's cell size or, equivalently, on the system's bandwidth efficiency.

14.3.3 Offset Correction in OFDMA UL

In OFDMA UL, there are, in principle, two ways of correcting offsets:

■ *Correction by users*: The synchronous timing scheme discussed in Section 14.3.1 can be extended to frequency offsets. Offsets in both time and frequency are corrected by users to achieve alignment at the BS. Offset estimates determined by the BS must thus be fed back to the users, which causes a bandwidth loss in the DL direction.

■ this is the most sophisticated and most challenging approach avoiding the feedback of parameter estimates.

In the following, we focus on dedicated low-complexity techniques to correct offsets. Most published work assumes a synchronous or quasi-synchronous timing scheme and focuses on the frequency offset. Again, structured subcarrier allocation simplifies the problem.

Subband allocation inherently separates users in the frequency domain to a certain degree. Figure 14.5 depicts a straightforward solution to frequency-offset compensation, hereinafter referred to as the *direct frequency-offset correction* method. Let r denote a receive block after purging the cyclic extension. For each user $k \in \{1, \ldots, K\}$, there is a dedicated processing path including the correction of the frequency offset in the time domain via multiplication by $e^{-j2\pi n\widehat{\epsilon}_k/N}$, where $\widehat{\epsilon}_k$ is the frequency-offset estimate, followed by an N-point FFT yielding R_k. In order to improve the

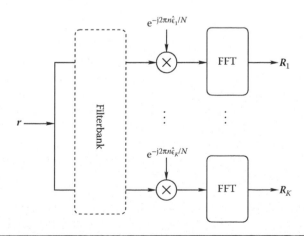

Figure 14.5 Direct frequency-offset correction.

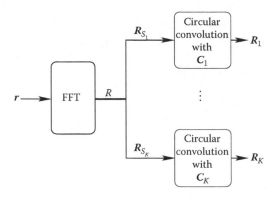

Figure 14.6   Carrier-frequency offset correction in FFT-domain (CLJL scheme [10]).

separation of users in the frequency domain, which is impaired by spectral leakage, a time-domain filterbank may be employed.

Direct frequency-offset correction requires one full-size FFT per user, which results in a computational complexity of $\mathcal{O}(KN \log_2 N)$ (excluding the filterbank). An approach to reduce complexity, hereinafter referred to as the *CLJL scheme* [10], is depicted in Figure 14.6. The idea of CLJL is to perform offset correction after FFT processing of r and thus use only a single FFT yielding R. At first glance, this does not seem to reduce complexity since the multiplication $r[n + 1]e^{-j2\pi n\widehat{\epsilon}_k/N}, n = 0, \ldots, N - 1$, in the discrete time domain (complexity $\mathcal{O}(N)$) corresponds to circular convolution of R and C_k in the frequency domain (complexity $\mathcal{O}(N \log_2 N)$), where $C_k[\ell] = \sum_{n=0}^{N-1} e^{-j2\pi n\widehat{\epsilon}_k/N} e^{-j2\pi(\ell-1)n/N}, \ell = 1, \ldots, N$, is the N-point DFT of $e^{-j2\pi n\widehat{\epsilon}_k/N}, n = 0, \ldots, N - 1$. However, using only the set $\mathcal{S}_k$ of subcarriers that are actually assigned to user k when computing the circular convolution yields a computational complexity of $\mathcal{O}(N \log_2 (N/K))$. In other words, instead of R, the vectors

$$R_{\mathcal{S}_k}[\ell] = \begin{cases} R[\ell], & \ell \in \mathcal{S}_k, \\ 0, & \text{otherwise} \end{cases}$$

are used to compute the circular convolutions. The rationale behind this approximation is that most of the energy corresponding to the kth user's signal should be concentrated in the kth user's subband $\mathcal{S}_k$, as long as the carrier-frequency offset is small compared to the subcarriers spacing. The CLJL scheme in fact outperforms the direct method without a filterbank in terms of bit error rate since processing only the subcarriers assigned to the corresponding user has a filter effect and thus improves separation of users.

Both the direct frequency-offset correction method and the CLJL scheme rely on subband allocation. In order to keep the ICI among users sufficiently low, frequency-domain guard bands may be necessary, which reduces spectral efficiency. An approach to correct frequency errors for interleaved and arbitrary assignment is *parallel interference cancelation* [11], a nonlinear multiuser-detection scheme that builds on the direct method or on the CLJL scheme. The basic idea is to iteratively improve estimates of users' receive signals. The direct offset correction method or the CLJL scheme can be used to obtain a set of K initial estimates for the separated receive signals. In each iteration, for each user $k \in \{1, \ldots, K\}$, a new receive signal estimate is computed as follows: first, the receive signal components of all the other users $\ell \in \{1, \ldots, K\} \backslash k$ impaired by their frequency offsets are reconstructed by imposing the corresponding frequency offset on each user's current (best available) frequency-corrected receive signal estimate; second, these receive components are subtracted from the aggregate receive signal, which yields a (hopefully) improved estimate of the kth user's receive signal (still containing the frequency offset); third, the frequency offset of the kth user is corrected (using one of the methods discussed above) to obtain a new estimate of the current user's frequency-corrected receive signal. Eventually, after a number of iterations, decisions on the data based on these estimates are taken.

Another approach to frequency-offset correction is *linear multiuser detection* [12]. The idea here is to start from the linear signal model

$$R = \boldsymbol{\epsilon}(\epsilon_1, \ldots, \epsilon_K) \underbrace{\begin{bmatrix} R'_1 \\ \vdots \\ R'_K \end{bmatrix}}_{R'} + n,$$

where $R'_k \in \mathbb{C}^{|\mathcal{S}_k| \times 1}$ denotes the data of the kth user multiplied by the corresponding frequency-domain channel coefficients (but excluding the impact of spectral leakage) and $\boldsymbol{\epsilon}(\epsilon_1, \ldots, \epsilon_K) \in \mathbb{C}^{N \times \sum_i |\mathcal{S}_i|}$ models the spectral leakage caused by the frequency offsets $\epsilon_k, k \in \{1, \ldots, K\}$. Then, parameter estimation is used to obtain frequency-offset-free receive signal estimates R'_k that are optimal in the sense of minimizing the least squares (LS) error or the minimum mean square error (MMSE). The LS estimate, given by

$$R'_{\mathrm{LS}} = \left(\boldsymbol{\epsilon}^{\mathrm{H}} \boldsymbol{\epsilon}\right)^{-1} \boldsymbol{\epsilon}^{\mathrm{H}} R,$$

is sometimes also referred to as linear decorrelating detector when used together with a decision device (and thus detecting data). The MMSE estimate, given by

$$R'_{\mathrm{MMSE}} = C_{R'} \boldsymbol{\epsilon}^{\mathrm{H}} \left(\boldsymbol{\epsilon} C_{R'} \boldsymbol{\epsilon}^{\mathrm{H}} + C_n\right)^{-1} R,$$

requires *a priori* knowledge about the noise and the parameter to be estimated in the form of their covariance matrices C_n and $C_{R'}$, respectively. In practice, the received data R' is often assumed to be uncorrelated, yielding $C_{R'} = I$. Also the noise n can often be assumed to be uncorrelated, yielding $C_n = \sigma_n^2 I$, where σ_n^2 is the per-sample noise variance. The advantage of the MMSE estimator over the LS estimator is the mitigation of the noise enhancement problem.

14.4 Example: Synchronization for WiMAX

As an example of an operational scheme, this section outlines the synchronization procedure in mobile WiMAX (IEEE 802.16e), a wireless metropolitan area network standard [13]. WiMAX embraces several of the principles discussed so far and is thus a well-suited example to round off this chapter. Synchronization for mobile WiMAX is described for DL first, followed by a discussion of the UL.

For WiMAX DL, each user needs to perform both timing synchronization and carrier frequency synchronization. Apart from finding the symbol boundary, a user also needs to identify the frame boundary (a frame consists of a number of symbols). For carrier frequency synchronization, the user synchronizes its carrier oscillator with the received carrier, whose frequency corresponds to the BS's carrier frequency plus the motion-incurred Doppler shift.

In WiMAX, DL acquisition can be based on a preamble inserted at the beginning of each frame. The preamble OFDMA symbol has the special property that only every third subcarrier is used, which results in a time-domain signal that has three identical parts within the useful part of an OFDMA symbol. These three identical parts can be used for both symbol timing synchronization and carrier frequency synchronization, based on the principle explained in Section 14.2. Since only the preamble OFDMA symbol exhibits these identical parts, symbol-timing acquisition identifies the frame boundary as well. With the initial acquisition based on the preamble OFDMA symbol, users can begin their FFT-processing with reasonably small ISI and ICI. Subsequently, users perform tracking using dedicated tracking-pilot subcarriers. Alternatively, CP-based tracking can be performed. The requirement for the symbol-timing error is 25% of the CP length according to the IEEE 802.16e standards and 6.25% of the CP length according to the WiMAX system-profile document that lists the requirements for WiMAX systems deployed in practice. The carrier-frequency offset must be below 2% of the subcarrier spacing. The offset after initial acquisition may be larger than 2%, but tracking can help to fulfill this requirement within acceptable time.

After DL synchronization is completed, a user is allowed to start transmitting signals to the BS in order to begin the process of UL synchronization. The main goal of UL synchronization is estimation and correction of path delay. Given the strict requirement for the symbol timing and carrier frequency for DL, both symbol timing and carrier frequency for UL are expected to be quite accurate. However, the user cannot estimate path delay based on the DL signal only. Thus, the path delay is estimated by the BS. In essence, WiMAX OFDMA employs a synchronous timing scheme. The synchronization parameters are estimated by the BS; the correction itself of both time and frequency offset is performed by the users. In addition, some residual symbol-timing error and carrier-frequency offset can be estimated by the BS. Parameter estimation is simplified through dedicated pilot symbols with periodic structure. Users employ individual pilot symbols with low mutual cross-correlation. Typically, one or a few users will try to establish UL synchronization simultaneously, which corresponds to a scenario that lies in between consecutive and joint estimation of UL parameters.

Before attempting to initiate UL communication, a user must first establish DL synchronization to be able to listen to control information broadcasted by the BS. In a time-division duplexed system, the frequency offset acquired in the process of DL synchronization can decrease the frequency offset in UL.* The first step is to acquire important UL transmission parameters, the so-called uplink channel descriptor (UCD) and uplink map (UL-MAP). With the UL transmission parameters, a user can figure out the uplink subframe structure. An exemplary UL subframe structure is shown in Figure 14.7. Then the so-called "initial ranging" procedure, which establishes time offset and transmit power, begins: the user randomly picks one out of a set of predefined PN sequences (so-called ranging codes), which is used to modulate the ranging subchannels (as defined in the UL-MAP). The so-designed symbol is transmitted (with proper prefix and postfix extension) in accordance with ranging-opportunity information (time slots available for ranging in the next UL frame as specified in the UCD) and repeated, immediately yielding a synchronization sequence

---

* Note that using the carrier frequency acquired during DL synchronization for UL transmission results in a UL frequency offset if the Doppler shift ΔF_c changes from DL reception to UL transmission.

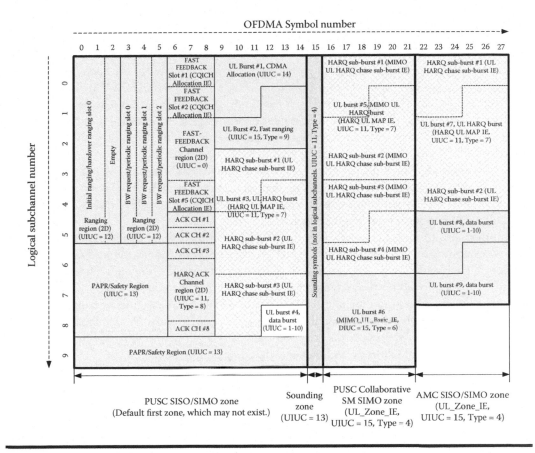

Figure 14.7 Example of a UL subframe in WiMAX showing ranging regions at the beginning of the subframe. Note: The number of OFDMA symbols in a UL subframe cannot exceed 21. Here, 28 OFDMA symbols are chosen for illustrative purposes.

that is two symbols long. In Figure 14.7, the initial ranging symbols are located in the first two OFDMA symbols. At the BS, it is thus guaranteed that out of three consecutive symbols, at least one (in the best, although very unlikely, case, two) symbol contains exclusively ranging information on the ranging subcarriers. Although in Figure 14.7, no data subcarriers are assigned at the same time as ranging subcarriers, ranging subcarriers can be multiplexed with data subcarriers before and after ranging-opportunity slots. PN sequences are used to cope with users simultaneously trying to establish UL communication (which is not too unlikely since ranging opportunities are limited). After successful identification of the PN sequence (or sequences, in case more than one user performs initial ranging) and estimation of the associated time offset(s), the BS broadcasts this information. Finally, aspiring users apply UL timing correction and continue with negotiating further physical layer parameters before eventually commencing their UL communication.

Users carrying out initial ranging are not time aligned yet and thus cause MAI disturbing synchronized users. In order to keep the MAI low, initial ranging begins with the lowest possible transmit power level. If synchronization fails (i.e., the user's PN-sequence identifier is not included in the broadcasted DL information), the initial ranging procedure is repeated with a larger transmit power level.

14.5 Literature

As a complement to the foray into selected synchronization techniques presented so far, this section aims at briefly summarizing the most important contributions available in the open literature.

Pilot-based single-user synchronization techniques include those in Refs. [1,14–19]. The technique in reference [14] is one of the first correlation-based timing-synchronization schemes for OFDM; however, it assumes that the impact of time dispersion is negligible. Moose [15] uses two successive identical OFDM blocks to estimate the frequency offset ($P = N$). In Ref. [16], a null symbol followed by a pilot symbol is used to detect the beginning of a frame. Lambrette et al. [17] use CAZAC (constant-amplitude zero-autocorrelation) sequences. Morelli and Mengali [18] and Coulson [19] investigate modifications of Ref. [1] avoiding the second pilot symbol. Morelli and Mengali [18] introduce a pilot symbol with more than two identical parts and Coulson [19,20] uses a pilot symbol based on a repeated maximum length sequence.

Another class of methods, here referred to as *semi-blind single-user synchronization techniques*, relies on exploiting inherent deterministic signal structure originally not introduced to aid synchronization. Signal repetition in the time domain introduced by the CP provides a great amount of structure to be exploited for timing synchronization [2,22–24]. Van de Beek et al. [2] present the maximum-likelihood solution for jointly estimating time and frequency offsets based on the CP. A very similar approach employing weighted moving average correlators is described in Ref. [22]. Lashkarian and Kiaei [23] improve the range of timing offsets using a modified likelihood function. A principal weakness of CP-based techniques is that they profit from an excess CP. Consequently, there is a tradeoff between CP-based synchronization performance and immunity to delay spread Landström et al. [24] present a CP-based time offset estimator that is immune to small frequency offsets and exhibits a certain degree of robustness to time dispersion. Another type of redundancy that can be exploited for synchronization are nulled subcarriers [25–28]. In a more rigorous approach, Bölcskei [29] exploits quasi-cyclostationarity (second order), introduced not only by a CP but also by frequency-domain guard bands or pulse shaping.

Truly *blind single-user synchronization methods* rely on higher-order statistical properties of the OFDM receive signal. Examples for kurtosis-based methods are those of reference [30], which measure distance from Gaussianity, and of reference [31], which exploit the spectral line corresponding to carrier frequency offset.

An excellent tutorial overview on *multiuser synchronization* in OFDMA UL in general and BS-based estimation and correction techniques in particular is provided in Ref. [32]. Parameter estimation in a consecutive fashion (one user at a time) is discussed in Ref. [4].

Multiuser parameter estimation techniques for systems with subband subcarrier allocation in combination with filterbank techniques can be borrowed from single-user setups [24,33]. A dedicated frequency estimation scheme for interleaved subcarrier allocation based on MUSIC [34] is presented in Ref. [5]. Synchronization schemes for unstructured subcarrier allocation in quasi-synchronous schemes using the EM-based [9] SAGE algorithm [8] are introduced in references [6,7]. Zhang et al. [35] consider estimation of both frequency and time offsets for synchronous OFDMA UL. Detection algorithms for initial ranging at the BS in WiMAX OFDMA are discussed in Ref. [36].

Multiuser offset correction by users for OFDMA systems was first mentioned in Ref. [33]. For subband subcarrier allocation, again filterbank techniques and individual correction can be applied [32]. A low-complexity correction scheme based on circular convolution is proposed in Ref. [10]. Offset correction for unstructured subcarrier allocation includes interference cancelation [11] and linear multiuser detection [12].

14.6 Summary and Conclusion

Synchronization in OFDMA is a challenge. In the DL direction, many concepts can be borrowed from single-user OFDM systems. In the UL direction, however, the time and frequency misalignment of different users' receive components destroys the orthogonality among users. Both estimation and correction of offsets are more complicated compared to the DL case, which renders the intuitively appealing multiple-access concept of user separation in the frequency domain less attractive. While research on the ultimate scheme of jointly estimating all parameters and correcting offsets at the BS with feasible complexity is ongoing, operational and emerging systems revert to estimating parameters of one or a few users at a time at the BS side and correcting offsets at the user side.

Symbols

ΔF_c	Carrier frequency offset in Hertz
ΔF_s	Sampling frequency offset in Hertz
F_c	Carrier frequency in Hertz
$F_s = 1/T_s$	Sampling frequency in Hertz
K	Number of users
N	OFDM symbol length in samples (excluding the cyclic prefix)
$T^{(CP)}$	Cyclic prefix length in seconds
$T^{(MC)} = NT_s$	OFDM symbol length (excluding cyclic prefix) in seconds
$T_s = 1/F_s$	Sampling period in seconds
$T^{(W)}$	Cyclic OFDM symbol extension for time-domain windowing in seconds
$\tau^{(CIR)}$	Channel impulse response length in seconds
$\tau_k^{(DL)}$	Downlink path delay of user k in seconds
$\tau_k^{(UL)}$	Uplink path delay of user k in seconds

Abbreviations

BS	Base station
CP	Cyclic prefix
DFT	Discrete Fourier transform
DL	Downlink
FFT	Fast Fourier transform
ICI	Intercarrier interference
ISI	Intersymbol interference
LS	Least squares
MAI	Multiple access interference
MMSE	Minimum mean square error
PN	Pseudo-random number
UL	Uplink

References

1. T. M. Schmidl and D. C. Cox, Robust frequency and timing synchronization for OFDM, *IEEE Transactions on Communications*, 45 (12), 1613–1621, 1997.
2. J.-J. van de Beek, M. Sandell, and P. O. Börjesson, ML estimation of time and frequency offset in OFDM systems, *IEEE Transactions on Signal Processing*, 45 (7), 1800–1805, July 1997.
3. A. M. Tonello, Multiuser detection and turbo multiuser decoding for asynchronous multitone multiple access systems, in *Proc. 56th IEEE Vehicular Technology Conference (VTC '02-Fall)*, 2002, 2, 970–974.
4. M. Morelli, Timing and frequency synchronization for the uplink of an OFDMA system, *IEEE Transactions on Communications*, 52 (2), 296–306, Feb. 2004.
5. Z. Cao, U. Tureli, and B Y. D. Yao, Deterministic multiuser carrier-frequency offset estimation for interleaved OFDMA uplink, *IEEE Transactions on Communications*, 52 (9), 1585–1594, Sept. 2004.
6. M. O. Pun, M. Morelli, and B C.-C. J. Kuo, Maximum-likelihood synchronization and channel estimation for OFDMA uplink transmissions, *IEEE Transactions on Communications*, 54 (4), 726–736, Apr. 2006.
7. M. O. Pun, M. Morelli, and C.-C. J. Kuo, Iterative detection and frequency synchronization for OFDMA uplink transmissions, *IEEE Transactions on Wireless Communications*, 6 (2), 629–639, Feb. 2007.
8. J. A. Fessler and A. O. Hero, Space-alternating generalized expectation-maximization algorithm, *IEEE Transactions on Signal Processing*, 42 (10), 2664–2677, Oct. 1994.
9. A. P. Dempster, N. M. Laird, and D. B. Rubin, Maximum likelihood from incomplete data via the EM algorithm, *Journal of the Royal Statistical Society*, 39, 1–38, 1997.
10. J. Choi, C. Lee, H. W. Jung, and B Y. H. Lee, Carrier frequency offset compensation for uplink of OFDM-FDMA systems, *IEEE Transactions on Communications*, 4 (12), 414–416, Dec. 2000.
11. D. Huang and K.B. Letaief, An interference-cancellation scheme for carrier frequency offsets correction in OFDMA systems, *IEEE Transactions on Communications*, 53 (7), 1155–1165, July 2005.
12. Z. Cao, U. Tureli, Y.-D. Yao, and P. Honan, Frequency synchronization for generalized OFDMA uplink in *Proc. IEEE Global Telecommunications Conference (GLOBECOM '04)*, December 2004, vol. 2, pp. 1071–1075.
13. IEEE standard for local and metropolitan area networks part 16: Air interface for fixed and mobile broadband wireless access systems amendment 2: Physical and medium access control layers for combined fixed and mobile operation in licensed bands and corrigendum 1, *IEEE Std 802.16e-2005 and IEEE Std 802.16-2004/Cor 1-2005 (Amendment and Corrigendum to IEEE Std 802.16-2004) Std.*, 2006.
14. W. D. Warner and C. Leung, OFDM/FM frame synchronization for mobile radio data communication, *IEEE Transactions on Vehicular Technology*, 42 (3), 302–313, Aug. 1993.
15. P. H. Moose, A technique for orthogonal frequency division multiplexing frequency offset correction, *IEEE Transactions on Communications*, 42 (10), 2908–2914, Oct. 1994.
16. H. Nogami and T. Nagashima, A frequency and timing period acquisition technique for OFDM systems, in *Proc 6th IEEE International Symposium on Personal, Indoor and Mobile Radio Communications (PIMRC '95)*, September 1995, vol. 3.
17. U. Lambrette, M. Speth, and H. Meyr, OFDM burst frequency synchronization by single carrier training data, *Communications Letters, IEEE*, 1 (2), 46–48, Mar 1997.
18. M. Morelli and U. Mengali, An improved frequency offset estimator for OFDM applications, *IEEE Communications Letters*, 3 (3), 75–77, Mar 1999.
19. A. J. Coulson, Maximum likelihood synchronization for OFDM using a pilot symbol: Algorithms, *IEEE Journal on Selected Areas in Communications*, 19 (12), 2486–2494, Dec. 2001.
20. A. J. Coulson, Maximum likelihood synchronization for OFDM using a pilot symbol: Analysis, *IEEE Journal on Selected Areas in Communications*, 19 (12), 2495–2503, Dec. 2001.
21. M. Schmidl and D. C. Cox, Blind synchronisation for OFDM, *Electronics Letters*, 33 (2), 113–114, Jan. 1997.
22. M.-H. Hsieh and C.-H. Wei, A low-complexity frame synchronization and frequency offset compensation scheme for OFDM systems over fading channels, *IEEE Transactions on Vehicular Technology*, 48, 1596–1609, Sept. 1999.

23. N. Lashkarian and S. Kiaei, Class of cyclic-based estimators for frequency-offset estimation of OFDM systems, *IEEE Transactions on Communications*, 48 (12), 2139–2149, Dec. 2000.
24. D. Landström, S. K. Wilson, J.-J. van de Beek, P. Ödling, and P. O. Börjesson, Symbol time offset estimation in coherent OFDM systems, *IEEE Transactions on Communications*, 50 (4), 545–549, Apr. 2002.
25. H. Liu and U. Tureli, A high-efficiency carrier estimator for OFDM communications, *IEEE Communication Letters*, 2, 104–106, Apr. 1998.
26. X. Ma, C. Tepedelenlioglu, G. B. Giannakis, and S. Barbarossa, Non-data-aided carrier offset estimators for OFDM with null subcarriers: Identifiability, algorithms, and performance, *IEEE Journal on Selected Areas in Communications*, 19, 2504–2515, Dec. 2001.
27. S. Barbarossa, M. Pompili, and G. B. Giannakis, Channel-independent synchronization of orthogonal frequency-division multiple-access systems, *IEEE Journal on Selected Areas in Communications*, 20, 474–486, Feb. 2002.
28. M. Ghogho and A. Swami, Blind frequency-offset estimator for OFDM systems transmitting constant-modulus symbols, *IEEE Communication Letters*, 6, 343–345, Aug. 2002.
29. H. Bölcskei, Blind estimation of symbol timing and carrier frequency offset in wireless OFDM systems, *IEEE Transactions on Communications*, 49, 988–999, June 2001.
30. Y. Yao and G. B. Giannakis, Blind carrier frequency offset estimation in SISO, MIMO, and multiuser OFDM systems, *IEEE Transactions on Communications*, 53 (1), 173–183, Jan. 2005.
31. M. Luise, M. Marselli, and R. Reggiannini, Low-complexity blind carrier frequency recovery for OFDM signals over frequency-selective radio channels, *IEEE Transactions on Communications*, 50 (7), 1182–1188, July 2002.
32. M. Morelli, C.-C.J. Kuo, and M.-O. Pun, Synchronization techniques for orthogonal frequency division multiple access (OFDMA): A tutorial review, *Proceedings of the IEEE*, 95 (7), 1394–1427, July 2007.
33. J.-J. van de Beek, P. O. Börjesson, M.-L. Boucheret, D. Landström, J. M. Arenas, P. Ödling, C. Östberg, M. Wahlqvist, and S. K. Wilson, A time and frequency synchronization scheme for multiuser OFDM, *IEEE Journal on Selected Areas in Communications*, 17 (11), 1900–1914, Nov. 1999.
34. R. Schmidt, Multiple emitter location and signal parameter estimation, *IEEE Transactions on Antennas and Propagation*, 34 (3), pp. 276–280, Mar. 1986.
35. Y. Zhang, R. Hoshyar, and R. Tafazolli, Timing and frequency offset estimation scheme for the uplink of OFDMA systems, *IET Communications*, 2 (1), 121–130, Jan. 2008.
36. H. A. Mahmoud, H. Arslan, and M. K. Ozdemir, Initial ranging for WiMAX (802.16e) OFDMA, in *Military Communications Conference*, 2006. MILCOM 2006, 1–7.

Chapter 15

Multiuser CFOs Estimation in OFDMA Uplink Systems

Yik-Chung Wu, Jianwu Chen, Tung-Sang Ng,
and Erchin Serpedin

Contents

15.1   Introduction..397
15.2   System Model..398
15.3   CFO Estimation in Subband-Based CAS Systems 400
15.4   CFO Estimation in Interleaved CAS Systems..401
15.5   CFO Estimation in Generalized CAS Systems 403
       15.5.1   ML Estimator 403
                15.5.1.1   Special Case: Asymptotic Decoupled ML Estimator..............404
       15.5.2   Iterative Algorithms for ML Estimator....................................405
       15.5.3   ML Estimation with Importance Sampling..................................407
                15.5.3.1   Choice of Importance Function $g(\omega)$..........................409
15.6   Complexity Analyses...410
15.7   Performance Comparisons...411
15.8   Concluding Remarks...414
References..414

15.1 Introduction

Since orthogonal frequency-division multiple access (OFDMA) is widely regarded as a promising technique for broadband wireless networks, it has attracted a lot of attention recently [1–4]. In OFDMA, all users transmit their data to the base station (BS) simultaneously by modulating exclusive sets of orthogonal subcarriers; thus the receiver at the BS can separate each user's signal easily in frequency domain.

Like other OFDM-based systems, the performance of OFDMA systems is very sensitive to carrier frequency offsets (CFOs). In particular, the CFO estimation problem is very challenging in uplink transmissions since different users have different CFOs, and thus in addition to intercarrier interference (ICI), multiple access interference (MAI) is also introduced in the received signal.

There are three different carrier assignment schemes (CASs) for OFDMA systems. Depending on the CAS employed, different algorithms for frequency estimation have been proposed for OFDMA uplink in the literature.

- For systems with *subband-based CAS*, bandpass filters are used to separate users before synchronization [5,6].
- To maximize frequency diversity and increase the capacity of OFDMA systems, *interleaved CAS* is required. For this kind of OFDMA systems, a frequency estimation scheme exploiting the periodic structure of the signal transmitted by each user was proposed in Ref. [7].
- Coping with the requirements for dynamic resource allocation and scheduling in future wireless systems, *generalized CAS* provides more flexibility than subband-based or interleaved schemes. For OFDMA systems with generalized CAS, a number of frequency estimation algorithms were proposed in Refs. [8–15]. Synchronization on one new user while assuming all existing users have already been synchronized was discussed in Ref. [8]. The more general case where all users need to be synchronized was also addressed [9–15], in which iterative techniques (e.g., alternating projection [9] and expectation Maximization [10–13]), importance sampling-based techniques [14] and asymptotic estimators [14,15] were proposed.

In this chapter, the details of the above algorithms will be reviewed and their relationships will be identified. Furthermore, their complexities and performances will be compared.

The following notations are used throughout this chapter. Superscripts $(\cdot)^{-1}$, $(\cdot)^{T}$, and $(\cdot)^{H}$ denote the inverse, the transpose, and the conjugate transpose operations, respectively. The $k \times k$ identity matrix is denoted by $\mathbf{I}_k$ and the zero matrix with dimension $k_1 \times k_2$ is denoted by $\mathbf{0}_{k_1 \times k_2}$. The operator diag$\{\mathbf{x}\}$ denotes a diagonal matrix with elements of $\mathbf{x}$ located on the main diagonal. Symbol $\|\mathbf{x}\|$ represents the L2 norm of the vector $\mathbf{x}$ and $\angle x$ is the angle of x. Notation $\mathbb{E}\{\cdot\}$ is the expectation of a random variable. Symbol vec$(\mathbf{Z})$ is the vec operator applied to the matrix $\mathbf{Z}$.

15.2 System Model

In the considered OFDMA system as shown in Figure 15.1, K users transmit different data streams simultaneously using exclusive sets of subcarriers to the BS. Before initiating the transmission, the timing for each user is acquired by using the downlink synchronization channel from BS. Consequently, the transmissions from all users can be regarded as quasi-synchronous. The total number of subcarriers is denoted as N and one block of frequency-domain symbols sent by the kth user is denoted as $\mathbf{d}_k = [d_k(0), \ldots, d_k(N-1)]^{T}$, where $d_k(i), i = 0, \ldots, N-1$ is nonzero if and only if the ith subcarrier is modulated by the kth user.

The frequency-domain data are first modulated onto different subcarriers by left multiplying an N-point inverse discrete Fourier transform (IDFT) matrix $\mathbf{F}^{H}$, where $\mathbf{F}$ is the discrete Fourier transform (DFT) matrix with $\mathbf{F}(p.q) = (1/\sqrt{N}) \exp(-j2\pi pq/N)$. After inserting a cyclic prefix (CP) of length L_{cp} into each block of the time domain signal (denoted as $\mathbf{F}^{H}\mathbf{d}_k$), the augmented block is serially transmitted through the multipath channel. Let the channel impulse

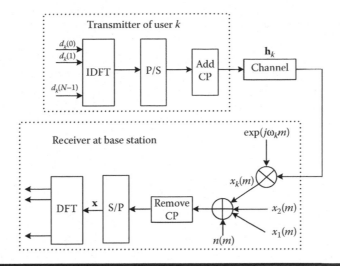

Figure 15.1   Baseband system diagram for OFDMA uplink. (From J. Chen, Y.-C.Wu, S. C. Chan, and T.-S. Ng, *IEEE Trans. Veh. Technol.*, 57 (6), 3462–3470, 2008. © 2008 IEEE. With permission.)

response (including all transmit/receive filtering effects) between the kth user and the BS be denoted as $\boldsymbol{\xi}_k = [\xi_k(0), \ldots, \xi_k(L_k - 1)]^{\mathrm{T}}$, where L_k is the channel length. Denoting the timing offset caused by propagation delay as μ_k, the compound channel response can be written as $\mathbf{h}_k \triangleq [\mathbf{0}^{\mathrm{T}}_{\mu_k \times 1} \quad \boldsymbol{\xi}^{\mathrm{T}}_k \quad \mathbf{0}^{\mathrm{T}}_{(L-\mu_k-L_k)\times 1}]^{\mathrm{T}}$, where L is the upper bound on the compound channel length. For user k, the normalized CFO is denoted as ε_k. At the BS, after timing synchronization and removal of CP, the received signal component from user k is given by

$$\mathbf{x}_k \triangleq [x_k(0), \ldots, x_k(N - 1)]^{\mathrm{T}} = \boldsymbol{\Gamma}(\omega_k)\mathbf{A}_k\mathbf{h}_k, \tag{15.1}$$

where

$$\boldsymbol{\Gamma}(\omega_k) \triangleq \mathrm{diag}\{1, \ldots, \exp(j(N - 1)\omega_k)\}, \tag{15.2}$$

$$\mathbf{A}_k \triangleq \sqrt{N}\mathbf{F}^{\mathrm{H}}\mathbf{D}_k\mathbf{F}_L, \tag{15.3}$$

$$\mathbf{D}_k \triangleq \mathrm{diag}\{\mathbf{d}_k\}, \tag{15.4}$$

$$\omega_k \triangleq \frac{2\pi\varepsilon_k}{N}, \tag{15.5}$$

with $\mathbf{F}_L$ denoting the first L columns of $\mathbf{F}$.

Since the received signal at the BS (denoted as $\mathbf{x}$) is a superposition of the signals from all the users plus noise, we have

$$\mathbf{x} = \sum_{k=1}^{K} \boldsymbol{\Gamma}(\omega_k)\mathbf{A}_k\mathbf{h}_k + \mathbf{n}, \tag{15.6}$$

where $\mathbf{n} = [n(0), \ldots, n(N - 1)]^{\mathrm{T}}$ is the complex white Gaussian noise vector with zero mean and covariance matrix $\mathbf{C_n} = \mathbb{E}\{\mathbf{nn}^{\mathrm{H}}\} = \sigma^2\mathbf{I}_N$.

In this chapter, only the case where the BS is equipped with one antenna is considered, but the above signal model can be easily generalized to the case where the BS is equipped with more than one antenna due to the fact that the multiple antennas equipped at the BS always share the same oscillator.

Remark: Although there is an implicit assumption that the upper bound of the sum of channel delay spread and propagation delay for each user (L) is less than the length of CP (L_{cp}), the considered system model is practical. The reason is that the timing offsets due to different propagation delays are limited to several samples only and in practical OFDM systems the CP is always longer than the exact channel order.

15.3 CFO Estimation in Subband-Based CAS Systems

If subcarriers are assigned to users in nonoverlapping contiguous groups, bandpass filters can be used to separate different users before synchronization. Interference from neighboring groups can be mitigated by the use of guard band between blocks of subcarriers assigned to different users. If the frequency offset of each user is smaller than the guard band, then conventional single user synchronization algorithms [16,17] can be applied to the separated signal. The idea is illustrated in Figure 15.2, where the output of the bandpass filter for the kth user is given by

$$\mathbf{r}_k = \mathbf{\Gamma}(\omega_k)\mathbf{A}_k\mathbf{h}_k + \mathbf{\Xi}_k + \mathbf{\eta}_k \tag{15.7}$$

with $\mathbf{\Xi}_k$ being the residual interference from other users to user k due to imperfect user separation, and $\mathbf{\eta}_k$ is the filtered noise. With $\mathbf{r}_k$, a number of conventional single user frequency estimation algorithms can be used. For example, in Ref. [5], the estimator exploiting CP correlation [16] is used:

$$\hat{\mu}_k = \arg \max_{\mu} \left\{ |\gamma_k(\mu)| - \frac{\text{SNR}_k}{1 + \text{SNR}_k} \Phi_k(\mu) \right\}, \tag{15.8}$$

$$\hat{\varepsilon}_k = -\frac{1}{2\pi} \angle \gamma_k(\hat{\mu}_k), \tag{15.9}$$

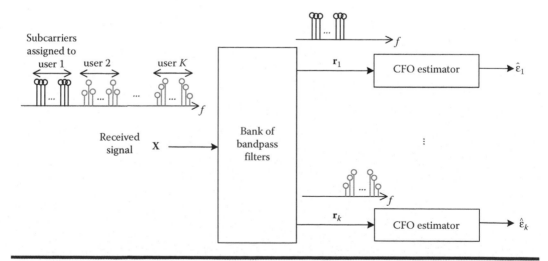

Figure 15.2   User separations and parameter estimations for an OFDMA system with subband-based CAS.

where

$$\gamma_k(\mu) = \sum_{n=\mu}^{\mu+L_{cp}-1} r_k^*(n)r_k(n+N), \tag{15.10}$$

$$\Phi_k(\mu) = \frac{1}{2}\left[\sum_{n=\mu}^{\mu+L_{cp}-1} |r_k(n)|^2 + |r_k(n+N)|^2\right], \tag{15.11}$$

with SNR_k being the signal-to-noise ratio of user k [16], and $r_k(n)$ is the nth element of $\mathbf{r}_k$. Note that the above CP-based estimator was originally derived for the AWGN channel and considerable performance degradation is expected when applied to frequency-selective fading channel.

To remedy this problem, other single user frequency estimators designed for frequency-selective fading channels can be applied instead. In Ref. [6], it was proposed that the estimator minimizing the received power at the virtual (null) subcarriers [17] of user k can be used:

$$\hat{\omega}_k = \arg\min_{\omega} \|\mathbf{W}_{-k}^H \mathbf{\Gamma}(\omega)^H \mathbf{r}_k\|^2, \tag{15.12}$$

where $\mathbf{W}_{-k}$ consists of a subset of columns of $\mathbf{F}^H$ that correspond to subcarriers that user k has not used. In fact, applications of other existing OFDM single user CFO estimators after signal separation are possible. Examples include the super-resolution subspace algorithm [18], cyclic-based estimator [19], and the algorithm exploiting the constant modulus property of data [20].

15.4 CFO Estimation in Interleaved CAS Systems

Subband-based CAS facilitates user separation, and makes synchronization easy. However, frequency diversity in frequency-selective fading channel is not fully exploited in subband-based CAS. If the coherent bandwidth of the channel is large, several consecutive subcarriers may be subject to the same fading at the same time. A deep fading may destroy the whole subband, leading to total loss of data for a user. Interleaved CAS avoids the above problem by providing maximum separation among the subcarriers assigned to a single user. Unfortunately, by interleaving subcarriers from different users, synchronization becomes more complicated since signals from different users cannot be separated simply by a bank of bandpass filters.

Without loss of generality, let $N = MR$ where both M and R are integers. Further, assume that the subcarrier indices assigned to user k are $\{i_k, i_k + R, i_k + 2R, \ldots, i_k + (M-1)R\}$ where i_k is any integer within $[0, R-1]$. Then, from Equation 15.1, the received samples from the user k can be written as

$$x_k(n) = \sum_{m=0}^{M-1} d_k(i_k+mR)\mathbf{H}_k(i_k+mR)e^{j2\pi(i_k+mR+\varepsilon_k)n/N}, \tag{15.13}$$

where $\mathbf{H}_k = \mathbf{F}_L\mathbf{h}_k$ and $\mathbf{H}_k(n)$ is the nth element of $\mathbf{H}_k$. Using Equation 15.13, it can be readily shown that

$$x_k(n+\nu M) = e^{j2\pi\nu\theta_k}x_k(n), \tag{15.14}$$

where ν is an integer and $\theta_k = (i_k + \varepsilon_k)/R$. This implies that the OFDMA time domain signal of user k has a periodic structure. Based on this periodic property, the received vector $\mathbf{x}$ in Equation 15.6

can be re-expressed in an $R \times M$ matrix as

$$Y = VS + Z, \tag{15.15}$$

where $\text{vec}(Y) = x$, $\text{vec}(Z) = n$,

$$V = \begin{bmatrix} 1 & 1 & \cdots & 1 \\ e^{j2\pi\theta^{(1)}} & e^{j2\pi\theta^{(2)}} & \cdots & e^{j2\pi\theta^{(K)}} \\ \vdots & \vdots & \ddots & \vdots \\ e^{j2\pi(R-1)\theta^{(1)}} & e^{j2\pi(R-1)\theta^{(2)}} & \cdots & e^{j2\pi(R-1)\theta^{(K)}} \end{bmatrix}, \tag{15.16}$$

$$S = \begin{bmatrix} x_1(0) & x_1(1) & \cdots & x_1(M-1) \\ x_2(0) & x_2(1) & \cdots & x_2(M-1) \\ \vdots & \vdots & \ddots & \vdots \\ x_K(0) & x_K(1) & \cdots & x_K(M-1) \end{bmatrix}. \tag{15.17}$$

In Ref. [7], a CFO estimator for interleaved CAS OFDMA systems was derived based on the equivalent system model in Equation 15.15.

The idea of the estimator in Ref. [7] is as follows. First, the correlation matrix of Y is given by

$$\Psi = \mathbb{E}[YY^H] = V\mathbb{E}[SS^H]V^H + \sigma^2 MI_R. \tag{15.18}$$

On the other hand, when $K < R$, eigendecomposition on Ψ gives

$$\Psi = U_s\Sigma_sU_s^H + \sigma^2 MU_zU_z^H, \tag{15.19}$$

where $\Sigma_s = \text{diag}\{\lambda_1, \lambda_2, \ldots, \lambda_K\}$ contains the eigenvalues of the signal subspace, U_s is composed of eigenvectors corresponding to $\lambda_1, \ldots, \lambda_K$, and U_z is composed of eigenvectors corresponding to the noise subspace. Since U_s and U_z are unitary and orthogonal to each other, multiplying Equations 15.18 and 15.19 by U_z from the right and subtracting the two resulting equations give

$$V\mathbb{E}[SS^H]V^HU_z = 0_{R\times(R-K)}. \tag{15.20}$$

Since V is a Vandermonde matrix, it is of full rank. Furthermore, assuming that signals from different users are independent, $\mathbb{E}[SS^H]$ is also of full rank. Therefore, we have $V^HU_z = 0_{K\times(R-K)}$.

Assume that the normalized CFO is in the range $\varepsilon_k \in (-0.5, 0.5)$, the range of θ_k is $((i_k - 0.5)/R, (i_k + 0.5)/R)$. Since different users employ distinct values of i_k, θ_k values for different users are in nonoverlapping ranges. Based on this fact and $V^HU_z = 0_{K\times(R-K)}$, a CFO estimator was proposed in Ref. [7] as locating the largest K local maximums of

$$\frac{1}{\|U_z^Ha(\theta)\|^2}, \tag{15.21}$$

where $a(\theta) = [1, e^{j2\pi\theta}, \ldots, e^{j2\pi(R-1)\theta}]^T$. As the ranges of θ_k for different k values are nonoverlapping, there will be one and only one $\hat{\theta}$ from Equation 15.21 falling in the range $((i_k - 0.5)/R,$

$(i_k + 0.5)/R)$. Denoting that estimate as $\hat{\theta}_k$, finally, the normalized CFO estimate for user k is given by

$$\hat{\varepsilon}_k = R\hat{\theta}_k - i_k. \tag{15.22}$$

Note that, in practice, $\mathbf{\Psi}$ is approximated by its sample correlation matrix $\hat{\mathbf{\Psi}} = \mathbf{YY}^H$, and $\mathbf{U}_z$ is approximated by $\hat{\mathbf{U}}_z$ obtained from the eigendecomposition of $\hat{\mathbf{\Psi}}$.

15.5 CFO Estimation in Generalized CAS Systems

Coping with the requirements of dynamic resource allocation and scheduling in future wireless systems, generalized CAS in OFDMA systems is very much desired. For generalized CAS, in addition to frequency diversity (also provided by interleaved CAS), multiuser diversity can also be obtained [4]. With the channel state information at the BS, users are assigned the best subchannels they experience, thus significant increases in throughput of OFDMA systems can be realized.

Unfortunately, CFOs estimation in generalized CAS is the most challenging the three CAS schemes, as signals from different users cannot be separated easily (as opposed to subband-based CAS) and there is no fixed structure in the OFDMA signal (as opposed to interleaved CAS). In Ref. [8], a CFO estimation scheme was proposed for a new user entering the OFDMA system, with other existing users already synchronized. However, this scheme may not be general enough for a practical system. In the following, we will first present the maximum likelihood (ML) estimator which jointly estimates the CFOs of all users; then we will discuss several efficient methods for obtaining the optimal solution. While previously introduced estimators for subband-based and interleaved CASs do not require training, in the following, it is assumed that a training OFDM symbol is transmitted at the beginning of the data packet by each user, to aid the CFO estimations.

15.5.1 ML Estimator

Denoting $\mathbf{h} = \left[\mathbf{h}_1^T \ \dots \ \mathbf{h}_K^T\right]^T$, the signal model in Equation 15.6 can be rewritten as

$$\mathbf{x} = \mathbf{Q}(\mathbf{\omega})\mathbf{h} + \mathbf{n}, \tag{15.23}$$

where $\mathbf{Q}(\mathbf{\omega}) = [\mathbf{\Gamma}(\omega_1)\mathbf{A}_1 \ \mathbf{\Gamma}(\omega_2)\mathbf{A}_2 \ \dots \ \mathbf{\Gamma}(\omega_K)\mathbf{A}_K]$. Based on the model in Equation 15.23, the ML estimate of parameters $\{\mathbf{h}, \mathbf{\omega}\}$ is given by maximizing [21]

$$\psi(\mathbf{x}; \tilde{\mathbf{h}}, \tilde{\mathbf{\omega}}) = \frac{1}{(\pi\sigma^2)^N} \cdot \exp\left\{-\frac{1}{\sigma^2}[\mathbf{x} - \mathbf{Q}(\tilde{\mathbf{\omega}})\tilde{\mathbf{h}}]^H[\mathbf{x} - \mathbf{Q}(\tilde{\mathbf{\omega}})\tilde{\mathbf{h}}]\right\} \tag{15.24}$$

or equivalently minimizing

$$\Lambda(\mathbf{x}; \tilde{\mathbf{h}}, \tilde{\mathbf{\omega}}) = [\mathbf{x} - \mathbf{Q}(\tilde{\mathbf{\omega}})\tilde{\mathbf{h}}]^H[\mathbf{x} - \mathbf{Q}(\tilde{\mathbf{\omega}})\tilde{\mathbf{h}}], \tag{15.25}$$

where $\tilde{\mathbf{h}}$ and $\tilde{\mathbf{\omega}}$ are trial values of $\mathbf{h}$ and $\mathbf{\omega}$, respectively. Due to the linear dependence of parameter $\mathbf{h}$ in Equation 15.23, the ML estimate for the channel vector $\mathbf{h}$ (when $\tilde{\mathbf{\omega}}$ is fixed) is given by

$$\hat{\mathbf{h}} = (\mathbf{Q}^H(\tilde{\mathbf{\omega}})\mathbf{Q}(\tilde{\mathbf{\omega}}))^{-1}\mathbf{Q}^H(\tilde{\mathbf{\omega}})\mathbf{x}. \tag{15.26}$$

Putting $\hat{\mathbf{h}}$ into Equation 15.25, the estimate of $\boldsymbol{\omega}$ can be obtained as

$$\hat{\boldsymbol{\omega}} = \arg \max_{\tilde{\boldsymbol{\omega}}} \left\{ \phi'(\tilde{\boldsymbol{\omega}}) \triangleq \|\mathbf{P}_\mathbf{Q}(\tilde{\boldsymbol{\omega}})\mathbf{x}\|^2 \right\}, \tag{15.27}$$

where $\mathbf{P}_\mathbf{Q}(\tilde{\boldsymbol{\omega}}) = \mathbf{Q}(\tilde{\boldsymbol{\omega}})(\mathbf{Q}^H(\tilde{\boldsymbol{\omega}})\mathbf{Q}(\tilde{\boldsymbol{\omega}}))^{-1}\mathbf{Q}^H(\tilde{\boldsymbol{\omega}})$.

A direct way maximizing Equation 15.27 is to use an exhaustive search over the multidimensional space spanned by $\tilde{\boldsymbol{\omega}}$. However, this approach is extremely computationally expensive and not suitable for implementation. In the next two subsections, several efficient methods for finding the maximum point in Equation 15.27 will be discussed.

As a remark, the ML estimator in Ref. [8] can be regarded as a special case of Equation 15.27, in which only the entering user needs to be synchronized and all other existing users' CFO and channel parameters are assumed to be known perfectly.

15.5.1.1 Special Case: Asymptotic Decoupled ML Estimator

From the definition of $\mathbf{A}_k$ in Equation 15.3 and noting that for OFDMA systems, only one user is allowed to transmit on each subcarrier, it is straightforward to show that $\mathbf{A}_i^H \mathbf{A}_j = \mathbf{0}_{L \times L}, \forall i \neq j$. Thus, for the OFDMA system with a large number of subcarriers, we have $\lim_{N \to \infty} \mathbf{A}_i^H \boldsymbol{\Gamma}(\omega_j - \omega_i)\mathbf{A}_j/N = \mathbf{0}_{L \times L}, \forall i \neq j$ [22], and therefore, if N is large enough, we can approximate $\mathbf{Q}^H(\boldsymbol{\omega})\mathbf{Q}(\boldsymbol{\omega})$ with its block diagonal version:

$$\mathbf{Q}^H(\boldsymbol{\omega})\mathbf{Q}(\boldsymbol{\omega}) = \begin{bmatrix} \mathbf{A}_1^H \mathbf{A}_1 & \cdots & \mathbf{A}_1^H \boldsymbol{\Gamma}(\omega_K - \omega_1)\mathbf{A}_K \\ \vdots & \ddots & \vdots \\ \mathbf{A}_K^H \boldsymbol{\Gamma}(\omega_1 - \omega_K)\mathbf{A}_1 & \cdots & \mathbf{A}_K^H \mathbf{A}_K \end{bmatrix} \tag{15.28}$$

$$\approx \begin{bmatrix} \mathbf{A}_1^H \mathbf{A}_1 & \cdots & \mathbf{0} \\ \vdots & \ddots & \vdots \\ \mathbf{0} & \cdots & \mathbf{A}_K^H \mathbf{A}_K \end{bmatrix} \triangleq \mathbf{B}. \tag{15.29}$$

Using the approximation in Equation 15.29, the CFO estimates for all users in Equation 15.27 can be decoupled as ($k = 1, \ldots, K$):

$$\hat{\omega}_k = \arg \max_{\tilde{\omega}_k} \left\{ \mathbf{x}^H \boldsymbol{\Gamma}(\tilde{\omega}_k)\mathbf{A}_k(\mathbf{A}_k^H \mathbf{A}_k)^{-1}\mathbf{A}_k^H \boldsymbol{\Gamma}^H(\tilde{\omega}_k)\mathbf{x} \right\}, \tag{15.30}$$

and the solution can be found by K one-dimensional searches [14]. Note that Equations 15.29 and 15.30 hold only when the number of subcarriers is infinite. For a practical system with finite subcarriers, Equation 15.30 only offers approximate solutions to the original estimation problem. In the "Performance Comparisons" section, we will see that the decoupled estimator in Equation 15.30 suffers significant performance loss when the number of subcarriers is not large enough. Thus, efficient algorithms that can find the exact solution of Equation 15.27 with affordable complexity are needed.

Recognizing that the approximation $\mathbf{Q}^H(\boldsymbol{\omega})\mathbf{Q}(\boldsymbol{\omega}) \approx \mathbf{B}$ is too coarse and leads to significant performance loss, Sezginer and Bianchi [15] propose using a better approximation of $\mathbf{Q}^H(\boldsymbol{\omega})\mathbf{Q}(\boldsymbol{\omega})$. They write $\mathbf{Q}^H(\boldsymbol{\omega})\mathbf{Q}(\boldsymbol{\omega}) = \mathbf{B} + \mathbf{E}$, where $\mathbf{E}/N \to \mathbf{0}_{KL \times KL}$ when $N \to \infty$. Then, in general, $(\mathbf{Q}^H(\boldsymbol{\omega})\mathbf{Q}(\boldsymbol{\omega}))^{-1}$ can be approximated by its Taylor series expansion truncated to order M_a, that is, $(\mathbf{Q}^H(\boldsymbol{\omega})\mathbf{Q}(\boldsymbol{\omega}))^{-1} = (\mathbf{B} + \mathbf{E})^{-1} \approx \sum_{m=0}^{M_a}(-1)^m(\mathbf{B}^{-1}\mathbf{E})^m\mathbf{B}^{-1}$. It was shown in Ref. [15] that with this approximation, the optimization problem (15.27) can be approximated as

$$\hat{\boldsymbol{\omega}} = \arg\max_{\tilde{\boldsymbol{\omega}}} \left\{ \mathcal{J}^{(M_a)}(\tilde{\boldsymbol{\omega}}) \triangleq (-1)^{M_a}\mathbf{x}^H[\mathbf{Q}(\tilde{\boldsymbol{\omega}})\mathbf{B}^{-1}\mathbf{Q}(\tilde{\boldsymbol{\omega}})^H - \mathbf{I}_N]^{M_a+1}\mathbf{x} \right\}. \tag{15.31}$$

Obviously, the larger the M_a, the better the approximation in Equation 15.31 with respect to Equation 15.27. Furthermore, when $M_a = 0$, Equation 15.31 reduces to the decoupled estimator in Equation 15.30. Note that although with better approximation, when $M_a \geq 1$, the multiple CFOs in Equation 15.31 are coupled. Therefore, while computation of the inverse $(\mathbf{Q}^H(\boldsymbol{\omega})\mathbf{Q}(\boldsymbol{\omega}))^{-1}$ can be avoided, we are still facing a multidimensional maximization problem.

In Ref. [23], another decoupled estimator was proposed, not by asymptotic argument, but by constraining the minimum distance between any two active subcarriers from different users, such that the MAI can be controlled to a small value. Then the multidimensional optimization problem can be approximately transformed into K one dimensional optimization problems. However, this scheme requires careful placement of pilots of different users.

15.5.2 Iterative Algorithms for ML Estimator

To reduce the computational burden brought by the multidimensional searches in the ML estimator, iterative methods for maximizing the likelihood function were proposed in Ref. [9] and Ref. [10]. In Ref. [9], the alternative projection algorithm is exploited. The alternative projection method reduces a K-dimensional maximization problem in Equation 15.27 into a series of one-dimensional maximization problems, by updating the CFO estimate of a single user at a time, while keeping the other CFOs fixed at the previous estimated values. Let $\hat{\omega}_k^{(i)}$ the estimate of ω_k at the ith iteration. Further, let

$$\hat{\boldsymbol{\omega}}_{-k}^{(i)} = [\hat{\omega}_1^{(i+1)} \ldots \hat{\omega}_{k-1}^{(i+1)} \ \hat{\omega}_{k+1}^{(i)} \ldots \hat{\omega}_K^{(i)}]^T. \tag{15.32}$$

Given the initial estimates $\{\hat{\omega}_k^{(0)}\}_{k=0}^K$, the ith ($i \geq 1$) iteration of the alternative projection algorithm for maximizing Equation 15.27 has the following form:

For $k = 1, \ldots, K$,

$$\hat{\omega}_k^{(i)} = \arg\max_{\tilde{\omega}_k} \left\{ \|\mathbf{P}_{\mathbf{Q}}(\tilde{\omega}_k, \hat{\boldsymbol{\omega}}_{-k}^{(i-1)})\mathbf{x}\|^2 \right\}. \tag{15.33}$$

Multiple iterations on i are performed until the CFO estimates converge to a stable solution.

For practical implementation, an equivalent but more computationally efficient scheme than Equation 15.33 was proposed in Ref. [9]. The idea is as follows. Observing that when updating $\hat{\omega}_k^{(i)}$, most of the columns of $\mathbf{Q}(\tilde{\omega}_k, \hat{\boldsymbol{\omega}}_{-k}^{(i-1)})$ are fixed and not related to $\tilde{\omega}_k$, thus there is no need to recompute these columns for different $\tilde{\omega}_k$. For example, when estimating $\hat{\omega}_1^{(i)}$, $\mathbf{Q}(\tilde{\omega}_1, \hat{\boldsymbol{\omega}}_{-1}^{(i-1)})$ is expressed as

$$\mathbf{Q}(\tilde{\omega}_1, \hat{\boldsymbol{\omega}}_{-1}^{(i-1)}) = [\underbrace{\boldsymbol{\Gamma}(\tilde{\omega}_1)\mathbf{A}_1}_{\mathbf{Q}_1(\tilde{\omega}_1)} \ \underbrace{\boldsymbol{\Gamma}(\hat{\omega}_2^{(i-1)})\mathbf{A}_2 \ \ldots \ \boldsymbol{\Gamma}(\hat{\omega}_K^{(i-1)})\mathbf{A}_K}_{\mathbf{Q}_2(\hat{\boldsymbol{\omega}}_{-1}^{(i-1)})}], \tag{15.34}$$

where $\mathbf{Q}_1(\tilde{\omega}_1)$ is the submatrix that depends on $\tilde{\omega}_1$ only, and $\mathbf{Q}_2(\hat{\boldsymbol{\omega}}_{-1}^{(i-1)})$ can be considered fixed with respect to $\tilde{\omega}_1$. In general, for every k, $\mathbf{Q}(\tilde{\omega}_k, \hat{\boldsymbol{\omega}}_{-k}^{(i-1)})$ can be decomposed into two parts: (1) $\mathbf{Q}_1(\tilde{\omega}_k)$ containing the columns related only to $\tilde{\omega}_k$ and (2) $\mathbf{Q}_2(\hat{\boldsymbol{\omega}}_{-k}^{(i-1)})$ containing the columns that are not related to $\tilde{\omega}_k$.

With this special structure of $\mathbf{Q}(\tilde{\omega}_k, \hat{\boldsymbol{\omega}}_{-k}^{(i-1)})$, it was shown in Ref. [9] that $\mathbf{P_Q}(\tilde{\omega}_k, \hat{\boldsymbol{\omega}}_{-k}^{(i-1)})$ can also be decomposed into two parts:

$$\mathbf{P_Q}(\tilde{\omega}_k, \hat{\boldsymbol{\omega}}_{-k}^{(i-1)}) = \mathbf{P_{Q_2}}(\hat{\boldsymbol{\omega}}_{-k}^{(i-1)}) + \mathbf{P_\Pi}(\tilde{\omega}_k, \hat{\boldsymbol{\omega}}_{-k}^{(i-1)}), \tag{15.35}$$

where

$$\mathbf{P_{Q_2}}(\hat{\boldsymbol{\omega}}_{-k}^{(i-1)}) = \mathbf{Q}_2(\hat{\boldsymbol{\omega}}_{-k}^{(i-1)})[\mathbf{Q}_2^H(\hat{\boldsymbol{\omega}}_{-k}^{(i-1)})\mathbf{Q}_2(\hat{\boldsymbol{\omega}}_{-k}^{(i-1)})]^{-1}\mathbf{Q}_2^H(\hat{\boldsymbol{\omega}}_{-k}^{(i-1)}), \tag{15.36}$$

$$\mathbf{P_\Pi}(\tilde{\omega}_k, \hat{\boldsymbol{\omega}}_{-k}^{(i-1)}) = \mathbf{\Pi}(\tilde{\omega}_k, \hat{\boldsymbol{\omega}}_{-k}^{(i-1)})[\mathbf{\Pi}^H(\tilde{\omega}_k, \hat{\boldsymbol{\omega}}_{-k}^{(i-1)})\mathbf{\Pi}(\tilde{\omega}_k, \hat{\boldsymbol{\omega}}_{-k}^{(i-1)})]^{-1}\mathbf{\Pi}^H(\tilde{\omega}_k, \hat{\boldsymbol{\omega}}_{-k}^{(i-1)}), \tag{15.37}$$

$$\mathbf{\Pi}(\tilde{\omega}_k, \hat{\boldsymbol{\omega}}_{-k}^{(i-1)}) = (\mathbf{I}_N - \mathbf{P_{Q_2}}(\hat{\boldsymbol{\omega}}_{-k}^{(i-1)}))\mathbf{Q}_1(\tilde{\omega}_k). \tag{15.38}$$

Since $\mathbf{P_{Q_2}}(\hat{\boldsymbol{\omega}}_{-k}^{(i-1)})$ is independent of $\tilde{\omega}_k$, Equation 15.33 can be rewritten as

$$\hat{\omega}_k^{(i)} = \arg \max_{\tilde{\omega}_k} \left\{ \|\mathbf{P_\Pi}(\tilde{\omega}_k, \hat{\boldsymbol{\omega}}_{-k}^{(i-1)})\mathbf{x}\|^2 \right\}. \tag{15.39}$$

Note that the computation of $\mathbf{P_\Pi}(\tilde{\omega}_k, \hat{\boldsymbol{\omega}}_{-k}^{(i-1)})$ involves only an $L \times L$ matrix inversion, as opposed to the the $KL \times KL$ matrix inversion in $\mathbf{P_Q}(\tilde{\omega}_k, \hat{\boldsymbol{\omega}}_{-k}^{(i-1)})$. Therefore, computational complexity is saved in Equation 15.39 with respect to Equation 15.33. An additional scheme that further reduces the complexity of Equation 15.39 was also proposed in Ref. [9] by approximating the inversion $[\mathbf{\Pi}^H(\tilde{\omega}_k, \hat{\boldsymbol{\omega}}_{-k}^{(i-1)})\mathbf{\Pi}(\tilde{\omega}_k, \hat{\boldsymbol{\omega}}_{-k}^{(i-1)})]^{-1}$ using truncated Taylor series expansion.

Apart from the alternating projection method, another iterative method for CFOs estimation was proposed based on space alternative generalized expectation-maximization (SAGE) algorithm [10,11], applied directly to the system model in Equation 15.6. Similar to alternative projection algorithm, SAGE algorithm also reduces the joint K-user multiple parameters estimation problem into a series of single user parameter estimation problems. Denoting $\hat{\mathbf{h}}_k^{(i)}$ as the estimated channel for the kth user at the ith iteration, the SAGE algorithm proceeds as follows. For $k = 1, \ldots, K$,

E-step: Compute an estimate of the kth user's signal

$$\hat{\mathbf{x}}_k^{(i)} = \mathbf{x} - \sum_{m=1}^{k-1} \mathbf{\Gamma}(\hat{\omega}_m^{(i)})\mathbf{A}_m\hat{\mathbf{h}}_m^{(i)} - \sum_{m=k+1}^{K} \mathbf{\Gamma}(\hat{\omega}_m^{(i-1)})\mathbf{A}_m\hat{\mathbf{h}}_m^{(i-1)} \tag{15.40}$$

with the notation $\sum_a^b$ is zero when $b < a$.

M-step: Compute an estimate of the CFO and channel for the user k

$$(\hat{\omega}_k^{(i)}, \hat{\mathbf{h}}_k^{(i)}) = \arg \min_{\tilde{\omega}_k, \tilde{\mathbf{h}}_k} \left\{ \|\hat{\mathbf{x}}_k^{(i)} - \mathbf{\Gamma}(\tilde{\omega}_k)\mathbf{A}_k\tilde{\mathbf{h}}_k\|^2 \right\}. \tag{15.41}$$

Using a similar derivation as the joint ML estimator in Section 15.5.1, the solution to the above minimization problem is given by

$$\hat{\omega}_k^{(i)} = \arg\max_{\tilde{\omega}_k} \left\{ (\hat{\mathbf{x}}_k^{(i)})^H \mathbf{\Gamma}(\tilde{\omega}_k) \mathbf{A}_k (\mathbf{A}_k^H \mathbf{A}_k)^{-1} \mathbf{A}_k^H \mathbf{\Gamma}^H(\tilde{\omega}_k) \hat{\mathbf{x}}_k^{(i)} \right\}, \qquad (15.42)$$

$$\hat{\mathbf{h}}_k^{(i)} = (\mathbf{A}_k^H \mathbf{A}_k)^{-1} \mathbf{A}_k^H \mathbf{\Gamma}^H(\hat{\omega}_k^{(i)}) \hat{\mathbf{x}}_k^{(i)}. \qquad (15.43)$$

Intuitively, the E-step is the MAI cancellation process, and the M-step is the estimation of CFO and channel for the kth user with the MAI-canceled signal. If the CFO and channel estimates of other users are accurate, then Equation 15.41 is in fact the ML estimator of CFO and channel for the kth user. Therefore, the SAGE algorithm can be viewed as a recursive approximation to the joint ML estimator in Equation 15.27. Interestingly, the estimation of CFO in Equation 15.42 is in the same form as the decoupled estimator in Equation 15.30 under asymptotic assumption. The only difference is that Equation 15.42 uses the MAI-reduced signal for estimation, while Equation 15.30 uses the original received data.

Variants of the above SAGE algorithms have also been proposed in Refs. [12,13]. Realizing that the CAS for the preamble and for the data symbols are not necessary to be the same, Fu et al. [12] propose a modification of SAGE by using an interleaved CAS at the preamble and incorporating the MAI cancellation in both time and frequency domains. The resulting algorithm is shown to have better convergence rate, lower complexity, and better robustness against number of users than the conventional SAGE. On the other hand, in Ref. [13], joint frequency synchronization and data detection based on EM-type algorithms are demonstrated.

Although the alternating projection and SAGE algorithms appear differently, their essences are actually the same. In particular, both algorithms are recursive and estimating the CFO of one user at a time. Also, both algorithms take care of the effect of MAI on the estimation process using the previous estimates of unknown parameters. Further discussion on the relationship between alternative projection and EM algorithms is given in Ref. [24].

In general, for iterative type algorithms, an initial estimate of $\{\hat{\omega}_k^{(0)}\}_{k=0}^K$ close to the optimal solution is required in order to avoid the algorithms locking into a local maximum. A simple method for obtaining the initial estimate is by setting $\hat{\omega}_k^{(0)} = 0$, since ω_k is typically a zero-mean random variable. Other possible initialization methods include the single user algorithm in Ref. [8] and the asymptotic decoupled ML estimator in Equation 15.30. However, note that no matter what initialization method is used, there is still no guarantee that an estimate obtained iteratively will be the global maximum. In the next subsection, we will review a computational method that guarantees is to obtain the global optimum of Equation 15.27 without the need of an initial estimate.

15.5.3 ML Estimation with Importance Sampling

In Ref. [25], Pincus showed that it is possible to obtain a closed-form solution for Equation 15.27 that guarantees to be the global optimum. Based on the theorem given by Pincus, the $\boldsymbol{\omega}$ that yields the global maximum of $\phi'(\tilde{\boldsymbol{\omega}})$ is given by

$$\hat{\omega}_k = \lim_{\rho_1 \to \infty} \frac{\int \cdots \int \omega_k \exp(\rho_1 \phi'(\boldsymbol{\omega})) \, d\boldsymbol{\omega}}{\int \cdots \int \exp(\rho_1 \phi'(\boldsymbol{\omega})) \, d\boldsymbol{\omega}}, \quad k = 1, \ldots, K \qquad (15.44)$$

where ρ_1 is a design parameter. If we denote $\phi(\omega) = \exp(\rho_1 \phi'(\omega))$, the normalized version of $\phi(\omega)$ can be obtained as

$$\bar{\phi}(\omega) = \frac{\phi(\omega)}{\int \cdots \int \phi(\omega)\, d\omega}. \tag{15.45}$$

Here, the function $\bar{\phi}(\omega)$ has all the properties of a probability density function (PDF), so it is referred to as the pseudo-PDF in ω. With this definition and Equation 15.44, the optimal solution of ω for Equation 15.27 is

$$\hat{\omega}_k = \int \cdots \int \omega_k \bar{\phi}(\omega)\, d\omega, \quad k = 1, \ldots, K \tag{15.46}$$

for some large value of ρ_1. Note that the multiple integral in Equation 15.46 is of the same form as the expectation of ω_k with respect to $\bar{\phi}(\omega)$. Suppose we can generate realizations of ω according to $\bar{\phi}(\omega)$, the value of the integral in Equation 45.46 can be found by the Monte Carlo approximation as [26]

$$\hat{\omega}_k \approx \frac{1}{T} \sum_{i=1}^{T} \omega_k^i, \quad k = 1, \ldots, K, \tag{15.47}$$

where T is the number of realizations and ω_k^i is the ith realization of ω_k generated according to the pseudo-PDF $\bar{\phi}(\omega)$. Unfortunately, generating samples from $\bar{\phi}(\omega)$ is difficult since it is a multidimensional PDF.

In Ref. [27], importance sampling is used to compute the multidimensional integral in Equation 15.46. This approach is based on the observation that integrals of the type $\int \cdots \int h(\omega)\bar{\phi}(\omega)\, d\omega$ can be expressed as

$$\int \cdots \int h(\omega)\bar{\phi}(\omega)\, d\omega = \int \cdots \int h(\omega)\frac{\bar{\phi}(\omega)}{\bar{g}(\omega)}\bar{g}(\omega)\, d\omega, \tag{15.48}$$

where

$$\bar{g}(\omega) = \frac{g(\omega)}{\int \cdots \int g(\omega)\, d\omega} \tag{15.49}$$

with $g(\omega) > 0$. The function $g(\omega)$ is called the importance function and its normalized version $\bar{g}(\omega)$ has all the properties of a PDF. Then, the right-hand side of Equation 15.48 can be expressed as the expected value of $h(\omega)(\bar{\phi}(\omega)/\bar{g}(\omega))$ with respect to the pseudo-PDF $\bar{g}(\omega)$. Supposing that generation of samples ω^i according to $\bar{g}(\omega)$ is relatively easy, the value of the integral in Equation 15.48 can be approximated as

$$\int \cdots \int h(\omega)\bar{\phi}(\omega)\, d\omega \approx \frac{1}{T} \sum_{i=1}^{T} h(\omega^i)\frac{\bar{\phi}(\omega^i)}{\bar{g}(\omega^i)}, \tag{15.50}$$

where T is the number of realizations.

Now recall the estimate of ω_k in Equation 15.46. By the importance sampling technique discussed above, and the fact that the frequency offset ω_k has the properties of a circular random variable, it was shown that Equation 15.46 can be approximated as [14]

$$\hat{\omega}_k \approx \frac{1}{2\pi} \angle \frac{1}{T} \sum_{i=1}^{T} \frac{\bar{\phi}(\omega^i)}{\bar{g}(\omega^i)} \exp(j2\pi\omega_k^i), \quad k = 1, \dots, K, \tag{15.51}$$

where samples ω^i are generated according to $\bar{g}(\omega)$. Note that $\hat{\omega}_k$ is only related to the angle of the complex value in Equation 15.51, the equivalent but simplified estimator is given by

$$\hat{\omega}_k = \frac{1}{2\pi} \angle \frac{1}{T} \sum_{i=1}^{T} \frac{\phi(\omega^i)}{g(\omega^i)} \exp(j2\pi\omega_k^i), \quad k = 1, \dots, K \tag{15.52}$$

in which the integrals needed for computing $\bar{\phi}(\omega)$ and $\bar{g}(\omega)$ can be avoided.

15.5.3.1 Choice of Importance Function $g(\omega)$

In general, the estimator in Equation 15.52 converges to Equation 15.44 by the Strong Law of Large Number regardless of the choice of the function $g(\omega)$. However, there are obviously some choices of $g(\omega)$ that are better than others in terms of computational complexity. For the problem under consideration, it was shown in Ref. [14] that the optimal choice of $g(\omega)$, which minimizes the variance of the estimator for a fixed number of realizations T, is proportional to $\phi(\omega)$. However, this choice is not practical, since if we can generate samples from $\phi(\omega)$, we do not need the importance function $g(\omega)$ to facilitate samples generation. In practice, the importance function $g(\omega)$ should be chosen according to the following two criteria [28]:

1. $g(\omega)$ should be a close approximation of $\phi(\omega)$,
2. $g(\omega)$ should be as simple as possible to facilitate sample generation.

From the asymptotic analysis in Equation 15.29, if we choose the importance function $g(\omega)$ as

$$g(\omega) = \exp(\rho_2 \mathbf{x}^H \mathbf{Q}(\omega) \mathbf{B}^{-1} \mathbf{Q}^H(\omega) \mathbf{x}), \tag{15.53}$$

$$= \exp\left(\sum_{k=1}^{K} \rho_2 \mathbf{x}^H \left\{ \mathbf{\Gamma}(\omega_k) \mathbf{A}_k (\mathbf{A}_k^H \mathbf{A}_k)^{-1} \mathbf{A}_k^H \mathbf{\Gamma}^H(\omega_k) \right\} \mathbf{x} \right), \tag{15.54}$$

$$= \prod_{k=1}^{K} \underbrace{\exp(\rho_2 \mathbf{x}^H \mathbf{\Gamma}(\omega_k) \mathbf{A}_k (\mathbf{A}_k^H \mathbf{A}_k)^{-1} \mathbf{A}_k^H \mathbf{\Gamma}^H(\omega_k) \mathbf{x})}_{\triangleq g_k(\omega_k)}, \tag{15.55}$$

where ρ_2 is a design parameter, then $g(\omega)$ resembles the asymptotic version of $\phi(\omega)$. Furthermore, the samples generation of ω is simplified as this choice of $g(\omega)$ reduces the process to the generation of K independent realizations of ω_k following the pseudo-PDFs

$$\bar{g}_k(\omega_k) = \frac{g_k(\omega_k)}{\int_{-2\pi\alpha/N}^{2\pi\alpha/N} g_k(\omega_k) \, d\omega_k}, \quad k = 1, \dots, K, \tag{15.56}$$

where α is the normalized estimation range of CFOs.

Now the problem becomes how to generate realization of ω_k from the pseudo-PDF $\bar{g}_k(\omega_k)$. There are a number of methods available in the literature [29–31] for random variable generation from one-dimensional PDF, including the inverse cumulative density function (CDF) and the ratio-of-uniform approaches. In particular, Chen et al. [14] propose a novel modification of the ratio-of-uniform method, which can maintain the low complexity of sample generation, irrespective of the estimation range α. Considerations of choosing ρ_1 and ρ_2 are also detailed in Ref. [14].

15.6 Complexity Analyses

In this section, the computational complexities of different CFO estimators will be investigated.

- In the subband-based CAS system, the computational complexity of the estimator exploiting correlation in CP (Equations 15.8 through 15.11) is dominated by the calculations of $\hat{\mu}_k$ in Equation 15.8. The CFO estimate in Equation 15.9 is of relatively low complexity after $\hat{\mu}_k$ is obtained. Computations of $\gamma_k(\mu)$ in Equation 15.10 and $\Phi_k(\mu)$ in Equation 15.11 are of the order of $\mathcal{O}(L_{cp})$. If the number of timing positions to be searched for each user in Equation 15.8 is N_μ, the total complexity for K users is $\mathcal{O}(L_{cp}KN_\mu)$. The complexity of bandpass filtering can be ignored since these bandpass filters are fixed, and thus can be implemented efficiently.

- In the subband-based CAS system, for the null subcarrier-based approach in Equation 15.12, the main complexity comes from the multiplication of $\mathbf{W}_{-k}^H$ with $\boldsymbol{\Gamma}(\omega)^H$, which is of the order of $\mathcal{O}(N^2)$. If the number of ω-values to be searched is N_ω, the total complexity of this scheme is $\mathcal{O}(KN^2N_\omega)$.

- For the frequency estimator for the interleaved CAS system in Equation 15.21, the main steps consist of computing $\hat{\boldsymbol{\Psi}}$ with complexity $\mathcal{O}(R^2M)$, eigendecomposition of $\hat{\boldsymbol{\Psi}}$ with complexity $\mathcal{O}(R^3)$ and the search of θ in Equation 15.21 with $\mathcal{O}(R^2N_\omega)$. Since N_ω is usually greater than M or R, the total complexity is given by $\mathcal{O}(R^2N_\omega)$.

- For the asymptotic decoupled ML estimator in Equation 15.30, assuming that the $N \times N$ matrix $\mathbf{A}_k(\mathbf{A}_k^H\mathbf{A}_k)^{-1}\mathbf{A}_k^H$ is precomputed and stored, the K one-dimensional searches result in complexity $\mathcal{O}(KN^2N_\omega)$.

- For the alternating projection method in Equation 15.39, the complexity is dominated by the calculation of $\mathbf{P}_\Pi(\tilde{\omega}_k, \hat{\boldsymbol{\omega}}_{-k}^{(i-1)})$ in Equation 15.37. With the assumption that N is usually much larger than L, for fixed $\tilde{\omega}_k$, k and i, the complexity order of this calculation is $\mathcal{O}(LN^2)$. Assuming the number of iterations needed in the estimation is N_c, the total complexity of the alternating projection algorithm is $\mathcal{O}(N_cLKN^2N_\omega)$.

- For the SAGE algorithm, the complexity is dominated by Equation 15.42. With $\mathbf{A}_k(\mathbf{A}_k^H\mathbf{A}_k)^{-1}\mathbf{A}_k^H$ precomputed and stored, the complexity order of Equation 15.42 per iteration per user is $\mathcal{O}(N^2N_\omega)$. Therefore, the overall complexity is $\mathcal{O}(N_cKN^2N_\omega)$.

- For the importance sampling-based estimator, the estimate in Equation 15.21 involves two main steps. In the first step, T samples of ω_k are generated according to $\bar{g}_k(\omega_k)$ in Equation 15.56. The complexity of this step depends on the method of sample generation. It was shown in Ref. [14] that, with the modified ratio-of-uniform method, the complexity order of this step is $\mathcal{O}(KN_\omega N^2 + KTN_aN^2)$, where N_a is the average PDF ($\bar{g}_k(\omega_k)$) evaluations for generating one realization of ω_k. In the second step, the T evaluations of the coefficient $\phi(\boldsymbol{\omega}^i)/g(\boldsymbol{\omega}^i)$ in Equation 15.52 have a complexity of order $\mathcal{O}(T((KL)^3 + N(KL)^2 + KN^2))$.

Summing the two steps and taking only the dominant order terms, the entire complexity is obtained as $\mathcal{O}(KTN_aN^2 + KN_\omega N^2 + T(KL)^3 + TN(KL)^2)$. Note that in general, the relative order of these terms can be determined only when the system specification is fixed. For the special case when the number of subcarriers N is much larger than the number of users K and the upper bound of channel length L, the complexity order is represented as $\mathcal{O}(KTN_aN^2 + KN_\omega N^2)$.

It can be seen that the complexities of estimators for generalized CAS are in general higher than that of subband-based and interleaved CASs.

15.7 Performance Comparisons

In the following, simulation results are presented to illustrate the properties and compare the performances of different algorithms. Since the algorithms for generalized CAS represent a more general and the largest class of estimator, we focus only on the algorithms for generalized CAS. For simplicity, in the simulation, it is assumed that the timing offset of different users $\mu_k = 0, \forall\ k$. The channel response for each user is generated according to the HIPERLAN/2 channel model with eight paths ($L = 8$) [32]. In detail, the channel taps are modeled as independent and complex Gaussian random variables with zero mean and an exponential power delay profile (normalized to unit power). Further, it is assumed that the total number of subcarriers is equally divided by all users. The subcarriers is randomly allocated to users with the generalized CAS (power of training symbols on each subcarrier is 1), and the users' signals arrive at the BS with equal average power. Without loss of generality, the normalized CFOs for all users ($\varepsilon_k, k = 1, \ldots, K$) are generated as random variables uniformly distributed in $[-0.5, 0.5]$. The signal-to-noise ratio (SNR) used in the simulation is defined as $\text{SNR} = 10 \log_{10}(1/\sigma^2)$. Since there are multiple CFOs to be estimated, the mean square error (MSE) is defined as

$$\text{MSE} = \frac{1}{J} \sum_{j=1}^{J} \sum_{k=1}^{K} (\hat{\omega}_k^j - \omega_k)^2,$$

where $\hat{\omega}_k^j$ is the estimate of ω_k at the jth simulation run and $J = 200$ is the number of simulation runs for each point in the figures. In all the figures, the Cramer–Rao bound (CRB) for multi-CFO estimation, which is given by [9]

$$\mathbf{CRB}(\boldsymbol{\omega}) = \frac{\sigma^2}{2} \left(Re \left\{ \mathbf{Z}^H \mathbf{P}_{\mathbf{Q}}^\perp \mathbf{Z} \right\} \right)^{-1},$$

where

$$\mathbf{Z} = \begin{bmatrix} \mathbf{Z}_1 & \mathbf{Z}_2 & \ldots & \mathbf{Z}_K \end{bmatrix},$$
$$\mathbf{Z}_k = \mathbf{M}\mathbf{\Gamma}(\omega_k)\mathbf{A}_k\mathbf{h}_k,$$
$$\mathbf{M} = \text{diag}\{0, 1, \ldots, N-1\},$$
$$\mathbf{P}_{\mathbf{Q}}^\perp = \mathbf{I}_N - \mathbf{Q}(\boldsymbol{\omega})[\mathbf{Q}^H(\boldsymbol{\omega})\mathbf{Q}(\boldsymbol{\omega})]^{-1}\mathbf{Q}^H(\boldsymbol{\omega}),$$

is also plotted for reference.

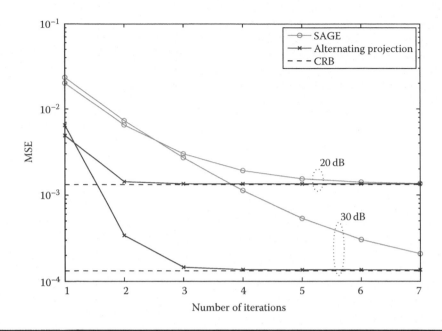

Figure 15.3   Convergence performances of alternating projection and SAGE algorithms.

Figure 15.3 shows the convergence performances of two iterative algorithms: alternating projection and SAGE algorithms. The considered OFDMA system has the following parameters: $N = 64$, $L_{cp} = 16$, and $K = 4$. The initial estimates of both algorithms are taken as $\hat{\omega}_k^{(0)} = 0$ for all k. It can be seen that in general the alternating projection converges faster than the SAGE algorithm. In particular, for SNR = 30 dB, the SAGE takes quite a long time to converge. Furthermore, although the solution from iterative algorithms cannot be theoretically guaranteed to be the global optimal, results from this figure show that their performance can reach the CRB after convergence. Additional simulations show that the convergence time also increases with both number of users and the range of CFO considered, but the figures are not presented here.

In Figure 15.4, the CRB and MSE of the CFO estimation ($K = 2$) using the ML estimator with importance sampling and the asymptotic decoupled estimator are presented versus different numbers of subcarriers N. The parameters for the importance sampling-based method are set as $\rho_1 = 2/K$ and $\rho_2 = 1/K$ and $T = 2000$ [14]. The SNRs in this simulation are 10 and 20 dB. From the figure, it can be seen that the MSE of the ML estimator meets the CRB quite well, which shows that the ML estimator with importance sampling is efficient. For the decoupled estimator, when the number of subcarriers is small, a significant performance loss occurs. On the other hand, the gap between the decoupled estimator and the ML estimator decreases when the number of subcarriers increases. This is especially true for SNR = 10 dB, in which the performance of the decoupled estimator is very close to the CRB when $N = 256$. However, for SNR = 20 dB, the asymptotic convergence rate is much slower, and the required value of N for the decoupled estimator to meet CRB is expected to be very large.

Finally, the MSE performances of the ML estimator with importance sampling, alternating projection, and SAGE algorithms are plotted against SNR in Figure 15.5. The OFDMA system has $N = 64$ subcarriers, $K = 4$ users and the number of iterations for alternating projection and SAGE algorithms taken as 3 and 10, respectively. The parameters for the importance sampling-based

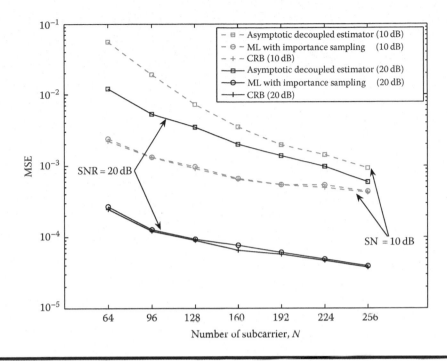

Figure 15.4 CRB and MSE of the decoupled estimator and ML estimator with importance sampling versus number of subcarriers.

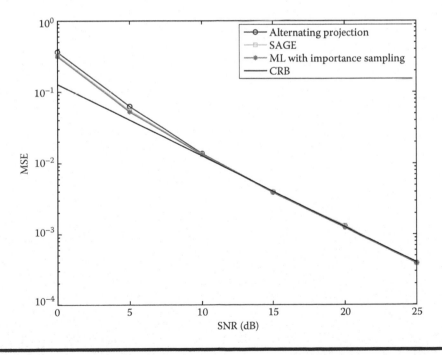

Figure 15.5 CRB and MSE of the ML estimator with importance sampling, alternating projection, and SAGE algorithms versus SNR.

estimator are the same as that in the previous figure. It is noted that for all the three methods considered, the MSEs always coincide with the CRB in medium to high SNR ranges, which means that they are efficient. They offer almost the same performance because they are optimizing the same cost function, but in different ways.

15.8 Concluding Remarks

In this chapter, a number of multi-CFO estimation schemes for OFDMA uplink systems have been reviewed, and the relationships between these estimators have been discussed. In particular, the challenging problem of CFOs estimation under generalized CAS has been addressed in detail. Due to its high complexity, the ML estimator for generalized CAS cannot be directly implemented. Advanced techniques such as alternating projection, SAGE algorithm and importance sampling are needed to locate the optimal solution. Complexity analysis and performance comparisons of these estimators have also been presented.

It is important to note that multi-CFO estimation represents only half of the process of frequency synchronization for OFDMA systems. After the estimates of CFOs of different users are obtained, these estimates are used to cancel or mitigate the ICI and MAI. This problem is not straightforward for interleaved and generalized CASs since signals from different users are mixed together. Compensating for the CFO with respect to one user causes misalignment of other users. An obvious way to solve this problem is to feedback the estimated CFOs to the users so that adjustments are made at the transmitters [5]. However, this involves additional transmission cost. More recently, the multi-CFO compensation problem has been viewed as a multiuser detection problem, where the linear multiuser detector [zero-forcing or minimum mean square error (MMSE)] [33] and iterative interference cancelation schemes [34,35] have been proposed. More details can be found in the corresponding references or another recent review article on synchronization techniques for OFDMA systems [36].

References

1. IEEE Std. 802.16a Standard for Local and Metropolitan Area Networks, Part 16: Air Interface for Fixed BroadbandWireless Access Systems- Amendment 2, IEEE Std. 802.16a, 2003.
2. K. Etemad, Overview of mobile WiMAX technology and evolution, *IEEE Commun. Mag.*, 46 (10), 31–40, 2008.
3. H. Yin and S. Alamouti , OFDMA: A broadband wireless access technology, in *Proc. IEEE Srnoff Symposium, 2006*, pp. 1-4, March. 2006.
4. P. Svedman, S. K. Wilson, L. J. Cimini, and B. Ottersten, Opportunistic Beamforming and Scheduling for OFDMA Systems, *IEEE Trans. Commun.*, 55 (5), 941–952, 2007.
5. J. J. van de Beek, P. O. Börjesson, M. L. Boucheret, D. Landstrom, J. M. Arenas, O. Odling, M. Wahlqvist, and S. K. Wilson, A time and frequency synchronization scheme for multiuser OFDM, *IEEE J. Sel. Areas Commun.*, 17 (11), 1900–1914, 1999.
6. S. Barbarossa, M. Pompili, and G. B. Giannakis, Channel-independent synchronization of orthogonal frequency division multiple access systems, *IEEE J. Sel. Areas Commun.*, 20 (2), 474–486, 2002.
7. Z. Cao, U. Tureli, and Y. D. Yao, Deterministic multiuser carrier-frequency offset estimation for interleaved OFDMA uplink, *IEEE Trans. Commun.*, 52 (9), 1585–1594, 2004.
8. M. Morelli, Timing and frequency synchronization for the uplink of an OFDMA system, *IEEE Trans. Commun.*, 52 (2), 296–306, 2004.
9. M.-O. Pun, M. Morelli, and C.-C. Jay Kuo, Maximum-likelihood synchronization and channel estimation for OFDMA uplink transmissions, *IEEE Trans. Commun.*, 54 (4), 726–736, 2006.

10. M.-O. Pun, S.-H. Tsai, and C.-C. Jay Kuo, An EM-based joint maximum likelihood estimation of carrier frequency offset and channel for uplink OFDMA systems, in *Proc. Vehicular Technology Conference*, 2004, pp. 598–602.

11. J.-H. Lee and S.-C. Kim, Time and frequency synchronization for OFDMA uplink system using the SAGE algorithm, *IEEE Trans. Wirel. Commun.*, 6 (4), 1176–1180, 2007.

12. X. Fu, H. Minn, and C. D. Cantrell, Two novel iterative joint frequency-offset and channel estimation methods for OFDMA uplink, *IEEE Trans. Commun.*, 56 (3), 474–484, 2008.

13. M.-O. Pun, M. Morelli, and C.-C. Jay Kuo, Iterative detection and frequency synchronization for OFDMA uplink transmissions, *IEEE Trans. Wireless Commun.*, 6 (2), 629–639, 2007.

14. J. Chen, Y.-C. Wu, S. C. Chan, and T.-S. Ng, Joint maximum-likelihood CFO and channel estimation for OFDMA uplink using importance sampling, *IEEE Trans. Veh. Technol.*, 57 (6), 3462–3470, 2008.

15. S. Sezginer and P. Bianchi, Asymptotically efficient reduced complexity frequency offset and channel estimators for uplink MIMO-OFDMA systems, *IEEE Trans. Signal Process.*, 56 (3), 964–979, 2008.

16. J. J. van de Beek, M. Sandell, and P. O. Borjesson, ML estimation of timing and frequency offset in OFDM systems, *IEEE Trans. Signal Process.*, 45 (7), 1800–1805, 1997.

17. H. Liu and U. Tureli, A high-efficiency carrier estimator for OFDM communications, *IEEE Commun. Lett.*, 2 (4), 104–106, 1998.

18. U. Tureli, H. Liu, and M. D. Zoltowski, OFDM blind carrier offset estimation: ESPRIT, *IEEE Trans. Commun.*, 48 (9), 1459–1461, 2000.

19. N. Lashkarian and S. Kiaei, Class of cyclic-based estimators for frequency-offset estimation of OFDM systems, *IEEE Trans. Commun.*, 48 (12), 2139–2149, 2000.

20. M. Ghogho and A. Swami, Blind frequency-offset estimator for OFDM systems transmitting constant-modulus symbols, *IEEE Commun. Lett.*, 6 (8), 343–345, 2002.

21. S. M. Kay, *Fundamentals of Statistical Signal Processing*, Englewood Cliffs, NJ: Prentice-Hall, 1993.

22. O. Besson and P. Stoica, On parameter estimation of MIMO flat-fading channels with frequency offsets, *IEEE Trans. Signal Process.*, 51, (3), 602–613, 2003.

23. Y. Zeng and A. Leyman, Pilot-based simplified ML and fast algorithm for frequency offset estimation in OFDMA uplink, *IEEE Trans. Veh Technol.*, 57 (3), 1723–1732, 2008.

24. P. Stoica and Y. Selen, Cyclic minimizers, majorization techniques, and expetation-maximization algorithms: A refresher, *IEEE Signal Process. Mag.*, 21 (1), 112–114, 2004.

25. M. Pincus, A closed form solution for certain programming problems, *Oper. Res.*, 16 (3), 690–694, 1962.

26. C. P. Robert and G. Casella, *Monte Carlo Statistical Methods*, New York: Springer, 2004.

27. S. Kay and S. Saha, Mean Likelihood Frequency Estimation, *IEEE Trans. Signal Process.*, 48 (7), 1937–1946, 2000.

28. A. F. M. Smith and A. Gelfand, *Bayesian Statistics without Tears: A Sampling-Resampling Framework.* New York: American Statistical Association, 1992.

29. H. Akima, A new method of interpolation and smooth curve fitting based on local procedures, *J. ACM*, 17, 589–602, 1970.

30. G. Marsaglia, The exact-approximation method for generating random variables in a computer, *J. Am. Stat. Assoc.*, 79, 218–221, 1984.

31. A. J. Kinderman and J. F. Monahan, Computer generation of random variables using the ratio of uniform deviates, *ACM Trans. Math. Softw.*, 3, 257–260, 1977.

32. ETSI. (2001). BRAN; HIPERLAN Type 2; Physical (PHY) Layer Specification (2nd ed.) Technical Specification 101 475. [Online]. Available: http://www.etsi.org

33. Z. Cao, U. Tureli, Y.-D. Tao, and P. Honan, Frequency synchronization for generalized OFDMA uplink, in *Proc. IEEE Globecom*, pp. 1071–1075, November 2004.

34. D. Haung and K. B. Letaief, An interference-cancellation scheme for carrier frequency offsets correction in OFDMA systems, *IEEE Trans. Commun.*, 53 (7), 1155–1165, 2005.

35. D. Marabissi, R. Fantacci and S. Papini, Robust multiuser interference cancellation for OFDM systems with frequency offset, *IEEE Trans. Wirel. Commun.*, 5 (11), 3068–3076, 2006.

36. M. Morelli, C.-C. J. Kuo, and M.-O. Pun, synchronization techniques for orthogonal frequency division multiple access (OFDMA): A tutorial review, *Proc. IEEE*, 95 (7), 1394–1427, 2007.

Chapter 16

Frequency Domain Equalization for OFDM and SC/FDE

Harald Witschnig

Contents

16.1	Introduction		418
16.2	Analysis of Frequency Domain Equalization		419
	16.2.1	Frequency Domain Equalization for Single Carrier Transmission	419
		16.2.1.1 Properties of the Fourier Transformation	420
		16.2.1.2 Linear Filtering in the Frequency Domain	421
		16.2.1.3 Linear Filtering Methods Based on the DFT	423
	16.2.2	The Concept of CP	424
		16.2.2.1 ZF Equalization	428
		16.2.2.2 MMSE Equalization	428
	16.2.3	Known Symbols versus Classical CP	429
16.3	FDE for Multicarrier Transmission		431
	16.3.1	Orthogonal Frequency Division Multiplexing—OFDM	433
		16.3.1.1 ISI and ICI in the Case of Multipath Propagation	434
16.4	Fundamental System Comparison		436
	16.4.1	Processing Load	437
		16.4.1.1 SC/FDE Based on a Classical CP	437
		16.4.1.2 SC/FDE based on UW	437
		16.4.1.3 OFDM	437
		16.4.1.4 Bandwidth Efficiency	438

　　　16.4.1.5　SC/FDE Based on CP..438
　　　16.4.1.6　SC/FDE Based on UW...438
　　　16.4.1.7　OFDM...438
　　　16.4.1.8　Loss of Signal-to-Noise Ratio due to a CP.......................439
　　　16.4.1.9　Front-End Requirements and RF-Related Issues..................439
16.5　Performance Comparison of SC/FDE versus OFDM.................................442
　　16.5.1　The OFDM System—Parameter Definition................................443
　　16.5.2　The SC/FDE System—Parameter Definition..............................445
　　16.5.3　Performance Comparison...447
16.6　Conclusion and Outlook...449
References...451

16.1 Introduction

Time dispersion caused by multipath propagation can severely distort the received signal and makes powerful equalization necessary in wireless communication systems. With increasing data rates (bandwidth), equalization becomes more and more of relevance and requires significant signal processing effort. Performing equalization in the frequency domain allows for a computationally efficient implementation because in that domain the effort grows significantly lower with the bandwidth compared to equalization in the time domain where the effort grows at least quadratically with bandwidth [2], as pointed out in Figure 16.1.

Therefore the concepts of frequency domain equalization represent one of the most powerful strategies to combat time dispersion caused by multipath propagation and are the basis for many broadband communication systems.

From a historical point of view, single carrier transmission might be the more obvious strategy, but it is the concept of multicarrier transmission that dominated and dominates for high-rate wireless communications. The interest frequency domain equalization for single carrier transmission (SC/FDE) has been growing rapidly only recently, when being discussed in the context of long term evolution (LTE). Finally, SC/FDMA was decided to be implemented for the uplink, based on basic

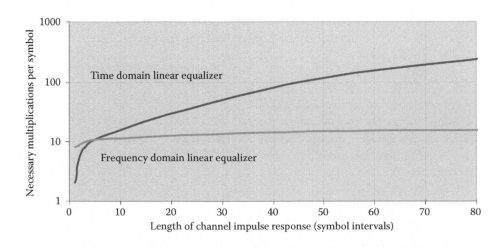

Figure 16.1　Implementation effort for time and frequency domain equalization.

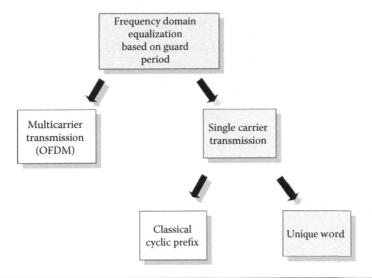

Figure 16.2   Concepts of frequency domain equalization.

principles of SC/FDE. It was realised that SC/FDE features advantages especially for the uplink transmission in mobile equipment, while OFDM shows advantages when it comes to flexibility.

A fundamental element for the correct and efficient implementation of the investigated frequency domain equalization is the so-called cyclic prefix (CP) (as introduced in a mathematical way in the following sections). Therefore this CP represents a central element of the underlying discussion and is used as a differentiator for the implementations in orthogonal frequency division multiplexing (OFDM) and SC/FDE.

Figure 16.2 anticipates this differentiation. While for OFDM almost only the *classical* CP structure is of practical relevance, a further differentiation makes sense for SC/FDE, namely between classical CP structure and a so-called *Unique Word (UW)* structure. This UW is based on known pilot symbols and shows significant advantages compared to the traditional CP Using the UW for equalization, synchronization, or channel estimation purposes makes this concept the most powerful one among the investigated strategies.

16.2 Analysis of Frequency Domain Equalization

It is the aim of this section to give an insight into the nature of frequency domain equalization and why it became such a fundamental factor for modern communications. It will be pointed out that the aim as well as the results due to frequency domain equalization are almost comparable for both single carrier—and multicarrier transmission but that the way these results are achieved is different.

16.2.1 Frequency Domain Equalization for Single Carrier Transmission

Despite the dominating position of the OFDM-like concepts, the concept of SC/FDE will be introduced first. SC/FDE is defined by the properties of the discrete Fourier transformation (DFT). Without going into details the main theorems of Fourier transformation that are relevant for this application will be reviewed [2–7].

16.2.1.1 Properties of the Fourier Transformation

As any investigation and implementation in this work are based on digital signal processing, only the DFT and its properties will be introduced.

Based on a continuous time signal $x(t)$, $x(n)^{T_a}$ describes the sequence achieved by sampling $x(t)$ based on the sampling time T_a

$$x^{T_a}(n) = x(t) \sum_{n=-\infty}^{\infty} \delta(t - nT_a). \tag{16.1}$$

To perform frequency analysis on a discrete-time signal, the time domain sequence is converted to an equivalent frequency domain representation by

$$X^{T_a}(f) = \sum_{n=-\infty}^{\infty} x^{T_a}(n)e^{-j2\pi nT_a f}. \tag{16.2}$$

Equation 16.2 describes the *discrete-time Fourier transformation* of the sequence $x(n)$, representing a periodic spectrum with the period $f_a = 1/T_a$. The *inverse discrete-time Fourier transformation* is given by

$$x^{T_a}(n) = \int_{-\pi/T}^{\pi/T} X^{T_a}(f)e^{j2\pi nT_a f} \, df. \tag{16.3}$$

Note that the application of the discrete-time Fourier transformation leads to a continuous spectrum and is therefore not a computationally convenient representation of the sequence $x^{T_a}(n)$. This drawback leads to the development of the so-called DFT.

For a given N-tupel of (time)values

$$x[0], x[1], \ldots, x[N-1] \tag{16.4}$$

the corresponding N-tupel of frequency values

$$X[0], X[1], \ldots, X[N-1] \tag{16.5}$$

is computed by applying the

$$X[k] = \sum_{n=0}^{N-1} x[n]e^{-j2\pi kn/N}, \quad k = 0, 1, 2, \ldots, N-1. \tag{16.6}$$

Of course this transformation is reversible, called *inverse discrete Fourier transformation* (IDFT), and is given by

$$x[n] = \frac{1}{N} \sum_{k=0}^{N-1} X[k]e^{j2\pi kn/N}, \quad n = 0, 1, 2, \ldots, N-1. \tag{16.7}$$

$X[k]$ as well $x[n]$ represent periodic functions in the frequency and time domains, respectively, based on the period N. The statement of Equation 16.6 or 16.7 is simply that a periodic signal $x[n]$

can be reconstructed from the periodic samples of $X[n]$ and vice versa. Equations 16.6 and 16.7 do not imply that a recovery of an aperiodic discrete-time signal $x^{T_a}(n)$ and its spectrum $X^{T_a}(f)$ is possible, making a more detailed characterization necessary.

Representing a continuous signal of duration T_0 by a sampled sequence of the form $x(n)$ for $n = 0, 1, \ldots N - 1$ and $T_a = T_0/N$, the discrete-time Fourier transformation is simplified to

$$X^{T_a}(f) = \sum_{n=0}^{N-1} x(n) e^{-j2\pi n T_a f}. \tag{16.8}$$

In comparison to this, the DFT is given by

$$X^{T_a}[k] = \sum_{n=0}^{N-1} x(n) e^{-j2\pi k n/N}. \tag{16.9}$$

The comparison of Equations 16.8 and 16.9 shows that if one takes $X^{T_a}(f)$ at the discrete frequency points $f_k = k/NT_a$, $X^{T_a}[k]$ results. So, for the case of $NT_a = T_0$ and $f_0 = 1/T_0$, the significant relationship

$$X^{T_a}[k] = X^{T_a}(k f_0) \tag{16.10}$$

follows. It is essential to point out the importance of this statement once more, namely that there is correspondence between the discrete-time Fourier transformed sequence $x(n)$ at the points $f = n f_0$ and the DFT of the sequence $x(n)$ within one period. This means that for sampling sequences of finite duration a sampled version of the discrete-time Fourier transformation is defined by the DFT without ambiguity.

It is this unambiguous correspondence between discrete-time Fourier transformation and DFT on the one hand and the existence of extremely efficient algorithms for computing the DFT (for the case of N being a power of two, they are well known as *fast Fourier transformation* algorithms) on the other hand that makes the Fourier transformation and further frequency domain equalization highly efficient compared to time domain equalization.

16.2.1.2 Linear Filtering in the Frequency Domain

In the time domain the equalization is carried out by a convolutional operation (linear finite impulse response [FIR] filtering [2]). Here it will be demonstrated how to carry out this convolutional operation in the frequency domain.

Of essential significance for any further investigation is the *Theorem of (linear) Convolution*, stating that the convolutional product of two functions $x(t)$ and $h(t)$ is transformed by the Fourier transformation into the algebraic product of the two Fourier transformed sequences $X(f)$ and $H(f)$

$$x(t) \star h(t) \quad \circ - \bullet \quad X(f) \cdot H(f). \tag{16.11}$$

The theorem of (linear) convolution allows one to replace the costly convolutional operation by a multiplication. Since the DFT provides a discrete frequency representation of a finite-duration sequence, it is essential to define its use also for linear system analysis, especially for linear filtering.

For that, two finite-duration sequences of length N, $x_1[n]$ and $x_2[n]$, are transformed by an N-point DFT to the frequency domain

$$X_1[k] = \sum_{n=0}^{N-1} x_1(n)e^{-j2\pi nk/N}, \quad k = 0, 1, \ldots, N-1, \tag{16.12}$$

$$X_2[k] = \sum_{n=0}^{N-1} x_2(n)e^{-j2\pi nk/N}, \quad k = 0, 1, \ldots, N-1, \tag{16.13}$$

Multiplying the two frequency domain sequences $X_1[k]$ and $X_2[k]$ element-wise results in a frequency domain sequence $X_3[k]$, which is the DFT of a sequence $x_3[n]$ of length N. Of significant meaning for the further steps is the relationship between $x_3[n]$ and the sequences $x_1[n]$ and $x_2[n]$. Due to Equation 16.11, $X_3[k]$ can be written as

$$X_3[k] = X_1[k]X_2[k], \quad k = 0, 1, \ldots, N-1. \tag{16.14}$$

The IDFT of $X_3[k]$ leads to

$$x_3[n] = \frac{1}{N} \sum_{k=0}^{N-1} X_3[k]e^{j2\pi kn/N} = \frac{1}{N} \sum_{k=0}^{N-1} X_1[k]X_2[k]e^{j2\pi kn/N}.$$

By substituting for $X_1[k]$ and $X_2[k]$ using the DFTs, one obtains

$$x_3[n] = \frac{1}{N} \sum_{k=0}^{N-1} \left[\sum_{r=0}^{N-1} x_1[r]e^{-j2\pi kr/N} \right] \left[\sum_{l=0}^{N-1} x_2[l]e^{-j2\pi kl/n} \right] e^{j2\pi kn/N},$$

$$= \frac{1}{N} \sum_{r=0}^{N-1} x_1[r] \sum_{l=0}^{N-1} x_2[l] \left[\sum_{k=0}^{N-1} e^{j2\pi k(n-r-l)/N} \right]. \tag{16.15}$$

Equation 16.15 will be simplified as the inner sum has the form

$$\sum_{k=0}^{N-1} a^k = \begin{cases} N & \text{for } a = 1, \\ \dfrac{1 - a^N}{1 - a} & \text{for } a \neq 1, \end{cases} \tag{16.16}$$

where a is defined as

$$a = e^{j2\pi(n-r-l)/N}. \tag{16.17}$$

Note that $a = 1$ if $n - r - l$ is a multiple of N. Besides this, $a^N = 1$ for any value of $a \neq 0$. As a result, Equation 16.16 reduces to

$$\sum_{k=0}^{N-1} a^k = \begin{cases} N & \text{for } l = n - r + pN = (n-r)_N, \\ 0 & \text{otherwise.} \end{cases} \tag{16.18}$$

In this equation p describes an integer value and $()_N$ describes a modulo operation. By substituting the result of Equation 16.18 into Equation 16.15, we obtain the required expression for $x_3[n]$ as follows:

$$x_3[n] = \sum_{n=0}^{N-1} x_1[r]x_2[(n-r)]_N, \quad n = 0, 1, \ldots, N-1. \tag{16.19}$$

Although this expression is similar to the form of a linear convolution, it is significantly different as it describes a periodic sequence indicated by the index $(n-r)_N$. Due to that periodicity, this convolution is called a *circular convolution*.

Comparable to the *Theorem of linear convolution* for the continuous Fourier transformation the *Theorem of cyclic convolution* for the DFT can be formulated as

$$x_1(n) \otimes x_2(n) \quad \circ - \bullet \quad X_1(k)X_2(k). \tag{16.20}$$

It is concluded that the cyclic convolution of two sequences is equivalently represented by the algebraic product of the discrete Fourier transformed sequences as

$$x[n] \otimes h[n] = \text{IDFT}_N\left\{\text{DFT}_N\{x[n]\} \cdot \text{DFT}_N\{h[n]\}\right\}. \tag{16.21}$$

Equation 16.21 represents the most essential relation in the context of SC/FDE, as it allows us to carry out an equivalent equalization in the frequency domain. Unfortunately, a direct use of Equation 16.21 is not possible as the distortion by the radio channel and the necessary equalization by an FIR filter are represented by a linear convolution but not a circular convolution.

16.2.1.3 Linear Filtering Methods Based on the DFT

Due to the previously mentioned problem, it is demonstrated in the following how to make use of the efficient DFT and to perform a correct linear filtering in the frequency domain.

The underlying investigations are based on a discrete-time finite duration sequence $x^{T_a}(n)$ of length L which excites an FIR filter $h^{T_a}(n)$ of length M based on a sampling time T_a. Without loss of generality, it is defined as

$$x^{T_a}(n) = 0, \quad n < 0 \quad \text{and} \quad n \geq L, \tag{16.22}$$

$$h^{T_a}(n) = 0, \quad n < 0 \quad \text{and} \quad n \geq M. \tag{16.23}$$

The output sequence $y^{T_a}(n)$ of the FIR filter is expressed in the time domain as the linear convolution of $x^{T_a}(n)$ and $h^{T_a}(n)$ by

$$y^{T_a}(n) = \sum_{k=0}^{M-1} h^{T_a}(k)x^{T_a}(n-k). \tag{16.24}$$

Since $h^{T_a}(n)$ and $x^{T_a}(n)$ are of finite duration, their linear convolution is also finite in duration and of length $L + M - 1$. Its frequency domain representation is given by

$$Y^{T_a}(f) = X^{T_a}(f)H^{T_a}(f). \tag{16.25}$$

If the sequence $y^{T_a}(n)$ is to be represented as unambiguous in the frequency domain by discrete samples of its spectrum $Y^{T_a}(f)$, the number of distinct samples must equal or exceed $L + M - 1$. Therefore, a DFT of size $N \geq L + M - 1$ is required.

From

$$Y(k) \equiv Y(f)|_{f=2\pi k/N} = X(f)H(f)|_{f=2\pi k/N}, \quad k = 0, 1, \ldots, N - 1, \tag{16.26}$$

it follows that

$$Y(k) = X(k)H(k), \quad k = 0, 1, \ldots, N - 1, \tag{16.27}$$

where $X(k)$ and $H(k)$ describe the N-point Fourier transform of the corresponding sequences. As the sequences $x(n)$ and $h(n)$ have a duration of less than N, they are simply zero padded. Note that the number of samples that represent these sequences in the frequency domain is increased to N, representing $x(n)$ of duration L and $h(n)$ of duration M. Due to Equation 16.10 there are N points of identity at the frequencies nf_0 between the discrete-time Fourier transformation and the DFT for the case of $NT_a = T_0$ and $f_0 = 1/T_0$, allowing a unique relation. Since an N-point DFT of the output sequence $y[n]$ is sufficient to represent this sequence in the frequency domain, it follows that the multiplication of the N-point DFTs $X[k]$ and $H[k]$ followed by the computation of the N-point IDFT must yield the sequence $y[n]$. This implies that the N-point circular convolution of $x[n]$ with $h[n]$ must be equivalent to the linear convolution of $x^{T_a}(n)$ with $h^{T_a}(n)$.

To conclude this consideration: By increasing the length of the sequences $x[n]$ and $h[n]$ to $N = L + M - 1$ points and then circularly convolving the resulting sequences, we obtain the same result as would have been obtained by a linear convolution. Thus, zero padding enables us to use a DFT to perform linear filtering.

16.2.2 The Concept of CP

A completely different approach to process a continual channel distorted datastream with the use of a DFT is the use of a CP that was introduced for the first time in Refs. [8,9], and was developed in detail by Huemer [10].

It is the aim of the CP to make the linear convolution due to the channel identical to a cyclic convolution by the use of a changed sent data-structure. In Ref. [8] it has been proposed for the first time to perform a blockwise transmission by inserting a guard interval (CP) between successive blocks. In order to mitigate interblock interference the duration T_G of the guard interval has to be longer than the duration T_h of the channel impulse response $h(t)$. The task of the frequency domain equalizer is to eliminate intersymbol interference (ISI) within the individual blocks. Figure 16.3 depicts the structure of one transmitted block that consists of the original sequence of N symbols of duration $T_{FFT} = NT$ and the cyclic extension duration of $T_G = N_G T$.

If $s_i(t)$ denotes the continuous-time representation of the original symbol sequence of the ith block with $s_i(t) = 0$ for $t \notin [0, T_{FFT}]$, then the extended block denoted by $\hat{s}_i(t)$ is given by

$$\hat{s}_i(t) = \begin{cases} s_i(t) & \text{for } t \in [0, T_{FFT}], \\ s_i(t + T_{FFT}) & \text{for } t \in [-T_G, 0], \\ 0 & \text{else.} \end{cases} \tag{16.28}$$

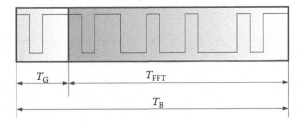

Figure 16.3 Sent data-structure using a CP.

The overall transmit signal can then be written as

$$s(t) = \sum_i \hat{s}_i(t - i(T_{\text{FFT}} + T_{\text{G}})).\tag{16.29}$$

The transmission of the cyclically extended sequence over a channel with the impulse response $h(t)$ is given by the linear convolution

$$\hat{s}_i(t) \star h(t) = \int\limits_{-\infty}^{\infty} \hat{s}(\tau)h(t - \tau)\, d\tau \tag{16.30}$$

and is defined for the interval $-T_{\text{G}}, T_{\text{FFT}} + T_h$. For the further steps, also the periodic sequences $s_{\text{p}}(t)$ and $h_{\text{p}}(t)$ are defined as

$$s_{\text{p}}(t) = \sum_{n=-\infty}^{\infty} s_i(t - nT_{\text{FFT}}),\tag{16.31}$$

$$h_{\text{p}}(t) = \sum_{n=-\infty}^{\infty} h(t - nT_{\text{FFT}}).\tag{16.32}$$

The transmission of $s_{\text{p}}(t)$ over the radio channel is given by

$$\int\limits_{-\infty}^{\infty} s_{\text{p}}(\tau)h(t - \tau)\, d\tau = \int\limits_{\Delta t}^{T_{\text{FFT}}+\Delta t} s_{\text{p}}(\tau)h_{\text{p}}(t - \tau)\, d\tau = s_i(t) \otimes h(t).\tag{16.33}$$

The choice of Δt is not of relevance as long as the integration is carried out over one period. By comparing Equation 16.30 with Equation 16.33 it becomes obvious that both convolution operations lead to the same result within the interval of $[-T_{\text{G}} + T_h, T_{\text{FFT}}]$. This behavior is characterized in Figure 16.4 additionally.

With the cyclic extension, the linear convolution of the i-th block with the channel impulse response becomes a circular convolution and the received block fulfills the condition

$$\hat{r}_i(t) = \hat{s}_i(t) \star h(t) = s_i(t) \otimes h(t) \quad \text{for} \quad t \in [-T_{\text{G}} + T_h, T_{\text{FFT}}].\tag{16.34}$$

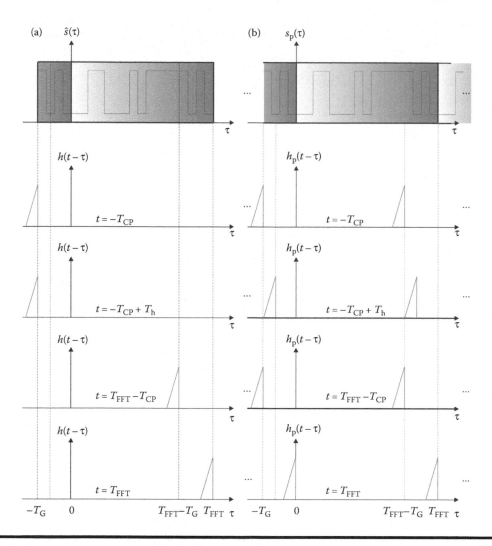

Figure 16.4 Linear (a) versus cyclic (b) convolution.

Considering only one block of the received block of duration T_{FFT}, this block includes one period of the cyclically extended signal $\hat{r}_i(t)$ and is denoted by $\tilde{r}_i(t)$. Based on the theorem of circular convolution one gets the essential relation

$$\tilde{R}_i(nf_0) = S_i(nf_0) \cdot H(nf_0) \quad f_0 = \frac{1}{T_{FFT}}, \quad n \in \mathbb{Z}. \tag{16.35}$$

Here the functions $\tilde{R}_i(f)$, $S_i(f)$, and $H(f)$ are related to the time domain signals $\tilde{r}_i(t)$, $s_i(t)$, and $h(t)$ by the continuous Fourier transform. Note that on the right-hand side of Equation 16.35 the Fourier transform of the original, noncyclically extended transmitted block $s_i(t)$ appears. This implies the important relation

$$\tilde{R}_i(nf_0) = R_i(nf_0), \quad n \in \mathbb{Z}, \tag{16.36}$$

where $R_i(f)$ denotes the Fourier transform of the received block $r_i(t)$ that would result from transmission of the original, noncyclically extended block $s_i(t)$ over the channel $h(t)$. The cyclic extension guarantees that every necessary information of a distorted and received block is concentrated within one period of the periodically extended structure of duration T_{FFT}.

It is essential to summarize the idea of the CP once more: It is our aim to use efficient DFT (FFT) operations to carry out the equalization in the frequency domain. But the DFT–multiplication–IDFT operation is equivalent to a cyclic convolution in the time domain, although the information sequence is influenced by the radio channel—represented by a linear convolution. The effects of this linear convolution shall be compensated for by a second linear convolution—the equalizer.

By using a CP, the linear convolution equals the cyclic convolution already at the channel and allows the instant application of the DFT–multiplication–IDFT operation due to the theorem of cyclic convolution. The price to be paid for these significant advantages is simply a reduced bandwidth efficiency of up to 20% as the guard period has to be inserted already at the transmitter. As a result, a correct blockwise equalization can be carried out.

Until now, only the equalization in the frequency domain as such has been presented, without defining the equalizer transfer-function or the equalization criterion itself. This will be compensate for in the following.

The equalizer error ε_k is defined as the difference between the transmit sequence $d(k)$ and the equalizer output $y(k)$

$$\varepsilon_k = d(k) - y(k). \tag{16.37}$$

This difference is caused by ISI as well as noise, making a detailed characterization of these interference factors necessary. The expectation ($\mathcal{E}$) of ϵ_k at the equalizer output is given by

$$\sigma^2 = \mathcal{E}|\varepsilon_k|^2, \tag{16.38}$$

where σ^2 denotes the variance, which can also be computed in the frequency domain as

$$\sigma^2 = \varphi_{\varepsilon\varepsilon}(0) = T \int_0^{1/T} \Phi_{\varepsilon\varepsilon}^T(f)\, df, \tag{16.39}$$

where $\varphi_{\varepsilon\varepsilon}(k)$ denotes the autocorrelation function and $\Phi_{\varepsilon\varepsilon}^T(f)$ the power spectral density of the error signal. From the block diagram shown in Figure 16.5, $\Phi_{\varepsilon\varepsilon}^T(f)$ can be derived as

$$\Phi_{\varepsilon\varepsilon}^T(f) = \sigma_d^2|Q^T(f)E^T(f) - 1|^2 + \Phi_{\nu\nu}^T(f)|E^T(f)|^2. \tag{16.40}$$

Figure 16.5 Transceive structure.

Equation 16.40 is based on the assumption of an uncorrelated data sequence $d(k)$ of variance σ_d^2. $\Phi_{vv}^T(f)$ denotes the power spectral density of the equalizer input noise sequence $v(k)$, colored by the matched filter. Inserting into Equation 16.39 yields

$$\sigma^2 = \varphi_{\varepsilon\varepsilon}(0) = T \int\limits_0^{1/T} \sigma_d^2 |Q^T(f)E^T(f) - 1|^2 + \Phi_{vv}(f)|E^T(f)|^2 \, df. \qquad (16.41)$$

While the first term describes the power due to residual ISI, the second term characterizes the output noise power. Two criteria have found widespread use in optimizing the equalizer coefficients. One is the zero forcing (ZF) criterion and the other is the minimum mean square error (MMSE) criterion.

16.2.2.1 ZF Equalization

The aim of a ZF equalizer is to satisfy the zero-ISI constraint. Based on that demand the ISI term in the integrand of Equation 16.41 has to be set to zero, resulting in the ZF equalizer given by

$$E_{ZF}^T(f) = \frac{1}{Q^T(f)}. \qquad (16.42)$$

The equalizer is simply the inverse filter to the linear filter model $Q^T(f)$. Consequently, $E_{ZF}^T(f)$ is a real valued function. The use of a ZF equalizer may result in significant noise enhancement if $Q^T(f)$ suffers from deep spectral fades. This can easily be seen by noting that in a frequency range where $Q^T(f)$ is small, the equalizer compensates for ISI by placing a large gain in that frequency range. Substituting Equation 16.42 into 16.41 yields the output noise power of the ZF equalizer:

$$\sigma_{ZF}^2 = T \int\limits_0^{1/T} \frac{\Phi_{vv}^T(f)}{|Q^T(f)|^2} \, df. \qquad (16.43)$$

The occurrence of noise enhancement may lead to severe degradations in terms of BER, making other, more powerful concepts necessary.

16.2.2.2 MMSE Equalization

MMSE equalizers alleviate the problem of noise enhancement by compromising the noise amplification and the ISI reduction. From Equation 16.41 the MMSE criterion leads the equation

$$\int\limits_0^{1/T} \sigma_d^2 |Q^T(f)E^T(f) - 1|^2 + \Phi_{vv}^T(f)|E^T(f)|^2 \, df \Rightarrow \min(\varepsilon^T(f)). \qquad (16.44)$$

Since the integrand is nonnegative, the criterion reduces to the frequency-wise criterion

$$\sigma^2 |Q^T(f)E^T(f) - 1|^2 + \Phi_{vv}^T(f)|E^T(f)|^2 \Rightarrow \min(\varepsilon^T(f)) \qquad (16.45)$$

at every frequency f. Applying the gradient method with respect to $E^T(f)$ yields the MMSE equalizer frequency response [10]:

$$E_{\mathrm{MMSE}}^T(f) = \frac{[Q^T(f)]^*}{|Q^T(f)|^2 + \Phi_{vv}^T(f)/\sigma_d^2}. \tag{16.46}$$

Again the equalizer transfer function is real valued, since all terms appearing at the right-hand side of Equation 16.46 are real valued as well. Thus, the complex conjugation in the numerator can be omitted. The additive term in the denominator of the expression appearing in Equation 16.46 protects against infinite noise enhancement. Substituting Equation 16.46 into Equation 16.41 yields the output noise power of the MMSE equalizer:

$$\sigma_{\mathrm{MMSE}}^2 = T \int_0^{1/T} \frac{\Phi_{vv}^T(f)}{|Q^T(f)|^2 + \Phi_{vv}^T(f)/\sigma_d^2} \, df. \tag{16.47}$$

Comparing Equations 16.47 and 16.43 it is obvious that σ_{ZF}^2 is always greater than or equal to σ_{MMSE}^2. At high SNRs the two equalizers become equivalent.

16.2.3 Known Symbols versus Classical CP

It has been stated that the concept of SC/FDE based on a CP suffers from a reduced bandwidth efficiency of up to 20% due to the CP. Additionally these 20% are simply chopped at the receiver without further use. The structure that is introduced now shows significant advantages compared to the classical CP as this structure allows one to use the overhead of the CP in many different ways.

The drawback of the classical CP is that it is chopped at the receiver and that the content of the CP changes with every processed block. To solve this problem, a restructuring of the classical CP structure is carried out. Figure 16.6 shows an obvious further development of the introduced CP concept [11–18]. In this diagram *UW* stands for *Unique Word*, describing a known pilot sequence. The first attempt would simply be the replacement of data symbols by known symbols at the end of every processed block, leading to exactly the same structure as for CP but having known symbols instead of arbitrary ones. Here the advantage of a known sequence is paid dearly at the cost of a further reduced bandwidth efficiency, as the UW carries no information data itself. This additional drawback can be compensated for using the structure shown in Figure 16.6c. It will be demonstrated that this structure fulfills the theorem of cyclic convolution in exactly, the same way as the classical CP structure.

The following mathematical description proves that the concept of UW fulfills the *theorem of cyclic convolution*. Figure 16.6b depicts the structure of one transmitted block, which consists of the original data sequence of N_s symbols and the sequence of the UW with N_G symbols. The overall duration of $N = N_s + N_G$ symbols is $T_{\mathrm{FFT}} = N_T$. Instead of the CP, a known sequence is part of every processed block. Let $s_{\mathrm{Data},i}(t)$ denote the continuous-time representation of the symbol sequence of the i-th transmitted block with $s_{\mathrm{Data},i}(t) = 0$ for $t \notin [0, T_{\mathrm{FFT}} - T_G]$, whereas $s_i(t)$

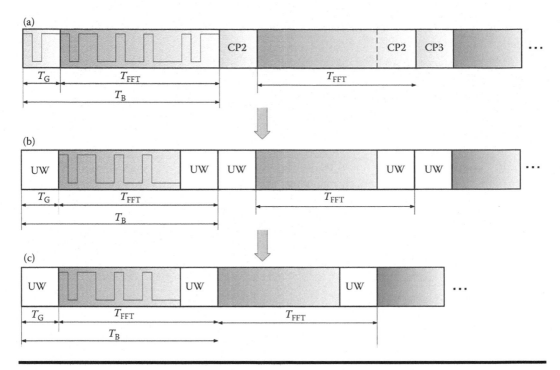

Figure 16.6 CP structure versus UW structures.

defines the extended block

$$s_i(t) = \begin{cases} s_{\text{Data},i}(t) & \text{for } t \in [0, T_{\text{FFT}} - T_{\text{G}}], \\ \text{uw}(t) & \text{for } t \in [T_{\text{FFT}} - T_{\text{G}}, T_{\text{FFT}}]. \end{cases} \tag{16.48}$$

For the further development $\hat{s}_i(t)$ is defined as

$$\hat{s}_i(t) = \begin{cases} s_i(t) & \text{for } t \in [0, T_{\text{FFT}}], \\ \text{uw}(t + T_{\text{FFT}}) & \text{for } t \in [-T_{\text{G}}, 0], \\ 0 & \text{elsewhere.} \end{cases} \tag{16.49}$$

Note that this representation includes not only the UW within T_{FFT} but also the UW from the previous block. With this extended block, the linear convolution of the i-th block with the channel impulse response becomes a circular convolution ($\otimes$), and the received block fulfills the condition

$$\hat{r}_i(t) = \hat{s}_i(t) * h(t) = s_i(t) \otimes h(t) \tag{16.50}$$

within the interval $[-T_{\text{G}} + T_h, T_{\text{FFT}}]$. Restricting the received block to the time interval $[0, T_{\text{FFT}}]$ and applying the theorem of circular convolution to Equation 16.50, we obtain the essential relation

$$\hat{R}_i(nf_0) = S_i(nf_0) \cdot H(nf_0) = R_i(nf_0) \tag{16.51}$$

for $f_0 = 1/T_{\mathrm{FFT}}$ and $n \in \mathbb{Z}$. $R_i(f)$ denotes the Fourier transform of the received block $r_i(t)$, which would result from transmission of the original block $s_i(t)$ over the channel $h(t)$. The frequency relation of Equation 16.51 shows that the concept of UW is identical to the concept of CP. The UW shows every characteristic of a CP, allowing the low complexity equalizer implementation on the one hand, but showing significant advantages on the other hand. These advantages are based on the following differences:

- The UW is not random as the CP but can be chosen in a preferred way.
- The UW is not removed at the receiver but is available after the equalization in the time domain.
- The UW does not change during transmission.

Although the step from the classical structure of a CP to the structure of the UW is a small one, the advantages are tremendous, as the UW will be used for

- Equalization
- Synchronization
- Channel estimation

It can be concluded that the concept of UW fulfills the theorem of cyclic convolution in exactly the same way as the traditional CP. Consequently, this concept allows a comparable low complexity equalization in the frequency domain as well as the implementation of an ideal, infinite long equalizer. The main and significant difference is that the UW will also be used for equalization, synchronization, and channel estimation.

16.3 FDE for Multicarrier Transmission

A different approach of frequency domain processing is the concept of multicarrier transmission. The underlying concept is a parallel transmission concept, where data are transmitted simultaneously over several spectral (adjacent placed) subcarriers as indicated in Figure 16.7 [19–24].

Due to the parallelization, the data rate of every individual subcarrier is reduced by the number of subcarriers N, which results in a symbol duration on every subcarrier that is enlarged by the same factor. That leads to a reduction of ISI compared to a single carrier transmission.

The principal multicarrier transmission concept will be introduced based on Figures 16.8 and 16.9, respectively. A multicarrier transmission starts with a serial/parallel conversion of length $N \cdot m$, where N describes the number of subcarriers, while m stands for the number of bits forming one symbol due to modulation. The mapping is followed by a pulse shaping filter $g(t)$, which may lead to a crosstalk between subcarriers. The heart of a multicarrier system is the spectral separation of every single subcarrier, by multiplying with $e^{j2\pi f_n t}$. Here f_n describes the specific center frequency. After a summation of all subcarriers, the transmitted signal is given by

$$s(t) = \sum_{i=-\infty}^{\infty} \sum_{n=0}^{N-1} d_n(i) \cdot g(t - iT) \cdot e^{j2\pi f_n t}. \tag{16.52}$$

In this equation, T describes the symbol duration and $d_n(i)$ describes the modulated symbol of the nth subcarrier at the time iT.

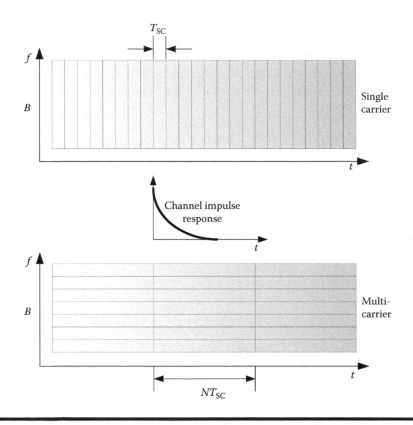

Figure 16.7   Single carrier versus multicarrier transmission [19].

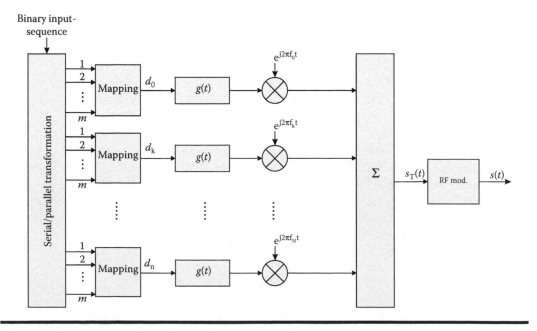

Figure 16.8   Transmitter for a multicarrier transmission.

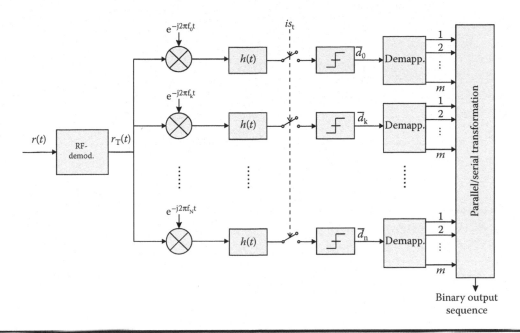

Figure 16.9   Receiver for a multicarrier transmission.

In the receiver, a spectral separation of every subcarrier is carried out by multiplication with $e^{-j2\pi f_n t}$. The recovery of symbols occurs by matched filtering and sampling on every subcarrier at a symbol duration of T.

It is now discussed how the concept of multicarrier transmission meets intercarrier interference (ICI) as well as ISI introduced by individual subcarriers on the one hand and by the radio channel on the other hand.

16.3.1 Orthogonal Frequency Division Multiplexing—OFDM

The concept of OFDM denotes a specific implementation of multicarrier transmission, which solves the problem of ICI as well as ISI [19–24]. OFDM shows a direct correlation between pulse shaping $g_n(t)$ and the subcarrier-center frequency f_n. By choosing $g_n(t)$ as a rectangular impulse response of duration T_s

$$|g_n(t)| = \begin{cases} 1 & \text{for } 0 \le t \le T_s, \\ 0 & \text{else,} \end{cases} \tag{16.53}$$

the spectrum results in a *Si*-function, whose center frequency varies with every subcarrier. The zeros of these subcarrier spectra have a distance of $1/T_s$—these zeros are of central importance for the entire system as they enable orthogonality between subcarriers and prevent ICI. It can easily be shown that this orthogonality is fulfilled for

$$|g_n(t)| = \begin{cases} e^{j2\pi f_n t} & \text{for } 0 \le t \le T_s, \\ 0 & \text{else,} \end{cases} \tag{16.54}$$

with $f_n = n/T_s$. Figure 16.10 shows a set of subcarriers, realizing orthogonality and enabling ICI-free transmission.

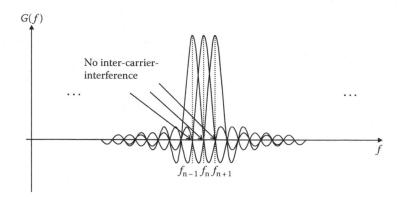

Figure 16.10 Spectrum of subcarriers.

For a correct demodulation in the receiver, every subcarrier needs a receive filter, adapted (matched) to the transmit filter. The receiver (matched) filter is thus given by

$$
h_n(t) = \begin{cases} \dfrac{1}{T_s} e^{j2\pi f_n t} & \text{for} \quad -T_s \le t \le 0, \\ 0 & \text{else.} \end{cases}
\tag{16.55}
$$

Under the assumption of an ideally synchronized symbol sampled at the time $T_0 = 0$ the convolution of every transmit- and receive-filter is given by

$$
g_m(t) \star h_n(t)|_{t=0} = \int_{-\infty}^{\infty} g_m(\tau) \cdot h_n(0 - \tau)\, d\tau,
\tag{16.56}
$$

$$
= \frac{1}{T_s} \int_{0}^{T_s} e^{j2\pi m\tau/T_s} \cdot e^{-j2\pi n\tau/T_s}\, d\tau,
\tag{16.57}
$$

$$
= \frac{1}{T_s} \int_{1}^{T_s} e^{j2\pi(m-n)\tau/T_s}\, d\tau.
\tag{16.58}
$$

For $n \ne m$ the integral vanishes and only the part for $n = m$ remains. It can be concluded that a proper choice of pulse shaping as well as subcarrier center frequency allows an ideal, ICI-free and ISI-free recovery of data.

16.3.1.1 ISI and ICI in the Case of Multipath Propagation

The previous description demonstrated that for ideal, undisturbed transmission an error-free recovery is possible. This statement is not valid in the case of multipath propagation, leading to ISI as well as ICI, making equalization concepts necessary. Figure 16.11 demonstrates that due to the dispersion of one symbol by the transmission over the mobile radio channel, ISI and ICI occur.

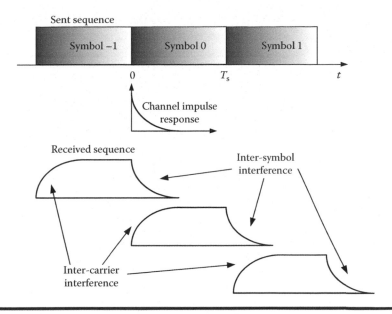

Figure 16.11 OFDM symbols due to the influence of a channel.

The solution of this problem is comparable to that of SC/FDE—actually it is to say that this method has been developed for OFDM earlier and has been adapted for SC/FDE. By using a guard period T_G between successive blocks, ISI will be prevented in the case of $T_G \geq T_h$ (T_h describes the length of the channel impulse response), indicated in Figure 16.12a. Nevertheless, a guard period is able to prevent ISI but cannot prevent the loss of orthogonality between subcarriers due to ICI. This degradation is to be prevented by using a CP as introduced in Section 16.2.2 and depicted in Figure 16.12b. Using a CP the necessary "settling" time, which is responsible for the ICI, is guaranteed.

Figure 16.13 shows that due to the use of a CP the settling time as well as the decaying time is not part of the processed OFDM symbol itself, having a duration of T_s. In that case exactly one period of duration T_s of a theoretical infinite long periodic signal can be processed at the receiver. Additionally it is mentioned that an easy and efficient use of the introduced UW structure as for

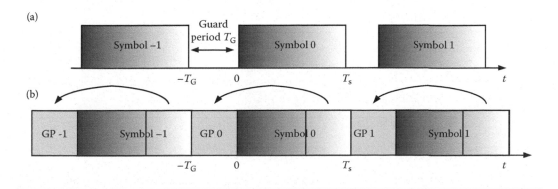

Figure 16.12 OFDM symbols by using (a) a guard period and (b) a CP.

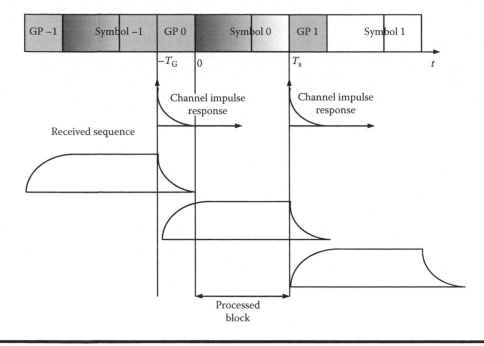

Figure 16.13   ICI and ISI in the case of using a CP.

SC/FDE is not as simple due to the use of the IFFT at the transmitter. Detailed investigations are the topic of actual research.

The above statements on ISI-free and ICI-free decisions allows for an extremely efficient equalization. As the channel at every subcarrier can be assumed to be flat, the equalization is reduced to a complex multiplication at every subcarrier, correcting for magnitude and phase. The implemented equalizer represents a one-tap filter fulfilling the ZF criterion, carried out before the demapping (see Figure 16.9). Based on the above statement it can be shown that the ZF criterion represents the optimum criterion for OFDM, an improvement based on other criteria is not given for the simple OFDM structure. The statement does not hold for (multiple input multiple output (MIMO) systems based on OFDM, where also the introduced MMSE criterion is used.

It can be concluded that the advantages as well as the dominant position of a multicarrier transmission or OFDM, respectively, are based on two facts:

■ By preventing ICI as well as ISI the equalization is reduced to a complex multiplication in the frequency domain for every subcarrier.
■ The spectral separation of subcarriers followed by a summation of these subcarriers is implemented very efficiently by a DFT (FFT) and IDFT (IFFT), respectively.

16.4 Fundamental System Comparison

To point out the specific differences and advantages of the following introduced concepts:

■ A single carrier system based on CP

- A single carrier system based on known pilot sequences (UW)
- A multicarrier system (OFDM)

a short comparison in terms of *processing load, bandwidth efficiency, and S/N loss* will be given in the following. Detailed comparisons in terms of performance, or RF-related issues will follow.

16.4.1 Processing Load

To make the introduced concepts comparable, the processing steps necessary to equalize a data block of size N will be compared.

It is to be mentioned that the SC/FDE concepts are based on a two-times oversampling to enable matched filtering. Furthermore, it has to be noted that this calculation of the processing load represents only a basic comparison as different forms of equalizer implementation exist.

16.4.1.1 SC/FDE Based on a Classical CP

- $2N$-point FFT to transform the processed block to the frequency domain
- $2N$ complex multiplications to carry out the equalization in the frequency domain
- N complex additions for a sampling rate reduction
- N-point IFFT to transform the processed block back to the time domain

16.4.1.2 SC/FDE Based on UW

- $2N$-point FFT to transform the processed block to the frequency domain
- $2N$ complex multiplications for the matched filtering in the frequency domain
- N complex additions for a sampling rate reduction
- N-point IFFT to transform the processed block back to the time domain

Although the processing loads for the CP and UW structure seem to be identical, the complexity of UW is of the order of $O(T_G/T_{FFT})$ higher than that of the CP as the UW itself carries no data information but has to be processed as any data information.

16.4.1.3 OFDM

While the concepts of single carrier transmission are based on a two-times oversampling, a symbol rate processing is sufficient for OFDM. Besides that, it is to be mentioned that FFT/IFFT, operations are separated between the transmitter and the receiver.

- N-point IFFT to transform the processed block to the time domain at the transmitter
- N-point FFT to transform the processed block to the frequency domain at the receiver
- N complex multiplications for the equalization

Based on the above investigation, it can be concluded that the implementation effort for SC/FDE is higher than that for OFDM—making twice as many operations necessary. In detail, the FFT size is 64 for OFDM and 128 for SC/FDE throughout this chapter. But, as especially for the FFT/IFFT operations, extremely efficient algorithms exist, this disadvantage is of reduced meaning. The essential point of this consideration is that the cost for the equalization does not grow quadratically with the bit rate as it does for a single carrier system with time domain equalization.

16.4.1.4 Bandwidth Efficiency

The demand for higher transmission rates on the one hand combined with the limited resource "bandwidth" on the other hand makes it inevitable to judge different communication systems by means of their (net) bandwidth efficiency.

A simplified calculation for the investigated systems will be given in the following. The simplification of this consideration is based on the fact that necessary preamble sequences are not taken into account yet. Furthermore, an exemplary coding rate (R) of 0.5 is taken into account.

For a single carrier system the bandwidth efficiency is dependent on the number of bits per symbol, indicated by M as well as by the roll-off factor (r) due to pulse shaping. For the case of quadrature phase shift keying (QPSK), M is set to 2 and the roll-off has been chosen 0.3.

$$\eta_{SC} = \frac{\text{Bitrate}}{\text{Bandwidth}} \cdot \text{Code rate} = \frac{M}{1+r} \cdot R = 0.77 \frac{\text{bit/s}}{\text{Hz}}. \tag{16.59}$$

16.4.1.5 SC/FDE Based on CP

To make the systems comparable, the symbol rate was chosen to be the same for the investigated systems, namely 12 Mbit/s. As a result the duration of the FFT window as well as the duration of the guard period differs slightly between SC/FDE and OFDM. For SC/FDE this leads to a duration of the FFT window of 4.49 ms and of 842 ns for the guard period. Based on that, the bandwidth efficiency is calculated by

$$\eta_{SC/FDE,CP} = \frac{M}{1+r} \left(\frac{T_{FFT}}{T_{FFT} + T_G} \right) \cdot R = 0.65 \frac{\text{bit/s}}{\text{Hz}}. \tag{16.60}$$

16.4.1.6 SC/FDE Based on UW

If using a UW the bandwidth efficiency will be further reduced a little bit if maintaining the block structure of $N = 64$ symbols, out of which 12 symbols (842 ns, respectively) represent the guard period. Allowing a processed block size of 76, this would not be the case.

$$\eta_{SC/FDE,UW} = \frac{M}{1+r} \left(\frac{T_{FFT} - T_G}{T_{FFT}} \right) \cdot R = 0.625 \frac{\text{bit/s}}{\text{Hz}}. \tag{16.61}$$

16.4.1.7 OFDM

For OFDM the parameters as defined by the 802.11a standard are used (number of total subcarriers: $N_{total} = 52$; number of pilots: 4; guard time: $T_G = 800$ ns; FFT integration time: $T_{FFT} = 3.2$ ms; number of bits/symbol: $M = 2$). Note that pilot symbols being multiplexed into the datastream further reduces, the bandwidth efficiency. Based on these parameters, the bandwidth efficiency is calculated by

$$\eta_{OFDM} = \frac{\text{bitrate}}{\text{bandwidth}} = \frac{M \cdot N_{data}/T_{FFT} + T_G}{N_{total}/T_{FFT}} \cdot R = 0.74 \frac{\text{bit/s}}{\text{Hz}}. \tag{16.62}$$

(In this equation, the system bandwidth has been approximated by $B = N_{total}/T_{FFT}$.)

16.4.1.8 Loss of Signal-to-Noise Ratio due to a CP

Besides a loss in terms of bandwidth efficiency the overhead due to the guard period additionally leads to a loss of signal-to-noise ratio, as the effective energy per data symbol is reduced by the symbols of the guard period. Based on a white Gaussian noise with the two-sided power spectral density of $N_0/2$, the S/N ratio results in

$$\frac{S}{N} = \frac{E_b}{N_0/2}\left(1 - \frac{T_G}{T_{FFT}}\right) \tag{16.63}$$

for the SC/FDE system using a CP. A summary of loss in terms of performance for the four cases is given in Table 16.1.

16.4.1.9 Front-End Requirements and RF-Related Issues

Other essential aspects that have not been taken into account within this comparison of OFDM and SC/FDE are front-end or RF-related issues. Especially OFDM suffers from high demands on the analog front-end. Maybe the most critical element of the front-end for OFDM is the power amplifier. The transmit signal envelope of a multicarrier signal shows a very high peak to average power ratio (PAPR). This is a problem when it comes to the use of nonlinear power amplifiers. Two major problems arise with nonlinear amplification:

1. Nonlinear amplification causes in-band distortions, which decrease the bit error performance
2. Nonlinear amplification causes out-of-band emission, making further filtering necessary

Of course this further filtering after power amplification is undesirable, because filters that can cope with high power are too expensive for the aspired solutions.

As a consequence, the power amplifier must be operated with significant back off, which means that the power efficiency drastically decreases. In comparison to OFDM, the demands on the power amplifier are reduced significantly for the SC/FDE concept. The peak-to-average power ratio is defined by the modulation scheme, enabling the use of a less cost-intensive power amplifier. Table 16.2 gives a comparison of the PAPR between SC/FDE and OFDM. It is obvious that OFDM shows a significantly higher PAPR especially for lower modulation schemes. Besides this, SC/FDE has the possibility to reduce this PAPR additionally by using modulation schemes with a constant envelope.

Table 16.1   S/N Loss due to Overhead

	S/N Loss Relative	S/N Loss Absolute (dB)
SC/FDE,CP	$\dfrac{T_{FFT}}{T_{FFT} + T_G}$	0.75
SC/FDE,UW	$1 - \dfrac{T_G}{T_{FFT}}$	0.90
OFDM	$\dfrac{T_{FFT}}{T_{FFT} + T_G}$	0.97

Table 16.2 Peak to Average Power Ratio

Modulation	OFDM	SC/FDE
QPSK	9.13	4.48
16-QAM	9.81	6.96
64-QAM	9.68	7.44

Note: QAM, quadrature amplitude modulation.

Of course it is to be mentioned that there exist a variety of algorithms to reduce the PAPR for OFDM, but in general these algorithms are adaptable for SC/FDE in an easy way, leading also for SC/FDE to a reduced PAPR. Additionally, it is to be mentioned that every algorithm for a reduced PAPR increases the signal processing complexity. To characterize the influence of this higher PAPR, the error vector magnitude (EVM) as well as the BER is given below, based on a nonlinear class A power amplifier, which is characterized by

- 1 dB compression point: $G_{1dB} = 22$ dBm
- Third order intercept point: IP3 = 34 dBm
- Linear amplification: $G = 38$ dB.

The EVM is a measure to describe distortions due to filters, amplifiers, or modulators in a constellation diagram. Therefore the EVM is defined as the ratio between error vector and reference vector as shown in Figure 16.14 and described by

$$\text{EVM(rms)} = \frac{\sqrt{\sum_{k=0}^{N-1} |E(k)|^2}}{\sqrt{\sum_{k=0}^{N-1} |S(k)|^2}}. \tag{16.64}$$

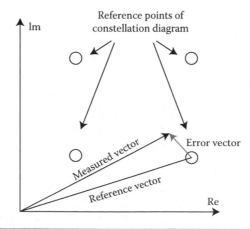

Figure 16.14   Definition of EVM.

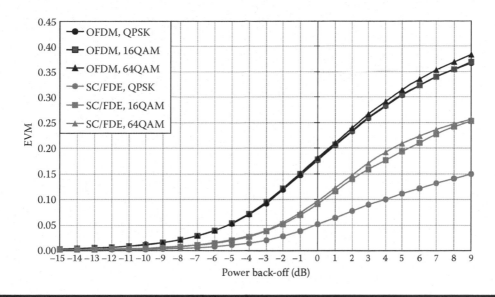

Figure 16.15   EVM for SC/FDE and OFDM.

Figure 16.15 depicts the EVM for the two investigated systems when operating the power amplifier at different levels. While the EVM is comparable for OFDM and SC/FDE, operating the amplifier at a significant power-back-off (PBO), this changes already at a PBO of about −6 dB. From that value SC/FDE shows a significantly reduced EVM. Although the EVM also grows significantly for SC/FDE with higher order modulation schemes, SC/FDE will always outperform OFDM.

As the IEEE 802.11a standard defines a maximal EVM for the different data rates, it is of interest to compare these values with the values achieved by simulation. The instructions given by the standard are given in Table 16.3. To fulfill the demands, especially for higher data rates, the amplifier would have to be operated at a high PBO. In comparison, it would be much easier for SC/FDE to fulfill the demands.

Table 16.3   Predefined EVM

Data Rate (Mbit/s)	Max. EVM (dB)	Max. EVM (%)
6	−5	47.3
9	−8	30.2
12	−10	22.3
18	−13	14.3
24	−16	9.1
36	−19	5.84
48	−22	3.71
54	−25	2.36

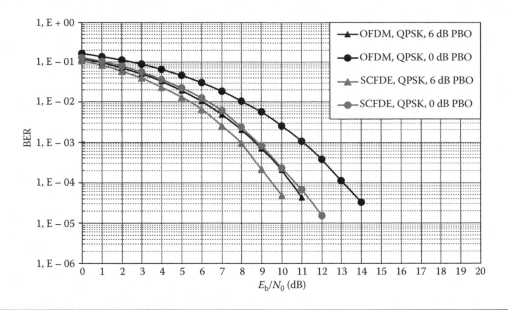

Figure 16.16   BER due to a nonlinear power amplifier (QPSK).

A high EVM leads inevitable to an additional loss in terms of BER, as characterized in Figure 16.16. The simulation results are based on additive white Gaussian noise (AWGN) without channel coding and the modulation scheme of QPSK. While the performance loss for SC/FDE is about 1 dB when operating the amplifier at the 1 dB compression point compared to the operation at a 6 dB PBO, the loss is about 2.5 dB for OFDM.

By carrying out a comparable simulation for the modulation scheme of 16-QAM, the performance loss gets significantly higher—for both systems (Figure 16.17). Especially OFDM shows already a strong saturation effect, operating the amplifier at the 1 dB compression point.

It should be noted that OFDM and SC/FDE take advantage or suffer from drawbacks of completely different elements of a transmission system, making these systems difficult to compare in terms of performance. Yet, it has been demonstrated that both strategies will show an overall comparable performance.

16.5 Performance Comparison of SC/FDE versus OFDM

The comparison of these two systems is quite difficult as they take advantage of or suffer from drawbacks by different elements as coding, equalizer structures, interleaving, channel-state information, front-end requirements, requirements in terms of synchronization, or others. From that point of view, the most fair comparison can only be the specific characterization of specific elements in an entire system and how they influence the actual performance. Nevertheless, although this comparison gives insight into the specific behavior of the investigated systems and the resulting performance, these investigations cannot give an all-embracing characterization.

Before going into details it is necessary to characterize the parameters that the actual investigations are based on.

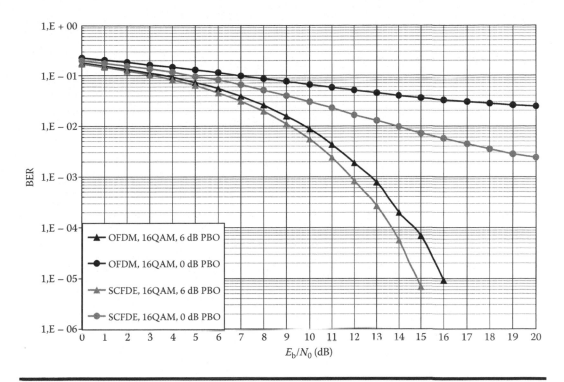

Figure 16.17   BER due to a nonlinear power amplifier (16-QAM).

16.5.1 The OFDM System—Parameter Definition

The following simulation results are exemplary based on the physical layer parameters of the IEEE 802.11a-like OFDM system, which are reviewed briefly in Refs. [25,26].

Figure 16.18 shows the main baseband processing steps of the investigated system. In the transmit path the binary input data are first scrambled (to ensure a small PAPR) and encoded by using the industry standard rate 1/2, constraint length 7 convolutional encoder with generator polynomials (133-171). By puncturing the rate 1/2 code, higher coding rates are achieved. An efficient coding scheme is a crucial requirement for OFDM since data symbols corresponding to carriers located in a region with deep spectral fades would otherwise be unreliable. After frequency domain interleaving, which ensures that adjacent coded bits are modulated on nonadjacent subcarriers, the binary values are converted to QAM values. The PHY layer offers different subcarrier modulation schemes and different coding rates, such that different data rates from 6 Mbit/s up to 54 Mbit/s can be provided. To facilitate coherent reception, four pilot values are added to each 48 data values, which yields a total of 52 QAM values per OFDM symbol. The heart of the OFDM system is the IFFT/FFT section— the 64-point IFFT/FFT algorithms perform the modulation and demodulation. In the transmitter the IFFT modulates a block of 52 QAM values onto 52 subcarriers. The DC subcarrier and the subcarriers at the band edges are not used for data transmission. To make the system robust against multipath propagation, the CP is added. The guard interval of length 800 ns ensures robustness to delay spreads up to 250 ns. To attain a narrow output spectrum, time domain windowing or digital filtering can be applied. Rectangular pulses are used in these investigations. Finally, the digital output signals are converted to analog signals, which are up-converted to the 5 GHz band.

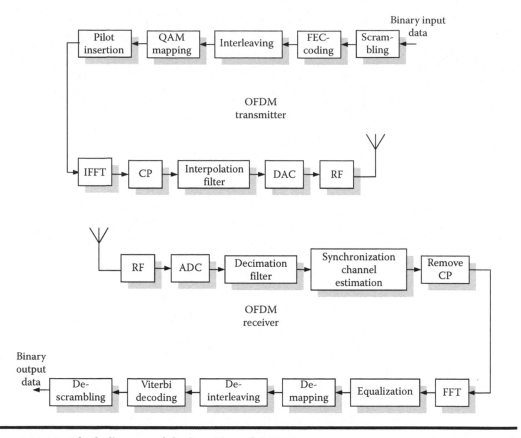

Figure 16.18   Block diagram of the investigated OFDM system.

In the receiver path, after passing the RF part and the analog-to-digital conversion, time and frequency synchronization is performed before starting the demodulation of the OFDM symbols. For each symbol the CP is removed, and an FFT is used to demodulate all subcarriers. The output of the FFT contains 48 distorted data values that are equalized by 48 complex multiplications. The pilot subcarriers are used to correct the remaining phase drift. The equalized QAM values are then demapped onto binary values and deinterleaved. Finally, decoding by a Viterbi decoder and descrambling produce the binary output data.

A summary of the introduced parameters are given in the following as well as in Table 16.4.

Number of subcarriers $N = 64$

Number of active subcarriers $N_{\text{active}} = 52$

Number of pilot symbols $N_{\text{pilot}} = 4$ (out of 52)

Duration of a processed block $T_{\text{FFT}} = 3.2 \, \mu s$

Duration of the guard period $T_{\text{G}} = 800 \, \text{ns}$

Sampling time (symbol spaced) $T_{\text{s}} = 50 \, \text{ns}$

Subcarrier spacing $B_{\text{subcarrier}} = 312.5 \, \text{kHz}$

Table 16.4 Modes of the IEEE 802.11a Standard

Data Rate (Mbit/s)	Modulation	Code-Rate	Bits/Processed Block
6	BPSK	1/2	24
9	BPSK	3/4	36
12	QPSK	1/2	48
18	QPSK	3/4	72
24	16-QAM	1/2	96
36	16-QAM	3/4	144
48	64-QAM	2/3	192
54	64-QAM	3/4	216

- ■ Entire bandwidth $B = 16.562\,\text{MHz}$
- ■ Modulation schemes   Binary phase shift keying (BPSK), quadrature phase shift keying (QPSK), 16-QAM, 64-QAM
- ■ Code rates $R = 1/2, 2/3, 3/4, (9/16)$
- ■ Resulting data rates   6, 9, 12, 18, 24, (27), 36, 48, 54 Mbit/s

16.5.2 The SC/FDE System—Parameter Definition

Figure 16.19 shows the block diagram of the considered SC/FDE system. The SC/FDE signal processing starts with encoding the binary input data using the same encoder as used for the OFDM system. After coding and time domain interleaving the binary data are converted to QAM values and a guard period is added between successive blocks. After pulse shaping (Root Raised Cosine Pulses) and digital-to-analog conversion the resulting I/Q signals are up-converted to the 5 GHz band. In the receiver path, after passing the RF part and the analog-to-digital conversion, time and frequency synchronization is performed first. For each symbol the CP has to be removed before the equalization can be performed. For the underlying investigations an MMSE equalizer has been used. The equalized QAM values are then demapped onto soft bipolar values and decoded by a Viterbi decoder.

Although the only obvious difference between OFDM and SC/FDE seems to be the placement of the IFFT, it is this placement that makes a significant difference.

The parameters of the SC/FDE system have been adapted in such a way that the same transmission rate in samples per second is reached. Based on that, the duration of a processed block and the guard period has been adapted, leading to slight deviations in terms of the duration of a processed block. This assumption leads to the following system parameters as well as Table 16.5:

- ■ Number of symbols per processed block $N = 64$
- ■ Number of symbols per guard period $N_G = 12$

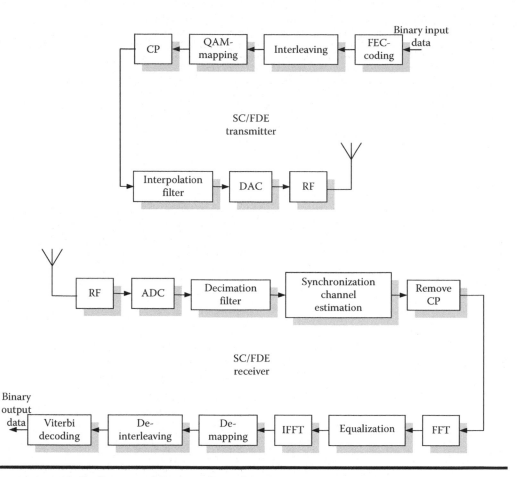

Figure 16.19   Block diagram of the investigated SC/FDE system.

**Table 16.5   Modes of the IEEE 802.11a Standard,
Adapted to SC/FDE**

Data Rate (Mbit/s)	Modulation	Code Rate	Bits/Processed Block
6	BPSK	1/2	32
9	BPSK	3/4	48
12	QPSK	1/2	64
18	QPSK	3/4	96
24	16-QAM	1/2	128
36	16-QAM	3/4	192
48	64-QAM	2/3	256
54	64-QAM	3/4	288

Duration of a processed block $T_{FFT} = 4.49\,\mu s$

Duration of the guard period $T_G = 842.1\,ns$

Sampling time (symbol spaced) $T_s = 70.18\,ns$

Pulse shaping (RRC)   Roll off: $r = 0.25$

Entire bandwidth $B = 17.81\,MHz$

Modulation schemes   BPSK, QPSK, 16-QAM, 64-QAM

Code rates $R = 1/2, 2/3, 3/4, (9/16)$

Resulting data rates   6, 9, 12, 18, 24, (27), 36, 48, 54 Mbit/s

16.5.3 Performance Comparison

It has already been mentioned that an all-embracing comparison of the two concepts is difficult. Due to that, the underlying simulation results are based on the simplification of an ideal synchronization and channel estimation. This assumption is justified by the fact that here only the performance of the basic concepts are discussed and compared.

Figure 16.20 depicts the BER behavior, based on the averaging of 250 randomly chosen indoor radio channel snapshots. The indoor radio channel impulse response (h) has been modeled as tapped delay line, each tap with random uniformly distributed phase and Rayleigh distributed magnitude and with the power delay profile decaying exponentially, according to IEEE 802.11a. In detail this means: It is a statistical model to investigate small-scale fading effects. The channel is seen as time-invariant for the duration of one burst. Based on this assumption, the channel can be modeled

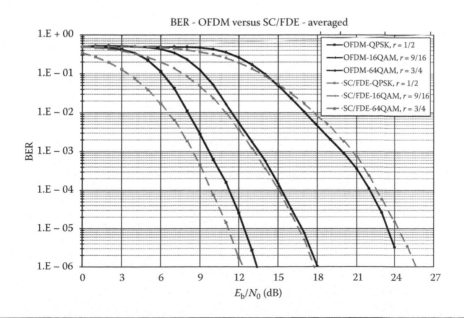

Figure 16.20   Performance comparison OFDM versus SC/FDE.

as a tapped delay line defined by

$$h(m) = \sum_{k=0}^{P-1} c_k \delta(m - k). \tag{16.65}$$

Here

$$c_k = a_k e^{-j\Theta_k}, \tag{16.66}$$

based on the following statistical model:

$$c_k = N\left(0, \frac{\sigma_k^2}{2}\right) + jN\left(0, \frac{\sigma_k^2}{2}\right),$$
$$\sigma_0^2 = 1 - e^{-\Delta\tau/\tau_{\mathrm{rms}}},$$
$$\sigma_k^2 = \sigma_0^2 e^{-\Delta\tau/\tau_{\mathrm{rms}}}. \tag{16.67}$$

In these equations $\Delta\tau$ describes the multipath resolution time, τ_{rms} stands for the channel delay spread, and $N(0, 1/2\sigma_k^2)$ is a zero mean Gaussian random variable with variance $1/2\sigma_k^2$. Based on that description the following characteristics result:

■ The amplitudes of echoes a_i follow a Rayleigh distribution, while the phase Θ_i follows a uniform distribution in the range of $[-\pi, \pi]$
■ The mean power decays exponentially
■ The mean receive power is independent of the actual channel impulse response because

$$\sum_k \sigma_k^2 = 1. \tag{16.68}$$

Criteria on how to choose $\Delta\tau$ as well as P are as follows

$$\Delta\tau \leq 1/B, \tag{16.69}$$
$$\Delta\tau \leq \tau_{\mathrm{rms}}/2, \tag{16.70}$$
$$P \approx 10\frac{\tau_{\mathrm{rms}}}{\Delta\tau}. \tag{16.71}$$

In Equation 16.69, B describes the overall system bandwidth. The chosen parameters are as follows:

$$P = 10\frac{\tau_{\mathrm{rms}}}{\Delta\tau},$$
$$\Delta\tau = 17.5 \text{ ns},$$
$$\tau_{\mathrm{rms}} = 100 \text{ ns}.$$

Exemplary three out of eight modes are given, representing the modulation schemes QPSK, 16-QAM, and 64-QAM, as well as the code rates 1/2, 9/16 (Hiperlan2), and 3/4. The simulation results show that for a low modulation scheme SC/FDE outperforms OFDM. This behavior changes for higher order modulation schemes such as 64-QAM. The underlying results for SC/FDE are based on an MMSE equalizer. It has been demonstrated in Ref. [1] that the performance gain due to MMSE equalization is about 2–3 dB for QPSK, while this gain is reduced to almost zero for the modulation scheme of 64-QAM. In comparison to this the simulations for OFDM are based on an equalizer, which in principle represents a ZF-equalizer (it has been mentioned in Section 16.3.1 already that the ZF criterion represents the optimum for OFDM). For low modulation schemes, the advantage of an MMSE equalizer for SC/FDE gets obvious, while it is not the case for higher modulation schemes. Nevertheless, the performance is comparable for the investigated concepts within 1 dB. These simulation results confirm that suitability for high data rate wireless communication is given for OFDM as well as for SC/FDE.

Nevertheless, the previous result is a general one and does not take into account specific features of the systems—for example, channel state information was not taken into acount, nor were interleaving strategies. Making use of these elements, OFDM improves its performance significantly. A comparable improvement is not achieved for SC/FDE in particular as the interleaving strategy fits well for OFDM but not in particular for SC/FDE. The interleaver for OFDM is defined by a two-step permutation. While the first permutation ensures that adjacent coded bits are mapped onto nonadjacent subcarriers, the second step ensures that adjacent coded bits are mapped alternately onto less and more significant bits of the constellation, preventing long runs of low reliability bits. The defined interleaver spans over one processed block and therefore its block size varies with the modulation scheme. This interleaving strategy represents a well-developed strategy for OFDM.

In comparison to this, the performance gain due to this specific interleaving strategy is diminishing for SC/FDE. As a better adapted interleaver strategy is not the topic of this work, it is only stated that a significantly enlarged size of the interleaver will lead to a better performance. A block interleaver of the size (128 × 32) leads to a performance gain of about 1 dB for QPSK, 2 dB for 16-QAM, and even 3 dB for 64-QAM for SC-FDE in comparison with OFDM. Of course it is to be mentioned that the effort for this introduced strategy in terms of necessary memory is high and additionally results in an unwanted time delay at the receiver for deinterleaving. Nevertheless, it is demonstrated that other concepts of interleaving are necessary for SC/FDE than for OFDM. Based on the above statements the performance in terms of BER is compared again, shown by Figure 16.21. Now OFDM outperforms SC/FDE by about 1 dB for QPSK and about 1.5 dB for 64-QAM, emphasizing the necessity of an adapted interleaving strategy for SC/FDE.

16.6 Conclusion and Outlook

The underlying investigations have demonstrated that both the concepts (OFDM and SC/FDE), which are based on the strategy of frequency domain equalization, are promising candidates for present and future broadband wireless communications. A comparison based on several different parameters (bandwidth efficiency, BER, RF-constraints) has been carried out for the two investigated concepts, leading to the statement of a comparable performance of the both. Of course it is to be mentioned that there are a variety of other aspects and parameters that have to be compared such as synchronization aspects, channel estimation, multiuser capabilities, or flexibility. Although the comparison based on specific and isolated elements will lead in general to advantages for the one or the other, the following points out the advantageous combination of OFDM and SC/FDE.

Referring to Figures 16.18 and 16.19 it is obvious that the two types of systems differ mainly in the placement of the IFFT. While OFDM uses the IFFT in the transmitter to multiplex the information onto subcarriers, the SC/FDE system uses the IFFT in the receiver to convert the equalized data back to the time domain. Although the overall signal processing complexity is

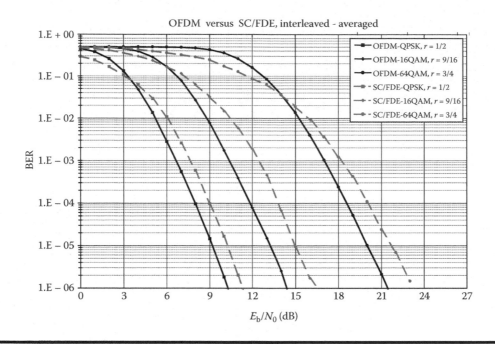

Figure 16.21 Performance comparison due to enhanced interleaving strategy for SC/FDE.

comparable for both systems, one can take advantage of the different placement of the IFFT by combining OFDM and SC/FDE. As depicted in Ref. [14] for the first time, and pointed out in Figure 16.22, a dual mode system, using OFDM for the downlink and SC/FDE for the uplink, would have significant advantages for the mobile station (subscriber end). These advantages can be described as follows:

■ It was stated that OFDM exhibits a high PAPR that forces the amplifier to be operated with a significant back-off, which decreases the power efficiency, while the PAPR depends only on the modulation scheme for SC/FDE. This leads to a higher efficiency in terms of power consumption due to the reduced PBO requirement, which is important for a long battery life in the mobile station.

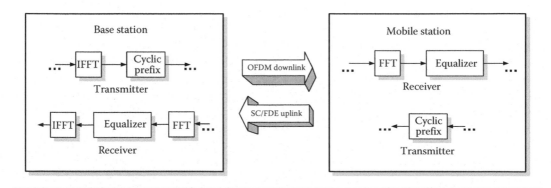

Figure 16.22 Combination of OFDM and SC/FDE.

■ The signal processing complexity is concentrated at the base station (three Fourier transformations at the base station versus one Fourier transformation at the mobile station).

Based on these basic statements it is concluded that this elementary idea was developed further and enhanced by the concepts of multiuser detection, multiantenna strategies or others, finally leading to the actual proposals and first implementations for the next generation mobile communication systems (LTE). An exemplary compendium of continuative literature is given in Refs. [27–34].

References

1. H. Witschnig, *Concepts of Frequency Domain Equalization—with Special Reference to Single Carrier Transmission*, VDM Verlag, Saarbrücken, ISBN: 978-3-8364-9285-0, 2008.
2. J.C. Proakis and D.G. Manolakis, *Digital Signal Processing*. 3. Edition, Prentice Hall, Upper Saddle River, NJ, 1996.
3. A. Papoulis, *The Fourier Integral and its applications*, McGraw-Hill Inc., New York, 1987.
4. A. Papoulis, *Signal Analysis*, McGraw-Hill Inc., New York, 1984.
5. A.V. Oppenheim and A.S. Willsky, *Signals and Systems*, Prentice-Hall Inc., Upper Saddle River, NJ, 1997.
6. H.K. Garg, *Digital Signal Processing Algorithms: Number Theory, Convolution, Fast Fourier Transforms, and Applications*, CRC Press, Boca Raton, FL, 1998.
7. V.K. Madisetti and D.B. Williams, *The Digital Signal Processing Handbook*, CRC Press, Boca Raton, FL, 1998.
8. H. Sari, G. Karam, and I. Jeanclaude, An analysis of orthogonal frequency-division multiplexing for mobile radio applications, *Proceedings of the IEEE Vehicular Technology Conference (VTC '94)*, Stockholm, Sweden, pp. 1635–1639, June 1994.
9. H. Sari, G. Karam, and I. Jeanclaude, Frequency-domain equalization of mobile radio and terrestrial broadcast channels, *Proceedings of the IEEE International Conference on Global Communications (GLOBE-COM '94)*, San Francisco, CA, pp. 1–5, 1994.
10. M. Huemer, *Frequenzbereichsentzerrung für hochratige Einträger-Übertragungs-systeme in Umgebungen mit ausgeprägter Mehrwegeausbreitung*, Dissertation, Institute for Communications and Information Engineering, University of Linz, Austria, 1999 (in German).
11. N. Benvenuto and S. Tomasin, On the comparison between OFDM and single carrier modulation with a DFE using frequency Domain feedforward filter, *IEEE Transactions on Communications*, 50(6), 947–955, 2002.
12. N. Benvenuto and S. Tomasin, Block iterative DFE for single carrier modulation, *IEEE Electronics Letters*, 38(19), 1144–1145, 2002.
13. N. Benvenuto and S. Tomasin, Frequency domain DFE: System design and comparison with OFDM, *Proceedings of the 8th Symposium on Communications and Vehicular Technology in the Benelux*, Delft, The Netherlands, October 2001.
14. D. Falconer, S.L. Ariyavisitakul, A. Benyamin-Seeyar, and B. Eidson, Frequency domain equalization for single-carrier broadband wireless systems, *IEEE Communications Magazine*, 40(4), 58–66, 2002.
15. H. Witschnig, T. Mayer, A. Springer, and R. Weigel, The advantages of a known sequence versus cyclic prefix for a SC/FDE system, *5th International Symposium on Wireless Personal Multimedia Communicationsm, (WPMC 2002)*, Honolulu HI, October 2002.
16. H. Witschnig, T. Mayer, A. Springer, L. Maurer, and M. Huemer, A different look on cyclic prefix for SC/FDE, *13th IEEE International Symposium on Personal, Indoor and Mobile Radio Communications*, 5 pages on CD, Lisbon, Portugal, September 2002.
17. H. Witschnig, T. Mayer, M. Petit, H. Hutzelmann, R. Weigel, and A. Springer, The advantages of a unique word for synchronisation and channel estimation in a SC/FDE system, *EPMCC03—European Personal Mobile Communications Conference*, Glasgow, Scotland, pp. 436–440, April 2003.

18. M. Huemer, H. Witschnig, and J. Hausner, Unique word based phase tracking algorithms for SC/FDE-systems, *GLOBECOM 2003*, San Francisco, CA, December 2003, 5 pages on CD.
19. H. Schmidt, OFDM für die drahtlose Datenübertragung innerhalb von Gebäuden, Dissertation, Fachbereich 1 (Physik/Elektrotechnik) der Universität Bremen, April 2001 (in German).
20. S.B. Weinstein, Data transmission by frequency-division multiplexing using the discrete Fourier transform, *IEEE Transactions on Communication Technologies*, 19(5), 628–634, 1971.
21. K. Fazel, A concept of digital terrestrial television broadcasting, *Wireless Personal Communications*, 2(1–2), 9–27, 1995.
22. R. Van Nee and R. Prasad, *OFDM for Wireless Multimedia Communications*, Artech House Publishers, Boston-London, 2000.
23. K. Fazel and S. Kaiser,(Ed.) *Multi-Carrier Spread-Spectrum & Related Topics*. Kluwer Academic Publishers, Norwell, MA, 2002.
24. M. Engels, *Wireless OFDM Systems*, Kluwer Academic Publishers, Boston, 2002.
25. IEEE Std 802.11a-1999: Part 11: Wireless LAN medium access control (MAC) and physical layer (PHY) specifications: High-speed physical layer in the 5 GHz band.
26. ETSI, TS 101 475: Broadband radio access networks (BRAN), HiperLAN Type 2, physical (PHY) layer.
27. H.G. Myung, J. Lim, and D.J. Goodman, Single carrier FDMA for uplink wireless transmission, *IEEE Vehicular Technology Magazine*, 1(3), 30–38, 2006.
28. B.E. Priyanto, H. Codina, S. Rene, T.B. Sorensen, and P. Mogensen, Initial performance evaluation of DFT-spread OFDM based SC-FDMA for UTRA LTE uplink, *IEEE Vehicular Technology Conference (VTC) 2007 Spring*, Dublin, Ireland, April 2007.
29. H. Ekstroem, A. Furuskär, J. Karlsson, M. Meyer, S. Parkvall, J. Torsner, and M. Wahlqvist, Technical solutions for the 3G long-term evolution, *IEEE Communications Magazine*, 44(3), 38–45, 2006.
30. 3rd Generation Partnership Project (3GPP): Requirements for evolved UTRA (E-UTRA) and evolved UTRAN (E-UTRAN), http://www.3gpp.org/ftp/Specs/html-info/25913.htm
31. 3rd Generation Partnership Project (3GPP): Technical Specification Group Radio Access Network; Physical Layer Aspects for Evolved UTRA, http://www.3gpp.org/ftp/Specs/html-info/25814.htm
32. U. Sorger, I. De Broeck, and M. Schnell, Interleaved FDMA—A new spread-spectrum multiple-access scheme, *Proc. IEEE ICC '98*, Atlanta, GA, June 1998, pp. 1013–1017.
33. H.G. Myung, J. Lim, and D.J. Goodman, Peak-to-average power ratio of single carrier FDMA signals with pulse shaping, *The 17th Annual IEEE International Symposium on Personal, Indoor and Mobile Radio Communications (PIMRC '06)*, Helsinki, Finland, September 2006.
34. J. Lim, H.G. Myung, K. Oh, and D.J. Goodman, Channel-dependent scheduling of uplink single Carrier FDMA systems, *IEEE Vehicular Technology Conference (VTC) 2006 Fall*, Montreal, Canada, September 2006.

Chapter 17

MIMO Beamforming Schemes for Multiuser Access in OFDM–SDMA

Ahmed Iyanda Sulyman and Mostafa Hefnawi

Contents

17.1   Introduction..454
17.2   OFDM–SDMA System Model...456
17.3   Capacity of OFDM–SDMA Systems..458
17.4   SER Performance of OFDM–SDMA Systems..............................458
    17.4.1   SINR Expression..459
    17.4.2   SER Expression...459
17.5   MIMO Beamforming Schemes in OFDM–SDMA..........................459
    17.5.1   MIMO–MRC System in OFDM–SDMA..........................459
        17.5.1.1   SER Analysis...460
    17.5.2   Capacity-Aware MIMO Beamforming System in OFDM–SDMA..........462
        17.5.2.1   SER Analysis...462
        17.5.2.2   OFDM–SDMA System Capacity with MIMO–MRC and
                  Capacity-Aware MIMO Beamforming Schemes..................464
    17.5.3   MIMO–MMSE Systems in OFDM–SDMA.........................465
        17.5.3.1   SER Performance......................................466
        17.5.3.2   Nonlinear MMSE Detectors...........................466
        17.5.3.3   Summary...467
17.6   Simulation Results..467
    17.6.1   Capacity Analysis...467

     17.6.2   SER Analysis..468
17.7 Conclusion...470
References..471

17.1 Introduction

A combined orthogonal frequency-division multiplexing/space-division multiple access (OFDM–SDMA) approach can effectively mitigate both the channel distortion due to multipath propagation and the low spectral efficiency due to the limited bandwidth that arise in the design of broadband wireless networks. OFDM mitigates the channel impairments by transforming a frequency-selective channel into a set of frequency-flat channels. SDMA on the other hand, helps to achieve higher spectral efficiency, by multiplexing multiple users on the same time–frequency resources [1–9]. In this chapter, we consider the uplink/downlink multiuser access in broadband wireless networks employing OFDM–SDMA access, where a multiple antenna-aided-base station (BS), receiver detects/transmits independent OFDM data streams from/to multiple SDMA users simultaneously on the same frequency band. We consider that access to the spectrum is slotted and controlled by the BS. We assume that L user terminals are assigned to each time slot, and are allowed to simultaneously transmit/receive streams of OFDM modulated symbols to/from the SDMA BS on the same frequency band. Thus, in a given time slot in OFDM–SDMA systems, multiple users share each OFDM subcarrier frequency simultaneously. Compared to the conventional access method where only one user is assigned a subcarrier, OFDM–SDMA access yields an L-fold system capacity increase as illustrated in Figure 17.1. Therefore, upcoming broadband wireless services like the WiMAX system will take advantage of the SDMA technology to further increase system capacity. However, for the SDMA technology to be effectively deployed in these applications, concerted efforts must be made to design appropriate beamforming schemes that allow the system to effectively separate the multiplexed user data without prejudice to the broadband nature of these applications. Therefore, a capacity-aware design method is highly necessary for the successful deployment of the SDMA technology in the upcoming broadband wireless services such as the WiMAX system.

Recently, several receive optimization schemes have been proposed for OFDM–SDMA systems and have been analyzed in terms of bit error rate (BER) performance improvements [1–4,10]. In Ref. [3], a minimum BER (MBER) detector that adaptively computes the optimal weights at the receiver was proposed. The scheme has better BER performance than the minimum mean square error (MMSE)-based algorithms presented in Ref. [4]. In Ref. [10] on the other hand, it was demonstrated that eigen-beamforming-based design can be used in an OFDM–SDMA system to provide improvement in system performance compared to the conventional systems. In Ref. [11], a new adaptive beamforming algorithm that targets the direct maximization of the OFDM–SDMA channel capacity, tagged here multiple input–multiple output (MIMO) capacity-aware as the beamforming scheme, was proposed. Using the steepest-ascent gradient of the OFDM–SDMA channel capacity, the algorithm adaptively seeks the transmitting weighting vectors that maximize the capacity of the OFDM–SDMA channel for each user, and thus enhance the overall system capacity. It is shown in Ref. [11] that the capacity of the multiantenna-aided OFDM–SDMA system whose transmitting weights have been computed using the proposed algorithm is significantly higher than that of the conventional systems such as eigen-beamforming in Ref. [10].

A number of works in the literature have also recently addressed the problem of joint design of transmit and receive weight matrices for parallel transmission of data over MIMO channels. These works are generally referred to as MIMO precoding systems [12]. In terms of symbol error rate

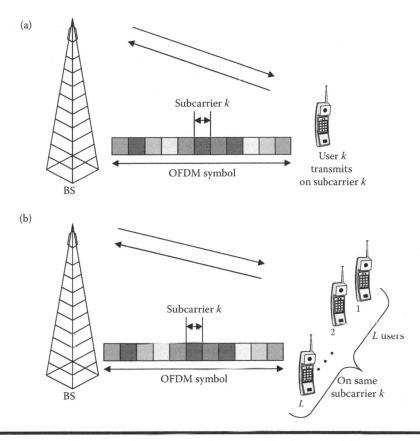

Figure 17.1   (a) Conventional access in BB wireless networks; (b) OFDM–SDMA access in broadband (BB) wireless networks.

(SER), and for the single-user case where the transmitter and receiver communicate in point-to-point mode, the MIMO precoding system is equivalent to multibeam MIMO beamforming as both jointly optimize the transmit and receive weights for optimum parallel data transmission through the channel that maximizes the SNR at the receiver. For the multiuser case, however, MIMO precoding systems do not traditionally mitigate multiuser interference. Every user in the system receives data transmitted to other users, and the desired user decodes its message using its receive weight matrix. There is no general solution for interference suppression in MIMO precoding systems [12]; however, a few works have recently considered adding such capability in the system. One such work can be found in Ref. [13] and is called successive MMSE precoding. For the multiuser MIMO beamforming system, however, multiuser access interference (MAI) suppression is an inherent criterion in the design of the transmit/receive weights. Thus, MAI is specifically mitigated in MIMO beamforming systems. In terms of capacity (or data rate per channel use), MIMO precoding has the same capacity as multibeam MIMO beamforming systems as both schemes allow parallel transmission of (possibly independent) data over the same time–frequency resources. The single-beam MIMO beamforming system, however, has lower capacity than MIMO precoding as the former allows only serial data transmission. Since single-beam beamforming and multibeam beamforming differ only in their implementations, the general term MIMO beamforming is commonly used for these schemes.

In this chapter, we focus on MIMO beamforming systems and present closed-form expressions for evaluating the SER performance of OFDM–SDMA, in the presence of the MAI encountered in OFDM–SDMA applications and correlated fading encountered in antenna array systems, when MIMO–MRC and the capacity-aware MIMO beamforming scheme are the underlying Tx/Rx beamforming weights optimization schemes. Using the derived expressions, we compare the SER performance of the capacity-aware adaptive MIMO beamforming scheme with the MIMO–MRC beamforming scheme in OFDM–SDMA applications. It is shown that in maximizing the capacity iteratively, the capacity-aware MIMO beamforming scheme also enhances the SINR, resulting in lower SER compared to conventional beamforming schemes. Then we present an overview of the MIMO–MMSE beamforming scheme in the OFDM–SDMA systems and discuss the prospects of linear and nonlinear MMSE detection methods in enhancing the SER and capacity of the system.

17.2 OFDM–SDMA System Model

Consider the uplink multiuser access in OFDM–SDMA systems, where a multiple-antenna-aided BS receiver detects independent OFDM data streams from multiple users simultaneously on the same time–frequency resources. The downlink multiuser access in OFDM–SDMA systems is similar to the uplink access; therefore, the transmitter/receiver (Tx/Rx) weight optimization schemes discussed here for OFDM–SDMA systems have applications both in the uplink and the downlink scenarios. Figure 17.2 displays the model for the uplink multiuser access in the OFDM–SDMA system. At any time instant, we consider that a BS receiver equipped with an N_r-element antenna array detects OFDM signals of L users, each of them equipped with an N_t-element transmit antenna array. The cascade of transmit beamforming weights of each user and their channel gains form a unique transfer function from each user's device to the BS antenna array, resulting in a unique spatial signature for each user. This can be exploited to effect the separation of the user data at the BS using appropriate multiuser detection weights [3,14]. Let $\mathbf{x}[k] = \{x[k]_1, x[k]_2, \ldots, x[k]_L\}$ denote the set

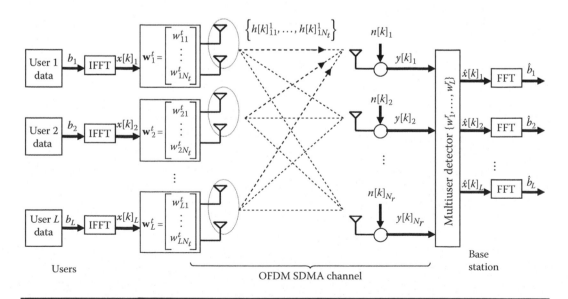

Figure 17.2　Multiple antenna aided OFDM–SDMA system.

of L user signals transmitted on each subcarrier, k, $k = 1, \ldots, N_c$, where N_c denotes the number of subcarriers per OFDM symbol in the system. The expression for the array output in Figure 17.2 can be written for each subcarrier as

$$\mathbf{y}[k] = \sum_{i=1}^{L} \mathbf{H}[k]_i \mathbf{w}[k]_i^t x[k]_i + \mathbf{n}[k], \tag{17.1}$$

where $\mathbf{y}[k] = [y[k]_1, y[k]_2, \ldots, y[k]_{N_r}]^T$ is the $N_r \times 1$ vector containing the outputs of the N_r-element array at the BS, with T denoting the transpose operation, and

$$\mathbf{H}[k]_i = \begin{bmatrix} h^i[k]_{11} & h^i[k]_{12} & \cdots & h^i[k]_{1N_t} \\ h^i[k]_{21} & h^i[k]_{22} & \cdots & h^i[k]_{2N_t} \\ \vdots & & & \vdots \\ h^i[k]_{N_r 1} & h^i[k]_{N_r 2} & \cdots & h^i[k]_{N_r N_t} \end{bmatrix}$$

is the $N_r \times N_t$ frequency-domain channel matrix representing the transfer functions from user i's N_t-element antenna array to the BS's N_r-element antenna array, on each subcarrier. Throughout this chapter, full-rate feedback of user channels is considered. Thus we assume that $\mathbf{H}[k]_i$ is perfectly known to the transmitter. Effects of limited feedback can be found in Ref. [15,16]. In most practical antenna array systems, the array elements are closely spaced such that the signals emanating from or arriving at the elements experience significant fading correlations. Therefore, we assume that $\mathbf{H}[k]_i$ consists of correlated elements and can be modeled as [17,18]

$$\mathbf{H}[k]_i = \mathbf{R}_r^{1/2} \mathbf{H}[k]_{w,i} \mathbf{R}_t^{1/2} \tag{17.2}$$

where $\mathbf{R}_r \in \mathcal{C}^{N_r \times N_r}$ and $\mathbf{R}_t \in \mathcal{C}^{N_t \times N_t}$ are positive definite Hermitian matrices specifying the receive and transmit fading correlations, respectively, and with eigenvalues $\beta_1 < \beta_2 < \cdots < \beta_{N_r}$, and $\omega_1 < \omega_2 < \cdots < \omega_{N_t}$ respectively. $\mathbf{H}[k]_{w,i} \in \mathcal{C}^{N_r \times N_t}$ consists of independent and identically distributed (iid) complex Gaussian entries, with zero mean and unit variance. $\mathbf{w}[k]_l^t = [w[k]_{l1}^t, w[k]_{l2}^t, \ldots, w[k]_{lN_t}^t]^T$ is the $N_t \times 1$ complex transmit weighting vector at the transmitter for user l, $l = 1, \ldots, L$, and $\mathbf{n}[k] = [n[k]_1, n[k]_2, \ldots, n[k]_{N_r}]^T$ is the $N_r \times 1$ complex additive white Gaussian noise (AWGN) vector at the receiver. The BS detects all L users simultaneously at the multiuser detection module of the OFDM–SDMA system, using the respective receiving weight vectors for each user l, $\mathbf{w}[k]_l^r = [w[k]_{l1}^r, w[k]_{l2}^r, \ldots, w[k]_{lN_r}^r]^T$, $l = 1, \ldots, L$. The detection of user l out of L users (with $L - 1$ interfering users) can thus be depicted as

$$\hat{x}[k]_l = \mathbf{w}[k]_l^{r^H} \mathbf{y}[k],$$

$$= \mathbf{w}[k]_l^{r^H} \mathbf{H}[k]_l \mathbf{w}[k]_l^t x[k]_l + \mathbf{w}[k]_l^{r^H} \sum_{i=1, i \neq l}^{L} \mathbf{H}[k]_i \mathbf{w}[k]_i^t x[k]_i + \mathbf{w}[k]_l^{r^H} \mathbf{n}[k],$$

$$= x[k]_l \underbrace{\mathbf{w}[k]_l^{r^H} \mathbf{H}[k]_l \mathbf{w}[k]_l^t + \mathbf{w}[k]_l^{r^H} \mathcal{I}[k]_l + \mathbf{w}[k]_l^{r^H} \mathbf{n}[k]}_{\text{OFDM–SDMA channel for user } l}, \tag{17.3}$$

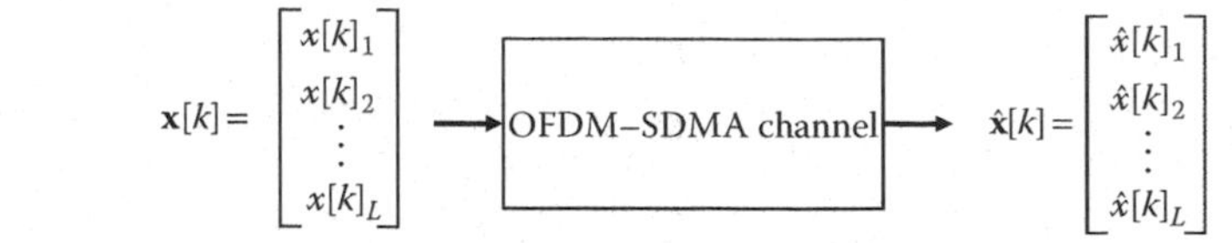

$$\mathbf{x}[k] = \begin{bmatrix} x[k]_1 \\ x[k]_2 \\ \vdots \\ x[k]_L \end{bmatrix} \longrightarrow \boxed{\text{OFDM–SDMA channel}} \longrightarrow \hat{\mathbf{x}}[k] = \begin{bmatrix} \hat{x}[k]_1 \\ \hat{x}[k]_2 \\ \vdots \\ \hat{x}[k]_L \end{bmatrix}$$

Figure 17.3   OFDM–SDMA channel model.

where H denotes the Hermitian transpose, and $\mathcal{I}[k]_l = \sum_{i=1, i \neq l}^{L} \mathbf{H}[k]_i \mathbf{w}[k]_i^t x[k]_i$ denotes the MAI for user l. Therefore, the OFDM–SDMA channel for all users can be represented in a compact form, for each subcarrier k, as shown in Figure 17.3, where the OFDM–SDMA channel represents the transfer function from input $x[k]_l$ to the output $\hat{x}[k]_l$ for the lth user data on subcarrier k, $l = 1, \ldots, L$.

17.3 Capacity of OFDM–SDMA Systems

Assuming $\mathbf{x}[k]$ is circularly symmetric complex Gaussian and equal power allocation to users, the Ergodic capacity of the OFDM–SDMA channel per subcarrier, for each user l, is given by [11,19,20]

$$\mathcal{C}\left(\mathbf{H}[k]_l, \mathbf{w}[k]_l^t\right) = E\left[\log_2\left\{1 + \frac{\rho_l}{N_t}|\mathbf{B}[k]_l^{-1/2}\mathbf{H}[k]_l\mathbf{w}[k]_l^t|^2\right\}\right], \quad (17.4)$$

where ρ_l is the total power transmitted by user l in a subcarrier, and $\mathbf{B}[k]_l$ is the covariance matrix of the interference-plus-noise, given by

$$\mathbf{B}[k]_l = \sum_{i=1, i \neq l}^{L} \mathbf{H}[k]_i \mathbf{w}[k]_i^t x[k]_i \left(x[k]_i \mathbf{w}[k]_i^t\right)^H \mathbf{H}[k]_i^H + \sigma_n^2 \mathbf{I}_{N_r}, \quad (17.5)$$

with σ_n^2 denoting the variance of the AWGN. Here we have assumed that for large L, the interference can be considered a Gaussian process [using the central limit theory (CLT)], and that the interference-plus-noise is accurately whitened by multiplying $\mathbf{y}[k]$ by $\mathbf{B}[k]_l^{-1/2}$, which happens when the user channels $\mathbf{H}[k]_l$, $l = 1, \ldots, L$, are known exactly. The capacity of the OFDM–SDMA channel for each user l (see Equation 17.3 and Figure 17.2) is therefore a function of $\mathbf{H}[k]_l$, $\mathbf{w}[k]_l^t$, and the interference-plus-noise, while the SER performance of the system is a function of the SINR, which depends on $\mathbf{w}[k]_l^r$, $\mathbf{H}[k]_l$, $\mathbf{w}[k]_l^t$, and the interference-plus-noise. Sulyman and Hefnawi [11] proposed an adaptive MIMO beamforming scheme that seeks to maximize the channel capacity for each user l iteratively, such that the overall system capacity is enhanced. The scheme is tagged here the capacity-aware MIMO beamforming scheme. In the following, we present an overview of the SER performance of the capacity-aware MIMO beamforming scheme in OFDM–SDMA applications, and compare its asymptotic behavior with that of MIMO–MRC systems, taking into account the prevailing MAI and fading correlations, for all the schemes.

17.4 SER Performance of OFDM–SDMA Systems

In this section, we present the error rate performance of Tx/Rx beamforming schemes suitable for OFDM–SDMA applications. First we write general expressions for the signal-to-intereference-plus-noise ratio (SINR) and SER performance of MIMO beamforming schemes in the presence of MAI experienced in OFDM–SDMA application, and correlated fading experienced in antenna array systems. Then we simplify the expressions for specific MIMO beamforming schemes examined in this work.

17.4.1 SINR Expression

Using the CLT [21], the instantaneous interference-plus-noise for user l in Equation 17.3 can be approximated as a zero-mean Gaussian random variable, with variance $\mathbf{w}[k]_l^{r^H}\mathbf{B}[k]_l\mathbf{w}[k]_l^r$. Therefore, the SINR for user l $\gamma[k]_l$ is given by

$$\gamma[k]_l = \frac{\|\mathbf{w}[k]_l^{r^H}\mathbf{H}[k]_l\mathbf{w}[k]_l^t x[k]_l\|^2}{(\mathbf{w}[k]_l^{r^H}\mathbf{B}[k]_l\mathbf{w}[k]_l^r)}. \tag{17.6}$$

17.4.2 SER Expression

The SER per subcarrier for MIMO beamforming schemes, in the presence of correlated fading and MAI in experienced OFDM–SDMA application, is derived next. Let $Pr(E, k)_l$ denote the probability of error associated with the kth subcarrier for user l, $l = 1, \ldots, L$. In this analysis, we consider modulation formats for which $Pr(E, k)_l$ can be expressed as [2]

$$Pr(E, k)_l = E_{\gamma[k]_l}\left[aQ\left(\sqrt{2b\gamma[k]_l}\right)\right], \tag{17.7}$$

where $E_{\gamma[k]_l}[\cdot]$ denotes the expectation with respect to the random variable $\gamma[k]_l$, the SINR on the kth subcarrier for user l, $Q(\cdot)$ denotes the Gaussian Q-function, and a and b are modulation-specific constants. For binary phase shift keying (BPSK), $a = 1$, $b = 1$, for binary frequency shift keying (BFSK) with orthogonal signaling $a = 1$, $b = 0.5$, while for M-ary phase shift keying (M-PSK) $a = 2$, $b = \sin^2(\pi/M)$. The average SER performance for user l, $l = 1, \ldots, L$, can be estimated as [23]

$$\text{SER}_l = \frac{1}{N_c}\sum_{k=0}^{N_c-1} Pr(E, k)_l. \tag{17.8}$$

Note from Equation 17.8 that if signals transmitted on all subcarriers employ the same modulation format, then $\text{SER}_l = Pr(E, k)_l$, $l = 1, \ldots, L$.

17.5 MIMO Beamforming Schemes in OFDM–SDMA

17.5.1 MIMO–MRC System in OFDM–SDMA

In the MIMO–MRC system, eigen beamforming is employed at the transmitter side, where the transmit beamforming vector for the lth user $\mathbf{w}[k]_l^t$ is chosen to be the eigen vector corresponding to the maximum eigen value of $\mathbf{H}[k]_l^H\mathbf{H}[k]_l$, on each subcarrier k. At the receiver side, the receive weights, $\mathbf{w}[k]_l^r$, that affect the separation of user l's data from the other interfering users, at the output of the antenna array, are computed using the MRC solution. Thus, the transmit and receive weights for the MIMO–MRC solution for each user l in the presence of OFDM–SDMA MAI are given by

$$\mathbf{w}[k]_l^t = \sqrt{P_l}\mathbf{u}_{\mathrm{max},l}, \tag{17.9}$$

$$\mathbf{w}[k]_l^r = g\mathbf{B}[k]_l^{-1}\mathbf{H}[k]_l\mathbf{w}[k]_l^t, \tag{17.10}$$

where $\mathbf{u}_{\mathrm{max},l}$ denotes the eigen vector corresponding to $\lambda_{\mathrm{max},l}$, the maximum eigen value of $\mathbf{H}[k]_l^H\mathbf{H}[k]_l$, and g is an arbitrary constant that does not alter the performance of the MIMO–MRC system. Without loss of generality, it can be assumed that $g = 1$ [11,24,25].

17.5.1.1 SER Analysis

For the MIMO–MRC system, the SINR expression above becomes

$$\gamma[k]_l = \frac{\| P_l \mathbf{u}_{\mathrm{max},l}^H \mathbf{H}[k]_l^H \mathbf{B}[k]_l^{-1^H} \mathbf{H}[k]_l \mathbf{u}_{\mathrm{max},l} \|^2}{\left(P_l \mathbf{u}_{\mathrm{max},l}^H \mathbf{H}[k]_l^H \mathbf{B}[k]_l^{-1^H} \mathbf{H}[k]_l \mathbf{u}_{\mathrm{max},l} \right)}. \tag{17.11}$$

We observe that in general, the off-diagonal elements of $\mathbf{B}[k]_l$ are nonzero, reflecting the color of the interference. However, in the asymptotic case of large L and given equal power transmitted by all users ($P_l = P$), the CLT can be invoked to show that [26]

$$\mathbf{B}[k]_l \rightarrow \left(\frac{(L-1)P}{\sigma_n^2} + 1 \right) \sigma_n^2 \mathbf{I}_{N_r}. \tag{17.12}$$

Thus, without loss of generality, we can express $\gamma[k]_l$ as

$$\gamma[k]_l = \frac{\| \left((L-1)P + \sigma_n^2 \right)^{-1} P \mathbf{u}_{\mathrm{max},l}^H \mathbf{H}[k]_l^H \mathbf{H}[k]_l \mathbf{u}_{\mathrm{max},l} \|^2}{\left(\left((L-1)P + \sigma_n^2 \right)^{-1} P \mathbf{u}_{\mathrm{max},l}^H \mathbf{H}[k]_l^H \mathbf{H}[k]_l \mathbf{u}_{\mathrm{max},l} \right)}.$$

$$= \frac{P}{(L-1)P + \sigma_n^2} \lambda_{\mathrm{max},l}. \tag{17.13}$$

Therefore, an alternative expression for Equation 17.7 can be written for the MIMO–MRC system, in the presence of fading correlation and MAI in OFDM–SDMA applications, as [27]

$$Pr(E, k)_l = \frac{a\sqrt{b}}{2\sqrt{\pi}} \int_0^{\infty} \frac{e^{-bu}}{\sqrt{u}} F_{\lambda_{\mathrm{max},l}} \left(\frac{(L-1)P + \sigma_n^2}{P} u \right) \, du \tag{17.14}$$

where $F_{\lambda_{\mathrm{max},l}}(u)$ is the cumulative distribution function (cdf) of $\lambda_{\mathrm{max},l}$. Note that Equation 17.14 gives the same average SER, $Pr(E, k)_l$, for all users ($l = 1, \ldots, L$). Therefore, in the sequel we use the notation $Pr(E, k)$ to denote the average SER for all users. For the case $N_t = 2$, $N_r \geq N_t$, for which a closed-form result is known for the SER of the MIMO–MRC system in correlated fading [27], a closed-form expression for $Pr(E, k)$ is given by

$$
Pr(E, k) = \frac{ab\left(P/[(L-1)P + \sigma_n^2]\right)\det(\mathbf{R_t})}{\Delta_2(\mathbf{R_t})\Delta_{N_r}(\mathbf{R_r})} \sum_{p=1}^{N_r} \sum_{t=1, t\neq p}^{N_r} (-1)^{p+\phi(t)} \left(\beta_p\beta_t\right)^{N_r-1} \Delta_{N_r-2}(\mathcal{A})
$$

$$
\times \left\{ \Psi\left(0, -\frac{[(L-1)P + \sigma_n^2]}{bP}\left(\frac{1}{\omega_2\beta_p} + \frac{1}{\omega_1\beta_t}\right)\right) \right.
$$

$$
- \sum_{l=0}^{N_r-1} \frac{1}{l!} \frac{\Psi\left(l, -[(L-1)P + \sigma_n^2]/bP\omega_2\beta_p\right)}{\left(-2b\left(P/[(L-1)P + \sigma_n^2]\right)\omega_1\beta_t\right)^l}
$$

$$
\left. - \sum_{l=0}^{N_r-1} \frac{1}{l!} \frac{\Psi\left(l, -[(L-1)P + \sigma_n^2]/bP\omega_1\beta_t\right)}{\left(-2b\left(P/[(L-1)P + \sigma_n^2]\right)\omega_2\beta_p\right)^l} \right\}, \tag{17.15}
$$

where

$$
\Psi(l, y) = (2l-3)!!(1-y)^{-(l-1/2)} - \sum_{k=0}^{2N_r-l-1} \left(\frac{y}{2}\right)^k \frac{(2(k+l)-3)!!}{k!}.
$$

For the case when one user is assigned per subcarrier ($L = 1$), the channel becomes noise limited and it is easily verified that Equation 17.15 reduces to its equivalent form in Equation A.1 in Ref. [24].

Asymptotic SER analysis: At high SNR, Equation 17.15 can be expressed asymptotically as [27]

$$
Pr(E, k)^\infty = \frac{a(4N_r - 1)!!}{b^{2N_r}2^{2N_r+1}N_r!(N_r + 1)!\det(\mathbf{R_t})^{N_r}\det(\mathbf{R_r})^2} \left(\frac{P}{\sigma_n^2}\right)^{-2N_r} \cdot \left((L-1)\frac{P}{\sigma_n^2} + 1\right)^{2N_r}. \tag{17.16}
$$

Therefore, compared to the single user MIMO–MRC system (case $L = 1$), MAI in the OFDM–SDMA system employing MIMO–MRC contributes asymptotically on SER degradation of $\left((L-1)(P/\sigma_n^2) + 1\right)^{2N_r} \equiv \left((L-1)(P/\sigma_n^2) + 1\right)^{N_t N_r}$ in each OFDM subcarrier, for each user. This represents an error floor in the SER plot for the desired user since increasing the desired user's SNR would not reduce this factor. This level of interference would easily dominate the SER performance for the desired user except when some additional mitigation techniques are employed. Thus, even though unique spatial signatures of each user, resulting from the combinations of $\mathbf{w}[k]_l^r$, $\mathbf{H}[k]_l$, and $\mathbf{w}[k]_l^t$, ensure that user data can be separated at the BS, the resulting error rate would be high except when additional pre/post-processing techniques are employed to mitigate the interference and reduce the resulting error floor. In Ref. [2], a subcarrier-hopping approach was proposed to reduce the proportion of time that users actually share the subcarriers. The error rate performance of the system was studied by simulation and a significant reduction in the error floor was reported. Equations 17.9 through 17.16 here provide a useful analytical framework for further works in this area.

17.5.2 Capacity-Aware MIMO Beamforming System in OFDM–SDMA

In the capacity-aware MIMO beamforming system, the transmitting weights are obtained using an iterative gradient-ascent search of the OFDM–SDMA channel capacity for each user, as [11]

$$\mathbf{w}(n+1)_l^t = \mathbf{w}(n)_l^t + \mu \nabla_{\mathbf{w}_l^t} \mathcal{C}\left(\mathbf{H}(n)_l, \mathbf{w}(n)_l^t\right), \tag{17.17}$$

where $\nabla_{\mathbf{w}_l^t} \mathcal{C}\left(\mathbf{H}(n)_l, \mathbf{w}(n)_l^t\right)$ is the gradient of $\mathcal{C}\left(\mathbf{H}(n)_l, \mathbf{w}(n)_l^t\right)$, the OFDM–SDMA channel capacity for user l, with respect to $\mathbf{w}(n)_l^t$, while μ is an adaptation constant (also known as the convergence factor). In practice, μ is typically chosen very small to ensure convergence of the adaptive algorithm [14,29]. Note that SDMA is subcarrier parallel [3,4] and that the update is done separately on each subcarrier. For brevity, therefore, we drop the frequency index $[k]$ and concentrate on the iteration index (n) in this recursion. For spatially white channel, the recursion can be expressed asymptotically as [11]

$$\mathbf{w}(n+1)_l^t = \mathbf{w}(n)_l^t + \mu \frac{1}{\ln 2}\left[\frac{\rho_l \mathbf{B}(n)_l^{-1} \mathbf{w}(n)_l^t}{\left(1 + \rho_l \mathbf{w}(n)_l^{t^H} \mathbf{B}(n)_l^{-1} \mathbf{w}(n)_l^t\right)}\right], \tag{17.18}$$

where ρ_l, $l = 1, \ldots, L$, is the total power transmitted by the lth SDMA user in an OFDM subcarrier, and $\mathbf{B}(n)_l^{-1}$ denotes $\mathbf{B}[k]_l^{-1}$ at iteration index n, with spatially white user channels. The computation of $\mathbf{B}(n)_l$ for each user requires knowledge of every other user's channels; thus we assume that the BS (which typically has all users' channel information) computes these transmit weight updates iteratively for all the users and exchanges the information with the users over the control channels. Users can initialize their transmit weights to the conventional schemes such as eigen-beamforming at the onset, and update their weights for higher system capacity adaptively as they receive updates using the proposed algorithm, from the BS. The update of the weights is continued until convergence of the algorithm is reached, after which the transmitting and receiving weights employed on the kth subcarrier for user l, $l = 1, \ldots, L$, is obtained as

$$\mathbf{w}[k]_l^t = \lim_{n \to \infty} \mathbf{w}(n+1)_l^t, \tag{17.19}$$

$$\mathbf{w}[k]_l^r = \mathbf{B}[k]_l^{-1} \mathbf{H}[k]_l \mathbf{w}[k]_l^t, \tag{17.20}$$

17.5.2.1 SER Analysis

Here, we examine the SER performance of the system, in comparison with conventional beamforming schemes. For example to quantify the enhancement provided by the system over MIMO–MRC, we initialize the weight updating equation to the MIMO–MRC solution. Thus, initializing Equation 17.18 to the eigen-beamforming weights, $\mathbf{w}(n)_{l,\text{EigBF}}^t$, at iteration index $n = 0$ and substituting the resulting expression into Equation 17.6, we can express Equation 17.6 at iteration index $n = 1$ as [24]

$$\gamma[k]_l = \frac{\left\{\|\mathbf{w}(n)_l^{r^H} \mathbf{H}(n)_l \mathbf{w}(n)_{l,\text{EigBF}}^t x(n)_l\|^2 + \alpha\right\}}{(\mathbf{w}(n)_l^{r^H} \mathbf{B}(n)_l \mathbf{w}(n)_l^r)}, \tag{17.21}$$

where

$$\alpha = \kappa^2\mu^2 \|\mathbf{w}(n)_l^{r^H}\mathbf{H}(n)_l\mathbf{B}(n)_l^{-1}\mathbf{w}(n)_{l,\text{EigBF}}^t x(n)_l\|^2$$

$$+ \kappa\mu \left(\mathbf{w}(n)_l^{r^H}\mathbf{H}(n)_l\mathbf{B}(n)_l^{-1}\mathbf{w}(n)_{l,\text{EigBF}}^t x(n)_l\left(x(n)_l\mathbf{w}(n)_{l,\text{EigBF}}^t\right)^H\mathbf{H}(n)_l^H\mathbf{w}(n)_l^r\right)$$

$$+ \kappa\mu \left(\mathbf{w}(n)_l^{r^H}\mathbf{H}(n)_l\mathbf{w}(n)_{l,\text{EigBF}}^t x(n)_l\left(x(n)_l\mathbf{w}(n)_{l,\text{EigBF}}^t\right)^H\mathbf{B}(n)_l^{-1^H}\mathbf{H}(n)_l^H\mathbf{w}(n)_l^r\right), \quad (17.22)$$

and $\kappa = (1/\ln2)\rho_l/\left(1 + \rho_l\mathbf{w}(n)_{l,\text{EigBF}}^{t^H}\mathbf{B}(n)_l^{-1}\mathbf{w}(n)_{l,\text{EigBF}}^t\right) = (1/\ln2)\rho_l/\left(1 + \rho_l(P/[(L-1)P +\sigma_n^2])\right)$, where we have used Equations 17.9 and 17.12 in writing the final expression for κ. Also using Equation 17.12, we can express Equation 17.22 as

$$\alpha = \left\{\frac{\kappa^2\mu^2}{\left[(L-1)P + \sigma_n^2\right]^2} + \frac{2\kappa\mu}{\left[(L-1)P + \sigma_n^2\right]}\right\} \|\mathbf{w}(n)_l^{r^H}\mathbf{H}(n)_l\mathbf{w}(n)_{l,\text{EigBF}}^t x(n)_l\|^2. \quad (17.23)$$

Using Equation 17.23 in Equation 17.21, we can write an expression for the SINR for the capacity-aware MIMO beamforming scheme, analogous to Equation 17.13 for the MIMO–MRC system, as

$$\gamma[k]_l = \left\{\frac{P}{\left[(L-1)P + \sigma_n^2\right]} + \frac{\kappa^2\mu^2 P}{\left[(L-1)P + \sigma_n^2\right]^3} + \frac{2\kappa\mu P}{\left[(L-1)P + \sigma_n^2\right]^2}\right\} \lambda_{\text{max},l}. \quad (17.24)$$

Therefore, analogous to Equation 17.14 for the MIMO–MRC system, we can derive the per subcarrier SER performance of the scheme as

$$Pr(E,k)_l$$

$$= \frac{a\sqrt{b}}{2\sqrt{\pi}} \int_0^\infty \frac{e^{-bu}}{\sqrt{u}} F_{\lambda_{\text{max},l}} \left(\frac{\left[(L-1)P + \sigma_n^2\right]^3 u}{\left\{\left[(L-1)P + \sigma_n^2\right]^2 P + \kappa^2\mu^2 P + 2\kappa\mu P\left[(L-1)P + \sigma_n^2\right]\right\}}\right) du,$$

$$(17.25)$$

which has a closed-form solution given by [35,36]

$$Pr(E,k) = \frac{ab\left(\left\{\left[(L-1)P + \sigma_n^2\right]^2 P + \kappa^2\mu^2 P + 2\kappa\mu P\left[(L-1)P + \sigma_n^2\right]\right\}/\left[(L-1)P + \sigma_n^2\right]^3\right)\det(\mathbf{R}_t)}{\Delta_2(\mathbf{R}_t)\Delta_{N_t}(\mathbf{R}_t)}$$

$$\times \sum_{p=1}^{N_t}\sum_{t=1,t\neq p}^{N_t} (-1)^{p+\phi(t)}\left(\beta_p\beta_t\right)^{N_t-1}\Delta_{N_t-2}(\mathcal{A})$$

$$\times \left\{\Psi\left(0, -\frac{\left[(L-1)P + \sigma_n^2\right]^3}{b\left\{\left[(L-1)P + \sigma_n^2\right]^2 P + \kappa^2\mu^2 P + 2\kappa\mu P\left[(L-1)P + \sigma_n^2\right]\right\}}\left(\frac{1}{\omega_2\beta_p} + \frac{1}{\omega_1\beta_t}\right)\right)\right.$$

$$-\sum_{l=0}^{N_r-1}\frac{1}{l!}\frac{\Psi\left(l,-\left[(L-1)P+\sigma_n^2\right]^3\Big/b\left\{\left[(L-1)P+\sigma_n^2\right]^2P+\kappa^2\mu^2P+2\kappa\mu P\left[(L-1)P+\sigma_n^2\right]\right\}\omega_2\beta_p\right)}{\left(-2b\left(\left\{\left[(L-1)P+\sigma_n^2\right]^2P+\kappa^2\mu^2P+2\kappa\mu P\left[(L-1)P+\sigma_n^2\right]\right\}\Big/\left[(L-1)P+\sigma_n^2\right]^3\right)\omega_1\beta_t\right)^l}$$

$$\left.-\sum_{l=0}^{N_r-1}\frac{1}{l!}\frac{\Psi\left(l,-\left[(L-1)P+\sigma_n^2\right]^3\Big/b\left\{\left[(L-1)P+\sigma_n^2\right]^2P+\kappa^2\mu^2P+2\kappa\mu P\left[(L-1)P+\sigma_n^2\right]\right\}\omega_1\beta_t\right)}{\left(-2b\left(\left\{\left[(L-1)P+\sigma_n^2\right]^2P+\kappa^2\mu^2P+2\kappa\mu P\left[(L-1)P+\sigma_n^2\right]\right\}\Big/\left[(L-1)P+\sigma_n^2\right]^3\right)\omega_2\beta_p\right)^l}\right\}.$$

$$(17.26)$$

Asymptotic SER analysis: In high SNR, Equation 17.26 can be expressed asymptotically as [27]

$$Pr(E,k)^\infty=\frac{a(4N_r-1)!!\left(P/\sigma_n^2\right)^{-2N_r}}{b^{2N_r}2^{2N_r+1}N_r!(N_r+1)!\det(\mathbf{R}_t)^{N_r}\det(\mathbf{R}_r)^2}\cdot\left((L-1)\frac{P}{\sigma_n^2}+1\right)^{2N_r}$$

$$\cdot\left(\frac{\left[(L-1)P+\sigma_n^2\right]^2}{\left\{\left[(L-1)P+\sigma_n^2\right]^2+\kappa^2\mu^2+2\kappa\mu\left[(L-1)P+\sigma_n^2\right]\right\}}\right)^{2N_r}$$

$$(17.27)$$

Comparing Equation 17.27 and 17.16, we observe that the additional factor in the SER expression of the capacity-aware adaptive MIMO beamforming scheme compared to MIMO–MRC when both are employed in OFDM–SDMA application is $([(L-1)P+\sigma_n^2]^2/\{[(L-1)P+\sigma_n^2]^2+\kappa^2\mu^2+2\kappa\mu[(L-1)P+\sigma_n^2]\})^{2N_r}\le1$, since $\mu>0$ and $\kappa>0$. Therefore the capacity-aware adaptive MIMO beamforming scheme also enhances the SINR (see Equation 17.13 and 17.24), and consequently it improves the SER performance of OFDM–SDMA systems compared to MIMO–MRC systems, albeit at an additional complexity increase as is apparent from Equation 17.18.

17.5.2.2 OFDM–SDMA System Capacity with MIMO–MRC and Capacity-Aware MIMO Beamforming Schemes

Here we compare the capacity of the OFDM–SDMA system when the capacity-aware MIMO beamforming scheme is employed with the case when MIMO–MRC is the underlying MIMO beamforming scheme. For example, initializing Equation 17.18 to the eigen-beamforming weights, $\mathbf{w}(n)_{l,\text{EigBF}}^t$, at iteration index $n=0$ and substituting the resulting expression into Equation 17.4, we can express Equation 17.4 at iteration index $n=1$ as

$$\mathcal{C}\left(\mathbf{H}(n)_l,\mathbf{w}(n)_l^t\right)=E\left[\log_2\left\{1+\frac{\rho_l}{N_t}\mid\mathbf{B}(n)_l^{-1/2}\mathbf{H}(n)_l\mathbf{w}(n)_{l,\text{EigBF}}^t\mid^2+\gamma\right\}\right],\qquad(17.28)$$

where

$$\gamma=\frac{\rho_l\mu}{N_t}\left(\nabla^{\text{H}}\mathbf{H}(n)_l^{\text{H}}\mathbf{B}(n)_l^{-1/2^{\text{H}}}\mathbf{B}(n)_l^{-1/2}\mathbf{H}(n)_l\mathbf{w}(n)_{l,\text{EigBF}}^t\right)$$

$$+\frac{\rho_l\mu}{N_t}\left(\mathbf{w}(n)_{l,\text{EigBF}}^{t^{\text{H}}}\mathbf{H}(n)_l^{\text{H}}\mathbf{B}(n)_l^{-1/2^{\text{H}}}\mathbf{B}(n)_l^{-1/2}\mathbf{H}(n)_l\nabla\right)+\frac{\rho_l\mu^2}{N_t}\left|\mathbf{B}(n)_l^{-1/2}\mathbf{H}(n)_l\nabla\right|^2,\text{ and}$$

$$\nabla=\alpha\mathbf{B}(n)_l^{-1}\mathbf{w}(n)_{l,\text{EigBF}}^t;\alpha=(1/\ln2)\,\rho_l\Big/\left(1+\rho_l\mathbf{w}(n)_{l,\text{EigBF}}^{t^{\text{H}}}\mathbf{B}(n)_l^{-1/2}\mathbf{B}(n)_l^{-1/2^{\text{H}}}\mathbf{w}(n)_{l,\text{EigBF}}^t\right).$$

Note that, since $\mu > 0$, and $\mathbf{u}^{\mathrm{H}}\mathbf{A}^{\mathrm{H}}\mathbf{A}\mathbf{u} = \| \mathbf{A}\mathbf{u} \|^2 \geq 0$, $\forall \mathbf{A}, \mathbf{u}$, we have $\gamma \geq 0$ [11]. The OFDM–SDMA channel capacity for user l is given by $\mathcal{C}_{l,\mathrm{OFDM-SDMA}} = 1/N_c \sum_{k=0}^{N_c-1} \mathcal{C}\left(\mathbf{H}[k]_l, \mathbf{w}[k]_l^t\right)$, and the overall system capacity for the OFDM–SDMA system can be obtained as [30] $\mathcal{C}_{\mathrm{OFDM-SDMA}} = \sum_{l=1}^{L} \mathcal{C}_{l,\mathrm{OFDM-SDMA}}$. Therefore, the capacity of the OFDM–SDMA system with the transmit weights obtained using the capacity-aware adaptive MIMO beamforming algorithm is greater than that obtained with the conventional schemes such as eigen-beamforming.

17.5.3 MIMO–MMSE Systems in OFDM–SDMA

Note that the signal model described by Equation 17.1 can be rewritten as [31]

$$\mathbf{y}[k] = \mathbf{H}[k]\mathbf{x}[k] + \mathbf{n}[k], \tag{17.29}$$

where $\mathbf{H}[k] = \left[\mathbf{H}[k]_1 \mathbf{w}[k]_1^t \ \mathbf{H}[k]_2 \mathbf{w}[k]_2^t \ \ldots \ \mathbf{H}[k]_L \mathbf{w}[k]_L^t\right]$ can be termed the total channel matrix, for all users, and $\mathbf{x}[k] = \left[x[k]_1 \ x[k]_2 \cdots x[k]_L\right]^{\mathrm{T}}$ is the vector of the transmitted user data.

In the MIMO–MMSE system, eigen-beamforming is used at the transmitter side, where the transmit beamforming vector for the lth user $\mathbf{w}[k]_l^t$ is chosen to be the eigen vector corresponding to the maximum eigen value of $\mathbf{H}[k]_l^{\mathrm{H}}\mathbf{H}[k]_l$, on each subcarrier k. At the receiver side, the system separates the signals of the simultaneous users sharing a subcarrier, using the MMSE filtering. In the MMSE solution, a linear filter with coefficient $\mathbf{w}[k]_l^{r,\mathrm{mmse}} = \left[w[k]_{l1}^{r,\mathrm{mmse}} \ w[k]_{l2}^{r,\mathrm{mmse}} \cdots w[k]_{lN_r}^{r,\mathrm{mmse}}\right]^{\mathrm{T}}$ combines the received signals $\mathbf{y}[k]$ to obtain soft estimates $\hat{x}[k]_l$ for the transmitted symbols $x[k]_l$ for the lth user, $l = 1, \ldots, L$.

$$\hat{\mathbf{x}}[k] = \mathbf{W}[k]_{\mathrm{MMSE}}^{\mathrm{H}}\mathbf{y}[k], \tag{17.30}$$

where $\hat{\mathbf{x}}[k] = [\hat{x}[k]_1 \ \hat{x}[k]_2 \ \cdots \ \hat{x}[k]_L]^{\mathrm{T}}$, and $\mathbf{W}[k]_{\mathrm{MMSE}} = [\mathbf{w}[k]_1^{r,\mathrm{mmse}}, \ \mathbf{w}[k]_2^{r,\mathrm{mmse}}, \ldots, \mathbf{w}[k]_L^{r,\mathrm{mmse}}]$. Unlike the MRC and the capacity-aware design methods, the receive weights (filter coefficients) in the MMSE solution, $\mathbf{w}[k]_l^{r,\mathrm{mmse}}$, are chosen as the unique vector minimizing the mean squared error (MSE) for each user l, expressed as

$$\begin{aligned} \mathrm{MSE}_l &= E\left[(x[k]_l - \hat{x}[k]_l)^2\right], \\ &= E\left[(x[k]_l - \hat{x}[k]_l)^*(x[k]_l - \hat{x}[k]_l)\right], \ l = 1, \ldots, L. \end{aligned} \tag{17.31}$$

Thus, MMSE combiner minimizes the expected variance of the error on the combined signal, reducing noise amplifications. MSE therefore makes a prudent tradeoff between multiuser interference mitigation and noise amplifications.

For a given noise power σ_n^2, and with $E\left[x[k]_l^* x[k]_l\right] = 1$, $l = 1, \ldots, L$, $\mathbf{W}[k]_{\mathrm{MMSE}}$ has to obey the L sets of linear equations

$$\left[\mathbf{H}[k]\mathbf{H}[k]^{\mathrm{H}} + \sigma^2 \mathbf{I}\right]\mathbf{W}[k]_{\mathrm{MMSE}} - \mathbf{H}[k] = 0,$$

$$\mathbf{W}[k]_{\mathrm{MMSE}} = \left[\mathbf{H}[k]\mathbf{H}[k]^{\mathrm{H}} + \sigma^2 \mathbf{I}\right]^{-1}\mathbf{H}[k]. \tag{17.32}$$

Using the MMSE weights computed by Equation 17.32 in Equation 17.30 yields the estimates of the transmitted data at the output of the linear MMSE detector, for the L users sharing each OFDM subcarrier.

17.5.3.1 SER Performance

The SINR for each user l can be estimated by substituting the respective MMSE weight, $\mathbf{w}[k]_l^{r,\mathrm{mmse}}$, into Equation 17.6, and the resulting expression used in the SER equation in Equation 17.8. For example for the downlink scenario, an alternate expression for the MMSE weight, $\mathbf{w}[k]_l^{r,\mathrm{mmse}}$, can be written for each user l as [32]

$$\mathbf{w}[k]_l^{r,\mathrm{mmse}} = \left[\sum_{i=1,i\neq l}^{L} \mathbf{H}[k]_l \mathbf{w}[k]_i^t \mathbf{w}[k]_i^{t^{\mathrm{H}}} \mathbf{H}[k]_l^{\mathrm{H}} + \sigma_n^2 \mathbf{I}_{N_r} \right]^{-1} \mathbf{H}[k]_l \mathbf{w}[k]_l^t. \tag{17.33}$$

If user channel coefficients are fairly similar in strength, this can be approximated as

$$\mathbf{w}[k]_l^{r,\mathrm{mmse}} \approx \mathbf{B}[k]_l^{-1} \mathbf{H}[k]_l \mathbf{w}[k]_l^t, \tag{17.34}$$

which is equivalent to Equation 17.10. Thus, the linear MMSE solution gives similar receive weights as the MRC system. Therefore, the SER performance of linear MIMO–MMSE detection is similar to the performance of MIMO–MRC in the presence of OFDM–SDMA MAI, presented earlier in this chapter.

17.5.3.2 Nonlinear MMSE Detectors

When some users, channel state information (CSI) is in deep fade, the average SER for the users concerned will be very high (see Equation 17.8 for average SER for each user). In such scenarios, linear MMSE filtering becomes less efficient because all users are detected simultaneously (see Equations 17.32 and 17.31) and thus the detection of the user with high SER cannot be isolated and treated in a special way that may enhance the user's SER performance. Another disadvantage of the linear MMSE detection is that its performance is heavily influenced by the number of users sharing a subcarrier. Therefore the use of nonlinear detection methods would be the focus of future research activities related to applications of MMSE in the OFDM–SDMA system. Nonlinear MMSE detection methods allow users to be detected successively one after another, offering the system the opportunity to isolate users with weak CSI and treat their detection in a way that enhances the SER. Inspired by its successful application in DS–CDMA systems [4,33,34] and the resulting enhancements in the error rate performance of the system, the application of this method in the OFDM–SDMA system is expected to significantly enhance the SER performance of OFDM–SDMA systems, albeit at an additional complexity increase. One such nonlinear detector is the ordered successive interference cancelation MMSE (OSIC-MMSE) multiuser detector, which orders and detects the "best" user's substream at the output of the MMSE linear decoder, reconstructs and subtracts its interference from the received signal and continues with the remaining user signals in the same way, successively. The "best" user substream can be selected using several criteria such as the MSE criterion, SNR/SINR, or the recently proposed capacity-aware criterion. The traditional criterion for ordering user substreams and computing the sequence for MMSE detection in nonlinear MMSE detectors are CSI, MSE, and SNR/SINR. Recently, the use of system capacity as a criterion for Tx/Rx weight designs in broadband wireless networks was proposed by Sulyman and Hefnawi

Table 17.1 Comparison between Different MIMO Beamforming (BF) Schemes for OFDM–SDMA Access in Broadband (BB) Wireless Networks

BF Schemes	Tx/Rx Weights	Capacity	SER Performance
MIMO–MRC	Tx: eigen BF Rx: MRC	Eigen BF capacity	Degrades by $\left((L-1)\frac{P}{\sigma_n^2}+1\right)^{N_tN_r}$, compared to 1-user MIMO–MRC
Capacity-aware	Tx: adaptively updated eigen BF Rx: MRC	Adaptively enhances eigen BF capacity	Enhancement of $\left(\dfrac{[(L-1)P+\sigma_n^2]^2}{\left\{[(L-1)P+\sigma_n^2]^2+\kappa^2\mu^2+2\kappa\mu[(L-1)P+\sigma_n^2]\right\}}\right)^{N_tN_r}$ compared to MIMO–MRC
MIMO–MMSE	Tx: eigen BF Rx: MMSE	Eigen BF capacity	*Linear:* similar to MIMO–MRC *Nonlinear:* outperforms MIMO–MRC

[11,24]. Motivation for this consideration is the fact that other criteria do not directly maximize the capacity, and for broadband transmission over wireless networks, the system needs design methods that would not prejudice the broadband nature of the underlying communications. Therefore, for MMSE design methods to be employed in broadband networks, it is highly necessary that the adopted MMSE scheme be capacity aware. Thus the use of capacity as a detection-sequence selection criterion in nonlinear MMSE schemes in OFDM–SDMA systems would be very attractive in future broadband wireless networks.

17.5.3.3 Summary

Table 17.1 presents a summary of comparisons among the various MIMO beamforming schemes examined in this section. In this table, we highlight the relative enhancements that each of the schemes provides, in terms of capacity and SER performance.

17.6 Simulation Results

17.6.1 Capacity Analysis

In our simulation setup we consider two OFDM–SDMA systems: one with four SDMA users per subchannel, $L = 4$, (for MIMO configuration $N_t = N_r = 4$), and the other with eight SDMA users per subchannel, $L = 8$ (for MIMO configuration $N_t = N_r = 8$). For OFDM configurations, we assume the 256-OFDM system widely deployed in broadband wireless access services, such as the WiMAX system. We generate the channel matrix $\mathbf{H}[k]_l$, $k = 1, \ldots, 256$, and compute the channel capacity per subcarrier for user l using Equation 17.4, with $\mathbf{w}[k]_l^t$ given by Equation 17.18 (initialized to the eigen-beamforming, similar to the approach in Ref. [3], and ran for 50 iterations with $\mu = 0.001$). The OFDM–SDMA channel capacity for user l is calculated as: $\mathcal{C}_{l,\text{OFDM–SDMA}} = 1/N_c \sum_{k=0}^{N_c-1} \mathcal{C}\left(\mathbf{H}[k]_l, \mathbf{w}[k]_l^t\right)$, $N_c = 256$, and the overall system capacity obtained as [29]: $\mathcal{C}_{\text{OFDM–SDMA}} = \sum_{l=1}^{L} \mathcal{C}_{l,\text{OFDM–SDMA}}$. Figure 17.4 displays the system capacity for the case when $\mathbf{w}[k]_l^t$ is obtained using the capacity-aware adaptive MIMO beamforming algorithm (with the algorithm parameter $\mu = 0.001$), as well as the eigen-beamforming-based design [10]. It is observed from the results that the adaptive algorithm achieves substantially higher

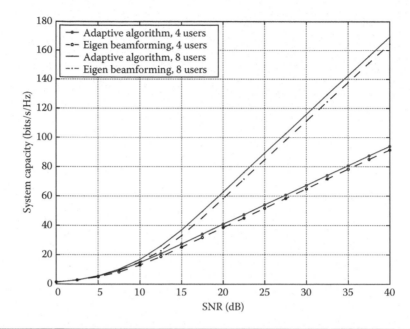

Figure 17.4 Capacity of OFDM–SDMA system for different numbers of SDMA users per OFDM subcarriers, case of 256-OFDM and $\mu = 0.001$.

system capacity compared to the eigen-beamforming design. For example for both cases shown in Figure 17.4, at high SNR ($\geq 15\,$dB), at least 3 bits/s/Hz enhancement in system capacity is observed with the proposed algorithm. For broadband wireless services like the WiMAX system with channel size up to 40 MHz, this translates to 120 Mbps capacity enhancement per channel in each cell, which can be very significant in a cellular deployment. This is due to the fact that the adaptive algorithm directly maximizes the channel capacity for each user iteratively. Consequently, it enhances the overall system capacity. In Figure 17.5, we display a similar result as in Figure 17.4, but with the algorithm parameter $\mu = 0.01$ for the adaptive scheme. It is observed from this figure that the capacity enhancements provided by the adaptive scheme over the eigen-beamforming-based design increases slightly, as the parameter μ (the step-size adaptation constant) increases.

17.6.2 SER Analysis

In our simulation setup we consider two OFDM–SDMA systems: (1) $N_t = N_r = 2$, with $L = 1, 2$, and 4, and (2) $N_t = 2, N_r = 4$, with $L = 1, 2$, and 4. We used $N_c = 256$ in the simulation; however, we assumed the same modulation format for all subcarriers. For fading correlation, we used

$$\mathbf{R}_t = \mathbf{R}_r = \begin{bmatrix} 1 & \rho \\ \rho & 1 \end{bmatrix},$$

for the 2×2 system, and

$$\mathbf{R}_t = \begin{bmatrix} 1 & \rho \\ \rho & 1 \end{bmatrix}, \mathbf{R}_r = \begin{bmatrix} 1 & \rho_1 & \rho_2 & \rho_2 \\ \rho_1 & 1 & \rho_1 & \rho_2 \\ \rho_2 & \rho_1 & 1 & \rho_1 \\ \rho_2 & \rho_2 & \rho_1 & 1 \end{bmatrix},$$

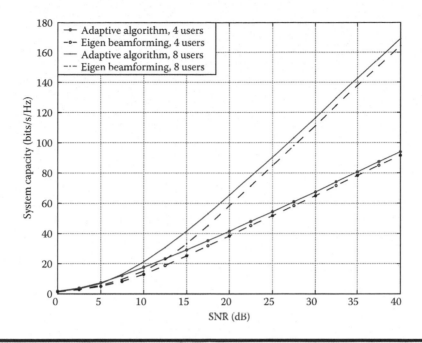

Figure 17.5 **Capacity of OFDM–SDMA system for different numbers of SDMA users per OFDM subcarriers, case of 256-OFDM and** $\mu = 0.01$**.**

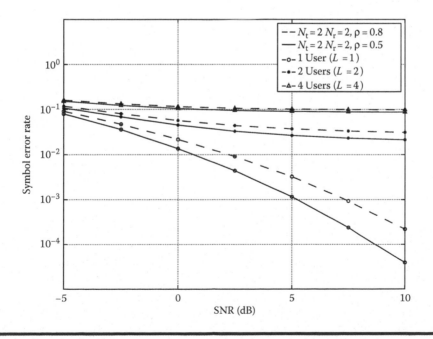

Figure 17.6 **SER of OFDM–SDMA system with the adaptive MIMO beamforming scheme for different level of fading correlations.**

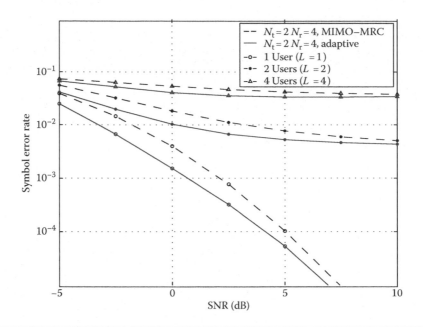

Figure 17.7 **SER of OFDM–SDMA systems for MIMO–MRC and capacity-aware adaptive MIMO beamforming schemes, 2 × 4 systems.**

for the 4×4 system, where ρ_1 denotes high-level correlation between adjacent antenna elements and ρ_2 denotes low-level correlations between nonadjacent ones.

In Figure 17.6, we plot the SER of OFDM–SDMA systems for 2×2 MIMO–MRC system and the capacity-aware adaptive MIMO beamforming scheme, with the algorithm parameter $\mu = 0.1$, and fading correlation parameters $\rho = 0.5$ and 0.8 representing respectively low- and high-level correlations. It is observed from the figure that the SER degrades as fading correlation between adjacent antenna elements increases, for all values of L. Figure 17.7 presents the SER of OFDM–SDMA systems for 2×4 MIMO–MRC and the capacity-aware adaptive MIMO beamforming scheme, with $\mu = 0.2$, $\rho_1 = 0.8$, and $\rho_2 = 0.5$. It is observed from the figure that the adaptive scheme has better SER performance than the MIMO–MRC system, especially at low SNR, which is of practical interest in OFDM–SDMA applications, where due to a high noise level (noise here denoting noise plus interference, after the pre-whitening in the MRC detection in Equation 17.10), the SNR of the operation is very low.

17.7 Conclusion

This chapter presents the performance evaluation of adaptive MIMO beamforming schemes employed for multiuser access in OFDM–SDMA systems. It is shown that the MIMO–MRC system enhances the SER performance of OFDM–SDMA by computing the Tx/Rx beamforming weights that enhance the SINR. On the other hand, the capacity-aware adaptive MIMO beamforming algorithm seeks the optimal transmit weight vectors that directly maximize the OFDM–SDMA channel capacity for each user iteratively and thus enhance the overall system capacity. We show that, asymptotically, the SER of OFDM–SDMA systems with the transmit weights obtained using the capacity-aware design method (and receiver weight by MRC) is better than that of the

MIMO–MRC system (eigen-beamforming at the transmitter with MRC receiver). This is due to the fact that the capacity-aware MIMO beamforming algorithm maximizes the capacity for each user iteratively; consequently, it enhances the SINR, resulting in lower SER. Finally, we present an overview of MIMO–MMSE schemes in OFDM–SDMA application and discuss their prospects in enhancing the performance of the system. While linear MMSE detection may give similar SER as MRC, nonlinear MMSE detection promises better performance. This work provides an useful analytical framework for evaluating the performance of OFDM–SDMA systems.

References

1. C. Ung and R. Johnston, A space division multiple access receiver, *Proc. IEEE-APS '01*, vol. 1, pp. 422–425, July 2001.
2. M. Jiang and L. Hanzo, Multiuser MIMO-OFDM systems using subcarrier hopping, *IEE Proc. Commun.*, 153(6), 802–809, 2006.
3. M. Y. Alias, A. K. Samingan, S. Chen, and L. Hanzo, Multiple antenna aided OFDM employing minimum bit error rate multiuser detection, *IEE Electron. Lett.*, 39(24), 1769–1770, 2003.
4. P. Vandenameele, L. Van Der Perre, M. G. E. Engels, B. Gyselinckx, and H. J. De Man, A combined OFDM/SDMA approach, *IEEE J. Sel. Areas Commun.*, 18(11), 2312–2321, 2000.
5. G. V. Tsoulos, Experimental and theoretical capacity analysis of space-division multiple access (SDMA) with adaptive antennas, *IEE Proc. Commun.*, 146(5), 307–311, 1999.
6. J. Gai and M. Hefnawi, A MIMO-beamforming scheme for frequency selective channels, *Proc. IEEE-MAPE 2005*, 2, 1311–1314, 2005.
7. M. Chryssomallis, Smart antennas, *IEEE Antennas Propag. Mag.*, 42(3), 129–136, 2000.
8. C.-H. Hsu, Uplink MIMO-SDMA optimization of smart antennas by phase-amplitude perturbations based on memetic algorithms for wireless and mobile communication systems, *IET Commun.*, 1(3), 520–525, 2007.
9. J. Li, K. B. Letaief, R. S. Cheng, and Z. Cao, Multi-stage low complexity maximum likelihood detection for OFDM/SDMA wireless LANs, *Proc. IEEE-ICC '01*, vol. 4, pp. 1152–1156, June 2001.
10. K.-K. Wong, R. S.-K. Cheng, K. B. Letaief, and R. D. Murch, Adaptive Antennas at the Mobile and Base Stations in an OFDM/TDMA System, *IEEE Trans. Commun.*, 49(1), 195–206, 2001.
11. A. I. Sulyman and M. Hefnawi, Adaptive MIMO beamforming algorithm based on gradient search of the channel capacity in OFDM–SDMA system, *IEEE Commun. Letters*, 12(9), 642–644, 2008.
12. V. Stankovic and M. Haardt, Generalized design of multi-user MIMO precoding matrices, *IEEE Trans. Wirel. Commun.*, 7(3), 953–961, 2008.
13. V. Stankovic and M. Haardt, Multi-user MIMO downlink precoding for users with multiple antennas, *Proc. 12th Meeting of the Wireless World Research Forum (WWRF)*, Toronto, Ontario, Canada, November 2004.
14. Q. H. Spencer, C. B. Peel, A. L. Swindlehurst, M. Haardt, An introduction to the multi-user MIMO downlink, *IEEE Commun. Mag.*, 60–67, 2004.
15. D. J. Love, R. W. Heath Jr, V. Lau, D. Gesbert, B. D. Rao, and M. Andrews, An overview of limited feedback in wireless communication systems, *IEEE J. Sel. Areas Commun.*, 26(8), 1341–1365, 2008.
16. S. Wang, H. Kim, B. K. Yi, and S. Kwon, On the feedback channel for MIMO beamforming, *Proc. IEEE-WCNC '08*, Las Vegas, NV, pp. 683–687, April 2008.
17. A. Paulraj, R. Nabar, and D. Gore, *Introduction to Space-Time Wireless Communications*, Cambridge University Press, Cambridge, 2003.
18. H. Bolcskei, M. Borgmann, and A. J. Paulraj, Impact of the propagation environment on the performance of space-frequency coded MIMO-OFDM, *IEEE J. Sel. Areas Commun.*, 21(3), 427–439, 2003.
19. E. Telatar, Capacity of multi-antenna gaussian channels, *Eur. Trans. Telecommun.*, 10(6), 585–595, 1999. [Published earlier in AT&T-Bell Lab memo, June 1995].

20. L. Liu and H. Jafarkhani, Novel transmit beamforming schemes for time-selective fading multiantenna Systems, *IEEE Trans. Signal Process.*, 54(12), 4767–4781, 2006.

21. A. Papoullis and S. U. Pillai, *Probability, Random Variables and Stochastic Processes*, 4th ed., New York: McGraw-Hill, Inc., 2002.

22. J. G. Proakis, *Digital Communications*, 4th ed., New York: McGraw-Hill, 2001.

23. A. Pascual-Iserte, A. I. Perez-Neira, and M. A. Lagunas, On power allocation strategies for maximum signal to noise and interference ratio in an OFDM-MIMO system, *IEEE Trans. Wirel. Commun.*, 3(3), 808–820, 2004.

24. A. I. Sulyman and M. Hefnawi, Performance evaluation of capacity-aware MIMO beamforming schemes in OFDM–SDMA systems, *IEEE Trans. Commun.*, 2008, to appear in (58)1, 2010.

25. M. Kang, L. Yang, and M.-S. Alouini, Certain computations involving complex Gaussian matrices with applications to the performance analysis of MIMO systems, in A. B. Gershman and N. D. Sidiropoulos (Eds) *Space-Time Processing for MIMO Communications*, Oxford, U.K.: John Wiley & Sons, Inc., 2005.

26. M. Webb, M. Beach, and A. Nix, Capacity limits of MIMO channels with co-channel interference, *Proc. IEEE-VTC 2004*, 2, 703–707, 2004.

27. M. R. Mckay, A. J. Grant, and I. B. Collings, Performance analysis of MIMO–MRC in double-correlated Rayleigh environments, *IEEE Trans. Commun.*, 55(3), 497–507, 2007.

28. S. Haykin, *Adaptive Filter Theory*, 3rd ed., Prentice-Hall, Englewood Cliffs, NJ, 1996.

29. A. I. Sulyman and A. Zerguine, Convergence and Steady-state Analysis of a variable step-size NLMS algorithm. *EURASIP Signal Process. J.*, 83(6), 1255–1273, 2003.

30. R. S. Blum, J. H. Winters, and N. R. Sollenberger, On the capacity of cellular systems with MIMO, *IEEE Commun. Lett.*, 6(6), 242–244, 2002.

31. S. Xi and M. D. Zoltowski, Transmit beamforming and detection design for uplink multiuser MIMO systems, *Proc. 40th Asilomar Conference on Signals, Systems, and Computers, AC SSC '06*, pp. 1593–1600, November 2006.

32. C. B. Peel, Q. H. Spencer, A. L. Swindlehurst, M. Haardt, and B. M. Hochwald, Linear and dirty-paper techniques for the multiuser MIMO downlink, in A. B. Gershman and N. D. Sidiropoulos (Eds) *Space-Time Processing for MIMO Communications*, Oxford, U.K.: Wiley, 2005.

33. S. Verdu, *Multi-User Detection*, Cambridge, U.K.: Cambridge University Press, 1998.

34. L. Hanzo and T. Keller, OFDM and MC-CDMA: A primer, Oxford, U.K.: John Wiley & Sons, Inc., 2006.

35. I. S. Gradshteyn and I. M. Ryzhik, *Table of Integrals, Series, and Products*, 4th ed., San Diego, CA: Academic, 1965.

36. M. Abramowitz and I. A. Stegun, *Handbook of Mathematical Functions with Formulas, Graphs, and Mathematical Tables*, 9th ed., New York: Dover, 1970.

## Chapter 18

Cooperative OFDMA in the Presence of Frequency Offsets

Zhongshan Zhang and Chintha Tellambura

Contents

18.1   Introduction...474
18.2   Cooperative OFDMA Uplink Signal Model 475
     18.2.1   Channel Model...476
     18.2.2   First Time Slot..477
     18.2.3   Second Time Slot..478
          18.2.3.1   AF Mode...478
          18.2.3.2   DF Mode 479
18.3   Channel Capacity for Cooperative OFDMA..479
     18.3.1   Capacity for Cooperative OFDMA without Considering
          the Frequency Offsets..479
     18.3.2   Capacity for Cooperative OFDMA with the Frequency Offsets 480
          18.3.2.1   Outage Information Rate in the AF Mode.......................480
          18.3.2.2   Outage Information Rate in the DF Mode 481
          18.3.2.3   Numerical Results 481
18.4   Frequency Offset Estimation in the Cooperative Scheme...........................483
     18.4.1   Point-to-Point Frequency Offset Estimation.................................484
     18.4.2   Cooperative Frequency Offset Estimation...................................486
          18.4.2.1   Frequency Offset Estimation in Space–Time
               Diversity System..486
          18.4.2.2   Relay Selection in Multiple Relay Cooperation..................487
          18.4.2.3   Alamouti-Coded Transmission to Mitigate the Effect of
               Frequency Offsets.......................................491

18.5   Channel Estimation in Cooperative OFDMA 493
     18.5.1 Signal Model 493
     18.5.2 Channel Estimation without Considering the Frequency Offsets 495
     18.5.3 Optimal Channel Estimation by Considering the
            Frequency Offsets 496
            18.5.3.1 Optimal Pilot Design to Eliminate MAI in the
                     Second Time Slot 497
            18.5.3.2 PEP Analysis 499
            18.5.3.3 Numerical Results 501
18.6   Conclusions 502
References 503

18.1 Introduction

Orthogonal frequency-division multiple access (OFDMA) is being considered for IEEE 802.16d/e [1,2]. In OFDMA, a unique set of subcarriers are allocated to each node, and the frequency-domain orthogonality can be guaranteed for different nodes. The use of orthogonal subcarrier sets for different nodes eliminates multiple access interference (MAI) under perfect conditions [3–8]. By adaptively allocating and modulating subcarriers for each node, the frequency diversity gain can be improved and, therefore, the OFDMA transmission can be optimized in terms of the bit error rate (BER) or the channel capacity [9,10].

Besides frequency diversity, spatial diversity gain can be realized by employing a space-time code and/or multiple co-located transmit antennas [11,12]. However, in cooperative diversity, a distributed virtual multiple antenna system is created to extract diversity gain [13–20]. A relay is used to improve the transmission quality between node S (the source) and node D (the destination), and the relay can operate in either the amplify-and-forward (AF) or the decode-and-forward (DF) mode. Joint optimization in the relay-precoders and decoders in a cooperative network is discussed in Ref. [21], with either full channel state information (CSI) or partial CSI being available at the destination and the relays. A hybrid forwarding scheme for cooperative relaying in orthogonal frequency-division multiplex (OFDM)-based networks is proposed in Ref. [22], and based on the current signal-to-noise ratio (SNR) of each subcarrier between the source, the relay and the destination, the relay can adaptively determine which relaying mode to use. Outage capacity of the slow-fading relay channel that focuses on the low SNR and low outage probability regime is discussed in Ref. [23]. It is shown that the Bursty Amplify-Forward protocol proposed in Ref. [23] achieves the optimal outage capacity per unit energy in the low SNR regime. An efficient algorithm for performing joint optimization in relay selection and resource allocation in a cellular system employing OFDMA is proposed in Ref. [24], where the optimization is performed not only by considering the power and bandwidth allocation, but also by selecting the best relay as well as the best relay strategy. The performance degradation of a cooperative OFDMA uplink due to the frequency offset is analyzed in Refs. [25,26].

In this chapter, the achievable performance improvement in an OFDMA uplink via cooperative relaying is discussed. As in Ref. [24], a unique subcarrier group is assumed to be allocated to each node. Symmetric cooperative transmission is assumed; that is, for each source with M relays, it also acts as a relay for the other M nodes simultaneously. The total transmit power used for transmitting each symbol is kept constant. As in Ref. [20], the relays can operate in either the AF or

DF mode.* In the AF mode, the relay simply retransmits its received signal, including the interference and its local additive noise. Although this mode simplifies the hardware design of the relay transceivers, the effective signal-to-interference-plus-noise ratio (SINR) is degraded due to the accumulated interference and noise. A DF relay decodes its received signal before retransmission. Since the interference generated in the source-to-relay transmission is eliminated, the DF mode improves the effective SINR considerably, but at the cost of an increased hardware complexity. We impose a constraint on each relay that does not permit us to receive and transmit signals simultaneously at the same time and the same frequency subband, that is, half-duplex constraint is assumed. The channel capacity in the cooperative OFDMA uplink transmission is first analyzed and then the performance of frequency offset estimation in cooperative OFDMA is also discussed. Finally, the channel estimation for the cooperative OFDMA is discussed, and the pairwise error probability (PEP) performance of the orthogonal space-time coded system in the presence of the frequency offset is also evaluated.

Cooperative OFDMA in the presence of the frequency offsets is studied. The ergodic or the outage channel information rate degraded by the frequency offsets is analyzed in Refs. [25,26]. It is proved that the DF mode always outperforms the AF mode in terms of the channel information rate. Relay selection in cooperative OFDMA frequency offset estimation is studied in Ref. [27]. A new relaying mode named DcF mode is proposed. Numerical results show that the AF mode outperforms DcF mode in terms of the frequency offset estimation. We also investigated the optimal pilot design for channel estimation in cooperative OFDM/OFDMA [28], and by using orthogonal space–time coding transmission, the DF mode always outperforms the AF mode in terms of PEP.

The remainder of this chapter is organized as follows. Section 18.2 introduces the cooperative OFDMA uplink signal model. Both the ergodic and the outage information rates are derived in Section 18.3. Section 18.4 discusses the frequency offset estimation in the cooperative OFDMA, and Section 18.5 proposes an optimal channel estimation algorithm in the presence of the frequency offset. The PEP performance of the orthogonal space–time coded system is also evaluated in Section 18.5. Finally, Section 18.6 concludes the chapter.

Notation: $(\cdot)^{\mathrm{T}}$ and $(\cdot)^{\mathrm{H}}$ denote the transpose and conjugate transpose of a matrix, respectively. $(\cdot)^{-1}$ represents the inverse of a matrix. The imaginary unit is $j = \sqrt{-1}.(\cdot)^*$ denotes the complex conjugate. A circularly symmetric complex Gaussian variable with mean a and variance σ^2 is denoted by $z \sim \mathcal{CN}(a, \sigma^2)$. $\mathbf{x}[i]$ represents the ith element of vector $\mathbf{x}$. $[\mathbf{A}]_{ij}$ represents the ijth element of matrix $\mathbf{A}$. The $N \times N$ identity matrix is $\mathbf{I}_N$, and the $N \times N$ all-zero matrix is $\mathbf{O}_N$. $\mathrm{diag}\{\mathbf{x}\}$ represents a diagonal matrix with its nnth element being equal to $\mathbf{x}[n]$. $\mathbb{E}\{x\}$ and $\mathrm{Var}\{x\}$ denote the mean and the variance x, respectively. $\theta \in \mathcal{A}$ means θ is an element of set $\mathcal{A}$.

18.2 Cooperative OFDMA Uplink Signal Model

Complex data symbols are mapped to a signal constellation such as phase-shift keying (PSK) or quadrature amplitude modulation (QAM) to modulate OFDMA subcarriers. Since only a part of the subcarriers are allocated to each node, when all the subcarriers allocated to one node are in a

---

* When discussing the frequency offset estimation, the relays can operate in either the AF or the decode-and-compensate-and-forward (DcF) mode. The DcF mode here is slightly different from the DF mode proposed in Ref. [20]: if the relay operates in the DcF mode, the relay should first estimate the frequency offset between the source node and itself and then demodulate and decode the received training sequence. After doing so, the relay will use the estimation result to pre-compensate for the frequency offset in the decoded training sequence and retransmit this regenerated training sequence to the destination.

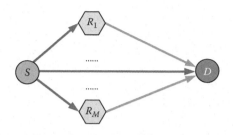

Figure 18.1 Cooperative transmission with *M* relays in the OFDMA uplink. (From Z. Zhang, W. Zhang, and C. Tellambura, *IEEE Trans. Wireless Commun.*, 8 (9), September 2009. With permission.)

deep-fading simultaneously, the transmission of that node will be degraded. Cooperative transmission can improve the transmission of each node.

Cooperative transmission with M relays is shown in Figure 18.1. Without loss of generality, we assume that the total number of subcarriers is N, and $\mathcal{M}$ is the total number of nodes (including source and relays), with each node being allocated N_u unique subcarriers, where $\mathcal{M}N_u \leq N$. Cooperative transmission takes place in two time slots (see Figure 18.2). In the first time slot, each mobile node transmits its first symbol to all its relays and the base station (node D) and simultaneously receives the first symbol of the other nodes. In the second time slot, each node will send its second symbol and forward the received symbols from the other nodes simultaneously. Without loss of generality, node S represents the source node and R_k, $k \in \{1, \ldots, M\}$, represents the kth relay of S.

18.2.1 Channel Model

We assume a frequency-selective quasi-static fading channel between any pair of nodes. The channel response between nodes a and b is $\tilde{\mathbf{h}}_{ab} = [h_{ab}(0), h_{ab}(1), \ldots, h_{ab}(L-1)]^{\mathrm{T}}$, where L is the maximum channel length for all nodes a and b. The corresponding frequency-domain channel attenuation matrix is given by $\mathbf{H}_{ab} = \mathrm{diag}\{H_{ab}(0), H_{ab}(1), \ldots, H_{ab}(N-1)\}$, with $H_{ab}(n) = \sum_{d=0}^{L-1} h_{ab}(d) e^{j2\pi nd/N}$ representing the channel attenuation at the nth subcarrier.

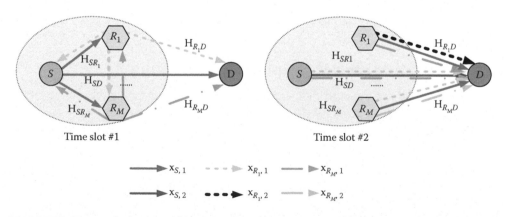

Figure 18.2 Cooperative transmission in the OFDMA uplink. (From Z. Zhang, W. Zhang, and C. Tellambura, *IEEE Trans. Wireless Commun.*, 8 (9), September 2009. With permission.)

18.2.2 First Time Slot

In the first time slot, each mobile node, including node S and the relays, transmits its first symbol to node D and the other mobile nodes. Let us use $\mathbf{x}_{S,1}$ and $\mathbf{x}_{R_k,1}$ ($1 \le k \le M$) to represent the first transmit vector of node S and node R_k, respectively, where both $\mathbf{x}_{S,1}$ and $\mathbf{x}_{R_k,1}$ are $N_u \times 1$ vectors, and without loss of generality, each entry of these two vectors is assumed to be an independent and identically distributed (i.i.d.) complex Gaussian random variable (RV) with zero mean and unit variance. We also use set G_z to represent the subcarrier group allocated to the zth node (the elements of G_z are the indexes of all the subcarriers allocated to the zth node, and the subcarriers allocated to each node need not be contiguous in the frequency domain), where $G_k \cap_{k \ne z} G_z = \emptyset$ and $\cup_{z=S,R_1,\ldots,R_M} G_2 \subseteq \{0,1,\ldots,N-1\}$.

The received signals at nodes $D, S,$ and each relay can be represented as

$$\mathbf{y}_{D,1} = \mathbf{E}_{SD}\mathbf{F}_S\mathbf{H}_{SD}\boldsymbol{\Phi}_{S,1}\mathbf{x}_{S,1} + \sum_{k=1}^{M}\mathbf{E}_{R_kD}\mathbf{F}_{R_k}\mathbf{H}_{R_kD}\boldsymbol{\Phi}_{R_k,1}\mathbf{x}_{R_k,1} + W_{D,1}, \quad (18.1)$$

$$\mathbf{y}_{S,1} = \sum_{k=1}^{M}\mathbf{E}_{R_kS}\mathbf{F}_{R_k}\mathbf{H}_{R_kS}\boldsymbol{\Phi}_{R_k,1}\mathbf{x}_{R_k,1} + \mathbf{w}_S, \quad (18.2)$$

and

$$\mathbf{y}_{R_k,1} = \mathbf{E}_{SR_k}\mathbf{F}_S\mathbf{H}_{SR_k}\boldsymbol{\Phi}_{S,1}\mathbf{x}_{S,1} + \sum_{l=1,l \ne k}^{M}\mathbf{E}_{R_lR_k}\mathbf{F}_{Rl}\mathbf{H}_{R_lR_k}\boldsymbol{\Phi}_{R_l,1}\mathbf{x}_{R_l,1} + \mathbf{w}_{R_k}, \quad (18.3)$$

respectively, where $\mathbf{y}_{D,1}$, $\mathbf{y}_{S,1}$, and $\mathbf{y}_{R_k,1}$ are $N \times 1$ vectors that represent the received time-domain signals at node D, node S and R_k, respectively. Furthermore,

$$\mathbf{E}_{zk} = \text{diag}\left\{e^{j\psi_{zk}}, e^{j((2\pi\varepsilon_{zk}/N)+\psi_{zk})}, \ldots, e^{j((2\pi\varepsilon_{zk}(N-1)/N)+\psi_{zk})}\right\} \quad (18.4)$$

with ψ_{zk} and ε_{zk} representing the initial phase and normalized carrier frequency offset (normalized to one subcarrier bandwidth) between nodes z and k, respectively.

Without loss of generality, we assume that each ψ_{zk} has been estimated and corrected, that is, $\psi_{zk} = 0$ is assumed in the following sections. We also assume that each ε_{zD}, $z \in \{S, R_1, \ldots, R_M\}$, is an i.i.d. RV with zero mean and variance σ_ϵ^2 (and is not necessarily Gaussian). Since $\varepsilon_{zD} = \varepsilon_{zk}+\varepsilon_{kD}$ is satisfied for each $z, k \in \{S, R_1, \ldots, R_M\}$, we have $\text{Var}\{\varepsilon_{zD}\} = \sigma_\epsilon^2$ and $\text{Var}\{\varepsilon_{zk}\} = \text{Var}\{\varepsilon_{zD} - \varepsilon_{kD}\} = 2\sigma_\epsilon^2$, $z, k \in \{S, R_1, \ldots, R_M\}$, $z \ne k$. $\mathbf{F}_z$ is the $N \times N_u$ inverse discrete Fourier transform (IDFT) matrix for the zth user. $\mathbf{F}_z$ can be generated from the $N \times N$ IDFT matrix $\mathbf{F}$ with $[\mathbf{F}]_{mn} = 1/\sqrt{N}e^{j2\pi mn/N}$, $0 \le m, n \le N - 1$, by deleting all the columns with the column indexes not belonging to G_z, where N is the IDFT length. $\boldsymbol{\Phi}_{z,1} = \text{diag}\{\sqrt{P_{z,1,i}} : i \in G_z, z \in \{S, R_1, \ldots, R_M\}\}$ is an $N_u \times N_u$ diagonal matrix with each diagonal entry representing the transmit power of one subcarrier allocated to node z in the first time slot. We assume that identical power allocation in each subcarrier is performed in either $S \rightarrow D$ or $R \rightarrow D$ transmission and, therefore, for each z and i, we have $\boldsymbol{\Phi}_{z,1} = \bar{P}\mathbf{I}_{N_u}$. The vectors $\mathbf{w}_z$ and $\mathbf{w}_{D,i}$ are additive white Gaussian noise (AWGN) with $\{\mathbf{w}_z[m], \mathbf{w}_{D,i}[m]\} \sim \mathcal{CN}(0, \sigma_w^2)$, where $z \in \{S, R_1, \ldots, R_M\}$, $0 \le m \le N - 1$, and $i = (1, 2)$. $\mathbf{H}_{zk} = \text{diag}\{H_{zk}^{(i)} : i \in G_z\}$ is an $N_u \times N_u$ diagonal matrix, where $H_{zk}^{(i)}$ denotes the channel attenuation between nodes z and k at the ith subcarrier.

18.2.3 Second Time Slot

In the second time slot, as in Ref. [20], the relays can operate in either the AF or the DF mode. In each transmission of node S, if the link between node S and another node is not in a deep fade, this node can be chosen as a relay for S. Since per subcarrier adaptive power allocation is not considered, the averaged channel attenuation between S and R_k can be quantified by using $\hbar_k = \text{trace}\{\mathbf{H}_{SR_k}\mathbf{H}_{SR_k}^H\}$, and a larger $\hbar_k$ implies less fading. The relays of node S can be chosen as follows: we first set a channel power threshold v where $v > 0$, and node k will act as a relay of S if $\hbar_k \geq vN_u$, $1 \leq k \leq M - 1$.

From Ref. [25], the probability that S has M relays is given by

$$P_{v,M} = \binom{M-1}{M} P_v^M (1 - P_v)^{M-M-1}, \tag{18.5}$$

where

$$P_v = e^{-N_u v} \sum_{m=0}^{N_u-1} \frac{v^m N_u^m}{m!}.$$

18.2.3.1 AF Mode

The received vector at node D in the second time slot is

$$
\begin{aligned}
\mathbf{y}_{D,2}^{AF} = {}& \mathbf{E}_{SD}\mathbf{F}_S\mathbf{H}_{SD}\mathbf{\Phi}_{S,2}^t\mathbf{x}_{S,2} + \sum_{k=1}^{M}\mathbf{E}_{R_kD}\mathbf{F}_{R_k}\mathbf{H}_{R_kD}\mathbf{\Phi}_{R_k,2}^t\mathbf{x}_{R_k,2} \\
& + \sum_{k=1}^{M}\varpi_{R_kS}\mathbf{E}_{SD}\mathbf{F}_{R_k}\mathbf{H}_{SD}\mathbf{\Phi}_{S,2}^r\mathbf{F}_{R_k}^H\mathbf{y}_{S,1} + \sum_{k=1}^{M}\varpi_{SR_k}\mathbf{E}_{R_kD}\mathbf{F}_S\mathbf{H}_{R_kD}\mathbf{\Phi}_{R_k,2}^r\mathbf{F}_S^H\mathbf{y}_{R_k,1} \\
& + \sum_{k=1}^{M}\sum_{l=1,l\neq k}^{M}\varpi_{R_lR_k}\mathbf{E}_{R_kD}\mathbf{F}_{R_l}\mathbf{H}_{R_kD}\mathbf{\Phi}_{R_k,2}^r\mathbf{F}_{R_l}^H\mathbf{y}_{R_k,1} + \mathbf{w}_{D,2}, \tag{18.6}
\end{aligned}
$$

where both $\mathbf{x}_{S,2}$ and $\mathbf{x}_{R_k,2}$ are $N_u \times 1$ vectors, which represent the second transmit vector of S and $R_k(1 \leq k \leq M)$, respectively. $\mathbf{y}_{D,2}^{AF}$ is an $N \times 1$ vector that represents the received time-domain signals at D in the second time slot. $\varpi_{R_kz} = (\theta_v\bar{P} + \sigma_w^2)^{-1/2}$, $z \in \{S, R_1, \dots, R_M\}$, $k \in \{R_1, \dots, R_M\}$, represents the amplify coefficients at R_k when it retransmits the signal for node z, where

$$\theta_v = 1 + \left(N_u! \sum_{m=0}^{N_u-1} \frac{v^{m-N_u}N_u^{m-N_u}}{m!}\right).$$

$\mathbf{\Phi}_{z,2}^t = \text{diag}\{\sqrt{P_{z,2,i}^t} : i \in G_z\}$ and $\mathbf{\Phi}_{z,2}^r = \text{diag}\{\sqrt{P_{z,2,i}^r} : i \in G_z\}(z \in \{S, R_1, \dots, R_M\})$ are $N_u \times N_u$ matrices, which represent the power allocation to node z to transmit its own signal and to retransmit signals for the other nodes in the second time slot, respectively. We assume that in the second time slot, the total transmit power $N_u\bar{P}$ is shared by the source node and its M relays. By allocating power $\alpha N_u\bar{P}$ to transmit its second symbol, the source node will use the remaining power $(1 - \alpha)N_u\bar{P}$ to retransmit signals for its M relays, where $0 \leq \alpha \leq 1$. Therefore,

$P^t_{z,2,i} = \alpha \bar{P}$ and $P^r_{z,2,i} = (1 - \alpha)\bar{P}/M$, and we have $\boldsymbol{\Phi}_{z,1}\boldsymbol{\Phi}^H_{z,1} = \bar{P}\mathbf{I}_{Nu}$, $\boldsymbol{\Phi}^t_{z,2}\boldsymbol{\Phi}^{tH}_{z,2} = \alpha\bar{P}\mathbf{I}_{Nu}$, and $\boldsymbol{\Phi}^r_{z,2}\boldsymbol{\Phi}^{rH}_{z,2} = ((1 - \alpha)P/M)\mathbf{I}_{N_u}$ for each z.

We can combine (18.1) and (18.6) into a compact equation as

$$\mathbf{y}^{AF}_D = \begin{bmatrix} \mathbf{y}^{AF}_{D,1} \\ \mathbf{y}^{AF}_{D,2} \end{bmatrix}. \tag{18.7}$$

18.2.3.2 DF Mode

Before retransmission, each DF relay node should first demodulate and decode the signal received from the other cooperative nodes. Here, we assume that the relay node can always demodulate and decode its relay signals correctly (this condition is met if and only if the transmit rate of S does not exceed the corresponding capacity of each of the $S \to R_k$ channels, $k = 1, \ldots, M$). The received vector at node D is

$$\mathbf{y}^{DF}_{D,2} = \mathbf{E}_{SD}\mathbf{F}_S\mathbf{H}_{SD}\boldsymbol{\Phi}^t_{S,2}\mathbf{x}_{S,2} + \sum_{k=1}^{M} \mathbf{E}_{R_kD}\mathbf{F}_{R_k}\mathbf{H}_{R_kD}\boldsymbol{\Phi}^t_{R_k,2}\mathbf{x}_{R_k,2} + \sum_{k-1}^{M} \mathbf{E}_{SD}\mathbf{F}_{R_k}\mathbf{H}_{SD}\boldsymbol{\Phi}^t_{S,2}\mathbf{x}_{R_k,1}$$

$$+ \sum_{k=1}^{M} \mathbf{E}_{R_kD}\mathbf{F}_S\mathbf{H}_{R_kD}\boldsymbol{\Phi}^r_{R_k,2}\mathbf{x}_{S,1} + \sum_{k=1}^{M}\sum_{l=1,l\neq k}^{M} \mathbf{E}_{R_kD}\mathbf{F}_{R_l}\mathbf{H}_{R_kD}\boldsymbol{\Phi}^r_{R_k,2}\mathbf{x}_{R_l,1} + \mathbf{w}_{D,2}, \tag{18.8}$$

where $\boldsymbol{\Phi}^t_{z,2}$ and $\boldsymbol{\Phi}^r_{z,2}$ are identical to those in the AF mode. Note that in Equation 18.8, M out of $\mathcal{M}$ nodes are chosen as relays of node S in the current transmission based on Equation 18.5, and both M and R_k may be changed for different transmissions.

We can represent Equations 18.1 and 18.8 into a compact equation as

$$\mathbf{y}^{DF}_D = \begin{bmatrix} \mathbf{y}^{DF}_{D,1} \\ \mathbf{y}^{DF}_{D,2} \end{bmatrix}. \tag{18.9}$$

18.3 Channel Capacity for Cooperative OFDMA

Channel capacity for point-to-point OFDMA connections has been considerably studied in Refs. [29–33], either with or without the frequency offsets being considered. In this section, we analyze the channel capacity of the cooperative OFDMA systems.

18.3.1 Capacity for Cooperative OFDMA without Considering the Frequency Offsets

The capacity for cooperative OFDMA is analyzed in Ref. [34], with both the AF and DF modes being considered. A relay network with multiple source nodes, multiple relays, and a single destination is considered in Ref. [34]. Let $\mathcal{S} = \{s_1, \ldots, s_K\}$, $\mathcal{R} = \{r_1, \ldots, r_M\}$, and d represent the set of sources, relays, and the destination, respectively. We also represent the channel gains of the nth subcarrier of $s_k \to d$, $s_k \to r_m$, and $r_m \to d$ as $c^{(n)}_k$, $a^{(n)}_{mk}$, and $b^{(n)}_m$, respectively. The total capacity over all the subcarriers of the proposed cooperative OFDMA in the AF mode can be derived as

$$C_{AF} = \frac{1}{2} \sum_{n=1}^{N_u} \log_2 \left(1 + \frac{\sum_{k=1}^{K} \|c_k^{(n)}\|^2}{\sigma_w^2} + \frac{\sum_{k=1}^{K} \left\| \sum_{m=1}^{M} \mu_m^{(n)} b_m^{(n)} a_{mk}^{(n)} \right\|^2}{\left(\sum_{m=1}^{M} \|\mu_m^{(n)}\|^2 \|b_m^{(n)}\|^2 + 1 \right) \sigma_w^2} \right), \qquad (18.10)$$

where $\mu_m^{(n)}$ represents the scaled power of each relay node in its retransmission, and we have $\|\mu_m^{(n)}\| = \left(\sum_{k=1}^{K} \|a_{mk}^{(n)}\|^2 + \sigma_w^2 \right)^{-1/2}$.

In the DF mode, the total capacity over all the subcarriers can be derived as

$$C_{DF} = \sum_{n=1}^{N_u} C_{DF}^{(n)}, \qquad (18.11)$$

where

$$C_{DF}^{(n)} = \frac{1}{2} \min \left\{ \sum_i \left(1 + \frac{\lambda_i}{\sigma_w^2} \right), \log_2 \left(1 + \frac{\sum_{m=1}^{M} \|b_m^{(n)}\|^2 + \sum_{k=1}^{K} \|c_k^{(n)}\|^2}{\sigma_w^2} \right) \right\}, \qquad (18.12)$$

and λ_i is the ith eigenvalue of $\mathbf{A}^{(n)}(\mathbf{A}^{(n)})^{\mathrm{H}}$ with $\mathbf{A}^{(n)} = \begin{bmatrix} a_{11}^{(n)} & \cdots & a_{1K}^{(n)} \\ \vdots & \ddots & \vdots \\ a_{M1}^{(n)} & \cdots & a_{MK}^{(n)} \end{bmatrix}$. Evidently, the

channel capacity is proportional to the number of relays.

18.3.2 Capacity for Cooperative OFDMA with the Frequency Offsets

Since OFDMA systems are very sensitive to the frequency offsets, the channel capacity must be degraded in the presence of the frequency offsets. In this subsection, the channel capacity for cooperative OFDMA in the presence of the frequency offsets is studied. Since the adaptive power and subcarrier allocation are not considered in this subsection, "outage information rate" rather than "channel capacity" will be used in this subsection.

18.3.2.1 Outage Information Rate in the AF Mode

As in Ref. [20], for a given M, the p $(0 \leq p \leq 1)$ outage information rate of the proposed cooperative transmission with AF relays, $\mathcal{I}_{p,\alpha,M}^{AF}$, is defined as $P_r\{\mathcal{I}_M^{AF}(\mathbf{x}_S; \mathbf{y}_D^{AF}) \leq \mathcal{I}_{p,\alpha,M}^{AF}\} = p$, where $\mathcal{I}_M^{AF}(\mathbf{x}_S; \mathbf{y}_D^{AF})$ denotes the mutual information between $\mathbf{x}_S$ and $\mathbf{y}_D^{AF}$, and $P_r\{\mathcal{I}_M^{AF}(\mathbf{x}_S; \mathbf{y}_D^{AF}) \leq \mathcal{I}_{p,\alpha,M}^{AF}\}$ represents the probability that event $\mathcal{I}_M^{AF}(\mathbf{x}_S; \mathbf{y}_D^{AF}) \leq \mathcal{I}_{p,\alpha,M}^{AF}$ happens.

When $0 \leq \alpha < 1$, cooperative relays are used, and for a given channel power threshold ν, the averaged outage information rate over M in the AF mode is derived as $\mathcal{I}_{out}^{AF} = \sum_{M=1}^{M-1} P_{\nu,M} \mathcal{I}_{p,\alpha,M}^{AF}$

[25], where

$$\mathcal{I}_{p,\alpha,M}^{\mathrm{AF}} = \frac{N_{\mathrm{u}}}{2} \log_2 \left(\mathcal{Z}_{\alpha,M}^{\mathrm{AF}} + \frac{\mathcal{L}_{\mathrm{u}} \cdot \theta_\nu (1-\alpha) \bar{P}^2 \beta_1 (p \cdot M!)^{1/M}}{M \sigma_{\xi_{S,2}^{\mathrm{AF}}}^2 (\theta_\nu \bar{P} + \sigma_w^2)} \right), \qquad (18.13)$$

$$\mathcal{Z}_{\alpha,M}^{\mathrm{AF}} = \left(1 + (\alpha \cdot \mathcal{L}_{\mathrm{u}} \cdot \bar{P} \cdot \beta_1 \cdot \ln(1/1-p)/\sigma_{\xi_{S,2}^{\mathrm{AF}}}^2) \right) \left(1 + (\mathcal{L}_{\mathrm{u}} \cdot \bar{P} \cdot \beta_1 \cdot \ln(1/1-p)/\sigma_{\xi_{S,1}^{\mathrm{AF}}}^2) \right), \sigma_{\xi_{S,1}^{\mathrm{AF}}}^2 =$$
$$(\mathcal{L}_{\mathrm{u}} M \pi^2 \sigma_\epsilon^2 \bar{P}/3M) + \sigma_w^2, \quad \sigma_{\xi_{S,2}^{\mathrm{AF}}}^2 = (2\mathcal{L}_{\mathrm{u}} \theta_\nu (M^2 - M + 1)(1 - \alpha) \pi^2 \sigma_\epsilon^2 \bar{P}^2/3M(\theta_\nu \bar{P} +$$
$$\sigma_w^2)) + (\mathcal{L}_{\mathrm{u}} \alpha M \pi^2 \sigma_\epsilon^2 \bar{P}/3M) + (\mathcal{L}_{\mathrm{u}}(M+1)(1-\alpha)\bar{P}\sigma_w^2/\theta_\nu \bar{P} + \sigma_w^2) + \sigma_w^2, \beta_1 \cong 1 - (\pi^2 \sigma_\epsilon^2/3) +$$
$$(\pi^4 \sigma_\epsilon^4/20), \text{ and } \bar{\gamma} = \bar{P}/\sigma_w^2 \text{ represents the average SNR of each subcarrier.}$$

18.3.2.2 Outage Information Rate in the DF Mode

Like in the AF mode, for a given M, the p outage information rate of the proposed DF mode, $\mathcal{I}_{p,\alpha,M}^{\mathrm{DF}}$, is also defined as

$$\mathbf{P}_{\mathrm{r}}\{\mathcal{I}_M^{\mathrm{DF}}(\mathbf{x}_S; \mathbf{y}_D^{\mathrm{DF}}) \leq \mathcal{I}_{p,\alpha,M}^{\mathrm{DF}}\} = p. \qquad (18.14)$$

When $0 \leq \alpha < 1$, the average outage information rate over M is given by $\mathcal{I}_{\mathrm{out}}^{\mathrm{DF}} = \sum_{M=1}^{\mathcal{M}-1} P_{\nu,M} \mathcal{I}_{p,\alpha,M}^{\mathrm{DF}}$ [25], where

$$\mathcal{I}_{p,\alpha,M}^{\mathrm{DF}} = \frac{N_{\mathrm{u}}}{2} \cdot \min \left\{ \log_2(\mathcal{Z}_{\alpha,M,\mathrm{relay}}^{\mathrm{DF}}), \log_2 \left(\mathcal{Z}_{\alpha,M}^{\mathrm{DF}} + \frac{\mathcal{L}_{\mathrm{u}}(1-\alpha)\bar{\gamma}\beta_1(p \cdot M!)^{1/M}}{(\mathcal{L}_{\mathrm{u}} M^2 \pi^2 \sigma_\epsilon^2 \bar{\gamma}/3M) + M} \right) \right\}, \qquad (18.15)$$

$$\mathcal{Z}_{\alpha,M,\mathrm{relay}}^{\mathrm{DF}} = \left(1 + \frac{(\alpha \cdot \mathcal{L}_{\mathrm{u}} \cdot \bar{\gamma} \cdot \beta_1 \cdot \ln(1/1-p)}{(\mathcal{L}_{\mathrm{u}} M \pi^2 \sigma_\epsilon^2 \bar{\gamma}/3M) + 1} \right) \left(1 + \frac{(\theta_\nu \cdot \bar{\gamma} \cdot \beta_2 \cdot \ln(1/1-p)}{(2(M-1)\pi^2 \sigma_\epsilon^2 \bar{\gamma}/3M) + 1} \right), \qquad (18.16)$$

$$\mathcal{Z}_{\alpha,M}^{\mathrm{DF}} = \left(1 + \frac{\alpha \cdot \mathcal{L}_{\mathrm{u}} \cdot \bar{\gamma} \cdot \beta_1 \cdot \ln(1/1-p)}{(\mathcal{L}_{\mathrm{u}} M \pi^2 \sigma_\epsilon^2 \bar{\gamma}/3M) + 1} \right) \left(1 + \frac{\mathcal{L}_{\mathrm{u}} \cdot \bar{\gamma} \cdot \beta_1 \cdot \ln(1/1-p)}{(\mathcal{L}_{\mathrm{u}} M \pi^2 \sigma_\epsilon^2 \bar{\gamma}/3M) + 1} \right), \qquad (18.17)$$

and $\beta_2 \cong 1 - (2\pi^2 \sigma_\epsilon^2/3) + (\pi^4 \sigma_\epsilon^4/5)$.

18.3.2.3 Numerical Results

In our simulation, a wireless channel of OFDMA uplink transmission with a bandwidth of 10 MHz and $N = 1024$ is considered. A cyclic-prefix (CP) of length 64 is added. A total of 64 users are assumed to access the base station.

From Figure 18.3 we also know that the DF mode does not always outperform the AF mode for each power allocation ratio α. For example, when $\alpha = 0$, the optimal outage information rate of 29.77 can be achieved at $\nu = 1.32$, but the optimum information rate for the DF mode is 29.22 (achieved at $\nu = 1.44$); that is, the AF mode outperforms the DF mode. We can explain it as follows: in an interference-limited environment, the transmission rates of both $S \to D$ and $S \to R$ links are functions of frequency offset variance σ_ϵ^2. When power allocation ratio α is very small, the transmission rate of each mode is dominated by $S \to R \to D$ link. Since $\mathrm{Var}\{\epsilon_{SR}\} = 2\mathrm{Var}\{\epsilon_{SD}\}$ for each R, the large demodulate errors at node R will degrade the information rate of the DF mode.

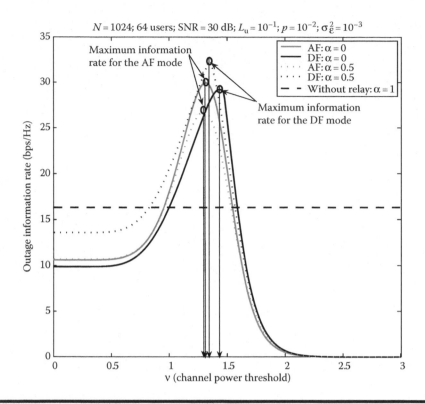

Figure 18.3   Outage information rate as a function of channel power threshold v. (From Z. Zhang, W. Zhang, and C. Tellambura, *IEEE Trans. Wireless Commun.*, 8 (9), September 2009. With permission.)

The ergodic information rate of the proposed cooperative transmission as a function of power allocation ratio α is evaluated in Figure 18.4. By approximating the interference as a Gaussian RV, the expectation operation in the ergodic information rate of both the AF and the DF mode can be simplified. To prove the validity of this approximation, we also perform a simulation by randomly generating an i.i.d. frequency offset for each user. The numerical results in Figure 18.4 show that the Gaussian approximation of the interference in either the AF or the DF mode is accurate enough, if frequency offset variance σ_ϵ^2 is not too large. Figure 18.4 also shows that the AF mode's ergodic information rate is a monotonically increasing function of power allocation ratio α, but that is not true for the DF mode. In this simulation, the maximum ergodic information rate is achieved at $\alpha = 0.9$. For a given SNR and $\alpha < 1$, the DF mode's ergodic information rate is always higher than that of the AF mode because of the noise and interference accumulation in the latter.

The outage information rate as a function of power allocation ratio α is illustrated in Figure 18.5. For different frequency offset variances σ_ϵ^2, the optimum power allocation ratio α for the DF mode will be different. For a given frequency offset variance σ_ϵ^2, the DF mode outperforms the AF mode only when power allocation ratio α is larger than a specified threshold. For example, when $\sigma_\epsilon^2 = 10^{-2}$, this threshold is $\alpha = 0.034$, and the AF mode outperforms the DF mode if $0 \leq \alpha < 0.034$. This threshold for $\sigma_\epsilon^2 = 10^{-3}$ is $\alpha = 0.037$. Numerical results in Figure 18.5 show that in an interference-limited environment, cooperative transmission with DF relays achieves a relatively better performance than cooperation with AF relays, if power allocation ratio α is larger than a specified threshold.

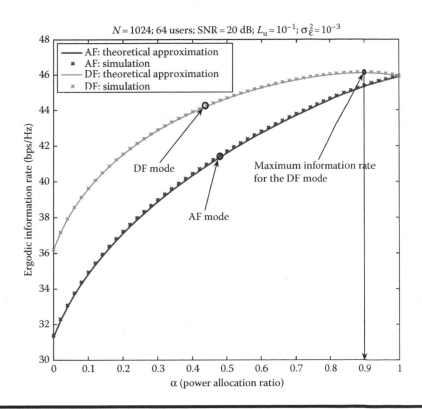

Figure 18.4   Ergodic information rate as a function of power allocation ratio α. (Adapted from Z. Zhang, C. Tellambura, and R. Schober, *Euro. Trans. Telecommun.*, DOI: 10.1002/eH.1385, 2009.)

The outage information rate of the proposed cooperative transmission as a function of frequency offset variance σ_ϵ^2 is illustrated in Figure 18.6. For a given outage probability p (e.g., $p = 10^{-2}$) and a finite SNR (e.g., 30 dB), the outage information rate of either the cooperative transmission or the conventional transmission is a monotonically decreasing function of frequency offset variance σ_ϵ^2. For each frequency offset variance σ_ϵ^2, the cooperative transmission with each mode will always outperform the conventional transmission in terms of the outage information rate, and a smaller frequency offset variance σ_ϵ^2 implies a higher performance improvement. For example, when SNR $= 20$ dB, the outage information rate improvement of the AF mode (or the DF mode) over the conventional transmission is about 6.89 bps/Hz (or 7.29 bps/Hz) when $\sigma_\epsilon^2 = 10^{-2}$, and this performance improvement will increase to about 8.76 bps/Hz (or 9.03 bps/Hz) when $\sigma_\epsilon^2 = 10^{-3}$.

18.4 Frequency Offset Estimation in the Cooperative Scheme

Since OFDMA systems are sensitive to the frequency offsets, accurate estimation of frequency offsets is critical. Frequency offset estimation in OFDMA has been widely studied [3–6,35–41]. However, all these existing algorithms perform frequency offset estimation based on a point-to-point connection, and the performance may be degraded due to the temporarily deep fading in the wireless channel between the source node and the base station. OFDMA uplink frequency offset estimation can be improved by using cooperative relays. By creating a duplicated link between the

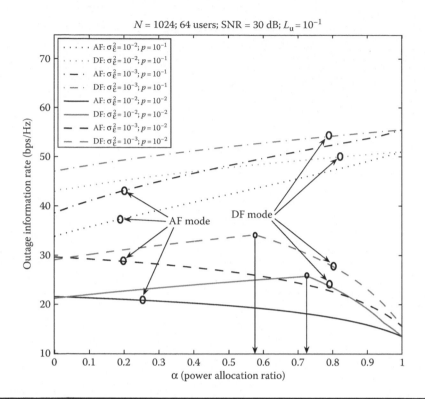

Figure 18.5 **Outage information rate as a function of power allocation ratio α. (Adapted from Z. Zhang, C. Tellambura, and R. Schober, *Euro. Trans. Telecommun.*, DOI: 10.1002/eH.1385, 2009.)**

source node and the destination through a third node, that is, the cooperative relay, to transmit a copy of the training sequence, the reliability of the transmission can be improved considerably. In cooperative OFDMA, the frequency offsets among the source node, the relay, and the destination will also degrade the cooperative transmission, and the frequency offset estimation in cooperative OFDMA is also a challenging work.

In this section, point-to-point OFDMA frequency offset estimation is first introduced and then we discuss the frequency offset estimation in cooperative transmission. Perfect time synchronization is assumed. Note that only the training-sequence and/or pilot-aided frequency offset estimation is considered in this chapter. Although the proposed cooperative scheme can improve the performance of any conventional training/pilot-based algorithm, the exact training/pilot design is beyond the scope of this chapter.

18.4.1 Point-to-Point Frequency Offset Estimation

All the algorithms proposed in Refs. [3–6,35–41] can perform well in a point-to-point OFDMA transmission. Reference [3] discusses CP-based synchronization for a multiuser OFDMA system. A reliable OFDMA uplink synchronization algorithm, which can be performed within one block by using periodically modulated pilots, is proposed in Ref. [5]. A high-performance maximum likelihood (ML) algorithm for both synchronization and channel estimation for an OFDMA uplink transmission is studied in Ref. [39], and the complexity is reduced by employing an

$N = 1024$; 64 users; $L_u = 10^{-1}$; $p = 10^{-2}$

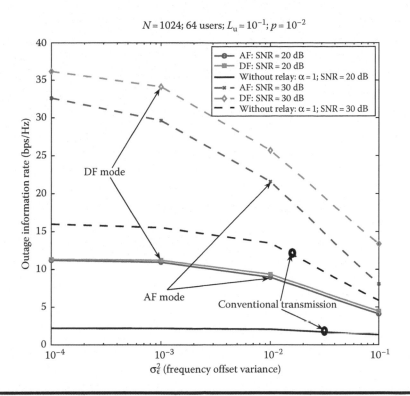

Figure 18.6 Outage information rate as a function of frequency offset variance σ_ϵ^2. (Adapted from Z. Zhang, C. Tellambura, and R. Schober, *Euro. Trans. Telecommun.*, DOI: 10.1002/eH.1385, 2009.)

alternating-projection method. An iterative time and frequency synchronization scheme using the space-alternating generalized expectation-maximization (SAGE) algorithm for an interleaved OFDMA uplink system is proposed in Ref. [6], where the MAI can be mitigated in case multiple asynchronous users are accessing the system. MAI cancellation in an OFDMA system is discussed in Ref. [42], and the MAI is eliminated correctly based on the perfect knowledge of the frequency offset of each user. Any conventional estimator, for example, [43], can be used for frequency offset estimation in Ref. [42].

However, the computational complexity of many existing algorithms is high. A simplified ML pilot-aided frequency offset estimation in OFDMA is proposed in Ref. [35] to reduce the complexity. Assuming that there are totally M users accessing the base station, the N subcarriers are divided into MQ subbands, and each subband has $P = N/(MQ)$ subcarriers. The subcarriers in the lth subband are $lP + m$, where $0 \leq l \leq MQ - 1$ and $m = 1, \ldots, P - 1$. Interleaved subcarriers are allocated to each user. In this algorithm, the distances between any two pilot subcarriers are required to be large enough.

By using $Y_k(n)$ to represent the post-discrete Fourier transform (DFT) signal of the kth user in the nth subcarrier, we can approximate $Y_k(n \pm 1/2)$ as

$$Y_k(n \pm 1/2) \cong |H_k^{(n)} X_k(n)| \frac{\cos(\pi \varepsilon_k)}{\pi(1/2 \mp \varepsilon_k)}, \tag{18.18}$$

where $H_k^{(n)}$ and $X_k(n)$ represent the channel attenuation and the pilot, respectively, of the kth user in the nth subcarrier, and ε_k is the frequency offset of the kth user. The frequency offset ε_k can be estimated as

$$\hat{\varepsilon}_k = \frac{\lambda_k/\rho_k - 1}{2(\lambda_k/\rho_k + 1)}, \tag{18.19}$$

where $\lambda_k = \sum_{n \in \mathcal{B}_k} |Y_k(n+1/2)|$ and $\rho_k = \sum_{n \in \mathcal{B}_k} |Y_k(n-1/2)|$, and $\mathcal{B}_k$ represents the subband allocated to the kth user.

Although the computational complexity is very low ($\mathrm{O}(N \log_2 N)$), the estimation accuracy is comparable to that in the existing algorithms, and a variance error of 10^{-3} can be obtained if the SNR is higher than 10 dB.

18.4.2 Cooperative Frequency Offset Estimation

Although many existing algorithms can achieve a high-performance frequency offset estimation in point-to-point OFDMA, temporarily deep fading may degrade their accuracy. In cooperative transmission, since the same training sequence is transmitted through different paths via cooperative relays, frequency offset estimation can be improved by exploiting the synchronization information from multiple copies of the training sequence at the destination. Frequency offset estimation for cooperative OFDM/OFDMA is discussed in Refs. [27,44–48].

18.4.2.1 Frequency Offset Estimation in Space–Time Diversity System

A space–time cooperative system based on OFDM is proposed in Ref. [44], and the proposed frequency synchronization algorithm can be easily adapted in OFDMA systems.

In Ref. [44], each frame is divided into two subframes, that is, the listening subframe and the cooperation subframe. Two-phase cooperative transmission is considered in this paper. In each subframe, the Preamble area is composed of a length-N_{sync} synchronization subblock, a CP, and a length-N_{ce} channel estimation subblock.

The synchronization subblock can be represented as

$$\mathbf{s}_{\mathrm{sync}} = \left\{ \underbrace{s_s}_{\text{length-}T_s}, s_s, \ldots, -s_s \right\} \tag{18.20}$$

$$\underbrace{\phantom{\left\{ s_s, s_s, \ldots, -s_s \right\}}}_{\text{length-}N_{\mathrm{sync}} \times T_s}$$

and the received synchronization subblock is represented as $\mathbf{r}_v = \{r_v[0], r_v[1], \ldots, r_v[N_{\mathrm{sync}} - 1]\}$, where $v = \{D, R\}$. By using the phase rotation between the received signals, the frequency offset can be estimated. In consideration of the timing offset Δ, the frequency offset ε_v can be estimated as

$$\hat{\varepsilon}_v = \frac{1}{2\pi L_{\mathrm{sync}} T_s} \arg \left\{ \sum_{n=0}^{n'-1} G[n] \right\}, \tag{18.21}$$

where $G[n] = \sum_{l=0}^{L_{\mathrm{sync}}-1} r_v^*[l + \Delta + nL_{\mathrm{sync}}] r_v[l + \Delta + (n+1)L_{\mathrm{sync}}]$, and $n' \geq 1$ represents the number of consecutive subblocks used to perform frequency offset estimation.

This algorithm is very simple to design, and as in Ref. [4], it is very robust to the fading channels. In addition, the proposed preamble structure can easily be adapted to a cooperative transmission with multiple relays.

18.4.2.2 Relay Selection in Multiple Relay Cooperation

In a cooperative transmission with multiple relays, the channel attenuation for some relays may be good, but some are not. In this case, the relays with a deep fading should not be used to perform retransmission, and as a result, the relay selection should be performed to optimize the cooperative transmission. Relay selection in multiple relay cooperative transmission is shown in Figure 18.7. Only pilot/training-based algorithms will be considered in this subsection.

When performing the frequency offset estimation, the cooperative transmission is also performed within two time slots, but the time slot here is conceptually interchangeable with the training sequence (each training sequence is composed of $T \geq 1$ OFDMA symbols). The forward operation depends on which relaying mode is applied. If the AF mode is applied, the relay simply amplifies and retransmits the received training sequence (including the additive noise) to D; if the DcF mode is applied, the chosen relay first decodes the received symbols and estimates the frequency offset between S and itself, and then the relay will use the estimated frequency offset to pre-compensate for the frequency offset between S and itself and then retransmit the regenerated training sequence to D.

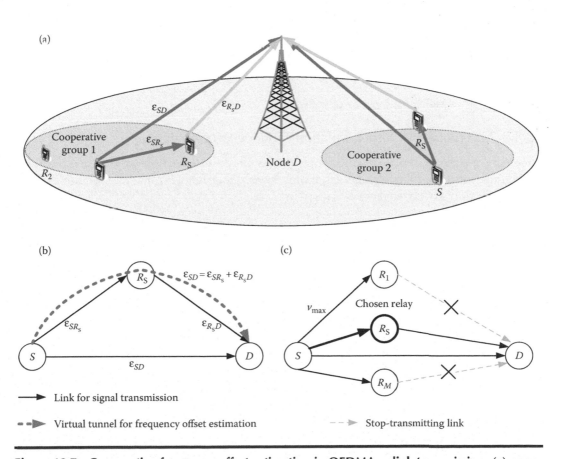

Figure 18.7 **Cooperative frequency offset estimation in OFDMA uplink transmission: (a) cooperative group organization, (b) virtual tunnel created by cooperative relay, (c) relay selection in cooperative OFDMA uplink transmission.**

18.4.2.2.1 Relay Selection

In the first time slot, by considering the channel coefficients and frequency offsets, we assume that the oscillators of the mobile nodes should be calibrated with the oscillator of the base station and, therefore, that each ε_{zD}, $z \in \{S, R_1, \ldots, R_M\}$, can be approximated as an i.i.d. RV with mean zero and variance σ_ε^2 (but not necessarily be Gaussian distributed). Since $\varepsilon_{SD} = \varepsilon_{SR_k} + \varepsilon_{R_kD}$ is satisfied for each k, we have $\mathrm{Var}\{\varepsilon_{zD}\} = \sigma_\varepsilon^2$ and $\mathrm{Var}\{\varepsilon_{SR_k}\} = \mathrm{Var}\{\varepsilon_{SD} - \varepsilon_{R_kD}\} = 2\sigma_\varepsilon^2$, $z \in \{S, R_1, \ldots, R_M\}$.

Based on the interference analysis in Ref. [25], the effective SINR at nodes D and R_k are given by $\gamma_{SD,1} = \alpha\bar{P}\beta_1 \cdot v_{SD}/(\mathcal{L}_u\pi^2 N_u\sigma_\varepsilon^2\alpha\bar{P}/3) + N_u\sigma_w^2)$ and $\gamma_{SR_k,1} = \alpha\bar{P}\beta_2 \cdot v_{SR_k}/(2\pi^2 N_u\sigma_\varepsilon^2\alpha\bar{P}/3) + N_u\sigma_w^2)$, respectively, where $v_{Sb} = \mathrm{trace}\{\mathbf{H}_{Sb}\mathbf{H}_{Sb}^H\}$, $\beta_1 = \left(1 - (\pi^2\sigma_\varepsilon^2/3) + (\pi^4\sigma_\varepsilon^4/20)\right)$ and $\beta_2 = \left(1 - (2\pi^2\sigma_\varepsilon^2/3) + (\pi^4\sigma_\varepsilon^4/5)\right)$.

In the second time slot, one of the M candidates, that is, $R_1, \ldots, R_M$, will be chosen as the relay of S. R_s is chosen as the relay of S if

$$R_s = \arg \max_{R_1,\ldots,R_M} \{v_{SR_1}, \ldots, v_{SR_M}\}, \tag{18.22}$$

as shown in Figure 18.7c.* The expectation of v_{DR_s} is derived in Ref. [27] as

$$\bar{v}_{\max} = \int_0^\infty v_{\max}f(v_{\max})\,dv_{\max} = M \sum_{k=0}^{M-1} \binom{M-k}{k} \sum_{n=0}^{N_u-1(M-k)} \frac{g_k^{(n)}(0)}{(M-k)^2(M-k-1)!}. \tag{18.23}$$

For a larger M, a higher $\bar{v}_{\max}$ can be obtained.

18.4.2.2.2 Frequency Offset Estimation Variance Error Analysis

If the relay operates in the AF mode, R_s simply retransmits the received training sequence, including the additive noise, by using a proper retransmit power. From Ref. [25], the received SINR in the second time slot in node reference D is given by

$$\gamma_2^{AF} = \frac{\mathcal{L}_u\alpha(1-\alpha)\bar{P}^2\beta_1\beta_2\bar{v}_{\max}v_{R_sD}}{(2\mathcal{L}_u\alpha\bar{P}\pi^2 N_u\sigma_\varepsilon^2/3) + N_u\sigma_w^2 + ((\mathcal{L}_u(1-\alpha)\bar{P}\pi^2 N_u\sigma_\varepsilon^2/3) + N_u\sigma_w^2)\xi_{SR_s}}, \tag{18.24}$$

where $\xi_{SR_s} = \mathcal{L}_u\alpha\bar{P}\beta_2\bar{v}_{\max} + (2\mathcal{L}_u\alpha\bar{P}\pi^2 N_u\sigma_\varepsilon^2/3) + N_u\sigma_w^2$.

If the relay operates in the DcF mode, the relay should first demodulate and decode the received training sequence, and then regenerate this training sequence and retransmit it in the second time slot. From Ref. [25], the average SINR in the second time slot in node D is

$$\gamma_2^{DcF} = \frac{(1-\alpha)\bar{P}\beta_1 \cdot v_{R_sD}}{(\mathcal{L}_u\pi^2 N_u\sigma_\varepsilon^2(1-\alpha)P/3) + N_u\sigma_w^2}. \tag{18.25}$$

The training sequences received in both time slots can be used to perform frequency offset estimation. From Refs. [4,43,50] we know that for an unbiased estimator, the Cramer–Rao lower

---

* It was proven by Bletsas et al. [49] that the opportunistic relaying strategy by using the "best" relay to perform retransmission is optimal in terms of outage probability.

bound (CRLB) is inversely proportional to the SINR and $\mathcal{A}_T$ (a positive coefficient specified by the structure of the training sequence $\mathbf{X}_S$. For example, by using the training sequence proposed in Ref. [43], we have $\mathcal{A}_T = 4\pi^2 N_u$). In the AF mode, for a given power allocation ratio α and (v_{SD}, v_{SR_s}, v_{R_sD}), the minimum variance of the estimation can be achieved as

$$\text{Var}\left\{e_{SD}|0 < \alpha < 1; v_{SD}, v_{SR_s}, v_{R_sD}\right\} \geq \frac{1}{\mathcal{A}_T(\gamma_{SD,1} + \gamma_2^{AF})}. \tag{18.26}$$

Similarly, the minimum variance for the DcF mode can be achieved as

$$\text{Var}\left\{e_{SD}|0 < \alpha < 1; v_{SD}, v_{SR_s}, v_{R_sD}\right\} \geq \frac{1}{\mathcal{A}_T(\gamma_{SD,1} + (\gamma_{SR_s,1}\gamma_2^{DcF}/\gamma_{SR_s,1} + \gamma_2^{DcF}))}. \tag{18.27}$$

Both Equations 18.26 and 18.27 are functions of power allocation ratio α and can be minimized by using the optimal power allocation ratio α. However, the optimal power allocation ratio α depends on whether or not the base station sends CSI to the mobile nodes. With feedback, the mobile nodes can adaptively optimize power allocation ratio α based on the current (v_{SD}, v_{SR_s}, v_{R_sD}) values. If the base station does not feedback CSI, then α can be optimized based on only the statistical information of (v_{SD}, v_{SR_s}, v_{R_sD}).

A hybrid cooperative scheme can be performed to adaptively optimize the transmission, as illustrated in Figure 18.8. For a given (M,σ_ϵ^2,SNR), if we can find a power allocation ratio α ($0 < \alpha < 1$) to make the cooperative transmission outperform the conventional transmission, the transmitter should be switched to "2" to perform the cooperative transmission; if not, the transmitter can be switched to "1" to perform the conventional transmission. The cooperative transmission may be performed in two cases: (1) if the base station does not feedback CSI, the second switch should be switched to "21" to perform "Without Feedback" cooperation; and (2) if the base station feedbacks the CSI to the mobile nodes, the second switch should be switched to "22" to perform "With Feedback" cooperation. An information-sharing scheme should be performed between S and R_s to guarantee that an identical power allocation ratio α will be used by them in the same transmission, but how to perform this information-sharing scheme is beyond the scope of this chapter.

18.4.2.2.3 Numerical Results

Our simulation considers an OFDMA uplink system with a DFT length of 1024. A CP of length 64 is used. Frequency offsets are assumed to be i.i.d. RVs with mean zero and variance σ_ϵ^2. The algorithm proposed in Ref. [43] is used here.

Figure 18.9 compares the variance errors of the proposed cooperative scheme and that of the conventional estimation as functions of power allocation ratio α (the latter is independent of power allocation ratio α), where SNR = 20 dB and $M = 16$. When power allocation ratio α is small, the proposed scheme may achieve a higher variance error than that achieved in the conventional estimation. When $\sigma_\epsilon^2 = 10^{-3}$, the ranges of power allocation ratio α that make the proposed AF and DcF schemes outperform the conventional scheme are $0.13 < \alpha < 1$ and $0.56 < \alpha < 1$, respectively. For each power allocation ratio α, the AF mode always outperforms the DcF mode.

Figure 18.10 evaluates the performance as a function of M when the base station feedbacks CSI to the mobile nodes. In either the AF or DcF mode, the variance error of the proposed cooperative

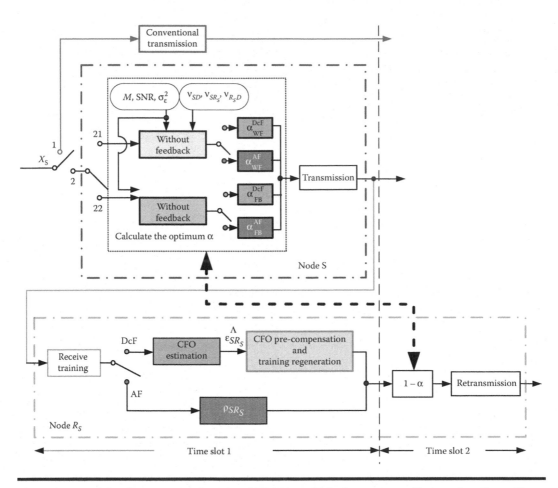

Figure 18.8 Adaptive cooperation in OFDMA uplink frequency offset estimation. (From Z. Zhang, W. Zhang, and C. Tellambura, *IEEE Trans. Wireless Commun.*, 8 (9), September 2009. With permission.)

scheme is always a monotonically decreasing function of M (compared to the proposed scheme, the conventional scheme is independent of M). The AF mode will still outperform the DcF mode for each M in this scenario. We can explain this finding as follows: In the interference-limited cooperative transmission, the interference due to the frequency offset in $S \rightarrow R_s$ link is twice that of either the $S \rightarrow D$ or $R_s \rightarrow D$ link. If the relay operates in the DcF mode, R_s should estimate ε_{SR_s}, and the estimation error will be accumulated and propagated to the final result.

The performance improvement by the proposed cooperative scheme over the conventional noncooperative scheme is shown in Tables 18.1 and 18.2. For a given SNR and frequency offset variance σ_ϵ^2, the proposed cooperative scheme with either the AF or DcF mode can achieve some performance advantage over the conventional one. Since the power allocation between S and R_s can be adaptively optimized in each transmission if the base station feedbacks CSI to the mobile nodes, a much higher performance advantage can be achieved than that achieved by the cooperative scheme without feedback. If the base station feedbacks CSI, the performance advantage by the proposed scheme can be increased to 4.95 and 1.75 dB for the AF and DcF modes, respectively.

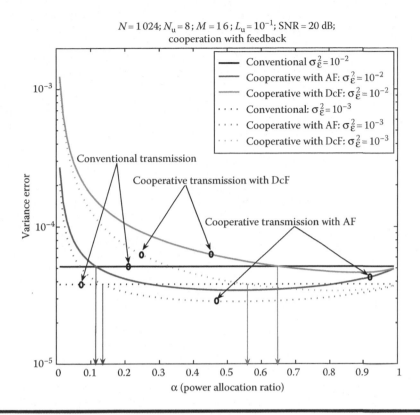

$N = 1\,024; N_u = 8; M = 16; L_u = 10^{-1}; \text{SNR} = 20 \text{ dB};$
cooperation with feedback

Figure 18.9 Cooperative frequency offset estimation as a function of power allocation ratio a without feedback from the base station. (From Z. Zhang, W. Zhang, and C. Tellambura, *IEEE Trans. Wireless Commun.*, 8 (9), September 2009. With permission.)

18.4.2.3 Alamouti-Coded Transmission to Mitigate the Effect of Frequency Offsets

Although many high-performance frequency offset estimation algorithms are proposed for cooperative OFDMA, the estimation errors are always nonzero in a noise-limited environment. In order to mitigate ICI caused by the frequency offset, an Alamouti-coded cooperative transmission is proposed in Ref. [45]. Although the proposed method is based on OFDM transmission, it is easily adapted into OFDMA transmissions.

Two-phase cooperative transmission is considered in Ref. [45]: in the first phase, the source node broadcasts signal to the two relays, and the relays forward the received signal to the destination in the second phase. The DF mode is considered in the relays. For a cooperative transmission with two relays, by considering a real-valued wireless channel, a spatial diversity order two can be achieved.

In each subcarrier, the Alamouti coding scheme is assumed in the relays. Let us denote the two relays as R_1 and R_2, respectively. We also denote two consecutive symbols on the kth subcarrier as $X_{k,1}$ and $X_{k,2}$, respectively. Based on the Alamouti code structure, during the first symbol period, R_1 transmits $X_{k,1}$ and R_2 transmits $X_{k,2}$ from the kth subcarrier, respectively. During the second symbol period, $-X_{k,2}^*$ is transmitted by R_1 and $X_{k,1}^*$ is transmitted by R_2. A quasi-static channel is assumed in this transmission.

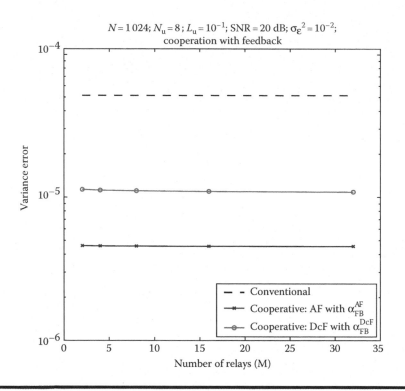

Figure 18.10 Cooperative frequency offset estimation with feedback from the base station. (From Z. Zhang, W. Zhang, and C. Tellambura, *IEEE Trans. Wireless Commun.*, 8 (9), September 2009. With permission.)

In order to eliminate the effect of ICI due to the frequency offset, a symmetric data-conjugate mapping is proposed: in the first OFDM symbol, we map $X_{N-k,1} = X_{k,1}^*$ at the first relay and map $X_{N-k,2} = X_{k,2}^*$ at the second relay; in the second OFDM symbol, we map $X_{N-k,2} = -(X_{k,2}^*)^* = -X_{k,2}$ at the first relay and map $X_{N-k,1} = (X_{k,1}^*)^* = X_{k,1}$ at the second relay, where $k = 1, 2, \ldots, N/2 - 1$.

For a real-valued channel, and assuming $\varepsilon_1 = \varepsilon_2$, where ε_1 and ε_2 are the frequency offsets for the first and the second relays, respectively,* the combined signal at the destination is proportional to $(|H_1|^2 + |H_2|^2)$, where H_1 and H_2 are the channel attenuations for the first and second relays, respectively, and as a result, the full diversity gain can be obtained when the frequency offset is small.

Since a spatial diversity of 2 can be obtained with two relays, the bit error rate (BER) performance of the proposed Alamouti-coded cooperative OFDM can be improved considerably. For example, when the frequency offset is ±0.25 and quadrature phase shift keying (QPSK) modulation is considered, the BER of 10^{-2} can be achieved when $E_b/N_0 = 20$ dB, and it will reduce to about 1.5×10^{-4} when the frequency offset is ±0.05.

---

* The equation $\varepsilon_1 = \varepsilon_2$ can be satisfied by estimating the frequency offsets in the relays firstly, and then adjusting these two offsets to make this equation hold.

Table 18.1 Performance Improvement in the Proposed Cooperative Scheme over the Conventional Noncooperative Algorithm without Feedback from the Base Station

Without Feedback, M = 16							
SNR	*20 dB*				*30 dB*		
σ_ϵ^2	10^{-4}	10^{-3}	10^{-2}	10^{-1}	10^{-4}	10^{-3}	10^{-3}	10^{-1}
AF	1.19	1.21	1.669	2.217	1.533	2.072	3.473	3.236
DcF	0.492	0.438	0.401	0.412	0.454	0.529	0.427	0.471

Source: Adapted from Z. Zhang, W. Zhang, and C. Tellambura, *IEEE Trans. Wireless Commun.,* 8 (9), September 2009.

Table 18.2 Performance Improvement in the Proposed Cooperative Scheme over the Conventional Noncooperative Algorithm with Feedback from the Base Station

With Feedback, M = 16							
SNR	*20 dB*				*30 dB*		
σ_ϵ^2	10^{-4}	10^{-3}	10^{-2}	10^{-1}	10^{-4}	10^{-3}	10^{-2}	10^{-1}
AF	25.35	20.17	10.48	4.95	19.97	10.58	5.56	3.48
DcF	15.86	15.55	6.66	1.75	15.56	6.65	1.73	0.6

Source: Adapted from Z. Zhang, W. Zhang, and C. Tellambura, *IEEE Trans. Wireless Commun.,* 8 (9), September 2009.

18.5 Channel Estimation in Cooperative OFDMA

Besides the frequency offset, the channel estimation errors can also significantly impact the system performance. The AF channel statistics, which are not Gaussian, are analyzed in Ref. [51]. Since relay networks are virtual multiple input–multiple output (MIMO) systems [20], many existing MIMO channel estimation algorithms, for example, those proposed in Refs. [52–55], can also be adapted. Channel estimations for cooperative networks are considerably discussed in Refs. [44,56, 57]. However, the optimal channel estimation for cooperative OFDM/OFDMA in the presence of the frequency offset is still an open issue.

18.5.1 Signal Model

In this section, we assume that the source node transmits its pilot and data in the first time slot, and the relays will forward the received symbol for the source node in the second time slot.

An $N \times 1$ vector $\tilde{\mathbf{X}}_S = [X_S[0], X_S[1], \ldots, X_S[N-1]]$ is used to represent these symbols sent by the node S. Pilot entries of $\tilde{\mathbf{X}}_S$ are given by the vector $\tilde{\mathbf{X}}_S^p$. A total of $\mathcal{N}_p$ pilots are allocated per node (including the source and the relays), and $\tilde{\mathbf{X}}_S^p$ is nonzero only at locations $(\theta_1, \cdots, \theta_{\mathcal{N}_p})$, where $0 \leq \theta_1 < \cdots < \theta_{\mathcal{N}_p} \leq N-1$. The transmit symbol vector $\tilde{\mathbf{X}}_S$ can be decomposed as $\tilde{\mathbf{X}}_S = \tilde{\mathbf{X}}_S^d + \tilde{\mathbf{X}}_S^p$, where $\tilde{\mathbf{X}}_S^d$ and $\tilde{\mathbf{X}}_S^p$ are $N \times 1$ data and pilot vectors.

In the first time slot, the source node will send the symbols to the relays and the receiver. In the second time slot, the relays will retransmit the received signal from the source S by using the power $(1 - \alpha)N\tilde{P}$, and an identical retransmit power is allocated to each relay.

In the AF mode, each relay simply retransmits the received signal, including MAI and noise, to the destination. The received signal at the destination D can be demodulated as

$$
\mathbf{r}_{D,2}^{AF} = \sum_{k=1}^{M} \underbrace{Q_1(\alpha) \mathbf{E}_{DR_{kS}}^{cir} \mathbf{X}_S^p \mathbf{F}_{(2L-1)}^{H}}_{\mathbf{P}_{DR_{ks}}^{AF} (N \times (2L-1))} \tilde{\mathbf{h}}_{DR_{kS}} + \sum_{k=1}^{M} Q_1(\alpha) \mathbf{E}_{DR_{kS}}^{cir} \mathbf{X}_S^d \mathbf{F}_{(2L-1)}^{H} \tilde{\mathbf{h}}_{DR_{kS}}
$$

$$
+ \underbrace{\sum_{k=1}^{M} Q_2(\alpha) \mathbf{E}_{DR_k}^{cir} \mathbf{W}_\eta \mathbf{F}_{(L)}^{H} \tilde{\mathbf{h}}_{DR_k} + \mathbf{F}^{H} \mathbf{w}_{D,2}}_{\eta_{D,2}^{AF}}, \tag{18.28}
$$

where $Q_1(\alpha) = \sqrt{\alpha(1-\alpha)N\tilde{P}^2/M(\alpha\tilde{P} + \sigma_w^2)}$, $Q_2(\alpha) = \sqrt{(1-\alpha)N\tilde{P}/M(\alpha\tilde{P} + \sigma_w^2)}$, $\mathbf{E}_{DR_{kS}}^{cir} = \mathbf{E}_{DR_k}^{cir}\mathbf{E}_{R_{kS}}^{cir} = \mathbf{E}_{DS}^{cir}$, $\mathbf{E}_{ab}^{cir} = \mathbf{F}^{H}\mathbf{E}_{ab}\mathbf{F}$ is a circulant matrix that is specified by the frequency offset ε_{ab}, $\mathbf{F}(L)$ is the first L rows of $\mathbf{F}$, and $\mathbf{X}_S = \mathbf{X}_S^d + \mathbf{X}_S^p$ is an $N \times N$ diagonal matrix with $\mathbf{X}_S^d = \text{diag}\{\mathbf{X}_S^d\}$ and $\mathbf{X}_S^p = \text{diag}\{\tilde{\mathbf{X}}_S^p\}$, $\mathbf{W}_\eta = \text{diag}\{\eta_{R_k,1}\}$, $\tilde{\mathbf{h}}_{DR_kS} = \left(\tilde{\mathbf{h}}_{R_kS}^{T} \otimes \tilde{\mathbf{h}}_{DR_k}^{T}\right)^{T}$, and $\otimes$ represents a convolutionary product operation.

In the DF mode, each relay first demodulates and decodes the received symbol, but only the relays without a decoding error will re-encode this symbol and retransmit it. We assume that m out of M relays can decode correctly,[*] where $0 \leq m \leq M$, and the received symbol at node D can be demodulated as

$$
\mathbf{r}_{D,2}^{DF} = \mathbf{F}^{H}\mathbf{y}_{D,2}^{DF} = \sum_{k=1}^{m} \underbrace{\sqrt{\frac{(1-\alpha)N\tilde{P}}{m}} \mathbf{E}_{DR_k}^{cir} \mathbf{X}_{R_k}^p \mathbf{F}_{(L)}^{H}}_{\mathbf{P}_{DR_k}^{DF} (N \times L)} \tilde{\mathbf{h}}_{DR_k}
$$

$$
+ \sum_{k=1}^{m} \sqrt{\frac{(1-\alpha)N\tilde{P}}{m}} \mathbf{E}_{DR_k}^{cir} \mathbf{X}_{R_k}^d \mathbf{F}_{(L)}^{H} \tilde{\mathbf{h}}_{DR_k} + \underbrace{\mathbf{F}^{H}\mathbf{w}_{D,2}}_{\eta_{D,2}^{DF}}. \tag{18.29}
$$

In the following subsections, we first introduce the channel estimation without considering the frequency offset, and then we discuss the optimal channel estimation in terms of the least-square (LS) variance errors.

---

[*] A correct decoding is assumed in one relay if and only if there is no decoding error in its received symbol in each subcarrier. The decoding error can be detected by using some error-detection techniques, for example, cyclic redundancy check (CRC) bits, which are beyond the scope of this chapter.

18.5.2 Channel Estimation without Considering the Frequency Offsets

Without considering the frequency offsets, the cooperative channel can be seen as a virtual MIMO channel, which is noise rather than interference limited, and many existing algorithms for MIMO channel estimation can be adapted here.

AF channel estimation in a single carrier (SC) cooperative channel is proposed in Ref. [58]. In multicarrier systems, a per-subcarrier channel estimation for AF OFDM without frequency offset is proposed in Ref. [57]. Only one relay is considered in Ref. [57]. By using the optimal preamble, the minimized CRLB for channel estimation with several power constraints can be obtained.

For the kth subcarrier, let $\mathbf{x}_{S,1}[k]$ and $\mathbf{x}_{S,2}[k]$ represent the signal transmitted by node S in the first and the second time slot, respectively. Also define

$$\mathbf{H} = [H_{DS}^{(k)} \; H_{DR}^{(k)}]^{\mathrm{T}},$$

$$\mathbf{X} = \begin{bmatrix} \mathbf{x}_{S,1}[k] & 0 \\ \mathbf{x}_{S,2}[k] & \frac{\sqrt{\rho R}}{\mathcal{L}_{\mathrm{u}}}\mathbf{x}_{S,1}[k] \end{bmatrix},$$

$$\mathbf{C} = \begin{bmatrix} \sigma_w^2 & 0 \\ 0 & \sigma_w^2(1 + \frac{1}{\mathcal{L}_u^2}) \end{bmatrix}, \quad \text{and}$$

$$\mathbf{Y} = [\mathbf{y}_{D,1}[k] \; \mathbf{y}_{D,2}[k]]^{\mathrm{T}},$$

where $\mathbf{y}_{D,1}[k]$ and $\mathbf{y}_{D,2}[k]$ represent the received vector in the first and second time slots, respectively, at node D. A minimum variance unbiased (MVU) estimator can be designed as

$$\hat{\mathbf{H}}_{\mathrm{MVU}} = (\mathbf{X}^{\mathrm{H}}\mathbf{C}^{-1}\mathbf{X})\mathbf{X}^{\mathrm{H}}\mathbf{C}^{-1}\mathbf{Y}. \tag{18.30}$$

Without considering the frequency offsets, from Ref. [59], the CRLB for the channel estimation can be derived as

$$\mathrm{CRLB}_{\mathbf{H}} = \sigma_w^2 \left[\frac{1}{\mathbf{x}_{S,1}^2[k]} \mathcal{L}_{\mathrm{u}} \times \frac{(1 + \rho_R^2)\mathbf{x}_{S,1}^2[k] + \mathbf{x}_{S,2}^2[k]}{\rho_R^2 \mathbf{x}_{S,1}^4[k]} \right]^{\mathrm{T}}. \tag{18.31}$$

In order to find the optimal preamble design to minimize the CRLB, there are two kinds of power constraints, that is, the sum power constraint (C1) and the individual power constraint (C2). The optimal preamble can be found by resolving the following problem:

$$\text{minimize} \quad \frac{1}{\mathbf{x}_{S,1}^2[k]} + \mathcal{L}_{\mathrm{u}} \times \frac{(1 + \rho_R^2)\mathbf{x}_{S,1}^2[k] + \mathbf{x}_{S,2}^2[k]}{\rho_R^2 \mathbf{x}_{S,1}^4[k]}$$

$$\text{s.t.} \quad \text{(C1)} \; \mathbf{x}_{S,1}^2[k] + \mathbf{x}_{S,2}^2[k] \leq 1;$$

$$\text{or}$$

$$\text{(C2)} \; \mathbf{x}_{S,1}^2[k] + \mathbf{x}_{S,2}^2[k] \leq \frac{1}{2}. \tag{18.32}$$

By resolving Equation 18.32, the optimal preamble for each constraint is given by

$$(\text{C1}) \; \mathbf{x}_{S,1}^2[k] = 1, \quad \mathbf{x}_{S,2}^2[k] = 0, \tag{18.33a}$$

$$(\text{C2}) \; \mathbf{x}_{S,1}^2[k] = \frac{1}{2}, \quad \mathbf{x}_{S,2}^2[k] = 0, \tag{18.33b}$$

and the mean square error (MSE) is

$$(\text{C1}) \; \text{MSE}_{\text{MVU}} = \frac{\sigma_w^2}{2} \left\{ 1 + \frac{\mathcal{L}_u(1 + \rho_R^2)}{\rho_R^2} \right\}, \tag{18.34a}$$

$$(\text{C2}) \; \text{MSE}_{\text{MVU}} = \sigma_w^2 \left\{ 1 + \frac{\mathcal{L}_u(1 + \rho_R^2)}{\rho_R^2} \right\}. \tag{18.34b}$$

Note that Equations 18.33 and 18.34 are based on a two-preamble case. In the case of N-preamble, we can derive the power constraints for the optimal preamble as

$$(\text{C1}) \; \sum_{n=1}^{N/2} \mathbf{x}_{S,2n-1}^2[k] = 1, \quad \mathbf{x}_{S,n}^2[k] = 0, \quad n \in \{2, 4, \ldots, N\} \tag{18.35a}$$

$$(\text{C2}) \; \mathbf{x}_{S,n}^2[k] = \frac{1}{N}, \quad \sum_{n=1}^{N/2} \mathbf{x}_{S,2n-1}^*[k], \mathbf{x}_{S,2n}[k] = 0, \tag{18.35b}$$

and the MSE is

$$(\text{C1}) \; \text{MSE}_{\text{MVU}} = \frac{\sigma_w^2}{2} \left\{ 1 + \frac{\mathcal{L}_u(1 + \rho_R^2)}{\rho_R^2} \right\}, \tag{18.36a}$$

$$(\text{C2}) \; \text{MSE}_{\text{MVU}} = \sigma_w^2(1 + \rho_R^2) \left\{ \frac{\mathcal{L}_u}{\rho_R^2} + \frac{1}{2 + \rho_R^2} \right\}. \tag{18.36b}$$

18.5.3 Optimal Channel Estimation by Considering the Frequency Offsets

Although the minimum MSE can be obtained in Ref. [57], the effect of the frequency offset is not considered yet. Moreover, only one relay is applied. Since the spatial diversity gain obtained by using relays is proportional to the number of relays, using more relays is helpful to improve the cooperative transmission. In this subsection, we will design the optimal pilot for cooperative channel estimation by considering the frequency offsets. Since the channel covariance matrix is assumed to be not available at the receiver, the optimal channel estimation in terms of LS variance error rather than minimum mean square error (MMSE) is studied.

The symbol frame structure is shown in Figure 18.11. Frame consists of T continuous symbols and a preamble. The preamble is transmitted within two time slots, with each time slot being equal to the preamble length. All the relays retransmit the received preamble simultaneously. Although a superposition of multiple copies of the preamble is received at the destination, it is possible for the destination to identify each retransmitted preamble if the preamble is designed properly.

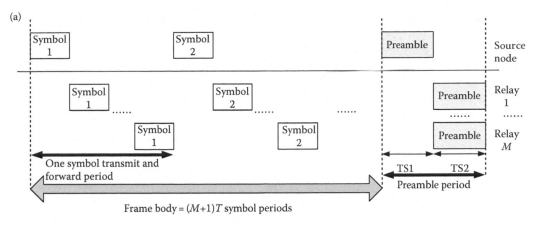

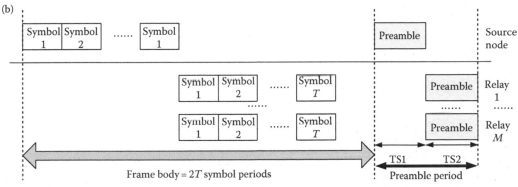

Figure 18.11   Preamble and data frame structure in the proposed cooperative transmission. (a) Relays retransmission in TDMA scheme and (b) relays retransmission in space-time block coding scheme. (From Z. Zhang, W. Zhang, and C. Tellambura, *IEEE Trans. Wireless Commun.*, 8 (9), September 2009. With permission.)

There are M relays. In time division multiple access (TDMA), for the data part, nodes take turns in symbol transmission and only one node transmits at a time. We also consider the spacetime block coding scheme, and the frame body is encoded into space-time block codes. The transmission of each data frame is also performed in two time slots, but the time slot here is conceptually interchangeable with the frame body. Since only $2T$ symbol periods are requested to transmit each space-time code, its transmission efficiency is much higher than that in the TDMA scheme.

Since the first time slot is MAI free and there is no interference among the relays and the receiver, a conventional LS estimator can be applied at each relay or the receiver. However, in the second time slot, the receiver will receive multiple copies of the same pilot from multiple relays, and MAI appears. MAI elimination is critical to the optimal LS channel estimation in the second time slot.

18.5.3.1 Optimal Pilot Design to Eliminate MAI in the Second Time Slot

In cooperative transmission, the received signal at D is a superposition of the retransmitted signal from all the relays. Since an identical pilot, that is, $\mathbf{X}_S^p$, is received at each relay in the first time slot, the received pilot at the destination D in the second time slot is also $\mathbf{X}_S^p$ if the relays simply retransmit the received signal without modifying it. In this case, the destination D does not know where the received pilot comes from and, therefore, node D will not be able to identify the channel related to each relay; that is, MAI appears.

With multiple relay transmissions, the MAI must be eliminated to minimize the MSE. Optimal pilot design in MIMO–OFDM systems in the presence of frequency offsets is discussed in Ref. [55], and the proposed pilots can be readily adapted for this case. Without loss of generality, we assume that the retransmitted pilot for the relay R_k is $\mathbf{X}_{R_k}^p$. The MAI between different relays can be eliminated if

$$(\mathbf{P}_{DR_kS}^{AF})^\dagger \mathbf{P}_{DR_iS}^{AF} = \mathbf{O}_{(2L-1)\times(2L-1)} \quad \text{(AF mode)}, \tag{18.37a}$$

$$(\mathbf{P}_{DR_k}^{DF})^\dagger \mathbf{P}_{DR_i}^{DF} = \mathbf{O}_{L\times L} \quad \text{(DF mode)} \tag{18.37b}$$

is satisfied for each $i \neq k$. In the DF mode, Equation 18.37 can be easily satisfied by modulating different pilots in different relays before their retransmission. However, in the AF mode, Equation 18.37 cannot be satisfied unless each relay does some modification to its received pilot.

18.5.3.1.1 AF Mode

In order to identify the retransmitted pilot from each relay in the receiver, the conventional AF mode should be modified. From Ref. [28], before the kth relay, R_k performs retransmission, its received signal $\mathbf{y}_{R_k,1}$ should be multiplied by a pre-modulation matrix $\mathbf{\Pi}_k$, that is, each relay uses a unique pre-modulation matrix $\mathbf{\Pi}_k$ to do some modification to its pilot subcarriers. We assume that $\mathbf{X}_{R_k}^p$ is a unique pilot of node k and different from $\mathbf{X}_S^p$.* From Ref. [28], $\mathbf{\Pi}_k$ can be resolved as

$$\mathbf{\Pi}_k = \mathbf{F}^H \mathbf{E}_{R_kS} \mathbf{F} \mathbf{\Lambda}_k \mathbf{F}^H \mathbf{E}_{R_kS}^{-1} \mathbf{F}, \tag{18.38}$$

where $[\mathbf{\Lambda}_k]_{\theta_i,\theta_i} = e^{[j2\pi\theta_i(k-1)(2L-1)]/N}$, and $[\mathbf{\Lambda}_k]_{ll} = 0$ for each $l \neq \theta_i$. The optimal pilot that satisfy Equation 18.37 for the kth node in the AF mode is

$$[\mathbf{X}_{R_k}^p]_{\theta_i,\theta_i} = e^{[j2\pi\theta_i(k-1)(2L-1)]/N}, \quad k = 1,\ldots,M, \ i = 1,\ldots,\mathcal{N}_p$$

$$\text{s.t. } (2L-1)M \leq \mathcal{N}_p \leq N, \quad \frac{N}{\mathcal{N}_p} = \text{integer};$$

$$\frac{\theta_i(k-1)(2L-1)}{N} \neq \text{integer}, \quad k \neq l;$$

$$\theta_2 - \theta_1 = \theta_3 - \theta_2 = \cdots = \theta_{\mathcal{N}_p} - \theta_{\mathcal{N}_p-1} = \frac{N}{\mathcal{N}_p}. \tag{18.39}$$

---

* In the following analysis, we assume that a unique pilot is allocated to each relay, and each relay knows its own pilot. A pilot-allocation scheme should be performed to mitigate the collision among pilots of different relays, but how to perform this pilot-allocation scheme is beyond the scope of this chapter. An example of pilot allocation is given by Zhang et al. [55].

In the AF mode, the channel order for all the $S \rightarrow R \rightarrow D$ channels is $2L - 1$. From Equation 18.39, the condition of $(2L - 1)M \leq \mathcal{N}_{\mathrm{p}} \leq N$ must be satisfied for the optimal pilot design, and consequently $M \leq \lfloor N/2L - 1 \rfloor$.

18.5.3.1.2 DF Mode

In the DF mode, each correct-decoding relay will retransmit in the second time slot by modulating the pilot subcarriers with its own pilot but without changing the data subcarriers. The optimal pilot for Rk in the DF mode is given by

$$[\mathbf{X}_{R_k}^p]_{\theta_i \theta_i} = e^{j 2\pi \theta_i (k-1)L/N}, \quad k = 1, \ldots, M, \quad i = 1, \ldots, \mathcal{N}_{\mathrm{p}}$$

$$\text{s.t. } LM \leq \mathcal{N}_{\mathrm{p}} \leq N, \quad \frac{N}{\mathcal{N}_{\mathrm{p}}} = \text{integer};$$

$$\frac{\theta_i(k-1)L}{N} \neq \text{integer}, \quad k \neq l;$$

$$\theta_2 - \theta_1 = \theta_3 - \theta_2 = \ldots = \theta_{\mathcal{N}_{\mathrm{p}}} - \theta_{\mathcal{N}_{\mathrm{p}}-1} = \frac{N}{\mathcal{N}_{\mathrm{p}}}. \tag{18.40}$$

In the DF mode, the channel order for each $R \rightarrow D$ channel is L. From Equation 18.40, the condition of $LM \leq \mathcal{N}_{\mathrm{p}} \leq N$ must be satisfied for the optimal pilot design, and we can easily conclude that $M \leq \lfloor \frac{N}{L} \rfloor$.

For a given L, the maximum number of active relays allowable in the DF mode is almost twice that of the AF mode. Thus, the DF spatial diversity gain is higher than that of the AF mode.

18.5.3.2 PEP Analysis

In this subsection, the PEP of cooperative OFDM by considering both the frequency offset and channel estimation errors is derived. An orthogonal space-time signal matrix $\tilde{\mathbf{X}}_S = [\tilde{\mathbf{X}}_S(1), \tilde{\mathbf{X}}_S(2), \ldots, \tilde{\mathbf{X}}_S(T)]$, which is $N \times T$ matrix, is assumed.

18.5.3.2.1 PEP for the AF Mode

The probability that $\bar{\mathbf{X}}_S$ will be mistaken for another code $\bar{\mathbf{L}}_S$ is upper bounded by [60]

$$\mathrm{P}_{\mathrm{r}}^{\mathrm{AF}}\left\{\bar{\mathbf{X}}_S \rightarrow \bar{\mathbf{L}}_S | 0 < \alpha < 1\right\} \leq \left(\prod_{n=0}^{2L-2} \frac{1}{1 + (\bar{\gamma}_{\mathrm{DRS},n}\ell_n/4)}\right)\left(\prod_{n=0}^{L-1} \frac{1}{1 + (\bar{\gamma}_{\mathrm{DS},n}\ell_n/4)}\right), \tag{18.41}$$

where $\bar{\gamma}_{\mathrm{DS},n}$ and $\bar{\gamma}_{\mathrm{DRS},n}$ represent the SINR of the $S \rightarrow D$ and $S \rightarrow R \rightarrow D$ channels, respectively, in the nth multipath tap, and ℓ_n is the nth eigenvalue of $(\bar{\mathbf{X}}_S - \bar{\mathbf{L}}_S)(\bar{\mathbf{X}}_S - \bar{\mathbf{L}}_S)^{\mathrm{H}}$. In a high SINR regime with $\sigma_e^2 \rightarrow 0$, $\bar{\gamma}_{\mathrm{DS},n}$ and $\bar{\gamma}_{\mathrm{DRS},n}$ can be approximated as $\lim\limits_{\substack{\sigma_e^2 \rightarrow 0 \\ \mathrm{SNR} \rightarrow \infty}} \bar{\gamma}_{\mathrm{DRS},n} \rightarrow \alpha \mathrm{SNR} \cdot |\tilde{\mathbf{h}}_{D,S}[n]|^2$ and

$\lim\limits_{\substack{\sigma_e^2 \rightarrow 0 \\ \mathrm{SNR} \rightarrow \infty}} \bar{\gamma}_{\mathrm{DRS},n} \rightarrow \alpha(1 - \alpha)\mathrm{SNR} \cdot \left|\sum_{k=1}^{M} \tilde{\mathbf{h}}_{D,R_k,S}[n]\right|^2 / M[(\mathcal{L}_u(M + 1)(1 - \alpha)/\alpha) + 1]$, where

$\text{SNR} = \bar{P}/\sigma_w^2$ denotes the average SNR, and Equation 18.41 can be rewritten as

$$
\lim_{\substack{\sigma_\epsilon^2 \to 0 \\ \text{SNR} \to \infty}} \mathbf{P}_r^{\text{AF}} \left\{ \bar{\mathbf{X}}_S \to \bar{\mathbf{L}}_S | 0 < \alpha < 1 \right\} \leq \underbrace{\left(\frac{4[(\mathcal{L}_u(M+1)(1-\alpha)/\alpha) + 1]}{\alpha(1-\alpha)\text{SNR}} \right)^{2L-1} \left(\frac{4}{\alpha\text{SNR}} \right)^L}_{\text{multipath diversity gain}}
$$

$$
\times \underbrace{\left(\prod_{n=0}^{2L-2} \frac{M}{\left| \sum_{k=1}^{M} \tilde{\mathbf{h}}_{D,R_k,S}[n] \right|^2 \ell_n} \right)}_{\text{multirelay diversity gain}} \left(\prod_{n=0}^{L-1} \frac{1}{\left| \tilde{\mathbf{h}}_{D,S}[n] \right|^2 \ell_n} \right). \tag{18.42}
$$

When L is small, the M-order multirelay diversity dominates the diversity gain in a high SINR regime, and a larger M implies a smaller PEP.

18.5.3.2.2 PEP for the DF Mode

In the DF mode, each relay decodes the received signal from S, and then retransmits it if there is no decoding error. The relays with decoding errors will not retransmit. By using P_{relay} to represent the average probability of decoding error at each relay, the probability that m out of M relays successfully decode the received signal is a binomial distribution, that is, $P_{\text{relay},m} = \binom{M}{m} (1 - P_{\text{relay}})^m P_{\text{relay}}^{M-m}$.
We also use $P_{S \to D}$ to represent the probability of decoding error at D in the first time slot.

In the second time slot, m relays with correct decoding will perform retransmission. The PEP that $\bar{\mathbf{X}}_S$ will be mistaken for another codeword $\bar{\mathbf{L}}_S$ is upper bounded by

$$
\mathbf{P}_{r,m}^{\text{DF}} \left\{ \bar{\mathbf{X}}_S \to \bar{\mathbf{L}}_S | 0 < \alpha < 1 \right\} \leq \prod_{n=0}^{L-1} \frac{1}{1 + \bar{\gamma}_{DR,m,n} \ell_n / 4}, \tag{18.43}
$$

where $\bar{\gamma}_{DR,m,n}$ represents the SINR of the $R \to D$ channel in the nth multipath tap. The averaged PEP of the DF mode is upper bounded by

$$
\overline{\text{PEP}^{\text{DF}}} \leq P_{S \to D} \cdot \sum_{m=0}^{M} P_{\text{relay},m} \mathbf{P}_{r,m}^{\text{DF}} \left\{ \bar{\mathbf{X}}_S \to \bar{\mathbf{L}}_S \, \middle| \, 0 < \alpha < 1 \right\}. \tag{18.44}
$$

In the high SINR regime with $\sigma_e^2 \to 0$, Equation 18.43 can be approximated as

$$
\lim_{\substack{\sigma_e^2 \to 0 \\ \text{SNR} \to \infty}} \mathbf{P}_r^{\text{DF}} \left\{ \bar{\mathbf{X}}_S \to \bar{\mathbf{L}}_S | 0 < \alpha < 1 \right\} \leq \underbrace{\left(\frac{4}{(1-\alpha)\text{SNR}} \right)^L}_{\text{multipath diversity gain}} \times \underbrace{\left(\prod_{n=0}^{L-1} \frac{m}{\left| \sum_{k=1}^{m} \tilde{\mathbf{h}}_{D,R_k}[n] \right|^2 \ell_n} \right)}_{\text{multirelay diversity gain}}
$$

$$
\tag{18.45}
$$

Comparison of Equations 18.42 and 18.45 shows that for a given M, in a high SINR regime with $\sigma_e^2 \to 0$, the AF mode outperforms the DF mode in terms of diversity gain. However, in

Table 18.3 Performance Comparison between AF and DF Modes

Relaying Mode	AF	DF
Complexity	Low	High
Channel estimation accuracy	Low	High
Maximum concatenated channel delay	$2L-1$	L
Maximum number of relays	$\lfloor\frac{N}{2L-1}\rfloor$	$\lfloor\frac{N}{L}\rfloor$
Multipath diversity gain	High	Low
Spatial diversity gain	Low	High
PEP	High	Low

Note: Cited from Table 1 in [28].

real systems, the DF mode usually outperforms the AF mode for the following reasons. First, for a given L, the maximum number of active relays used in the DF mode is almost twice that of the AF mode, and the achievable cooperative diversity gain in the DF mode is much higher than that obtained in the AF mode. Second, by considering the frequency offset, the OFDM transmission is usually interference limited, and the diversity gain obtained in the AF mode may be deteriorated by the interference and noise accumulated in the relays. The interference-mitigation capability in the DF mode provides a performance advantage over the AF mode in the low-SINR regime. A brief performance comparison between the AF and DF modes is shown in Table 18.3.

18.5.3.3 Numerical Results

Uniform power-delay profiles are used between any pair of nodes; that is, $\mathbb{E}\{|h_{R_k,S}(l)|^2\} = 1/L$ and $\mathbb{E}\{|h_{D,R_k}(l)|^2\} = \mathbb{E}\{|h_{D,S}(l)|^2\} = \mathcal{L}_u/L$, where $l = 0, 1, 2, \ldots, L-1$.* We also assume that $N = 128$. The SNR of the pilot subcarriers SNR $= \bar{P}/\sigma_w^2$. The average power of each pilot subcarrier is assumed to be identical to that of each data subcarrier, unless otherwise stated. An i.i.d. frequency offset estimation error is assumed for each of $S \to R$, $R \to D$, and $S \to R \to D$ channels with a variance σ_e^2.

The PEP performance of the proposed cooperative transmission in the presence of both the frequency offset and channel estimation errors is illustrated in Figures 18.12 and 18.13. Since the channel estimation MSE is a function of the variance of the frequency offset estimation error (i.e., σ_e^2), we use σ_e^2 as the only parameter of impairment. $\mathcal{L}_u$ is set to 10^{-1}. All the eigenvalues of $(\bar{X}_S - \bar{L}_S)(\bar{X}_S - \bar{L}_S)^H$ are set to 1.

The PEP performance as a function of σ_e^2 is shown in Figure 18.12, which considers ($L = 4$, $M = 16$) and ($L = 8$, $M = 8$). Both the AF and DF mode PEPs monotonically increase with σ_e^2. For each number of relays, the DF mode always outperforms the AF mode for a small σ_e^2, although both modes approach the same PEP performance for large σ_e^2 (i.e., the system is interference limited rather than noise limited, and the diversity gain cannot be improved through cooperation).

---

* Although other power profiles such as the exponential power profile can be readily considered, our discussion is limited to the uniform profile. The performance analysis in the proposed cooperative transmission based on the uniform profile is also valid in other profiles.

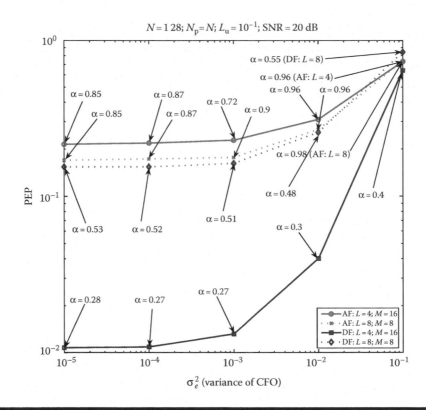

Figure 18.12 PEP of the proposed cooperative transmission as a function of σ_e^2 with $L = 4, 8$. (From Z. Zhang, W. Zhang, and C. Tellambura, *IEEE Trans. Wireless Commun.*, 8 (9), September 2009. With permission.)

The PEP performances as functions of SNR are shown in Figures 18.13 for $L = 4, M = 16$, and $\sigma_e^2 = 10^{-2}$ and 10^{-3}. Since the system is noise limited for a small σ_e^2 ($\sigma_e^2 = 10^{-3}$), the relays realize the spatial diversity gain. The PEP performance of the DF mode is about 9 dB better than that of the AF mode at an error rate of 5×10^{-3}. As σ_e^2 increases to 10^{-2}, the system becomes interference limited, and an error floor appears in both the AF and DF modes. In this environment, the performance increases to about 11.3 dB; that is, the DF mode has a higher interference-mitigation capability than the AF mode.

18.6 Conclusions

This chapter discussed cooperative OFDMA uplink transmission with AF or DF relays. Since the accumulation of interference and noise in the relays in the AF mode degraded the effective SINR, the DF mode outperformed the AF mode in terms of the outage information rate, with an additional relay complexity being required in the former. Cooperative transmission also improved the performance of the frequency offset estimation. In the multiple-relay scenario, relay selection should be performed to optimize the transmit efficiency, and the AF mode outperformed the DcF mode in terms of estimation accuracy. The channel estimation in the cooperative networks was also studied. In the optimal LS channel estimation in the presence of the frequency offsets, the DF mode

$N = 128; N_p = N; L_u = 10^{-1}; L = 4; M = 16$

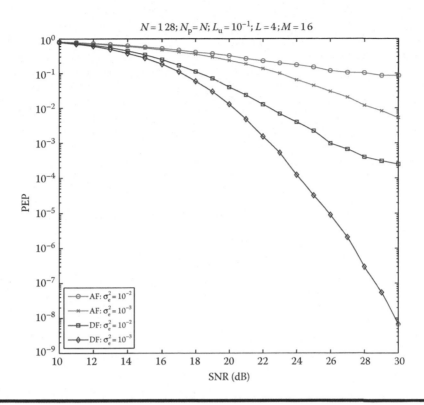

Figure 18.13 **PEP of the proposed cooperative transmission as a function of SNR with $L = 4$, $M = 16$ and $\sigma_e^2 = 10^{-2}, 10^{-3}$. (From Z. Zhang, W. Zhang, and C. Tellambura, *IEEE Trans. Wireless Commun.*, 8 (9), September 2009. With permission.)**

had a better performance than the AF mode in terms of PEP, because the maximum number of the active relays in the DF mode could be as many as twice that in the AF mode.

References

1. IEEE Std 802.16-2004, Part 16: Air interface for fixed broadband wireless access systems, IEEE standard for local and metropolitan area networks, October 2004.
2. Draft 802.16e/D9, Part 16: Air interface for fixed and mobile broadband wireless access systems, IEEE standard for local and metropolitan area networks, June 2005.
3. J.-J. van de Beek, P. Borjesson, M.-L. Boucheret, D. Landstrom, J. Arenas, P. Odling, C. Ostberg, M. Wahlqvist, and S. Wilson, A time and frequency synchronization scheme for multiuser OFDM, *IEEE J. Sel. Areas Commun.*, 17 (11), 1900–1914, 1999.
4. M. Morelli, Timing and frequency synchronization for the uplink of an OFDMA system, *IEEE Trans. Commun.*, 52 (2), 296–306, 2004.
5. Z. Cao, U. Tureli, and Y.-D. Yao, Deterministic multiuser carrier-frequency offset estimation for interleaved OFDMA uplink, *IEEE Trans. Commun.*, 52 (9), 1585–1594, 2004. [Online]. Available: http://dx.doi.org/10.1109/TCOMM.2004.833183
6. J.-H. Lee and S.-C. Kim, Time and frequency synchronization for OFDMA uplink system using the SAGE algorithm, *IEEE Trans. Wirel. Commun.*, 6 (4), 1176–1181, 2007.

7. I. Koffman and V. Roman, Broadband wireless access solutions based on OFDM access in IEEE 802.16, *IEEE Commun. Mag.*, 40 (4), 96–103, 2002.

8. A. Tonello, N. Laurenti, and S. Pupolin, Analysis of the uplink of an asynchronous multi-user DMT OFDMA system impaired by time offsets, frequency offsets, and multi-path fading, in *Proc. IEEE Vehicular Technology Conf. (VTC)*, vol. 3, September 2000, pp. 1094–1099.

9. C. Pan, Y. Cai, and Y. Xu, Adaptive subcarrier and power allocation for multiuser MIMO-OFDM systems, in *Proc. IEEE Int. Conf. Communications (ICC)*, vol. 4, May 2005, pp. 2631–2635.

10. J. M. Choi, J. S. Kwak, H. S. Kim, and J. H. Lee, Adaptive subcarrier, bit, and power allocation algorithm for MIMO-OFDMA system, in *Proc. IEEE Vehicular Technology Conf. (VTC)*, vol. 3, May 2004, pp. 1801–1805.

11. V. Tarokh, N. Seshadri, and A. R. Calderbank, Space-time codes for high data rate wireless communication: performance criterion and code construction, *IEEE Trans. Inform. Theory*, 44, (2), 744–765, 1998.

12. S. M. Alamouti, A simple transmit diversity technique for wireless communications, *IEEE J. Sel. Areas Commun.*, 16 (8), 1451–1458, 1998.

13. A. Sendonaris, E. Erkip, and B. Aazhang, Increasing uplink capacity via user cooperation diversity, in *Proc. IEEE Int. Symp. on Infor. Theory (ISIT)*, Cambridge, MA, August 1998.

14. T. Cover and A. E. Gamal, Capacity theorems for the relay channel, *IEEE Trans. Inform. Theory*, 25 (5), 572–584, 1979.

15. J. N. Laneman and G. W. Wornell, Distributed space-time-coded protocols for exploiting cooperative diversity in wireless networks, *IEEE Trans. Inform. Theory*, 49 (10), 2415–2425, 2003.

16. M. Gastpar and M. Vetterli, On the capacity of large gaussian relay networks, *IEEE Trans. Inform. Theory*, 51 (3), 765–779, 2005.

17. K.-D. Lee and V. C. M. Leung, Capacity planning for group-mobility users in OFDMA wireless networks, *EURASIP J. Wirel. Commun. Net.*, Volume 2006, pp. 1–12, 2006.

18. V. Sreng, H. Yanikomeroglu, and D. D. Falconer, Relayer selection strategies in cellular networks with peer-to-peer relaying, in *Proc. IEEE Vehicular Technology Conf. (VTC)*, vol. 3, October 2003, pp. 1949–1953.

19. M. Yu and J. Li, Is amplify-and-forward practically better than decode-and-forward or vice versa? in *Proc. IEEE Int. Conf. Acoustics, Speech, and Signal Processing (ICASSP)*, vol. 3, March 2005.

20. R. U. Nabar, H. Bolcskei, and F. W. Kneubuhler, Fading relay channels: Performance limits and space-time signal design, *IEEE J. Select. Areas Commun.*, 22 (6), 1099–1109, 2004.

21. Z. Yi and I.-M. Kim, Joint optimization of relay-precoders and decoders with partial channel side information in cooperative networks, *IEEE J. Select. Areas Commun.*, 25 (2), 447–458, 2007.

22. B. Can, H. Yomo, and E. De Carvalho, Hybrid forwarding scheme for cooperative relaying in OFDM based networks, in *Proc. IEEE Int. Conf. Communications (ICC)*, Istanbul, vol. 10, June 2006, pp. 4520–4525.

23. A. S. Avestimehr and D. N. C. Tse, Outage capacity of the fading relay channel in the low-SNR regime, *IEEE Trans. Inform. Theory*, 53 (4), 1401–1415, 2007.

24. T. C.-Y. Ng and W. Yu, Joint optimization of relay strategies and resource allocations in cooperative cellular networks, *IEEE J. Sel. Areas Commun.*, 25 (2), 328–339, 2007.

25. Z. Zhang, C. Tellambura, and R. Schober, Improved OFDMA uplink transmission via cooperative relaying in the presence of frequency offsets—Part I: Ergodic information rate analysis, *Euro. Trans. Telecommun.*, DOI: 10.1002/eH.1384, 2009.

26. Z. Zhang, C. Tellambura, and R. Schober, Improved OFDMA uplink transmission via cooperative relaying in the presence of frequency offsets—Part II: Ergodic information rate analysis, *Euro. Trans. Telecommun.*, DOI: 10.1002/eH.1385, 2009.

27. Z. Zhang, W. Zhang, and C. Tellambura, OFDMA uplink frequency offset estimation via cooperative relaying, *IEEE Trans. Wireless Commun.*, 8 (9), September 2009.

28. Z. Zhang, W. Zhang, and C. Tellambura, Cooperative OFDM channel estimation in the presence of frequency offsets, *IEEE Trans. Veh. Technol.*, 58 (7), September 2009.

29. S. Pfletschinger, Achievable rate regions for OFDMA with link adaptation, in *IEEE Int. Symposium on Personal, Indoor and Mobile Radio Commun. (PIMRC)*, Athens, September 2007, pp. 1–5.

30. S. Gault, W. Hachem, and P. Ciblat, Performance analysis of an OFDMA transmission system in a multicell environment, *IEEE Trans. Commun.*, 55 (4), 740–751, 2007.
31. Z. Zhang and C. Tellambura, The effect of imperfect carrier frequency offset estimation on OFDMA uplink transmission, in *Proc. IEEE Int. Conf. Communications (ICC)*, Glasgow, June 2007.
32. S. E. Elayoubi, B. Fourestie, and X. Auffret, On the capacity of OFDMA 802.16 systems, in *Proc. IEEE Int. Conf. Communications (ICC)*, Istanbul, vol. 4, June 2006, pp. 1760–1765.
33. S. Ko and K. Chang, Capacity optimization of a 802.16e OFDMA/TDD cellular system using the joint allocation of sub-channel and transmit power, in *Advanced Communication Technology, The 9th International Conference on*, Gangwon-Do, vol. 3, February 2007, pp. 1726–1731.
34. G. Li and H. Liu, On the capacity of broadband relay networks, in *Signals, Systems and Computers, 2004. Conference Record of the Thirty-Eighth Asilomar Conference on*, vol. 2, November 2004, pp. 1318–1322.
35. Y. Zeng and A. R. Leyman, Pilot-based simplified ML and fast algorithm for frequency offset estimation in OFDMA uplink, *IEEE Trans. Veh. Technol.*, 57 (3), 1723–1732, 2008.
36. S. Barbarossa, M. Pompili, and G. Giannakis, Channel-independent synchronization of orthogonal frequency division multiple access systems, *IEEE J. Sel. Areas Commun.*, 20 (2), 474–486, 2002.
37. J. Choi, C. Lee, H. W. Jung, and Y. H. Lee, Carrier frequency offset compensation for uplink of OFDM-FDMA systems, *IEEE Commun. Lett.*, 4 (12), 414–416, 2000.
38. H. Bolcskei, Blind high-resolution uplink synchronization of OFDM-based multiple access schemes, in *IEEE Workshop on Signal Processing advances in Wireless Commun.*, Annapolis, MD, 1999, pp. 166–169.
39. M.-O. Pun, M. Morelli, and C.-C. Kuo, Maximum-likelihood synchronization and channel estimation for OFDMA uplink transmissions, *IEEE Trans. Commun.*, 54 (4), 726–736, 2006.
40. X. Dai, Carrier frequency offset estimation and correction for OFDMA uplink, *IET Commun.*, 1, 273–281, 2007.
41. Z. Zhang, W. Zhang, and C. Tellambura, Robust OFDMA uplink synchronization by exploiting the variance of carrier frequency offsets, *IEEE Trans. Veh. Technol.*, 57 (5), 3028–3039, 2008.
42. D. Huang and K. Letaief, An interference-cancellation scheme for carrier frequency offsets correction in OFDMA systems, *IEEE Trans. Commun.*, 53 (7), 1155–1165, 2005.
43. P. Moose, A technique for orthogonal frequency division multiplexing frequency offset correction, *IEEE Trans. Commun.*, 42 (10), 2908–2914, 1994.
44. O. S. Shin, A. M. Chan, H. T. Kung, and V. Tarokh, Design of an OFDM cooperative space-time diversity system, *IEEE Trans. Veh. Technol.*, 56, 2203–2215, 2007.
45. Z. Li and X.-G. Xia, An alamouti coded OFDM transmission for cooperative systems robust to both timing errors and frequency offsets, *IEEE Trans. Wirel. Commun.*, 7, 1839–1844, 2008.
46. X. Li, F. Ng, and T. Han, Carrier frequency offset mitigation in asynchronous cooperative OFDM transmissions, *IEEE Trans. Signal Process.*, 56 (2), 675–685, 2008.
47. N. Benvenuto, S. Tomasin, and D. Veronesi, Multiple frequency offsets estimation and compensation for cooperative networks, in *IEEE Wireless Commun. and Networking Conf.*, Kowloon, March 2007, pp. 891–895.
48. K. S. Woo, H. I. Yoo, Y. J. Kim, H. Lee, H. K. Chung, and Y. S. Cho, Synchronization and channel estimation for OFDM systems with transparent multi-hop relays, in *Proc. IEEE Vehicular Technology Conf. (VTC)*, Dublin, April 2007, pp. 2414–2418.
49. A. Bletsas, H. Shin, and M. Z. Win, Cooperative communications with outage-optimal opportunistic relaying, *IEEE Trans. Wirel. Commun.*, 6 (9), 3450–3460, 2007.
50. Z. Zhang, W. Jiang, H. Zhou, Y. Liu, and J. Gao, High accuracy frequency offset correction with adjustable acquisition range in OFDM systems, *IEEE Trans. Wirel. Commun.*, 4 (1), 228–237, 2005.
51. C. Patel, G. Stuber, and T. Pratt, Statistical properties of amplify and forward relay fading channels, *IEEE Trans. Veh. Technol.*, 55 (1), 1–9, 2006.
52. I. Barhumi, G. Leus, and M. Moonen, Optimal training design for MIMO-OFDM systems in mobile wireless channels, *IEEE Trans. Signal Process.*, 51 (6), 1615–1624, 2003.
53. H. Minn, N. Al-Dhahir, and Y. Li, Optimal training signals for MIMO-OFDM channel estimation in the presnece of frequency offset and phase noise, *IEEE Trans. Commun.*, 54 (6), 1081–1096, 2006.
54. M. Ghogho and A. Swami, Training design for multipath channel and frequency-offset estimation in MIMO systems, *IEEE Trans. Signal Process.*, 54 (10), 3957–3965, 2006.

55. Z. Zhang, W. Zhang, and C. Tellambura, MIMO-OFDM channel estimation in the presence of frequency offsets, *IEEE Trans. Wirel. Commun.*, 7 (6), 2329–2339, 2008.
56. R. A. Iltis, S. Mirzaei, R. Kastner, R. E. Cagley, and B. T. Weals, GEN05-4: Carrier offset and channel estimation for cooperative MIMO sensor networks, in *IEEE Global Telecommn. Conf. (GLOBECOM)*, San Francisco, CA, USA, November 2006, pp. 1–5.
57. K. Kim, H. Kim, and H. Park, OFDM channel estimation for the amply-and-forward cooperative channel, in *Proc. IEEE Vehicular Technology Conf. (VTC)*, Dublin, April 2007, pp. 1642–1646.
58. C. Patel and G. Stuber, Channel estimation for amplify and forward relay based cooperation diversity systems, *IEEE Trans. Wirel. Commun.*, 6 (6), 2348–2356, 2007.
59. S. M. Kay, *Fundamentals of Statistical Signal Processing: Estimation Theory*. Englewood Cliffs, NJ: Prentice Hall, 1993.
60. H. Mheidat, M. Uysal, and N. Al-Dhahir, Equalization techniques for distributed space-time block codes with amplify-and-forward relaying, *IEEE Trans. Signal Process.*, 55, 1839–1852, 2007.
61. Z. Zhang, W. Zhang, and C. Tellambura, Improved OFDMA uplink frequency offset estimation via cooperative relaying: AF or DcF? in *Proc. IEEE Int. Conf. Communications (ICC)*, Beijing, May 2008.

Chapter 19

Performance and Optimization of Relay-Assisted OFDMA Networks

Wern-Ho Sheen and Shiang-Jiun Lin

Contents

19.1 Introduction.. 508
19.2 System Models... 509
    19.2.1 Multicell Architecture... 509
    19.2.2 Subcarrier Allocation and Permutation............................ 509
    19.2.3 Propagation Models.. 511
    19.2.4 Power Setup.. 512
    19.2.5 Frequency Reuse over RS–MS Links................................ 513
19.3 Downlink Optimization... 513
    19.3.1 Objective Function... 513
    19.3.2 Optimization Algorithm... 515
    19.3.3 Multicell Optimization.. 517
19.4 Uplink Optimization.. 517
    19.4.1 Minimization of Transmit Power 518
    19.4.2 Maximization of Spectral Efficiency 519
19.5 System Performance... 521
    19.5.1 Downlink Performance 521
        19.5.1.1 Single-Cell... 521

19.5.1.2 Multicell 526
19.5.2 Uplink Performance.. 526
19.6 Conclusions and Open Issues 528
Appendix: Proofs of Lemmas 1 and 2 532
References... 533

19.1 Introduction

Next-generation mobile communication is envisaged to support high-rate multimedia services in a wide variety of environments: indoors, outdoors, low-mobility, high-mobility, and so on. In Ref. [1], International Telecommunication Union-Radiocommunication Sector (ITU-R) has targeted the next-generation International Mobile Telecommunications (IMT)-advanced system to provide 100 Mbps and 1 Gbps data rates in the high-mobility and stationary/nomadic environments, respectively. Meanwhile, very high spectral efficiency, at least one order of magnitude higher than the current systems, is required to best utilize the very scarce radio spectrum in the future [1]. One key design challenge of the system is to overcome inter-symbol interference (ISI) incurred by high-rate transmission. Furthermore, technology advancements in multiple access, cellular architecture, multiple antenna systems, and others have to be fully explored to achieve the targeted spectral efficiency [2,3].

Orthogonal frequency-division multiplexing (OFDM) is an effective modulation scheme to combat ISI in a high-rate environment. By using parallel orthogonal subcarriers along with cyclic-prefix, ISI can be removed completely as long as the cyclic-prefix is larger than the maximum delay spread of the channel [4]. OFDM can also be used as an effective multiple access scheme [4,5]. In particular, orthogonal frequency-division multiple-access (OFDMA) has been widely considered as one of the most promising multiple access schemes for the next-generation systems [4,5]. OFDMA has been adopted in the 3GPP-LTE (long-term evolution) [6] and IEEE 802.16e [7] specifications.

Cooperative communication with relay stations (RSs) is an emerging technology to improve the performance of a wireless communication system and has been a topic of extensive research both in academia and industry [8–32]. At the link level, by exploiting cooperative diversity, RSs can be used to improve user throughput, outage probability, and error rate in a faded environment [8–17]. At the system level of cellular systems, RSs can be deployed to improve system capacity, extend cell coverage, save transmit power (in the uplink) and provide more uniform data rates to users who are scattered over a cell [18–32]. In addition, a relay-assisted cellular system allows flexible and fast network roll-out and is easy to adapt to traffic load, which is particularly important at the initial stage of system roll-out, where extensive base station (BS) deployment might not be economically viable [20,29,31]. Very recently, the first commercial relay-assisted OFDMA network has been standardized by IEEE 802.16j [32].

The performance of relay-assisted cellular systems has been a topic of research interest [23–27]. In Ref. [23], the uplink performance of a relay-assisted code division multiple access (CDMA) network was analyzed in a single-cell environment, where six RSs are deployed at fixed locations to improve system performance. Pole capacity and cell coverage were investigated for different frequency allocation methods. In Ref. [24], a non-CDMA relay-assisted network was investigated in the multicell environment also with six RSs in a cell, where a frequency reuse arrangement in the relaying link was proposed to improve spectral efficiency. In Ref. [25], the issues of RS positioning and spectrum partitioning were investigated with RSs located on the lines connecting BS and the six vertices of a hexagonal cell. Again, they considered the case of using six RSs in a cell. The RSs' positions are optimized (along the line) to maximize the user throughput at the cell boundary. And, in Refs. [26,27], the performance of a relay-assisted OFDMA network was evaluated for the specific setup of three RSs in a cell with and without intracell frequency reuse; numerical results showed that the relay-assisted system significantly outperforms the conventional cellular system with respect to system capacity and coverage.

So far, the performance investigation of the relay-assisted network has been limited to very specific system setups: with fixed RS numbers and locations in a cell and/or a fixed reuse pattern. In this work, the theoretical performance of a general relay-assisted OFDMA network is investigated in the multicell environment. In the downlink, a genetic algorithm (GA)-based, two-step approach is proposed for the joint multicell optimization of system parameters including RSs' positions, reuse pattern, path selection, and resource allocation to maximize the system spectral efficiency. In the uplink, given the RSs' positions determined in the downlink, the transmit power of mobile stations (MSs) or the system spectral efficiency is optimized by using optimal path selection and resource allocation. Performance improvement provided by RSs is illustrated under different system setups.

The rest of the chapter is organized as follows. Section 19.2 describes the system model including the multicell architecture, subcarrier allocation, propagation model, power setup criterion and reuse pattern. In Sections 19.3 and 19.4, the downlink and uplink optimizations are detailed under different criteria and system setups. Section 19.5 gives the numerical results for both downlink and uplink. Finally, the chapter is concluded in Section 19.6, where potential topics for future research are also discussed.

19.2 System Models

19.2.1 Multicell Architecture

We investigate a general multicell OFDMA network that consists of BSs, fixed RSs, and MSs. Figure 19.1 is such a network with the cluster size (reuse factor) K equal to 3; that is, three frequency bands A, B, and C are allocated to and shared by the BSs denoted by $\boxed{A}$, $\boxed{B}$, and $\boxed{C}$, respectively. Obviously, only the co-channel cells need to be considered in the design.

Figure 19.2a is a more detailed layout of a cell, where R is the cell radius, BS is located at the center of the cell, N RSs are deployed to improve the system performance, $\vec{r}_j$ is the position vector of the jth RS, R_j, and $\vec{m}$ is the MS's position vector. MSs are assumed to be uniformly distributed over the cell region Ω.

The RSs operate in the decode-and-forward mode in which the received signal is decoded and forwarded to the destination using the same transmission technology as the BS at the same frequency band. All stations (BS, RS, and MS) are equipped with one radio-frequency (RF) transceiver and an omni-directional antenna. In addition, MS can communicate with the BS through either a direct path or a 2-hop path (via a RS) that constitutes the BS–RS link and the RS–MS link.

Figure 19.2b is the radio resource allocation that is applicable to both downlink and uplink; orthogonal radio resources are allocated to a direct path and a 2-hop path, respectively. The radio resource for the 2-hop path is further divided into ones for the BS–RS link and the RS–MS link. With this type of allocation, RS does not need to transmit and receive at the same time (half-duplex relaying).

19.2.2 Subcarrier Allocation and Permutation

In the OFDMA systems, the frequency band is divided into orthogonal subcarriers, and subcarriers are grouped to form a channel, which is the basic resource unit to serve users. In real systems, the

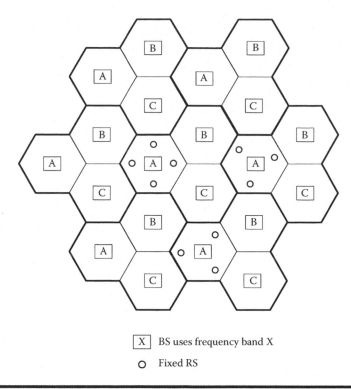

$\boxed{X}$ BS uses frequency band X

○   Fixed RS

Figure 19.1   Multicell architecture with cluster size $K = 3$.

subcarriers of channels in a cell/sector are not overlapping so as to avoid intracell/sector interference [6,7]. Thus, the co-channel intercell/sector interference, called multiple access interference (MAI) in the following, becomes the major factor that limits the system capacity in a multicell deployment.

Two types of subcarrier allocation schemes are popular in real OFDMA systems [6,7]. One is diversity allocation, where subcarriers in a channel are distributed over the frequency band, and the other is adjacent allocation, where subcarriers are adjacent to each other. Diversity allocation is designed to obtain frequency diversity at the link level, and adjacent allocation is designed to exploit multiuser diversity (through frequency-selective scheduling) for low-mobility users at the system level [33]. In this work, we are concerned only with the diversity allocation, where MAI can be accurately modeled as additive white Gaussian noise (AWGN), as will be discussed in the following.*

In diversity allocation, allocating subcarriers into a channel is often permuted randomly among cells/sectors so as to randomize the intercell/sector MAI. For example, in the PUSC (partial usage of subcarriers) mode of the IEEE 802.16e OFDMA system, subcarrier allocation is done with a subcarrier mapping process, called subcarrier permutation [7], where permutation is done pseudo-randomly and determined by the cell/sector ID [7]. To facilitate our analysis, nevertheless, subcarrier allocation will be assumed to be fully random from one cell/sector to another. With this arrangement, we say that the intercell/sector MAI has been randomized completely and, therefore, each

---

* The model may not be as accurate in the case of adjacent subcarrier allocation.

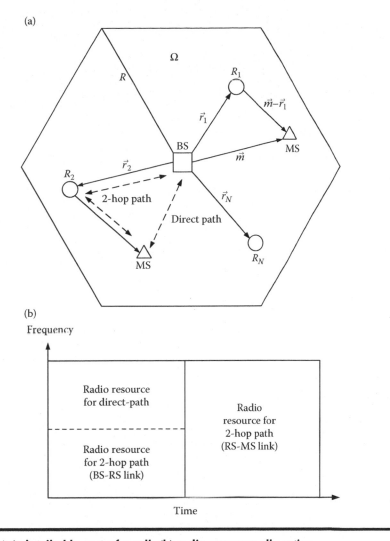

Figure 19.2   (a) A detailed layout of a cell, (b) radio resource allocation.

subcarrier/channel experiences the same MAI characteristics. It has been shown that for diversity allocation, MAI can be accurately modeled as a white Gaussian process [34,35].

19.2.3 Propagation Models

The path and shadowing losses are the two propagation impairments considered in this work.* The line-of-sight (LOS) and non-line-of-sight (NLOS) path-loss models in Ref. [36] for the suburban

---

* The effects of small-scale fading can be easily included in the formulation if the formula of spectral efficiency in Equation 19.4 is modified accordingly.

macrocell environment are adopted, which are given in Equations 19.1 and 19.2, respectively.

$$L_{\text{path}}^{\text{LOS}}(d) = 23.8 \log_{10}(d) + 41.9 \text{ dB}, \tag{19.1}$$

$$L_{\text{path}}^{\text{NLOS}}(d) = 40.2 \log_{10}(d) + 27.7 \text{ dB}, \tag{19.2}$$

where d is the separation (in meters) between the transmitter and the receiver. The LOS model is used for the BS–RS link, whereas the NLOS one is for the BS–MS link and RS–MS links because of the high altitude of RSs antenna. Note that the NLOS model is also used for the calculation of MAI.

For the shadowing loss, the simplified model in Equation 19.3 is adopted, mainly for verifying the effectiveness of the proposed method. Other shadowing models can be used as well.

$$L_{\text{shadow}}(\vec{m}) = \begin{cases} \delta \text{ dB} & \text{if } \vec{m} \text{ in a shadowed area,} \\ 0 \text{ dB} & \text{otherwise.} \end{cases} \tag{19.3}$$

In particular, the shadowed environment in Figure 19.3, which consists of four obstacles, will be used in Section 19.5 for numerical results. A MS is said to be in a shadowed area if the LOS between BS (RS) and MS passes through the obstacle. No shadowing loss is imposed on the BS–RS link.

19.2.4 Power Setup

A pre-specified spectral efficiency for users at the cell boundary is used to set up the transmitted power of BSs and RSs. Under the effects of the path and shadowing losses, the link spectral efficiency

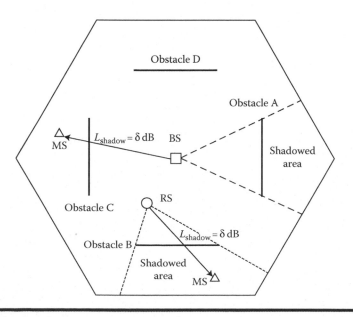

Figure 19.3 A simplified shadowing model.

is given by the equation [37]

$$S(d) = \log_2\left(1 + \frac{p_{\mathrm{T}} \cdot L_{\mathrm{all}}^{-1}(d)}{N_0 + I_0}\right) \text{ bps/Hz,} \tag{19.4}$$

where p_{T} is the transmit power spectral density (PSD), $L_{\mathrm{all}}(d) = L_{\mathrm{path}} \cdot L_{\mathrm{shadow}}$ is the overall propagation loss (in linear scale), and N_0 and I_0 are the PSD of AWGN and MAI, respectively. Recall that for diversity allocation, MAI can be modeled as a white Gaussian process.

Let S_{t} be the targeted link spectral efficiency at the cell edge; according to Equation 19.4, the transmit PSD of BS is set as

$$p_{\mathrm{B}} = \left(2^{S_{\mathrm{t}}} - 1\right) \cdot (N_0 + I_0) \cdot L_{\mathrm{all}}^{\mathrm{NLOS}}(R) \text{ W/Hz.} \tag{19.5}$$

Likewise, the transmit PSD of $R_j, j = 1, \ldots, N$, is

$$p_{R_j} = \left(2^{S_{\mathrm{t}}} - 1\right) \cdot (N_0 + I_0) \cdot L_{\mathrm{all}}^{\mathrm{NLOS}}(R - \|\vec{r}_j\|) \text{ W/Hz,} \tag{19.6}$$

where $\|\vec{r}_j\|$ denotes the Euclidean norm of the vector $\vec{r}_j$. It is commonplace to set $I_0 = N_0$ during the power setup, although some other values may be used.

19.2.5 Frequency Reuse over RS–MS Links

In Figure 19.2, it can be seen that radio resource may be reused over the RS–MS links in a cell to increase spectral efficiency, thanks to the spatial separation between RSs. To exploit this advantage, the RSs will be divided into L groups, and each group shares the same radio resource, where $L \le N$. The reuse pattern then can be specified conveniently by the set $\mathbf{G} \doteq \{G_i\}_{i=1}^{L}$, where $G_i = \{R_{j_i}\}$ is the set of RSs in the ith group. Obviously, $\sum_{i=1}^{L} |G_i| = N$, where $|G_i|$ is the cardinality of the set G_i.

19.3 Downlink Optimization

In the downlink, the optimum RSs' positions, reuse pattern, path selection, and bandwidth allocation of the 2-hop path will be sought to maximize the cell spectral efficiency. Two QoE criteria are investigated: one is the *fixed-bandwidth allocation* (FBA), where a fixed bandwidth is allocated to MSs, and the other is the *fixed-throughput allocation* (FTA), where a fixed throughput (data rate) is supported no matter where the MS is located (uniform data rate coverage). These two criteria represent two extreme ends of QoE, and a practical system's QoE will lie between them. In the following, optimization for the single-cell case will be presented first and then extended to the multicell case.

19.3.1 Objective Function

Let $w_{B \to M}(\vec{m})$, $w_{B \to R_j}(\vec{m})$, and $w_{R_j \to M}(\vec{m})$ denote the bandwidth per unit area (Hz/m^2) allocated to the BS–MS, BS–R_j, and R_j–MS links, respectively. For a frequency reuse pattern $\mathbf{G}$, the aggregate bandwidth (Hz) of the cell is obtained as

$$W_{\text{cell}} = W_{\text{direct}} + W_{\text{2-hop}}, \tag{19.7}$$

where

$$W_{\text{direct}} = \int\limits_{\vec{m} \in \Omega_B} w_{B \to M}(\vec{m}) \, dA, \tag{19.8}$$

and

$$W_{\text{2-hop}} = \sum_{j=1}^{N} W_{B \to R_j} + \sum_{l=1}^{L} W_{R \to M, l}, \tag{19.9}$$

with

$$W_{B \to R_j} = \int\limits_{\vec{m} \in \Omega_{R_j}} w_{B \to R_j}(\vec{m}) \, dA, \tag{19.10}$$

$$W_{R \to M, l} = \max_{R_j \in G_l} \{W_{R_j \to M}\}, \tag{19.11}$$

and

$$W_{R_j \to M} = \int\limits_{\vec{m} \in \Omega_{R_j}} w_{R_j \to M}(\vec{m}) \, dA. \tag{19.12}$$

Here, Ω_B and Ω_{R_j} are the coverage areas of BS (direct-path) and R_j, respectively, and $\{\Omega_B, \Omega_{R_1}, \ldots, \Omega_{R_N}\}$ is a partition of the cell region Ω.

Furthermore, assume that

$$S_{B \to M}(\vec{m}) = \log_2 \left(1 + \frac{p_B \cdot (L_{\text{all}}^{\text{NLOS}}(\|\vec{m}\|))^{-1}}{N_0 + I_{B \to M}(\vec{m})} \right), \tag{19.13}$$

$$S_{B \to R_j}(\vec{r}_j) = \log_2 \left(1 + \frac{p_B \cdot (L_{\text{all}}^{\text{LOS}}(\|\vec{r}_j\|))^{-1}}{N_0 + I_{B \to R_j}(\vec{r}_j)} \right), \tag{19.14}$$

and

$$S_{R_j \to M}(\vec{m}) = \log_2 \left(1 + \frac{p_{R_j} \cdot (L_{\text{all}}^{\text{NLOS}}(\|\vec{r}_j - \vec{m}\|))^{-1}}{N_0 + I_{R_j \to M}(\vec{m})} \right) \tag{19.15}$$

are the spectral efficiencies of the BS–MS, BS–R_j, and R_j-MS links, respectively, where $I(\vec{x})$ is the PSD of MAI appearing at the position $\vec{x}$. Then, the theoretical throughput per area (bps/m^2) supported by the BS–MS, BS–R_j, and R_j–MS links is given, respectively, by

$$t_{B \to M}(\vec{m}) = w_{B \to M}(\vec{m}) \cdot S_{B \to M}(\vec{m}), \tag{19.16}$$

$$t_{B \to R_j}(\vec{m}) = w_{B \to R_j}(\vec{m}) \cdot S_{B \to R_j}(\vec{r}_j), \tag{19.17}$$

and

$$t_{R_j \to M}(\vec{m}) = w_{R_j \to M}(\vec{m}) \cdot S_{R_j \to M}(\vec{m}). \tag{19.18}$$

Let $t_{B \to R_j \to M}(\vec{m})$ denote the effective throughput per unit area of a 2-hop path via R_j. It is clear that $t_{B \to R_j \to M}(\vec{m}) = \min\{t_{B \to R_j}(\vec{m}), t_{R_j \to M}(\vec{m})\}$. From Lemmas 1 and 2 given in the Appendix, $t_{B \to R_j}(\vec{m}) = t_{R_j \to M}(\vec{m})$ gives the highest spectral efficiency. Using this, the aggregate throughput (bps) of the cell is obtained as

$$T_{\text{cell}} = T_{\text{direct}} + \sum_{j=1}^{N} T_{\text{2-hop},R_j}, \tag{19.19}$$

where

$$T_{\text{direct}} = \int_{\vec{m} \in \Omega_B} t_{B \to M}(\vec{m}) \, dA \tag{19.20}$$

and

$$T_{\text{2-hop},R_j} = \int_{\vec{m} \in \Omega_{R_j}} t_{B \to R_j \to M}(\vec{m}) \, dA. \tag{19.21}$$

Note that for FBA, both a direct path and a 2-hop path are allocated a fixed bandwidth w_t, that is $w_t = w_{B \to M}(\vec{m}) = w_{B \to R_j}(\vec{m}) + w_{R_j \to M}(\vec{m})$, whereas for FTA, $w_{B \to M}(\vec{m})$, $w_{B \to R_j}(\vec{m})$, and $w_{R_j \to M}(\vec{m})$ are allocated to achieve the targeted user throughput $t_t = t_{B \to M}(\vec{m}) = t_{B \to R_j \to M}(\vec{m})$ $\forall \, \vec{m}$.

With these formulations, our objective is to find the RSs' positions $\Upsilon \doteq \{\vec{r}_1, \vec{r}_2, \ldots, \vec{r}_N\}$ and the reuse pattern $\mathbf{G}$ to maximize the system spectral efficiency,

$$S_{\text{cell}} \doteq \frac{T_{\text{cell}}}{W_{\text{cell}}}. \tag{19.22}$$

Apparently, S_{cell} depends on how a direct path or a 2-hop path is selected for the communication between BS and MSs. Two path selection schemes are investigated: one is the received signal to interference plus noise ratio (SINR) based, and the other is spectral efficiency based. For the SINR-based path selection, the direct path is selected if $\text{SINR}_{B \to M}(\vec{m}) \geq \max_{R_j}\{\text{SINR}_{R_j \to M}(\vec{m})\}$; otherwise, a 2-hop path via R_i is selected, where $R_i = \arg\{\max_{R_j}\{\text{SINR}_{R_j \to M}(\vec{m})\}\}$, and $\text{SINR}_{B \to M}(\vec{m})$ and $\text{SINR}_{R_j \to M}(\vec{m})$ are the received SINR at MS through a direct path and a 2-hop path via R_j, respectively. For the spectral efficiency-based selection, for FBA, a direct path is selected if $t_{B \to M}(\vec{m})/w_t \geq t_{\text{2-hop}}(\vec{m})/w_t$, where $t_{\text{2-hop}}(\vec{m}) = \max_{R_j}\{t_{B \to R_j \to M}(\vec{m})\}$; otherwise a 2-hop path via R_i is selected, where $R_i = \arg\{\max_{R_j}\{t_{B \to R_j \to M}(\vec{m})\}\}$. For FTA, on the other hand, a direct path is selected if $t_t/w_{B \to M}(\vec{m}) \geq t_t/w_{\text{2-hop}}(\vec{m})$, where $w_{\text{2-hop}}(\vec{m}) \doteq \min_{R_j}\{w_{B \to R_j}(\vec{m}) + w_{R_j \to M}(\vec{m})\}$; otherwise a 2-hop path via R_i is selected, where $R_i = \arg\{\min_{R_j}\{w_{B \to R_j}(\vec{m}) + w_{R_j \to M}(\vec{m})\}\}$.

19.3.2 Optimization Algorithm

The spectral efficiency, S_{cell}, is a highly nonlinear function of Υ and $\mathbf{G}$, and, generally, an analytic solution for the optimum Υ and $\mathbf{G}$ is not possible. In this section, a two-step optimization algorithm is proposed. First, a GA is employed to find the optimum RSs' positions Υ under a reuse pattern $\mathbf{G}$. Then, the optimization is carried out over the reuse patterns with an exhaustive search.

GA is a stochastic search technique and has been successfully applied to a wide range of optimization problems that involve a large number of variables [38–41]. Inspired by Darwin's theory on evolution, GA is started with a generation of initial population that consists of a set of chromosomes (possible solutions), and each chromosome is evaluated against an objective function such as the one in Equation 19.22. A *selection* is then performed over the population to select chromosomes based on the criterion of *survival of the fittest*, which means chromosomes with less fitness will be eliminated and the ones with better fitness will survive. Next, the algorithm goes through the *genetic evolution* including the operations of *crossover* and *mutation*. The *crossover* (recombination) is carried out over the selected chromosome mates to reproduce offspring who inherit partial genes from parents. *Mutation*, on the other hand, is performed for a few offspring whose genes may be altered in order to avoid the solutions to be stuck in the local optimum. The evolution process repeats from one generation to another until a termination condition is met, for example, the number of iterations.

The GA used in this work can be described in more detail as follows:

- *Chromosome representation:* The Cartesian coordinate of a RS ($\vec{r}_j$) is encoded as a gene, and the set of N RSs' positions Υ is a chromosome. In our formulation, the cell region is discretized into grids with $\vec{r}_j$ located at a vertex of the grids.
- *Initial population:* At the beginning, N_{pop} chromosomes are randomly generated and used as the population involved in the evolution of each generation. Each randomly generated coordinate is rounded to the nearest grid.
- *Fitness evaluation:* Each chromosome Υ is evaluated with the objective function S_{cell} under different strategies of radio resource allocation, path selection and reuse patterns. Chromosomes with better fitness survive and will be involved in the evolution in the next generation, whereas those with less fitness are discarded. The selection rate is denoted as β, so only the best $N_{sur} = \beta \cdot N_{pop}$ chromosomes survive after evaluation, and the rest are discarded to make rooms for the new offspring.
- *Mate selection:* Two mates (chromosomes) are selected from the mating pool of N_{sur} chromosomes to produce two new offspring. We apply the most commonly used mate selection scheme in GA applications: Roulette wheel selection [39], where the survival chromosome is chosen to be one of the parents with probability given in Equation 19.23,

$$P(\Upsilon_k) = \frac{S_{cell}(\Upsilon_k)}{\sum_{\Upsilon_i} S_{cell}(\Upsilon_i)}, \tag{19.23}$$

where $S_{cell}(\Upsilon_k)$ is the fitness value for the particular chromosome Υ_k. Roulette wheel selection emulates the survival of the fittest mechanism in nature, where the fitter chromosome (to have good genes to survive) will have higher chances of being selected to create offspring. The offspring that inherits good genes from parents may have higher probability of survival in the subsequent generation.

- *Crossover:* After two distinct chromosomes (mates) are selected, a crossover is performed to produce two new offspring. The uniform crossover is adopted in this work, where each gene of an offspring randomly decides whether or not to interchange information between the two mates [39,40]. In addition, in a real-valued encoding scheme a small zero-mean random perturbation is suggested to add to each gene of the offspring [41] in order to prevent the evolutions from being dominated by a few genes. In our method, the Gaussian variable with the variance equal to a grid length, which is specified in Section 19.5, is used as the perturbation. The newly generated offspring together with the survival population in the previous generation forms a new generation for the next step of evolution.

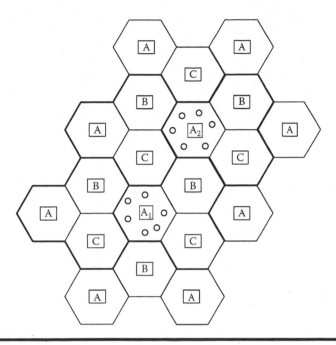

Figure 19.4 An example of multicell optimization (two-cell).

■ *Mutation:* Besides the crossover, mutation is another way to force an algorithm to explore other areas of the solution space in the search process. A mutation probability P_{mut} is set to determine whether a gene is mutated or not. Once a mutation is performed, the coordinate of the corresponding RS will be regenerated. The purpose of mutation is to avoid the GA from converging overly fast into a local optimum.

19.3.3 Multicell Optimization

In the multicellular environment, the deployment of RSs in the co-channel cells can be jointly optimized to maximize the overall spectral efficiency. The optimization method developed above can be applied to the multicell case as well, provided that Ω is considered as the overall cell region, N as the total number of RSs in the co-channel cells that an involved in the joint optimization, and $\vec{m}$ and $\vec{r}_j$ are the position vectors to their own BS. Figure 19.4 is such an example where the RSs in cells A_1 and A_2 are to be deployed jointly in an optimal way.

19.4 Uplink Optimization

In addition to system spectral efficiency, the average transmit power of MS is also an important performance measure in the uplink. In this section, given a fixed bandwidth, system parameters are optimized to minimize the MS's average transmit power for a specified throughput or to maximize throughput (spectral efficiency) under a fixed transmit power. Let w_t (Hz/m^2) be the bandwidth allocated to MS for both direct path and 2-hop path at any location $\vec{m}$;* then, $w_t = w_{M \to B}(\vec{m})$

---

* In this section, notations are kept the same as those used in the downlink, although they represent different values.

$= w_{M \to R_j}(\vec{m}) + w_{R_j \to B}(\vec{m})$. In the uplink optimization, we use the RSs' positions determined in the downlink.

Following the notations in the previous section, the spectral efficiency for the MS–BS, R_j–BS, and MS–R_j links is given, respectively, by

$$S_{M \to B}(\vec{m}) = \log_2 \left(1 + \frac{P_{M,\text{direct}}(\vec{m})(L_{\text{all}}^{\text{NLOS}}(\|\vec{m}\|))^{-1}}{(N_0 + I_{M \to B})\, w_t} \right), \tag{19.24}$$

$$S_{R_j \to B} = \log_2 \left(1 + \frac{P_{R_j}(\vec{m})(L_{\text{all}}^{\text{LOS}}(\|\vec{r}_j\|))^{-1}}{\left(N_0 + I_{R_j \to B}\right) w_{R_j \to B}(\vec{m})} \right), \tag{19.25}$$

and

$$S_{M \to R_j}(\vec{m}) = \log_2 \left(1 + \frac{P_{M,\text{2-hop}}(\vec{m})(L_{\text{all}}^{\text{NLOS}}(\|\vec{r}_j - \vec{m}\|))^{-1}}{\left(N_0 + I_{M \to R_j}\right) w_{M \to R_j}(\vec{m})} \right), \tag{19.26}$$

where $P_M(\vec{m})$ is the transmit power of MS, $P_{R_j}(\vec{m})$ is the transmit power of R_j, and $I_{M \to B}$, $I_{R_j \to B}$, and $I_{M \to R_j}$ are the PSDs of average MAI appearing over the MS-BS, R_j-BS, and MS-R_j links, respectively. $I_{M \to B}$ and $I_{R_j \to B}$ contain only intercell interference, whereas $I_{M \to R_j}$ contains both intercell and intracell interferences if intracell frequency reuse is considered. Note that the intercell and intracell interferences are averaged over all MSs who are distributed uniformly over the co-channel interfering area and transmit at the maximum allowable power. Since the maximum MAI is considered, the performance obtained here is a lower bound of the real performance. In addition, given Υ, $S_{R_j \to B}$ is fixed because the transmit PSD of R_j is already determined in the downlink at the power setup stage to provide the targeted spectral efficiency to users at the cell edge.

Furthermore, the throughput expressions are given by

$$t_{M \to B}(\vec{m}) = w_t \cdot S_{M \to B}(\vec{m}), \tag{19.27}$$

$$t_{R_j \to B}(\vec{m}) = w_{R_j \to B}(\vec{m}) \cdot S_{R_j \to B}, \tag{19.28}$$

and

$$t_{M \to R_j}(\vec{m}) = w_{M \to R_j}(\vec{m}) \cdot S_{M \to R_j}(\vec{m}). \tag{19.29}$$

19.4.1 Minimization of Transmit Power

In this criterion, given a targeted throughput t_t, the transmit power of MS is minimized by the optimization of path selection and bandwidth allocation of a 2-hop path. From Equations 19.24 and 19.27, the MS's transmit power for the direct path is

$$\hat{P}_{M,\text{direct}}(\vec{m}) = \left(2^{t_t/w_t} - 1 \right) \cdot (N_0 + I_{M \to B})\, w_t \cdot L_{\text{all}}^{\text{NLOS}}(\|\vec{m}\|). \tag{19.30}$$

For a 2-hop path, similar to Lemma 2 in the Appendix, the optimal bandwidth allocation is

$$\hat{w}_{R_j \to B} = \frac{t_t}{S_{R_j \to B}} \tag{19.31}$$

$$\hat{w}_{M \to R_j} = w_t - \hat{w}_{R_j \to B}, \tag{19.32}$$

which achieve the condition $t_t = t_{M \to R_j}(\vec{m}) = t_{R_j \to B}(\vec{m})$. With this optimal bandwidth allocation, the MS's transmit power is

$$\hat{P}_{M,R_j}(\vec{m}) = \left(2^{t_t/\hat{w}_{M \to R_j}} - 1\right) \cdot \left(N_0 + I_{M \to R_j}\right) \hat{w}_{M \to R_j} \cdot L_{\text{all}}^{\text{NLOS}}(\|\vec{r}_j - \vec{m}\|). \tag{19.33}$$

The optimal path selection is then to select a direct path if $\hat{P}_{M,\text{direct}}(\vec{m}) \leq \hat{P}_{M,\text{2-hop}}(\vec{m})$, where $\hat{P}_{M,\text{2-hop}}(\vec{m}) = \min_{R_j}\{\hat{P}_{M,R_j}(\vec{m})\}$; otherwise a 2-hop path via R_i is selected, where $R_i = \arg\{\min_{R_j}\{\hat{P}_{M,R_j}(\vec{m})\}\}$. Finally, the average MS's transmit power is given by

$$P_{M,\text{avg}} = \frac{1}{A_{\Omega_{\text{direct}}}} \int_{\Omega_B} \hat{P}_{M,\text{direct}}(\vec{m}) \, dA + \frac{1}{A_{\Omega_{\text{2-hop}}}} \int_{\Omega_{\text{2-hop}}} \hat{P}_{M,\text{direct}}(\vec{m}) \, dA, \tag{19.34}$$

where A_ψ is the area of ψ.

19.4.2 Maximization of Spectral Efficiency

In this criterion, given a fixed transmit power of MS P_M, the system spectral efficiency is maximized by performing optimization on path selection and bandwidth allocation of the 2-hop path.

For the direct-path, the throughput is

$$\hat{t}_{M \to B}(\vec{m}) = w_t \cdot \log_2\left(1 + \frac{P_M(L_{\text{all}}^{\text{NLOS}}(\|\vec{m}\|))^{-1}}{(N_0 + I_{M \to B}) w_t}\right). \tag{19.35}$$

For the 2-hop path, similar to Lemma 1 in Appendix, it can be shown that the optimal bandwidth allocation is that

$$w_{R_j \to B}(\vec{m}) = \frac{S_{M \to R_j}(\vec{m})}{S_{R_j \to B} + S_{M \to R_j}(\vec{m})} \cdot w_t \tag{19.36}$$

and

$$w_{M \to R_j}(\vec{m}) = \frac{S_{R_j \to B}}{S_{R_j \to B} + S_{M \to R_j}(\vec{m})} \cdot w_t. \tag{19.37}$$

On the other hand,

$$S_{M \to R_j}(\vec{m}) = \log_2\left(1 + \frac{P_M(\vec{m})(L_{\text{all}}^{\text{NLOS}}(\|\vec{r}_j - \vec{m}\|))^{-1}}{\left(N_0 + I_{M \to R_j}\right) w_{M \to R_j}(\vec{m})}\right). \tag{19.38}$$

Therefore, we need to solve the following nonlinear equation

$$S_{M \to R_j}(\vec{m}) = \log_2 \left(a S_{M \to R_j}(\vec{m}) + b \right), \tag{19.39}$$

for the optimal $\hat{S}_{M \to R_j}(\vec{m})$, where

$$a = \frac{P_M(\vec{m})(L_{\text{all}}^{\text{NLOS}}(\|\vec{m}\|))^{-1}}{\left(N_0 + I_{M \to R_j} \right) S_{R_j \to B} \cdot w_t} \tag{19.40}$$

and

$$b = \frac{P_M(\vec{m})(L_{\text{all}}^{\text{NLOS}}(\|\vec{m}\|))^{-1}}{\left(N_0 + I_{M \to R_j} \right) w_t} + 1. \tag{19.41}$$

As a result,

$$\hat{w}_{M \to R_j}(\vec{m}) = \frac{S_{R_j \to B}}{S_{R_j \to B} + \hat{S}_{M \to R_j}(\vec{m})} \cdot w_t, \tag{19.42}$$

and the maximum attainable throughput for the 2-hop path is

$$\hat{t}_{M \to R_j \to B}(\vec{m}) = \hat{w}_{M \to R_j}(\vec{m}) \cdot \hat{S}_{M \to R_j}(\vec{m}). \tag{19.43}$$

The optimal path selection is then to the select a direct path if $\hat{t}_{M \to B}(\vec{m}) \geq \hat{t}_{M \to B, \text{2-hop}}(\vec{m})$, where $\hat{T}_{M \to B, \text{2-hop}}(\vec{m}) = \max_{R_j}\{\hat{t}_{M \to R_j \to B}(\vec{m})\}$; otherwise, a 2-hop path via R_i is selected where $R_i = \arg \max_{R_j}\{\hat{t}_{M \to R_j \to B}(\vec{m})\}$. Finally, the system spectral efficiency is given by

$$S_{\text{cell}} \doteq \frac{T_{\text{cell}}}{W_{\text{cell}}} \text{ bps/Hz}, \tag{19.44}$$

where

$$T_{\text{cell}} = T_{\text{direct}} + \sum_{j=1}^{N} T_{\text{2-hop},R_j}, \tag{19.45}$$

where

$$T_{\text{direct}}(\vec{m}) = \int_{\vec{m} \in \Omega_B} \hat{t}_{M \to B}(\vec{m}) \, dA, \tag{19.46}$$

$$T_{\text{2-hop},R_j}(\vec{m}) = \int_{\vec{m} \in \Omega_{R_j}} \hat{t}_{M \to R_j \to B}(\vec{m}) \, dA, \tag{19.47}$$

and

$$W_{\text{cell}} = W_{\text{direct}} + W_{\text{2-hop}}, \tag{19.48}$$

where

$$W_{\text{direct}} = \int_{\vec{m} \in \Omega_B} w_t \, dA, \tag{19.49}$$

and

$$W_{2\text{-hop}} = \sum_{j=1}^{N} W_{R_j \to B} + \sum_{l=1}^{L} W_{M \to R,l}, \tag{19.50}$$

with

$$W_{R_j \to B} = \int_{\vec{m} \in \Omega_{R_j}} w_{R_j \to B}(\vec{m}) \, dA, \tag{19.51}$$

$$W_{M \to R,l} = \max_{R_j \in G_l} \{W_{M \to R_j}\}, \tag{19.52}$$

and

$$W_{M \to R_j} = \int_{\vec{m} \in \Omega_{R_j}} w_{M \to R_j}(\vec{m}) \, dA. \tag{19.53}$$

19.5 System Performance

19.5.1 Downlink Performance

In our numerical results, the cell radius is 2000 m, the cell region is divided into grids with each side equal to 20 m, and all stations (BS, RSs, and MS) are located at the grid vertices. Four cases of system setup will be considered, including FBA with spectral efficiency-based path selection (FBA-SE), FBA with SINR-based path selection (FBA-SINR), FTA with SE-based path selection (FTA-SE) and FTA with SINR-based path selection (FTA-SINR). $S_t = 0.5$ bps/Hz, and $N_{\text{pop}} = 200$, $\beta = 0.5$, and $P_{\text{mut}} = 0.01$ are adopted in GA. In addition, the number of generations depends on the number of RSs involved; more generations are needed for a larger number of RSs. For convenience, all results are obtained with $w_t = 1$ Hz for FBA and $t_t = 1$ bps for FTA.

19.5.1.1 Single-Cell

Figure 19.5 shows how RSs are placed in the optimal way and what they can do to improve the user throughput (spectral efficiency) for FBA–SE. No frequency reuse and shadowing are considered. Figure 19.5a shows the case of no RS ($N = 0$), where 25.43 bps throughput is achieved at the locations close to BS and 0.5 bps at the cell edge (as planned at the power setup of BS). Clearly, the achievable throughput decreases rapidly from the cell center to the edge due to path loss. Figure 19.5b shows the case of two RSs, where they are placed on the lines connecting the cell center and two vertices of the hexagonal cell, at a distance of 960 m from BS. As can be seen, user throughput is improved significantly in the close proximity of RSs. Note that the optimal placement of RSs is not unique in this case; as long as RSs are placed on the lines connecting the cell center and any two of the vertices at the same distance from BS, the system performance is the same. Figure 19.5c illustrates the case of six RSs. Again, RSs are placed on the lines connecting the cell center and vertices, and all of them are located on the ring with a distance of 960 m from BS. As the number of RS increases, however, the optimal RS locations can be split into more than one ring so as to maximize the system spectral efficiency, as one might expect. Figure 19.5d is such an example with 14 RSs. Table 19.1 summarizes the important system parameters including service areas of BS and RSs, bandwidth allocation ratio for the BS–MS link, the BS–RS link and the RS–MS link, and the system spectral

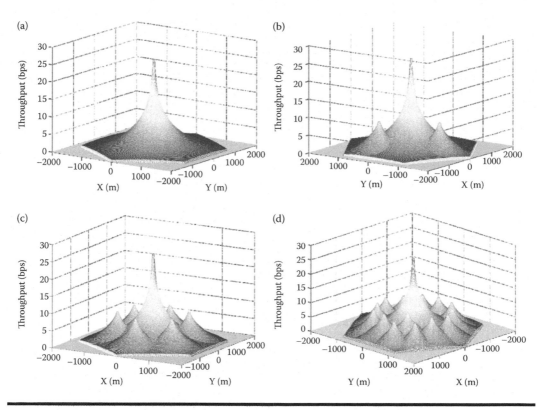

Figure 19.5 Optimal RSs' positions and throughput distribution for FBA–SE. (a) $N = 0$; (b) $N = 2$; (c) $N = 6$; (d) $N = 14$.

efficiency. As can be seen, the system spectral efficiency is improved significantly with the deployment of RSs, around 4.4% for each addition of RS when $N \leq 6$. The improvement becomes smaller, however, as the number of RS is larger than 6 (also see Figure 19.8), where RSs begin to compete with each other for serving MS rather than to compete with BS. Figure 19.6 shows the cumulative distribution function (CDF) of user throughput for different N. It is shown that the percentage

Table 19.1 Important Parameters for FBA–SE

N	0	2	4	6	8	10	12	14	16
Ω_B (%)	100	76	53	29	23	22	21	20	18
Ω_R (%)	–	24	47	71	77	78	79	80	82
$W_{B \to M}$ (%)	100	76	53	29	23	22	21	20	18
$W_{B \to R}$ (%)	–	5	9	14	16	17	18	19	20
$W_{R \to M}$ (%)	–	19	38	57	61	61	61	61	62
S_{cell} (bps/Hz)	2.7372	2.9793	3.2214	3.4634	3.6786	3.8414	3.9606	4.0524	4.1386
S_{cell} gain (%)	–	8.84	17.69	26.53	34.39	40.34	44.70	48.05	51.20

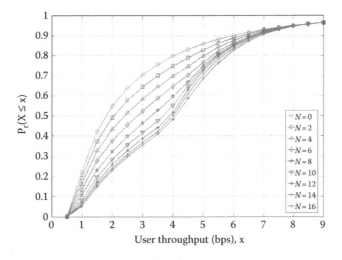

Figure 19.6   CDF of user throughput for FBA–SE.

of user with low throughput decreases as N increases. In other words, the achievable data rate is distributed more uniformly over the cell. Further, we see that the improvement becomes saturated when $N \geq 12$. Finally, the cumulative function approaches 1 as the throughput is increased to the maximum throughput of 25.43 bps, where all lines merge together at that region.

As to FBA–SINR, the optimal RSs are placed in a way similar to FBA–SE except that generally the RSs are a bit far from BS than that of FBA–SE. Table 19.2 is a summary of important system parameters. Somewhat surprisingly, FBA–SINR performs very similarly to that of FBA-SE with respect to system spectral efficiency after performing system optimization with GA. That is because under different system criteria, GA always searches the optimum RS locations to maximize the system spectral efficiency. SINR-based path selection is much simpler to implement than the SE-based one.

Figure 19.7 shows the complement CDF of bandwidth consumption for FTA–SE. As it can be seen, the percentage of high bandwidth consumption is reduced as the number of RSs increases.

Table 19.2   Important parameters for FBA–SINR

N	0	2	4	6	8	10	12	14	16
Ω_B (%)	100	75	50	25	18	18	19	19	17
Ω_R (%)	–	25	50	75	82	82	81	81	83
$W_{B \to M}$ (%)	100	75	50	25	18	18	19	19	17
$W_{B \to R}$ (%)	–	5	10	15	17	18	19	19	21
$W_{R \to M}$ (%)	–	20	40	60	65	64	62	62	62
S_{cell} (bps/Hz)	2.7372	2.9705	3.2033	3.4369	3.6563	3.8224	3.9436	4.0356	4.1160
S_{cell} gain (%)	–	8.52	17.03	25.56	33.58	39.65	44.07	47.44	50.37

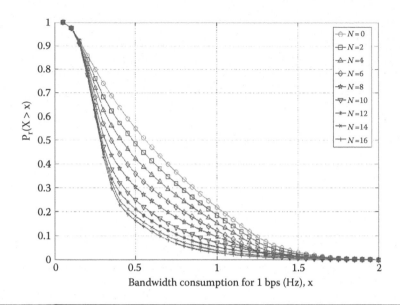

Figure 19.7 Complement CDF of bandwidth consumption for FTA–SE.

Table 19.3 summarizes the important system parameters for FTA–SE. Figure 19.8 compares the system spectral efficiency of the four system setups. Clearly, the system spectral efficiency of FBA is much smaller than that of FTA because a lot more bandwidth is required to maintain a fixed data rate for MSs who are in bad channel conditions. On the other hand, there is little difference between SE-based path selection and the SINR-based one after system optimization. Similarly, there is little difference between SE-based path selection and the SINR-based one after system optimization in the FTA case.

To further utilize the radio resource, frequency reuse over the RS–MS link is explored. The case of six RSs is taken as an example with the following reuse patterns, $\mathbf{G} = \{G_1\}$, $\mathbf{G} = \{G_1, G_2\}$,

Table 19.3 Important Parameters for FTA–SE

N	0	2	4	6	8	10	12	14	16
Ω_B (%)	100	82	64	46	37	30	30	32	32
Ω_R (%)	–	18	36	54	63	70	72	68	68
$W_{B \to M}$ (%)	100	84	65	41	31	21	17	20	20
$W_{B \to R}$ (%)	–	2	5	8	10	12	14	14	15
$W_{R \to M}$ (%)	–	14	30	51	59	67	69	66	65
$\overline{W}$ (Hz)	0.6428	0.5903	0.5379	0.4854	0.4475	0.4106	0.3789	0.3613	0.3460
$\overline{W}$ saving (%)	–	8.17	16.32	24.49	30.38	36.12	41.05	43.79	46.17
S_{cell} (bps/Hz)	1.5557	1.6939	1.8591	2.0601	2.2347	2.4356	2.6390	2.7678	2.8900
S_{cell} gain (%)	–	8.88	19.50	32.42	43.65	56.56	69.63	77.91	85.77

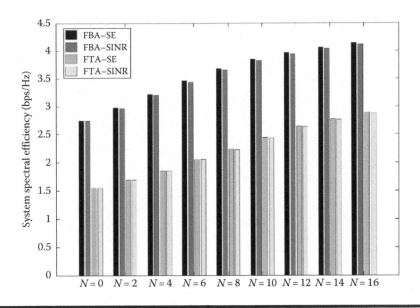

Figure 19.8 System spectral efficiency of four system setups.

$\mathbf{G} = \{G_1, G_2, G_3\}$, and $\mathbf{G} = \{G_1, G_2, G_3, G_4, G_5, G_6\}$, where each reuse group has an equal number of RSs. Figure 19.9 depicts the throughput distribution of FBA–SE. Circles with the same letter indicate the service area of RSs in the same reuse groups. Again, RSs are placed on the line connecting the cell center to six vertexes of the cell, with RSs in the same reuse group being pulled away as far as possible. In the full reuse case, as is shown in Figure 19.9d, severe interference leads to a shrinking of the service area of RSs. With respect to system spectral efficiency, the reuse cases always outperform the no reuse cases, as is shown in Figure 19.10, where the reuse pattern $\mathbf{G} = \{G_1, G_2\}$ gives the highest spectral efficiency. Also, it is very interesting to see that SINR-based path selection has a higher spectral efficiency than the SE-based path selection for both FBA and FTA in the frequency-reuse cases. This can be attributed to that RSs always have a larger service area in the SINR-based path selection and that leads to a larger bandwidth saving in the frequency-reuse case. Table 19.4 summarizes the important system parameters for four different system setups. Note that in this example, the reuse pattern $\mathbf{G} = \{G_1, G_2, G_3\}$ gives the largest service area of RS, while $\mathbf{G} = \{G_1, G_2\}$ gives the highest system spectral efficiency; there is a tradeoff between MAI and frequency reuse.

Figure 19.11 is the result considering the shadowing effect with four RSs in the cell to improve the system performance. The simplified shadow model in Figure 19.3 is adopted with four obstacles (of length 1000 m) placed at $(0, \pm1050)$ and $(\pm1050, 0)$. Two values of shadowing loss $\delta = 10$ and 20 dB are considered. Only FBA–SE is shown here as an example. In Figures 19.11a and c, as expected, the achievable throughput in the shadowed area is very low if there is no RS in the cell. With the deployment of RSs, on the other hand, downlink optimization suggests that the optimal RSs' positions should be placed at $(0, \pm1060)$ and $(\pm1060, 0)$ for both cases, which are right behind the four obstacles. The throughput enhancements are shown in Figures 19.11b and d for $\delta = 10$ and 20, respectively. In Figure 19.11b, in addition to serving the severely shadowed area, RSs also provide service to small fractions of the area across the obstacle where the throughput provided by RSs is higher than that by BS. In Figure 19.11d, we can see that this area is reduced as the shadowing loss increases.

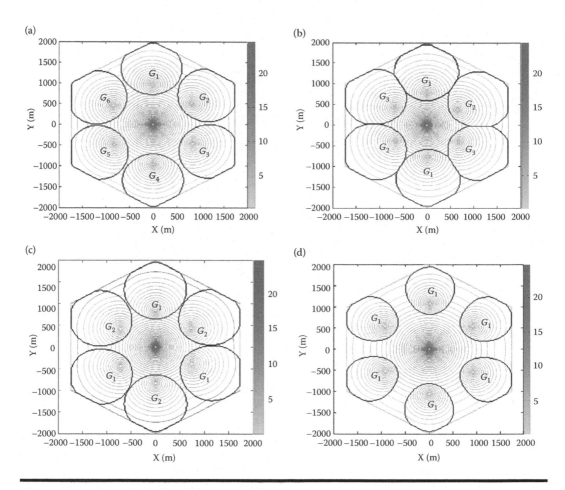

Figure 19.9 Throughput distribution of different frequency-reuse patterns for FBA–SE ($N = 6$): (a) $G = \{G_1, G_2, G_3, G_4, G_5, G_6\}$; (b) $G = \{G_1, G_2, G_3\}$; (c) $G = \{G_1, G_2\}$; (d) $G = \{G_1\}$.

19.5.1.2 Multicell

In this part, we just use an example to illustrate how the joint optimization is done over multiple cells. Figure 19.4 is such an example, where a total of 12 RSs are jointly deployed in the cells A_1 and A_2 with six RSs per cell. The reuse factor K equals 3, and the first tier co-channel interference is taken into account for both cells in the optimization. No intracell frequency-reuse among RSs is considered. Figure 19.12 shows the optimal placement of RSs and the distribution of bandwidth consumption for FTA–SE. As expected, cells A_1 and A_2 have almost the same layout, and RSs in the same group are placed as far as possible in order to reduce the co-channel interference.

19.5.2 Uplink Performance

In this subsection, the uplink performance is presented in the single-cell environment, where six RSs are employed to improve system performance. No intracell frequency reuse is considered, and the optimal RSs' positions are determined in the downlink according to the FBA–SE setup. In this case, $S_{R \rightarrow B} = 10.7$ bps/Hz. In the following, the bandwidth allocated to a MS is fixed at 14.2 KHz/m^2,

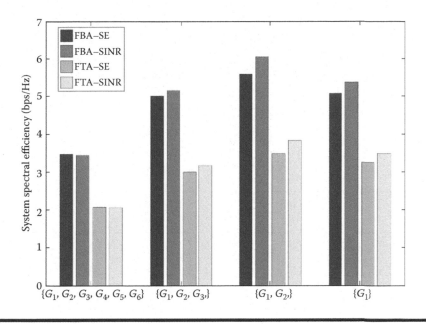

Figure 19.10 System spectral efficiency of different frequency-reuse patterns (N = 6).

and for the 2-hop path, the bandwidth is further partitioned into $w_{M \to R}$ and $w_{R \to B}$, as given in Equations 19.31 and 19.32.

In the uplink, the transmit power of MS is an important performance measure due to the limited battery capacity of MS. Firstly, we look at the issue as to how RSs can be used to reduce the average transmit power of MS. Figure 19.13 is the result for the target rate of 28.4 Kbps/m². Figure 19.13a is the distribution of transmit power (in dBm) for the no RS case ($N = 0$). Clearly, more transmit power is needed for a more distant MS. The average transmit power is 20.77 dBm in this case. Figure 19.13b is the distribution of power consumption for the case of $N = 6$, where the optimal path determined in the downlink is also used in the uplink (symmetrical path usage). In this scenario, the service areas of BS and RSs in the uplink are exactly the same as those in the downlink. As can be seen, the uplink transmit power is reduced significantly in the area served by RSs. The average transmit power is 13.56 dBm in this case, which is 7.21 dB lower than that of no RS. The transmit power can be reduced further, nevertheless, if MS is allowed to select the most power efficient uplink path of its own (optimal uplink or asymmetrical path usage). Figure 19.13c is such a result, where the average transmit power is 7.27 dBm: a gain of 13.5 and 6.29 dB as compared to $N = 0$ and 6 with symmetrical path usage, respectively. Note that in Figure 19.13c as the RS–BS link gets better and better, the MS tends to choose the nearest serving station for uplink transmission. In other words, the service area of BS and RS will be determined by the perpendicular bisector of any two serving stations if the RS–BS link is perfect. Figure 19.14 shows the complement CDFs of power consumption for all three cases. As can be seen, the percentage of large power consumption area is reduced significantly with the help of RSs. In the symmetric uplink path selection, however, there is still about 10% area where the power consumption exceeds 15 dBm; in fact, the area is close to cell boundary and yet served by BS through direct path due to the fact that the path selection is done in the downlink. On the other hand, as expected, the area can be reduced largely if the optimal uplink path is employed.

Table 19.4   Important Parameter for Frequency-Reuse Patterns, $N = 6$

	G	$d_{B \to R}$ (m)	Ω_B (%)	Ω_R (%)	$W_{B \to M}$ (%)	$W_{B \to R}$ (%)	$W_{R \to M}$(%)	S_{cell} (bps/Hz)
FBA, SE-based	$\{G_1, G_2, G_3,$ $G_4, G_5, G_6\}$	958	29	71	29	14	57	3.4634
	$\{G_1, G_2, G_3\}$	770	20	80	30	22	48	5.0005
	$\{G_1, G_2\}$	845	27	73	45	22	33	5.5899
	$\{G_1\}$	1066	48	52	74	15	11	5.0693
FBA, SINR-based	$\{G_1, G_2, G_3,$ $G_4, G_5, G_6\}$	990	25	75	25	15	60	3.4369
	$\{G_1, G_2, G_3\}$	798	14	86	21	25	54	5.1483
	$\{G_1, G_2\}$	808	16	84	30	28	42	6.0553
	$\{G_1\}$	1041	42	58	70	17	13	5.3705
FTA, SE-based	$\{G_1, G_2, G_3,$ $G_4, G_5, G_6\}$	1168	46	54	41	8	51	2.0601
	$\{G_1, G_2, G_3\}$	878	25	75	32	15	53	3.0013
	$\{G_1, G_2\}$	892	28	72	43	17	40	3.4860
	$\{G_1\}$	1109	49	51	75	12	13	3.2445
FTA, SINR-based	$\{G_1, G_2, G_3,$ $G_4, G_5, G_6\}$	1181	42	58	37	8	55	2.0524
	$\{G_1, G_2, G_3\}$	862	16	84	19	18	63	3.1725
	$\{G_1, G_2\}$	888	19	81	29	21	50	3.8324
	$\{G_1\}$	1103	44	56	70	14	16	3.4887

Secondly, we investigate the spectral efficiency improvement offered by RSs, given a fixed MS transmit power. Again, both symmetrical and asymmetrical (optimal uplink) path selections are considered. The transmit power is fixed at 25 dBm. Figure 19.15 shows the CDF of spectral efficiency. The low-spectral-efficiency area is improved by RSs. Lastly, $S_{cell} = 2.27$, 4.24, and 4.5 bps/Hz for $N = 0$ and $N = 6$ with the symmetrical uplink path, and $N = 6$ with optimal uplink path, respectively.

19.6 Conclusions and Open Issues

In this work, the performance of relay-assisted OFDMA networks is investigated in the multicell environment with optimized system parameters. A GA based approach is proposed in the downlink for joint multicell optimization of system parameters including RSs' positions, path selection, reuse pattern, and resource allocation to maximize the system spectral efficiency. Four different system

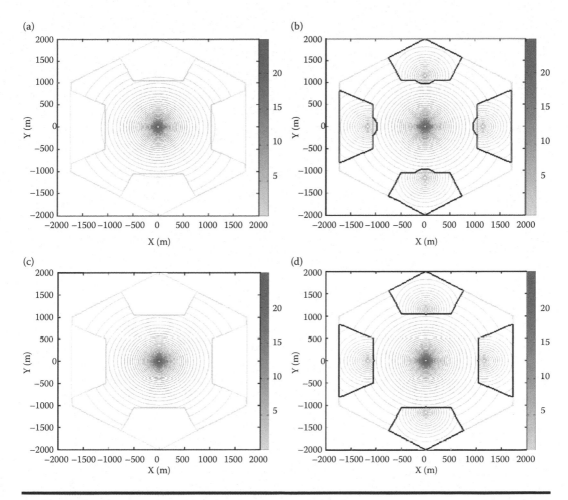

Figure 19.11 **Throughput distribution for FBA–SE under shadowing effects: (a)** $\delta = 10$, $N = 0$; **(b)** $\delta = 10$, $N = 4$; **(c)** $\delta = 20$, $N = 0$; **(d)** $\delta = 20$, $N = 4$.

setups are investigated including (FBA–SE) fixed-bandwidth allocation with spectral efficiency-based path selection, FBA–SINR (fixed-bandwidth allocation with SINR-based path selection), FTA–SE (fixed-throughput allocation with spectral efficiency-based path selection), and FTA–SINR (fixed-throughput allocation with SINR-based path selection). In the downlink, numerical results show that (i) RSs provide significant improvement with respect to system spectral efficiency and user throughput over the traditional cellular systems, (ii) uniformity of user data rate comes at the expense of a large loss in system spectral efficiency when FTA is employed, and (iii) somewhat surprisingly, the low-complexity SINR-based path selection performs nearly as good as the SE-based one for the no reuse case, while slightly better in the frequency-reuse case. In the uplink, the minimum transmit power of MS or system spectral efficiency is optimized with the RSs' positions determined in the downlink optimization. Numerical results show that the average transmit power and system spectrum efficiency are significantly improved over the traditional OFDMA networks with the deployment of RSs. In addition, it is shown that asymmetrical path selection between the downlink and the uplink is particularly helpful to those users close to the cell boundary.

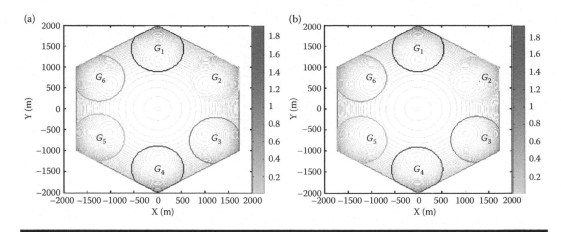

Figure 19.12 **RSs' positions of jointly designed cells and distribution of bandwidth consumption for FTA–SE: (a) Cell A_1; (b) Cell A_2.**

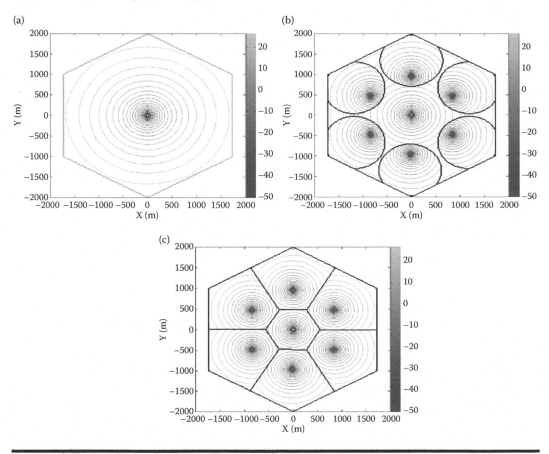

Figure 19.13 **Distribution of uplink transmit power: (a) $N = 0$; (b) Symmetrical uplink path; (c) Optimal uplink path.**

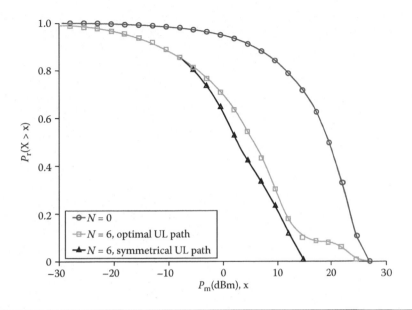

Figure 19.14   Complement CDFs of uplink transmit power.

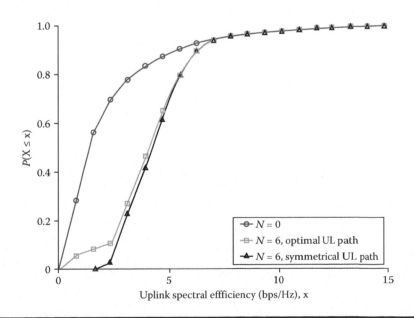

Figure 19.15   CDF of uplink spectral efficiency.

In this work, we only focus on the theoretic performance of the relay-assisted OFDMA systems from the information theoretical viewpoint, where all stations are equipped with a single antenna. There is much more to be done in this area of research including the following topics.

- *Performance evaluation with finite MCS (modulation and coding schemes) and signaling overhead:* In the practical systems, the achievable spectral efficiency is limited to the used MCSs rather than the theoretic one predicted in Equation 19.5. In addition, overhead will be imposed for signaling messages between BS and RS, RS and MS, and BS and MS. All these factors significantly affect the system performance of a practical system and should be taken into account for performance evaluation.
- *Performance improvement with multiple transmit and receive antennas:* Using multiple transmit and receive antennas in a wireless communication system, a.k.a. the multiple-input multiple-output (MIMO) system, is capable of providing diversity gain, array gain (power gain), and/or degree-of-freedom gain over the single-input single-output (SISO) systems. Space-time coding, beam-forming, and spatial multiplexing are modes of operations to exploit the diversity gain, array gain, and degree-of-freedom gain, respectively. It will be interesting to see what the MIMO technology can do to improve the system performance of the relay-assisted OFDMA networks, especially if the multiuser MIMO technology is applied.
- *Performance improvement with cooperative diversity:* In this work, only the simple 2-hop relaying protocol is investigated. More sophisticated relaying protocols such as transmit diversity, receive diversity, and others can be used to further improve the system performance. It would be interesting to see how these protocols perform in a practical system, where the effects of MCS and signaling overhead are also taken into consideration.

Appendix: Proofs of Lemmas 1 and 2

In this appendix, we prove the following two lemmas which state the conditions for the best bandwidth allocation in a 2-hop path.

Lemma 1

For the FBA, the highest spectral efficiency of a 2-hop path is achieved with the following bandwidth allocation,

$$w_{B \to R_j}(\vec{m}) = \frac{S_{R_j \to M}(\vec{m})}{S_{R_j \to M}(\vec{m}) + S_{B \to R_j}(\vec{r_j})} \cdot w_t \qquad (19.54)$$

and

$$w_{R_j \to M}(\vec{m}) = \frac{S_{B \to R_j}(\vec{m})}{S_{R_j \to M} + S_{B \to R_j}(\vec{r_j})} \cdot w_t \qquad (19.55)$$

where $S_{R_j \to M}(\vec{m})$ and $S_{B \to R_j}(\vec{r_j})$ are the spectral efficiencies of the BS–R_j link and the R_j–MS link, respectively.

Proof: Recall that for the FBA, $w_t = w_{B \to R_j}(\vec{m}) + w_{R_j \to M}(\vec{m})$ for a 2-hop path. Let

$$w_{B \to R_j}(\vec{m}) = \left(\frac{S_{R_j \to M}(\vec{m})}{S_{R_j \to M}(\vec{m}) + S_{B \to R_j}(\vec{r_j})} + \alpha \right) \cdot w_t \qquad (19.56)$$

be the bandwidth allocated to the BS–R_j link, where $0 \leq \alpha \leq 1$. Then, the bandwidth for the R_j–MS link is

$$w_{R_j \to M}(\vec{m}) = \left(\frac{S_{B \to R_j}(\vec{r_j})}{S_{R_j \to M}(\vec{m}) + S_{B \to R_j}(\vec{r_j})} - \alpha \right) \cdot w_t. \tag{19.57}$$

In addition, the throughput for the BS–R_j link and the R_j–MS link is given, respectively, by

$$t_{B \to R_j}(\vec{m}) = \left(\frac{S_{R_j \to M}(\vec{m})}{S_{R_j \to M}(\vec{m}) + S_{B \to R_j}(\vec{r_j})} - \alpha \right) w_t \cdot S_{B \to R_j}(\vec{r_j}) \tag{19.58}$$

and

$$t_{R_j \to M}(\vec{m}) = \left(\frac{S_{B \to R_j}(\vec{r_j})}{S_{R_j \to M}(\vec{m}) + S_{B \to R_j}(\vec{r_j})} + \alpha \right) w_t \cdot S_{R_j \to M}(\vec{m}). \tag{19.59}$$

Since $t_{B \to R_j \to M}(\vec{m}) = \min\{t_{B \to R_j}(\vec{m}), t_{R_j \to M}(\vec{m})\}$,

$$t_{B \to R_j \to M}(\vec{m}) \leq \left(\frac{S_{B \to R_j}(\vec{r_j}) S_{R_j \to M}(\vec{m})}{S_{B \to R_j}(\vec{r_j}) + S_{R_j \to M}(\vec{m})} \right) \cdot w_t. \tag{19.60}$$

The right side of Equation 19.60 is obtained by setting $\alpha = 0$ in Equations 19.54 and 19.55. With $\alpha = 0$, we have $t_{B \to R_j \to M}(\vec{m}) = t_{B \to R_j}(\vec{m}) = t_{R_j \to M}(\vec{m})$.

Lemma 2
For the FTA, the highest spectral efficiency of a 2-hop path is achieved with the following bandwidth allocation

$$w_{B \to R_j}(\vec{m}) = \frac{t_t}{S_{B \to R_j}(\vec{r_j})} \tag{19.61}$$

and

$$w_{R_j \to M}(\vec{m}) = \frac{t_t}{S_{R_j \to M}(\vec{m})}. \tag{19.62}$$

Proof: Since $t_{B \to R_j \to M}(\vec{m}) = \min\{t_{B \to R_j}(\vec{m}), t_{R_j \to M}(\vec{m})\}$, one needs to have

$$t_{B \to R_j}(\vec{m}) = S_{B \to R_j}(\vec{r_j}) \cdot w_{B \to R_j}(\vec{m}) \geq t_t \tag{19.63}$$

and

$$t_{R_j \to M}(\vec{m}) = S_{R_j \to M}(\vec{m}) \cdot w_{R_j \to M}(\vec{m}) \geq t_t. \tag{19.64}$$

The proof follows immediately from Equations 19.63 and 19.64.

References

1. ITU-R M.1645, Framework and overall objectives of the future development of IMT-2000 and systems beyond IMT-2000, 2003.
2. IEEE 802.16m-07/002r5, IEEE 802.16m system requirements, August 2008.

3. 3GPP TR 36.913, 3rd Generation Partnership Project; technical specification group radio access network; requirements for further advancements for E-UTRA (LTE-Advanced) V8.0.0, June 2008.
4. R. V. Nee and R. Prasad, *OFDM for Wireless Multimedia Communications*, Norwood, MA: Artech House Publisher, 2000.
5. H. Yang, A road to future broadband wireless access: MIMO-OFDM-based air interface, *IEEE Commun. Mag.*, 43, 53–60, 2005.
6. 3GPP TS 36.201 V1.0.0, Technical specification group radio access network; LTE physical layer–general description LTE, March 2007.
7. IEEE 802.16e-2005, IEEE standard for local and metropolitan area networks, part 16: Air interface for fixed and mobile broadband wireless access systems, amendment for physical and medium access control layers for combined fixed and mobile operation in licensed bands, February 2006.
8. E. C. van der Meulen, Three-terminal communication channels, *Adv. Appl. Prob.*, 3, 120–154, 1971.
9. T. M. Cover and A. El Gamal, Capacity theorems for the relay channel, *IEEE Trans. Info. Theory*, 25, 572–584, 1979.
10. A. Host-Madsen and J. Zhang, Capacity bounds and power allocation for wireless relay channel, *IEEE Trans. Info. Theory*, 51, 2020–2040, 2005.
11. A. Host-Madsen, On the capacity of wireless relaying, in *Proc. IEEE Veh. Technol. Conf.*, 3, 1333–1337, 2002.
12. R. U. Nabar, H. Bölcskei, and F. W. Kneubühler, Fading relay channels: Performance limits and space-time signal design, *IEEE J. Sel. Areas Commun.*, 22, 1099–1109, 2004.
13. J. N. Laneman, D. N. C. Tse, and G. W. Wornell, Cooperative diversity in wireless networks: Efficient protocols and outage behavior, *IEEE Trans. Info. Theory*, 50, 3062– 3080, 2004.
14. H. Ochiai, P. Mitran, and V. Tarokh, Variable rate two phase collaborative communication protocols for wireless networks, *IEEE Trans. Info. Theory*, 52, 4299–4313, 2006.
15. W. Su, A. K. Sadek, and K. J. Ray Liu, Cooperative communication protocols in wireless networks: Performance analysis and optimum power allocation, *Wireless Pers. Commun.*, 44, 181–217, 2008.
16. A. Ribeiro, X. Cai, and G. B. Giannakis, Symbol error probabilities for general cooperative links, *IEEE Trans. Wireless Commun.*, 4, 1264–1273, 2005.
17. N. Ahmed and B. Aazhang, Throughput gains using rate and power control in cooperative relay networks, *IEEE Trans. Commun.* 55, 656–660, 2007.
18. A. Sendonaris, E. Erkip, and B. Aazhang, User cooperation diversity–Part I: System description, *IEEE Trans. Commun.* 51, 1927–1938, 2003.
19. A. Sendonaris, E. Erkip, and B. Aazhang, User cooperation diversity–Part II: Implementation aspects and performance analysis, *IEEE Trans. Commun.* 51, 1939–1948, 2003.
20. R. Pabst, et al., Relay-based deployment concepts for wireless and mobile broadband radio, *IEEE Commun. Mag.*, 80–89, 2004.
21. A. Nosratinia, T. E. Hunter, and A. Hedayat, Cooperative communiaction in wireless networks, *IEEE Commun. Mag.*, 74–80, 2004.
22. Ö, Oyman, J. N. Lanaman, and S. Sandhu, Multihop relaying for broadband wireless mesh networks: From theory to practice, *IEEE Commun. Mag.*, 116–122, 2007.
23. D. R. Basgeet and Y. C. Chow, Uplink performance analysis for a relay based cellular system, in *Proc. IEEE VTC-2006 Spring*, pp. 132–136, May 2006.
24. H. Hu, H. Yanikomeroglu, D. D. Falconer, and S. Periyalwar, Range extension without capacity penalty in cellular networks with digital fixed relays, in *Proc. IEEE Global Telecommunications Conference*, pp. 3053–3057, November, 2004.
25. P. Li, et al., Spectrum partitioning and relay positioning for cellular system enhanced with two-hop fixed relay nodes, *IEICE Trans. Commun.*, vol. E90-B, pp. 3181-3188, Novermber 2007.
26. E. Weiss, S. Max, O. Klein, G. Hiertz, and B. Walke, Relay-based vs. conventional wireless networks: Capacity and spectrum efficiency, in *Proc. IEEE PIMRC'2007*, pp. 1–5, September 2007.
27. R. Pabst, D. C. Schultz, and B. H. Walke, Performance evalaution of a relay-based 4G network deployment with combined SDMA/OFDMA and resource partitioning, in *Proc. IEEE VTC 2008-Spring*, pp. 2001–2005, May 2008.

28. A. Adinoyi and H. Yanikomeroglu, Cooperative relaying in multi-antenna fixed relay networks, *IEEE Trans. Wireless Commun.*, 6, 533–544, 2007.
29. D. Soldani and S. Dixit, Wireless relays for broadband access, *IEEE Commun. Mag.*, 58–66, 2008.
30. Y. Yu, S. Murphy, and L. Murphy, Planning base station and relay station locations in IEEE 802.16j multi-hop relay networks, in *Proc. IEEE CCNC'2008*, pp. 922–926, January 2008.
31. https://www.ist-winner.org
32. IEEE 802.16j-06/026r4, Part 16: Air interface for fixed and mobile broadband wireless access systems: Multihop relay specification, June, 2007.
33. WiMAX White Paper, Mobile WiMAX-Part I: A technical overview and performance evaluation, http://www.wimaxforum.org/technology/downloads/, February 2006.
34. G. Auer, S. Sand, A. Dammann, and S. Kaiser, Analysis of cellular interference for MC-CDMA and its impact on channel estimation, *Eur. Trans. Telecommun.*, 15, 173–184, 2004.
35. G. Auer, A. Dammann, S. Sand, and S. Kaiser, On modeling cellular interference for multi-carrier based communication systems including a synchronization offset, in *Proc. International Symposium on Wireless Personal Multimedia Communication*, October 2003.
36. A. Adinoyi, et al., Description of identified new relay based radio network deployment concepts and first assessment by comparison against benchmarks of well known deployment concepts using enhanced radio interface technology, IST-2003-507581 WINNER D3.1, November 2004.
37. T. Cover and J. Thomas, *Elements of Infomation Theory*. Wiley, New York, 1991.
38. D. E. Goldberg, *Genetic Algorithms in Search, Optimization, and Machine Learning*. Addison-Wesley, Reading, MA, 1989.
39. R. L. Haupt and S. E. Haupt, *Practical Genetic Algorithms*. Wiley, New York, 2004.
40. K. F. Man, K. S. Tang, and S. Kwong, Genetic algorithms: Concepts and applications, *IEEE Trans. Industrial Electronics*, 43, 519–534, 1996.
41. E. K. P. Chong and S. H. Zak, *An Introduction to Optimization*. Wiley, New York, 2001.

Chapter 20

OFDM–MIMO Applications for High Altitude Platform Communications

Abbas Mohammed and Tommy Hult

Contents

20.1 Introduction to HAPs and the Proposed OFDM–MIMO HAP System 537
20.2 A Novel STP Channel Model. 541
20.3 The OFDM–MIMO System 546
20.4 The Space-Polarization Domain 547
20.5 Simulation Results 550
20.6 A Novel Orthogonal Momentum Division Multiplexing (OMDM) Technique. 555
20.7 Conclusions. 558
References. 558

20.1 Introduction to HAPs and the Proposed OFDM–MIMO HAP System

We are beginning to witness an exciting era for researchers and developers of advanced future generation multimedia telecommunication systems. High altitude platform (HAP) systems are among these novel new technologies [1] and are starting to attract considerable attention worldwide. Research and development activities include the EU FP6 CAPANINA Project [2] and the COST 297 Action [3] in Europe, along with government-funded projects in Japan, Korea, and the United States. Commercial projects are also under way in Switzerland, United States, People's Republic of China, and the United Kingdom.

HAPs are airships or planes, operating in the stratosphere, at altitudes of typically 17–22 km (around 75,000 ft). At this altitude (which is well above the commercial aircraft height), they can support payloads to deliver a range of services: principally communication and remote sensing. Communication services, including broadband, WiMAX, 3G, and emergency communications, as well as broadcast services, are under consideration. A recent HAP trial in Sweden has successfully tested the usage of a HAP at 24 km altitude, operating in the mm-wave band to send data via Wi-Fi (802.11b) to a coverage area 60 km in diameter, to demonstrate the potential of this new and novel technology.

A HAP can provide the best features of both terrestrial masts (which may be subject to planning restrictions and/or related environmental/health constraints) and satellite systems (which are usually highly expensive) [4]. This makes HAP a viable competitor/complement to conventional terrestrial infrastructures and satellite systems. In particular, HAPs permit rapid deployment, and highly efficient use of the radio spectrum (largely through intensive frequency re-use). The relatively close range of HAPs compared to satellites means that data rates can be significantly higher for the same size antennas, and imaging and remote sensing are highly effective, offering low cost and high resolution. A variety of hybrid applications may also be envisaged, such as traffic management, navigation, security management, and so on. There are two fundamental types of platform technology capable of stratospheric flight: manned and unmanned aircraft and unmanned airships. Other platform technologies at lower altitudes including manned aircraft and tethered aerostats and unmanned aerial vehicles (UAVs) may also play a developmental role in HAPs and their applications [4].

Many countries have made significant efforts in the research on the HAP system and its applications. Some well-known projects are (1) HeliNet and CAPANINA of the European Union (EU) [5]; (2) SkyNet project in Japan [6]; (3) a project managed by ETRI and KARI in Korea [7]; (4) in USA, Sanswire Technologies Inc., Sky Station, and Angel Technologies carried out a series of researches and demonstrations HAP's on practical applications [6]; (5) engineers from Japan have demonstrated that HAPs can be a new platform to provide high-definition television (HDTV) services and international mobile telecommunications (IMT-2000) wideband code division multiple access (WCDMA) service successfully. These projects mainly focus on IMT-2000 services, IEEE802.1x services, and fixed broadband wireless access (FBWA) in the Ka/Ku band and carry out various feasibility studies for using the HAP system and research a HAP applications [4,8–12].

Despite the advantages of HAPs as listed above, the mobility of the terminals and platforms and the characteristics of the stratospheric channel may cause strong attenuation, multipath fading, and Doppler shift, which might strongly impact their performance. A particularly promising candidate for next generation wireless mobile communication systems is the combination of orthogonal frequency division multiplexing (OFDM) and multiple-input and multiple-output (MIMO) technology [13–17]. OFDM is a technique that has gained tremendous interest in recent years because of its *high spectral efficiency* and robustness against intersymbol interference (ISI) and multipath fading. The use of multiple antennas at both the transmitter and the receiver (MIMO technology) holds the potential to drastically improve the *spectral efficiency* and link reliability of future wireless communications systems. OFDM combines very naturally with MIMO. Thus, in this chapter, we propose a combined OFDM–MIMO technique in order to *increase the capacity* of HAP communication links [18–27] and combat the channel impairments.

The proposed system consists of virtual MIMO spatial channels (created by HAP diversity) in conjunction with the OFDM subcarriers [18–27], and the antenna polarization/pattern diversity (formed via special compact MIMO antenna arrangements) [28–31] as shown in Figure 20.1. This can be viewed as a means of achieving independent channels over polarization space as well as frequency. In essence, the MIMO technique achieves orthogonality over space by employing the singular value decomposition (SVD), and the OFDM technique achieves orthogonality over frequency by employing the discrete Fourier transform (DFT) with a cyclic prefix (CP).

It is worth mentioning here that the application of MIMO to HAP communications is quite scarce in the open literature. However, the application of MIMO to satellite communication links is reported in some recent papers [32–36]. The concept of polarization diversity is of particular interest due to its potential application satellite and stratospheric systems as stressed in these papers.

The number of relaying HAPs used in the system is dependent on the number of spatial channels created by the employed antenna arrays. The antenna arrays investigated in this chapter divide the scenarios into two groups, which utilize a platform diversity consisting of three or six relaying HAP platforms. Figure 20.1 shows the diversity setup for scenario 2 (Table 20.1) where six HAPs are employed.

In scenarios 1 and 2 we assume that the HAPs are located along a fixed line with the HAPs uniformly distributed and separated by the angles θ. In these scenarios the wave propagation channel $\mathbf{H}_{ch}$ from the N transmitter antenna array elements to the M receiver antenna array elements is modeled as a free space loss (FSL) model [17], and is defined as a transformation containing the space-dependent decaying values of the signal being transmitted

$$H_{mn}(\mathbf{r}, f) = \frac{c}{4\pi f |\mathbf{r}_m - \mathbf{r}_n|}, \qquad (20.1)$$

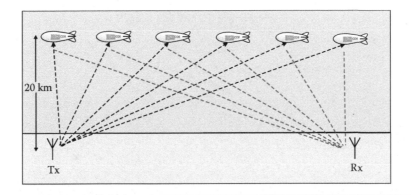

Figure 20.1   The MIMO-HAP diversity system with six relaying HAPs (scenario 2) and the channel paths from the transmitter to the receiver.

Table 20.1   Platform Scenarios

Scenario	Platform	Number of Platforms	Type of Channel Model
1	HAP	3	FSL
2	HAP	6	FSL
3	HAP	3	Multiple STP
4	HAP	6	Multiple STP

where c is the speed of light and $|\mathbf{r}_m - \mathbf{r}_n|$ is the distance along the path between transmitter array element m and receiver array element n. In this propagation channel an uncorrelated Gaussian noise is assumed and atmospheric or multipath interference is not taken into account.

For scenarios 3 and 4, a novel space-time-polarization (STP) multichannel simulator model is utilized. Full details of the STP model are presented in Section 20.3. In these scenarios, a signal is transmitted through a propagation channel with physical objects that generate multipath components of the transmitted signal, each with its own attenuation, polarization, correlation, phase shift, delay, and direction of propagation. Depending on the material of the objects in the channel, frequency-dependent stochastic scattering effects can also affect the received signal. Each object in the propagation channel is modeled as a cluster of microscatterers [37], see Figure 20.2, in the local area (within a few hundreds wavelengths) centered about the center position of the cluster. This cluster is denoted as a macroscatterer. Furthermore, both the transmitter and the receiver can respectively be positioned inside a local area surrounded by microscatterers.

In order to model the atmospheric angle spreading effect due to scintillation phenomena and the depolarization effects due to precipitation, clusters are added in the direct line-of-sight (LOS) path as well [21–23]. In the presence of different polarizations, attenuations, phase shifts, and polarizations generated by the multipath cluster interaction with the signal can differ for different

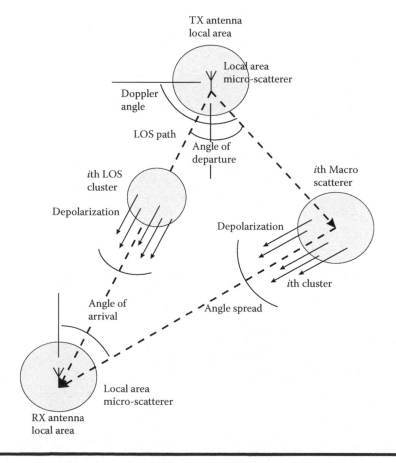

Figure 20.2 The cluster geometry for the space-time-polarization simulator.

incident polarization states and will therefore also generate a cross-polarization effect. The relationship between these polarization components of the electromagnetic wave is measured by the cross-polar discrimination (XPD) defined as [38,39]

$$XPD = 20 \log_{10} \left(\frac{E_{\|}}{E_{\perp \to}} \right), \tag{20.2}$$

where $E_{\|}$ is the amount of signal that remains in the same polarization state, and $E_{\perp \to}$ is the amount of signal that has scattered out into the opposite orthogonal polarization state.

Depolarization from rain is strongly correlated with the rain attenuation and can be calculated from the following empirical formula [38,39]

$$XPD = a - b \log(L), \tag{20.3}$$

where $a = 35.8$ and $b = 13.4$ are reasonable values for frequencies below 10 GHz, and L is the rain attenuation [38,39].

20.2 A Novel STP Channel Model

The spatial properties of wireless communication channels are extremely important in determining the performance of future wireless communication systems. In light of the vast interest in MIMO systems and in multicarrier transmission employing OFDM and compact MIMO antenna array applications, new and improved channel models are necessary to evaluate the parameters and performance of these systems. Thus, in this section, we propose a novel multiple antenna channel model and simulator [21–23] that takes into account the *STP* properties affecting signal transmission in wireless communications, and we test its performance for the proposed OFDM–MIMO HAP system [21–23]. The proposed channel model is an extension of Ref. [37], in which only the spatial and temporal (ST) properties of the signals were implemented.

The signals carried by the different spatial subchannels are written in a vector format as

$$\mathbf{i}(t) \triangleq \left[i^{(1)}(t), i^{(2)}(t), \ldots, i^{(N_p)}(t) \right]^{\mathrm{T}}, \tag{20.4}$$

where $i^{(p)}(t)$ is the information signal transmitted with the pth mode. The transmitting antenna is assumed to be a linear transformation of the input signal $\mathbf{i}(t)$ into the independent radio channels $\mathbf{s}(t)$, according to

$$\mathbf{s}(t) \triangleq \mathbf{M}_{tx}^{\mathrm{T}} \mathbf{i}(t). \tag{20.5}$$

At the receiver we have a similar transformation from the signals in the independent radio subchannels $\mathbf{s}_{rx}(t)$ into the output signal $\mathbf{r}(t)$ of the system, that is, the vector of the elementary signals received in each mode

$$\hat{\mathbf{r}}(t) \triangleq \left[\hat{r}^{(1)}(t), \hat{r}^{(2)}(t), \ldots, \hat{r}^{(N_p)}(t) \right]^{\mathrm{T}} = \mathbf{M}_{rx} \mathbf{s}_{rx}(t), \tag{20.6}$$

where $\mathbf{s}(t)$ and $\mathbf{s}_{rx}(t)$ are vectors containing the independent spatial-polarization modes that are active for a specific antenna type and they are related through

$$\mathbf{s}_{rx}(t) = \mathbf{H}_{ch} * \mathbf{s}(t), \tag{20.7}$$

where the symbol $*$ indicates matrix convolution, as defined in Ref. [37]. The channel $\mathbf{H}_{ch}$, from the transmitter mode vector $\mathbf{s}(t)$ to the receiver mode vector $\mathbf{s}_{rx}(t)$, is the simulated propagation channel. Thus, the mode signal vector $\hat{\mathbf{r}}(t)$ at the receiver can be rewritten using Equations 20.5 through 20.7 as a linear combination of all the transmitted modes, each of which is subjected to the effects of the propagation channel

$$\hat{\mathbf{r}}(t) = \mathbf{M}_{rx}\mathbf{H}_{ch} * \mathbf{M}_{tx}^{T}\mathbf{i}(t). \tag{20.8}$$

The coefficients of the linear combination represent the effect of the fading, cross-correlation and depolarization induced by the channel.

In general, it is possible to assume that each mode is transmitted by an antenna array that uses a specific spatial processor (transmitter beamformer) for each specific mode. The pth transmitter beamforming weight vector is then denoted as $\mathbf{v}^{(p)}$. The signals propagating through the wireless channel are affected by several factors (e.g., ST fast fading, shadowing, etc.). In particular, the signals are affected by reflection and diffraction phenomena, both at the macroscopic and the microscopic (scattering) scale that generates attenuations, phase shifts, and time delays. It also causes the signals to arrive at the receiver from different direction of arrivals (DOAs) other than that of the direct LOS path. This is characterized by a certain angular spread of the signals. There is also the possibility of mobility at both the transmitter and the receiver side, which contributes to the scattering phenomenon and gives rise to temporal fading processes. However, in the presence of different polarizations, the attenuations and phase shifts generated by the microscopic interactions with the scattering structures may differ for different polarizations and could generate cross-polar effects.

From Equation 20.8 the elementary signal received in the pth mode can be written as

$$\hat{r}^{p}(t) = \mathbf{w}^{(p)H}\mathbf{a}_{rx}(\theta)\sum_{q=1}^{N_{p}}\gamma_{qp}\mathbf{a}_{tx}^{H}(\xi)\mathbf{v}^{(q)}h(\zeta;t) * i^{(q)}(\zeta), \tag{20.9}$$

where $\mathbf{w}^{(p)}$ is the receiver beamforming weight vector used for the pth mode, $\mathbf{a}_{tx}(\xi)$ and $\mathbf{a}_{rx}(\theta)$ are the transmitter and receiver array steering vectors, respectively. ξ and θ are the direction of departure (DOD) and DOA expressed with respect to the transmitter and receiver reference frames. $h(\zeta;t)$ is the time-domain elementary path impulse response defined in Ref. [37], $*$ is the convolution operator, and γ_{qp} is the real coefficient representing the fraction of the qth polarization being scattered into the pth part

$$\gamma_{qp} = \frac{E_{\parallel}}{E_{\perp\rightarrow}}, \tag{20.10}$$

$$\gamma_{pq} = \frac{E_{\parallel}}{E_{\perp\leftarrow}}, \tag{20.11}$$

assuming that $p \triangleq \perp$ and $q \triangleq \parallel$. Furthermore,

$$\mathbf{a}_{tx}^{T}(\xi)\mathbf{v}^{(p)} = \left[\mathbf{M}_{tx}^{T}\right]_{(q,q)}, \tag{20.12}$$

$$\mathbf{w}^{(p)H}\mathbf{a}_{rx}(\theta) = [\mathbf{M}_{rx}]_{(p,p)}, \tag{20.13}$$

since from Equations 20.5 and 20.6 it is possible to set the relationships

$$\mathbf{M}_{tx} \triangleq \mathbf{D}_{v} \otimes \mathbf{a}_{tx}(\xi), \tag{20.14}$$

$$\mathbf{M}_{rx} \triangleq \mathbf{D}_{w} \otimes \mathbf{a}_{rx}(\theta), \tag{20.15}$$

where

$$\mathbf{D}_{v} \triangleq \begin{bmatrix} \mathbf{v}^{(1)} & 0 & \cdots & 0 \\ 0 & \mathbf{v}^{(2)} & & \vdots \\ \vdots & & \ddots & \\ 0 & \cdots & & \mathbf{v}^{(N_p)} \end{bmatrix} \in \mathbb{C}^{NN_p,N_p}. \tag{20.16}$$

$\mathbf{D}_{w} \in \mathbb{C}^{MN_p,N_p}$ is defined in a similar way and $\otimes$ indicates the Kronecker product. The expression in Equation 20.9 can then be rewritten in matrix notation as

$$\hat{r}^{p}(t) = \mathbf{w}^{(p)H}\mathbf{a}_{rx}(\theta)\left(\boldsymbol{\gamma}^{(p)T} \otimes \mathbf{a}_{tx}^{T}(\xi)\right)\mathbf{D}_{v}h(\varsigma;t) * \mathbf{i}(\varsigma), \tag{20.17}$$

where

$$\boldsymbol{\gamma}^{(p)} \triangleq \left[\boldsymbol{\gamma}_{11}, \gamma_{21}, \ldots, \gamma_{N_p,1}\right]^{T}. \tag{20.18}$$

The vector of the received elementary modes, that is, the vector of the elementary signals received in each mode, can now be expressed as

$$\begin{aligned}
\hat{\mathbf{r}}(t) &\triangleq \left[\hat{r}^{(1)}(t), \hat{r}^{(2)}(t), \ldots, \hat{r}^{(N_p)(t)}\right]^{T} \\
&= \begin{bmatrix} \mathbf{w}^{(1)H}\mathbf{a}_{rx}(\theta)\left(\gamma^{(1)T} \otimes \mathbf{a}_{tx}^{T}(\xi)\right)\mathbf{D}_{v}h(\varsigma;t) * \mathbf{i}(\varsigma) \\ \vdots \\ \mathbf{w}^{(N_p)H}\mathbf{a}_{rx}(\theta)\left(\gamma^{(N_p)T} \otimes \mathbf{a}_{tx}^{T}(\xi)\right)\mathbf{D}_{v}h(\varsigma;t) * \mathbf{i}(\varsigma) \end{bmatrix} \\
&= \mathbf{D}_{w}^{H}\boldsymbol{\Gamma}^{T} \otimes \left(\mathbf{a}_{rx}(\theta)\mathbf{a}_{tx}^{T}(\xi)\right)\mathbf{D}_{v}h(\varsigma;t) * \mathbf{i}(\varsigma),
\end{aligned} \tag{20.19}$$

where

$$\begin{aligned}
\boldsymbol{\Gamma} &\triangleq \left[\gamma^{(1)}, \gamma^{(2)}, \ldots, \gamma^{(N_p)}\right] \\
&= \begin{bmatrix} \gamma_{11} & \gamma_{12} & \cdots & \gamma_{1N_p} \\ \gamma_{21} & \gamma_{22} & & \vdots \\ \vdots & & \ddots & \\ \gamma_{N_p 1} & \cdots & & \gamma_{N_p N_p} \end{bmatrix} \in \mathbb{R}^{N_p,N_p}
\end{aligned} \tag{20.20}$$

is the cross-polarization matrix. Thus, it is possible to define the multispace-polarization elementary path channel matrix as

$$\hat{\overline{\mathbf{H}}}(\zeta, t) \triangleq \mathbf{\Gamma}^{\mathrm{T}} \otimes \mathbf{a}_{\mathrm{rx}}(\theta)\mathbf{a}_{\mathrm{tx}}^{\mathrm{T}}(\xi)h(\zeta, t) \in \mathbb{C}^{MN_p, N_p}. \tag{20.21}$$

This is also depicted as a block diagram in Figures 20.3 and 20.4, where we can identify the elementary-path channel matrix $\hat{\mathbf{H}}(\zeta, t) = \mathbf{a}_{\mathrm{rx}}(\theta)\mathbf{a}_{\mathrm{tx}}^{\mathrm{T}}(\xi)h(\zeta, t)$ obtained in Ref. [37]. Thus,

$$\hat{\overline{\mathbf{H}}}(\zeta, t) = \mathbf{\Gamma}^{\mathrm{T}} \otimes \hat{\mathbf{H}}(\zeta, t), \tag{20.22}$$

and the received elementary modes vector becomes

$$\hat{\overline{\mathbf{r}}}(t) = \mathbf{D}_{\mathrm{w}}^{\mathrm{H}}\hat{\overline{\mathbf{H}}}(\zeta, t) * \mathbf{D}_{\mathrm{v}}\mathbf{i}(\zeta), \tag{20.23}$$

which allows the equivalence, $\mathbf{D}_{\mathrm{w}}^{\mathrm{H}}\hat{\overline{\mathbf{H}}}(\zeta, t)\mathbf{D}_{\mathrm{v}} \equiv \mathbf{M}_{\mathrm{rx}}\mathbf{H}_{\mathrm{ch}}\mathbf{M}_{\mathrm{tx}}^{\mathrm{T}}$, as derived from Equation 20.8.

Following the analogy developed in Ref. [37], it is possible to identify the lth cluster multispace-polarization channel matrix for each cluster of microscopic scattering elements in the physical channel. Thus, let us indicate with $\mathcal{D}(\boldsymbol{\chi})$ the multidimensional domain of the vector of the

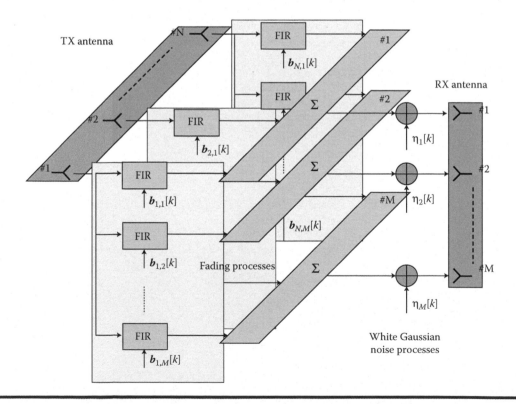

Figure 20.3 The block diagram of the complete simulator where the signal from each transmitting antenna element is filtered by a separate STP FIR filter.

independent channel variables $\chi \triangleq [\xi, \theta, \psi_{tx}, \psi_{rx}]^T$, where ψ_{tx} and ψ_{rx} are the Doppler shift of the transmitter and receiver, respectively, with the nominal value χ_l for each cluster $l = 0, 1, \ldots, L_S$. The lth cluster channel matrix can then be computed as

$$\overline{\mathbf{H}}_l(\zeta, t) \triangleq \int_{\mathcal{D}(\chi_l)} \hat{\overline{\mathbf{H}}}(\zeta, t)\, d\chi. \tag{20.24}$$

The global multispace-polarization channel matrix can now be written as

$$\overline{\mathbf{H}}(\zeta, t) \triangleq \sum_{l=0}^{L_S} \overline{\mathbf{H}}_l(\zeta, t). \tag{20.25}$$

where $\overline{\mathbf{H}}(\zeta, t)$ is the total STP multichannel simulator model utilized in scenarios 3 and 4.

We can now write the received signal vector $\hat{\mathbf{r}}(t)$ using Equations 20.5 through 20.7, where we define $\mathbf{H}_{ch} \triangleq \overline{\mathbf{H}}(\zeta, t)$, as a linear combination of all the transmitted modes:

$$\hat{\mathbf{r}}(t) = \mathbf{M}_{rx}\mathbf{H}_{ch} * \mathbf{M}_{rx}^T\mathbf{i}(t). \tag{20.26}$$

The coefficients of the linear combination represent the effect of the spatio-temporal fading, cross-correlation, and depolarization induced by the channel.

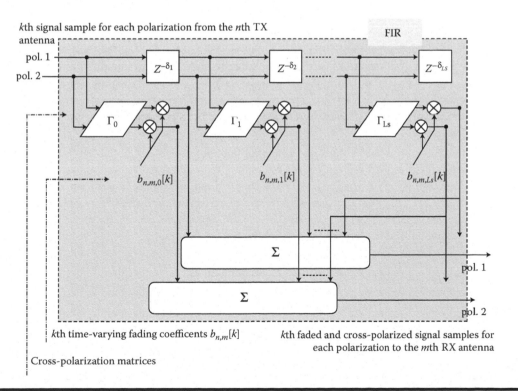

Figure 20.4 The block diagram of the separate STP FIR filter for each cluster.

20.3 The OFDM–MIMO System

In this section, we discuss the combined OFDM–MIMO system. Through the combination of MIMO and OFDM we can also achieve orthogonality over the frequency domain by using the DFT with a CP to mitigate the effects of frequency selective channels. Assuming $CP \geq \tau_d f_s$, where $f_s = T_s^{-1}$ and T_s is the symbol time, we can write the received signals as a circular convolution ⊛ over the DFT frame as

$$r_m(t) = \sum_{n=1}^{N} h_{mn}(t) \circledast s_n(t) + v_m(t), \tag{20.27}$$

where N_{DFT} is the size of the DFT, $m = 1, \ldots, M$ are the receiving antennas and $v_m(t)$ is the additive white Gaussian noise (AWGN). If we have a MIMO antenna system with N transmitting antennas and M receiving antennas, we can then write the signal for subchannel k in the frequency domain between any pair of transmitting and receiving antennas as

$$r_m(k) = \sum_{n=1}^{N} h_{mn}(k)s_n(k) + v_m(k), \tag{20.28}$$

where k denotes each separate subcarrier. If we write Equation 20.28 in vector notation, we get

$$\mathbf{r}(k) = \mathbf{H}(k)\mathbf{s}(k) + \mathbf{v}(k), \tag{20.29}$$

where $\mathbf{r}(k)$ is an $(M \times 1)$ vector and $\mathbf{s}(k)$ is an $(N \times 1)$ vector of the received and transmitted signals, respectively. $\mathbf{H}(k)$ is the $(M \times N)$ frequency response matrix $\mathbf{H}(k) = h_{mn}(k), m = 1, \ldots, M$ and $n = 1, \ldots, N$ in Equation 20.27 of the channel between N transmitters and M receivers. The noise in the system $\mathbf{v}(k)$ is an $(M \times 1)$ vector assumed to be AWGN. Thus, the correlation matrix of the noise vector $\mathbf{v}(k)$ is $\mathrm{E}\{\mathbf{v}(k)\mathbf{v}^{\mathrm{H}}(k)\} = \sigma_v^2 \cdot \mathbf{I}_M$, where σ_v^2 is the variance of the noise and $\mathbf{I}_M$ is the $(M \times M)$ identity matrix. Since we are using the SVD technique, we can now write the channel matrix $\mathbf{H}(k)$ as

$$\mathbf{H}(k) = \mathbf{U}(k)\Sigma(k)\mathbf{V}^{\mathrm{H}}(k), \tag{20.30}$$

where $\Sigma(k)$ is an $(M \times N)$ matrix containing singular values that are larger than zero $\sigma_1(k) \geq \sigma_2(k) \ldots \geq \sigma_r(k) > 0$, where r is the rank of the matrix $\mathbf{H}(k)$, and the $(M \times M)$ matrix $\mathbf{U}(k)$ and the $(N \times N)$ matrix $\mathbf{V}(k)$ contain the corresponding eigenvectors as matrix column vectors. To obtain a diagonalized system, we then define

$$\mathbf{y}(k) = \Sigma(k)\mathbf{x}(k) + \mathbf{n}(k), \tag{20.31}$$

where

$$\begin{cases} \mathbf{y}(k) = \mathbf{U}^{\mathrm{H}}(k)\mathbf{r}(k), \\ \mathbf{s}(k) = \mathbf{V}(k)\mathbf{x}(k), \\ \mathbf{n}(k) = \mathbf{U}^{H}(k)\mathbf{v}(k). \end{cases} \tag{20.32}$$

Since the OFDM–MIMO channels in Equation 20.30 are uncorrelated and the correlation of the noise $\mathbf{n}(k)$ is $\mathrm{E}\{\mathbf{n}(k)\mathbf{n}^{\mathrm{H}}(k)\} = \sigma_n^2 \cdot \mathbf{I}_M$, we can write the theoretical maximum information capacity of the system [16] as

$$C = \sum_{k=0}^{N_{\text{DFT}}-1} \sum_{m=1}^{r} \log_2 \left(1 + \sigma_{x_m}^2(k) \frac{\sigma_m^2(k)}{\sigma_n^2} \right),$$ (20.33)

where $\sigma_{x_m}^2$ is the variance of the separate uncorrelated input signals in vector $\mathbf{x}(k)$. The capacity in Equation 20.33 is constrained by the total radiated power from the transmitting antennas, defined as

$$P_{\text{rad}} = \sum_{k=0}^{N_{\text{DFT}}-1} \sum_{m=1}^{r} \sigma_{x_m}^2(k).$$ (20.34)

To maximize the total sum of capacities in all the subchannels, we use the so-called *water-filling* technique in which we allocate more power to the subchannels with high eigenvalues. The optimal *water-filling* solution [40] is then given by

$$\begin{cases} \sigma_{x_m}^2(k) = \gamma - \dfrac{\sigma_n^2}{\sigma_m^2(k)} & \text{if } \gamma - \dfrac{\sigma_n^2}{\sigma_m^2(k)} \geq 0, \\[3mm] \sigma_{x_m}^2(k) = 0 & \text{if } \gamma - \dfrac{\sigma_n^2}{\sigma_m^2(k)} \leq 0, \end{cases}$$ (20.35)

where $k = 0, \ldots, N_{\text{DFT}} - 1$, $m = 1, 2, \ldots, r$, and γ is a pre-defined threshold level of the signal-to-noise ratio (SNR) in the system and it is dependent on the total transmitted power P_{rad} in Equation 20.34. The average signal-to-noise ratio, SNR_{avg}, in the simulations is calculated as

$$\text{SNR}_{\text{avg}} = 10 \log_{10} \left(\frac{1}{N_{\text{active}}} \sum_{k=0}^{N_{\text{DFT}}-1} \sum_{m=1}^{r} \sigma_{x_m}^2(k) \frac{\sigma_m^2(k)}{\sigma_n^2} \right),$$ (20.36)

where N_{active} is the number of active channels used for which the variance of the input signals $\sigma_{x_m}^2 > 0$.

20.4 The Space-Polarization Domain

In the proposed system, each transmit and receive antenna arrangement of the system consists of a special compact MIMO antenna array configuration, which possesses a structure of different design and complexity as shown in Figure 20.5. The tested antennas shown in this figure are: the vector element antenna, MIMO-cube, MIMO-tetrahedron, and the proposed novel MIMO-octahedron [28–31].

The polarization and antenna radiation pattern of the electromagnetic field can be expressed as a multipole expansion [41,42] of the field emanating from a virtual sphere enclosing the antenna that is being analyzed. This series expansion consists of weighted orthogonal base functions on the

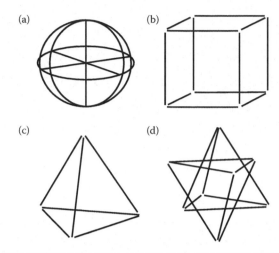

Figure 20.5 The structure of various compact MIMO antenna array configurations: (a) vector element antenna, (b) MIMO-cube, (c) MIMO-tetrahedron, and (d) MIMO-octahedron.

surface of the virtual sphere and allows for a solution to Maxwell's equations that can be written as

$$
\begin{cases}
\vec{E} = \sum_{l,m} \left[\dfrac{j}{k} a_E(l,m)(\nabla \times f_l(kr)\vec{X}_{lm}) + a_M(l,m)g_l(kr)\vec{X}_{lm} \right], \\[3mm]
\vec{H} = \dfrac{1}{\eta_0} \sum_{l,m} \left[a_E(l,m)f_l(kr)\vec{X}_{lm} - \dfrac{j}{k} a_M(l,m)(\nabla \times g_l(kr)\vec{X}_{lm}) \right],
\end{cases}
\tag{20.37}
$$

where η_0 is the intrinsic impedance of free space. The base functions $\vec{X}_{lm}(\varphi,\theta) = \vec{L}Y_{lm}(\varphi,\theta)$ are orthogonal vector functions of the spherical field in the φ and θ directions when the far-field of the antenna is projected onto the virtual sphere. Y_{lm} is the scalar spherical harmonic function and $\vec{L} = j\hat{r} \times \nabla$ is the angular momentum operator. The radial functions g_l and f_l in Equation 20.37 are spherical Hankel functions representing an outgoing (transmitted) wave or an incoming (received) wave. The weights a_E and a_M are the corresponding real valued coefficients and will give the gain of each orthogonal function (mode) for a particular electromagnetic far-field pattern, as shown by

$$
\begin{cases}
a_E(l,m) \approx \dfrac{1}{\eta_0} \cdot \dfrac{k}{\sqrt{l(l+1)}} f_l(kr) \displaystyle\int Y_{lm}^*(\theta,\phi)\,\vec{r} \cdot \vec{E}\, d\Omega, \\[3mm]
a_M(l,m) \approx -\dfrac{k}{\sqrt{l(l+1)}} g_l(kr) \displaystyle\int Y_{lm}^*(\theta,\phi)\,\vec{r} \cdot \vec{H}\, d\Omega.
\end{cases}
\tag{20.38}
$$

By using Equation 20.38, we can calculate which modes are active on any arbitrary antenna enveloped by a virtual sphere by knowing the current distribution $\vec{J}$, the charge distribution ρ and the intrinsic magnetization $\vec{M}$ of the antenna structure. These modes are theoretically orthogonal to each other and therefore represent independent ports of the antenna. The vector element antenna and MIMO-Tetrahedron have theoretically six independent ports, while the MIMO-cube and MIMO-octahedron have 12 independent ports.

The transmitting channel $\mathbf{H}_{tx}$ is then assumed to be the linear transformation of the input signal $\mathbf{x}$ into the mode domain $\mathbf{a}_{tx}$ according to $\mathbf{a}_{tx} = \mathbf{H}_{tx}\mathbf{x}$. For the receiving channel we have a similar transformation from the mode domain $\mathbf{a}_{rx}$ into the output signal $\mathbf{y}$ of the system following $\mathbf{y} = \mathbf{H}_{rx}\mathbf{a}_{rx}$, where $\mathbf{a}_{tx}$ and $\mathbf{a}_{rx}$ are vectors containing the mode gains for a specific antenna type. In order to separate these modes into uncorrelated channels, we use SVD defined as

$$\mathbf{H} = \mathbf{U}\Sigma\mathbf{V}^{H}, \tag{20.39}$$

where $\mathbf{U}$ and $\mathbf{V}$ are matrices containing the eigenvectors of the antenna and matrix Σ comprise of the antenna gains for independent MIMO subchannels.

The orthogonalization of the channel into the independent subchannels is done by multiplying the signal to be transmitted $\mathbf{x}$ with the matrix $\mathbf{V}$ on the transmitter side of the channel and multiply the received signal $\mathbf{y}$ on the receiver side of the channel with the matrix $\mathbf{U}^{H}$

$$\mathbf{y} = \mathbf{U}^{H}\mathbf{H}\,(\mathbf{V}\mathbf{x}), \tag{20.40}$$

where $\mathbf{y}$ is a vector containing the decoded signal from the independent MIMO subchannels and $\mathbf{x}$ is a vector of the separate signals to be transmitted through the MIMO channel.

If spatial correlation and mutual coupling effects are taken into account, the channel matrix is given by

$$\mathbf{H} = \frac{\mathbf{Z}_{rx}\,(\mathbf{R}_{rx})^{1/2}\,\mathbf{H}_0\,(\mathbf{R}_{tx})^{1/2}\,\mathbf{Z}_{tx}}{C_{rx}C_{tx}}, \tag{20.41}$$

where $\mathbf{H}_0$ is the channel response without spatial correlation and mutual coupling. $\mathbf{R}_{rx}$ and $\mathbf{R}_{tx}$ are the spatial correlation matrices on the receiving and transmitting sides, respectively. The mutual coupling between the elements of the compact antenna array is denoted $\mathbf{Z}_{rx}$ and $\mathbf{Z}_{tx}$ normalized by C_{rx} and C_{tx}, respectively. The mutual coupling is calculated from

$$\begin{cases} \mathbf{Z}_{tx} = \mathbf{Z}_0^{tx}\left(\mathbf{Z}_0^{tx} + \mathbf{Z}_L\right)^{-1}, \\[2mm] \mathbf{Z}_{rx} = \mathbf{Z}_L\left(\mathbf{Z}_0^{rx} + \mathbf{Z}_L\right)^{-1}, \end{cases} \tag{20.42}$$

where $\mathbf{Z}_0$ represents the impedance matrix of the transmitting and receiving compact arrays, $\mathbf{Z}_L$ is a diagonal matrix containing the source impedance that has been chosen as the complex conjugate of the self-impedance given by the diagonal impedance matrix $\mathbf{Z}_{0,ii}$, and C_{rx} and C_{tx} are normalizing factors [43].

The spatial correlation matrices $\mathbf{R}_{tx}$ and $\mathbf{R}_{rx}$ are calculated between the antenna element positions and polarization states according to [44]

$$\rho_{p,q} = \frac{\int_\varphi \int_\theta \int_\zeta \int_\kappa \mathbf{a}_q(\Theta)\mathbf{a}_p^*(\Theta)\sin\theta\mathbf{p}(\Theta)\,d\kappa\,d\zeta\,d\theta\,d\varphi}{\sqrt{\int_\varphi \int_\theta \int_\zeta \int_\kappa |\mathbf{a}_q(\Theta)|\sin\theta\mathbf{p}(\Theta)\,d\kappa\,d\zeta\,d\theta\,d\varphi}} \\ \times \frac{1}{\sqrt{\int_\varphi \int_\theta \int_\zeta \int_\kappa |\mathbf{a}_p(\Theta)|\sin\theta\mathbf{p}(\Theta)\,d\kappa\,d\zeta\,d\theta\,d\varphi}}, \tag{20.43}$$

where the element p, q of matrix $\mathbf{R}$ is the correlation between antenna elements p and q. φ and θ are the spherical coordinates expressing the spatial domain, and ζ and κ are the polarization angle and phase difference, respectively, that account for the effects in the polarization domain [44]. In Equation 20.43, $\mathbf{a}(\Theta)$ is the steering vector and $\mathbf{p}(\Theta)$ is the joint probability distribution function of the parameter vector $\Theta = \left[\theta \; \varphi \; \zeta \; \kappa \right]^{\mathrm{T}}$, where T denotes a transpose operator. It is assumed that all the parameters are independent of each other and that $p(\zeta) = u\{0, \pi\}$ and $p(\kappa) = u\{-\pi, \pi\}$ are uniformly distributed.

20.5 Simulation Results

In this section, we present the simulation results of the various parameters affecting the performance of the proposed OFDM–MIMO HAP system. The simulation results are divided according to the number of platforms and type of channel model employed (referring to Table 20.1) of the different scenarios and the various compact MIMO antenna arrays.

First we compare the capacity for different ideal (no mutual coupling or spatial correlation) compact antenna arrays using scenarios 1 and 2 for a system of HAPs operating at an altitude of 20 km and with a separation angle of 20°, and plotted against the average SNR of the system. The elevation angle is held at 10° throughout this investigation. It is evident from Figure 20.6 that the multiple HAP diversity system provides superior performance as compared to the single HAP case. It is also apparent that the MIMO-cube antenna or MIMO-octahedron antenna provides a better capacity than both the MIMO-tetrahedron and the vector element antenna due to the higher number of acquired platforms (independent channels). For example, if we compare the performance for the multiple platform systems (using three and six platforms) taken together with that of the

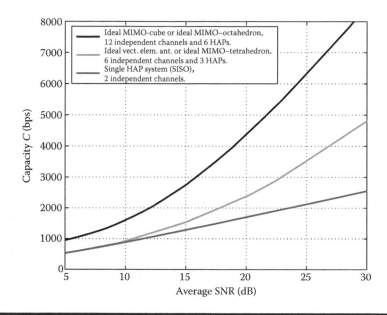

Figure 20.6   The capacity of the multiple HAP system for the ideal (no mutual coupling or spatial correlation) compact MIMO arrays in Figure 20.5.

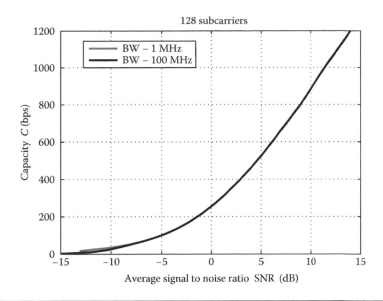

Figure 20.7   The capacity of the multiple HAP system for different bandwidths of the 128 OFDM carrier waves.

single platform case at an SNR of 20 dB, we can observe that in the multiple HAP system the capacity of six HAPs is 158% higher than that of the single HAP system and 85% higher than that of the three HAP system, which in turn is 39% higher than the single HAP system.

Next we investigate the influence of OFDM parameters on the system performance. In figure 20.7 we show the effect of changing between a narrow and a wide OFDM subchannel bandwidth, ranging from 1 to 100 MHz for an OFDM system of 128 subcarriers. It can be observed that this increase in the bandwidth has a negligible influence on the capacity of the system. On the other hand, if we increase the number of OFDM subcarrier waves from a single subcarrier to an N subcarrier OFDM system, we obtain an N-fold increase in system capacity, as can be clearly seen from Figure 20.8.

Next we investigate the effect of different separation angles between platforms on the capacity for different compact antenna arrays. The maximum capacity, calculated in accordance with Equation 20.33, is achieved when there is a spatial-polarization alignment between the modes of the transmitting array and the receiving array. Since the spatial-polarimetric radiation pattern is dependent on the geometrical shape of the compact array, the separation angle where the maximum capacity of the system occurs is also dependent on the geometry of the array and therefore it differs for the different arrays depicted in Figure 20.5. Figure 20.9 show the effect of the separation angles between platforms (ranging from 0 to 35°) on the capacity for the various compact antenna arrays at a fixed SNR of 20 dB. It is clear from this figure that the separation angle between HAPs has a great impact on the system capacity. It is also evident that the MIMO-Cube and the vector element antenna array have their maximum capacity at separation angles of 15 and 20°, respectively, while both the MIMO-tetrahedorn and MIMO-octahedron have their maximum capacity at 10°. The agreement in the optimal separation angles of the MIMO-tetrahedorn and MIMO-octahedron arrays is due to the similar geometry of the arrays, since the MIMO-octahedron can be seen as two co-located MIMO-tetrahedron arrays. It is also clear from Figure 20.9 that both the vector element antenna and the MIMO-cube array are more robust in situations where the platforms are widely

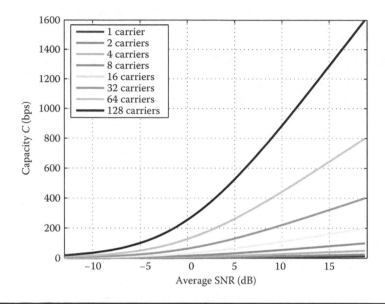

Figure 20.8 The capacity of the multiple HAP system for different numbers of OFDM carrier waves and for a bandwidth of 1 MHz.

spread, and that the MIMO-tetrahedron and MIMO-octahedron arrays have better performance in situations where the platforms are positioned close together. We can also observe in Figure 20.9 that if the separation angle tends toward 0° the spatial diversity will collapse into a SISO (single platform system).

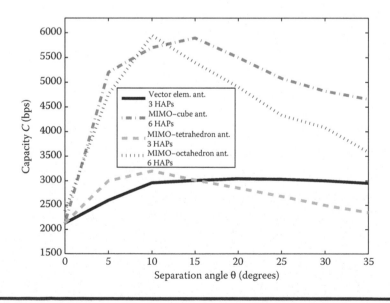

Figure 20.9 The capacity of the multiple HAP system for different separation angles using the compact MIMO antenna arrays in Figure 20.5, at an SNR of 20 dB.

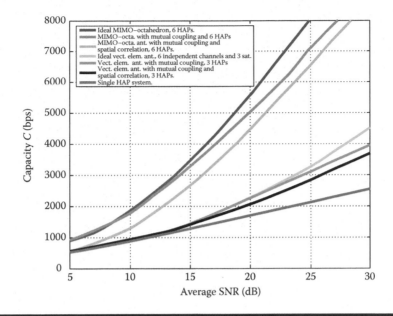

Figure 20.10 The effect imposed on capacity by mutual coupling and spatial-polarization correlation for the MIMO-octahedron and the vector element antenna arrays. The single antenna system is included in the graph for comparison purposes.

Next, we show the effects of mutual coupling and spatial-polarization correlation on the capacity of the system, and the results are plotted in Figures 20.10 and 20.11 for the various compact antennas. It is evident from these figures that although the capacity is degraded by correlation and mutual coupling, we still achieve significant gain compared to the single antenna case. For example, if we compare the performance in Figure 20.10 for an average SNR of 20 dB, we get a 146% higher capacity from the ideal (no correlation and mutual coupling) MIMO-octahedron antenna than the ideal vector element antenna, and a 229% higher capacity compared with the nondiversity single HAP system. If correlation and the mutual coupling are taken into account, the MIMO-octahedron antenna still gives a 115% higher capacity than the vector element antenna and a 161% higher than the single HAP system.

So far, the investigations have been performed with a FSL channel model. In scenarios 3 and 4, we will use the proposed STP channel model simulator described in Section 20.3 to analyze the atmospheric propagation effects on the performance of the multiple platform diversity system. In the following simulations, we set up a multichannel model dominated by LOS components and with weak NLOS components, with a Rice factor of 10, which corresponds to a rural type of environment. For scenarios 3 and 4, the platforms are randomly positioned according to a uniform distribution at at altitudes between 17 and 22 km and within an angular sector from $-60°$ to $+60°$. Figure 20.12 shows the impact on the capacity for various XPD values (see Section 20.1). XPD = ∞ represents a channel with no depolarization effects and is comparable to scenarios 1 and 2. For example, fairly harsh precipitation in the troposphere (XPD = 20 dB) would result in a 30% drop in capacity for an average SNR of 20 dB. A worse precipitation (XPD 10 dB) would result in a 59% drop in capacity. Thus, it is clear that precipitation in the troposphere can also cause severe degradation in system performance due to the loss of several communication subchannels. Extreme cases, XPD of 0 dB, for example, will yield a total loss of the signal.

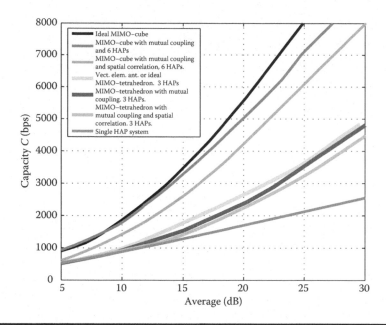

Figure 20.11 **The effect imposed on capacity by mutual coupling and spatial-polarization correlation for the MIMO-tetrahedron and MIMO-cube antenna arrays. The single antenna system is included in the graph for comparison purposes.**

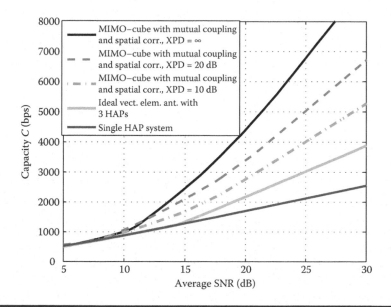

Figure 20.12 **The effect of depolarization on the MIMO-cube HAP diversity system compared to the vector element antenna array and the single HAP system.**

20.6 A Novel Orthogonal Momentum Division Multiplexing (OMDM) Technique

In this section, we propose a novel combination of OFDM and orbital angular momentum (OAM) of the electromagnetic field to produce a novel technique we denote as *orthogonal momentum division multiplexing (OMDM)* [45]. This OMDM technique is applied to a HAP diversity system employing three HAPs. We will show that the proposed OMDM–HAP diversity system can achieve a theoretic N_{OAM}-fold increase in system capacity as compared with a system utilizing OFDM only, where N_{OAM} denotes the number of helical wave modes.

In the past two decades, there has been a great interest in the research and applications of OAM [45–52] particularly in the optics field (e.g., rotating optical tweezers [48] and quantum communications [49]), where it was shown that the spin angular momentum (polarization) and the OAM of a paraxial light beam can be controlled separately. The OAM of the electromagnetic field can be expressed as a multipole expansion [50] of the field emanating from a rotating beam around the z-axis. This series expansion that consists of an angular distribution of the radiated field can then be expressed as [53]

$$
\begin{aligned}
\Omega_l^m(\theta, \varphi) =& \frac{1}{2}\left(1 - \frac{m(m+1)}{l(l+1)}\right)|Y_{l(m-1)}(\theta, \varphi)|^2 \\
&+ \frac{m^2}{l(l+1)}|Y_{lm}(\theta, \varphi)|^2 + \frac{1}{2}\left(1 - \frac{m(m+1)}{l(l+1)}\right)|Y_{l(m+1)}(\theta, \varphi)|^2,
\end{aligned}
\tag{20.44}
$$

where the functions $Y_{lm}(\theta, \varphi)$ are the normalized scalar spherical harmonics [41]. These functions have a null direction aligned with the z-axis for all $m \neq 0$. To create these field modes we use a circular array of electric dipoles with an interelement separation distance $d = \pi(D/N)$, where D is the diameter of the array and N is the number of array elements. The rotating beam is produced by progressively increasing the signals phase angle to each array element by $\varphi_n = 2\pi m$. Figure 20.13 shows the radiated mode beam for $m = 1$ using $N = 8$ array elements.

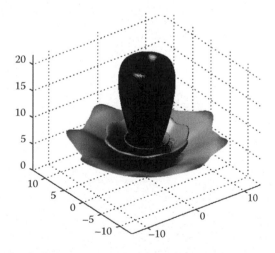

Figure 20.13 An example of the radiated OAM mode for $m = 1$. This radiated electromagnetic mode is produced by an array of eight santenna elements.

These radiated modes will create clockwise/counterclockwise helical wave fronts [52] (a radiowave vortex) around the z-axis. The direction of rotation depends on the sign of the mode number m. Then, by using an array of several concentric rings of array elements we can create combinations of superimposed modes.

The OMDM technique we propose is a novel joint 2D multiplexing method [45] using an OFDM of N_{DFT} number of carrier waves in combination with N_{OAM} number of helical wave modes, which can be perceived as a second frequency domain of orbital frequencies that are orthogonal to the normal frequency domain of translational frequencies. Figure 20.14 shows the block diagram of the proposed OMDM system [45].

Assuming that we have an OAM system with N_{OAM} modes, we can then write the signal for each subchannel in the translational frequency domain between the transmitting and receiving antenna arrays for each OAM mode as

$$r_m(k) = h_m(k)s_m(k) + n_m(k), \quad k \in \{0, \ldots, N_{DFT} - 1\}, \tag{20.45}$$

where k denotes each separate subcarrier and m denotes each OAM mode. If we write Equation 20.45 in vector notation form, we get

$$\mathbf{r}(k) = \mathbf{h}(k)\mathbf{s}(k) + \mathbf{n}(k), \tag{20.46}$$

where $\mathbf{r}(k)$ is an ($N_{OAM} \times 1$) vector and $\mathbf{s}(k)$ is an ($N_{OAM} \times 1$) vector of the received and transmitted signals, respectively. $\mathbf{h}(k)$ is the ($N_{OAM} \times 1$) frequency response vector of the N_{OAM} mode channels. The noise in the system $\mathbf{n}(k)$ is an ($N_{OAM} \times 1$) vector assumed to be AWGN. The correlation matrix of the noise vector $\mathbf{n}(k)$ is $\mathrm{E}\{\mathbf{n}(k)\mathbf{n}^H(k)\} = \sigma_n^2 \cdot \mathbf{I}_{N_{OAM}}$, where σ_n^2 is the variance of the noise and $\mathbf{I}_{N_{OAM}}$ is the ($N_{OAM} \times N_{OAM}$) identity matrix.

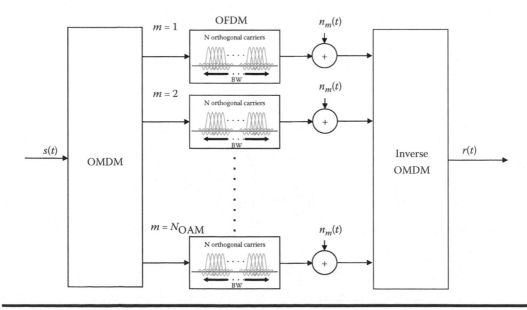

Figure 20.14 The block diagram of the proposed OMDM system [45]. Each of the N_{OAM} channels, $m = 0, \ldots, N_{OAM} - 1$, uses N_{DFT} OFDM subcarriers.

Since the OMDM channels in Equation 20.46 are assumed to be uncorrelated and the correlation of the noise $n(k)$ is $\sigma_n^2 \cdot \mathbf{I}_{N_{OAM}}$, we can then write the theoretical maximum information capacity of the system [15,45] as

$$C = \sum_{k=0}^{N_{DFT}-1} \sum_{m=0}^{N_{OAM}-1} \log_2 \left(1 + \frac{\sigma_{s_m}^2(k)}{\sigma_n^2} \right), \qquad (20.47)$$

where $\sigma_{s_m}^2(k)$ is the variance of the separate uncorrelated input signals in vector $\mathbf{s}(k)$. The capacity in Equation 20.47 is constrained by the total radiated power from the transmitting antennas, defined as [45]

$$P_{rad} = \sum_{k=0}^{N_{DFT}-1} \sum_{m=0}^{N_{OAM}-1} \sigma_{s_m}^2(k). \qquad (20.48)$$

The capacity results shown in Figure 20.15 are obtained for a system of HAPs operating at an altitude of 20 km and plotted against the average SNR of the system. The elevation angle is held at 10° throughout the simulations. The array has eight elements placed along a circle with the radius of two wavelengths. In this simulation, we use $N_{OAM} = 3$ OAM modes in combination with $N_{DFT} = 128$ OFDM subcarriers. It is evident from Figure 20.15 that the OMDM–HAP system provides superior performance as compared to the HAP system utilizing only OFDM. For example, for an average SNR of 10 dB, we get a 190% increase in capacity from the three HAP OMDM systems as compared to the conventional OFDM HAP system, which is approximately an N_{OAM}-fold increase of the total system capacity.

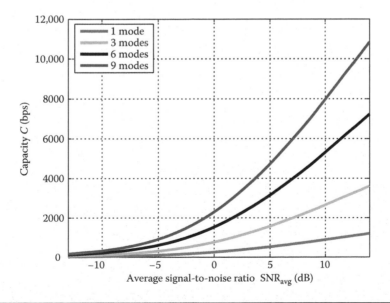

Figure 20.15 The capacity of the multiple HAP system for different numbers of OAM modes and with 128 OFDM carrier waves and a bandwidth of 1 MHz.

20.7 Conclusions

In this chapter, a combined OFDM–MIMO technique to enhance the capacity of HAP communication relaying links using various compact MIMO antenna array configurations was proposed. Simulation results show that multiple platform diversity systems utilizing compact MIMO antenna arrays outperform a single platform system. It was also shown that the MIMO-cube and MIMO-octahedron antenna arrays are superior to the MIMO-tetrahedron antenna array and the vector element antenna since they have twice the number of independent channels, which will result in a higher capacity.

Simulation results have also shown that a small degradation in capacity results because of the effects of spatial correlation and mutual coupling between the separate antenna array elements of the compact antenna arrays. We have also shown the effects of the separation angle between platforms on system performance and determined the optimal separation angle that maximizes the total capacity of the system. It has also been shown that MIMO-cube and vector element antenna arrays are preferred when the platforms are widely separated ($>15°$) and that the novel MIMO-octahedron and MIMO-tetrahedron are preferred when the platforms are closely positioned ($<15°$).

We have also presented a novel multichannel simulator that takes into account the temporal, spatial, and polarization properties affecting signal transmission in HAP systems. In addition, we have also determined the impact on the total capacity of the proposed platform diversity system. Simulation results have shown that the depolarization will have a severe impact on the performance.

Finally, we proposed a novel combination of OFDM and OAM, which we denoted here as the *OMDM*. This technique is used in conjunction with a HAP diversity technique in order to further enhance the capacity of HAP communication systems. Simulation results show that the performance of the OMDM–HAP system outperforms the conventional HAP system utilizing OFDM only. We have also shown that the proposed OMDM technique gives approximately an N_{OAM}-fold increase in the theoretical capacity as compared to the standard OFDM system.

Further HAP trials are ongoing and more are expected to be conducted in the coming years. The measurement trials and results would be extremely valuable to provide a deeper understanding of the propagation environment experienced by HAPs and would allow for better parameterizing of the proposed channel model and simulator. In addition, the measurement data would be particularly useful in evaluating the performance of the proposed OFDM–MIMO system and comparing it with the simulation results. It should be stressed here that the results of the OMDM technique are preliminary and thus further research is necessary in order to evaluate the performance limits and the practical aspects of using OAM technology in radio communications.

References

1. G. M. Djuknic, J. Freidenfelds, and Y. Okunev, Establishing wireless communications services via high-altitude aeronautical platforms: A concept whose time has come?, *IEEE Communications Magazine*, 35, 128–135, 1997.
2. The CAPANINA Project, CAPANINA—Communications from aerial platform networks delivering broadband information for all, http://www.capanina.org/
3. The European Community COST 297 Action, HAPCOS—high altitude platforms for communications and other services, http://www.cost297.org/
4. A. Mohammed, S. Arnon, D. Grace, M. Mondin, and R. Miura, Advanced communication techniques and applications for high altitude platforms, Editorial for a special issue, *EURASIP Journal on Wireless*

Communications and Networking, 2008, doi:10.1155/2008/934837. (See also the Call for Papers at: http://www.hindawi.com/journals/wcn/si/hap.html)

5. D. Grace, M. Mohorcic, M. Oodo, M. H. Capstick, M. B. Pallavicini, and M. Lalovic, CAPANINA— Communications from aerial platform networks delivering broadband information for all, IST Mobile Communications Summit, Dresden, Germany, 2005.

6. T. C. Hong, B. J. Ku, J. M. Park, D. S. Ahn, and Y. S. Jang, Capacity of the WCDMA system using high altitude platform stations, Springer Netherlands, *International Journal of Wireless Information Networks*, 13(1), 5–17, 2005.

7. J. M. Park, B. J. Ku, Y. S. Kim, and D. S. Ahn, Technology development for wireless communications system using stratospheric platform in korea, IEEE PIMRC 2002, 2002.

8. T. Hult, A. Mohammed, and D. Grace, WCDMA uplink interference assessment from multiple high altitude platforms configurations, *EURASIP Journal on Wireless Communications and Networking*, Special Issue on Advanced Communications Techniques and Applications for High Altitude Platforms, 2008, 2008. doi: 10.1155/2008/182042.

9. Z. Yang, A. Mohammed, and T. Hult, Performance evaluation for wimax broadband from high altitude platforms cellular systems and terrestrial coexisting capability, *EURASIP Journal on Wireless Communications and Networking*, Special issue on advanced communications techniques and applications for high altitude platforms, 2008, 2008. doi: 10.1155/2008/348626.

10. Z. Yang, A. Mohammed, T. Hult, and D. Grace, Downlink coexistence performance assessment and techniques for WiMAX services from high altitude platform and terrestrial deployments, *EURASIP Journal on Wireless Communications and Networking*, Special Issue on Advanced Communications Techniques and Applications for High Altitude Platforms, 2008, 2008. doi: 10.1155/2008/291450.

11. Z. Yang and A. Mohammed, High altitude platforms for wireless sensor network applications, *IEEE 5th International Symposium on Wireless Communication Systems (ISWCS'08)*, Reykjavik, Iceland, October 2008.

12. Z. Yang and A. Mohammed, A study of multiple access schemes for wireless sensor network applications via high altitude systems, *IEEE 69th Semiannual Vehicular Technology Conference*, Barcelona, Spain, 26–29 April 2009.

13. H. Bölcskei, Principles of MIMO-OFDM wireless systems, in *Signal Processing for Mobile Communications Handbook,* M. Ibnkahla (Ed)., CRC Press, Boca Raton, FL, 2004.

14. H. Schulze and C. Luder, *Theory and Applications of OFDM and CDMA*, Wiley, 2005.

15. I. E. Telatar, Capacity of multi-antenna Gaussian channels, *European Transactions on Telecommunication*, 10(6), 585–595, 1999.

16. G. Foschini and M. Gans, On limits of wireless communications in a fading environment when using multiple antennas, *Wireless Personal Communications*, 6, 311–335, 1998.

17. M. Martone, *Multiantenna Digital Radio Transmission*, Artech House Inc., 2002.

18. T. Hult and A. Mohammed, MIMO for HAPs: An idea whose time has come, EU COST 297 - HAPCOS meeting and workshop, Oberpfaffenhofen, Germany, April 2006.

19. T. Hult, A. Mohammed, Z. Yang, and D. Grace, *Performance of a multiple HAP system employing multiple polarization, Wireless Personal Communications Journal,* Special Issue J. No. 11277; 2008. Invited Paper as one of the Highest Quality Papers of Wireless Personal Multimedia Communications 2006 Conference (WPMC06).

20. T. Hult, A. Mohammed, D. Grace, and Z. Yang, Performance of a multiple HAP system employing multiple polarization, *Wireless Personal Multimedia Communications* 2006, WPMC06, San Diego, CA, 17–20 September 2006.

21. T. Hult and A. Mohammed, Theoretical analysis and assessment of depolarization effects on the performance of high altitude platforms, EU COST 297–HAPCOS Meeting and Workshop, Nicosia, Cyprus, 7–9 April 2008.

22. T. Hult, E. Falletti, A. Mohammed, and F. Sellone, Multi-antenna multi-HAP channel model for space-polarization, EU COST 297–HAPCOS Meeting and Workshop, Nicosia, Cyprus, 7–9 April 2008.

23. T. Hult, E. Falletti, A. Mohammed, and F. Sellone, Multi-channel model for space-polarization systems, European Union Conference on Antennas and Propagation, EUCAP2007, Edinburgh, U.K., 11–16 November 2007.

24. T. Hult and A. Mohammed, Assessment of a HAP diversity system employing compact MIMO-tetrahedron antenna, EU COST 297–HAPCOS Meeting and Workshop, Prague, Czhech Repulic, 29–30 March 2007.

25. T. Hult and A. Mohammed, Compact MIMO antennas and HAP diversity for enhanced data rate communications, *IEEE 65th Semiannual Vehicular Technology Conference*, Dublin, Ireland, 22–25 April 2007.

26. T. Hult and A. Mohammed, Capacity of multiple HAP system employing multiple polarizations, European Union Conference on Antennas and Propagation, EUCAP2006, Nice, France, 6–9 November 2006.

27. T. Hult and A. Mohammed, Compact MIMO antennas and HAP diversity for high data rate communications, IEEE 64th Semiannual Vehicular Technology Conference, Montreal, Canada, 25–28 September 2006.

28. M. R. Andrews, P. P. Mitra, and R. deCarvalho, Tripling the capacity of wireless communications using electromagnetic polarization, *Nature*, 409, 316–318, 2001.

29. S. Nordebo, A. Mohammed, and J. Lundbck, On the use of polarization channels in satellite communications. Proc. of Radio Vetenskap och Kommunication 02, Stockholm, Sweden, 10–13 June, 2002.

30. J. B. Andersen and B. N. Getu, The MIMO cube–A compact MIMO antenna, *IEEE 5th International Symposium on Wireless Personal Multimedia Communications*, vol. 1, pp. 112–114, 2002.

31. M. Gustafsson and S. Nordebo, Characterization of MIMO antennas using spherical vector waves, *IEEE Transactions on Antennas and Propagation*, 54(9), 2679–2682, 2006.

32. I. Frigyes and P. Horvath, Polarization-time coding in satellite links, *Satellite and Space Communications*, 15(1), 6–8, 2005.

33. P. Horvath and I. Frigyes, Application of the 3D polarization concept in satellite MIMO systems, *Proceedings of the 49th Annual IEEE Global Telecommunications Conference (Globecom '06)*, San Francisco, CA, December 2006.

34. P. Horvath, G. Karagiannidis, P. King, S. Stavrou, and I. Frigyes, Investigations in satellite MIMO channel modeling: Accent on polarization, *EURASIP Journal on Wireless Communications and Networking*, 2007, 2008, doi: 10.1155/2008/934837.

35. T. Hult and A. Mohammed, MIMO antenna applications for LEO satellite communications, *Proceedings of the 3rd European Space Agency International Workshop of the European COST 280*, Prague, Czech Republic, June 2005.

36. A. Mohammed and T. Hult, Performance evaluation of a MIMO satellite diversity system, European Space Agency 10th International Workshop on Signal Processing for Space Communications (SPSC 2008), Rhodes Island, Greece, October 2008.

37. E. Falletti and F. Sellone, A matrix channel model for transmit and receive smart antennas systems, *Vehicular Technology Conference, VTC 2006-Spring, IEEE 63rd*, 6, 2967–2971, 2006.

38. ITU-RECOMMENDATION, ITU-R P.618-9: Propagation data and prediction methods required for the design of Earth-space telecommunication systems, *International Telecommunication Union/ITU Radio Communication Sector*, 2007.

39. S. R. Saunders and A. Aragon-Zavala, *Antennas and Propagation for Wireless Communication Systems*, ISBN:975-0-470-84879-1, John Wiley & Sons, New York, 1999.

40. M. Khalighi, J. Brossier, G. Jourdain, and M. Raoof, Water filling capacity of Rayleigh MIMO channels, 12th IEEE Symposium on Personal, Indoor and Mobile Radio Communications, vol. 1, pp. 155–158, 1998.

41. J. D. Jackson. *Classical Electrodynamics*, 3rd ed., John Wiley & Sons, New York, pp. 107–110, 1998.

42. Fitzpatrick, R. *Lecture notes: advanced classical electromagnetism*. University of Texas, Austin, 1996.

43. R. Janaswamy, Effect of Element Mutual Coupling on the Capacity of Fixed Length Linear Arrays, *IEEE Antennas and Wireless Propagation Letters*, 1, 157–160, 2002.

44. S. K. Yong and J. S. Thompson, Three-dimensional spatial fading correlation models for compact MIMO receivers. *IEEE Transactions Wireless Communications*, 4(6), 2005.

45. T. Hult and A. Mohammed, Performance evaluation of a satellite diversity system, Institution of Engineering and Technology (IET) Workshop on Wideband, Multiband, Antennas and Arrays for Civil or Defence Applications, March 2008.

46. L. Allen, M. W. Beijersbergen, R. J. C. Spreeuw, and J. P. Woerdman, Orbital angular momentum of light and the transformation of Laguerre-Gaussian modes, *Physical Revie A*, 45, pp. 8185–8189, 1992.
47. A. Mair, A. Vaziri, G. Weihs, and A. Zeilinger, Entanglement of the orbital angular momentum states of photons, *Nature*, 412, pp. 313–316, 2001.
48. N. B. Simpson, K. Dholakia, L. Allen, and M. J. Padgett, Mechanical equivalence of spin and orbital angular momentum of light: An optical spanner, *Optical Letters* 22, 52–54, 1997.
49. G. Molina-Terriza, J. P. Torres, and L. Torner, Management of the angular momentum of light: Preparation of photons in multidimensional vector state of angular momentum, *Physical Reviews Letters*, 88, 2002.
50. M. Harwit, Photon orbital angular momentum in astrophysics, *The Astrophysical Journal*, 597, 1266–1270, 2003.
51. B. Thide, H. Then, J. Sjöholm, K. Palmer, J. Bergman, T. D. Carozzi, Y. N. Istomin, N. H. Ibragimov, R. Khamitova, Utilization of photon orbital angular momentum in the low-frequency radio domain, *Physical Review Letters*, 99(8), 2007.
52. L. Marrucci, C. Manzo, and D. Paparo, Optical spin-to-orbital angular momentum conversion in inhomogenous anisotropic media, *Physical Review Letters*, 96(16), 163905, 2006.
53. J. M. Blatt and V. F. Weisskopf, *Theoretical Nuclear Physics*, Wiley, New York, 1952.

Chapter 21

OFDMA Systems and Applications

André Noll Barreto and Robson Domingos Vieira

Contents

21.1 Introduction...564
21.2 WLANs—IEEE 802.11 Family..565
    21.2.1 IEEE 802.11a/g PHY 565
    21.2.2 IEEE 802.11n..568
21.3 Mobile WiMAX (IEEE 802.16e)...568
    21.3.1 The OFDMA PHY...570
        21.3.1.1 OFDM Parameters.................................571
        21.3.1.2 Subchannelization Schemes 572
        21.3.1.3 Burst Profiles.....................................574
        21.3.1.4 Duplexing and Framing...........................574
        21.3.1.5 HARQ...576
        21.3.1.6 Advanced Antenna Systems 576
21.4 Evolved Universal Terrestrial Radio Access (E-UTRA)...............577
    21.4.1 E-UTRA Requirements...578
    21.4.2 E-UTRAN Architecture..578
    21.4.3 E-UTRAN Multiple Access...579
        21.4.3.1 Frame Structure...................................580
        21.4.3.2 OFDMA...581
        21.4.3.3 Downlink Transport Channels 582
        21.4.3.4 Downlink Physical Channels......................582
        21.4.3.5 E-UTRAN Uplink Multiaccess: SC-FDMA....................584
        21.4.3.6 Uplink Transport Channel.........................585

           21.4.3.7   Uplink Physical Channel 585

      21.4.4   E-UTRA: Advanced Multiantenna Techniques.............................587

21.5   Open Research Challenges: IMT-A Next Cellular Network Generation...............589

      21.5.1   LTE Advanced..590

           21.5.1.1   WiMAX: 802.16m 591

21.6   Summary..593

References..593

21.1 Introduction

Many wireless communications systems rely on orthogonal frequency division multiplexing (OFDM), and orthogonal frequency division multiple access (OFDMA) to guarantee reliable transmission and high capacity in broadband frequency-selective channels. OFDMA is employed for instance both in mobile WiMAX (IEEE 802.16e) and in evolved universal terrestrial radio access network (E-UTRAN), which are the main contenders for beyond 3G broadband mobile wireless systems. More recently, International Mobile Telecommunications-Advanced (IMT-A) next generation mobile networks (4G) requirements have been specified for operation in cellular licensed bands. The IMT-A networks will be designed to provide significantly improved performance compared with other high-rate broadband cellular network systems. Hence, new challenges for OFDMA cellular networks are expected. OFDM with other multiple access schemes is also the transmission technology of choice for other wireless systems, such as 802.11-based wireless local area networks (WLANs) (WiFi) and digital television [digital video broadcasting (DVB) and integrated services digital broadcasting (ISDB)].

Even though OFDM is the common underlying technology for all these systems, each one of them has a different purpose, with different applications and propagation environments. Therefore, different OFDM(A) configurations are required.

For instance, OFDMA can be employed in both time division duplexing (TDD) and frequency division duplexing (FDD) systems. In TDD systems, downlink and uplink transmissions are made in the same frequency band, but in different time intervals. In FDD the opposite is done, and downlink and uplink transmissions are simultaneously done in different frequency bands [1]. TDD has a higher flexibility in the resource allocation between the uplink and the downlink, and is therefore more appropriate for services with asymmetric data services, besides allowing an adequacy to different spectrum allocations, as there is no need for a separate band for each link. TDD also has some other advantages, such as lower user terminal complexity, because only one modem is needed; channel reciprocity, making the use of precoding techniques easier; and higher frequency diversity, because the whole available spectrum is used in any single direction. However, TDD requires synchronisation among all base stations (BSs) in a WiMAX system and all cells must use the same downlink and uplink frame sizes, in order to avoid interference across links. For instance, a mobile station (MS) transmitting in the uplink may interfere with another MS who is receiving in the downlink, if the frames are not synchronised. In OFDM systems, FDD can also provide a better link budget, because subcarrier power can be higher since the transmission bandwidth can be half as wide as that in TDD.

OFDM parameters have a large influence on the system behavior. For instance, the guard interval determines the system robustness in the presence of multipath, but introduces some overhead, and must, therefore be carefully chosen. The overhead will be less significant for long symbols, and, hence, narrow subcarrier spacing. However, systems with narrow subcarrier spacing are more affected by the frequency offset. A certain number of pilot symbols is fundamental for guaranteeing reliable transmission, but on the other hand consumes valuable resources and transmit power. The choice of the minimum allocation units is also a trade-off between allocation flexibility and signaling needs. Understanding OFDM/OFDMA parameters is therefore essential if we want to understand the capabilities and limitations of each technology.

In this chapter, we briefly describe the main OFDM and OFDMA parameters of the three above-mentioned wireless systems, WiFi, WiMAX, and long-term evolution (LTE). We discuss relevant OFDM parameters, such as carrier frequencies, subcarrier spacing, fast Fourier transform (FFT) size, guard interval length, the use of cyclic prefix (CP), frame structure, including pilots and preambles, modulation and coding schemes (MCSs, OFDMA frequency) and time allocation chunks. We focus here on the physical layer (PHY) aspects of the different technologies. The advanced transmission techniques applied at each system are also addressed, like, for instance, multiple-input multiple-output (MIMO), and hybrid automatic repeat request (HARQ).

In Section 21.2, we briefly describe the OFDM-based WLANs. Overviews of the two main contenders for broadband wireless, mobile WiMAX, and E-UTRAN, are given in Sections 21.3 and 21.4, respectively. Finally, in Section 21.5, we analyze the main trends of OFDMA technologies for 4G systems that fall under the IMT-A umbrella, such as IEEE 802.16m and LTE-A.

21.2 WLANs—IEEE 802.11 Family

Wireless local area networks (WLANs) based on the IEEE 802.11 family of standards have been one of the greatest market successes of the past decade. The technology is promoted by the WiFi Alliance [2], an industrial consortium, which is responsible for certification of WiFi products. The first WiFi devices, based on the IEEE 802.11b standards, employed spread-spectrum techniques and provided data rates of up to 11 Mbps. In order to provide increased data rates of up to 54 Mbps, OFDM was the technique of choice for 802.11 evolution, in both IEEE 802.11a* and 802.11g standards. IEEE 802.11a works in the 5 GHz unlicensed frequency band, whereas IEEE 802.11g is backwards compatible to IEEE 802.11b and operates in the same 2.4 GHz band, also unlicensed.

21.2.1 IEEE 802.11a/g PHY

Apart from the carrier frequencies, both IEEE 802.11a and IEEE 802.11g have very similar PHY. The main difference between them is in the packet preamble and header, which for IEEE 802.11g must guarantee backwards compatibility to legacy IEEE 802.11b terminals, through the Physical Layer Convergence Protocol (PLCP). Knowing about these similarities, we focus here on a brief description of the IEEE 802.11a PHY [3].

OFDM transmission in IEEE 802.11a uses a bandwidth of approximately 20 MHZ, and the main OFDM parameters are listed in Table 21.1. As we can see, only 52 subcarriers are effectively used for data and pilots; the DC subcarrier and 11 subcarriers at the spectrum extremes remain unused. Pilot symbols are transmitted in every OFDM symbol and always at the same subcarriers. The guard interval between consecutive OFDM symbols occupies 20% of the total OFDM symbol duration and is filled up by a cylic prefix. Because WiFi is targeted mostly at indoor and short-range applications, with channels with a low delay spread, the guard interval can be relatively short. Therefore, the guard interval overhead can be kept proportionally low even with short symbols and,

---

* Standardization work began earlier at 802.11a than at 802.11b, but it was released about four years later.

Table 21.1   Main OFDM Parameters for IEEE 802.11a/g

Parameter	Value
(I)FFT size	$K = 64$
Data subcarriers	$K_d = 48$
Pilot subcarriers	$K_p = 4$
Subcarrier spacing	$\Delta_f = 312.5\,\text{kHz}$
Effective symbol duration	$T_S = 1/\Delta_f = 3.2\,\mu s$
Sampling rate	$f_s = \Delta_f * K = 20\,\text{MHz}$
Guard interval	$T_G = 800\,\text{ns}$
Symbol duration	$T_{OFDM} = T_S + T_G = 4\,\mu s$

consequently, with a large subcarrier spacing. This makes a small inverse FFT (IFFT) size possible, which in turn makes implementation less complex.

Different MCSs can be used, as described in Table 21.2. Bit-interleaved coded modulation (BICM) [4] is employed, consisting of a convolutional encoder followed by a bit interleaver before the modulation symbol mapper. BICM is known to provide excellent performance in fast fading channels, as is the case across the different OFDM subchannels. Different code rates are achieved by puncturing of a convolutional code (CC) with a rate $R = 1/2$ and a constraint length $k = 7$.

Table 21.2   MCSs in IEEE 802.11a

Data Rate (Mbps)	Modulation Scheme	Coding Rate	Code Bits/ OFDM Symbol	Data Bits/ OFDM Symbol
6	BPSK[a]	1/2	48	24
9	BPSK	3/4	48	36
12	QPSK[b]	1/2	96	48
18	QPSK	3/4	96	72
24	16-QAM[c]	1/2	192	96
36	16-QAM	3/4	192	144
48	64-QAM	2/3	288	192
54	64-QAM	3/4	288	216

[a] BPSK, binary phase shift keying.
[b] QPSK, quaternary phase shift keying.
[c] QAM, quadrature amplitude modulation.

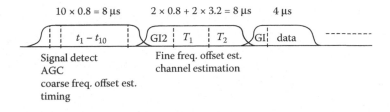

Figure 21.1 IEEE 802.11a preamble.

Because of the various available MCSs, link adaptation is essential for a good performance of IEEE 802.11a systems. Because of their low redundancy and high-order modulation schemes, high-rate transmission modes require high levels of signal-to-interference-plus-noise ratio (SINR). However, 802.11 standards do not specify any feedback channel for reporting channel quality information (CQI). In order to perform link adaptation, the transmitter must rely only on the mandatory acknowledgment packets (ACK), which must be transmitted every time a packet is correctly received. If an ACK is not received within a given period, then the transmitter may assume that the packet was not correctly detected, and may choose to retransmit it. This also indicates a possible degradation in the link quality and may trigger a reduction in the transmission data rate. On the other hand, if several consecutive packets are correctly detected and, consequently, ACKs are received at the transmitter side, then the transmitter may attempt to increase the data rate and exploit a good link quality. In Ref. [5], we can see how ACKs can be exploited to implement effective link adaptation in IEEE 802.11a/g.

In the beginning of each PHY frame, a known preamble is transmitted as a training sequence. Initially, a short symbol is built modulating with a known sequence 12 subcarriers, which are spaced four times higher (1.25 MHz) than the subcarrier spacing of other OFDM symbols. This 0.8 μs symbol is repeated 10 times, and these repeated symbols are used to detect the signal, adjust the automatic gain control (AGC), and synchronize and perform a course channel and a frequency offset estimation. After that, a training symbol employing all the 52 subcarriers is also formed, repeated twice, and transmitted with a long guard interval of 1.6 μs. These long symbols are used for a finer estimation of channel response and frequency offset. The complete preamble lasts for 16 μs and is depicted in Figure 21.1. Remember that data symbols last for 4 μs each. The pilot subcarriers included in the data symbols may be used to keep the estimates up-to-date during longer packets.

The frame structure is shown in Figure 21.2. After the preamble, the SIGNAL field is transmitted, conveying information about the length of the PHY service data unit (PSDU) and the used MCS. Without this information, the packet cannot be demodulated and decoded, and, therefore, the SIGNAL field is always transmitted with the most robust MCS, BPSK with a rate $R = 1/2$. The rest of the frame is transmitted with the MCS defined in the SIGNAL field. The first 16

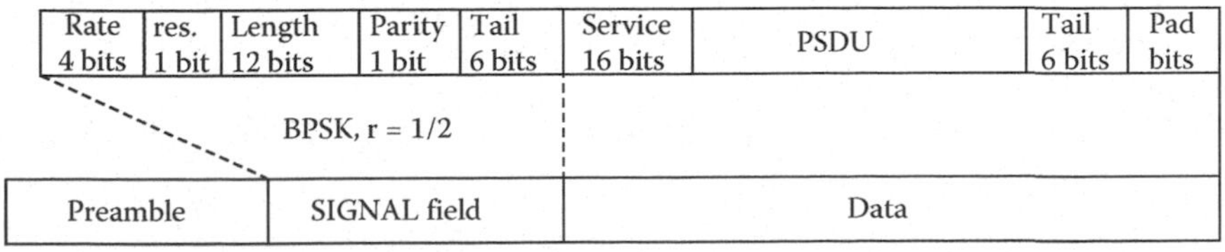

Rate 4 bits	res. 1 bit	Length 12 bits	Parity 1 bit	Tail 6 bits	Service 16 bits	PSDU	Tail 6 bits	Pad bits
		BPSK, r = 1/2						

Preamble	SIGNAL field	Data

Figure 21.2 IEEE 802.11a PHY frame.

bits are the SERVICE field, whose 7 first bits are used to initialize the scrambler. The scrambler is a pseudo-random bit sequence that is XORed with the data sequence, and has the purpose of avoiding long all-zero bit sequences, which would have unfavorable peak-to-average power ratio (PAPR) properties. The signal from the preamble to the SERVICE field forms the PLCP header. After the header, the PSDU containing the user data is transmitted. Each PSDU corresponds to one MAC packet data unit (MPDU), of up to 4095 bytes. The code is terminated with 6 zero tail bits to bring the trellis to the initial state. Some padding bits may be needed at the end to fill up an integer number of OFDM symbols.

The MAC protocol is the same for all PHYs of the IEEE 802.11 family. A contention-free point coordination function (PCF) for centralized networks is specified in the standards, but a contention-based distributed coordination function (DCF), adequate for *ad hoc* networks, is more commonly used in practice. The DCF relies on the CSMA/CA (carrier sensing multiple access/collision avoidance) protocol, which is described and analyzed in Ref. [6].

21.2.2 IEEE 802.11n

The evolution of WiFi toward higher data rates of up to 600 Mbps is currently under standardization in the IEEE 802.11n amendment.* Work in IEEE 802.11n started in 2003, and commercial products based on a draft version of the standards are already available.

Several advanced techniques are going to be needed to allow IEEE 802.11n to reach such higher date rates [7] and to increase the transmission range. Among the most relevant ones we can mention:

- The use of MIMO techniques with up to 4×4 antennas, including both space-time (ST) codes and spatial multiplexing
- Provisioning of a feedback channel for CQI, what will improve the efficiency of link adaptation
- The possibility of using a reduced guard interval, if the channel delay spread so allows
- Low-density parity codes (LDPC) instead of CCs for better error-correcting capability
- Wider bandwidth, by bonding two 20 MHz channels into one 40 MHz wide channel
- Aggregation of more than one MAC PDU into a PHY SDU, thus reducing the relative overhead

21.3 Mobile WiMAX (IEEE 802.16e)

Among the many emerging wireless broadband (WiBro) technologies, mobile WiMAX deserves some special attention. The name WiMAX stands for Worldwide Interoperability for Microwave Access, and, as the name says, WiMAX was devised in order to provide an internationally accepted standard for broadband wireless access, as opposed to the several proprietary systems under development at the beginning of the century. Mobile WiMAX is based on the IEEE 802.16e standard, which was released in 2005. Large-scale implementation of WiMAX is already under way in several countries all around the world, but being a new and complex technology, many of its characteristics and features are still not well understood.

The IEEE 802.16 standard was originally thought as a WiBro system for fixed and nomadic applications, as a competitor of cabled broadband access, such as ADSL. The initial IEEE 802.16

---

* Standardization work was still going on at the time of writing, but the final version is expected to be available in 2009.

version was approved in 2001, specifying a packet-switched WiBro system for line-of-sight (LOS) transmission at high frequencies (10–66 GHz), and, after some amendments (IEEE 802.11a/b/c, later withdrawn), the specifications for nomadic and fixed broadband wireless access were released in October 2004 (IEEE 802.16d or 802.16-2004) [8]. IEEE 802.16-2004 defines the PHY and MAC layers for broadband wireless access with and without LOS, also for frequencies below 11 GHz. The following four different PHYs are specified:

- *WirelessMAN-SC PHY:* Single-carrier modulation transmission targeted at LOS operation at frequencies between 10 and 66 GHZ, using time division multiple access (TDMA) and DAMA (Demand Assigned Multiple Access).
- *WirelessMAN-SCa PHY:* Single-carrier modulation scheme adapted to transmission without LOS at frequencies below 11 GHz, using TDMA.
- *WirelessMAN-OFDM PHY:* An OFDM-based transmission scheme for non-line-of-sight transmission at frequencies below 11 GHz, employing TDMA.
- *WirelessMAN-OFDMA PHY:* Also for transmission at frequencies below 11 GHz without LOS, employing OFDMA, which is the focus of our contribution. This is also the PHY that is most likely to be widely employed in real networks.

A set of advanced techniques for providing data transmission with high spectral efficiency was considered in IEEE 802.16-2004, including fast link adaptation, frequency-domain scheduling, automatic repeat request (ARQ), and HARQ. Particular emphasis is given to multiple antenna techniques, such as beamforming, ST coding (STC) and spatial multiplexing MIMO, with multiple antennas both at the base station (BS) and MS sides. These techniques are part of what is called advanced antenna systems (AAS) in the standards.

IEEE 802.16 also specifies authentication and encryption procedures to guarantee secure wireless communications. These procedures are however out of the scope of this contribution.

The specification of a mobile WiBro IP-based network was originally a task of the mobile broadband wireless Access (MBWA) Working Group in IEEE 802.20, started in 2002. However, in parallel, the IEEE 802.16 Working Group proceeded with the extension of its standards to support mobility, with similar goals as IEEE 802.20, and completed the amendment IEEE 802.16e in the end of 2005, becoming the de facto IEEE standard for mobile broadband wireless communications. IEEE 802.16e is a differential amendment, correcting and complementing the previous IEEE 802.16-2004 standard, and both standards must be read together to be understood. IEEE 802.16e has full mobility support, including handovers between cells and roaming between different networks. Mobile PHY is based on SOFDMA (scalable OFDMA), a multicarrier modulation technique with subchannelization, which will be described in some more detail in Section 21.3.1.

The WiMAX forum is an industrial consortium, formed in 2001, with currently over 400 members, among network operators, equipment suppliers, and electronic component manufacturers. Its goal is to promote the worldwide adoption of IEEE 802.16 compatible network equipment and mobile terminals. In order to achieve this goal, WiMAX forum is responsible for defining and running certification tests, for guaranteeing interoperability among WiMAX equipment from different suppliers, for making large-scale manufacturing of WiMAX products economically feasible.

As mentioned before, IEEE 802.16e is a very complex standard, with four different PHYs and many possible combinations of transmission features and parameters. Therefore, in order to make the implementation and interoperability easier, the WiMAX forum also defines some profiles, containing a subset of the features defined in the standards. It is based on these profiles that

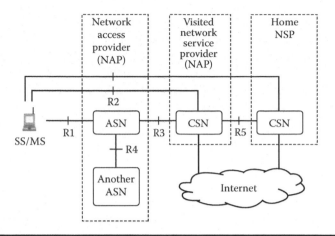

Figure 21.3   Reference WiMAX architecture.

certification tests are run. The first fixed WiMAX equipment was certified at the end of 2005, and the first mobile equipment at the end of 2007.

IEEE 802.16 specifies only the two lowest layers, PHY and MAC. Because of that, it is also the task of the WiMAX forum to recommend an end-to-end architecture, which is based solely on existing technologies and standards. A network reference model defines the interfaces among the different network entities. This architecture should be adopted by different network operators to ease the roaming procedures. The R1 to R5 interfaces in the reference architecture can be seen in Figure 21.3. The user terminal can be either a subscriber station (SS) or an MS. In the context of mobile WiMAX, we are considering only MSs. R1 is the air interface between the MS and the BS, which is specified in IEEE 802.16e. The access service network (ASN) consists of the WiMAX BSs and network gateways. The connectivity service network (CSN) provides actual internet access and can be built using off-the-shelf transmission control protocol/internet protocol (TCP/IP) equipment, such as authentication, authorization and accounting (AAA) servers, content providers, application and IMS servers, and billing and operation support systems.

The WiMAX forum is also in charge of defining policies for sharing of industrial property.

Parallel to the development of IEEE 802.16, and in close cooperation with it, work was conducted in Europe under the high performance radio metropolitan area networks (HIPERMAN) standard [9]. HIPERMAN was harmonized with IEEE 802.16 and is concerned with spectrum access rules for the 2–11 GHz in Europe. Some parallel work in mobile broadband wireless was also done by South Korean suppliers, which was marketed under the name of WiBro and has been in commercial service in South Korea since 2006. WiBro was also integrated into the IEEE 80.16e standard and is now officially recognized as the commercial name of mobile WiMAX in South Korea [10].

In 2007, WiMAX has also been officially selected by the International Telecommunications Union (ITU) as one of the third-generation technologies in the IMT-2000 program.

In the following section, we describe the main characteristics of mobile WiMAX OFDMA-PHY and MAC, based on the IEEE 802.16e standard.

21.3.1 The OFDMA PHY

Mobile WiMAX can operate either in licensed or in unlicensed bands. Different licensed bands are allocated by the various national telecommunications regulators, but typical frequency bands for broadband wireless communications can be found at around 2.5 and 3.5 GHz. According to IEEE 802.16, bandwidth is flexible, varying from 1.25 to 20 MHz, allowing the technology to be adapted to different spectrum allocations. However, the WiMAX forum has defined so far only profiles with 5 or 10 MHz bandwidth.

As mentioned before, IEEE 802.16 has support for different transmission schemes in the PHY, but for WiMAX transmission it is based on OFDMA. The OFDMA PHY will be described in some details in the following subsections.

21.3.1.1 OFDM Parameters

WiMAX employs SOFDMA. With this scheme, different bandwidths are achieved by using different numbers of subcarriers (corresponding to different IFFT sizes), maintaining a fixed interval between adjacent subcarriers. This interval is $\Delta f = 10.937$ kHz, and, consequently, the useful OFDM symbol duration is always $T_S = 91.43\,\mu s$. By having always the same subcarrier spacing, we guarantee that transmission performance, in terms of robustness against multipath and carrier frequency offset, is independent of the chosen bandwidth, making network planning easier, besides simplifying the modem implementation. The guard interval in IEEE 802.16e is filled up with a CP and may have different sizes. The ratio of the guard interval length to the useful symbol length can be any of the following values: $T_G/T_S = 1/32, 1/16, 1/8,$ or $1/4$, which allows an adequacy for various channel conditions. However, current WiMAX forum profiles specify only $T_G = T_S/8$. With this CP length, the main OFDM parameters of WiMAX are listed in Table 21.3. WiBro has a single bandwidth of 8.75 MHz and some slightly different parameter values at the PHY, which are specified in the last column of Table 21.3.

Table 21.3 Main OFDM Parameters for IEEE 802.16e OFDMA PHY

Parameter	*Value*				
Bandwidth W (MHz)	1.25	5^a	10^a	20	$8,75^b$
Sampling factor n_s	28/25				8/7
Sampling frequency (MHz) $f_s = W * n_s$	1.4	5.6	11.2	22.4	10
FFT size K	128	512	1024	2048	1024
Subcarrier spacing (kHz) Δ_f	10.937				9.766
Useful symbol time (μs) $T_s = \frac{1}{\Delta_f}$	91.4				102.4
Guard time $T_G = T_s/8$ (μs)	11.4				12.8
OFDM symbol duration (μs) $T_{OFDM} = T_s + T_G$	102.9				115.2

[a] Approved by WiMAX Forum.
[b] WiBro.

21.3.1.2 Subchannelization Schemes

The OFDM subcarriers can be divided into three different types:

- *Pilot subcarriers:* Transmit a known sequence and are used for channel estimation and synchronization
- *Data subcarriers:* Carry either user data or signaling information
- *Null subcarriers:* Set to zero at both spectrum extremes for spectrum shaping and at the central subcarrier to avoid DC offset, specially in direct-conversion receivers

The subcarriers in each OFDMA symbol are divided into logical subchannels, and IEEE 802.16e specifies several different subchannelization schemes, that is, several different ways to map subcarriers into logical subchannels. The subcarriers in a subchannel are in general nonadjacent, and are pseudo-randomly spread over the whole available spectrum, which provides larger frequency and interference diversity. This is called a distributed permutation. Optionally, subchannels with adjacent subcarriers can also be formed. Different subchannels can be then allocated to different users.

The main subchannelization schemes (or permutation schemes) are

- *Downlink PUSC (partial use of subchannels):* It is a distributed permutation scheme with support for sectorization, that is, the subcarriers are pseudo-randomly divided into groups, which are assigned to up to six different sectors. The subchannels are formed from the subcarriers of a single group, where each group contains subcarriers spread all over the spectrum. PUSC will be described in some details later.
- *Uplink PUSC:* In the uplink, PUSC also supports sectorization, but its algorithm is slightly different from the downlink. We will also give some more details about it further down the text.
- *FUSC (full use of subchannels):* It can be used only in the downlink, and, differently from PUSC, has no support for sectorization. It is also a pseudo-random distributed permutation. In FUSC, the pilots are separated at the beginning and only the data subcarriers are pseudo-randomly separated into subchannels. There are two different options of FUSC, with different numbers of pilots and data subcarriers.
- *TUSC (downlink tile usage of subchannels):* It is a similar algorithm as in the uplink PUSC, but applied in the downlink.
- *OPUSC (optional PUSC):* It is for the uplink only, and is a different type of PUSC with less pilots, allowing the use of more data subcarriers.
- *AMC (adaptive modulation and coding):* In this scheme, subchannels are formed by adjacent subcarriers. It allows frequency-dependent allocation and a better use of beamforming techniques with multiple antennas.

PUSC, both in the downlink and in the uplink, is the most widely used subchannelization scheme in WiMAX and is used for the signaling information, thus being mandatory. In the current WiMAX forum profiles, only PUSC, FUSC, and AMC are supported.

In permutation-based subchannelization schemes, such as PUSC and FUSC, the subcarriers are spread all over the wideband spectrum, and, thus, a high level of frequency diversity can be achieved. The mapping algorithm from subcarriers into subchannels is based on the permutation base, determined by DL_PermBase and UL_PermBase parameters. These parameters are selected by the network planner and should be different for neighboring cells. Therefore, within each subchannel, interference typically comes from different users in different subcarriers, providing us with interference diversity.

In downlink PUSC, with a 10 MHz bandwidth, 840 subcarriers (out of 1024) are effectively used for either pilot or data. We initially form clusters of 14 adjacent subcarriers, including two pilot subcarriers. The pilots positions vary, whether it is an odd symbol or an even symbol. Again in 10 MHz, 60 clusters can be formed. These clusters are shuffled and divided into six groups, where three groups have 12 clusters and three groups have eight clusters each. When sectorization is employed, the groups are assigned to different segments, corresponding to the BS sectors. Subchannels are now formed by choosing 24 data subcarriers among the ones assigned to a given group, according to a permutation formula based on DL_PermBase. Up to 30 subchannels are thus available in a 10 MHz bandwidth. The main PUSC parameters for different bandwidths are listed in Table 21.4.

In uplink PUSC, we start by building tiles, instead of clusters. Each tile consists of four adjacent subcarriers over three OFDM symbols, totalling 12 modulation symbols. Out of the 12 modulation symbols eight are data symbols and four are pilot symbols, where two pilot symbols are located in the first OFDM symbol and two in the third. Therefore, in a 10 MHz bandwidth, we have 210 tiles for each three OFDM symbols. The tiles are grouped into subchannels, each of them consisting of six nonadjacent tiles, following a pseudo-random permutation formula based on UL_PermBase. There are 35 uplink PUSC subchannels in a 10 MHz bandwidth.

If we have reliable information about the SINR over different chunks of the wideband spectrum, we can significantly increase the transmission capacity by selecting in each part of the spectrum the user with the highest SINR. WiMAX has support for a feedback channel with frequency-related CQI, and, provided the users have low mobility, up-to-date reliable information can be made available to the BS scheduler. The AMC permutation can be used to exploit the advantages of frequency-domain scheduling. Nevertheless, the resource allocation based on frequency-domain CQI must also take into account transmission fairness for all the users, as well as satisfy the quality of service (QoS) requirements of each individual user. Of critical importance is the transmission delay, because for most services we cannot delay the transmission for too long by waiting for the channel to become favorable. Scheduling algorithms like proportional fair [11] may help overcome this problem. Capacity gains with AMC are promising, but practical issues, such as choosing the optimal frequency resource allocation in real time and obtaining accurate channel estimation in a

Table 21.4 Main PUSC Parameters

Parameters	Values			
Bandwidth (MHz)	1.25	5	10	20
FFT size	128	512	1024	2048
Number of used subcarriers	84	420	840	1680
Number of data subcarriers	72	360	720	1440
Number of pilot subcarriers	12	60	120	240
Number of clusters	6	30	60	120
Number of subchannels	3	15	30	60

mobile environment make the use of permutation schemes, such as PUSC, initially more likely in WiMAX.

21.3.1.3 Burst Profiles

WiMAX supports different combinations of modulation and coding. Each possible combination is called in IEEE 802.16e a burst profile. The allowed modulation schemes are QPSK, 16-QAM, and 64-QAM, this latter optional in the uplink. Error-correcting coding with a rate $R = 1/2$ is also specified, with higher code rates achievable by means of puncturing. The standard defines CC, convolutional turbo codes (CTC) and LDPC, whereas the latter is optional in the WiMAX forum profiles. If lower code rates are required, the code words can be repeated up to six times.

Link adaptation is therefore essential for a good link performance. In the downlink, the most adequate burst profile for each transmission can be chosen based on CQI transmitted in an uplink feedback channel. In the uplink, the BS measures the CQI and decides which burst profile shall be used by the MS.

21.3.1.4 Duplexing and Framing

IEEE 802.16e has support for both FDD and TDD, but the WiMAX forum initially supported only the TDD mode, which is likely to be the most widespread mode.

One example of a WiMAX TDD frame can be seen in Figure 21.4, where in the x-axis we have the time dimension, represented by the OFDMA symbol number, and in the y-axis we have the frequency domain, represented by the logical subchannel index. Note that the logical subchannel may consist of several nonadjacent subcarriers.

IEEE 802.16e defines several TDD frame sizes, ranging from 2 to 20 ms, but the WiMAX forum supports only a 5 ms frame, in which the lengths of the downlink frame and uplink frames are variable. Between a downlink frame and an uplink frame an interval, the transmit/receive transition gap (TTG), is required, in order to allow the BS to ramp down its transmit power

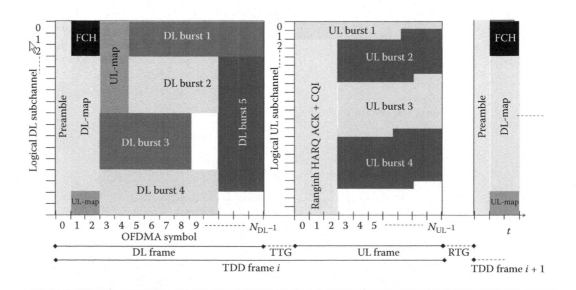

Figure 21.4　WiMAX TDD frame.

before switching to reception in the uplink. A similar interval is required between the uplink and the downlink frames called the receive/transmit transition gap (RTG). Those gaps also serve the purpose of dealing with the propagation delay. Already subtracting the TTG and RTG, each 5 ms frame consists of 48 OFDM symbols.

A half-duplex FDD (HFDD) mode is also proposed, combining advantages of both TDD and FDD modes. In the HFDD mode, DL and UL are transmitted in different frequencies, as in FDD, but, for any given user, they are also done in different time intervals. Thus, each MS still requires just one single modem, because transmission and reception do not have to occur simultaneously for a single connection. The BS, on the other hand, is required to transmit and receive at the same time. In WiMAX, the minimum allocation resource is called a slot. The slot definition depends on the permutation scheme, but a slot always contains 48 data modulation symbols. For instance, in downlink PUSC a slot consists of one subchannel (24 data subcarriers) over two OFDMA symbols, remembering that PUSC subchannels are not formed by contiguous subcarriers. In uplink PUSC, a slot consists of one subchannel over three OFDMA symbols, that is, six tiles.

The time–frequency region is divided into bursts, corresponding to contiguous slot allocations. In the downlink, the first transmitted symbol is always a known preamble, which allows the MS to synchronize with the BS and perform an initial channel and signal quality estimation. The preamble consists of a pseudo-random code transmitted with a boosted BPSK modulation. There are 32 different preamble sequences for each of three segments. (Segments are sets of disjoint subcarriers. Each segment can be allocated to a different sector.) The used code is determined by the IDCell parameter, and, when planning a network, one must take care not to let neighboring cells use the same code. In the first subchannels of the next two symbols, the frame control header (FCH) is transmitted, which is a broadcast channel containing important information about the cell configuration, including the MAP parameters. The MAP follows the FCH, and, among other signaling information, it contains information about the position of the user data bursts. There is one MAP for the downlink and one for the uplink, where the uplink MAP in the i-th frame refers to the uplink frame of the $i + 1$th frame.

In the downlink, user data bursts are always rectangular regions of several slots. Each burst may contain subbursts of different users, provided they all employ the same burst profile. The downlink frame allocation scheme offers the WiMAX equipment manufacturer a great amount of flexibility. Nevertheless, the best way to schedule bursts and subbursts in a frame is still a challenging research subject.

As we can see in Figure 21.4, we can leave some slots unused. However, all pilot symbols must be transmitted in all subcarriers to facilitate channel estimation and synchronization in the downlink.

In the uplink, some slots are reserved for a ranging channel. This is a contention channel, which serves for various purposes, such as bandwidth requests and handover signaling by the MSs. Remembering that carrier frequency offset is a serious issue in OFDMA systems, and that each MS transmits using a different oscillator, it is fundamental that the MSs adjust their frequencies before transmitting in the uplink. This is also done using the ranging channel. Some other slots are usually also reserved for the CQI feedback channel, already mentioned in Section 21.3.1 and for HARQ subburst acknowledgment.

In the uplink, the burst allocation is no longer rectangular. Instead, the slots are filled up sequentially, initially in the time-domain, and when a subchannel is full, the allocation moves to the first available slot in the next subchannel. This is done in order to improve the uplink transmission range by minimizing the number of subcarriers $N_{\mathrm{UL},u}$ used by transmission of user u. Supposing the MS has an available transmit power P_{MS}, the transmit power at each subcarrier is $P_{\mathrm{SC}} = P_{\mathrm{MS}}/N_{\mathrm{UL},u}$.

Hence, the less the subcarriers we use, the higher the transmit power per subcarrier, which is what ultimately impacts the transmission range.

If AMC subchannelization is employed in the uplink, a few uplink OFDMA symbols may be reserved for a sounding channel, in which the MSs may transmit a training sequence. This channel is needed for the BS to estimate the uplink channel on different frequency bands, and use this information to optimize uplink frequency-domain scheduling.

Each frame may be divided into several permutation zones. Each permutation zone consists of an adjacent number of OFDMA symbols, and each zone may employ a different permutation scheme. However, the first downlink permutation zone must always employ PUSC, which is used for the signaling messages in FCH, as well as DL- and UL-MAP.

21.3.1.5 HARQ

Hybrid ARQ (HARQ) is a fundamental technique to guarantee low packet error rates in harsh interference-limited wireless channels. HARQ [12] is a combination of packet retransmissions with error-correcting codes. Even though IEEE 802.16e specifies two types of HARQ, namely incremental redundancy (IR) and Chase combining, only the latter is mandatory. With Chase combining, in case of a detection error, the received packet is not discarded, and the received signal for each modulation symbol is stored. The transmitter is informed about the detection through the ACK channel and retransmits exactly the same packet, using the same burst profile. The receiver may then combine the received signal with the previously stored packets, thus substantially improving the probability of a correct packet detection.

HARQ has to deal, however, with memory constraints on the receiver and imposes an increase in the transmission delay when retransmissions are required. To address these problems, the number of retransmissions is limited by a parameter, after which the packets are discarded. Furthermore, the number of simultaneous HARQ outstanding packets is also limited.

HARQ can be seen as a cross-layer technique, since it requires support from both the PHY and MAC layers.

21.3.1.6 Advanced Antenna Systems

Multiple antenna techniques may be employed in WiMAX to achieve either a higher transmission rate and a higher robustness under low SINR values, both of which have a positive impact in the user throughput. OFDMA systems facilitate the use of multiple antenna techniques by making the channel flat in each subcarrier. The multiple antenna techniques employed in WiMAX can be divided into the following categories:

■ *Diversity schemes* exploit the different instant channel realizations at the different antennas to improve the link performance, particularly in low SINR conditions. Diversity schemes may employ multiple antennas both at the BS and at the MS. However, because of the higher cost of transmitters, it is expected that, at least at an early phase, the MSs will employ multiple antennas for reception only. Among others, the following diversity techniques can be used:
  – *MRC (maximal ratio combining)* is a well-known signal processing technique, in which the received signals at each antenna are weighted by their individual SINR and coherently combined. MRC is the optimum linear combining technique in the presence of white noise [13]. With MRC only one antenna is used for transmission.
  – *IRC (interference rejection combining):* MRC maximizes the postcombining SINR only when the noise is uncorrelated among the different antennas. In interference dominated

scenarios, however, there is a correlation among the interference in the different antennas. If the noise-plus-interference covariance matrix can be estimated, then IRC is the best combining technique [14]. The advantage of IRC over MRC is particularly significant when the interference is caused by one dominant interferer. Nevertheless, IRC is usually only feasible in the downlink. In the uplink the estimation of the covariance matrix is usually unfeasible, because transmission is made in short bursts, whereas in the downlink the BS always transmits the preamble and pilots.

 - *CDD (cyclic delay diversity)* can be employed with multiple transmit antennas in the downlink. With CDD, a virtual delay can be applied to one or several transmit antennas, by applying different phase shifts on each subcarrier [15]. This virtual delay acts like an extra multipath and thus increases the frequency diversity of the system.
 - *STBC (space-time block code).* The STC scheme known as the Alamouti [16] code is supported in IEEE 802.16e. In this scheme, two modulation symbols are transmitted over two antennas in two consecutive OFDMA symbols. This is thus a coding scheme with a rate $R = 1$, and as two antennas are used at the same time, spatial diversity can be obtained. As long as the channel remains constant over two OFDMA symbols, this code is orthogonal, and the symbols can be detected by a simple linear combination at the receiver. When two receive antennas are available, both MRC and IRC can be employed. The standards also support four transmit antennas, however, with a reduction in the code rate.

■ *Beamforming:* By using knowledge of the channel frequency response and applying adequate weights on each antenna, it tries to optimize the antenna radiation pattern by increasing the antenna gain in the direction of the desired user, while minimizing it in the direction of the main interferers. It requires correlated antennas with small spacing.

■ *MIMO with spatial multiplexing (SM):* In a system where both transmitter and receiver have multiple antennas, N_{tx} and N_{rx}, respectively, we can transmit $N_{SM} = \min(N_{tx}, N_{rx})$ parallel data flows over the different antennas [17]. In the downlink, this is used to increase the transmission data rate N_{SM}-fold. In the uplink, the collaborative MIMO approach is employed, where each MS is considered an antenna in a MIMO system, and different users are spatially multiplexed in the same time–frequency resources.

STBC, SM, and beamforming require some modifications in the pilot structures of the permutation schemes, to allow for channel estimation over the different antennas, and thus require a different zone. All single-transmit antenna schemes and CDD can be employed in the normal PUSC zone.

21.4 Evolved Universal Terrestrial Radio Access (E-UTRA)

The user penetration of third generation wideband code division multiple access (WCDMA) radio access technology and the usage of mobile broadband services, supported by high speed packet access (HSPA), have been increasing recently. It seems that HSPA systems will satisfy operators' and users' data requirements, including higher bit rates, lower delays, and higher capacity, in the near and mid-term future. However, in a longer time perspective, the Third Generation Partnership Project (3GPP) proposed a study item called Evolved UTRA and UTRAN [18], in order to be prepared for further increasing user demands and for competition from new radio-access technologies such as WIMAX. The objective of the study item was to develop a framework for the evolution of the 3GPP radio-access technology toward a high-data-rate (around 100/50 Mbps for downlink/uplink), low-latency, and packet-optimized radio-access technology, with higher spectral efficiency and coverage and acceptable complexity and cost [18]. As a basis for this study item work, 3GPP has concluded a set of targets and requirements for this LTE [19].

During the feasibility study a selection of the multiple access schemes and the basic radio access network architecture were defined. The 3GPP considered different multiple access options, but OFDMA and single carrier frequency division multiple access (SC-FDMA) were chosen for downlink and uplink transmission directions.

Nowadays, the standardization of LTE is well advanced and this chapter summarizes the main characteristics based on E-UTRAN Release 8.

21.4.1 E-UTRA Requirements

For the LTE, clearly more ambitious goals were necessary to make it worth the effort, when compared with what can be achieved with current 3G systems. For instance, from the radio-interface point of view, in the HSPA system, 3GPP Releases 5 and 6, peak data rates up to 14.4 Mbps downlink and 5.7 Mbps uplink can be achieved. Thus, the following targets were defined [19]:

- Instantaneous downlink peak data rate of 100 Mbps within a 20 MHz downlink spectrum allocation (5 bps/Hz)
- Instantaneous uplink peak data rate of 50 Mbps (2.5 bps/Hz) within a 20 MHz uplink spectrum allocation
- Average downlink user throughput per MHz three to four times higher than Release 6 high speed downlink packet access (HSDPA)
- Average uplink user throughput per MHz two to three times higher than Release 6 Enhanced Uplink
- In a loaded network, target for downlink spectrum efficiency (bits/s/Hz/site) three to four times higher than Release 6 high speed downlink packet access (HSDPA)
- In a loaded network, target for uplink spectrum efficiency (bits/s/Hz/site) two to three times higher than Release 6 Enhanced Uplink
- Cell-edge user throughput improved by a factor of two for the uplink and the downlink
- Radio network user plane latency below 10 ms round trip time (RTT) with 5 MHz or higher spectrum allocation
- E-UTRA shall operate in spectrum allocations of different sizes, including 1.4, 3.0, 5, 10, 15, and 20 MHz in both the uplink and the downlink.
- Support for interworking with existing 3G systems and non-3GPP specified systems
- Support of packet-switched domain only (including VoIP)
- Reduced cost for operator and end user

21.4.2 E-UTRAN Architecture

The E-UTRAN has a flat architecture where the radio-related functionalities are located in the BS. Hence, similarly to HSUPA and HSDPA, more intelligence was added to the BS.

The E-UTRAN consists of enhanced NodeBs (eNodeB) (see Figure 21.5), which provide the E-UTRA user plane and control plane protocol terminations towards the user equipment (UE) [20]. The eNodeBs are interconnected with each other, to the mobility management entity (MME) and to the serving gateway (S-GW). The evolved packet core (EPC) is compounded by MME and S-GW. Figure 21.6, obtained from Ref. [20], presents the main functionalities of eNodeB, MME, and S-GW. More details can be found in Ref. [21].

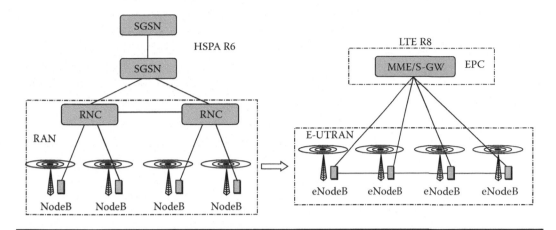

Figure 21.5   E-UTRAN architecture evolution.

21.4.3 E-UTRAN Multiple Access

The multiple access scheme for the LTE PHY is based on OFDM with a CP in the downlink, and on SC-FDMA with a CP in the uplink [22]. To support transmission in paired and unpaired spectra, two duplex modes are supported: FDD, supporting full duplex and half duplex operations, and TDD.

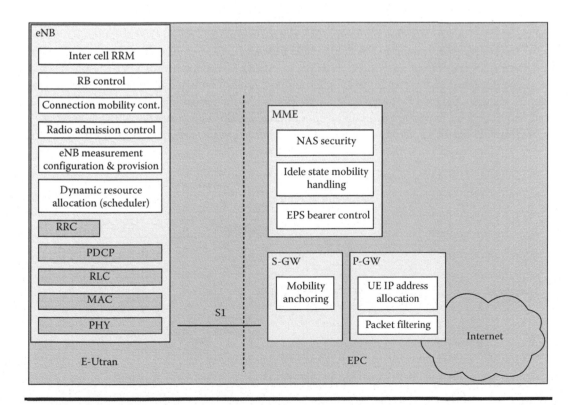

Figure 21.6   Functional split between E-UTRAN and EPC.

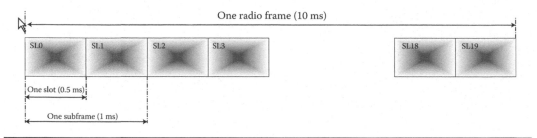

Figure 21.7   E-UTRAN FDD frame structure. *Abbreviation:* SL, slot.

21.4.3.1 Frame Structure

Two radio frame structures are supported for DL/UL E-UTRAN transmissions: Type 1, applicable to FDD and Type 2, applicable to TDD [23].

The Type 1 frame structure has a 10 ms ($T_f = 307,200 \times T_s$) duration, where T_s is a basic time unit defined in standard, given by $T_s = 1/(15000 \times 2048)$s. Each Type 1 frame has 20 slots, numbered from 0 to 19, with a 0.5 ms duration ($T_{slot} = 15,360 \times T_s$). A subframe is defined as two consecutive slots. Figure 21.7 presents the Type 1 frame structure. As it is well known, in case of FDD (paired spectrum), all subframes of a carrier are used either for downlink transmission or for uplink transmission.

The Type 2 frame structure, the TDD mode or the unpaired spectrum, has 10 ms length as well. However, each radio frame is divided in two half-frames of 5 ms of duration for downlink and uplink directions. Each half-frame consists of five subframes of 1 ms of duration. The subframes can be reserved for uplink, downlink, and special transmissions. The special subframes have three fields: downlink pilot time slot (DwPTS), guard period (GP) and uplink pilot time slot (UpPTS) and their total joint duration is equal to 1 ms. In Ref. [23], the lengths of DwPTS, GP, UpPTS are defined. Figure 21.8 presents the Type 2 frame structure. The first and sixth subframes of each frame are always assigned for downlink transmission while the remaining subframes can be flexibly assigned to be used for either downlink or uplink transmission. The main reason for the predefined assignment is that these subframes have synchronization signals. The synchronization signals are

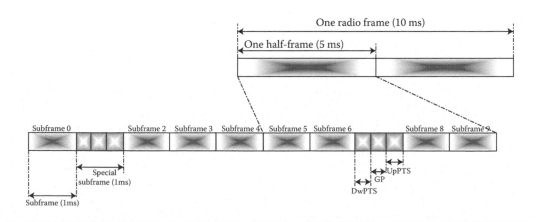

Figure 21.8   E-UTRA TDD frame structure.

Table 21.5   Transmission Configuration and RBs in E-UTRAN Channel Bandwidths

Normal channel bandwidth BWChannel MHz	1.4	3	5	10	15	20
Transmission bandwidth configuration N_{sc}^{RB}	6	15	25	50	75	100
FFT size		128	256	512	1024	1536	2048

transmitted on the downlink of each cell and are used for initial cell search as well as for neighbor-cell search. This flexible assignment of subframes allows for different proportion of radio resources for downlink and uplink transmissions.

In Ref. [23], six different uplink–downlink configurations are defined. Two downlink-to-uplink switch-points periodicities are allowed in a TDD frame: 5 and 10 ms. When downlink-to-uplink switch-point periodicity is 5 ms, the special subframe exists in both half-frames and the subframes are defined as two slots of length. On the other hand, the special subframe exists only in the first half-frame for downlink-to-uplink switch-points periodicity of 10 ms.

21.4.3.2 OFDMA

OFDMA has been selected as the radio access scheme for E-UTRAN downlink [23]. It is well known that in high data rate, OFDM has superior performance in frequency-selective fading channels. Besides, the complexity of the baseband receiver is much lower, when compared with equalized single carrier transmission. Another advantage of OFDM is its easy combination with MIMO systems. High capacity and spectral efficiency can be achieved with MIMO–OFDM systems when the time, space, and frequency domains are explored. In layer 2, better and flexible radio resource management algorithms can be developed with OFDM in order to achieve frequency-domain diversity and scheduling.

The E-UTRAN system has a scalable bandwidth of up to 20 MHz, as we can see in Table 21.5 [24]. In order to allow the operation in differently sized spectrum allocations, a fixed subcarrier spacing of 15 kHz was defined.

In the downlink, the smallest time–frequency unit for transmission is defined as a resource element. Each resource element is represented by one frequency and one OFDM symbol in a slot.

Resource blocks are defined as N_{sc}^{RB} consecutive subcarriers in the frequency domain and N_{symb}^{DL} consecutive OFDM symbols in the time domain. In other words, a resource block represents one slot in the time domain and 180 kHz in the frequency domain. The quantity of resource blocks will depend on the downlink transmission bandwidth, and the number of symbols in a slot will depend on the CP length and the subcarrier spacing given by Table 21.6, which presents the physical resource blocks parameters.

Table 21.6   Physical Resource Blocks Parameters for Downlink

Configuration	N_{sc}^{RB}	N_{symb}^{DL}
Normal CP $\Delta F = 15$ kHz	12	7
Extended CP $\Delta F = 15$ kHz	12	6
Extended CP $\Delta F = 7.5$ kHz	24	6

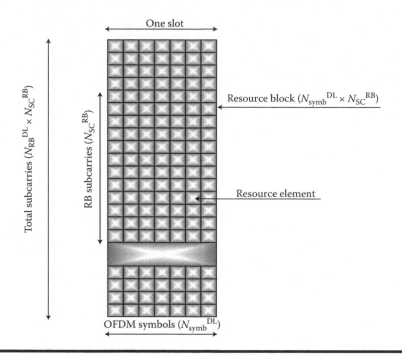

Figure 21.9 Downlink resource grid.

Resource blocks are employed to map the physical channels into resource elements and they can be classified as physical and virtual. There are two types of virtual resource blocks: virtual resource block of the localized type and virtual resource block of the distributed type [23].

Figure 21.9 presents a resource grid for downlink transmission. When multiple antennas are used, one resource grid must be defined per antenna.

21.4.3.3 Downlink Transport Channels

The interface between the PHY and higher layers, more specifically to the MAC layer, is carried out through the transport channels. E-UTRAN has adopted the same basic principle of WCDMA/HSPA in which data are delivered to the PHY in the form of transport blocks of a certain size. The following transport channels are defined for E-UTRAN:

Broadcast channel (BCH) broadcasts information in the entire coverage area of the cell.
Downlink shared channel (DL-SCH) transmits HARQ, dynamic link adaptation by varying the modulation, coding and transmit power, dynamic and semi-static resource allocations, and UE discontinuous reception information.
Paging channel (PCH) is also used for broadcast information.
Multicast channel (MCH) is characterized by support for MBSFN combining MBMS transmission on multiple cells.

21.4.3.4 Downlink Physical Channels

The information originated from higher layers are transported through some resource elements in the downlink physical channel. The following downlink physical channels are defined in the

E-UTRAN standard [23]: *The Physical Downlink Shared Channel (PDSCH)* is the physical channel available for the transmit user data. The PDSCH channel supports QPSK, 16QAM, and 64QAM, modulations and may use up to 20 MHz bandwidth [23]. The channel coding for the user data is a parallel concatenated convolutional code (PCCC) with two 8-state constituent encoders combined with turbo code internal interleaver. The defined coding rate of the turbo encoder is 1/3. This channel coding is similar to the turbo coding in Release 99 WCDMA, but with a new turbo interleaver.

Physical control format indicator channel (PCFICH) informs the UE about the number of OFDM symbols used for transmission of the PDCCHs and it is transmitted in every subframe.

Physical downlink control channel (PDCCH) carries the scheduling assignments and other control information.

Physical hybrid ARQ indicator channel (PHICH) carries the hybrid ARQ ACK/NACK information.

Physical broadcast channel (PBCH) broadcasts system information, and the coded BCH transport block is mapped to four subframes within a 40 ms interval.

Figure 21.10 depicts the mapping between transport and physical channels.

While the signaling physical channels PDCCH, PBCH, and PCFICH use QPSK modulation, the BPSK modulation is defined for PHICH. The channel coding schemes for control signaling are a tail-biting CC with a constraint length 7 and a coding rate 1/3 for PDCCH and PBCH, a block code with a coding rate 1/16 for PCFIH and a repetition code with a coding rate 1/3 for PHICH.

The E-UTRAN standard still defines two types of signals: the synchronization signal, to allow cell search, and the reference signal to facilitate channel estimation and channel quality measurements.

The downlink reference signals are known symbols inserted in the first and third last OFDM symbols of each slot and with a frequency-domain spacing of six subcarriers. Hence, within each resource block, consisting of 12 subcarriers during one slot, there are four reference symbols. When multiple antennas are present, the mobile terminal must be able to estimate the downlink channel corresponding to each transmit antenna. Thus, there is one reference signal transmitted per downlink antenna port. For example, in case of two transmit antennas, the reference symbols of the first and second antennas are frequency multiplexed of three subcarriers.

The throughput-enhancing mechanisms in E-UTRAN include AMC, based on link adaptation, and HARQ based on fast L1 retransmissions. Link adaptation, with various modulation schemes and channel coding rates is applied to the shared data channel. Downlink HARQ is based on IR. Chase combining is a special case of IR and is thus implicitly supported as well.

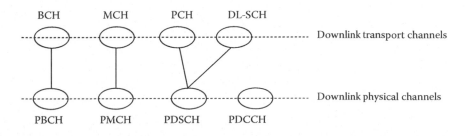

Figure 21.10 Mapping between downlink transport channels and downlink physical channels.

21.4.3.5 E-UTRAN Uplink Multiaccess: SC-FDMA

OFDM is known to provide good performance in frequency-selective environments. OFDM can be easily implemented by means of an IFFT at the transmitter and an FFT at the receiver. A similar result can be obtained if both FFT and IFFT are performed at the receiver, with single-tap equalization done after the FFT. This technique is called single carrier with frequency domain equalization (SC-FDE). In OFDMA, the subcarriers in an OFDM system can be allocated to different users. The same can be done in an SC-FDE system, which is called single carrier-frequency domain multiple access (SC-FDMA). Differently from the downlink, for the uplink, E-UTRAN adopted SC-FDMA (Figure 21.11).

SC-FDMA has some advantages over OFDMA, which makes it a good choice, particularly in the uplink of wireless systems. Perhaps the main advantage of SC-FDMA is its lower PAPR. The great amplitude range of OFDM signals is a big drawback when nonlinear amplifiers are employed, because large backoffs are required to minimize signal distortion and out-of band radiation, thus reducing amplifier's efficiency and possibly, on account of lower transmit powers, also coverage. This is specially a problem in the uplink, where power-efficient transmission with low-cost amplifiers is required. SC-FDMA can help overcome this problem. Figure 21.12 presents a comparison between SC-FDMA and OFDMA for different number of subcarriers. As we can see, SC signals have a much lower PAPR.

Another advantage of SC-FDMA is that it is less sensitive to carrier frequency offset, which in the uplink of OFDMA may be a problem, because each user has a different oscillator.

It is also well known that uncoded OFDMA may suffer in frequency-selective fading channels, as some subchannels may be in a deep fade, and thus their information may be lost. In wideband single carrier systems, each symbol occupies a wide range of frequencies and is thus inherently resilient to frequency-selective fading.

In coded systems, the performances of SC-FDMA and OFDMA are similar, but, nevertheless, SC-FDMA still allows operation with coding rates close to one, but at the cost of complexity in the receiver.

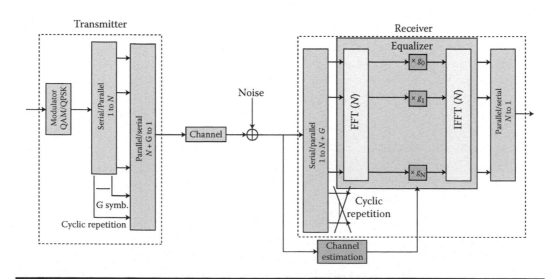

Figure 21.11 Single carrier with frequency-domain equalization (FDE).

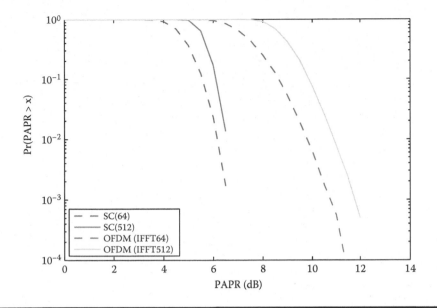

Figure 21.12 Comparison of SC and OFDM-PAPR.

Similarly to the downlink, in the uplink the smallest time–frequency unit for transmission is defined as resource element. Besides, the same definition of downlink resource block can also be applied for uplink. Table 21.7 presents the physical RB parameters.

Figure 21.13 presents a resource grid for uplink transmission.

21.4.3.6 Uplink Transport Channel

For the uplink, some transport channels are also defined:

Uplink shared channel (UL-SCH): Allows transmission of HARQ, dynamic link adaptation by varying power, modulation and coding, and dynamic and semi-static resource allocation information.

Random access channel(s) (RACH): Carries limited control information.

21.4.3.7 Uplink Physical Channel

For the uplink, three physical channels are defined:

The physical uplink shared channel (PUSCH) is a physical channel available for transmitting uplink user data. The same downlink set of modulation (64QAM is optional for devices) and channel coding schemes is supported for the PUSCH channel [23].

Table 21.7 Physical Resource Blocks Parameters for Uplink

Configuration	N_{sc}^{RB}	N_{symb}^{UL}
Normal CP $\Delta F = 15\,\text{kHz}$	12	7
Extended CP $\Delta F = 15\,\text{kHz}$	12	6

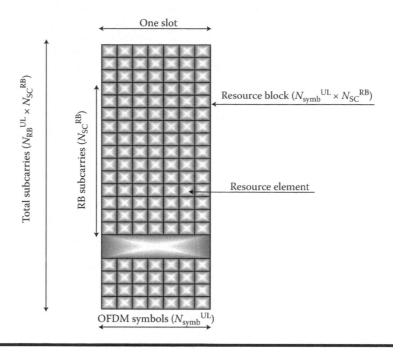

Figure 21.13   Uplink resource grid.

Physical uplink Control Channel (PUCCH) is a signaling control channel. It carries HARQ ACK/NAKs in response to downlink transmission, carries scheduling requests and CQI reports.

Physical Random Access Channel (PRACH) is another control channel and carries the random access preamble.

Figure 21.14 depicts the mapping between transport and physical channels. More details can be found in Ref. [23]. Similarly to the downlink, reference signals for channel estimation are also needed for the uplink and are time multiplexed with uplink data transmitted in the fourth block of each slot.

Two types of uplink reference signals are supported:

■ Demodulation reference signal, associated with transmission of PUSCH or PUCCH
■ Sounding reference signal

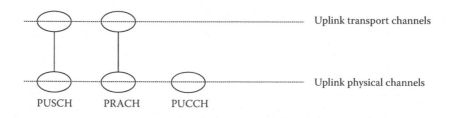

Figure 21.14   Mapping between uplink transport channels and uplink physical channels.

21.4.4 E-UTRA: Advanced Multiantenna Techniques

Multiple antenna systems known as MIMO systems have been proposed as an effective way to satisfy the user demand for high data rate applications in wireless systems. This is especially important in systems where the capacity attained with traditional techniques is limited by the adverse characteristics of the propagation environment. MIMO systems are also especially adequate for transmission of high data rates that require large bandwidth in which frequency selectiveness generally occurs.

Hence, advanced multiantenna techniques will play an important role in fulfilling the E-UTRAN requirements, as they increase capacity, coverage, and reliability, all of them being important requirements of this technology. This potential is large and not always fully exploited in existing radio access technologies. Although MIMO technology improves reliability and transmission rates, its cost-effective implementation remains a major challenge.

Increasing data rates can be achieved by transmitting multiple parallel streams or layers to a single user. This multilayer transmission is referred to as spatial multiplexing. The benefits of spatial multiplexing are achieved in conditions with favorable signal-to-noise ratio and rich scattering environments.

Beamforming implies that multiple antennas are used to form the transmission or reception beam, which increase the signal-to-noise ratio at the receiver. This technique can both be used to improve coverage of a particular data rate and to increase the system spectral efficiency. The increased signal-to-noise ratio is not only due to a larger gain in the direction of the desired user, but also due to a reduction of interference in the cell.

The concept of transmit diversity, more specifically STC was proved effective in combating fading and enhancing data rates. Exploiting the presence of spatial diversity offered by multiple transmit and/or receive antennas, ST coding relies on simultaneous coding across space and time to achieve diversity gain without necessarily sacrificing precious bandwidth.

Diversity, spatial multiplexing, and beamforming are the multiple antennas techniques proposed for E-UTRAN [23]. For the downlink, two or four transmit antennas and two or four receive antennas are the proposed configurations and allow for multilayer transmissions with up to four streams.

Antenna mapping occurs in two separated stages. Initially, the symbols are mapped in layers followed by a precoding [23] (Figure 21.15). In the layer mapping stage, the complex-valued modulation symbols obtained from a codeword are mapped on one or several layers, which are mapped onto the transmit antennas ports. Streams are referenced as spatial layers in the E-UTRAN.

A maximum of two codewords is allowed, even when four antennas are used in the downlink transmission. Figure 21.16 presents the codeword-to-layer mapping for spatial multiplexing. Note that the number of codewords is always less than the number of layers or antennas.

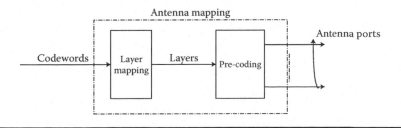

Figure 21.15 Antenna mapping: layer mapping followed by a precoding.

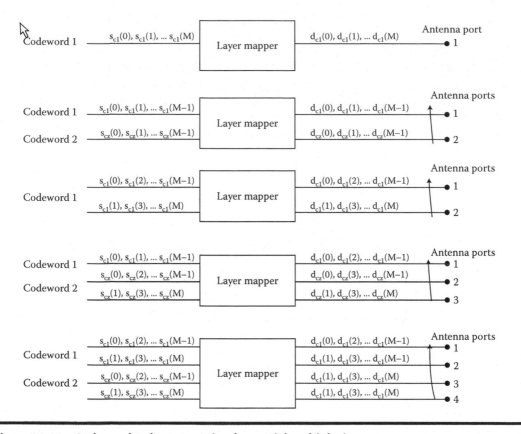

Figure 21.16   Codeword to layer mapping for spatial multiplexing.

Actually, multiple codewords could be used in the layer mapping. However, in order to reduce the amount of control signaling, due to CQI reporting and HARQ ACK/NACK, a reduced number of codewords was chosen.

Open-loop and closed-loop spatial multiplexing modes were proposed for the LTE system. Based on the type of transmission mode, different codebooks were defined in order to form the transmitted layers [23]. In the second antenna mapping stage, the layers are precoded based on the chosen transmission mode.

For the closed-loop spatial multiplexing mode on the basis of the measurements on the downlink reference signals of the different antennas, the UE decides the predefined precoder matrix and feeds it back to the eNodeB. Maximizing the capacity based on the receiver capabilities is the chosen metric for defining the precoder matrix.

For the open-loop spatial multiplexing transmission mode, CDD precoding was defined. In this transmission mode, the OFDM symbol is transmitted from the first antenna and a cyclically delayed replica of the same in the second antenna. In other words, a spatial diversity is converted in frequency diversity, because a time delay is identical to applying a phase shift in the frequency domain.

Transmit diversity is only defined for two and four transmit antennas and one codeword as illustrated in Figure 21.17.

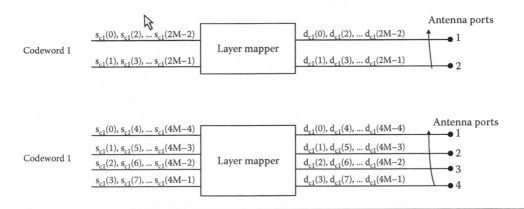

Figure 21.17 Codeword to layer mapping for transmit diversity.

When diversity is required and eNodeB has two antennas, space-frequency block code (SFBC) is used [23]. SFBC is a frequency domain version of the well-known STBC, also known as Alamouti codes [16]. This family of codes uses orthogonal structures to design a STBC, which is of much interest since it requires a low complexity decoding scheme, where maximum-likelihood decoding can be performed in a very simple way with linear processing at the receiver.

For the case of four transmit antennas, frequency switched transmit diversity (FSTD) and its combination with SFBC were defined for the LTE system (see precoding matrix below)

$$
\begin{matrix}
s_1 & s_2 & 0 & 0 \\
0 & 0 & s_3 & s_4 \\
-s_2^* & s_1^* & 0 & 0 \\
0 & 0 & -s_3^* & s_4^*
\end{matrix}
$$

In the first phase of E-UTRAN specifications, multiantenna uplink transmission is not specified.

21.5 Open Research Challenges: IMT-A Next Cellular Network Generation

International Mobile Telecommunications-2000 (IMT-2000) is the global standard for Third generation (3G) wireless communications. 3G systems have been evolving during the past years and evolution will continue. Nevertheless, 3G systems will eventually reach their limits. For the sake of evolution, ITU recommendation ITU-R M.1645 describes the user trends and needs for systems beyond IMT-2000 [25], which are referenced as IMT-A.

The key features of IMT-A can be defined as

- A high degree of commonality of design worldwide, while retaining the flexibility to support a wide range of services and applications in a cost-efficient manner
- Compatibility of services within IMT and with fixed networks
- Capability of interworking between IMT and other access systems
- High-quality mobile services
- User equipment suitability for worldwide use

- User-friendly applications, services, and equipment
- Worldwide roaming capability
- Enhanced peak user data rates to support new services and applications such as multimedia

The new IMT-A systems need to provide high data rates with high mobility, covering high speed on highways or fast trains, and data rates of up to approximately 100 Mbps for high-mobility mobile access and up to approximately 1 Gbps for low-mobility nomadic/local wireless access.

21.5.1 LTE Advanced

In 3GPP, work on LTE Advanced has started recently [21]. However, LTE Release 8 fulfils most of the wide-area requirements for IMT-A. Future LTE releases will consider enhancements in order to fulfil all IMT-A requirements. It is likely that major enhancements to LTE-A will be introduced in Release 10 and shall satisfy the 4G requirements defined by the ITU.

The requirements for LTE Advanced are as follows:

- Instantaneous downlink and uplink peak data rates of 1 Gbps and 500 Mbps, respectively
- Transmission bandwidth wider than approximately 70 MHz (up to 100 MHz) in DL and 40 MHz in UL
- *Latency:* C-plane from Idle (with IP address allocated) to Connected in less than 50 ms and U-plane latency shorter than 5 ms one way in RAN taking into account 30% retransmissions
- Cell-edge user throughput two times higher than that in LTE Release 8
- Average user throughput three times higher than that in LTE Release 8
- Capacity (spectrum efficiency) three times higher than that in LTE Release 8
- Peak spectrum efficiency in the DL of 30 bps/Hz, and in the UL of 15 bps/Hz
- *Spectrum flexibility:* Support of scalable bandwidth and spectrum aggregation
- Same mobility as that in LTE Release 8
- Coverage should be optimized for deployment in local areas/microcell environments with site distance up to 1 km
- Backward compatibility and interworking with LTE Release 8 with 3GPP legacy systems

According to Ref. [26], a new spectrum in the lower and higher bands will be required. Lower frequency bands must be used to increase coverage, and higher frequency bands to obtain broadband experience for small cell sizes and low mobility. However, not all bands will be available on a global scale, making global deployment and roaming a challenge.

In the ITU World Radiocommunication Conference (WRC07), the following new spectrum bands were proposed to the IMT 2000 and IMT-A:

- The 450 MHz band
- The UHF band (698–960 MHz)
- The 2.3 GHz band
- The C-band (3400–4200 MHz)

The 3GPP working groups have started working on the PHY in order to achieve the IMT-A requirements. The research topics could roughly be categorized as follows:

- Relay concepts
- More antennas in UE

- Scalable system bandwidth exceeding 20 MHz, potentially up to 100 MHz
- Local area optimization of air interface
- Nomadic/local area network and mobility solutions
- Spectrum sharing
- Cognitive radio concepts
- Automatic and autonomous network configuration and operation
- Enhanced precoding and forward error correction
- Interference management and suppression
- Asymmetric bandwidth assignment for FDD
- Hybrid OFDMA and SC-FDMA in the uplink
- UL/DL inter eNodeB coordinated MIMO
- New RRM algorithms for spectrum aggregation

21.5.1.1 WiMAX: 802.16m

IEEE 802.16m technology has been proposed as the 802.16e evolution. The general requirements for IEEE 802.16m systems are defined in Ref. [27]. These requirements are intended to supplement the requirements specified by the ITU-R for IMT-A systems.

The requirements for 802.16m are

- Interoperability for legacy WirelessMAN-OFDMA equipment
- Minimize complexity of the architecture and protocols
- Support existing services more efficiently as well as facilitate the introduction of new/emerging types of services
- Flexibility in order to support services required for next generation mobile networks
- Coexisting with other IMT-A technologies
- Support for scalable bandwidths from 5 to 20 MHz. Other bandwidths shall be considered as necessary to meet operator and ITU-R requirements
- Support for MIMO and beamforming operation
- Instantaneous downlink and uplink peak data rates of 1 Gbps and 500 Mbps, respectively
- Average user throughput twice as high as IEEE 802.16e (DL/UL)
- Cell edge user throughput twice as high as IEEE 802.16e (DL/UL)
- Peak spectrum efficiency in the DL larger than 6.5 bps/Hz, in the UL larger than 2.8 bps/Hz
- Maximum data latency downlink/uplink of 10 ms
- Support for interference mitigation schemes
- Support for flexible frequency reuse schemes
- Optimized handover with legacy IEEE 802.16e systems
- Support for multi radio access technology (RAT) operation
- Suitability for deployment both in spectrum already identified for IMT RATs and for any additional spectrum identified for IMT RATs by ITU (e.g., at WRC 2007)
- Enhancements to enable multihop relays

The IEEE 802.16m standard is in the process of development and some drafts can be found in Ref. [28]. In Ref. [29], the modifications and new features proposed for IEEE 802.16m are presented.

One main modification proposed is related to the frame structure, which was completely modified when compared with the IEEE 802.16e frame. The IEEE 802.16m basic frame structure is illustrated in Figure 21.18.

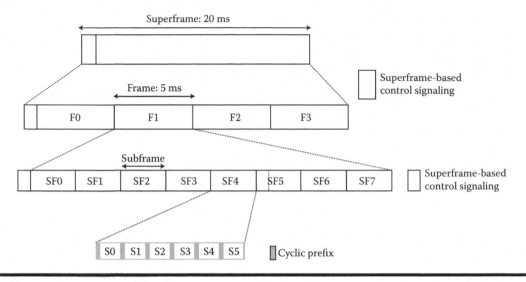

Figure 21.18 Basic 802.16m frame structure.

In the 802.16m frame structure, a superframe of 20 ms duration is divided into four 5 ms radio frames. Each frame consists of eight subframes. There are two types of subframes depending on the size of the CP:

- *Type 1:* Subframe consists of six OFDMA symbols
- *Type 2:* Subframe consists of seven OFDMA symbols

Table 21.8 OFDMA Parameters for IEEE 802.16m

Parameter	Value				
Bandwidth W (MHz)	5	7	8.75	10	20
Sampling factor n_s	28/25	8/7	8/7	28/25	28/25
Sampling frequency (MHz)	5.6	8	10	11.2	22.4
FFT size K	512	1024	1024	1024	2048
Subcarrier spacing (kHz)	10.937500	7.812500	9.765625	10.937500	10.937500
Useful symbol time (μs)	91.429	128	102.4	91.429	91.429
Symbol Time (μs) T_s (CP $= 1/8T_u$)	102.857	144	115.2	102.857	102.857
Number of OFDM symbols per frame (CP $= 1/8T_u$)	48	34	43	48	48
Symbol time (μs) T_s (CP $= 1/16T_u$)	97.143	136	108.8	97.143	97.143
Number of OFDM symbols per frame (CP $= 1/16T_u$)	51	36	45	59	59

Table 21.9 Comparison Among WiFi, WiMAX, and LTE

	WiFi	*WiMAX*	*LTE*
Bandwidth	20 MHz	5–20 MHz	1.4–20 MHz
Duplexing	TDD	TDD/FDD/HFDD (mainly TDD)	TDD/FDD/HFDD (mainly FDD)
Downlink multiple access	CSMA/CA	OFDMA	OFDMA
Uplink multiple access	CSMA/CA	OFDMA	SC-FDMA
Subcarrier spacing	312.5 kHz	10.9375 kHz	15 kHz
Useful symbol length	3.2 μs	91.4 μs	67 μs
CP	800 ns	11.4 μs	4.7 μs
Frame length (TTI)	Variable	1 ms	5 ms
Coding scheme	Convolutional	Turbo	Turbo
Modulation	BPSK, QPSK, 16-QAM, and 64-QAM	QPSK, 16-QAM, and 64-QAM	QPSK, 16-QAM, and 64-QAM
MIMO schemes	Only in 11 n	Yes	Yes

A subframe can be be assigned for either DL or UL transmission and it can be applied for both FDD and TDD duplexing schemes.

The OFDMA parameters for the IEEE 802.16m are specified in Table 21.8.

21.6 Summary

We have described in this chapter some of the main aspects of WiFi, WiMAX, and LTE, which all employ OFDM/OFDMA with different parameters. To sum up our contribution, we have listed some of the main parameters of the different technologies in Table 21.9.

References

1. P. W. C. Chan et al., The evolution path of 4G networks: FDD or TDD? *IEEE Commun. Mag.*, 42–50, December 2006.
2. http://www.wi-fi.org.
3. IEEE Std 802.11a-1999, Wireless LAN medium access control (MAC) and physical layer (PHY) specifications–High-speed physical layer in the 5 GHz band.
4. G. Caire, G. Taricco, and E. Biglieri, Bit-interleaved coded modulation, *IEEE Trans. Information Theory*, 44, 927–946, 1998.
5. J. Jelitto, A. N. Barreto, and H. L. Truong, A dynamic link adaptation algorithm for IEEE 802.11a wireless LANs, in *Proc. Int'l. Conf. on Communications* (ICC), Anchorage, AK, May 2003.
6. G. Bianchi, Performance analysis of the IEEE 802.11 distributed coordination function, *IEEE J. Sel. Areas Commun.*, 18, 535–547, 2000.

7. T. Paul and T. Ogunfunmi, Wireless LAN comes of age: Understanding the IEEE 802.11n amendment, *IEEE Circuits Syst. Mag.*, 1st Quarter, 2008, 28–54.
8. IEEE, Air interface for fixed broadband wireless access systems, *IEEE STD 802.16 - 2004*, October 2004.
9. http://www.etsi.org/
10. http://www.wibro.org.kr
11. H.J. Kushner and P.A. Whiting, convergence of proportional-fair sharing algorithms under general conditions, *IEEE Trans. Wirel. Commun.*, 3, 1250–1259, 2004.
12. J.-F. Cheng, Coding performance of hybrid ARQ schemes, *IEEE Trans. Commun.*, 54(6), 1017–1029, 2006.
13. A. N. Barreto and R. D. Vieira, A critical analysis of receiver diversity with multiple antennas with spatially coloured noise, in *Proc. Workshop on Signal Processing Advances in Wireless Communications*, 2008, Recife, Brazil, July 2008.
14. J. Karlsson and J. Heinegard, Interference rejection combining for GSM, in *Proc. of ICUPC*, 1996.
15. A. N. Barreto, Antenna transmit diversity for wireless OFDM systems, in *Proc. Vehicular Tech. Conf. (VTC-Spring)*, Birmingham, AL, May 2002.
16. S. Alamouti, A simple transmit diversity technique for wireless communications, *IEEE J. Sel. Areas Commun.*, 16(8), 1451–1458, 1998.
17. G. Foschini, Layered space-time architecture for wireless communication in a fading environment when using multi-element antennas, *Bell Lab. Tech. J.*, 1996.
18. 3GPP TD RP-040461, Proposed study item on evolved UTRA and UTRAN.
19. 3GPP TR 25.913 V7.2.0 (2005-12), Requirements for evolved UTRA (E-UTRA) and evolved UTRAN(E-UTRAN).
20. 3GPP TS 36.300 V8.6.0 (2008-09), Evolved universal terrestrial radio access (E-UTRA) and evolved universal terrestrial radio access network (E-UTRAN); overall description; Stage 2 (Release 8).
21. 3GPP TS 36.401 V8.3.0 (2008-09), Evolved universal terrestrial radio access network (E-UTRAN); Architecture description(Release 8).
22. H. G. Myung, J. Lim, and D. J. Goodman, Single carrier FDMA for uplink wireless transmission, *Veh. Tech. Mag.*, 30–38, September 2006.
23. 3GPP TS 36.211 V8.3.0 (2008-05), Evolved universal terrestrial radio access (E-UTRA), physical channels and modulation (Release 8).
24. 3GPP TS 36.104 V8.3.0 (2008-09), Evolved universal terrestrial radio access (E-UTRA), base station (BS) radio transmission and reception (Release 8).
25. Rec. ITU-R M.1645, Framework and overall objectives of the future development of IMT-2000 and systems beyond IMT-2000 (2003).
26. 3GPP TR 36.913 V8.0.0 (2008-06), Requirements for further advancements for E-UTRA (LTE-Advanced) (Release 8).
27. IEEE 802.16 Broadband Wireless Access Working Group, Draft IEEE 802.16m Requirements (2007-06-08).
28. http://ieee802.org/16/
29. IEEE 802.16 Broadband Wireless Access Working Group, Draft IEEE 802.16m system description document (2008-07-29).

Chapter 22

OFDMA-Based Mobile WiMAX

Jinsong Wu and Pei Xiao

Contents

22.1 Introduction 596
22.2 OFDMA Frame Structure and Subchannelization 597
    22.2.1 Elements of the OFDMA Frame Structure and Subchannelization 597
        22.2.1.1 OFDMA Symbol 597
        22.2.1.2 Cluster 598
        22.2.1.3 Slot 598
        22.2.1.4 Subchannel 598
        22.2.1.5 Several Elements with Similar Definitions in IEEE
                802.16-2004 and IEEE 802.16e-2005 598
        22.2.1.6 OFDMA Data Mapping in IEEE 802.16e-2005 599
        22.2.1.7 Frame, Subframe, and Zone 599
        22.2.1.8 Subburst 601
    22.2.2 Subchannelization 601
    22.2.3 Construction of Permutation Zones 603
        22.2.3.1 DL Preamble 603
        22.2.3.2 DL PUSC for a Single Transmit Antenna 604
        22.2.3.3 DL PUSC for STC Using Two Transmit Antennas 606
        22.2.3.4 UL PUSC for a Single Transmit Antenna 607
        22.2.3.5 UL PUSC for Space–Time Coding Using Two Transmit
                Antennas and Collaborative SM Using a Single Transmit
                Antenna per MS 608
        22.2.3.6 AMC for a Single Transmit Antenna 610

          22.2.3.7   AMC for Two Transmit Antennas...............................612

22.3   Power Saving Mode 614

      22.3.1   Sleep Mode 614

      22.3.2   Idle Mode 616

22.4   Handover..617

      22.4.1   Introduction..617

      22.4.2   Decision and Initiation of MDHO and FBSS.......................618

            22.4.2.1   Diversity Set and Anchor BS.....................618

            22.4.2.2   MDHO Decision....................................618

            22.4.2.3   FBSS Decision......................................618

            22.4.2.4   Required Conditions to Enable MDHO/FBSS....................618

      22.4.3   Diversity Set Update for MDHO/FBSS...............................619

      22.4.4   Anchor BS Update for MDHO/FBSS.................................619

            22.4.4.1   HO MAC Management Message Method.........................619

            22.4.4.2   Fast Anchor BS Selection Feedback Mechanism.................620

22.5   Related Research and Investigations...620

22.6   Conclusion...621

References...621

22.1 Introduction

The Worldwide Interoperability for Microwave Access (WiMAX) Forum, an industry-led, not-for-profit organization, is formed to certify and promote the compatibility and interoperability of broadband wireless products based upon the harmonized IEEE 802.16 [1,2]/ETSI HiperMAN standard [3–5]. The WiMAX Forum officially declare their multiple access control (MAC) and physical layer (PHY) specifications as system profiles. Currently, the WiMAX Forum has two different system profiles: one based on IEEE 802.16-2004 orthogonal frequency division multiple (OFDM) PHY, called the fixed system profile, and the other one based on IEEE 802.16e-2005 scalable OFDMA PHY, called the mobility system profile [6,7]. Mobile WiMAX was adopted as one of the ITU (International Telecommunication Union) IMT-2000 standard technologies in November 2007 [8].

Mobile WiMAX is not the same as IEEE 802.16e-2005, rather it is a subset of the IEEE STD 802.16 standard features and functionalities [9,10]. Both IEEE 802.16-2004 and IEEE 802.16e-2005 have multiple physical-layer options: single-carrier-based Wireless-MAN-SC and WirelessMAN-SCa, an orthogonal frequency division multiple (OFDM)-based WirelessMAN-OFDM, and an orthogonal frequency division multiple access (OFDMA)-based Wireless-OFDMA [1,2]. However, mobile WiMAX based on IEEE 802.16e-2005 currently supports only OFDMA. IEEE 802.16-2004 and IEEE 802.16e-2005 support both time division duplexing (TDD) and frequency division duplexing (FDD), as well as a half-duplex FDD, while the current mobile WiMAX system profile Release 1.0 [6,7] supports only TDD [6,11]. TDD holds its advantages: (1) flexibility in choosing uplink-to-downlink data rate ratios, (2) ability to exploit channel reciprocity, (3) ability to implement in a nonpaired spectrum, and (4) less complex transceiver design. WiMAX products may simultaneously support both fixed WiMAX and mobile WiMAX through duo-mode base stations (BS) or mobile stations (MS).* It is worth mentioning that future mobile WiMAX system

---

* MS are stations in the mobile service intended to be used while in motion or during halts at unspecified points. An MS is always a subscriber station (SS) unless specified otherwise in the standard [1,2]. An SS is a generalized equipment set providing connectivity between subscriber equipment and a BS [1,2].

profile Release 1.5 may further support both FDD and half-duplex FDD [12,13], and all required fixes and minor enhancements to support Release 1.5 are included in IEEE 802.16 REV2, which is a unified specification document embracing the IEEE 802.16-2004 as well as IEEE 802.16e/f/g amendments and related corrigenda [12].

Note that the WiMAX Forum only chooses the 256 subcarriers fast Fourier transformation (FFT)-based OFDM physical layer within the 802.16-2004 standard for the first-generation fixed WiMAX products, while the mobile WiMAX air interface utilizes OFDMA as the radio access method for improved multipath performance in the non-line-of-sight environment.

This chapter focuses on discussions and comparisons of several MAC and PHY aspects of TDD-based OFDMA between IEEE 802.16-2004 and IEEE 802.16e-2005 standards. It gives readers a systematic and panoramic view of the physical and MAC layer aspects of the OFDMA-based mobile WiMAX systems without digging deeply into the lengthy specifications. This chapter provides a thorough survey on OFDMA subchannelization, power saving mechanisms, and handovers for OFDMA-based mobile WiMAX. We explain complicated terminologies of the OFDMA frame structure and subchannelization in a systematic manner and let readers familiarize themselves with the mobile WiMAX system by comparing it to the fixed broadband wireless standard IEEE 802.16-2004.

22.2 OFDMA Frame Structure and Subchannelization

This section explains OFDMA frame construction with the aid of comparisons between IEEE 802.16-2004 and IEEE 802.16e-2005, and this is to help readers understand some differences between fixed WiMAX and mobile WiMAX and the standard evolvement from IEEE 802.16-2004 to IEEE 802.16e-2005. In both IEEE 802.16-2004 and IEEE 802.16e-2005, the OFDMA frame structure may be different in point-to-multipoint (PMP) and mesh modes [1,2]. This section only discusses and makes comparisons of the PMP mode OFDMA frame structure and subchannelization in IEEE 802.16-2004 and IEEE 802.16e-2005.

22.2.1 Elements of the OFDMA Frame Structure and Subchannelization

22.2.1.1 OFDMA Symbol

1. *Resemblance between IEEE 802.16-2004 and IEEE 802.16e-2005:* An OFDMA symbol is made of subcarriers, the number of which determines the FFT size. An OFDMA symbol consists of data, pilot, and null [guard band and direct-current (DC)] subcarriers. The following values for the ratio of cyclic prefix (CP) time to "useful" time shall be supported: 1/32, 1/16, 1/8, and 1/4 [1,2].

2. *Difference between IEEE 802.16-2004 and IEEE 802.16e-2005 [1,2]:*
   (a) *IEEE 802.16-2004:* FFT size is only 2048 [3,8].
   (b) *IEEE 802.16e-2005:* FFT sizes include 2048 (backward compatible to IEEE 802.16-2004), 1024, 512, and 128, which is called scalable OFDMA. This facilitates the support of various channel bandwidths. The MS may implement a scanning and search mechanism to detect the downlink (DL) signal when performing initial network entry and this may include dynamic detection of the FFT size and the channel bandwidth employed by the BS.

3. In the current mobile WiMAX [6,7], the only mandatory ratio of CP is 1/8, and the only FFT sizes specified in the band class index table are 1024 and 512.

22.2.1.2 Cluster

1. *Resemblance between IEEE 802.16-2004 and IEEE 802.16e-2005 [1,2]:* A physical cluster contains 14 adjacent subcarriers, and is made of data and pilot subcarriers. The index of each physical cluster may have different logical indexes.
2. *Difference between IEEE 802.16-2004 and IEEE 802.16e-2005 [1,2]:* The data and pilot subcarrier allocations within a cluster for even and odd OFDMA symbols are different between IEEE 802.16-2004 and IEEE 802.16e-2005.

22.2.1.3 Slot

1. *Resemblance between IEEE 802.16-2004 and IEEE 802.16e-2005 [1,2]:* A slot requires both a time and subchannel dimension for completeness, and is the minimum possible data allocation unit. The definition of an OFDMA slot depends on the OFDMA symbol structure, which varies for uplink (UL) and DL, for full usage of subchannels (FUSC) and partial usage of subchannels (PUSC), and for the distributed subcarrier permutations and the adjacent subcarrier permutation. The number of data subcarriers per slot is 48.
   (a) *Under distributed subcarrier permutations:* For DL PUSC using the distributed subcarrier permutation, one slot is one subchannel by two OFDMA symbols. For DL FUSC and DL optional FUSC using the distributed subcarrier permutation, one slot is one subchannel by one OFDMA symbol. For UL PUSC and UL optional PUSC using the distributed subcarrier permutations, one slot is one subchannel by three OFDMA symbols.
   (b) *Under adjacent subcarrier permutations:* In the adaptive modulation and coding (AMC) zone, one slot consists of six contiguous bins in one OFDMA symbol, where a bin is defined in Section 22.2.1.
2. *Difference between IEEE 802.16-2004 and IEEE 802.16e-2005:* In IEEE 802.16e-2005, for AMC zone, one slot may additionally consist of two bins by three OFDMA symbols, three bins by two OFDMA symbols, or one bin by six OFDMA symbols. Optional DL tile usage of the subchannels (TUSC), TUSC1 and TUSC2, are newly introduced in IEEE 802.16e-2005. One slot of TUSC is one subchannel by three OFDMA symbols.

22.2.1.4 Subchannel

1. *Resemblance between IEEE 802.16-2004 and IEEE 802.16e-2005:* Subchannels consist of data subcarriers in slots.
2. *Difference between IEEE 802.16-2004 and IEEE 802.16e-2005:* There are some allocation and subcarrier permutation differences for slots between IEEE 802.16-2004 and IEEE 802.16e-2005, so the allocation of data subcarriers in the subchannel may be different. Since there are newer types of slots introduced, there are newer constructions of subchannels in IEEE 802.16e-2005.

22.2.1.5 Several Elements with Similar Definitions in IEEE 802.16-2004 and IEEE 802.16e-2005

1. *Segment:* A segment is a subdivision of available subchannels.
2. *Data region:* A data region is a two-dimensional allocation of a group of contiguous sub-channels, in a group of contiguous OFDMA symbols. All the allocations refer to logical subchannels. A two-dimensional allocation may be visualized as a rectangle.

3. *Bin:* A bin is a set of nine contiguous subcarriers within an OFDMA symbol.
4. *Tile:* Tiles are frequency–time blocks. There are two types of tiles:
   (a) Optional PUSC, 3×3 frequency–time block containing nine subcarriers—one pilot/null tone and eight data tones.
   (b) PUSC, 4×3 frequency–time block containing 12 subcarriers—four pilot tones (or two pilot tones and two null/data tones) and eight data tones.
5. *Burst:* Burst is an OFDMA data region to which a MAC data should be mapped. A burst consists of slots. Note that a burst may consist of slots from multiple transmit antennas [for instance, in the case of vertical encoding of spatial multiplexing (SM) mode, as discussed in Section 22.2.3].

22.2.1.6 OFDMA Data Mapping in IEEE 802.16e-2005

After channel coding, interleaving, and complex symbol mapping, the data symbols shall be mapped to an OFDMA data region for DL and UL:

1. *DL:*
   (a) Partition of the data into blocks sized to fit into one OFDMA slot.
   (b) Each slot shall span one subchannel in the subchannel axis and one or more OFDMA symbols in the time axis. Map the slots so that the lowest numbered slot occupies the lowest numbered subchannel in the lowest numbered OFDMA symbol.
   (c) Continue the mapping allocated slots to the corresponding bursts along increasing sub-channel indices. When the edge of the data region is reached, continue the mapping from the lowest numbered subchannel in the next available OFDMA symbol.
2. *UL:*
   (a) The OFDMA slots allocated to each burst are selected.
      i. The initial procedure is similar to the first two steps of DL slot allocation except using different slot structures.
      ii. Continue allocating slots (skipping allocations made with UIUC (uplink interval usage code) = 0, 12, 13, where UIUC = 0 refers to fast feedback, UIUC = 12 code division multiple access (CDMA) ranging and bandwidth request (BR) allocations, UIUC = 13 PAPR (peak-to-average-power-ratio)/safety zone allocations). When the edge of the UL zone (which is marked with Zone_IE) is reached, continue allocating from the lowest numbered OFDMA symbol in the next available subchannel.
      iii. An UL allocation is created by selecting an integer number of contiguous slots, according to the previously determined ordering.
   (b) The allocated slots are mapped to the burst.
      i. Map the slots so that the lowest numbered slot occupies the lowest numbered sub-channel in the lowest numbered OFDMA symbol.
      ii. Continue the mapping allocated slots to the corresponding bursts along increasing subchannel indices. When the last subchannel is reached, continue the mapping from the lowest numbered subchannel in the next OFDMA symbol that belongs to the UL allocation.

22.2.1.7 Frame, Subframe, and Zone

Each OFDMA frame is a frequency–time plane, including multiple OFDMA symbols, and is divided into DL and UL subframes, separated by transmit/receive and receive/transmit transition

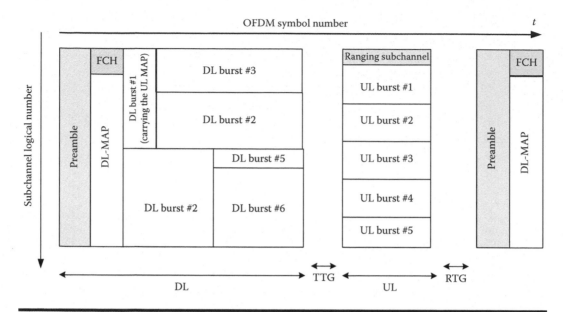

Figure 22.1 An example of the OFDMA frame (with only a mandatory zone) in TDD mode. (From IEEE Std 802.16e™ -2005 and IEEE Std 802.16™-2004/Cor1-2005, February 2006. With permission.)

gaps (TTG and RTG), respectively, where

1. TTG is a gap between the DL burst and the subsequent UL burst in a TDD transceiver. During TTG, BS switches from transmit to receive mode and SSs switch from receive to transmit mode.
2. RTG is a gap between the uplink burst and the subsequent DL burst in a TDD transceiver. During RTG, BS switches from receive to transmit mode and SSs switch from transmit to receive mode.

TTG and RTG in the mobile WiMAX system profile Release 1.0 [6,7] support a maximum cell size of approximately 20.7 km for 3.5 or 7 MHz bandwidth, and 8.4 km for 5 or 10 MHz bandwidth mobile WiMAX systems [8].

As illustrated in Figure 22.1, a TDD OFDMA frame may include a preamble for synchronization, a Frame control header (FCH) for providing the frame configuration information,* DL-MAP,† and UL-MAP‡ for offering subchannel allocation and other control information for the DL and UL subframes,§ respectively, UL ranging for MSs to perform closed-loop time, frequency, and power

---

* FCH is the first message at the beginning of the second DL OFDM symbol. FCH provides the required information to decode the subsequent DL-MAP message. The FCH provides information like the subchannels being used by the sector in the current frame, coding, and size of the DL-MAP [8].
† The DL-MAP message defines the usage of the DL intervals for a burst mode PHY.
‡ The UL-MAP message defines the UL usage in terms of the offset of the burst relative to the Allocation Start Time (units PHY-specific).
§ Each MAP message embraces several information elements (IEs), which may include a fixed part and a variable part. The size of the variable part proportionates to the number of DL and UL users scheduled in that frame [8].

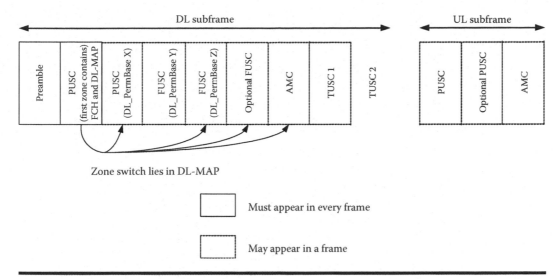

Figure 22.2 Illustration of the OFDMA frame with multiple zones. (From IEEE Std 802.16e™-2005 and IEEE Std 802.16™-2004/Cor1-2005, February 2006. With permission.)

adjustment as well as BRs, and UL CQICH (channel quality information channel) channel for the MS to feedback DL HARQ (hybrid automatic-repeat-request) acknowledgment.

As shown in Figure 22.2, the OFDMA frame may include multiple zones (such as PUSC and FUSC), and the transition between zones is indicated in DL-MAP by Zone_switch information element (IE). No DL-MAP or UL-MAP allocations can span over multiple zones. As noted in Ref. [8], mobile WiMAX may support variable frame sizes (e.g., 2, 2.5, and 5 ms).

22.2.1.8 Subburst

Subburst is defined in IEEE 802.16e-2005 instead of IEEE 802.16-2004. A burst allocation is further partitioned into subbursts, by allocating a specified number of slots to each subburst. All subbursts of a data region shall only support one of the HARQ modes. The slots are allocated in a frequency first order, starting from the slot with the smallest symbol number and smallest subchannel number, and continuing to slots with increasing subchannel number. When the edge of the allocation is reached, the symbol number is increased by a slot duration. Each subburst is separately encoded. Note that a subburst may consist of slots from multiple transmit antennas, for example in the case of vertical encoding of SM mode.

22.2.2 Subchannelization

1. Resemblance between IEEE 802.16-2004 and IEEE 802.16e-2005:
   (a) For both DL and UL, the used subcarriers are allocated to pilot subcarriers and data subcarriers. However, there is a difference between different possible zones.
   (b) Subchannel allocation may be performed in the following ways: PUSC where some of the subchannels are allocated to the transmitter and FUSC where all subchannels are allocated to the transmitter. For DL FUSC, the pilot tones are allocated first; what remains are data subcarriers, which are divided into subchannels that are used exclusively for data.

For UL PUSC, the set of used subcarriers is first partitioned into subchannels, and then the pilot subcarriers are allocated within each subchannel. Thus, in FUSC, there is one set of common pilot subcarriers.

(c) In UL PUSC and UL optional PUSC, partition all the tiles to multiple groups. Each slot consists of six tiles, each of which is chosen from different tile groups.

(d) A group of bins is called a logical band. As optional adjacent subcarrier permutations for AMC, band-AMC allocations use a logical band with six contiguous bins in one OFDMA symbol as an AMC slot.

2. Difference between IEEE 802.16-2004 and IEEE 802.16e-2005:

(a) *IEEE 802.16-2004:* For DL PUSC, the set of used subcarriers is first partitioned into subchannels, and then the pilot subcarriers are allocated within each subchannel. In DL PUSC, each subchannel contains its own set of pilot subcarriers.

(b) *IEEE 802.16e-2005:*

i. For DL PUSC, the pilot tones are allocated first; what remains are data subcarriers. In DL PUSC, there is one set of common pilot subcarriers in each major group, but in UL PUSC, each subchannel contains its own set of pilot subcarriers.

ii. There are two kinds of optional DL tile usage of subchannels, TUSC1 and TUSC2: The symbol structure of TUSC1 is similar to that of UL PUSC [2], while the symbol structure of TUSC2 is similar to that of UL optional PUSC [2], and thus TUSC1 and TUSC2 are of different frequency–time block sizes. For both TUSC1 and TUSC2, each transmission uses 48 data subcarriers as the minimal block of processing; the active subchannels shall be renumbered consecutively starting from 0; the pilots in permutation are regarded as part of the allocation and as such shall be beamformed in a way that is consistent with the transmission of the allocated data subcarriers. Both TUSC1 and TUSC2 permutation shall only be used within an adaptive antenna system (AAS) zone. TUSC1/TUSC2 may be used in the support for spatial division multiple access (SDMA). In this case, the pilots in an AAS zone with TUSC1 or TUSC2 permutation are regarded as part of the allocation, and as such shall be beamformed in a way that is consistent with the transmission of the allocated data subcarriers. In an SDMA region, the pilots of each allocation may correspond to a different pilot pattern.

iii. Additional AMC adjacent subcarrier permutations using additional three kinds of slot construction as discussed in Section 22.2.1.

iv. Two optional permutations for PUSC adjacent subcarrier allocation (PUSC-ASCA) are introduced: one using adjacent clusters and the other using distributed clusters.

3. In the current mobile WiMAX system profile Release 1.0 [7], mandatory support of permutation zones includes

(a) For the single transmit antenna

i. Preamble (BS required)

ii. *DL:* PUSC, PUSC with all subchannels, PUSC with dedicated pilots (BS required for IO-BF,* MS required), FUSC, AMC 2 × 3, AMC 2 × 3 with dedicated pilots (BS required for IO-BF, MS required)

iii. *UL:* PUSC, AMC 2 × 3 with dedicated pilots (BS required for IO-BF, MS required), AMC 2 × 3

---

* IO-BF: Group of inter-operable option features related to beam forming (BF) operation, required for "WiMAX certified with BF capability" [6].

(b) For space–time coding (STC)/multiple-input multiple-output (MIMO).
   i. Preamble (BS required)
   ii. *DL:* Two-antenna matrix A* using PUSC (IO-MIMO[†] for BS, required for MS), two-antenna matrix B[‡] with vertical encoding using PUSC (BS required for IO-MIMO, BS required)
   iii. *UL:* Collaborative SM for two MSs with a single transmit antenna using PUSC (BS required for IO-MIMO, MS required),

where permutation zones without following bracketed comments are required for both BS and MS.

22.2.3 *Construction of Permutation Zones*

This subsection introduces several typical permutation zones: DL preamble, DL PUSC, UL PUSC, and AMC for both single and two transmit antennas. This section mainly focuses on PUSC of FFT sizes 1024 and 512, since those two FFT sizes are parts of band classes in Ref. [6] and are currently widely adopted in industry.

22.2.3.1 *DL Preamble*

The DL can be divided into a three-segment structure and includes a preamble that begins the transmission. The preamble subcarriers are divided into three carrier-set. There are three possible groups, each consisting of a carrier-set, that may used by any segment.

The first symbol of the DL transmission is the preamble. There are three types of preamble carrier-sets, which are defined by allocation of different subcarriers for each one of them. Those subcarriers are modulated using a boosted BPSK modulation with a specific pseudo-noise (PN) code. The preamble carrier-sets are defined using the following equation

$$\mathbf{P}_n = \{n + 3k\}$$

where $\mathbf{P}_n$ is the subcarrier index set for the n-th segment of the preamble, $n = 0, \ldots, 2$, k is the running subcarrier index within each segment,

1. FFT of size 2048, $k = 0, \ldots, 567$
2. FFT of size 1024, $k = 0, \ldots, 283$
3. FFT of size 512, $k = 0, \ldots, 142$
4. FFT of size 128, $k = 0, \ldots, 35$

---

* For two transmit antennas, Matrix A is referred to space–time codes or space–frequency codes (frequency hopping diversity coding or FHDC) based on orthogonal space–time block codes (OSTBC) [2,14]. For three or four transmit antennas, Matrix A is referred to space–time–frequency codes (STFC) based on the structure of OSTBC [14,15].

† IO-MIMO: Group of interoperable option features related to MIMO operation, required for "WiMAX certified with MIMO capability" [6].

‡ For two transmit antennas, Matrix B is mainly referred to SM [2,16], while Matrix B is referred to a hybrid switching scheme between OSTBC and SM in closed-loop MIMO DL enhanced mode. For three or four transmit antennas, Matrix B may be also referred to quasi-orthogonal STFC (with a higher data rate than OSTBC-based STFC) using OSTBC as component codes [2].

Each segment uses a preamble composed of one carrier-set out of the three available carrier-sets in the following manner:

1. Segment 0 uses preamble carrier-set 0
2. Segment 1 uses preamble carrier-set 1
3. Segment 2 uses preamble carrier-set 2

Therefore, each segment eventually modulates one-third of the subcarriers of each preamble symbol. In the case of segment 0, the DC carrier will not be modulated at all, where a DC carrier is defined as the subcarrier whose frequency would be equal to the radio frequency (RF) center frequency of the station, and the appropriate PN will be discarded. Therefore, the DC carrier shall always be zeroed. The number of guard band subcarriers for different FFT sizes is

1. *1024-FFT size:* 86 guard band subcarriers on the left side and the right side of the spectrum
2. *512-FFT size:* 42 guard band subcarriers on the left side and 41 guard band subcarriers on the right side of the spectrum
3. *128-FFT size:* 10 guard band subcarriers on the left side and the right side of the spectrum

The PN series modulating the preamble carrier-set in a hexadecimal format is defined in Table 309a-309c of Ref. [2]. The series modulated depends on the segment used and IDcell parameter, where segment index may refer to sector index of each cell and IDcell index may refer to cell or BS index in cellular systems. The defined series shall be mapped onto the preamble subcarriers in ascending order. The value of the PN is obtained by converting the series to a binary series (W_k) and then mapping the PN from the MSB of each symbol to the LSB (0 mapped to +1 and 1 mapped to 1; for example for Index = 0, segment = 0, $W_k = 110000010010\ldots$, and the mapping shall follow: $-1 - 1 + 1 + 1 + 1 + 1 + 1 - 1 + 1 + 1 - 1 - 1\ldots$).

22.2.3.2 DL PUSC for a Single Transmit Antenna

The DL PUSC symbol structure is constructed using pilots, data, and zero (null) subcarriers:

1. Guard band and DC carriers are allocated.
2. The rest of the subcarriers are divided into the number of physical clusters containing 14 adjacent subcarriers each (starting from carrier 0), and pilots and data subcarriers are allocated within each physical cluster. The number of physical clusters per OFDMA symbol, N clusters, varies with FFT sizes:
   (a) N clusters = 60 for 1024 FFT
   (b) N clusters = 30 for 512 FFT
   The permutation of pilots and data subcarriers within each physical cluster depends on the OFDMA symbol indices counted from the beginning of the current zone, and the first symbol in the zone is even. The preamble shall not be counted as part of the first zone. For the case of a single transmit antenna, the permutation of pilots and data subcarriers within physical clusters for even and odd OFDMA symbols is illustrated in Figure 22.3.
3. The physical clusters are renumbered into logical clusters. If the physical clusters are placed in the first DL zone or "Use All SC Indicator" = 0 in the STC_CL_Zone_IE, then *LogicalCluster = RenumberingSequence(PhysicalCluster)*. Otherwise, *LogicalCluster = RenumberingSequence*[(*PhysicalCluster*) + 13 × *DL_PermBase*] mod N clusters, where *DL_PermBase* is an integer ranging from 0 to 31, which is set to preamble IDCell in the

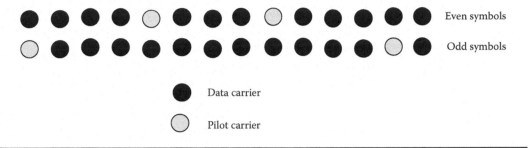

Figure 22.3 Cluster structure. (From IEEE Std 802.16e™-2005 and IEEE Std 802.16™-2004/Cor1-2005, February 2006. With permission.)

first zone and determined by the DL-MAP for other zones. When "Use All SC indicator" is set to 1, this indicator indicates transmission on all available subchannels for PUSC; for other permutations, this field shall be ignored. The *RenumberingSequence* of 60 clusters for 1024 FFT is 6, 48, 37, 21, 31, 40, 42, 56, 32, 47, 30, 33, 54, 18, 10, 15, 50, 51, 58, 46, 23, 45, 16, 57, 39, 35, 7, 55, 25, 59, 53, 11, 22, 38, 28, 19, 17, 3, 27, 12, 29, 26, 5, 41, 49, 44, 9, 8, 1, 13, 36, 14, 43, 2, 20, 24, 52, 4, 34, 0. The *RenumberingSequence* of 30 clusters for 512 FFT is 12, 13, 26, 9, 5, 15, 21, 6, 28, 4, 2, 7, 10, 18, 29, 17, 16, 3, 20, 24, 14, 8, 23, 1, 25, 27, 22, 19, 11, 0.

4. *Allocate logical clusters to major groups:* The allocation algorithm varies with FFT sizes. Table 22.1 illustrates how to construct major groups from logical clusters for FFT size 1024 and 512. These groups may be allocated to up to three segments, which may refer to sectors of a cell in cellular systems. If a segment is being used, then at least one group shall be allocated to it.

5. Allocating the subcarriers to subchannels in each major group is performed separately for each OFDMA symbol, that is to say, one subchannel does not contain subcarriers in more than one major group. First, we allocate the pilot carriers within each cluster, and then allocate all the remaining data carriers within each major group of each OFDMA symbol into subchannels according to

$$
\begin{aligned}
\text{subcarrier}(k, s) = {}& N_{\text{subchannels}} \cdot n_k \\
&+ \left\{ p_s \left[n_k \bmod N_{\text{subchannels}} \right] + DL\_PermBase \right\} \bmod N_{\text{subchannels}}, \quad (22.1)
\end{aligned}
$$

where

(a) $s = 0, \ldots, N_{\text{subchannels}} - 1$ and $k = 0, \ldots, N_{\text{subcarriers}} - 1$, where $N_{\text{subchannels}}$ is the number of subchannels in the current major group, and $N_{\text{subcarriers}}$ is the number of subcarriers per subchannel per OFDMA symbol

(b) subcarrier(k, s) is the logical data subcarrier index in the current major group for the kth logical data subcarrier within the s-th subchannel in the current major group, and subcarrier$(k, s) = 0, \ldots, N_{\text{subchannels}} \cdot N_{\text{subcarriers}}$

(c) $n_k = (k + 13s) \bmod N_{\text{subcarriers}}$

(d) $p_s[a]$ is the ath element of the series obtained by rotating the basic permutation sequence cyclically to the left s times

The parameters in Equation 22.1 vary with FFT sizes:

1. For FFT size = 1024

Table 22.1 PUSC Major Group Permutation

FFT Size	Group Index	Group 0	Group 1	Group 2	Group 3	Group 4	Group 5
1024	Logic cluster indices	0–11	12–19	20–31	32–39	40–51	52–59
	Number of logic clusters	12	8	12	8	12	8
	Number of data subcarriers	144	96	144	96	144	96
	Number of subchannels per group	6	4	6	4	6	4
	Number of major groups	6					
512	Logic cluster indices	0–9	N/A	10–19	N/A	20–29	N/A
	Number of logic clusters	10	N/A	10	N/A	10	N/A
	Number of data subcarriers	120	N/A	120	N/A	120	N/A
	Number of subchannels per group	5	N/A	5	N/A	5	N/A
	Number of major groups	3					

(a) For even-numbered major groups, $N_{subchannels} = 6$, $N_{subcarriers} = 24$, p_0 basic permutation sequence, *PermutationBase6*, is 3,2,0,4,5,1

(b) For odd-numbered major groups, $N_{subchannels} = 4$, $N_{subcarriers} = 24$, p_0 basic permutation sequence, *PermutationBase4*, is 3,0,2,1

2. For FFT size = 512, all major groups are even numbered, $N_{subchannels} = 5$, $N_{subcarriers} = 24$, p_0 basic permutation sequence, *PermutationBase6*, is 4,2,3,1,0

22.2.3.3 DL PUSC for STC Using Two Transmit Antennas

In IEEE 802.16, the term "space time coding" (STC) refers to both orthogonal space-time codes, also called space–time transmit diversity (STTD), and SM. Transmitters having two transmit antennas may perform SM using either horizontal coding or vertical coding. For horizontal coding, which has evolved from the concept of H-BLAST [16], two bursts are first individually modulated and then transmitted one per antenna (first burst on antenna #0 and, the second burst on antenna #1). For vertical coding, which has evolved from the concept of V-BLAST [16–18], a single burst is modulated and then transmitted according to the mapping order.

Unlike the physical cluster structure of size 14 subcarriers by two OFDMA symbols for single transmit antenna DL PUSC, the physical cluster structure for STC using two transmit antennas is of size 14 subcarriers by four OFDMA symbols as shown in Figure 22.4. For the same OFDMA symbol period, the data subcarrier allocation of pair clusters in two transmit antennas is the same, while the rest of subcarriers, pilot and null subcarriers, are differently allocated for pair clusters

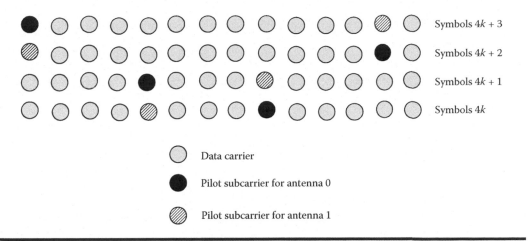

Figure 22.4   Cluster structure for STC PUSC using two antennas. (From IEEE Std 802.16e™-2005 and IEEE Std 802.16™-2004/Cor1-2005, February 2006. With permission.)

in two transmit antennas, and the pilot locations change in a period of four OFDMA symbols. OFDMA symbols are counted from the beginning of the current zone. The first symbol in the zone is even. STC encoding is performed on each pair of symbols $2n, 2n + 1 (n = 0, 1, \ldots)$.

22.2.3.4 UL PUSC for a Single Transmit Antenna

The UL follows the DL model; therefore it also supports up to three segments. Unlike DL PUSC, tile instead of cluster becomes the minimal subcarrier allocation block. The subchannel is constructed using six UL tiles. As shown in Figure 22.5, each tile has four successive active subcarriers.

The following is the construction procedure of UL PUSC symbol structure.

1. Guard band and DC carriers are allocated.
2. The usable subcarriers in the allocated frequency band shall be divided into *Ntiles* physical tiles as illustrated in Figure 22.5.

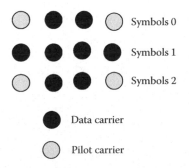

Figure 22.5   Illustration of a UL tile of size 3 × 4. (From IEEE, IEEE 802.16-2004 standard for local and metropolitan area networks Part 16: Air interface for fixed broadband wireless access systems, October 2004. With permission.)

3. Given a subchannel index and a logical tile index, the corresponding physical tile index can be calculated using

$$Tiles(s, n) = N_{\text{subchannels}} \cdot n$$
$$+ \left\{ p_t \left[(s + n) \bmod N_{\text{subchannels}} \right] + UL\_PermBase \right\} \bmod N_{\text{subchannels}}, \quad (22.2)$$

where

(a) n is the logical tile index, $n = 0, \ldots, 5$, in the current s-th subchannel, where $s = 0, \ldots, N_{\text{subchannels}} - 1$, and $N_{\text{subchannels}}$ is the total number of subchannels, depending on FFT sizes;

(b) *Tiles*(s, n) is the physical tile index in the FFT with tiles being ordered consecutively from the most negative to the most positive used subcarrier (0 is the starting tile index);

(c) p_t is the tile permutation;

(d) *UL_PermBase* is an integer value in the range $0, \ldots, 69$, which is assigned by a management entity.

4. The data subcarriers per slot are enumerated by the following process:

(a) After the pilot carriers are allocated in each tile, the data subcarriers within each slot are indexed starting from the first symbol at the lowest indexed subcarrier of the lowest indexed tile, continuing in an ascending manner through the subcarriers in the same symbol, then going to the next symbol at the lowest indexed data subcarrier, and so on. Data subcarriers shall be indexed from 0 to 47.

(b) Map data symbols onto subcarriers according to the permutation

$$\text{subcarrier}(n, s) = (n + 13 \cdot s) \bmod N_{\text{subcarriers}} \quad (22.3)$$

i. n is the data logical subcarrier index in the current s-th indexed subchannel, where $n = 0, \ldots, 47$

ii. $N_{\text{subcarriers}} = 48$ is the number of subcarriers per slot

iii. subcarrier(n, s) is the permutated subcarrier index corresponding to data subcarrier n in subchannel s

The parameters in Equation 22.2 vary with FFT sizes:

1. For FFT size $= 1024$, $N_{\text{subchannels}} = 35$, p_t permutation sequence, *TilePermutation*, is 11,19,12,32,33,9,30,7,4,2,13,8,17,23,27, 5,15,34,22,14,21,1,0,24,3,26,29,31,20,25, 16, 10,6,28,18. The total number of tiles is 210.

2. For FFT size $= 512$, all major groups are even numbered, $N_{\text{subchannels}} = 17$, p_t permutation sequence, *TilePermutation*, is 11,15,10,2,12,9,8,14,16,4,0,5,13,3,6,7,1. The total number of tiles is 102.

Note that UL optional PUSC as well as TUSC2 uses 3×3 tile structure, as shown in Figure 22.6.

22.2.3.5 UL PUSC for Space–Time Coding Using Two Transmit Antennas and Collaborative SM Using a Single Transmit Antenna per MS

The data subcarrier allocation in UL PUSC physical tile structure per antenna for STC using two transmit antennas is the same as that in UL PUSC physical tile structure for a single transmit

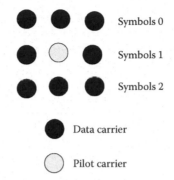

Figure 22.6 Illustration of an uplink tile of size 3 × 3. (From IEEE, IEEE 802.16-2004 standard for local and metropolitan area networks Part 16: Air interface for fixed broadband wireless access systems, October 2004. With permission.)

antenna as shown in Figure 22.7. However, the pilots in each tile shall be split between the two antennas for STC.

Two MSs, MS #1 and MS #2, using a single transmit antenna can perform collaborative SM onto the same subcarrier. In this case, MS #1 shall use the UL tile with pattern A, and MS #2 shall use the UL tile with pattern B. The pilot patterns are as shown in Figure 22.7. Transmit data for both MS #1 and MS #2 shall be coded, interleaved, modulated, and mapped to time/frequency as in the non-MIMO case.

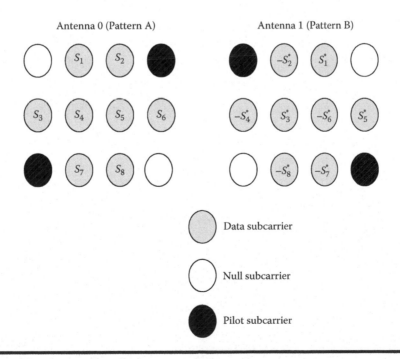

Figure 22.7 Mapping of data subcarriers in STTD tile mode. (From IEEE Std 802.16e™-2005 and IEEE Std 802.16™-2004/Cor1-2005, February 2006. With permission.)

To perform SM with either vertical or horizontal coding, an MS needs to signal both of its antennas. To this end, the subscriber uses both pilot patterns A and B. Antenna #0 shall be signaled using pattern A, and antenna #1 using pattern B. For non-MIMO transmissions, only antenna #0 shall be used.

22.2.3.6 AMC for a Single Transmit Antenna

In AAS zones, AMC subchannels based on contiguous subcarriers are mainly considered. Compared with distributed subcarrier permutation, contiguous subcarriers may exploit much less frequency diversity, while using subchannel with contiguous subcarriers may greatly facilitate multiuser diversity approaches to significantly improve the overall system capacity and throughput. Multiuser diversity may take advantage of the dynamic nature of the wireless channels, which leads to different AMC subchannels for different users with different SNRs in time.

In an AMC-based AAS zone, symbol data within a subchannel are assigned to adjacent subcarriers and the pilot and data subcarriers are assigned fixed positions in the frequency domain within an OFDMA symbol. This permutation is the same for both UL and DL. Table 22.2 shows the AMC subcarrier allocations for 1024-FFT, 512-FFT, and 128-FFT, respectively. Pilots are allocated first and separated from the data subcarriers. The indexing of pilot tones is done using the equation [2]: pilot subcarrier index = $9k + 3m + 1$ with m = [symbol index] mod 3, where

1. k is the pilot index within each OFDMA symbol,
   (a) for 1024 FFT, $k = 0, \ldots, 95$
   (b) for 1024 FFT, $k = 0, \ldots, 47$
   (c) for 1024 FFT, $k = 0, \ldots, 11$
2. OFDMA symbol of index 0, used in pilot subcarrier index calculation, should be the first symbol of the current zone, m is incremented only for data symbols, excluding preambles, safety zones, sounding symbols, midambles, and so on [2]
3. DC subcarrier is excluded when the pilot subcarrier index is calculated by the equation

Table 22.2   OFDMA AMC Subcarrier Allocations

FFT size	1024	512	128
Number of DC subcarriers	1	1	1
Number of left guard subcarriers	80	40	10
Number of right guard subcarriers	79	39	9
Number of pilot subcarriers	96	48	12
Number of data subcarriers	768	384	96
Number of physical bands	24	12	3
Number of bins per physical band	4	4	4
Number of data subcarriers per slot	48	48	48

An AMC subchannel of type $N \times M$ (where $N \times M = 6$) is defined as six contiguous bins (a slot consists of N bins by M symbols). AMC subchannel allocations are determnined by two mechanisms, where subchannels are differently indexed:

1. Regular/normal AMC allocations made by the DL-MAP or UL-MAP:
   (a) *Nonband AMC for normal DL/UL-MAP:* The subchannels are numbered from the lowest (0) to the highest frequency so that subchannel k ($k = 0,\ldots,(N_{\text{bins}}/N)$) consists of bins with indices $N \times k$ to $(N \times k) + N - 1$. In this case, there is only one type of AMC subchannel, which consists of two bins by three symbols.
   (b) *Band-AMC for normal DL/UL-MAP:* In this case, a logical band is a group of the physical AMC bands. In general, if $K = $ Max Logical Bands, then logical band $J = 0,\ldots,(K-1)$ contains physical bands $(M/K) \times J, ((M/K) \times J) + 1, \cdots, ((M/K) \times (J+1)) - 1$, where M is the number of physical AMC bands.
       i. For FFT sizes of 1024, the number of Max Logical Bands is defined as 12, and logical band 0 consists of AMC bands (0,1), logical band 1 consists of AMC bands (2,3), and logical band 2 consists of AMC bands (4,5).
       ii. For FFT sizes of 512 and 128, the number of Max Logical Bands is the same as the number of physical AMC bands, that is 12 and 3, respectively.
2. *Band AMC allocations made by HARQ MAP message:* In this case, an AMC slot consists of six contiguous bins in the same logical band. There are four types of AMC subchannels, which are different in the collection of six bins in a band. In the first type (default type in Ref. [2]) of AMC subchannel, a slot consists of six consecutive bins, and the enumeration of the available bins in a band starts from the lowest bin in the first symbol to the last bin in the symbol, then going to the lowest bin in the next symbol, and so on. The rest of the three type slots have been discussed in Section 22.2.1. In the last three types of AMC subchannel, enumeration of bins in a slot is the same as in the first type.

The logical index of the data subcarriers within an AMC slot can be numbered from 0 to 47. The index of the first data subcarrier in the first bin is 0, the next one is 1, and so on. The index of the data subcarriers increases along the subcarriers within a bin first and then along subcarriers within the next bin. The jth symbol of the 48 symbols where a band AMC slot is allocated is mapped onto the $S_{\text{per}}^{\text{off}}(j)$th subcarrier of a slot, where $j = 0,\ldots,47$, and $S_{\text{per}}^{\text{off}}(j)$ is as follows

$$S_{\text{per}}^{\text{off}}(j) = \begin{cases} P_{\text{per}}(j) + \text{off}, P_{\text{per}}(j) + \text{off} \neq 0, \\ \text{off}, P_{\text{per}}(j) + \text{off} = 0, \end{cases} \tag{22.4}$$

where
1. $P_{\text{per}}(j)$ is the jth element of the left cyclic shifted version of basic sequence P_0 by per
2. Basic sequence P_0 is defined in GF(7^2): 01, 22, 46, 52, 42, 41, 26, 50, 05, 33, 62, 43, 63, 65, 32, 40, 04, 11, 23, 61, 21, 24, 13, 60, 06, 55, 31, 25, 35, 36, 51, 20, 02, 44, 15, 34, 14, 12, 45, 30, 03, 66, 54, 16, 56, 53, 64, 10 in hepta-notation
3. per $= PermBase \bmod 48$
4. off $= (\lfloor PermBase/48 \rfloor) \bmod 49$
5. In the DL, *PermBase* shall be set to *DL_PermBase* specified in the preceding STC DL zone IE, while in the UL, *PermBase* shall be set to *UL_PermBase* specified in the preceding UL Zone IE

S_0	$-S_1^*$	S_{32}	$-S_{33}^*$	S_{64}	$-S_{65}^*$
Pilot		S_{34}	$-S_{35}^*$	S_{66}	$-S_{67}^*$
S_2	$-S_3^*$	S_{36}	$-S_{37}^*$	S_{68}	$-S_{69}^*$
S_4	$-S_5^*$	S_{38}	$-S_{39}^*$	S_{70}	$-S_{71}^*$
S_6	$-S_7^*$	S_{40}	$-S_{41}^*$	Pilot	
S_8	$-S_9^*$	S_{42}	$-S_{43}^*$	S_{72}	$-S_{73}^*$
S_{10}	$-S_{11}^*$	S_{44}	$-S_{45}^*$	S_{74}	$-S_{75}^*$
S_{12}	$-S_{13}^*$	Pilot		S_{76}	$-S_{77}^*$
S_{14}	$-S_{15}^*$	S_{46}	$-S_{47}^*$	S_{78}	$-S_{79}^*$
S_{16}	$-S_{17}^*$	S_{48}	$-S_{49}^*$	S_{80}	$-S_{81}^*$
Pilot		S_{50}	$-S_{51}^*$	S_{82}	$-S_{83}^*$
S_{18}	$-S_{19}^*$	S_{52}	$-S_{53}^*$	S_{84}	$-S_{85}^*$
S_{20}	$-S_{21}^*$	S_{54}	$-S_{55}^*$	S_{86}	$-S_{87}^*$
S_{22}	$-S_{23}^*$	S_{56}	$-S_{57}^*$	Pilot	
S_{24}	$-S_{25}^*$	S_{58}	$-S_{59}^*$	S_{88}	$-S_{89}^*$
S_{26}	$-S_{27}^*$	S_{60}	$-S_{61}^*$	S_{90}	$-S_{91}^*$
S_{28}	$-S_{29}^*$	Pilot		S_{92}	$-S_{93}^*$
S_{30}	$-S_{31}^*$	S_{62}	$-S_{63}^*$	S_{94}	$-S_{95}^*$

☐ Data subcarrier

☐ Pilot subcarrier

▨ Null subcarrier

Figure 22.8 Data mapping in the optional AMC Zone with a 2 Tx antenna using matrix A. (From IEEE Std 802.16e™-2005 and IEEE Std 802.16™-2004/Cor1-2005, February 2006. With permission.)

Note that the addition between two elements in $GF(7^2)$ is component-wise addition modulo 7 of two representations. For example, $(56) + (34)$ in $GF(7^2) = (13)$.

22.2.3.7 AMC for Two Transmit Antennas

In an AMC zone for two transmit antennas

1. All pilots in the even symbols shall be allocated for antenna 0, and Pilot Location for Antenna #0 $= 9k + 3[m \bmod 3] + 1$, where m is even
2. All pilots in the odd symbols shall be allocated for antenna 1, and Pilot Location for Antenna #1 $= 9k + 3[(m - 1) \bmod 3] + 1$, where m is odd

where m is the OFDMA symbol index, symbol index 0 is the the first symbol (except midamble) in which the STC zone is applied, and k is the pilot index defined in Section 22.2.3. The positions of pilots in the odd OFDMA symbols are further switched with those of data subcarriers whose locations coincide with pilots in the previous OFDMA symbol.

For the optional AMC permutation in the STC zone, the data subchannels shall adopt 2×6 (two bins for six symbols) format. The subcarrier permutation shown by Equation 22.4 shall not be applied for the optional AMC permutation within STC zones.

Antenna #0
with vertical encoding

S_0	S_{32}	S_{64}	S_{96}	S_{128}	S_{160}
Pilot		S_{66}	S_{98}	S_{130}	S_{162}
S_2	S_{34}	S_{68}	S_{100}	S_{132}	S_{164}
S_4	S_{36}	S_{70}	S_{102}	S_{134}	S_{166}
S_6	S_{38}	S_{72}	S_{104}	Pilot	
S_8	S_{40}	S_{74}	S_{106}	S_{136}	S_{168}
S_{10}	S_{42}	S_{76}	S_{108}	S_{138}	S_{170}
S_{12}	S_{44}	Pilot		S_{140}	S_{172}
S_{14}	S_{46}	S_{78}	S_{110}	S_{142}	S_{174}
S_{16}	S_{48}	S_{80}	S_{112}	S_{144}	S_{176}
Pilot		S_{82}	S_{114}	S_{146}	S_{178}
S_{18}	S_{50}	S_{84}	S_{116}	S_{148}	S_{180}
S_{20}	S_{52}	S_{86}	S_{118}	S_{150}	S_{182}
S_{22}	S_{54}	S_{88}	S_{120}	Pilot	
S_{24}	S_{56}	S_{90}	S_{122}	S_{152}	S_{184}
S_{26}	S_{58}	S_{92}	S_{124}	S_{154}	S_{186}
S_{28}	S_{60}	Pilot		S_{156}	S_{188}
S_{30}	S_{62}	S_{94}	S_{126}	S_{158}	S_{190}

 Data subcarrier

 Pilot subcarrier

 Null subcarrier

Figure 22.9   Data mapping in the optional AMC zone with a 2 Tx antenna using matrix B with vertical encoding. (From IEEE Std 802.16e™-2005 and IEEE Std 802.16™-2004/Cor1-2005, February 2006. With permission.)

The subcarrier mappings in the STC zone for two transmit antennas are described as follows:

1. *Matrix A:* Using the pattern of space time block codes [14], STC-encoded data symbols shall be time mapped starting from the first two OFDMA symbols. The subcarrier-to-zone mapping starts at the lowest numbered subcarriers of the lowest slot and continues in an ascending manner in subchannels first and then proceeds to the next two symbols in time. Figure 22.8 illustrates the mapping rule for antenna #0, assuming 2 Tx with Matrix A for a block of two slots.
2. *Matrix B:*
   (a) *Vertical encoding:* In this case, modulated data symbols shall be sequentially mapped for two transmit antennas along the subcarriers of the first symbol. The mapping continues in an ascending manner in subchannels first and then proceeds to the next symbol in time. For a block of two slots, Figure 22.9 illustrates the mapping rule for antenna #0.
   (b) *Horizontal encoding:* There are two separately encoded streams, and each encoded stream is separately mapped to the corresponding antenna. Figure 22.10 illustrates the mapping rule for antenna #0.

Antenna #0
with horizontal encoding

S_0^0	S_{16}^0	S_{32}^0	S_{48}^0	S_{64}^0	S_{80}^0
Pilot	▨	S_{33}^0	S_{49}^0	S_{65}^0	S_{81}^0
S_1^0	S_{17}^0	S_{34}^0	S_{50}^0	S_{66}^0	S_{82}^0
S_2^0	S_{18}^0	S_{35}^0	S_{51}^0	S_{67}^0	S_{83}^0
S_3^0	S_{19}^0	S_{36}^0	S_{52}^0	Pilot	▨
S_4^0	S_{20}^0	S_{37}^0	S_{53}^0	S_{68}^0	S_{84}^0
S_5^0	S_{21}^0	S_{38}^0	S_{54}^0	S_{69}^0	S_{85}^0
S_6^0	S_{22}^0	Pilot	▨	S_{70}^0	S_{86}^0
S_7^0	S_{23}^0	S_{39}^0	S_{55}^0	S_{71}^0	S_{87}^0
S_8^8	S_{24}^0	S_{40}^0	S_{56}^0	S_{72}^0	S_{88}^0
Pilot	▨	S_{41}^0	S_{57}^0	S_{73}^0	S_{89}^0
S_9^0	S_{25}^0	S_{42}^0	S_{58}^0	S_{74}^0	S_{90}^0
S_{10}^0	S_{26}^0	S_{43}^0	S_{59}^0	S_{75}^0	S_{91}^0
S_{11}^0	S_{27}^0	S_{44}^0	S_{60}^0	Pilot	▨
S_{12}^0	S_{28}^0	S_{45}^0	S_{61}^0	S_{76}^0	S_{92}^0
S_{13}^0	S_{29}^0	S_{46}^0	S_{62}^0	S_{77}^0	S_{93}^0
S_{14}^0	S_{30}^0	Pilot	▨	S_{78}^0	S_{94}^0
S_{15}^0	S_{31}^0	S_{47}^0	S_{63}^0	S_{79}^0	S_{95}^0

☐ Data subcarrier

☐ Pilot subcarrier

▨ Null subcarrier

Figure 22.10   Data mapping in the optional AMC zone with a 2 Tx antenna using matrix B with horizontal encoding. (From IEEE Std 802.16e™-2005 and IEEE Std 802.16™-2004/Cor1-2005, February 2006. With permission.)

22.3 Power Saving Mode

The normal power operating mode is awake mode, in which the MSs and BSs shall continuously process DL and/or UL traffics. In 802.16-2004, fixed wireless communications are mainly considered, and thus only awake mode is available and no power saving mechanisms are provided. To efficiently utilize battery resources of MSs, two power-efficient enhancement modes of IEEE 802.16e-2005 over IEEE 802.16-2004—sleep mode and idle mode [2] are introduced in IEEE 802.16e-2005. Both power-saving methods are considered in current Mobile WiMAX System Profile [6,7].

22.3.1 Sleep Mode

In the current mobile WiMAX system profile Release 1.0, sleep mode is a required feature for MSs [6,7]. Sleep mode is intended to minimize MS power usage and the usage of the serving BS air interface resources. The sleep mode also offers flexibility when the MS scans other BSs to collect information to assist handover during the sleep mode.

In the sleep mode, several concepts are listed as follows:

1. A sleep window is the predetermined amount of time for which an MS with active connections using one or more communication identifiers (CIDs) communicates with the BS to temporarily disrupt its connection over the air interface.
2. A listen window is the predetermined amount of time during which the MSs maintain an active connection with the BSs.
3. An unavailability interval is the period of time when all the MS connections are in their sleep windows, during which the MS cannot receive any DL transmission or send any UL transmission. During the unavailability interval, no transmission is done from the BS to the MS; therefore, the MS may power down one or more physical operation components or perform other activities without the need of communications with the BS.
4. An availability interval is a time interval that does not overlap with any unavailability interval. During the availability interval, the MS is able to receive all DL transmissions in the same way as in the state of normal operations. In addition, the MS shall examine the DCD and UCD change counts and the frame number of the DL-MAP PHY synchronization field to verify synchronization with the BS.

Each sleep window is followed by a listen window, and the MS goes through alternating sleep and listen windows for each connection.

IEEE 802.16e-2005 defined power-saving classes, based on the manner in which sleep mode is executed. In mobile WiMAX, Power-Saving Classes (PSCs) of type I are mandatory. If there is a connection at the MS, which is not associated with any active power-saving class, the MS shall be considered available on a permanent basis. The length of each sleep and listen window is negotiated between the MS and BS and is dependent on the current power-saving class of the sleep-mode operation. These three different classes are suitable for different service flows, as each flow has different requirements and methods for requesting bandwidth and sending data:

1. In PSC of type I, sleep windows are interleaved with listening windows of fixed size duration. The length of a sleep window is twice the length of the previous sleep window but no greater than a specified final value.

$$sleepwin = \min \left(2 \cdot prevSleepWin \cdot finalSleepWinBase \cdot 2^{finalSlpWinExp} \right), \quad (22.5)$$

where *sleepWin* is the sleep window, *prevSleepWin* is the previous sleep window, *finalSleepWinBase* is the final-sleep window base, and *finalSlpWinExp* is the final-sleep window exponent.

   The sleep window size is first calculated using the initial sleep window size and the final sleep window size. Once the sleep window size achieves the final sleep window size, all the subsequent sleep windows are of the same length. At any time during the sleep-mode operation, the BS can reset the window size to the initial sleep window size, and the process of doubling sleep window sizes is repeated. This class is typically recommended when the MS transmits best-effort or non-real-time traffic.

2. PSC of type II uses both fixed-length sleep windows and listening windows recommended for UGS service. Opposite to Power Saving Class type I, during listening windows of Power Saving Class type II, the MS may send or receive any MAC service data units (SDUs) or their

fragments at connections comprising the Power Saving Class as well as acknowledgments to them.

3. PSC of type III only uses a single sleep window. The start time, "Start frame number for first sleep window" and the length of the sleep window, specified as base/exponent, are indicated by the BS before entering this mode. After the expiration of the sleep window, power-saving class automatically becomes inactive. For multicast service, BS may conjecture when the next portion of data will appear. Then the BS allocates a sleep window for all time when it does not expect the multicast traffic to arrive. After expiration of the sleep window, available multicast data may be transmitted to relevant MSs. After that, the BS may choose to reactivate the power-saving class.

22.3.2 Idle Mode

Mobile WiMAX requires the support of general idle mode functionality [6,7]. Idle mode is to allow the MS to periodically turn off and become available for DL broadcast traffic messaging without registration at a specific BS as the MS traverses an air link environment populated by multiple BSs, typically over a large geographical area.

Idle mode enjoys even more power saving than sleep mode, since, during idle mode

1. An MS does not have the active requirement for handover (HO), and all normal operation requirements, and the only activity of the MS is to scan at discrete intervals, which support the MS conserving power and operational resources.
2. The network and BS obtain a simple and timely method for alerting the MS to pending DL traffic directed toward the MS, by eliminating the air interface and network HO traffic from essentially inactive MS.

In idle mode, the BSs are divided into logical groups called paging groups. Each paging group offers a contiguous coverage region in which the MS does not need to transmit in the UL, yet can be paged in the DL if there is traffic targeted at it. Without the concept of the paging area, the network would need to page the MSs in all the BSs within the entire network. The paging groups should be large enough so that most MSs will remain within the same paging group most of the time and small enough so that the paging overhead is reasonable. A BS is allowed to be a member of one or more paging groups, which provides support for not only the geographic requirements of idle mode operation but may also support differentiated and dynamic quality of service requirements and scalable load-balancing distribution.

In idle mode, an MS is assigned to a paging group by the BS before going into idle mode, and the MS periodically monitors the DL transmission of the network to determine the paging group of its current location. When the MS has decided to move to a new paging group, the MS periodically wakes up to perform a paging group update, during which it informs the network of the current paging group.

There are two concepts about MS paging periods:

1. *MS paging unavailable interval:* During this period, the MS may power down, scan neighbor BSs, reselect a preferred BS, conduct ranging, or perform other activities for which the MS will not guarantee availability to any BS for DL traffic. Should the MS reselect a preferred BS during this period, then the MS shall return to the MS Broadcast Paging message time synchronization stage.

2. *MS paging listening interval:* During this period, the MS shall scan, decode the DL channel descriptor (DCD) and DL-MAP, and synchronize on the DL for the preferred BS in time for the MS to begin decoding any BS Broadcast Paging message during the entire BS paging interval. At the end of the MS paging listening interval,

    (a) if the MS is paged in the broadcast paging message, the MS responds to the page and terminates its idle-mode operation;

    (b) if the MS does not elect to terminate the MS idle mode, the MS may return to the MS paging unavailable interval.

22.4 Handover

22.4.1 Introduction

IEEE 802.16-2004 was established for fixed broadband wireless systems and may allow slow movement over cells; however, it does not provide any support of handover [1,10]. Mobile WiMAX based on IEEE 802.16e-2005 specifies handovers for portability, simple mobility, and full mobility, of the users [6,7].

Handover (HO) is the process in which an MS migrates from the air interface provided by one BS to the air interface provided by another BS [2]. Two HO variants are defined in IEEE 802.16e-2005 [2]:

1. *Break-before-make HO:* An HO where service with the target BS starts after a disconnection of service with the previous serving BS, which is also called hard handover.
2. *Make-before-break HO:* An HO where service with the target BS starts before disconnection of the service with the previous serving BS, which is also called soft handover.

Mobile WiMAX and IEEE 802.16e-2005 accept three types of handover: hard handover (HHO), macro diversity handover (MDHO), and fast base station switching (FBSS) [2,6,7]. HHO, a type of break-before-make HO, is mandatory in mobile WiMAX systems [6,7], while MDHO and FBSS, two types of make-before-break HO, are optional [6,7].

Several handover-related concepts are introduced here [2]:

1. *Serving BS:* For any MS, the serving BS is the BS with which the MS has most recently completed registration at initial network entry or during a handover.
2. *Active BS:* An active BS is informed of the MS capabilities, security parameters, service flows, and full MAC context information. For MDHO, the MS transmits/receives data to/from all active BSs in the diversity set.
3. *Anchor BS:* For MDHO or FBSS supporting MSs, this is a BS where the MS is registered, synchronized, performs ranging, and monitors the DL for control information. For FBSS supporting MS, this is the serving BS that is designated to transmit/receive data to/from the MS at a given frame.
4. *Diversity set:* The diversity set contains a list of active BSs to the MS.
5. *Neighbor BS:* For any MS, a neighbor BS is a BS (other than the serving BS) whose DL transmission can be received by the MS.
6. *Scanning interval:* A time period intended for the MS to monitor neighbor BSs to determine the suitability of the BS as targets for handover.
7. *Target BS:* The BS that an MS intends to be registered with at the end of a handover.

22.4.2 Decision and Initiation of MDHO and FBSS

22.4.2.1 Diversity Set and Anchor BS

For an MS and a BS that supports MDHO/FBSS, both the MS and the BS shall maintain a diversity set. One of the BSs in the diversity set is defined as anchor BS. Regular HHO operation is a particular case of MDHO/FBSS with a diversity set consisting of a single BS, the anchor BS.

22.4.2.2 MDHO Decision

In MDHO mode, the MS communicates with all BSs in the diversity set for UL and DL unicast messages and traffic. An MDHO begins with a decision for an MS to transmit/receive unicast messages and traffic from multiple BSs at the same time interval. For DL MDHO, two or more BSs provide synchronized transmission of MS DL data such that diversity combining can be performed by the MS. For UL MDHO, the transmission from an MS is received by multiple BSs such that selection diversity of the information received by multiple BSs can be performed.

There are two ways for MS to monitor DL control elements and DL broadcast messages:

1. The MS monitors only the anchor BS for DL control information and DL broadcast messages.
2. The MS monitors all the BSs in the diversity set for DL control information and DL broadcast messages. In this case, the DL-MAP and UL-MAP of any active BS may contain burst allocation information for the other active BSs.

In MDHO, a decision is made for an MS to transmit/receive unicast messages and traffic from multiple BSs at the same time interval:

1. *DL:* Two or more BSs offer synchronized transmission of MS DL data such that the MS can perform diversity combining.
2. *UL:* The transmission from an MS is received by multiple BSs such that multiple BSs can carry out selection diversity of the received information.

22.4.2.3 FBSS Decision

In FBSS, an MS only communicates with the anchor BS for UL and DL messages including management and traffic connections, and transition from one anchor BS to another ("switching") is performed without invocation of the HO procedure. A FBSS handover begins with a decision for an MS to receive/transmit data from/to the anchor BS that may change within the diversity set.

22.4.2.4 Required Conditions to Enable MDHO/FBSS

There are several similar required conditions to enable MDHO/FBSS between MS and a group of BSs:

1. BSs involved in MDHO/FBSS
   (a) are synchronized based on a common time source
   (b) have a synchronized frame structure
   (c) have the same frequency assignment
   (d) are required to share or transfer MAC context
2. The frames sent by the involving BSs at a given frame time arrive at the MS within the prefix interval.

Since MDHO involves multiple BSs synchronized transmission, MDHO requires stronger extra system conditions than FBSS:

1. BSs involved in MDHO shall use the same set of CIDs for the connections that are established with the MS.
2. The same MAC/PHY PDUs shall be sent by all the BSs involved in MDHO to the MS.

22.4.3 Diversity Set Update for MDHO/FBSS

In the DCD message BS broadcasts for supporting MDHO/FBSS, there are two thresholds for updating the diversity set: H_Add and H_Delete. In FBSS/MDHO, an MS uses these thresholds to determine if diversity set update request message, MOB_MSHO-REQ, should be sent:

1. When long-term/mean carrier-interference-to-noise-ratio (CINR) of a serving BS is less than H_Delete threshold, the MS shall send MOB_MSHO-REQ to request dropping this serving BS from the diversity set.
2. When long-term/mean CINR of a neighbor BS is higher than H_Add threshold, the MS shall send MOB_MSHO-REQ to require adding this neighbor BS to the diversity set.

The decision to update the diversity set starts with a notification by the MS through the MOB_MSHO-REQ message or by the BS through the MOB_BSHO-REQ management message.

When MOB_MSHO-REQ is sent by an MS, the MS may provide a possible list of BSs to be included in the MS's diversity set. The MS may evaluate the possible list of BSs, and previously performed signal strength measurement, propagation delay measurement, scanning, ranging, and association activity. When MOB_BSHO-RSP is sent by the anchor BS or BSs in the MS's current diversity set, the BSs may provide a list of BSs recommended for incorporation into the MS's diversity set. When MOB_BSHO-REQ is sent by the anchor BS or BSs in the MS's current diversity set, the BSs may provide a recommended list of BSs to be included in the MS's diversity set. The criteria for the recommendation may be based on expected quality of service (QoS) performance to MS requirements and list of BSs that can be involved in MDHO/FBSS.

22.4.4 Anchor BS Update for MDHO/FBSS

Two mechanisms for the MS and BS to perform anchor BS update are provided in Ref. [2]: one using the HO messages, the handover MAC management method, and the other using the fast anchor BS selection feedback, the fast anchor BS selection mechanism.

22.4.4.1 HO MAC Management Message Method

With this method, the MS starts the anchor BS update process by reporting the preferred anchor BS with the MOB_MSHO-REQ message. The BS informs the MS of the anchor BS update through the MOB_BSHO-REQ or MOB_BSHO-RSP message with the estimated switching time, and then the MS shall update its anchor BS. The BS may reconfigure the anchor BS list and retransmit the MOB_BSHO-RSP or MOB_BSHO-REQ message to the MS. After an MS or BS has initiated an anchor BS update using MOB_MSHO/BSHO-REQ, the anchor BS update can be canceled by MS at any time. When switching to a new anchor BS within the MS's diversity set, the network entry procedures are not required and shall not be performed by the MS.

22.4.4.2 Fast Anchor BS Selection Feedback Mechanism

With this method, when the MS has more than one BS in its diversity set, the MS shall transmit fast anchor BS selection information to the current anchor BS using fast-feedback channel. If the MS wants to transmit anchor BS selection information, it transmits the codeword corresponding to the selected anchor BS via its fast-feedback channel. Fast-feedback channel shall be allocated by one of the following three methods:

1. Pre-allocated by MOB_BSHO-RSP or MOB_BSHO-REQ when a BS is added to the diversity set
2. Allocated through Anchor_Switch_IE during anchor switching operation
3. Allocated by UL-MAP of the new anchor BS after the switching period

The current anchor BS may send the Anchor_Switch_IE prior to the expiry of the switching timer to do one of the following:

1. Acknowledge the MS's switch indication and/or assign a CQICH at the new anchor BS, and/or specify a new action time when the switch shall occur, and/or specify a new anchor BS to switch to
2. Cancel the MS switching event

The responses of the MS may be based on the following conditions:

1. Whether or not the MS does not receive an Anchor_BS_switch_IE prior to the expiry of the switching timer
2. Whether or not new action time is specified
3. Whether or not the MS receives an Anchor_BS_Switch_IE with cancelation prior to the expiry of the switching timer

Based on those conditions, the MS may decide to switch to the new Anchor BS after the expiry of the switching timer or cancel the switching operation prior to the expiry of the switching timer. If the MS successfully decodes an Anchor_BS_Switch_IE, the MS shall acknowledge the reception of the IE using the allocated codeword over the CQICH. Prior to the expiry of the switching timer, the MS shall report the CQI of the current anchor BS and anchor switch indication on alternate frames.

22.5 Related Research and Investigations

As a hot topic in wireless communications, mobile WiMAX have attracted a lot of research and investigation in both industrial and academic communities. Instead of a thorough survey, this section only provides a brief review on some investigations and contributions to the OFDMA-based mobile WiMAX.

The authors of Ref. [19] focused on the major technical comparisons between IEEE 802.16e-2005-based Mobile WiMAX and 3G/WCDMA/HSDPA for mobile/portable broadband data services. Etemad offers a high-level review of mobile WiMAX technology and its evolution roadmap from both radio and network perspectives [12]. Wang et al. [8] provided an overview of the current mobile WiMAX system, its performance (under various configurations, channel conditions, and types of data traffic), and future mobile WiMAX evolution. Balachandran et al. [20], investigated the link and system-level performance on the DL of an IEEE 802.16e-2005-based OFDMA communication system and provided recommendations on high-performance IEEE 802.16e-2005 system design and deployment configurations. To improve system throughput as compared to the conventional subchannelization scheme such as IEEE 802.16e-2005 FUSC in various channel environments, Kim et al. [19] proposed a dynamic subchannelization scheme under multipath fading channels, which vary with a user's speed and leads. Das et al. [21] considered approaches to better support mobility in IEEE 802.16e-2005 and proposed hard handover with MAC context reset for a low mobility scenario and a layered migration procedure using fast base station switching (FBSS) with MAC context transfer for a high-mobility scenario and Internet Protocol (IP) packet forwarding for seamless mobility, and their handover solution is applicable to both flat and centralized network architectures [22]. Han et al. [23] and Kim et al. [24] have studied the energy efficiency design using power-saving mechanisms of IEEE 802.16e-2005. To shorten the frame delay and deal with the energy of the node effectively in IEEE 802.16e-2005 systems [25], Jung et al. provided an algorithm that efficiently reduces the next sleep window interval to the medium between the minimum and the maximum sleep window interval after the current sleep window reaches the maximum sleep window interval. For power-saving mode for a VoIP connection in the context of IEEE 802.16e-2005 standard, Lee and Cho [26], based on the network delay model, have developed an algorithm to find the most energy-efficient length of the sleep interval while satisfying the given delay constraint of the end-to-end VoIP connection.

22.6 Conclusion

This chapter introduces several PHY and MAC aspects of TDD OFDMA-based mobile WiMAX, including OFDMA frame structure and subchannelization, power-saving enhancements, and handover mechanisms. Through comparisons between IEEE 802.16-2004 and IEEE 802.16e-2005, this chapter gives a systematic survey of IEEE 802.16 standard updates from fixed broad wireless and mobile broad wireless communications. A brief review on research investigations and contributions to the OFDMA technology-based mobile WiMAX is also provided in this chapter. Mobile Wimax is an industrial standard with state-of-art technologies and a bright future. Mobile WiMAX Release 1.0 systems enable broadband internet access on a vehicle traveling at 120 km/h. After the mobile system profile Release 1.5, the next major Release of mobile WiMAX, Release 2.0, based on the next generation of IEEE 802.16 being developed in the 16m technical group (TGm) of 802.16, will target major enhancements in spectrum efficiency, latency, and scalability of the access technology to wider bandwidths in challenging spectrum environments [12]. It is expected that IEEE 802.16m and WiMAX Certification of Release 2 products would be formally completed in early 2010 and early 2011, respectively [12]. Release 2.0 systems may support vehicles with moving speed up to 350 km/h, and achieve 100 Mbit/s for high mobility and 1 GBit/s for stationary applications.

References

1. IEEE, IEEE 802.16-2004 standard for local and metropolitan area networks Part 16: Air interface for fixed broadband wireless access systems, October 2004.
2. IEEE Std 802.16e^{TM}-2005 and IEEE Std 802.16^{TM}-2004/Cor1-2005, IEEE 802.16e-2005 standard for local and metropolitan area networks Part 16: Air interface for fixed and mobile broadband wireless

access systems, amendment for physical and medium access control layers for combined fixed and mobile operation in licensed bands and corrigendum 1, February 2006.

3. ETSI, Broadband Radio Access Networks (BRAN); HiperMAN; System profiles, ETSI TS 102 210 version 1.2.1, January 2005.

4. ETSI, Broadband Radio Access Networks (BRAN); HiperMAN; Physical (PHY) layer, ETSI TS 102 177 version 1.4.1, November 2007.

5. ETSI, Broadband Radio Access Networks (BRAN); HiperMAN; Data Link Control (DLC) layer, ETSI TS 102 178 version 1.4.1, November 2007.

6. WiMAX Forum, *WiMAX Forum*^{TM} mobile system profile Release 1.0 approved specification (Revision 1.4.0), May 2007.

7. WiMAX Forum, *WiMAX Forum*^{TM} mobile system profile Release 1.0 approved specification (Revision 1.7.1), November 2008.

8. F. Wang, A. Ghosh, C. Sankaran, P. Fleming, F. Hsieh, and S. Benes, Mobile WiMAX systems: Performance and evolution, *IEEE Commun. Mag.*, 46 (10), 41–49, October 2008.

9. WiMAX Forum, Mobile WiMAX-Part I: A technical overview and performance evaluation, March 2006.

10. WiMAX Forum, Mobile WiMAX-Part II: A comparative analysis, April 2006.

11. WiMAX Forum, A comparative analysis of mobile WiMAX deployment alternatives in the access network, May 2007.

12. K. Etemad, Overview of mobile WiMAX technology and evolution, *IEEE Commun. Mag.*, 46 (10), 31–40, October 2008.

13. WiMAX Forum, The *WiMAX Forum Certified*^{TM} program driving the adoption of interoperable wireless broadband worldwide, September 2008.

14. S. Alamouti, A simple transmit diversity technique for wireless communications, *IEEE J. Selected Areas in Commun.*, 16 (8), 1451–1458, 1998.

15. V. Tarokh, H. Jafarkhani, and A.R. Calderbank, Space-time block codes from orthogonal designs, *IEEE Trans. Inform. Theory*, 45 (5), 1456–1467, July 1999.

16. G.J. Foschini, Layered space-time architecture for wireless communication in a fading environment when using multi-element antennas, *Bell Labs Tech. J.*, 1 (2), 41–59, 1996.

17. P.J. Wolniansky, C.J. Foschini, G.D. Golden, and R.A. Valenzuela, V-BLAST: An architecture for realizing very high data rates over the rich-scattering wireless channel, in *Proc. IEEE ISSSE*, 1998.

18. G.D. Golden, G.J. Foschini, R.A. Valenzuela, and P.W. Wolniansky, Detection algorithm, and initial laboratory results using the V-BLAST space-time communication architecture, *Electron. Letter.*, 35 (1), 14–16, January 1999.

19. Intel Corporation, Understanding WiMAX and 3G for portable/mobile broadband wireless—technical overview and comparison of WiMAX and 3G technologies, December 2004.

20. K. Balachandran, D. Calin, F.-C. Cheng, et al., Design and analysis of an IEEE 802.16e-Based OFDMA Communication System, *Bell Labs Tec. J.*, 11 (4), 53–73, November 2007.

21. J.-C. Kim, J.-Y. Hwang, and Y. Han, Dynamic subchannelization for adaptation to multipath fading in mobile WiMAX, in *Proc. IEEE Int. Symposium on Personal, Indoor and Mobile Radio Commun.*, September 2007.

22. S. Das, T. Klein, A. Rajkumar, et al., System aspects and handover management for IEEE 802.16e, *Bell Labs Tech. J.*, 11 (1), 123–142, May 2006.

23. Y.-H. Han, S.-G. Min, and D. Jeong, Performance comparison of sleep Mode Operations in IEEE 802.16e terminals, Lecture Notes in Computer Science, Springer Berlin/Heidelberg, July 2007.

24. B. Kim, J. Park, Y.-H. Choi, Power saving mechanisms of IEEE 802.16e: Sleep mode vs. idle mode, Lecture Notes in Computer Science, Springer Berlin/Heidelberg, November 2006.

25. W.J. Jung, H.J. Ki, T.-J. Lee, and M.Y. Chung, Adaptive sleep mode algorithm in IEEE 802.16e, in *Proc. Asia-Pacific Conf. Communications*, October 2007.

26. J. Lee and D. Cho, An optimal power-saving class II for VoIP traffic and its performance evaluations in IEEE 802.16e, *Comput. Commun.*, 31 (14), 3204–3208, September 2008.

27. Senza Fili Consulting on behalf of the WiMAX Forum, Fixed, nomadic, portable and mobile applications for 802.16-2004 and 802.16e WiMAX networks, November 2005.

Index

1G. *See* First-generation (1G)
"2A32PSK", 279, 280, 285
2D. *See* Two-dimensional (2D)
3G. *See* Third generation (3G)
3GPP. *See* Third Generation Partnership Project (3GPP)
3GPP-LTE, 358, 359, 366, 366, 508
"4 A16PSK," 278, 285
   modulation scheme, 279
   PCM1, 279, 280
   subconstellations, 279
4G. *See* Fourth-generation (4G)
64–APSK modulation, 278. *See also* APSK modulation
   "APSK1", 281
   "APSK2", 282
   "APSK3", 283
   "4A16PSK", 278–279
   "2A32PSK", 279–281

A

AAA. *See* Authentication, authorization and accounting (AAA)
AAS. *See* Advanced antenna systems (AAS)
Access service network (ASN), 570
ACK. *See* Acknowledgement (ACK)
Acknowledgement (ACK), 203
   packets, 567
Acquisition, 363, 364, 368, 376, 390. *See* Synchronization—procedure
   DL, 390
   frequency, 368, 382
   range, 381
   symbol-timing, 378–381
   timing, 380
Adaptation constant (μ), 462, 468
Adaptive modulation, 42, 191, 193, 302–303, 306, 307, 309–314, 318, 324
   adaptive schemes, 310
   algorithms, 309

approaching capacity, 303
   multiuser diversity, 305
Adaptive modulation and coding (AMC), 219, 572, 598
   allocations, 611
   AMC slot, 502
   band-AMC, 211, 602, 611
   link adaptation, 583
   nonband AMC, 611
   *opt*ional, 613
   permutation, 573
   single transmit antenna, 610
   slot, 602
   subchannel, 611
   subchannelization, 576
   two transmit antenna, 612
Adaptive power allocation (APA), 102, 111, 112
   greedy power allocation algorithm, 112
   Levin–Campello algorithm, 113
Adaptive transmission. *See* Adaptive modulation
Adaptive transmission concept, 240. *See also* Resource allocation algorithm
Additive Gaussian noise (AGN), 30
Additive white Gaussian noise (AWGN), 239, 303, 363, 442, 477, 510, 546
   channels, 74, 363
   waterfilling distribution, in, 304
Advanced antenna systems (AAS), 569, 572, 576–577
Advanced mobile phone system (AMPS), 2
AF mode. *See* Amplify-and-forward (AF) mode
AGC. *See* Automatic gain control (AGC)
AGN. *See* Additive Gaussian noise (AGN)
AM/AM. *See* Amplitude-to-amplitude conversion (AM/AM)
AMC. *See* Adaptive modulation and coding (AMC)
Amplify-and-forward (AF) mode, 474, 478
Amplitude and phase shift keying (APSK) modulation, 272
   factors influence, 283
   results for tested, 284
Amplitude-to-amplitude conversion (AM/AM), 55
Amplitude-to-phase conversion (AM/PM), 55
AM/PM. *See* Amplitude-to-phase conversion (AM/PM)

AMPS. *See* Advanced mobile phone system (AMPS)
Angular momentum operator, 548
Antenna, 236
    BS, 257
    channel matrix, 549
    compact antenna, 550–551
    independent ports, 548
    mapping, 587
    MIMO-cube, 548
    MIMO-octahedron, 548
    MIMO-tetrahedron, 548
    mutual coupling, 549
    orthogonalization, 549
    receiver, 32, 254
    spatial correlation, 549
    vector element, 547, 548, 550, 551, 553, 554
APA. *See* Adaptive power allocation (APA)
Application Working Group (AWG), 193
a priori, 366
"APSK1", 281. *See also* "4A16PSK"
"APSK2", 282
"APSK3", 283
APSK modulation. *See* Amplitude and phase shift keying
        (APSK) modulation
ARIB. *See* Association of Radio Industries and
        Broadcasting (ARIB)
ARQ. *See* Automatic Repeat Request (ARQ)
asCRB. *See* asymptotic Cramer–Rao bound (asCRB)
ASN. *See* Access service network (ASN)
Association of Radio Industries and Broadcasting (ARIB), 45
asymptotic Cramer–Rao bound (asCRB), 330, 332, 334, 335
    frequency offset estimation, 336
    joint frequency and channel estimation, 339
    normalized FIM, 334
    subcarriers vs, 345, 346
Authentication, authorization and accounting (AAA)
Automatic gain control (AGC), 331, 567
Automatic Repeat Request (ARQ), 192, 203
    feedback types, 203
    sliding window technique, 203
Automatic repeat request (ARQ), 569
Average signal-to-noise ratio, 547
AWG. *See* Application Working Group (AWG)
AWGN. *See* Additive white Gaussian noise (AWGN)
AWGN channel, 273, 304
    64–APSK modulation results, 285
    system performance, 285
Axioms, 72–73

B

BABS. *See* Bandwidth assignment based on SNR (BABS)
Band-AMC allocations, 611
Bandwidth, 418
    allocation, 204–205, 219–228, 513, 518–521, 529, 532
    assignment, 108
    calculation, 514–515
    channel, 13, 33, 199, 581
    coherence, 17, 18, 29, 32, 74, 401
    consumption, 523–526
    correlation, 22

    cost, 165
    efficiency, 26, 116, 165, 268, 387, 429, 437, 438
    excess, 359–360
    exploitable, 166, 169
    fair share, 107
    loss, 2, 439, 587
    OFDM basics, 3
    on-demand technology, 43
    request, 204, 209, 215–218, 224
    signal, 254
    stealing, 205
    subcarrier, 14, 28, 29
    subchannels, 1
    tone, 158
    transmission, 14
Bandwidth allocation on rate estimation (BARE)
        algorithm, 108
Bandwidth assignment based on SNR (BABS), 108
Bandwidth efficiency (BE), 268. *See* Spectral efficiency
Bandwidth request (BR), 599
Bandwidth request (BWR), 215
    aggregate, 215
    comparisons of mechanisms, 217
    explicit mechanisms, 216
    implicit mechanisms, 216
    incremental, 215
    types, 215
BARE. *See* Bandwidth allocation on rate estimation (BARE)
Bargaining, 72, 87
    axioms, 72–73
    comparison, 94–97
    cooperative, 71
    disagreement point, 72
    Egalitarian solution, 73
    feasible outcome set, 72
    game, 72, 86–88
    Kalai–Smorodinsky solution, 73
    locations, 95
    Nash bargaining solution, 73, 87, 91, 92
    utilitarian solution, 73
Base station (BS), 6, 74, 103, 303, 331, 351, 397, 190
        237, 596
    active, 617
    anchor, 617
    neighbor, 617
    serving, 617
    target, 617
Bayesian
    CFO estimation, 370
    framework, 369
    probabilities, 369
BC. *See* Block code (BC); Broadcast channel (BC)
BCH. *See* Broadcast channel (BCH)
BE. *See* Bandwidth efficiency (BE); Best effort (BE)
Beamforming, 587
BER. *See* Bit error rate (BER); Bit error ratio (BER)
Best effort (BE), 117, 204
BFSK. *See* Binary frequency shift keying (BFSK)
BICM. *See* Bit-interleaved coded modulation (BICM)
Binary frequency shift keying (BFSK), 459
Binary phase shift keying (BPSK), 317, 566

Binary series (W_k), 604
Bin group, 602
    IEEE difference, 602
    IEEE resemblance, 601
    partition, 602
    pilot, 602
Bins, 306, 316, 599
Bit and power loading, 303. *See also* Adaptive modulation
Bit error rate (BER), 74, 104, 268, 304, 355, 454, 474, 492
Bit error ratio (BER), 14
Bit-interleaved coded modulation (BICM), 566
BLER. *See* Block error rate (BLER)
Block code (BC), 273
Block error rate (BLER), 194
Block turbo coding (BTC), 203
BPSK. *See* Binary phase shift keying (BPSK)
BR. *See* Bandwidth request (BR)
Break-before-make HO, 617
Broadband communication, wireless, 42
    ISM bands, 43
    OFDM, 42–43
    spectrum efficiency, 43
    spectrum regrowth, 43–44
    WLAN systems target data rates, 43
Broadband wireless access (BWA), 190
Broadcast channel (BC), 236
Broadcast channel (BCH), 582
BS. *See* Base station (BS)
BTC. *See* Block turbo coding (BTC)
Burst, 598
BWA. *See* Broadband wireless access (BWA)
BWR. *See* Bandwidth request (BWR)

C

C1. *See* Criterion 1 (C1)
C2. *See* Criterion 2 (C2)
CA. *See* SubChannel Allocation (CA)
Cable TV (CATV), 3
CAC. *See* Connection admission control (CAC)
Capacity evaluation, 210, 214
    data workload, 206
    mobile TV workload, 206
    traffic models, 206
    triple play services, for, 205
    VoIP workload, 206
    WiMAX application classes, 207
    workload characteristics, 208
Carrier allocation schemes (CAS), 5
    generalized, 5
    interleaved, 5
    subband, 5
    subcarrier, 312, 512
Carrier assignment schemes (CASs), 398
    generalized, 403
    interleaved, 401–403
    performance comparisons, 411
    subband-based, 400–401
Carrier-frequency acquisition, 381
    CP-based, 382
    pilot-based, 381

Carrier frequency offset (CFO), 7, 46, 330, 351, 390, 398, 571
    DA ML joint estimate, 365
    effect, 352
    estimation, 362–364, 390
    feedback loop, 364
    FFT-domain, 388
    localized subcarrier allocation, 364–365
    normalized, 477
    performance decay, 358
    tracking, 364, 365
Carrier-interference-to-noise-ratio (CINR), 619
Carrier sense multiple access/collision avoidance (CSMA-CA), 192
Carrier to interference and noise ratio (CINR), 225
CAS. *See* Carrier allocation schemes (CAS)
CASs. *See* Carrier assignment schemes (CASs)
CATV. *See* Cable TV
CBR. *See* Constant bit rate (CBR)
CC. *See* Convolutional code (CC)
CCI. *See* Cochannel interference (CCI)
CDD. *See* Cyclic delay diversity (CDD)
CDF. *See* Cumulative density function (CDF); Cumulative distribution function (CDF)
CDM (F). *See* Code division multiplexing in frequency-domain (CDM (F))
CDMA. *See* Code division multiple access (CDMA)
CDMA2000. *See* Code-division multiple access 2000 (CDMA2000)
Central limit theory (CLT), 458
CFO estimation, 330–331, 335, 337, 338, 362–372
    CAS systems, 400–403
    complexity analyses, 410
    fine, 363–364
    importance function $g(\omega)$, 409
    ML estimator, 403
    performance comparisons, 411–414
    rough, 362–363
CFO estimators, 403, 410. *See also* CFO estimation
CFO. *See* Carrier frequency offset (CFO)
Channel aware scheduling, 132
Channel coherence time, 13, 36, 355
Channel estimation in OFDM/A, 36
    2-D Wiener filtering, 36
    MMSE filtering, 36
    pilot symbols, 36
    "return" channel model, 36
Channel feedback, 314
Channel impulse response (CIR), 15
    correlation function, 21
    correlation in US case, 21
Channel matrix, 546
Channel modeling, 13–14, 33, 34–35
    affected signal/system design parameters, 15
    BER floor, 14
    channel-counteracting techniques, 14
    characterization functions, 23
    latency, 14
Channel parameters, 17, 18, 30
    coherence bandwidth, 17
    CTF equation, 19

Channel parameters (*Continued*)
    definitions, and, 18
    Doppler spread, 17
    fading amplitude functions, 20
    guard time, 30
    multipath delay spread, 17
    OFDM/A signal parameters, 30
    persistence process, 19
    rules, 31
    slow fading, 31
    TDL channel model, 20
    time-varying CIR, 19
Channel quality identifiers (CQI), 194, 236
Channel quality indicator (CQI), 194
Channel quality information (CQI), 160, 567, 601
Channel quality information channel, (CQICH), 601
Channel State-dependent Round Robin (CSD-RR), 192
Channel state information (CSI), 102, 236, 269,
        302, 466, 474
Channel transfer function (CTF), 16
    correlation function, 20–21
    correlation in US case, 21–23
Chunks, 307. *See also* Bins; Cluster
    FDD mode, for, 308
    non-overlapping frequency–time, 132
CID. *See* Connection IDentifier (CID)
CINR. *See* Carrier-interference-to-noise-ratio (CINR)
CIR. *See* Channel impulse response (CIR)
Circular convolution, 423, 426. *See also*
        Cyclic convolution theorem
Classical scheduling algorithm, 118. *See also* Scheduling
    EDF scheduler, 119
    EXP rule, 121
    GPS scheduler, 118
    M-LWDF algorithm, 120–121
    PF scheduler, 120
    WFQ scheduler, 119
    WRR scheduler, 118–119
Closed-loop spatial multiplexing mode, 588
CLT. *See* Central limit theory (CLT)
Cluster, 306, 316. *See also* Bins; Chunks
    channel matrix, 545
CNA. *See* SubChannel Number Assignment (CNA)
Coalition algorithm, pair-bargaining, 88
    comparison, 97
    grouping strategies, 89
    overall multiuser algorithm, 88–89
    two-user bargaining algorithm, 90–92
Co-channel intercell/sector interference. *See* Multiple
        access interference (MAI)
Cochannel interference (CCI), 237
Code division multiple access (CDMA), 1, 2, 68,
        121, 132, 508, 599
Code-division multiple access 2000 (CDMA2000), 190
Code division multiplexing in frequency-domain
        (CDM(F)), 344
Codeword, 215, 587–589, 620
Codeword, layer mapping
    spatial multiplexing, 588
    transmit diversity, 589
Coherence bandwidth. *See* Correlation bandwidth

Common public radio interface (CPRI), 256
Communication identifiers (CIDs), 615
Compact array geometry, 551
Comparison, WiFi, WiMAX, and LTE, 592
Compression, feedback data
    algorithms, 314, 315, 320–324
    compressed, 315, 318–319
    data-limited, 314, 315–317
    error impact, 319
    fairness and QoS remarks, 319–320
    quantized, 314, 317–318
Connection admission control (CAC), 218
Connection IDentifier (CID), 192
Connectivity service network (CSN), 570
Constant bit rate (CBR), 205
Convergence factor. *See* Adaptation constant (μ)
Convergence performances
    alternating projection, 412, 413
    performance comparisons, 411–414
    SAGE algorithm, 412
Convolutional code (CC), 566
Convolutional turbo codes (CTC), 203, 574
Convolution coding (CC), 203
Convolution theorem. *See* Linear convolution
        theorem
Cooperative communication, 508
Cooperative game
    bargaining, 72, 86, 87
    convex hull, 87
    Egalitarian solution, 73, 87
    existence and uniqueness theorem, 88
    Kalai–Smorodinsky solution, 73, 88
    Nash bargaining solution, 73, 87
    strategy set, 86
    utilitarian solution, 73, 87
    utility function, 86
Cooperative OFDMA, channel estimation, 493
    AF mode, 498
    DF mode, 499
    frequency offsets, 495–496
    numerical results, 501
    optimal pilot design, 497
    PEP analysis, 499–501
    signal model, 493
Cooperative OFDM PEP, 499
    AF Mode, 499
    DF mode, 500
    performance, 501
Cooperative transmission, 487
    Alamouti-coded, 491
    PEP performance, 501, 502
    performing, 489
    preamble and data frame, 496
    two-phase, 491
Correlation bandwidth, 22, 23
Correlation function
    CIR, 21
    CTF, 21
    delay-Doppler, 21, 22
    double Fourier transform, 21
    output Doppler spread, 21

RMS-DS, 22
SFST, 22
CP. *See* Cyclic prefix (CP)
CPRI. *See* Common public radio interface (CPRI)
CQI. *See* Channel quality identifiers (CQI); Channel quality
          indicator (CQI); Channel quality
          information (CQI)
CQICH. *See* Channel quality information channel
          (CQICH)
Cramer–Rao bound (CRB), 330, 332, 411, 413
   FIM, 333
   finite sample, 333
   matrix, 333
Cramer–Rao lower bound (CRLB), 488, 495
CRC. *See* Cyclic redundancy check (CRC)
Criterion 1 (C1), 286
Criterion 2 (C2), 286
CRLB. *See* Cramer–Rao lower bound (CRLB)
Crossover, 516
Cross-polar discrimination (XPD), 541
Cross-polarization matrix, 543
CS. *See* Cyclic suffix (CS)
CSD-RR. *See* Channel State-dependent Round
          Robin (CSD-RR)
CSI. *See* Channel state information (CSI)
CSMA-CA. *See* Carrier sense multiple
          access/collision avoidance (CSMA-CA)
CSN. *See* Connectivity service network (CSN)
CTC. *See* Convolutional turbo codes (CTC)
CTF. *See* Channel transfer function (CTF)
CUM. *See* subChannel User Matching (CUM)
Cumulative density function (CDF), 410
Cumulative distribution function (CDF), 245, 259, 260
   inverse channel condition number, 259
   SNR statistics, 259
Cyclic convolution theorem, 423. *See also* Linear
          convolution theorem
   DFT, 424, 427
   UW concept, 429, 431
Cyclic delay diversity (CDD), 577
Cyclic extensions, 26, 171, 176, 180
   FFT, use of, 26
   guard interval, 26
   OFDM with CP, 27
   OFDM with guard interval, 26
Cyclic prefix (CP), 4, 25, 30, 170, 175, 331, 351, 359, 360,
          361, 376, 398, 419, 508, 539, 565, 597
   classical, 429
   concept, 424
   sent data-structure, 425
   transceive structure, 427
   vs UW, 430–431
Cyclic redundancy check (CRC), 198, 494
Cyclic suffix (CS), 170

D

DA. *See* Data—aided (DA)
DAB. *See* Digital—audio broadcasting (DAB)
DAPSK schemes. *See* Differential amplitude and phase shift
          keying (DAPSK) schemes

Data
   aided (DA), 350
   carrier, 605
   region, 308, 598. *See also* Slot
Data mapping, 599
   2 Tx antenna with horizontal encoding, 614
   2Tx antenna with vertical encoding, 613
   optional AMC Zone, 612
   subcarriers, 609
DC. *See* Direct-current (DC)
DC carrier, 604, 607
DCD. *See* Downlink—channel descriptor (DCD)
DcF. *See* Decode-and-compensate-and-forward (DcF)
DCF. *See* Distributed coordination function (DCF)
DCT. *See* Discrete cosine transform (DCT)
Decode-and-compensate-and-forward (DcF),
          474n, 475
Decode-and-forward (DF) mode, 474, 479
Deficit fair priority queuing (DFPQ), 224
Deficit round robin (DRR), 198
Deficit weighted round robin (DWRR), 223
Delay-based utility function, 125
   queueing system stability region, 125
   resource assignment goal, 125
Delay Threshold Priority Queuing (DTPQ), 224
DF mode. *See* Decode-and-forward (DF) mode
DFPQ. *See* Deficit fair priority queuing (DFPQ)
DFT. *See* Discrete Fourier transform (DFT)
Differential amplitude and phase shift keying (DAPSK)
          schemes, 273. *See also* APSK modulation
Differential space frequency block codes (DSFBCs), 274
   differential modulation, 274–276
   incoherent demodulation, 275
Differential space time block code (DSTBC), 268, 272, 277
   description, 273
   evolution, 272
   modulation schemes, 278
   new class, 277
   proposed structure, 277
Differential space time frequency block codes (DSTFBCs),
          276. *See also* DSFBCs
   equal gain combining, 295
   maximum ratio combining, 296
   modulation schemes, 278. *See* 64–APSK modulation
   power control mechanism, 284, 285
   system performance, 285, 288
   technique with receive diversity, 293
Differential unitary space time modulations, 272
Digital
   audio broadcasting (DAB), 42
   crystal oscillator (DXO), 352
   signal processing (DSP), 42
   subscriber lines (DSLs), 303
   video broadcast (DVB), 3, 564
   video broadcasting-terrestrial (DVBT), 42
Direct-current (DC), 597
Direction of arrivals (DOAs), 542
Direction of departure (DOD), 542
Dirty paper coding (DPC), 236, 314
Discrete cosine transform (DCT), 319

Discrete Fourier transform (DFT), 37, 42, 170, 240, 351, 376, 398, 419, 420, 485, 485
    cyclic convolution theorem, 423
    discrete-time Fourier transformation, 420
    fast Fourier transformation, 421
    linear convolution theorem, 421
    linear filtering methods, 423
    properties, 242, 420
Distributed coordination function (DCF), 568
Diversity set, 617
DL. *See* Downlink (DL)
DL-MAP. *See* Downlink (DL)—map (DL-MAP)
DL/UL-MAP, band-AMC, 611
DOAs. *See* Direction of arrivals (DOAs)
DOD. *See* Direction of departure (DOD)
Doppler
    effect, 354
    shift, 354, 545
    spectrum, 29, 354
    spread, 17, 29–30, 354–355
Downlink (DL), 166, 191, 237, 257, 378, 597, 599
    channel descriptor (DCD), 617
    map (DL-MAP), 192
    MAP message, 600n
    overhead, 209, 210–211
    pilot time slot (DwPTS), 580
    preamble carrier-sets, 603, 604
    subframe, 601
DPC. *See* Dirty paper coding (DPC)
DRR. *See* Deficit round robin (DRR)
DSA. *See* Dynamic subchannel assignment (DSA)
DSFBCs. *See* Differential space frequency block codes (DSFBCs)
DSLs. *See* Digital—subscriber lines (DSLs)
DSP. *See* Digital—signal processing (DSP)
DSTBC. *See* Differential space time block code (DSTBC)
DSTFBCs. *See* Differential space time frequency block codes (DSTFBCs)
DTPQ. *See* Delay Threshold Priority Queuing (DTPQ)
DVB. *See* Digital—video broadcast (DVB)
DVBT. *See* Digital—video broadcasting-terrestrial (DVBT)
DwPTS. *See* Downlink (DL)—pilot time slot (DwPTS)
DWRR. *See* Deficit weighted round robin (DWRR)
DXO. *See* Digital—crystal oscillator (DXO)
Dynamic resource allocation, 69
Dynamic subchannel assignment (DSA), 102, 107, 109, 111
    ACG algorithm, 110–111
    bandwidth assignment, 108
    cost minimization, 109
    FI calculation, 107
    Hungarian algorithm, 109–110
    Jang and Lee algorithm, 107
    RCG algorithm, 110
    Rhee and Cioffi max–min algorithm, 107
    Zhang and Letaif algorithm, 111, 112

E

Earliest deadline first (EDF), 119, 220
EDF. *See* Earliest deadline first (EDF)
EDGE. *See* Enhanced data rates for global evolution (EDGE)

Egalitarian solution, 73
Egress reduction techniques
    comparison, 184–186
    filtering, 175
    intrasymbol windowing, 176–177, 178–179
    nyquist windowing, 180–181
    power loading, 175–176
    run-time complexity, 185
    single subcarriers PSDs, 178
    spectral efficiency R/N, 184, 185
    transmit windowing, 176
EGT. *See* Equal gain—transmission (EGT)
EM algorithm. *See* Expectation-maximization (EM) algorithm
Enhanced data rates for global evolution (EDGE), 190
enhanced NodeBs (eNodeB), 578, 579
eNodeB. *See* enhanced NodeBs (eNodeB)
EPC. *See* Evolved packet core (EPC)
Equal gain, 295
    combination module, 295, 296, 297
    results, 296
    transmission (EGT), 239
Error vector magnitude (EVM), 440–442
ertPS. *See* Extended real-time polling service (ertPS)
ETSI. *See* European Telecommunications Standards Institute (ETSI)
European Telecommunications Standards Institute (ETSI), 42
E-UTRA. *See* Evolved Universal Terrestrial Radio Access (E-UTRA)
E-UTRAN. *See* Evolved universal terrestrial radio access network (E-UTRAN)
EVM. *See* Error vector magnitude (EVM)
Evolved packet core (EPC), 578
Evolved Universal Terrestrial Radio Access (E-UTRA), 577–589
    AMC, 583
    downlink channels, 582–583
    downlink resource grid, 582
    E-UTRAN architecture, 578–579, 579–586
    E-UTRAN FDD, 580–581
    HARQ, 583
    multiantenna techniques, 587
    OFDMA, 581–582
    requirements, 578
    resource blocks, 581, 582, 585
    SC and OFDM-PAPR, 585
    SC-FDE, 584
    SC-FDMA, 584
    turbo coding, 583
    uplink physical channel, 585, 586
    uplink reference signals, 586
    uplink resource grid, 586
    uplink transport channel, 585, 586
Evolved universal terrestrial radio access network (E-UTRAN), 564
Excess bandwidth, 359
Expectation-maximization (EM) algorithm, 366
Exponential (EXP) rules, 121
EXP rules. *See* Exponential (EXP) rules
Extended real-time polling service (ertPS), 204

F

Fairness, 218, 220, 222, 226
  channel-aware schedulers, 221, 225
  PFS, 226–227
  proposals, 227
  scheduler design factors, 219
Fairness index (FI), 107
Fast base station switching (FBSS), 617, 621
  anchor BS update, 619–620
  decision, 618
  diversity set update, 619
  fast-feedback channel, 620
Fast-feedback channel, 620
Fast Fourier transform (FFT), 37, 42, 166, 331, 376, 597
  algorithms, 421
  guard band subcarriers, 604
FBA. *See* Fixed-bandwidth allocation (FBA)
FBSS. *See* Fast base station switching (FBSS)
FBWA. *See* Fixed broadband wireless access (FBWA)
FCC. *See* Federal Communications Commission (FCC)
FCH. *See* Frame control header (FCH)
FDD. *See* Frequency division duplex (FDD)
FDE. *See* Frequency domain equalization (FDE)
FDE system comparison, 436
  bandwidth efficiency, 438
  front-end requirements, 439
  OFDM, 437, 438
  processing load, 437
  SC/FDE based, 437, 438
  signal-to-noise ratio loss, 439
FDM. *See* Frequency division multiplexing (FDM)
FDMA. *See* Frequency division multiple access (FDMA)
Feasible Earliest Due Date (FEDD), 192
FEC. *See* Forward error control (FEC); Forward error
    correction (FEC)
FEDD. *See* Feasible Earliest Due Date (FEDD)
Federal Communications Commission (FCC), 46
FFT. *See* Fast Fourier transform (FFT)
FI. *See* Fairness index (FI)
Field programmable gate array (FPGA), 257
FIFO. *See* First in first out (FIFO)
File transfer protocol (FTP), 117, 205
FIM. *See* Fisher information matrix (FIM)
finalSleepWinBase, 615
Finite impulse response (FIR), 421, 423
Finite period of time. *See* Time slots
FIR filtering. *See* Finite impulse response (FIR)
First-generation (1G), 2
First in first out (FIFO), 219
Fisher information matrix (FIM), 333
Fixed-bandwidth allocation (FBA), 513
Fixed broadband wireless access (FBWA), 538
Fixed system profile, 596
Fixed-throughput allocation (FTA), 513
Forward error control (FEC), 193
Forward error correction (FEC), 45, 257
Fourth-generation (4G), 42, 330
FPGA. *See* Field programmable gate array (FPGA)
Frame control header (FCH), 193, 575, 600
Frank–Wolfe algorithm, 135

Free space loss (FSL), 539
Frequency division duplex (FDD), 116, 138, 191,
    236, 307, 352, 564, 596
Frequency division multiple access (FDMA), 1–2, 68,
    193, 312
Frequency division multiplexing (FDM), 3, 42, 342
Frequency domain equalization (FDE), 7
  analysis, 419
  concepts, 419
  CP, 424, 429
  Fourier transformation, 420
  implementation effort, 418
  linear filtering, 421–424
  MMSE equalization, 428
  multicarrier transmission, 431
  multipath propagation, 434
  OFDM, 433
  single carrier transmission, 419, 584
  ZF equalization, 428
Frequency offset, 6, 7, 27–29, 46, 63, 169, 351, 389, 400
  estimation, 336, 483–502
  modeled as random process, 28
  time variation, 28
  time-varying channel effect, 27
Frequency offset estimation, 483, 486
  frequency offsets, 479–480
  numerical results, 481–483, 489
  OFDMA uplink transmission, 487
  outage information rate, 480–481
  performance improvement, 493
  point-to-point, 484
  relay selection, 487
  space–time diversity system, 486
  variance error analysis, 488
  with feedback, 492
Frequency synchronization algorithm, 486
FSL. *See* Free space loss (FSL)
FTA. *See* Fixed-throughput allocation (FTA)
FTP. *See* File transfer protocol (FTP)
Full usage of subchannels (FUSC), 219, 308, 598, 601
FUSC. *See* Full usage of subchannels (FUSC)

G

GA. *See* Genetic algorithm (GA)
Game theoretical framework, 69, 98
  approaches to OFDMA, 70
  distributed optimization, 70–71
  inherent risks, 71
  motivations, 70
  net utility method, 70
  novel allocation scheme, 70
  open issues, 99
  rational decision making, 70
Generalized weighted fairness (GWF), 226
General processor sharing (GPS), 118, 223
Genetic algorithm (GA), 509
  chromosome representation, 516
  crossover, 516
  fitness evaluation, 516
  initial population, 516

Genetic algorithm (GA) (*Continued*)
    mate selection, 516
    mutation, 517
Global multispace-polarization channel matrix, 545
Global systems for mobile communications (GSM), 2
GM. *See* Grant management (GM)
GoB. *See* Grid of beams (GoB)
GP. *See* Guard—period (GP)
GPC. *See* Grant per connection (GPC)
GPS. *See* General processor sharing (GPS)
GPSS. *See* Grant per subscriber station (GPSS)
Gradient-based algorithms, 132
Grant management (GM), 193
Grant per connection (GPC), 215
Grant per subscriber station (GPSS), 215
Greedy bandwidth assignment, 108
    average user gain, 108
    BABS algorithm, 108
    BARE algorithm, 108–109
Grid of beams (GoB), 237
    concept, 239
    DFT beams, 240
Grouping strategies, pair-bargaining, 89
    Hungarian method, 89–90
    random method, 89
GSM. *See* Global systems for mobile communications (GSM)
Guard
    band subcarriers, 175, 604
    period (GP), 580
GWF. *See* Generalized weighted fairness (GWF)

H

Handover (HO), 617
    concepts, 617
    types, 617
    variants, 617
HAP. *See* High altitude platform (HAP)
Hard handover (HHO), 617
HARQ. *See* Hybrid ARQ (HARQ); Hybrid automatic-repeat-request (HARQ)
HDTV. *See* High-definition television (HDTV)
Head of line (HOL), 219
Helical wave fronts, 555–556
Heterogeneous traffic
    suboptimal solution, 126
    utility-based scheduling, 126
HHO. *See* Hard handover (HHO)
High altitude platform (HAP), 537, 538
    advantages, 538
    limitations, 538
    MIMO-HAP diversity system, 539
    OFDM–MIMO technique, 538
    platform scenarios, 539
    polarization diversity, 539
    STP, 540–541
    stratospheric flight, 538
    wave propagation channel, 539, 540
    wireless systems, 538
High-definition television (HDTV), 538

High-power amplifier
    amplifier's output, 55–56
    IP_3 calculation, 56
    IP_5 determination, 56
    mathematical model, 55
    Taylor model, 55
    two-tone test, 56, 57
High speed downlink packet access (HSDPA), 578
HO. *See* Handover (HO)
HOL. *See* Head of line (HOL)
Horizontal encoding, 613, 614
HSDPA. *See* High speed downlink packet access (HSDPA)
HTTP. *See* Hypertext transfer protocol (HTTP)
Hungarian
    algorithm, 109
    method, 89–90
Hybrid ARQ (HARQ), 192, 203
    retransmissions, 203, 204
    types, 203
Hybrid automatic-repeat-request (HARQ), 601
Hypertext transfer protocol (HTTP), 117, 206

I

ICAS. *See* Interleaved CAS (ICAS)
ICI. *See* Intercarrier interference (ICI); Inter-channel interference (ICI); Inter-subcarrier interference (ICI)
Ideal compact antenna, 550–551
IDFT. *See* Inverse discrete Fourier transform (IDFT)
Idle mode, 616
    MS paging periods, 616
    power saving, 616
IEEE. *See* Institute of Electrical and Electronics Engineers (IEEE)
IEs. *See* Information elements (IEs)
IFFT. *See* Inverse fast Fourier transform (IFFT)
i.i.d. *See* independent and identically distributed (i.i.d.)
Importance function $g(\omega)$, 408, 409
IMT. *See* International Mobile Telecommunications (IMT)
IMT-2000. *See* International Mobile Telecommunications—2000 (IMT-2000)
IMT-A. *See* International Mobile Telecommunications—Advanced (IMT-A)
Incremental redundancy (IR), 203, 576
independent and identically distributed (i.i.d.), 477
Independent ports, 548
Industrial, scientific, and medical (ISM) frequency bands, 43
Information elements (IEs), 195, 212, 600n
In-phase and quadrature (IQ) component, 256
Institute of Electrical and Electronics Engineers (IEEE), 38
    802.11a systems, 44, 566
    802.11 family, 565
    802.11n amendment, 568n
    802.16. *See* WiMax, 190, 568
    802.16e, 569
    data mapping, 599
    data region, 309
    difference, 597, 598, 602
    downlink FUSC, 308–309

downlink PUSC, 309
FEC with coding rates, 46
guard interval, 45
OFDM, 45, 46, 566, 592
PHYs types, 44–45, 307
resemblance, 598, 601
slots, 308
SNR loss minimization, 45, 46
spectrum allocation, 45
transmitter and receiver block, 47–48
U-NII bands, 45, 46–47
Integer programming (IP), 229
Integrated services digital broadcasting (ISDB), 564
Intercarrier interference (ICI), 4, 330, 351, 379, 398, 433
Inter-channel interference (ICI), 27
Interference rejection combining (IRC), 576
Interleaved CAS (ICAS), 5
International Mobile Telecommunications (IMT), 508
   2000 (IMT-2000), 589–590
   Advanced (IMT-A), 564
International Telecommunication
      Union-Radiocommunication Sector
      (ITU-R), 508
Inter-subcarrier interference (ICI), 27
Intersymbol interference (ISI), 4, 50, 303, 330,
      356, 379, 424, 508, 538
Inverse discrete Fourier transform (IDFT), 171, 398,
      420, 477
   time Fourier transformation, 420
Inverse fast Fourier transform (IFFT), 3, 15
IP. *See* Integer programming (IP)
IP$_3$. *See* Third-order interception point (IP$_3$)
IQ component. *See* In-phase and quadrature (IQ) component
IR. *See* Incremental redundancy (IR)
IRC. *See* Interference rejection combining (IRC)
ISDB. *See* Integrated services digital broadcasting (ISDB)
ISI. *See* Intersymbol interference (ISI)
ISM. *See* Industrial, scientific, and medical (ISM)
      frequency bands
ITU-R. *See* International Telecommunication
      Union-Radiocommunication Sector (ITU-R)

J

Jang and Lee algorithm, 107
Joint 2D multiplexing method, 556

K

Kalai–Smorodinsky solution, 73
Karush–Kuhn–Tucker (KKT), 77, 318
KKT. *See* Karush–Kuhn–Tucker (KKT)

L

Largest weighted delay first (LWDF), 220
Largest-weighted-throughput (LWT), 228
LDPC. *See* Low-density parity codes (LDPC)
Least squares (LS), 389, 494
Lempel–Ziv–Welch (LZW) coding, 318
Levin–Campello algorithm, 113

Linear combination coefficients, 542
Linear convolution, 25, 26, 170, 421, 423,
      424, 425, 426, 427, 430
   theorem, 421
Linear decorrelating detector. *See* Least squares (LS)
Linear minimum mean square error (LMMSE), 346
Line of sight (LOS), 38, 192, 253, 512, 540, 569
Link adaptation, 302, 567, 568, 574, 582, 585.
         See Adaptive modulation
LMMSE. *See* Linear minimum mean square error
      (LMMSE)
Localized subcarrier allocation, 365
Long term evolution (LTE), 237, 307, 350, 418, 565
   frame structure, 307
   synchronization phase, 367
LOS. *See* Line of sight (LOS)
Low-density parity codes (LDPC), 203, 245, 568
Lower layer overhead, 209. *See also* Upper layer overhead
   downlink overhead, 209
   MAC overhead, 209
   uplink overhead, 211
LS. *See* Least squares (LS)
LTE. *See* Long term evolution (LTE)
LWDF. *See* Largest weighted delay first (LWDF)
LWT. *See* Largest-weighted-throughput (LWT)
LZW coding. *See* Lempel–Ziv–Welch (LZW) coding

M

MAC. *See* Multiple access control (MAC)
MAC layer. *See* Medium access layer (MAC layer)
MAC overhead, 206, 209
   bandwidth request, 209
   capacity evaluation, 210
   downlink fragmentation, 208, 209
   packing subheaders, 208, 209
   UL preamble and MPDU, 209
MAC packet data unit (MPDU), 568
MAC protocol data units (MPDUs), 192
Macro diversity handover (MDHO), 617
   anchor BS update, 619–620
   conditions to enable, 618–619
   decision, 618
   diversity set update, 619
   fast-feedback channel, 620
MAI. *See* Multiple access interference (MAI); Multiuser
      access interference (MAI)
Make-before-break HO, 617
MA optimization. *See* Margin adaptive (MA) optimization
MAP. *See* maximum *a posteriori* probability (MAP)
Margin adaptive (MA) optimization, 103–104
   conditions, 104
   MCS system, 104
   M-QAM schemes, 104–105
   Shannon capacity expression, 104
M-ary phase shift keying (M-PSK), 304, 459
M-ary quadrature amplitude modulation (M-QAM),
      104, 304
Maximal ratio combining (MRC), 576
Maximum *a posteriori* probability (MAP), 383
Maximum information capacity, 547, 557

Maximum likelihood (ML)
  algorithm, 484
  estimator, 332
Maximum ratio combining (MRC), 239, 295, 296
  "combination module", 297
  results, 298
Maximum sustained (MST) rate, 205
Maximum throughput (MT) scheduling, 249
max-SNR scheduling, 315
MB. *See* Multiband (MB)
MBER. *See* Minimum BER (MBER)
MBWA. *See* Mobile broadband wireless Access (MBWA)
MCH. *See* Multicast channel (MCH)
MCS. *See* Modulation and coding scheme (MCS)
MDHO. *See* Macro diversity handover (MDHO)
Mean square error (MSE), 411, 413, 496
Medium access control (MAC), 38, 44, 102
Medium access layer (MAC layer), 256
MIMO. *See* Multiple input multiple output (MIMO)
MIMO antenna array, 541, 548
  configurations, 548
  independent ports, 548
  MIMO-cube antenna, 548
  MIMO-octahedron antenna, 548
  MIMO-tetrahedron antenna, 548
  vector element antenna, 548
MIMO-HAP diversity system, 539
MIMO–MRC system, OFDM–SDMA
  asymptotic SER analysis, 461
  SER analysis, 460
MIMO–OFDM system, 268, 271. *See also* MIMO system
  capacity scaling, 249, 254
  channel evaluation, 241
  concept, 237
  experimental results, 255–261
  GoB, 239
  implementation, 271
  link level, 244
  MIMO channel models, 269
  model extensions, 251
  resource allocation algorithm, 240
  resource scheduling, 243
  scores determination, 242
  setup, 244, 252
  system level, 251
  test-bed parameter set, 256
  throughput performance, 245
MIMO-OFDM system isolated cell, 244
  capacity scaling, 249
  setup, 244
  throughput performance, 245
MIMO-OFDM system multicell, 251
  capacity scaling, 254
  setup, 252
  simulation assumptions, 253
  system model extensions, 251
  triple-sectored hexagonal cell, 252
MIMO system, 268, 272
  channel, 270
  decoder, 270
  encoder, 270

features, 269
  scheme, 269
  system representation, 271
  techniques classification, 269
MIMO techniques
  MB models, 35–36
  OFDM, 35
  uses, 35
Minimum BER (MBER), 454
Minimum mean square error (MMSE), 37, 239,
      319, 389, 428, 496
  equalization, 428
  equalization vector, 242
  equalizers, 429
  estimate, 389
  estimator, 389
Minimum variance unbiased (MVU), 495
MISO. *See* Multiple-input single-output (MISO)
ML. *See* Maximum likelihood (ML)
ML estimator, 403, 413
  asymptotic decoupled, 404
  importance sampling, 407
  iterative algorithms, 405. *See also* Space-alternating
      generalized expectation-maximization (SAGE)
  performance comparisons, 411–414
M-LWDF. *See* Modified largest weighted delay first
      (M-LWDF)
MMAC. *See* Multimedia Mobile Access Communications
      (MMAC)
MME. *See* Mobility management entity (MME)
MMSE. *See* Minimum mean square error (MMSE)
Mobile broadband wireless Access (MBWA), 569
Mobile station (MS), 74, 190, 596
  initialization, 190, 204
  paging periods, 616–617
Mobile WiMAX (IEEE 802.16e), 568–577
  antenna category, 576–577
  basis, 568
  beamforming, 577
  burst profile, 574
  data subcarriers, 572
  downlink slots, 575
  MIMO with SM, 577
  null subcarriers, 572
  OFDM Parameters, 571
  pilot subcarriers, 572
  ranging channel, 575
  SOFDMA, 571
  uplink slots, 575
Mobility management entity (MME), 578
Mobility system profile, 596
Modified largest weighted delay first (M-LWDF),
      120–121, 227
Modulation and coding scheme (MCS), 104, 192, 302
Moving picture experts group (MPEG), 205
MPC. *See* Multipath component (MPC)
MPDUs. *See* MAC protocol data units (MPDUs)
MPEG. *See* Moving picture experts group (MPEG)
M-PSK. *See* M-ary phase shift keying (M-PSK)
M-QAM. *See* M-ary quadrature amplitude modulation
      (M-QAM)

MRC. *See* Maximal ratio combining (MRC);
        Maximum ratio combining (MRC)
MS. *See* Mobile station (MS)
ms. *See* multistream (ms)
MSE. *See* Mean square error (MSE)
MST. *See* Maximum sustained (MST) rate
MT scheduling. *See* Maximum throughput (MT) scheduling
Multiband (MB), 31
Multicarrier transmission, FDE
    concept, 431
    ICI and ISI, 436
    multipath propagation, 434
    OFDM concept, 433
    OFDM symbols, 435
    receiver, 433
    single carrier vs, 432
    subcarriers spectrum, 434
    transmitter, 432
Multicast channel (MCH), 582
Multicell
    achievable rate, 75
    downlink, 139
    efficiency, 261
    link spectral efficiency, 513
    OFDMA, 70, 74, 76, 79
    power setup, 512–513
    scenario, 75, 98
    shadowing model, 512
    SINR, 75
    system model, 251–254
    total rate, 75
    transmit PSD, 513
Multicell architecture, 509, 510
    2-hop path, 509
    cell layout, 511
    cluster size, 509
    direct path, 509
    omni-directional antenna, 509
    radio resource allocation, 509, 511
    RF transceiver, 509
Multicell downlink
    bandwidth consumption distribution, 530
    cell region, 517
    FBA, 513
    FTA, 513
    optimization, 513–517
    performance, 517, 526
    RSs' positions, 530
Multicell propagation models, 511–512
    line-of-sight (LOS), 512
    non-line-of-sight (NLOS), 512
    shadowing loss, 512
Multicell simulation. *See* MIMO-OFDM system multicell
Multicell system models, 509–513
    cluster size, 509, 510
    frequency reuse over RS–MS Links, 513
    multicell architecture, 509
    power setup, 512–513
    propagation models, 511–512
    subcarrier allocation and permutation, 509–511
Multimedia Mobile Access Communications (MMAC), 45

Multipath component (MPC), 21
Multipath delay spread, 17, 22, 30, 269
Multiple access control (MAC), 596
    layer, 256
    overhead, 209
    PHY block diagram, 117
    protocol, 568
    scheduler, 118
Multiple access interference (MAI), 2, 330, 357, 384, 398,
        405, 455, 474, 485, 498, 510
    AWGN, 510
    cancellation process, 407
    elimination, 498
    OFDM–SDMA system, 461
    reduced signal, 407
    suppression, 455
Multiple access method, 1
Multiple antenna(s), 115, 236, 237, 400, 538, 583
    clustered S-best criterion, 116
    CSI estimation, 116
    systems, 587. *See also* MIMO system
    Zhang–Letaief algorithm, 115
Multiple HAP system, 550
    capacity, 550–552, 557
Multiple-input multiple-output (MIMO), 7, 38, 191, 314,
        330, 436, 454, 456, 565, 603
    link, 236
    MAC, 257
    system, 268, 272
Multiple-input single-output (MISO), 273, 314
Multiple physical-layer options, 596
    WirelessMANOFDM, 596
    Wireless-MAN-SC, 596
    WirelessMAN-SCa, 596
    Wireless-OFDMA, 596
Multiple signal classification (MUSIC), 385
Multipole expansion, 547–548
Multispace-polarization elementary path, 544
multistream (ms), 237
Multiuser access interference (MAI), 455
Multiuser MIMO (MU-MIMO), 236, 246, 255, 257, 261
    channel-adaptive transmission, 238
    DL, 237
Multiuser OFDMA system, 310
    CFO, 330
    channel estimation, 330
    constant power allocation, 311
    maximizing rate, 310
    minimizing power, 310
    optimal training condition, 330, 335
    packet-based, 331–332
    SNR gap approximation, 311
    two-step allocation, 312–313
Multiuser OFDM systems training design
    CDM(F) sequences, 343
    channel-dependent, 336
    channel estimation, 335, 339
    channel-independent, 337
    FDM sequences, 342
    frequency-domain realization, 341
    frequency offset estimation, 336

Multiuser OFDM systems training design (*Continued*)
    joint CFO, 339
    minimax approach, 339–340
    optimal condition 335
    performance comparisons, 344–346
    time-domain realizations, 340–341
Multiuser orthogonal frequency-division multiplex
    (OFDM), 330
MU-MIMO. *See* Multiuser MIMO (MU-MIMO)
MUSIC. *See* Multiple signal classification (MUSIC)
Mutation, 516
Mutual coupling, 549
Mutually exclusive groups, 5. *See also* Subband
MVU. *See* Minimum variance unbiased (MVU)

N

Nash bargaining solution, 73
Nash equilibrium, 71, 72
    discrete actions, 71
    fundamental elements, 71
    mixed strategy, 71
    pure strategy, 71
    stable operating point, 72
    strategic form, 71
    strategy profile, 71
    theorems, 72
NDA. *See* Non-data-aided (NDA)
Next-generation
    systems, 508
    wireless systems, 538
NLOS. *See* Non-line of sight (NLOS)
Nonband AMC, 611
Noncooperative game
    assignment strategy set, 77
    formulation, 76
    game formulation, 76
    iterative best-response, 77
    Nash equilibrium, 71
    net utility, 76
    numerical results, 83
    Pareto optimal strategies, 72
    power minimization, 76–77
    power minimization game, 80–81
    power strategy set, 77
    rate as utility, 76
    strategy identification, 77
    system, 83
    utility function, 76
    virtual referee, with, 79, 80, 81
Noncooperative game, iterative best-response, 77
    algorithm, 79
    assume constant power, 77
    channel allocation results, 77–78
    generic game break down, 77
    interference scenario, 84
    lemma, 79
    maximization problem, 78
    Nash equilibrium, 78–79, 84
    oscillating utility, 85
    random cell utility, 84

Non-data-aided (NDA), 350
Non-deterministic polynomial time (NP), 105
Nonfrequency adaptive scheme, 310
Nonlinearity
    results, 44
    RF power amplifier, 44
    WLAN OFDM systems, 48
Non-line of sight (NLOS), 192, 253n, 512
Non-real-time polling service (nrtPS), 117, 204, 205, 218
Nonreturn-to-zero (NRZ) function, 54
Nonstationarity (NS) models, 23
Nonstationary channel modeling, 34
    campus area NLOS parameters, 34
    two-state persistence process, 34, 35
NP. *See* Non-deterministic polynomial time (NP)
nrtPS. *See* Non-real-time polling service (nrtPS)
NRZ. *See* Nonreturn-to-zero (NRZ) function
NS models. *See* Nonstationarity (NS) models

O

OAM. *See* Orbital angular momentum (OAM)
OFDM. *See* Orthogonal frequency division
        multiplexing (OFDM)
OFDMA. (Orthogonal frequency division multiple access), 1,
        3, 4–5, 68, 132, 166, 190, 256, 302, 330, 351,
        397, 474, 508, 564. *See also* Channel feedback
    802.16m frame structure, 592
    adaptive modulation, 302
    AF mode, 478
    Alamouti-coded transmission, 491
    algorithms, 311–313
    application, 302–303
    capacity, 479–481
    capacity region, 139–140
    channel capacity, 479
    channel estimation, 493
    channel model, 476
    complexity and performance, 313
    cross-layer visibility, 142
    DF mode, 479
    downlink transmission, 6
    dual function, 144–145
    duality theory, 148–149
    dual of problem, 141
    dual variables, 141, 142, 148
    FDD, 564
    feasibility, 149–150
    feasible solution construction, 141
    first time slot, 477
    frame structure, 194
    frequency offset estimation, 483
    generalized CAS, 5
    general power constraint, 139
    general rate regions, 136
    gradient-based wireless, 135
    ICAS, 5
    joint subcarrier, bit, and power allocation, 313
    Lagrangian marked constraints, 149
    limitations, 151–152
    LTE advanced, requirements, 590

maximizing Lagrangian, 143
maximum rate constraints, 140
MCSs, 199
methods, 148
minimizing Lagrangian, 150
multicell considerations, 313
multiuser diversity, 305
optimal algorithms, 140
optimal power, 144
optimization problem, 140–141
optimizing dual function, 145–146, 147–148
orthogonalization, 549
parameters, 200, 202
parameters for IEEE 802.16m, 592
PHY research topics, 590–591
physical layers (PHYs), 192
primal optimal solution, 148
problem, 149
problem formulation, 135
rate regions, 136–137
receiver, 6
related work, 133–134
research challenges, 589
SCAS, 5
scheduling, 132, 134
second time slot, 478
self-noise, 137
signal model, 493
single cell problem, 139
SMUD algorithm, 315
SNR, 137–138
spectrum proposed, 590
subcarrier allocation, 5
subChannel allocation, 150–151
subframes, 592
TDD, 564
time–frequency structure, 302
transmitter, 5, 6
two-step allocation, 312–313
uplink signal model, 475
weaknesses, 7
WiMAX: 802.16m, 591
WiMAX networks, 191
OFDM/A, 12
  arbitrary frequency band, 12
  canonical, 31–32
  channel estimation, 36
  channel model(s), 32
  channel modeling, 15, 16
  CIR length, 15
  conditions in time-domain, 32
  Doppler spreading, 32, 33
  fading, 33
  features, 15
  flat fading approximation, 31
  frequency-domain model, 33
  IEEE 802.11a WLAN, 12
  indoors, 12
  interference effects, 13
  noise effects, 13
  nonterrestrial environments, 12

OFDMA "profile", 31
  open issues, 37
  parameters, 34
  practical, 32–34
  Rayleigh/Ricean fading, 31
  signal model, 23–24
  SUI types, 32–33
  transmissions, 15
  Wi-Fi and WiMax systems, 13
  WINNER models, 33
OFDMA synchronization
  advanced methods, 366
  case study, 366
  CFO effects, 357
  CFO estimation, 362–366
  classical approach, 360
  dynamic effects, 353
  offsets origin, 352
  origin, 352
  parameters, 351–352
  performance impacts, 355
  recovery, 359
  static effects, 352
  STO effects, 355–357
  STO estimation, 360–362
  techniques, 367
  uplink, 358–359
OFDMA system, 35, 68, 74, 169, 398, 412, 480, 483, 509
  assumptions, 170
  framework, 132, 134
  frequency-domain, 171
  Gaussian channel, 170
  gradient-based scheduling, 132
  matrix, 170, 171
  multi-cell scenario, 75, 133
  parameter estimations, 400
  resource allocation, 132
  single-cell scenario, 74, 133
  transmit symbol, 169–170
  user separations, 400
OFDMA uplink
  asynchronous system, 383
  baseband system, 399
  carrier-frequency, 388
  "chicken–egg problem", 383
  correction by users, 387
  direct frequency, 387
  frequency estimation, 398
  linear multiuser detection, 389
  multi-CFO estimation, 400–403, 410, 411
  offset correction, 387
  offset estimation, 385
  parallel interference cancellation, 388
  parameters consecutive estimation, 385
  quasi-synchronous system, 383
  quasi-synchronous timing, 386
  structured subcarrier allocation, 385
  synchronization, 383, 384, 385
  transmission, 6
OFDM–MIMO HAP
  compact array geometry, 551

OFDM–MIMO HAP (*Continued*)
   ideal compact antenna, 550–551
   separation angles, 551
   subcarrier wave effect, 551
   system capacity, 551–554
   system performance, 551
OFDM–MIMO technique, 538
OFDM–SDMA, MIMO beamforming, 459
   capacity-aware, 462–465
   MIMO–MMSE systems, 465
   MIMO–MRC system, 459
   nonlinear MMSE detectors, 466–467
OFDM–SDMA system, 454
   access, 455
   capacity, 458
   channel model, 458
   MIMO beamforming schemes, 459–464
   MIMO–MMSE systems, 465
   model, 456
   multiple antenna aided, 456
   nonlinear MMSE detectors, 466
   SER expression, 459
   SER performance, 458, 466
   simulation results, 467–470
   SINR expression, 459
OFDM signal model, 49
   16-QAM encoding table, 52
   amplified PSD, 57
   baseband data signal, 53
   constellation bit encoding, 52
   Fourier transform, 54
   guard interval time, 50
   mathematical model, 49, 53, 55
   modulation schemes, 51
   normalization factor, 52, 53
   NRZ function, 54
   OFDM frame, 51
   OFDM symbols concatenation, 50
   PHY modes, 51
   PSD, 54
   sent and received signals, 50
   time-windowing function, 50
OFDM signal PSD, 57, 60, 64
   comparison with simulations, 63–64
   emission power level, 62–63
   Fourier transform, 57–59
   IP_3 and IP_5 calculation, 59, 60–62
   observations, 62
   Wiener–Khintchine theorem, 57
OMDM. *See* Orthogonal momentum division
      multiplexing (OMDM)
OMDM–HAP diversity system, 555
Opportunistic user scheduling, 305
Optimization algorithm, 515–517
Orbital angular momentum (OAM), 555
Orthogonal frequency division multiplexing
      (OFDM), 42, 190, 377
Orthogonal momentum division multiplexing
      (OMDM), 555
   block diagram, 556
   helical wave fronts, 555–556
   joint 2D multiplexing method, 556
   maximum information capacity, 557
   multiple HAP capacity, 551
   OMDM–HAP diversity system, 555
   radiated OAM mode, 555
   series expansion, 555
   total radiated power, 557
Out-of-band emission, 166
Overall multiuser algorithm, 88–89
Overhead analysis, 206
   lower layer, 209
   upper layer, 206

P

Paging channel (PCH), 582
Pairwise error probability (PEP), 475, 499, 501, 502
PA phase. *See* Power Allocation (PA) phase
PAPR. *See* Peak-to-average power ratio (PAPR)
Parallel concatenated convolutional code (PCCC), 583
Partial usage of subchannels (PUSC), 196, 308, 598, 601
   clusters, 604
      DL PUSC symbol, 604
      major group permutation, 606
      OFDMA symbols, 605
Payload header suppression (PHS), 206
PBO. *See* Power-back-off (PBO)
PCCC. *See* Parallel concatenated convolutional code (PCCC)
PCF. *See* Point coordination function (PCF)
PCFICH. *See* Physical control format indicator channel
      (PCFICH)
PCH. *See* Paging channel (PCH)
PCM. *See* Power control mechanism (PCM)
PDCCH. *See* Physical downlink—control channel
      (PDCCH)
PDF. *See* Probability density function (PDF)
PDP. *See* Power delay profile (PDP)
PDSCH. *See* Physical downlink—shared channel (PDSCH)
PDUs. *See* Protocol data units (PDUs)
Peak-to-average power ratio (PAPR), 43, 357, 439,
      440, 568, 599
PEP. *See* Pairwise error probability (PEP)
Perfect sequence, 341
Permutation schemes, 572
   AMC, 572
   downlink PUSC, 572, 573, 575
   FUSC, 572
   OPUSC, 572
   TUSC, 572
   uplink PUSC, 572, 573, 575
Permutation zones, 603
PF. *See* Proportional Fairness (PF)
PFS. *See* Proportional fairness scheme (PFS)
Phase-shift keying (PSK), 475
PHICH. *See* Physical hybrid ARQ indicator channel
      (PHICH)
PHS. *See* Payload header suppression (PHS)
PHY frame initiation, 567
PHY layer. *See* Physical layer (PHY)
PHY service data unit (PSDU), 567
Physical control format indicator channel (PCFICH), 583
Physical downlink

control channel (PDCCH), 583
shared channel (PDSCH), 583
Physical hybrid ARQ indicator channel (PHICH), 583
Physical layer (PHY), 38, 42, 132, 192, 256, 565, 596
channel coding, 203
IEEE (802.16–2004), 569
OFDMA, 193
OFDM-TDMA, 192–193
WirelessMAN-OFDM, 569
WirelessMAN-OFDMA, 569
WirelessMAN-SC, 192, 193
WirelessMAN-SC, 569
WirelessMAN-SCa, 569
Physical Layer Convergence Protocol (PLCP), 565
header, 568
Physical Random Access Channel (PRACH), 586
Physical resource block (PRB), 307
Physical uplink
control channel (PUCCH), 586
shared channel (PUSCH), 585
Pilot
carrier, 605
index, 610
subcarrier, 607
Pilot index, 610
single transmit antenna, 610
subchannel allocations, 611
Pilot symbol. *See* Synchronization—symbol
Pilot symbol-assisted modulation (PSAM), 330
PLCP. *See* Physical Layer Convergence Protocol (PLCP)
PLMN. *See* Public land mobile network (PLMN)
PMP. *See* Point-to-multipoint (PMP)
PN. *See* Pseudo-noise (PN); Pseudo-random number (PN)
Point coordination function (PCF), 568
Point-to-multipoint (PMP), 597
Polarization, 547, 555
angle, 550
antenna, 538
diversity, 539
Power
Allocation (PA) phase, 152, 158
back-off (PBO), 441
constraint, 228–229
control mechanism (PCM), 278, 279, 284–288
delay profile (PDP), 22
detection process, 350
DSTBCs new class, 285
IP approach, 229
LWT algorithm, 228–229
minimization game, 80–81
PCM criteria, 286
PCM selection procedures, 286
PCM swapping technique, 229
PSC types, 229
saving class (PSC), 229, 615–616
saving mode, 614
spectral density (PSD), 54, 166, 513
PRACH. *See* Physical Random Access Channel (PRACH)
PR algorithm. *See* Priority-based (PR) algorithm
PRB. *See* Physical resource block (PRB); *Bins*
prevSleepWin, 615

PRF. *See* Priority function (PRF)
Primary synchronization sequences (PSS), 351, 368
Priority-based (PR) algorithm, 224
Priority function (PRF), 122
Probability density function (PDF), 408
Processing load, FDE system comparison, 437
bandwidth efficiency, 438
front-end requirements, 439
OFDM, 437, 438
SC/FDE based on CP, 437, 438
SC/FDE based on UW, 437, 438
signal-to-noise loss, 439
Progressive subchannel allocation, 152–154
Proportional fairness (PF), 132, 218
algorithm, 320
polling, 218
scheduler, 120, 236
score-based scheduler, 236
Proportional fairness scheme (PFS), 226
Proportional fair scheduler, 120
Protocol data units (PDUs), 209
PSAM. *See* Pilot symbol-assisted modulation (PSAM)
PSC. *See* Power saving class (PSC)
PSD. *See* Power spectral density (PSD)
PSDU. *See* PHY service data unit (PSDU)
Pseudo-noise (PN), 603
Pseudo-random number (PN), 382
PSK. *See* Phase-shift keying (PSK)
PSS. *See* Primary synchronization sequences (PSS)
Public land mobile network (PLMN), 367
PUCCH. *See* Physical uplink Control Channel (PUCCH)
PUSC. *See* Partial usage of subchannels (PUSC)
PUSCH. *See* Physical uplink shared channel (PUSCH)

Q

QAM. *See* Quadrature amplitude modulation (QAM)
QoS. *See* Quality of service (QoS)
QPSK. *See* Quadrature phase shift keying (QPSK);
Quaternary phase shift keying (QPSK)
Quadrature amplitude modulation (QAM), 45, 199, 475, 566
Quadrature phase shift keying (QPSK), 199, 492
Quality of service (QoS), 69, 102, 116, 117, 120, 132, 191, 192, 313
BE, 117
CAC in assurance, 218–219
guarantee, 225, 227
IEEE 802.16 standard, 117
mechanism, 192
nrtPS, 117
parameters, 220
rtPS, 117
scheduling, 219
service classes, 204
UGS, 117
user data rate, 102
users' requirements, 226
utility function, 102
Quaternary phase shift keying (QPSK), 566

R

Radiated OAM mode, 555
Radiated power, 547, 557
Radio frequency (RF), 68
    application, 42, 509, 604
    resource problem, 68
Radio resource map (RRM), 257
Random
    method, 89
    variable (RV), 477
Ranging codes, 390
RA optimization. *See* Rate adaptive (RA) optimization
Rate adaptive (RA) optimization, 103, 105
Rate-based utility function, 123
    closed-form expression, 124, 125
    general utility function, 124
    optimal subcarrier assignment, 123, 124
    optimization problem, 124–125
    sorting–search suboptimal algorithm, 124
Rate craving greedy (RCG) algorithm, 110
RB. *See* Resource block (RB)
RBPP. *See* Resource block—pre-processor (RBPP)
RCG. *See* Rate craving greedy (RCG) algorithm
Real-time polling service (rtPS), 117, 204, 205
Real-time transport protocol (RTP), 206
Received elementary modes vector, 543, 544
Receive diversity, 293
Receiver (Rx), 38
Receiver mode vector, 542
Receive–transmit transition gap (RTG), 194, 600
Reference symbols (RS), 368
Relay-assisted
    cellular system, 508, 509
    OFDMA systems, 532
Relay stations (RSs), 508
Resource allocation, 102, 103
    ACG algorithm, 114
    algorithms, 103, 105, 106, 113, 240
    APA optimization algorithms, 106
    BABS algorithm, 115
    capacity distribution, 115
    channel evaluation, 241
    Cioffi algorithm, 114
    CQI-based feedback, 243
    DSA, 106
    FI, 113, 114
    frequency domain, 103
    frequency selective channels, 103
    Jang–Lee algorithm, 114
    MA optimization, 103
    possible approaches, 103
    problem solving, 106
    RA optimization, 103
    rate maximization optimization, 106
    received signal, 103
    resource scheduling, 243
    Rhee–Cioffi algorithm, 114
    scores determination, 242
    simulations, 113
    structure, 241

sum capacity comparison, 114
system model, 103
system parameters, 113
Zhang–Letaif algorithm, 114
Resource block (RB), 237, 238, 307, 581, 582. *See* Bins
    pre-processor (RBPP), 257
    resources partitioning, 238
Resource management, 102
    issues, 68
    multiple access, 68
    open problems, 126–127
    parts, 102
    QoS, 69, 102
    resource allocation, 102
    resources sharing, 68–69
    scheduling, 102
    system resources, 68
RF. *See* Radio frequency (RF)
RF power amplifier, 44
    amplified WLAN OFDM signal, 48–49
    IP_3, 49
    nonlinearity, 44, 48
    nonlinearity results, 44
    out-of-band emission levels, 49
    third-order intermodulation, 44
R_H, 240
Rhee and Cioffi max–min algorithm, 107
RMS-DS. *See* Root-mean-square delay spread (RMS-DS)
Robust header compression (ROHC), 206
ROHC. *See* Robust header compression (ROHC)
Root-mean-square delay spread (RMS-DS), 22
Rotating optical tweezers, 555
Round robin (RR), 219
    bargaining solutions result, 93
    carrier assignment algorithm, 92, 93–94
    comparison, 97
    complexity, 93
    Kalai–Smorodinsky solution, 93
    theoretical evaluation, 92–93
    theoretical Nash solution, 93
Round trip time (RTT), 578
RR. *See* Round robin (RR)
RRM. *See* Radio resource map (RRM)
RS. *See* Reference symbols (RS)
RSs. *See* Relay stations (RSs)
RTG. *See* Receive–transmit transition gap (RTG)
RTP. *See* Real-time transport protocol (RTP)
rtPS. *See* Real-time polling service (rtPS)
RTT. *See* Round trip time (RTT)
RV. *See* Random variable (RV)
Rx. *See* Receiver (Rx)

S

SAGE. *See* Space-alternating generalized expectation-maximization (SAGE)
Sampling clock offset (SCO), 351
Satisfaction Oriented Resource Allocation (SORA), 126
SB scheduling strategy. *See* Score-based (SB) scheduling strategy

SC. *See* Single carrier (SC)
Scalable OFDMA (SOFDMA), 569
SCAS. *See* Subband CAS (SCAS)
SC/FDE, 418, 419, 446
   based on CP, 437, 438
   based on UW, 437, 438
   OFDM, 442, 443–445, 447, 450–451
   parameters, 445
   system, 445
SC-FDMA. *See* Single-carrier—frequency division multiple
      access (SC-FDMA); Single carrier—frequency
      domain multiple access (SC-FDMA)
Scheduling, 116
   aim, 116
   algorithms, 118, 122
   application, 116
   classical, 118
   cross-layer, 116
   PRF, 122
   properties, 117–118
   QoS, 116, 117
   rules, 121–122
   scheduler, 117, 122
   utility-based, 122, 126
SCME. *See* Spatial channel model extended (SCME)
SCO. *See* Sampling clock offset (SCO)
Score-based (SB) scheduling strategy, 239
SDMA. *See* Space-division multiple access (SDMA);
      Spatial—division multiple access
      (SDMA)
SDUs. *See* Service data units (SDUs)
Secondary synchronization sequence (SSS), 366
Sector index, 604
Segment, 575, 598
Selective Multi User Diversity (SMUD), 315
Separation angles, 551
SER. *See* Symbol error rate (SER)
Service data units (SDUs), 219, 615
SERVICE field, 568
Serving gateway (S-GW), 578
SFST. *See* Spaced-frequency spaced-time (SFST)
S-GW. *See* Serving gateway (S-GW)
SIGNAL field, 567
Signal to interference and noise ratios (SINRs), 75, 239, 303,
      458, 475, 567
   FBA, 521
   FTA, 521
   postequalization, 242
Signal-to-noise ratio (SNR), 29, 45, 74, 132, 194, 237, 258,
      268, 303, 356, 411, 474
"sinc-*F*" function, 173
Single carrier (SC), 190, 192, 495
   frequency division multiple access (SC-FDMA), 358
   frequency domain multiple access (SC-FDMA), 584
   transmission, 418
Single-cell downlink
   aggregate bandwidth, 514
   cell region, 514
   FBA–SE, 522, 523, 526
   FBA–SINR, 523
   frequency-reuse patterns, 528

FTA–SE, 524
full reuse, 525, 526
optimization, 513–515
path loss, 521
performance, 521–526
RSs' positions, 522, 525
shadowing effect, 525
spectral efficiency, 515, 525, 527
Single cell link-level simulation. *See* MIMO-OFDM
      system isolated cell
Single-cell scenario, 74, 98
   achievable rate, 74
   AWGN channels, 74
   BS, 74
   channel estimation, 74
   MS, 74
   power constraint, 74
   SNR, 74
   total rate, 75
Single-cell uplink
   performance, 526–528
   service area determination, 527
   spectral efficiency, 527, 529, 531
   transmit power, 527, 529, 530, 531
Single-input single output (SISO), 249, 271, 369
single stream (ss), 237
Single transmit antenna, 602
   subcarrier allocation, 602
   UL PUSC symbol procedure, 607
   UL tile, 608
Single-user (SU), 351
Single-user MIMO (SU-MIMO), 236
Singular value decomposition (SVD), 539, 549
SINRs. *See* Signal to interference and noise ratios
      (SINRs)
SISO. *See* Single-input single-output (SISO)
Sleep mode, 614–615
sleepWin, 615
Slot, 307. *See also* Bins
   IEEE 802.16e, 308
   time, 2
SM. *See* Spatial—multiplexing (SM)
SMUD. *See* Selective Multi User Diversity (SMUD)
SMUX mode. *See* Spatial multiplexing (SMUX) mode
SNR. *See* Signal-to-noise ratio (SNR)
SNR gap, 311
SOAs. *See* Suboptimal algorithms (SOAs)
SOAs, low complexity, 151
   algorithm, 153, 154
   CA in SOA, 152, 154
   choices for step (4), 152, 153–154
   CNA step, 154
   CUM step, 154
   optimal tone allocation, 152
   PA phase, 152
SOFDMA. *See* Scalable OFDMA (SOFDMA)
SORA. *See* Satisfaction Oriented Resource Allocation
      (SORA)
Space-alternating generalized expectation-maximization
      (SAGE), 386
   algorithm, 386, 406–407, 413, 485

Space-alternating generalized expectation-maximization (SAGE) (*Continued*)
 convergence performances, 412
 E-step (compute expectation), 386
 M-step (maximize expectation), 386
 performance comparisons, 411–414
Spaced-frequency spaced-time (SFST), 22
Space-division multiple access (SDMA), 1, 2, 454
Space-polarization domain, 547–550
 angular momentum operator, 548
 channel matrix, 549
 independent ports, 548
 multipole expansion, 547–548
 mutual coupling, 549
 orthogonalization, 549
 scalar spherical harmonic function, 548
 spatial correlation matrices, 549
 spherical Hankel functions, 548
 SVD, 549
Space–time (ST), 568
 block code (STBC), 577
 cluster structure, 607
 coding (STC), 603, 606
 polarization (STP), 540–541
 simulator, 540
 subcarrier mappings, 613–614
 transmit diversity (STTD), 606
 UL PUSC, 608
 XPD, 541
Spatial
 channel model extended (SCME), 252
 correlation matrices, 549
 division multiple access (SDMA), 116
 multiplexing (SM), 599, 610
 polarization modes, 542
 processor, 542
 subchannels, 541
Spatial diversity, 293. *See also* Multiple input multiple output (MIMO) system; Receive diversity
Spatial multiplexing (SM) mode
 closed-loop, 588
 layer mapping, 588
 open-loop, 588
Spatial multiplexing (SMUX) mode, 236, 248
Spatial transmission modes, 237
 adaptive mode switching system, 246
 multistream (ms), 237
 selection probabilities, 246, 247, 255
 single-mode systems, 248
 single stream (ss), 237
Spectral compensation, 181, 186
 active methods, 182
 frequency-domain interpretations, 182–184
 passive methods, 182
 shaping matrix, 181–182
 transmit block, 181
Spectral efficiency, 125, 165–166, 184, 185, 359, 517, 519–521, 525, 529, 532–533
 high, 166
 lower, 166
 open issues, 186

 spectral leakage, 166, 167, 168
 worst-case interference, 167–169
Spectrum regrowth. See Nonlinearity
Spherical Hankel functions, 548
Spin angular momentum. *See* Polarization
ss. *See* single stream (ss)
SS. *See* Subscriber station (SS)
SSS. *See* Secondary synchronization sequence (SSS)
ST. *See* Space-time (ST)
"Stanford University Interim" (SUI), 24
Statistical channel, 16
 coherence bandwidth, 17
 coherence time, 17
 discrete impulses, 17
 Doppler spread, 17
 large-scale fading, 16
 multipath delay spread, 17
 parameters, 17
 propagation path loss, 16
 small-scale fading, 16
 statistical models, 16–17
 wireless channel, 17
STBC. *See* Space-time—block code (STBC)
STC. *See* Space–time—coding (STC)
STO. *See* Symbol timing offset (STO)
STP. *See* Space-time—polarization (STP)
STP channel model, 541–545
 AWGN, 546
 channel matrix, 545, 546
 cross-polarization, 543
 Doppler shift, 545
 linear combination coefficients, 542
 maximum information capacity, 547
 path channel, 544
 radiated power, 547
 receiver mode vector, 542
 signal vector, 543, 544
 SNR, 547
 spatial-polarization, 542
 spatial processor, 542
 spatial subchannels, 541
 transmitter mode vector, 542
 water-filling, 547
STP FIR filter, 544
Stratospheric flight, 538
STTD. *See* Space–time—transmit diversity (STTD)
SU. *See* Single-user (SU)
Subband, 5, 35, 167, 174, 237, 365, 385, 475. *See* Mutually exclusive groups
 allocation, 387, 388, 392
 CAS, 398, 400–403, 410
Subband CAS (SCAS), 5
Subburst, 601
Subcarrier
 allocation and permutation, 509–511
 ASCA, 510
 AWGN, 510
 diversity SCA, 510
 frequency band, 510
 mapping, 613, 510

permutation, 510
wave effect, 551
Subchannel, 307
allocation, 601, 602
assignment improvement stage, 156
basic assignment stage, 154–156
bin groups, 602
processing, 601
Subchannel Allocation (CA), 150–152, 601
Subchannel Number Assignment (CNA), 154
Subchannel User Matching (CUM), 154, 156
algorithm(3), 157
Hungarian algorithm, 157
permutation matrix, 157–158
Suboptimal algorithms (SOAs), 152
achievable rate, 158
algorithms, 159
baseline, 159
integer-dual, 159
resource allocation, 159–160
SOA1, 159
SOA2, 159
worst-case complexity, 159
Subscriber station (SS), 570
Successive MMSE precoding, 455
SUI. *See* "Stanford University Interim" (SUI)
SU-MIMO. *See* Single-user MIMO (SU-MIMO)
SVD. *See* Singular value decomposition (SVD)
Symbol error rate (SER), 311, 454
Symbols, 186–187, 393, 398
Symbol-timing acquisition, 378
channel dispersion, and, 378
CP-based, 380
pilot-based, 380
Symbol timing offset (STO), 351, 352
downlink estimation, 360–361
uplink estimation, 361
Synchronization
carrier-frequency acquisition, 381
clock offset, 376
frequency offset, 376
OFDM, 378
procedure, 351, 378, 389
recovery, 359
symbol, 380
timing acquisition, 376
tracking, 382
System
AMC subcarrier allocations, 610
capacity, 551–554
cluster, 598
data mapping, 599
elements, 598
exponential rule, 228
frame structure, 597, 599–600, 601
multi-cell downlink, 526
profile, 596
single-cell downlink, 521–526
slot, 598
subchannel, 597–598
subchannel allocation, 228

symbol, 597–598
throughput maximization, 228
uplink performance, 526–528

T

Tapped-delay line (TDL), 15, 19
TCP. *See* Transmission control protocol (TCP)
TDD. *See* Time division—duplex (TDD)
TDL. *See* Tapped-delay line (TDL)
TDM. *See* Time-division multiplex (TDM)
TDMA. *See* Time division—multiple access (TDMA)
Third generation (3G), 43
Third Generation Partnership Project (3GPP), 33, 34, 206,
324, 350, 577, 578, 590
Third-order interception point (IP$_3$), 49
Tile, 200, 599
index, 608
subchannels, 573
"tiling" problem, 195
UL, 607
uplink PUSC, 573, 602
usage of the subchannels (TUSC), 598
Time division
duplex (TDD), 116, 138, 191, 352n, 564, 596
multiple access (TDMA), 1, 2, 68, 132, 192, 497
multiplex (TDM), 134, 351
Time–frequency domain, 24–25
channel models, 24
circular convolution, 25
CIR function, 24
CTF function, 24
discrete frequency-domain, 25
frequency domain vector, 25
SUI models, 24
time-domain model, 25
Time–frequency grid, 305, 306
description, 306
downlink FUSC, 308–309
downlink PUSC, 309
IEEE 802.16e, 307
LTE, 307
WINNER project, 307
Time slots, 2
Timing offset, 28
decrease SNR, 29
OFDM symbol, 28
Tones, 137n
Tracking, 378
Transmission control protocol (TCP), 203
Transmission equation, 239
Transmit
diversity, 587, 588–589
power, 518–519, 527, 529, 530, 531
receive transition gap (TTG), 194, 574, 600
Transmit spectra, 171, 172
aggregate PSD, 175
autocorrelation, 173, 175
continuous-time version, 171
discrete-time, 174, 175
Fourier transform, 172, 174

Transmit spectra (*Continued*)
    IDFT matrix, 171
    out-of-band region, 174
    PSDs, 174
    "sinc-F " function, 173
    spectral measures, 172
    transmit PSD, 173
    "true" out-of-band emission, 174
Transmitter (Tx), 38
Transmitter mode vector, 542
"true" out-of-band emission, 174
TTG. *See* Transmit–receive transition gap (TTG)
TUSC. *See* Tile usage of the subchannels (TUSC)
Two-dimensional (2D), 77, 306
Two-step algorithm. *See* Resource allocation
        algorithm
Two transmit antennas, 606
    AMC, 612
    Matrix A, 603n
    Matrix B, 603n
Two-user bargaining algorithm, 90
    Lagrangian multiplier, 92
    two-band partitioning algorithm, 92
    waterfilling problem, 91
Tx. *See* Transmitter (Tx)

U

UAVs. *See* Unmanned aerial vehicles (UAVs)
UCD. *See* Uplink (UL)—channel descriptor (UCD)
UEPS. *See* Urgency and efficiency-based packet
        scheduling (UEPS)
UE. User equipment (UE)
UGS. *See* Unsolicited grant service (UGS)
UHF. *See* Ultra high frequency (UHF)
UIUC. *See* Uplink (UL)—interval usage code (UIUC)
UL. *See* Uplink (UL)
UL-MAP. *See* Uplink (UL)—map (UL-MAP)
UL timing scheme
    asynchronous, 166
    synchronous, 166
Ultra high frequency (UHF), 12
Uncorrelated scattering (US) channel, 21
U-NII. *See* Unlicensed National Information
        Infrastructure (U-NII)
Unique Word (UW) structure, 419, 429
    advantages, 431
    CP structure vs, 430
Universal variable length code (UVLC), 318
Unlicensed National Information Infrastructure (U-NII)
        bands, 45
Unmanned aerial vehicles (UAVs), 538
Unsolicited grant service (UGS), 117, 204
Uplink (UL), 166, 191, 383, 598, 599
    bandwidth mechanisms, 215
    BE consideration, 218
    burst allocation, 575
    BWR, 215
    channel descriptor (UCD), 390
    delay requirements, 215, 216
    ertPS consideration, 217–218

interval usage code (UIUC), 599
    map (UL-MAP), 192, 390
    MAP message, 600n
    nrtPS consideration, 218
    optimization, 517–521
    overhead, 211
    pilot time slot (UpPTS), 580
    rtPS consideration, 216, 217
    spectral efficiency, 519–521
    subframe, 601
    system spectral efficiency, 520
    tile, 607, 609
    transmit power, 518–519
    UGS consideration, 215
    zone, 599
Upper layer overhead, 206. *See also* Lower layer overhead
    PHS, 206
    ROHC-RTP mode, 206, 208
    ways to reduce, 206
UpPTS. *See* Uplink—pilot time slot (UpPTS)
Urgency and efficiency-based packet scheduling (UEPS), 227
US. *See* Uncorrelated scattering (US)
User equipment (UE), 578
User terminals (UTs), 237
Utilitarian solution, 73
Utility function, 122, 123, 124
    delay-based, 125
    high-priority user, for, 124
    rate-based, 123
UTs. *See* User terminals (UTs)
UVLC. *See* Universal variable length code (UVLC)
UW structure. *See* Unique Word (UW) structure

V

Variable bit rate (VBR), 205
VBR. *See* Variable bit rate (VBR)
VCTCXO. *See* Voltage controlled temperature compensated
        crystal oscillator (VCTCXO)
Vertical encoding, 613
Very large scale integration (VLSI), 3, 42
Virtual referee game, 81
    algorithm, 82
    channel removal, 81
    mathematical optimization, 81
    optimal solution, 82–83
    power and rate allocation, 85, 86
    rate adjustment, 81
    rate maximization game, 81
    waterfilling problem, 81, 82
VLSI. *See* Very large scale integration (VLSI)
Voice over IP (VoIP), 102
VoIP. *See* Voice over IP (VoIP)
Voltage controlled temperature compensated crystal oscillator
        (VCTCXO), 352

W

Waterfilling
    distribution, 304, 311
    technique, 547

water-filling solution, 547
Wave propagation channel, 539, 540
WCDMA. *See* Wideband code division multiple access
  (WCDMA)
Weighted Fair Queuing (WFQ), 119, 223
Weighted multicarrier proportional
  fair algorithm (WMPF), 120
Weighted round robin (WRR), 118–119, 222
WF²Q. *See* Worst-case fair weighted fair queueing (WF²Q)
WFQ. *See* Weighted Fair Queuing (WFQ)
WiBro. *See* Wireless broadband (WiBro)
Wideband code division multiple access (WCDMA), 538, 577
Wide-sense stationary (WSS), 21
  CIR, 22
  relative mathematical simplicity, 22–23
Wide-sense stationary with uncorrelated scattering
  (WSSUS) channel, 288
  "2L-APSK" performance, 293, 294, 295
  "4A16PSK PCM2" performance, 291, 292, 295
  simulation parameters, 289
  system performance, 288
  technique improvement, 292
WiFi. *See* Wireless local area network (WLAN)
WIM. *See* WINNER channel model (WIM)
WiMAX *See* Worldwide interoperability
  for microwave access (WiMAX)
WiMAX BS, 190
WiMAX coding, mobile, 203
  FEC coding methods, 203
  interleaving, 203
  modulation, 203
  PHY channel coding, 203
  randomization, 203
  reducing and correcting steps, 203
  repetition, 203
WiMAX configuration, 202
  OFDMA parameters, 200, 202
  slot capacity, 201, 202
  subcarriers in PUSC, 201, 202
WiMAX forum, 190–191, 192, 193, 206
WiMAX frame structure, 193
  2D criterion, 195
  bin packing problem, 195
  BLER curves, 195
  bursts, 195
  CQI, 194
  eOCSA, 196
  fragmentation and packing, 198
  issues, 194
  mapping for downlink, 197
  mapping problem, 195–196
  MCS, 193–194
  MPDU, 193
  MPDU frame format, 194
  normalized over-allocations, 196
  normalized unused space, 198
  scheduler designing, 194
  "tiling" problem, 195
  types of bursts, 196
  unused slots, 196
WiMAX Interclass scheduling, 224

DFPQ, 224
DTPQ, 224
negative effect of priority, 224
PR algorithm, 224, 225
two-step scheduler, 225
WiMAX Intra-class scheduling, 222
  delay-based algorithms, 224
  DWRR, 223
  fragmentation mechanism, 224
  RR algorithm, 222
  weight calculation, 223–224
  WF2Q, 223
  WFQ, 223
  WRR, 222, 223
WiMAX networks, 192, 196, 224, 225
  configuration, 202
  deployment scenarios, 191
  frame structure, 193
  key features, 191
  mapping problem, 195–196
  mobile coding schemes, 203
  OFDMA MCSs, 199
  OFDMA parameters, 200
  pitfalls, 211
  QoS, 192, 204
  scheduling algorithms, 192
WiMAX OFDMA MCSs, 199
  definition of slots, 199
  FDD, 199
  fields, 199
  modulation methods, 199
  slot formation for PUSC, 200
  subcarriers allocation, 199
  TDD, 199
  users allocation, 199
WiMAX pitfalls, 211
  enhanced scheduler, 213
  MAC header, 212
  number of DL users, 212, 213
  number of users, 211–214
  problem with analysis, 213
  subheaders, 212
WiMAX QoS service classes, 204
  comparison, 204
  ertPS, 205
  nrtPS, 205
  rtPS, 205
  traffic priority, 205
  types, 204
  UGS, 204, 205
WiMAX scheduler, 194, 218, 225
  allocation, 219
  CAC role in QoS assurance,
    218–219
  channel-aware schedulers, 221, 225
  channel-unaware schedulers, 221
  comparison, 222, 223
  compensation issue, 226
  component schedulers, 220
  energy consumption, 220
  fairness, 220, 226

WiMAX scheduler (*Continued*)
   FIFO technique, 219
   four classes, 225
   GWF concept, 226
   implementation complexity, 221
   other users' QoS requirements, 226
   permutation, 219
   power constraint, 228
   power control, 220
   QoS guarantee, 219, 220, 227
   RR technique, 219
   scalability, 221
   scheduling, 218, 221, 222, 224
   throughput, 220, 228
WINNER channel model (WIM), 244
WINNER project. *See* Wireless World Initiative
        New Radio (WINNER) project
Wireless
   broadband communication, 42–44
   communication development, 41–42
   MANOFDM, 596
   MAN-SC, 596
   MAN-SCa, 596
   metropolitan area network (WMAN), 2, 303
   OFDMA, 596
   personal area networks (WPANs), 303
   system, 1
Wireless broadband (WiBro), 568
   AAS, 569, 576–77
   burst profiles, 575
   duplexing and framing, 574–576
   HARQ, 576
   OFDM subcarriers, 572
   subchannelization schemes, 572
Wireless local area network (WLAN), 2, 12, 42, 303,
        330, 350, 564, 565
   IEEE 802.11a/g PHY, 565
   IEEE 802.11a PHY frame, 567
   IEEE 802.11a preamble, 567
   IEEE 802.11n, 565
   link adaptation, 567
WirelessMAN, 190
WirelessMAN-SC, 192, 193
Wireless World Initiative New Radio (WINNER) project,
        244
W_k. *See* Binary series (W_k)
WLAN. *See* Wireless local area network (WLAN)
WMAN. *See* Wireless—metropolitan area network (WMAN)

WMPF. *See* Weighted multicarrier proportional
        fair algorithm (WMPF)
World Radio communication Conference (WRC07), 590
Worldwide interoperability for microwave access (WiMAX),
        3, 12, 190, 237, 330, 376, 568, 570, 596
   antenna category, 576–577
   beamforming, 577
   CP, 597
   DL synchronization, 390
   FFT, 597
   first fixed equipment, 570
   first mobile equipment, 570
   fixed system profile, 596
   forum, 569
   handover types, 617
   "initial ranging", 390
   MIMO with SM, 577
   mobile, 568
   mobility system profile, 596
   reference architecture, 570
   single transmit antenna, 602
   synchronization, 389
   UL subframe in, 391
   UL synchronization, 390
   variable frame sizes, 601
   WiMAXTDD frame, 574
Worst-case fair weighted fair queueing (WF^2Q), 223
WPANs. *See* Wireless—personal area networks (WPANs)
WRC07. *See* World Radio communication Conference
        (WRC07)
WRR. *See* Weighted round robin (WRR)
WSS. *See* Wide-sense stationary (WSS)
WSSUS channel. *See* Wide-sense stationary with uncorrelated
        scattering (WSSUS) channel

X

XPD. *See* Cross-polar discrimination (XPD)

Z

Zadoff–Chu (ZC) sequences, 360, 367–368
ZC sequences. *See* Zadoff–Chu (ZC) sequences
Zero forcing (ZF), 115, 428
ZF. *See* Zero forcing (ZF)
Zhang and Letaif algorithm, 111, 112
Zone_IE, 599. *See* Uplink (UL)—zone